T0331120

Applied Computational Aerodynamics

This computational aerodynamics (CA) textbook is written at the undergraduate level, based on years of teaching focused on developing the engineering skills required to become an intelligent user of aerodynamic codes, unlike most available books which focus on learning how to write codes. This is done by taking advantage of CA codes that are now freely available and doing projects to learn the basic numerical and aerodynamic concepts required.

The authors believe that new and vibrant ways to interact with CA are important in order to improve understanding of aerodynamics. This book includes a number of unique features to make studying computational aerodynamics more enjoyable. These include:

- The computer programs used in the book's projects are all open source and accessible to students and practicing engineers alike on the book's website, www.cambridge.org/aerodynamics.
- CA Concept Boxes appear throughout the book to make material more relevant and to provide interesting asides from the material at hand.
- Flow Visualization Boxes are used throughout the book to give readers the opportunity to "see" fluid dynamic flows first hand.
- Profiles of both experienced and beginning practitioners of CA are included throughout to add a more personal dimension to the practice of numerical simulations of aerodynamics.
- Best Practices summaries are included at the end of most chapters to provide real-world guidelines for how CA is typically used.
- The website includes access to images, movies, programs, CA codes, additional material, and links to a variety of resources vital to the discussions contained within the book (www.cambridge.org/aerodynamics).

Russell M. Cummings is a professor of aeronautics at the U.S. Air Force Academy, where he teaches and conducts research in fluid mechanics, aerodynamics, and computational aerodynamics.

William H. Mason is a professor emeritus of aerospace engineering at Virginia Polytechnic Institute and State University.

Scott A. Morton is a researcher at the University of Dayton Research Institute and the principal software developer for the Kestrel Fixed Wing Aircraft Navier-Stokes flow solver.

David R. McDaniel is an associate research professor at the University of Alabama at Birmingham where he works on the Kestrel Fixed Wing Aircraft Navier-Stokes flow solver.

Cambridge Aerospace Series

Editors: Wei Shyy and Vigor Yang

1. J. M. Rolfe and K. J. Staples (eds.): *Flight Simulation*
2. P. Berlin: *The Geostationary Applications Satellite*
3. M. J. T. Smith: *Aircraft Noise*
4. N. X. Vinh: *Flight Mechanics of High-Performance Aircraft*
5. W. A. Mair and D. L. Birdsall: *Aircraft Performance*
6. M. J. Abzug and E. E. Larrabee: *Airplane Stability and Control*
7. M. J. Sidi: *Spacecraft Dynamics and Control*
8. J. D. Anderson: *A History of Aerodynamics*
9. A. M. Cruise, J. A. Bowles, C. V. Goodall, and T. J. Patrick: *Principles of Space Instrument Design*
10. G. A. Khoury (ed.): *Airship Technology*, Second Edition
11. J. P. Fielding: *Introduction to Aircraft Design*
12. J. G. Leishman: *Principles of Helicopter Aerodynamics*, Second Edition
13. J. Katz and A. Plotkin: *Low-Speed Aerodynamics*, Second Edition
14. M. J. Abzug and E. E. Larrabee: *Airplane Stability and Control: A History of the Technologies that Made Aviation Possible*, Second Edition
15. D. H. Hodges and G. A. Pierce: *Introduction to Structural Dynamics and Aeroelasticity*, Second Edition
16. W. Fehse: *Automatic Rendezvous and Docking of Spacecraft*
17. R. D. Flack: *Fundamentals of Jet Propulsion with Applications*
18. E. A. Baskharone: *Principles of Turbomachinery in Air-Breathing Engines*
19. D. D. Knight: *Numerical Methods for High-Speed Flows*
20. C. A. Wagner, T. Hüttl, and P. Sagaut (eds.): *Large-Eddy Simulation for Acoustics*
21. D. D. Joseph, T. Funada, and J. Wang: *Potential Flows of Viscous and Viscoelastic Fluids*
22. W. Shyy, Y. Lian, H. Liu, J. Tang, and D. Viieru: *Aerodynamics of Low Reynolds Number Flyers*
23. J. H. Saleh: *Analyses for Durability and System Design Lifetime*
24. B. K. Donaldson: *Analysis of Aircraft Structures*, Second Edition
25. C. Segal: *The Scramjet Engine: Processes and Characteristics*
26. J. F. Doyle: *Guided Explorations of the Mechanics of Solids and Structures*
27. A. K. Kundu: *Aircraft Design*
28. M. I. Friswell, J. E. T. Penny, S. D. Garvey, and A. W. Lees: *Dynamics of Rotating Machines*
29. B. A. Conway (ed): *Spacecraft Trajectory Optimization*
30. R. J. Adrian and J. Westerweel: *Particle Image Velocimetry*
31. G. A. Flandro, H. M. McMahon, and R. L. Roach: *Basic Aerodynamics*
32. H. Babinsky and J. K. Harvey: *Shock Wave–Boundary-Layer Interactions*
33. C. K. W. Tam: *Computational Aeroacoustics: A Wave Number Approach*
34. A. Filippone: *Advanced Aircraft Flight Performance*
35. I. Chopra and J. Sirohi: *Smart Structures Theory*
36. W. Johnson: *Rotorcraft Aeromechanics*
37. W. Shyy, H. Aono, C. K. Kang, and H. Liu: *An Introduction to Flapping Wing Aerodynamics*
38. T. C. Lieuwen and V. Yang: *Gas Turbine Emissions*
39. P. Kabamba and A. Girard: *Fundamentals of Aerospace Navigation and Guidance*
40. R. M. Cummings, W. H. Mason, S. A. Morton, and D. R. McDaniel: *Applied Computational Aerodynamics*

APPLIED COMPUTATIONAL AERODYNAMICS

A Modern Engineering Approach

Russell M. Cummings

United States Air Force Academy

William H. Mason

Virginia Polytechnic Institute and State University

Scott A. Morton

University of Dayton Research Institute

David R. McDaniel

University of Alabama at Birmingham

CAMBRIDGE
UNIVERSITY PRESS

CAMBRIDGE
UNIVERSITY PRESS

University Printing House, Cambridge CB2 8BS, United Kingdom

One Liberty Plaza, 20th Floor, New York, NY 10006, USA

477 Williamstown Road, Port Melbourne, VIC 3207, Australia

314-321, 3rd Floor, Plot 3, Splendor Forum, Jasola District Centre, New Delhi - 110025, India

79 Anson Road, #06-04/06, Singapore 079906

Cambridge University Press is part of the University of Cambridge.

It furthers the University's mission by disseminating knowledge in the pursuit of education, learning and research at the highest international levels of excellence.

www.cambridge.org
Information on this title: www.cambridge.org/9781107053748

First published 2015
Reprinted 2017

A catalogue record for this publication is available from the British Library

Library of Congress Cataloging in Publication data
Cummings, Russell M. (Russell Mark), author.
Applied computational aerodynamics : a modern engineering approach / Russell M. Cummings, United States Air Force Academy, William H. Mason, Virginia Polytechnic Institute and State University, Scott A. Morton, United States Air Force, David R. McDaniel, University of Alabama at Birmingham.
  pages   cm. – (Cambridge aerospace series)
Includes bibliographical references and index.
ISBN 978-1-107-05374-8 (hardback)
1. Air flow – Mathematical models.   2. Aerofoils – Mathematical models.   3. Aerodynamics, Supersonic – Data processing.   I. Morton, Scott A., author.   II. Mason, William H. (William Henry), 1947– author.   III. McDaniel, David R., author.   IV. Title.
TL574.F5C86   2015
629.132′300151–dc23   2014020402

ISBN 978-1-107-05374-8 Hardback

Contents

Color plates follow page 338

Preface

Aren't there already plenty of excellent books on the topic of CFD? Yes, there are ... if you are a graduate student who wants to learn the intricacies of numerical methods applied to solving the fundamental equations of fluid dynamics. However, we believe that a paradigm shift has taken place in CFD, where the development of algorithms and codes has largely been replaced by people applying well-established codes to real-world applications. While this is a natural progression in any field of science and engineering, we do not believe that the paradigm shift has filtered into the academic world. In academia, undergraduates learning about aerodynamics are still going through theories and applications that were being taught 40 or 50 years ago. We believe that it is time to write a book for people who want to be "intelligent users" of CA, not for those who want to continue developing CA tools. We strongly endorse the perspective of David Darmofal and Earll Murman of MIT (see AIAA Paper 2001–0870):

> Within aerodynamics, the need for re-engineering the traditional curriculum is critical. Industry, government, and (to some extent) academia has seen a significant shift away from engineering science and highly specialized research-oriented personnel toward product development and systems-thinking personnel. While technical expertise in aerodynamics is required, it plays a less critical role in the design of aircraft than in previous generations. In addition to these influences, aerodynamics has been revolutionized by the development and maturation of computational methods. These factors cast significant doubt that a traditional aerodynamics curriculum with its largely theoretical approach remains the most effective education for the next generation of aerospace engineers. We believe that change is in order.

We agree completely and believe that CA needs to be brought into the undergraduate classroom as soon as possible. That is why we have written this book!

The target audience for *Applied Computational Aerodynamics* is advanced undergraduates in aerospace engineering who want (or need) to learn CA in the broad context of learning to do computational investigations, while also learning engineering methods and aerodynamics. In addition, we believe that working engineers who need to apply CA methods, but who have no CA background, will also find the book valuable.

The educational objectives of the book include: (1) providing a context for computational aerodynamics within aeronautical engineering; (2) learning how to approach and solve computational problems; and (3) providing an entry into the literature by including numerous references and trying to put them in some sort of relevant context. Our overall goal, as mentioned previously, is to educate competent and intelligent users (or even observers) of CA, which will be accomplished through the use of well-defined projects where students will learn how to use the available tools within the context of understanding aerodynamics.

The contents of the book include: a brief history of computational aerodynamics and computers (why and how CA is used); engineering problem solving with emphasis on using the computer, but in the broad context of experimental, analytical, and engineering methods; review of the governing equations used in CA; an introduction to aerodynamic concepts; "classical" linear computational aerodynamics methods; the central idea of CFD – the numerical solution of PDEs; geometry and grids; viscosity and turbulence models; the art of CA, including rules of thumb, overall approaches to the simulation of aerodynamics, grid generation, convergence, grid studies, and flow visualization; and projects illustrating both CA and aerodynamics. The book and accompanying website (www.cambridge.org/aerodynamics) will provide access to CA software so that anyone within an academic or industrial environment will be able to accomplish the various projects with readily available computer resources. Finally, we want to reiterate what we are doing and what we are not doing. This is a book designed to teach aerodynamics through the use of modern computational tools. This is not a book for CFD algorithm developers (those books have already been written).

The presentation of material in this book presumes that the reader previously completed a course in fluid mechanics or aerodynamics, although readers without any background in aerodynamics will be able to work through the concepts presented (with a little extra effort). The typical student using this book may still be learning about basic engineering and aerodynamics, but probably does not have well-developed skills in computational problem solving. The book was written with these students in mind. It contains several unique features, including biographies of people who work in computational aerodynamics, as well as concept boxes that help explain certain ideas more clearly. Projects are included for most chapters, as well as some traditional problems where appropriate. We have found that projects work well with the material in this book, and suggest that you perform projects rather than just complete homework problems. We hope that you enjoy our approach to computational aerodynamics while you learn! Sample course outlines are also available to aid instructors in presenting the material.

There are a number of biographical sketches throughout this book which we hope will be interesting and possibly inspirational as you look forward to your career. We have included a wide variety of profiles, including those from some "up and coming" young researchers. The response to our requests for profiles was so positive, we actually received many more than we could publish. All of the profiles, however, are available at www.cambridge.org/aerodynamics.

Unique Features

Computational Aerodynamics is such a new and vibrant field, and we believe that new and vibrant ways to interact with the subject matter are important. To that end, we have included a number of unique features that should make studying Computational Aerodynamics more enjoyable. These include:

- Various computer programs are used within the projects contained in the book, all of which are open source and accessible to students and practicing engineers alike.
- CA Concept Boxes appear throughout the book to make material more relevant and to provide interesting asides from the material at hand.
- Flow Visualization Boxes are used throughout the book to give readers the opportunity to "see" fluid dynamic flows first hand.
- Profiles of both experienced and beginning practitioners of Computational Aerodynamics are included throughout the book to give a more personal dimension to the study of this material.
- Summaries of Best Practices are included at the end of most chapters to provide real-world guidelines for how Computational Aerodynamics is typically used.
- Access to a website with images, movies, programs, additional material, and links to a variety of resources vital to the discussions contained within the book (www.cambridge.org/aerodynamics).

Acknowledgments

As is true with any undertaking of this magnitude, there are many people who have helped us along the way, and we are extremely grateful. First of all we want to thank all the cadets who took the computational aerodynamics course at the U.S. Air Force Academy (USAFA) – they gave us a great deal of valuable feedback and advice about the course and the book (whether we wanted to hear it or not). We are also very grateful to the many people who have taught Computational Aerodynamics at USAFA since its inception. As each new group of people teaches the course, they make observations and suggestions for improvement, many of which have been implemented here. Included in the growing list of people who have taught the course are: Scott Morton, Barrett McCann, Robert Van Dyken, Jacob Freeman, Robert Decker, Charlie Hoke, Marc Riviere (as a visiting officer from École de l'air), Martiqua Post, Bill Mason (as a visiting professor from Virginia Tech), Roger Greenwood, Christopher Coley, Andrew Lofthouse, and Russell Cummings.

Perhaps the first person to recognize a need for this book (besides our students) was Ray Cosner of The Boeing Company – Ray encouraged us to write the book, and to keep in mind that it would be beneficial for students as well as practicing engineers; we thank him for the nudge! Also, two early supporters and constant fans were John J. Bertin of the U.S. Air Force Academy and John McMasters of The Boeing Company – we miss their witty and intelligent interaction. We are also greatly indebted to Doug Blake and Jim Forsythe, who wrote early versions of Chapters 3 and 8, respectively.

The book contains a number of biographical sketches of people who work in the field of computational aerodynamics – we thank W. Mark Saltzman of Yale University for this wonderful idea. Among those who provided sketches are: Ken Badcock of the University of Liverpool; Tracie Barber of the University of New South Wales; Marsha Berger of New York University; Tuncer Cebeci of Long Beach State University (who provided the material for the sketch of A.M.O. Smith); Kozo Fujii of the Japan Aerospace Exploration Agency; Karen Gundy-Burlet, Larry Erickson, Scott Murman, and Tom Pulliam of NASA Ames Research Center; Zach Hoisington of The Boeing Company; Kerstin Huber of the German Aerospace Center (DLR); Mark Lewis of the University of Maryland; Bob MacCormack and

Antony Jameson of Stanford University; Dimitri Mavriplis of the University of Wyoming; Earll Murman of MIT; Bryan Richards of the University of Glasgow; Christopher Roy of Virginia Tech; Christopher Rumsey of NASA Langley Research Center; Joe Thompson of Mississippi State University; Wei Shyy and Bram Van Leer of the University of Michigan; and David Vallespin of Airbus. The response to our requests for biographical sketches was overwhelming, and unfortunately there was not enough room in the book for all of the sketches, but they are all included on the book website at www.cambridge.org/aerodynamics.

We would also like to thank the following people who supplied technical information or feedback for the book in one form or another: Vedat Akdag of Metacomp Technologies; Ken Badcock, David Vallespin, Simao Marques, Lucy Schiavetta, and George Barakos of the University of Liverpool; Tim Baker and Luigi Martinelli of Princeton University; Tracie Barber of the University of New South Wales; Wolf Bartelheimer of BMW; John J. Bertin, Keith Bergeron, Mehdi Ghoreyshi, and Tiger Jeans of the U.S. Air Force Academy; Doug Blake and Miguel Visbal of the U.S. Air Force Research Laboratory; Okko Boelens and Koen de Cock of the Dutch National Aerospace Labs (NLR); Patrick Champigny of ONERA; Bill Dawes of the University of Cambridge; James DeSpirito of the U.S. Army Research Laboratory; Scott Eberhardt of Analytical Methods, Inc.; Lars-Erik Eriksson of Chalmers University of Technology; Jim Forsythe of the Office of Naval Research; Mike Giles of Oxford University; Reynaldo Gomez of NASA Johnson Space Center; Pres Henne of Gulfstream Aerospace Corporation; Jean Hertzberg of the University of Colorado; Colin Johnson of Desktop Aeronautics, Inc. John Lamar, Chris Rumsey, Neal Frink, and Pieter Buning of NASA Langley Research Center; José Longo of the European Space Agency; Samantha Magill and Kathleen Bangs of Honda Aircraft Company; Dimitri Mavriplis of the University of Wyoming; Heather McCoy of Pointwise, Inc.; Rob McDonald, Brian J. German, and Alejandro Ramos of Cal Poly; John McMasters, Ray Cosner, Ed Tinoco, John C. Vassberg, and Zach Hoisington of The Boeing Company; Scott Murman, Tom Pulliam, Karen Gundy-Burlet, Neal Chaderjian, Terry Holst, and Larry Erickson of NASA Ames Research Center; Gary J. Page of Loughborough University; Cori Pasinetti of SGI; Adrian Pingstone; Max Platzer of the Naval Postgraduate School; Frits Post of the TU Delft Visualization Group; Mark Potsdam of the U.S. Army Aeroflightdynamics Center at NASA Ames Research Center; Sjaak Priester; Daniel Reckzeh and Klaus Becker of Airbus; Art Rizzi of the Royal Institute of Technology (KTH); Chris Roy of Virginia Tech; Neil Sandham of the University of Southampton; William S. Saric of Texas A&M University; Andreas Schütte, Stefan Görtz, Ralf Heinrich, and Andreas Krumbein of the German Aerospace Center (DLR); Brian R. Smith of Lockheed Martin; Richard Smith of Symscape; Fred Stern of the University of Iowa; Lei Tang of D & P LLC; John Tannehill of Iowa State University; Ken Taylor of the Mercer Engineering Research Center; Kunihiko Taira and Tim Colonius of the California Institute of Technology; Joe Thompson of Mississippi State University; the members of the AIAA Fluid Dynamics Technical Committee's Discussion Group on CFD in Undergraduate Education; and the members of NATO RTO/STO Task Groups 113, 161, 189, and 201. We also want to thank the book evaluators for their insightful and helpful comments. Earl Duque and Steve Legensky of Intelligent Light were invaluable in adding the Flow Visualization boxes throughout the book. A very special thank you goes to our wonderful editor, Peter Gordon, who was supportive and attentive beyond all of our expectations. We also want to thank Wei Shyy and Vigor Yang, the editors of the Cambridge Aerospace Series for Cambridge University Press, for including our book in that series. Great appreciation is reserved for Patricia Bowen of the U.S. Air Force Academy for proofreading the manuscript. Finally, we would not have been able to accomplish many of the computations and projects over the years that gave us the background and ability to write this book without the support of the U.S. Department of Defense (DoD) and the U.S. Air Force, specifically the computational support from the DoD High Performance Computing Modernization Program and the High Performance Computing Research Center at the U.S. Air Force Academy.

Abbreviations

ACSYNT	AirCraft SYNThesis design program
ADI	Alternating Direction Implicit
AF	Approximate Factorization
AFLR	Advancing-Front/Local-Reconnection
AFM	Advancing Front Method
AGARD	Advisory Group for Aeronautics Research and Development (NATO, later replaced by RTO)
AGPS	Aero Grid and Paneling System
AIAA	American Institute of Aeronautics and Astronautics
ALM	Advancing Layer Method
AMR	Adaptive Mesh Refinement
ARC2D	Ames Research Center 2D flow solver
ARC3D	Ames Research Center 3D flow solver
ASME	American Society of Mechanical Engineers
BCFD	Boeing CFD flow solver
CA	Computational Aerodynamics
CAD	Computer Aided Design
CAGD	Computer Aided Geometry Design
CALSPAN	Cornell Aeronautical Laboratory Corporation
CAM	Computer Aided Manufacturing
CART3D	CARTesian 3D flow solver
CASI	Canadian Aeronautics and Space Institute
$C^2A^2S^2E$	Center for Computer Applications in AeroSpace Science and Engineering
CDC	Control Data Corporation
CFD	Computational Fluid Dynamics
CFL	Courant-Friedrichs-Lewy number
CFL3D	Computational Fluids Laboratory 3D flow solver
CGI	Computer Generated Imagery
CGNS	CFD General Notation System
CPU	Central Processing Unit
CRM	Common Research Model

CUBRC	CALSPAN/University of Buffalo Research Center
DATCOM	USAF Stability and Control DATa COMpendium
DDES	Delayed Detached-Eddy Simulation
DES	Detached-Eddy Simulation
DLR	Deustches Zentrum für Luft– und Raumfahrt (German Aerospace Center)
DNS	Direct Numerical Simulation
DNW	Deutsch-Niederländisches Windkanal (German-Dutch Wind Tunnels)
DOD	Department of Defense
DOE	Department of Energy
DPW	Drag Prediction Workshop
EADS	European Aeronautic Defence and Space Company
EASM	Explicit Algebraic Stress Model
EARSM	Explicit Algebraic Reynolds Stress Model
EFD	Experimental Fluid Dynamics
ENSICA	École Nationale Supérieure d'Ingénieurs de Constructions Aéronautiques (French National Higher School of Aeronautical Construction)
ENSOLV	Euler Navier-Stokes SOLVer
ESDU	Engineering Science Data Unit
EVM	Eddy Viscosity Model
FAST	Flow Analysis Software Toolkit
FDE	Finite Difference Equation
FLOP	Floating Point OPeration
FLOMANIA	FLOw physics Modelling – AN Integrated Approach
FOI	Totalförsvarets Forskningsinstitut (Swedish Defence Research Agency)
FORTRAN	FORmula TRANslation program language
FTF	Flap Track Fairing
FUN3D	Full Unstructured Navier-Stokes 3D flow solver
GASP	General Aerodynamic Simulation Program
GIGO	Garbage In/Garbage Out
GIS	Grid-Induced Separation
GMGG	Geometry Modeling and Grid Generation
GPU	Graphics Processor Unit
GUI	Graphics User Interface
HPC	High Performance Computing
HSCT	High Speed Civil Transport
IBM	International Business Machines
ICASE	Institute for Computer Applications in Science and Engineering
IDDES	Improved Delayed Detached-Eddy Simulation
I/O	Input/Output

IGES	Initial Graphics Exchange Specification
ISAE	Institut Supérieur de l'Aéronautique et de l'Espace (merged institute consisting of ENSICA and SUPAERO)
ISAS	Institute of Space and Aeronautical Sciences
JATO	Jet-Assisted Take Off
JAXA	Japan Aerospace Exploration Agency
KTH	Kungliga Tekniska Högskolan (Royal Institute of Technology)
LES	Large Eddy Simulation
LEX	Leading-Edge EXtension
LHS	Left Hand Side
LIC	Line Integral Convolution
LINAIR	LINear AIR vortex lattice code
LU	Lower/Upper
MDO	Multidisciplinary Design Optimization
MTVI	Modular Transonic Vortex Interaction
N+1, N+2, N+3	Next Generation, Second Generation, Third Generation aircraft technology
NACA	National Advisory Committee for Aeronautics
NAL	National Aerospace Laboratory (Japan)
NAS	NASA Advanced Supercomputer division (formerly Numerical Aerodynamic Simulation)
NASA	National Aeronautics and Space Administration
NLR	Nationaal Lucht– en Ruimtevaartlaboratorium (Dutch National Aerospace Lab)
NRC	National Research Council
NSF	National Science Foundation
NSU3D	Navier-Stokes Unstructured 3D flow solver
NTF	National Transonic Facility
ODE	Ordinary Differential Equation
ONERA	Office National d'Etudes et Recherches Aérospatiales (French Aerospace Lab)
OVERFLOW	OVERset grid FLOW solver
PANDA	Program for ANalysis and Design of Airfoils
PC	Personal Computer
PDE	Partial Differential Equation
PIV	Particle Image Velocimetry
PMARC	Panel Method Ames Research Center
PMB	Parallel Multi-Block
PNS	Parabolized Navier-Stokes
PSB	Periodic Suction and Blowing
PSP	Pressure Sensitive Paint
PSC	Personal SuperComputer
RAE	Royal Aeronautical Establishment

RANS	Reynolds-Averaged Navier-Stokes equations
RHS	Right Hand Side
RISC	Reduced Instruction Set Computing
RMS	Root Mean Square
RTO	Research and Technology Organization (NATO, follow on to AGARD, later replaced by STO)
SLOR	Successive Line Over-Relaxation
SA	Spalart-Allmaras
SARC	Spalart-Allmaras with Rotation Correction
SGS	Sub-Grid Scale
SOR	Successive Over-Relaxation
SSBD	Shaped Sonic Boom Demonstrator
SST	Shear-Stress Transport
STO	Science and Technology Organization (NATO, follow on to RTO)
SUPAERO	École Nationale Supérieure de l'Aéronautique et de l'Espace (French National Higher School of Aeronautics and Space)
TE	Truncation Error
TetrUSS	Tetrahedral Unstructured Software System
TLNS	Thin-Layer Navier-Stokes
TOGW	Take-Off Gross Weight
TSDE	Transonic Small Disturbance Equation
TVD	Total Variation Diminishing
UCAV	Unmanned Combat Air Vehicle
UPACS	Unified Platform for Aerospace Computational Simulation
URANS	Unsteady RANS
USAF	U.S. Air Force
USAFA	U.S. Air Force Academy
USC	University of Southern California
VLES	Very Large Eddy Simulation
VLM	Vortex Lattice Method
VORLAX	VORtex LAttice code
WMLES	Wall Modeled LES

Nomenclature

a	speed of sound (acoustic speed)
$\vec{a}$	acceleration vector
A	axial force *or* area
AR	wing aspect ratio, $\equiv b^2 / S$
b	wing span
c	chord *or* wave speed
c_p	specific heat at constant pressure
c_v	specific heat at constant volume
C_d	section (airfoil) drag coefficient, $\equiv d / q_\infty c$
C_D	drag coefficient, $\equiv D / q_\infty S$
C_{D_0}	zero-lift drag coefficient
C_{D_i}	induced drag coefficient
C_{DES}	constant in DES turbulence model
C_f	local skin friction coefficient, $\equiv \tau / q_\infty$
C_F	total skin friction coefficient, $\equiv D_f / q_\infty S$
C_l	section (airfoil) lift coefficient, $\equiv l / q_\infty c$
$C_{l_{MAX}}$	section (airfoil) maximum lift coefficient
C_{l_α}	section (airfoil) lift curve slope, $\equiv \partial C_l / \partial \alpha$
C_L	lift coefficient, $\equiv L / q_\infty S$
$C_{L_{MAX}}$	maximum lift coefficient
C_{L_α}	lift curve slope, $\equiv \partial C_L / \partial \alpha$
C_m	section (airfoil) pitching moment coefficient, $\equiv m / qc^2$
C_{m_α}	section (airfoil) pitching moment curve slope, $\equiv \partial C_m / \partial \alpha$
C_M	pitching coefficient, $\equiv M / q_\infty Sc$
C_{M_α}	pitching moment curve slope, $\equiv \partial C_M / \partial \alpha$
C_p	pressure coefficient, $\equiv (p - p_\infty) / q_\infty$
d	section (airfoil) drag *or* distance to the wall
D	drag
e	wing span efficiency factor *or* specific energy
E	total energy
f_x, f_y, f_z	forces in x, y, z directions
$\vec{F}$	force vector, $\equiv f_x \hat{i} + f_y \hat{j} + f_z \hat{k}$
F, G, H	conserved flux quantities in x, y, z directions

g	gravity
G	amplification factor
h	enthalpy, $\equiv e + p / \rho$
H	helicity density, $\equiv \vec{V} \cdot \vec{\omega}$
i, j, k	structured grid indices
$\hat{i}, \hat{j}, \hat{k}$	unit vectors in x, y, z directions
J	structured grid transformation Jacobian
k	thermal conductivity coefficient *or* turbulent kinetic energy
k_t	turbulent thermal conductivity
l	section (airfoil) lift *or* characteristic length
$\vec{l}$	vector along a curve
L	lift
m	mass *or* section (airfoil) pitching moment *or* doublet strength
M	Mach number, $\equiv V / a$ *or* pitch moment
M_{crit}	critical Mach number
M_{DD}	drag divergence Mach number
n	time index
N	normal force
p	pressure
Pr	Prandtl number, $\equiv \mu c_p / k$
q	pitch rate or dynamic pressure, $\equiv \rho V^2 / 2 = \gamma p M^2 / 2$
Q	Q criterion, $\equiv \left(\left\| \vec{\xi} \right\|^2 - \left\| \vec{S} \right\|^2 \right)$ *or* conserved flow variables *or* heat
r	radius
R	universal gas constant, $= c_p - c_v$
Re	Reynolds number, $\equiv \rho V l / \mu$
S	reference area *or* surface area *or* strain rate
$\vec{S}$	rate of shearing
t	thickness *or* physical time
T	temperature
u, v, w	velocity components in x, y, z directions
V	velocity magnitude
$\vec{V}$	velocity vector, $\equiv u\hat{i} + v\hat{j} + w\hat{k}$
Ψ	volume
W	work
$\vec{x}$	position vector
x, y, z	Cartesian coordinates
y^+	wall unit distance normal to surface

Greek

α	angle of attack
$\dot{\alpha}$	plunge rate
β	angle of sideslip
δ	thickness

δ_{ij}	Kronecker delta
Δ	grid spacing in DES turbulence model
ε	turbulent dissipation
ϕ	linearized velocity potential
Φ	velocity potential *or* viscous dissipation
γ	ratio of specific heats, $\equiv c_p / c_v$ *or* local circulation
Γ	circulation *or* vortex strength
η	Kolmogorov length microscale
λ	taper ratio, c_t / c_r *or* stability parameter
Λ	wing sweep *or* source strength
μ	fluid viscosity *or* Mach angle
μ_t	turbulent viscosity
ν	kinematic viscosity, $\equiv \mu / \rho$ *or* Courant number, $\equiv c\Delta t / \Delta x$
ν_t	turbulent eddy viscosity
ρ	fluid density
τ	shear stress *or* computational domain time *or* Kolmogorov temporal microscale
ω	vorticity
ξ	rotational velocity
ξ, η, ζ	computational domain spatial coordinates
ψ	stream function

Subscripts or Superscripts

f	friction
LE	leading edge
o	stagnation/total property
r	root
ref	reference value
t	tip *or* turbulent
w	wave
wet	wetted (in contact with the fluid)
$*$	nondimensional quantity
∞	freestream condition

1 Introduction to Computational Aerodynamics

Codes do not produce results, people produce results using codes.

—Dave Whitfield,
a longtime CFD code developer and user[1]

CFD simulation of a modern commercial transport (courtesy of Airbus; a full color version of this image is available in the color insert pages of this text).

LEARNING OBJECTIVE QUESTIONS

After reading this chapter, you should know the answers to the following questions:

- What is computational aerodynamics?
- What is computational fluid dynamics?
- What are the major goals of computational aerodynamic simulation efforts?
- What does it mean to be an intelligent user of computational aerodynamics?
- How have computational aerodynamic simulation capabilities changed during the past fifty years? Why?
- Can you list five ways that computational aerodynamics could be used on the design of a new airplane?
- Why is computational aerodynamics needed? Why not just determine aerodynamic characteristics with flight testing or wind tunnel testing?
- What is the difference between a real flow and a model of a real flow?
- What are the four steps of the computational fluid dynamics process?
- What is one of the largest sources of error in computational aerodynamics?

1.1 Introduction

Welcome to the world of computational aerodynamics! If you want to learn to apply what you know about fluid dynamics and aerodynamics to computationally simulate the flow around real aircraft, such as the F-16 shown in Fig. 1.1, please read on. Having the ability to predict the aerodynamics of a real aircraft can be exciting, and we hope you decide to join us as we learn the information required to perform such a simulation.

Not too many years ago, *computational aerodynamics* (CA) was reserved for those who performed numerical methods research or perhaps were program developers, or even for those developing new ways to apply computational aerodynamics to real-world analysis and design. Aerodynamicists, numerical methods researchers, and graduate students spent countless hours developing programs to apply computational aerodynamics to real-world analysis and design problems. While you may do some programming to arrive at numerical solutions in the course of your career, chances are you will be applying the techniques, programs, and models developed by others to new and interesting aircraft and aerodynamic concepts, like the X-31 flying at a high angle of attack as shown in Fig. 1.2. The vortices developing on the aircraft's canards, strake, and wing are clearly seen in this simulation.

Figure 1.1 F-16 fighter in formation flight with a computational aerodynamic simulation showing strake vortices, surface pressures, and the exhaust of the engine (courtesy of Stefan Görtz and the USAFA High Performance Computing Research Center; a full color version of this image is available in the color insert pages of this text as well as on the website: www.cambridge.org/aerodynamics).

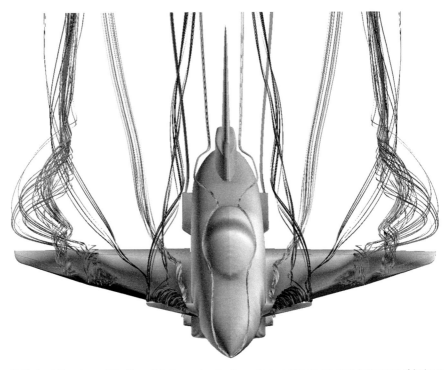

Figure 1.2   X-31 simulation at $\alpha = 16°$ with particle traces showing the various vortices on the aircraft (courtesy of Andreas Schütte of DLR; a full color version of this image is available in the color insert pages of this text as well as on the website: www.cambridge.org/aerodynamics).

Being able to adequately predict these vortices is essential to understanding how the airplane flies.

The ready and affordable availability of fast and efficient CA programs and the ever-increasing speed and memory of modern computers have made CA an integral component of modern aerospace design and analysis. You may perform applied research in your career, but it is more likely that you will be called on to analyze and design real and proposed aircraft components using the tools of computational aerodynamics. For example, perhaps you are designing an aircraft like the AV-8B Harrier that is required to hover above the ground, and you use CA tools to better understand how the vehicle will fly in this situation, such as for the simulation shown in Fig. 1.3. Having a detailed understanding of how the hovering vehicle interacts with the ground is essential to being able to design and control the aircraft in this situation, and CA tools can be extremely useful in determining these characteristics. Performing this type of simulation requires you to be able to properly understand and utilize the tools of computational aerodynamics. Since you are among the first undergraduate engineers to embark upon this world as practitioners of what has only recently become a well-established field of engineering, you have a special responsibility to take CA beyond where it has been.

The goal of this book is to make you an *intelligent user* of computational aerodynamics. What kind of user will you be? Will you know how to use the

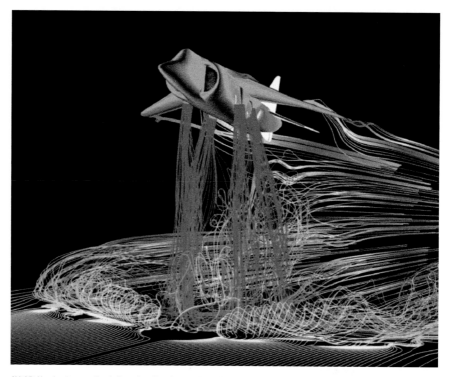

Figure 1.3   AV-8B Harrier hover simulation showing interaction of various jets with the ground plane; a full color version of this image is available in the color insert pages of this text as well as on the website www.cambridge.org/aerodynamics.

tools of CA? Will you know which tools are appropriate for which applications, or will you blindly input numbers into a program and get results that may not be meaningful? As an intelligent user, you will want to have enough knowledge and experience to understand what you are doing and why you are doing it. As someone learning to become an intelligent user, we hope you enjoy finding out about, and using, CA to accomplish aerodynamic analysis and design.

If you are going to learn about CA, you should probably know what CA is. Different people may have different opinions about a formal definition of computational aerodynamics, but a simple working definition should serve us well here: whenever you use a computer to model an aerodynamic flow-field, you are using some form of computational aerodynamics. Whether you are solving a potential flow problem using computer algebra software on a laptop, or you are using the most powerful supercomputer to perform a direct numerical simulation of the Navier-Stokes equations, you are using computational aerodynamics.

In order to better understand what all of this means, take a look at the F-16 fighter in flight as shown in Fig. 1.1 and 1.4. Fig. 1.1 shows the F-16 cruising at a constant speed and altitude, and Fig. 1.4 shows the same airplane while maneuvering. The differing colors shown on the solid surfaces of the airplane indicate the variation of pressure over the surfaces at a specific Mach number and angle of attack – these images are the result of a computational

Figure 1.4 F-16 fighter with surface colored to show the pressures plus surfaces of constant vorticity showing various vortex flow structures (courtesy of Stefan Görtz and the USAFA High Performance Computing Research Center; a full color version of this image is available in the color insert pages of this text as well as on the website: www.cambridge.org/aerodynamics).

aerodynamics simulation. Notice that the vortices and separated flow regions are visualized by surfaces of constant vorticity, which also show the shear layer created by the engine exhaust. Each one of these flow features makes an important contribution to the airplane's ability to fly well, and understanding each of these features is the job of an aerodynamicist. Because of the great power of such simulations, giving the aeronautical engineer the ability to simulate various aircraft components and the performance of the entire airplane, computational aerodynamics plays a critical role in current aerodynamic analysis and design, especially in the development process for advanced vehicles. That is why it is important to have a working knowledge of the appropriate roles and limitations of CA and to understand how and when CA can be used.

Another important distinction is how computational aerodynamics differs from *computational fluid dynamics*, and how you can tell the difference between the two. Computational fluid dynamics (CFD) involves using numerical methods and computers to solve the governing equations of fluid dynamics directly (the governing equations are a set of coupled nonlinear, second-order, partial differential equations that we will derive and discuss in Chapter 3). While CFD is a partial subset of computational aerodynamics, computational aerodynamics is broader and more encompassing than CFD, and certain applications of CFD are not directly related to aerodynamics. In the Venn diagram shown in Fig. 1.5, CFD is shown as an overlapping set with computational aerodynamics. Some of the CFD space lies outside of

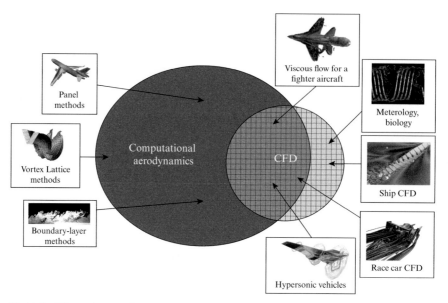

Figure 1.5   What is the difference between Computational Aerodynamics and CFD? (Don't worry if some of the concepts mentioned are not familiar – we will discuss most of them in later chapters.) Boundary layer image reprinted with permission © 1977 AIP Publishing, LLC; vortex lattice methods image courtesy of Phil Beran of the Air Force Research Laboratory; MD-11 courtesy of Analytical Methods, Inc.; F-16 courtesy of U.S. Air Force Academy; meteorology image courtesy of the U.S. Air Force; ship CFD courtesy of Fred Sterns of the University of Iowa; race car CFD courtesy of Richard Smith of Symscape © 2013; hypersonic vehicle courtesy of NASA Dryden Flight Research Center.

the domain of computational aerodynamics (including CFD applied to biological applications such as blood flow,[2] or geophysical applications such as weather prediction,[3] or for the design of ships and myriad other applications). But much of the CFD world lies inside the larger domain of computational aerodynamics – using computers to solve aerodynamic problems (including flow over airplanes, hypersonic vehicles, and even sports applications such as race car aerodynamics[4]). Computational aerodynamics also includes non-CFD applications, such as the so-called panel methods and vortex lattice methods (which will be developed in Chapter 5) used to predict flowfields assuming the flow is inviscid and irrotational, and boundary layer methods, among a host of other approximate but still valuable techniques.

1.2 The Goals of Computational Aerodynamics

Understanding the difference among these various CA approaches to analyzing problems, and knowing their strengths and weaknesses, is crucial to being a good user of CA methods. For example, the design and analysis of airplanes using CA is quite pervasive today, but the analysis can use a wide variety of tools, from low-fidelity (more approximate) models such as panel methods to high-fidelity models such as those that can predict the aerodynamics for full aircraft including viscous effects. Understanding how and why CA is being used is an important aspect of knowing how to use CA correctly. Some typical CA usages for airplanes include:

- Airframe design and analysis, i.e., development of airfoils, wings, and entire configurations, including the determination of loads for structural design
- Aircraft/propulsion system integration, i.e., the design of inlets, diffusers, and nozzles
- Performance prediction: estimation of lift and drag characteristics for determination of takeoff, climb, cruise, descent, and landing performance of aircraft
- Vehicle stability, control, and handling characteristics, i.e., to provide the mathematical model for flight simulation, and use of simulations to resolve any related flight problems
- Aeroelastic analysis, including flutter and divergence – requiring coupling with structural analysis and control system design analysis methodology
- Multidisciplinary optimization, i.e., simultaneously computing and optimizing multiple facets of an aircraft design, such as aerodynamics, structures, propulsion, performance, and mission objectives

When performing these types of calculations, it will become apparent that there are many differences between CFD and CA, which leads to the need to be able to intelligently choose which computational aerodynamics tool to use for a given task. Do you need to use a full Navier-Stokes code[i] to solve for the pressure distribution over an airfoil at low angles of attack (where the flow is attached) and low speeds (where the flow is essentially incompressible)? Can you use a potential flow code to analyze a jet fighter that is flying at very high angles of attack? The answers to these questions forms the purpose of this textbook – anyone using computational aerodynamics should be an *intelligent user* of the programs and methods available, as we will describe.

1.3 The Intelligent User

Skilled engineers can obtain valuable results using computational aerodynamics with knowledge, ingenuity, and judgment. The computer power available to every engineer today is greater than the total computing power available to the engineers who put men on the moon in the Apollo program in the 1960s, and even to those who designed the Space Shuttle during the 1970s. Significant responsibility accompanies the use of this computational capability! Unfortunately, it is possible for an engineer using this large computational power to make errors and not catch them, especially when that engineer puts too much faith and not enough skepticism in the untested results

---

[i] A *code* (or *source code*) is a collection of statements written to run on a computer.

of a massive computation. As Stan Lee, the creator of Spider-Man, said, "With great power there must also come great responsibility."[5] We believe that becoming an "intelligent user" of CA is very important and comes with responsibility; we therefore want to spend some time helping you to understand what is involved in using the great power of computers.

While numerous books are available that describe CA, guidance on the application of the methods and approaches for CA is often scarce. Many books concentrate on how to develop CA tools. It is more likely, however, that you, along with the vast majority of engineers working in computational aerodynamics today, will be *applying* existing methods and tools, not *developing* new ones. Successful practitioners must, however, understand the underlying algorithms and assumptions employed in CA methods to be *effective users*. The ability to approach aerodynamics problems using computational methods, assess the results, and make good decisions is a critical engineering skill.

An analogy may help you understand what it means to be an intelligent user of computational aerodynamics software. Traditionally there are at least three "levels" of carpenters: the apprentice, the journeyman, and the master – each level requiring increasingly more sophisticated abilities and tools. When building a house a variety of skills are required which use various tools, including hand tools and power tools. The apprentice carpenter, new to the trade, learns the basics of carpentry: using tools, joinery, planning, etc. A journeyman carpenter has mastered all of the manual skills required to perform basic carpentry and is capable of working with others to complete every aspect of a project. The master carpenter has gone beyond the level of the journeyman by showing that all aspects of using tools and building requirements are known and understood. A master carpenter has many more tools on his tool belt than an apprentice carpenter and a great deal more knowledge about how to use those tools. However, just knowing how to use a tool does not make you a master carpenter; it is the ability to apply that knowledge toward solving a problem (such as working with an architect or contractor to suggest more economical, efficient, or structurally superior methods of building a house) that makes one a skilled carpenter. Being a master carpenter means knowing how to use all the tools, knowing the right tool to use for each task, and seeing where each task fits within the project as a whole.

How does all of that relate to computational aerodynamics? Most of the history of computational aerodynamics has been devoted to developing the tools that are needed to perform computations (numerical algorithms, grid generators, flowfield visualization tools, etc., all of which will be explained in detail in later chapters). This would be equivalent to a modern carpenter spending time making his own tools, which is something that carpenters had to be able to do a long time ago. But carpenters today no longer have to make their own tools; there are many well-designed and affordable tools already available. Computational aerodynamics has reached a level of maturity where analysis can be performed by those of us who do not want to develop tools, but want to apply the already available tools. We want to build the house! This book is designed to help you become an intelligent user of computational aerodynamics tools, rather than a developer of computational aerodynamic tools. If you want to learn to develop the tools, there are numerous excellent books available, including those by Blazek,[6] Pletcher, Tannehill, and Anderson,[7] Hoffmann and Chiang,[8] and Löhner.[9] However, if your goal is to learn how to use the existing tools effectively, then please continue reading.

To become a "master user" of computational aerodynamics tools, you will find that you need to have at least a rudimentary understanding of aerodynamics, numerical methods, experimental procedures, and computer systems. This "entry-level" knowledge has traditionally forced computational aerodynamics into the realm of graduate studies at universities, since the average undergraduate student would not be able to become "expert" enough in all these areas to make any meaningful headway. Now that computer capabilities have advanced significantly, and various codes are becoming mature, undergraduate students should be able to take well-developed CA tools and apply them. The danger is that these easy-to-use codes can be used by people who do not understand enough about them, leading to errors that may be costly (or even dangerous). Knowing how much you need to know, however, is quite difficult (especially if you do not know what you need to know!). Finding the balance between the two is one of the important goals of this book, which will show you what you need to know about the basic physics and numerical methods involved in CA, code usage (other than how to make it run), and computer-related issues that you should understand in order to use the code efficiently and effectively. Also, we will make you aware of experimental and theoretical aspects of aerodynamics. Maybe you will not have to know all of these things in your career, but we believe having a basic knowledge is essential to being an intelligent user. These are the types of issues you should resolve before you "blindly" begin using computational aerodynamics programs.

Computational Aerodynamics Concept Box

The Garbage In/Garbage Out (GIGO) Syndrome of Computer Usage

"On two occasions I have been asked, 'Pray, Mr. Babbage, if you put into the machine wrong figures, will the right answers come out?' ... I am not able rightly to apprehend the kind of confusion of ideas that could provoke such a question."

– Charles Babbage, *Life of a Philosopher*

Actually, we have heard this question from our students for years! We all have seemingly developed a belief that computers are smart and capable of doing anything. This had led the average person to somehow believe that computers are smarter than the people who program them or the codes they run. Simply put, however, computers are really quite dumb, since they only do *exactly* what you tell them to do. If you put Garbage In you will get Garbage Out, which is known as the GIGO syndrome.

This not only applies to using computers, it also applies to programming computers. Students and practicing engineers alike may believe that the computer knows everything, even things that they have not "told" the computer (including correct input information or unknown information not supplied by the user). Over the years, the syndrome has spread to include those who make decisions based on the computer results. Supervisors, managers, accountants, and politicians have all been guilty of basing decisions on faulty, incomplete, or imprecise data. These perceptions are often so pervasive that a new GIGO syndrome has developed called Garbage In, Gospel Out. This is based on our mistaken tendency to believe and trust computerized results and blindly accept whatever the computer says. Computers are very good on certain things, such as processing large amounts of data very quickly, but they are not very good at knowing what we want unless we tell them completely and correctly.

A very troubling example of GIGO happened to Wall Street bankers during the financial collapse of 2008. Federal regulations required that each bank and investment firm have computers perform risk management analysis for every investment. These computer programs are supposed to raise a red flag when the bank is over-exposed to certain risky investments, and then the bank is supposed to reduce the amount that is invested in the risky enterprise. "In other words, the computer is supposed to monitor the temperature of the party and drain the punch bowl as things get hot," according to *New York Times* reporter Saul Hansell.[10] Unfortunately, "the people who ran the financial firms chose to program their risk-management systems with overly optimistic assumptions and to feed them oversimplified data. This kept them from sounding the alarm early enough.... Lying to your risk management computer is like lying to your doctor. You just aren't going to get the help you really need." Bad data were put into the computers (Garbage In) and the risk management programs did not warn the bankers that they were in trouble (Garbage Out).

1.4 A Bit of Computational Aerodynamics History

Computational aerodynamics is a relatively young field of engineering. While computers have been around for many decades, the capacity of the computers to perform high-level calculations (the kind required for many CA applications) has only recently become sufficient to perform the calculations and display the results for full aircraft of interest. In fact, it would be safe to say that in the past CA has helped to push the development of larger, faster computers. A fuller understanding of this will be evident once you work your way through the next few chapters, but for now we can see the progress that CA has made by looking at some examples from the past; this should help to put things in perspective.

The original use of computers for CA predictions occurred during the 1960s primarily for various potential flow methods. These methods included the panel method and the vortex lattice method, both of which will be described in greater detail in Chapters 4 and 5. Panel methods allowed researchers to represent an airplane by a series of panels, where each panel

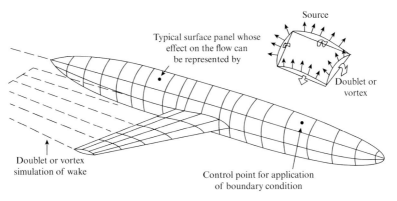

Source

Typical surface panel whose
effect on the flow can
be represented by

Doublet or
vortex

Doublet or vortex
simulation of wake

Control point for application
of boundary condition

Figure 1.6 Schematic showing how an airplane is represented by panels and boundary conditions; initially used for design in
the 1960s (Ref. 11).

had a potential flow function (such as a source, vortex, or doublet), and then
boundary conditions were applied (see Fig. 1.6 for a schematic representation of a panel method[11]). This required the solution of a large system
of linear equations that could be put into matrix form. This is where the
computer comes in! Matrix algebra hand calculations for even the smallest number of equations can be quite tedious and take a great deal of time.
When the number of equations gets into the hundreds or even thousands,
the work becomes far too intense and lengthy. Solving these matrices was
the first computational aerodynamics problem, and the computers of the
day were taxed to their limits working to obtain solutions. Very quickly after
being able to find solutions for airfoils and wings, engineers extended these
methods to full aircraft calculations – these methods are still useful today for
defining cruise aerodynamics where flow separation is not important.

As computer memory and processing speeds grew larger and faster, the
potential flow methods became easier and easier to use, and researchers
began trying to solve more difficult CA problems. In the late 1960s and early
1970s, many researchers started to develop numerical methods to calculate
the flowfields for airfoils and wings using inviscid CFD methods, and once
again, these methods required the full capacity of existing computers. By
the mid- to late 1970s, these methods had also been extended to full aircraft
and inverse design approaches (using the code to design the wing), and once
again CA was looking for even faster computers. An example of the transonic design of a wing using inviscid methods is shown in Fig. 1.7.[12]

Finally, since CA practitioners were always attempting to solve more and
more challenging problems, methods were developed to find the viscous flow
solutions for airfoils and wings. Computers of the 1960s could not have performed these calculations very efficiently, but by the mid-1970s, solutions
were being obtained for airfoils and wings. Viscous solutions for full aircraft
(using turbulence models, which we will discuss in Chapter 8) were finally
being obtained during the mid-1980s (see Fig. 1.8 for an example of an F-16
at transonic conditions).[13] Once these initial solutions were obtained, more

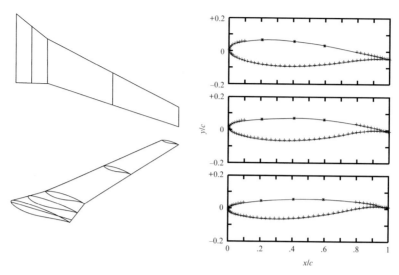

Figure 1.7 Cessna Citation III wing geometry for inviscid transonic calculations; design done in 1986 (Ref. 12, Courtesy of Terry Holst of NASA Ames Research Center).

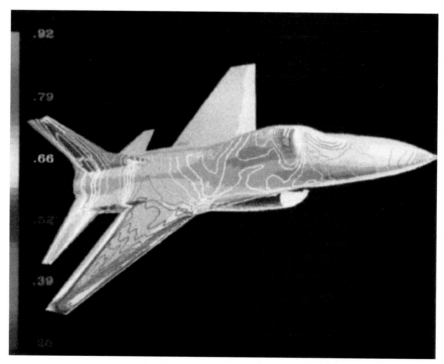

Figure 1.8 Pressure coefficient contours on the upper surface of the F-16A for $M = 0.9$, $\alpha = 6.0$ deg, and $Re = 4.5 \times 10^6$; prediction made in 1990 (Ref. 13; reprinted by permission of the American Institute of Aeronautics and Astronautics, Inc.).

and more complex configurations were attempted with ever more complex turbulence models, and some researchers even started performing unsteady calculations for problems such as aileron buzz or aeroelasticity applications. It has only been within the past decade or so that these full aircraft simulations could be performed for aircraft that were undergoing a maneuver,

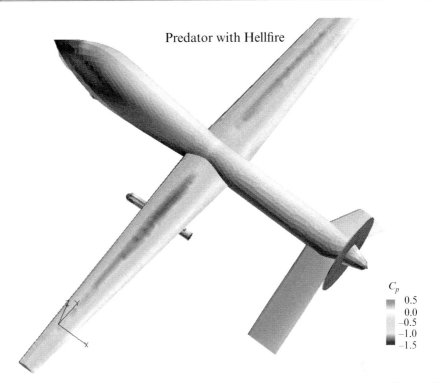

Predator with Hellfire

C_p
0.5
0.0
−0.5
−1.0
−1.5

Figure 1.9   Pressure coefficient variation on the upper surface of the Predator with an under-wing missile; prediction made in 2002 (Ref. 14; reprinted by permission of the American Institute of Aeronautics and Astronautics, Inc.).

again because the computer capabilities had to continuously improve before researchers could obtain solutions.

In addition to these groundbreaking applications of CA to ever more complex aircraft configurations using ever more complex governing equations, other researchers were finding ways to apply these methods to more practical aerodynamic applications; the technical literature from over the past few decades is filled with examples of CA being used in analysis and design, such as for the Predator drone aircraft shown in Fig. 1.9.[14] Sometimes, however, these researchers found that although we could perform certain calculations, we did not always get the best results. Bradley and Bhateley[15] reviewed the status of CA in 1983 and proposed a classification scheme in terms of the types of flowfield being predicted. They divided flows into seven categories and defined the capability to compute each type of flow over a variety of geometries of increasing complexity. These categories included attached flows, separated flows, vortex flows, and other more complex flow types. Their assessment at the time was that CA had a long way to go before being able to accurately predict most flowfields, and they were right!

An AIAA Progress Series volume edited by Henne described the state of the art in 1990 through many aerodynamic applications.[16] For example, "normal" two-dimensional airfoil analysis and design could be done reliably using the computational methods of that time, without wind tunnel validation, but

Figure 1.10 Shock wave visible against clouds on the Lockheed L-1011 wing (Ref. 17; Courtesy of NASA Dryden Flight Research Center; photo by Carla Thomas).

other complex flows required a great deal of experience to predict and analyze (such as the transonic flow over an airliner wing where shock waves are present – see Fig. 1.10 for an example on an L-1011 wing[17]). The need for engineers capable of exercising judgment when using CA results was important then, and it is still important.

For example, the three-stage, air-launched winged space booster Pegasus was designed using computational and engineering methods alone, with no wind tunnel tests being conducted (see Fig. 1.11).[18] The initial launches were successful, and it appeared that the accuracy of the aerodynamic analysis was adequate for this unmanned vehicle. However, after a subsequent launch failure, a dispute arose over whether the aerodynamics had been adequately predicted or whether the control system was too sensitive to imperfections in the aerodynamic model (the problem was determined to be in the lateral-directional aerodynamic characteristic of the vehicle, something which the methods of the day predicted to be slightly stable, while the actual vehicle turned out to be slightly unstable). Understanding the uncertainties in the analysis required careful evaluation, since no prediction is exact. Subsequently, wind tunnel testing had to be done to clarify the situation. The important lesson here is to remember that we cannot do everything equally well, and you should understand when and why CA tools might not perform up to your expectations.

Figure 1.11 Pegasus space booster after launch from B-52 with a T-38 chase plane (Courtesy of NASA Dryden Flight Research Center).

Profiles in Computational Aerodynamics: Russell Cummings

"I have loved airplanes since I was about 12 years old while growing up in Santa Cruz, California – I can thank my uncle who was in the Air Force for that. He came to visit us on his way to serve in the Vietnam War, and during his stay he bought me just about every model of every Air Force fighter that he could find. Putting together those models and hanging them from my bedroom ceiling meant that I could look at those airplanes (like the F-100, F-101, and F-104) every night while I was going to sleep, and dream about either flying or designing them someday.

"That eventually took me to Cal Poly to study aeronautical engineering in 1973, which was a wonderful place to study about airplanes – the department's hangar was full of them! Nearly limitless laboratory and project experiences meant that I got to study just about everything that went into making an airplane, including aerodynamics, structures, propulsion, and flight dynamics, but it was aerodynamics that really caught my attention. After completing my BS and MS degrees at Cal Poly, I went to work at Hughes Aircraft Company in Los Angeles in 1979, working in the Aerodynamic Design Department. Hughes Aircraft Company had been started by Howard Hughes in the 1930s, but really came to maturity as a major aerospace defense contractor in the 1950s. Hughes was a wonderful place to work, with a great deal of freedom to explore interests and to find where you wanted to 'fit in.' I was fortunate to work in the Aerodynamics Section, where we not only predicted and evaluated aerodynamics, but we also worked closely with wind tunnel testing and flight performance. One of the reasons I went to Hughes was because I wanted to have that level of broad experience in my first job. During my time at Hughes, among many other things, I was exposed to a wide variety of aerodynamic prediction capabilities, including semi-empirical methods (which we will discuss in Chapter 2), panel methods (which we will discuss in Chapter 5), and computational fluid dynamics (CFD, which we will discuss starting in Chapter 6). Hughes had an incredibly supportive educational program known as the Hughes Fellowship Program – I immediately applied to the program and started studying aerodynamics at the University of Southern California, eventually receiving a PhD in aerospace engineering. While I truly enjoyed working at Hughes, my true desire was to return to academia to teach someday, so in 1986 I accepted a position back at Cal Poly.

"Barely having become acquainted with life as a professor, I was offered a two-year position as a National Research Council post-doctoral researcher in CFD at NASA Ames Research Center, and I spent a wonderful time doing research on high angle of attack aerodynamics with Lew Schiff in the Applied CFD Branch. The CFD groups at NASA Ames during the late 1980s were a vibrant and exciting place to work, with many of the leaders of CFD development working there. Walking down the halls when I first arrived meant seeing the offices of people I had studied in school, including: Bill Ballhaus (who was the Ames Research Center director by that time), Harvard Lomax, Ron Bailey, Joe Steger, Paul Kutler, Terry Holst, Peter Goorjian, Denny Chausee, Richard Beam, Robert Warming, Marcel Vinokur, Bill McCroskey, Dochan Kwak, and Tom Pulliam (and those were just the more experienced researchers – the younger CFD researchers at the time were making major contributions to CFD as well). I felt like my CFD education really happened while I was at NASA Ames working with and around these great people.

"Upon my return to Cal Poly I started a long and productive research program with NASA Ames, which allowed me to work with many talented graduate students who were interested in CFD. Helping each of them get started in this exciting field was truly rewarding. In 2001 I left Cal Poly and came to the U.S. Air Force Academy to work with Scott Morton, Jim Forsythe, and Doug Blake, who were conducting equally exciting applied CFD research, as well as with John Bertin, who was the aerodynamics discipline director at the time. Our collaboration led to the creation of an undergraduate computational aerodynamics (CA) course at the Academy, and this book is a direct outgrowth of that course – we could not find a good CA book for undergraduates, so we started writing our own! I hope your experiences with CA are as rewarding and interesting as mine have been."

Russ Cummings lives on a ranch in Larkspur CO with his wife Signe, daughter Cornelia, and son Carl, as well as horses, sheep, alpacas, and other assorted animals.

Case studies also provide another way to understand computational aerodynamics capabilities. In 1986, several aerodynamic design problems ("anomalies") that arose in flight on various transport aircraft were identified, as shown in Table 1.1.[19] These problems were examined to determine if the use of the computational aerodynamic tools of the time would have

Table 1.1 Aerodynamic "anomalies" on Douglas airliners that CA would not have been able to solve at the time (Ref. 19)

Year	Airplane	Aerodynamic Problem
1950s	DC-8	Drag rise prior to drag divergence Mach number
1963	DC-8	Engine pylon compressibility drag increase
1965	DC-9	T-tail design to combat deep stall
1968	DC-10	Nacelle strake design to increase high lift C_{Lmax}
1970s	MD-80	Fuselage nose strake design

(a) DC-9 with T-tail

(b) KC-10 nacelle strake

(c) MD-83 fuselage strake (visible below cockpit)

Figure 1.12 Three examples of airplane design features that were not originally well modeled with CA (Ref. 19; DC-9 is a NASA photo, KC-10 is a U.S. Air Force photo by Airman 1st Class Meghan Geis). (a) DC-9 with T-tail; (b) KC-10 nacelle strake; (c) MD-83 fuselage strake.

predicted them. The conclusion at that time was that uninformed computational aerodynamics (true predictions with no a priori knowledge of the anomaly) would not likely have prevented these problems. They included subtle aspects of airfoil and wing aerodynamics, such as the ability to compute deep stall characteristics of T-tail aircraft (such as the DC-9 shown in Fig. 1.12a), the use of nacelle strakes to improve high-lift characteristics (such as the KC-10 shown in Fig. 1.12b, where the strake vortex is visible), and fuselage strakes to improve high angle of attack directional stability (such as the

MD-83 shown in Fig. 1.12c). Though starting to become dated, these results provide valuable insight into the way aerodynamicists think about design. Since CA is playing an ever-increasing role in vehicle design, readers should become familiar with the experiences of those who have pioneered the use of computational aerodynamic tools, including Tinoco,[20] who reviewed the computational aerodynamics situation at Boeing in 1998; and Busch,[21] who provides a good review of the use of CFD in the design of the YF-23.

Computational Aerodynamics Concept Box

The Advent of Commercial CA Codes

As computers matured through the 1980s and 1990s, most CA codes were developed in-house (at aerospace companies or government labs) or by individual users (at universities). These codes were often limited in scope and designed for specific applications, rarely capable of being applied to a wide variety of applications. Over time, however, a number of small start-up companies began to develop commercially available software for CA simulations and led to the advent of codes that could be purchased for use. Companies such as ANSYS Fluent, Metacomp, Cobalt Solutions, Aerosoft, and CFD Research Corporation (among others) developed codes that could be easily used by non-experts.

SR-71 flow solution from Metacomp Technologies using the CFD++ flow solver (a full color version of this image is available on the website: www.cambridge.org/aerodynamics)

These codes are often driven by Graphical User Interfaces (GUIs) and are relatively simple and straightforward to run (assuming you have the computer power to use them). The codes are very powerful and are often written for a wide variety of applications. The development of these commercial codes represented a paradigm shift in CA, since individual users no longer had to write and apply their own software in order to get results.

While some people still write their own codes (typically for academic research), the majority of applications today use these commercial codes. These commercial codes are relatively easy to use, but with that ease of use comes responsibility (see the GIGO Box)!

1.5 What Can Computational Aerodynamics Do Today and Tomorrow?

Now that we have briefly discussed the background of CA usage from the 1960s to about the year 2000, we can begin to gain a fuller understanding of what CA can do today and what we hope it will be able to do tomorrow. Again, we will see that CA is constantly trying to do more and more detailed applications, which constantly requires newer and better computer systems. The paradigm of CA requiring better computers to perform more challenging applications will be with us for decades to come!

1.5.1 Commercial Aircraft Applications

Modern commercial aircraft, like the Boeing 787 and the Airbus A380, used CA extensively in the design phase of their development. The high level of CA usage during the design of these airplanes has led to significant reductions in design time, as well as improvements in aerodynamic efficiency. There are many examples of how CA application has greatly improved the design of modern aircraft, but specifically there are two flight conditions that are the most critical: high subsonic/transonic cruise conditions and high-lift configurations for takeoff and landing. Table 1.2 describes how the application of CA has greatly improved aircraft design, including the reduction of drag, the design of flaps and leading-edge devices, and the design of wing-tip devices.

The first critical flight condition occurs at transonic speeds, when shock waves start to appear on the airplane. These shock waves cause an increase in drag due to the losses associated with shocks and the disturbance to the boundary layer imposed by the shock wave pressure gradient. This is a limiting case for transonic cruise aircraft, and it is critical to be able to predict the drag rise conditions accurately. Transonic airfoils that were developed in wind tunnels during the 1960s and 1970s are now designed reliably using the computational codes of today.

Table 1.2. Commercial aircraft design improvements due to CA usage

Aircraft	Year Reported	Aerodynamic Improvement from CA
Airbus A320	1986	Used inviscid/viscous methods coupled with extensive wind tunnel testing to design lighter, more affordable high-lift system[22]
Boeing 777	1995	Used viscous prediction methods (mainly 2D sections) to obtain some reduction in high-lift design cycle, but still required extensive experimental testing[23]
Airbus A380	2003	Full-span high-lift system design with single-slotted flap[24]
Airbus Long Range Configuration	2006	Wing-tip "shark" shape with about 5% reduction in drag, including weight and operating cost considerations[25]
Boeing 787	2008	2%–3% aerodynamic improvement; cut wind tunnel testing time in half since design of Boeing 767; uses single-slotted flap system[26]

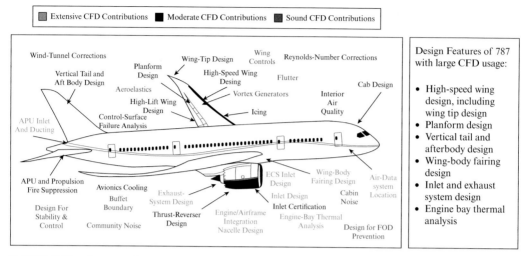

The other important flight condition occurs at low speeds, where takeoff and landing requirements lead to the need to predict the maximum lift coefficient. In this case the use of elaborate high-lift systems leads to complicated geometries and complicated flowfields. Both of these key flight conditions (transonic cruise and low speed, high lift, flight) are difficult to predict and demand the use of high-fidelity methods. Figure 1.13 shows an example of how CA has been used on the design of the Boeing 787,[26] which represents a large increase in CA usage for aircraft design over earlier aircraft. Michael Garrett of The Boeing Company says that "the use of CFD tools has allowed Boeing engineers to address a wide range of design challenges, including traditional wing design, the even distribution of cabin air and reduction in overall airplane noise."[27]

Specifically, the 787 design included use of CA tools in areas that were considered "mature" (the application was previously well understood), such as providing wind tunnel corrections, the design of wing planforms and vertical tails, and the design of the fuselage. Other areas of application are still considered "immature" (not as well understood), such as using CA for determining stability and control characteristics, determining flutter problems with wings, and understanding the noise levels produced by the aircraft.

Airbus has also greatly expanded the use of CA in evaluating the designs for the Airbus A380 and A350 aircraft, including running entire aircraft CFD simulations for use in performance and handling quality predictions. "Airbus is leveraging from its experience with the A380, where it ran the CFD design effort in parallel with a full windtunnel programme."[28] Their experiences with the A380 allowed them to reduce the amount of time spent

Table 1.3. Future improvements in computational aerodynamics for various applications

Year	Application/Observer(s)	Future CA Improvements Required
2002	High lift/Rumsey and Ying[29]	Quantification of requirements for accurate prediction of C_{Lmax} Additional experimental databases for CA validation Improved slat wake predictions Improved turbulent shear stress predictions Exploration of unsteady effects on high-lift prediction
2002	High lift/van Dam[30]	Fast grid generation Improved turbulence models Boundary-layer transition models Improved prediction of separation bubbles
2005	Aircraft Design/Fujii[31]	Improved prediction of scaling effects (low to high Reynolds numbers) Rapid turn-around predictions using CFD for conceptual design
2006	Wind Tunnel/Bosnyakov, et al.[32]	Improvement of numerical methods for more accurate prediction Improved representation of turbulence models in codes Better representations of boundary conditions for wind tunnels
2006	Scaling/Bushnell[33]	Improved prediction of separated flows Better predictions of transonic flow, high-lift aerodynamics, spin/stall, and buffet
2007	Aeroelasticity/Henshaw, et al.[34]	Usable nonlinear aeroelastic methods are required for accurate predictions Aeroelastic evaluation can be done in stages with varying levels of fidelity Affordable grid movement schemes are crucial Improved fluid/structure coupling schemes
2008	Maneuvering Aircraft/Cummings, et al.[35]	Accurate representation of unsteadiness via grid/time-step convergence Improved turbulence models for highly separated flows

conducting wind tunnel tests on the A350 by 40 percent and also allowed them to conduct their design cycles faster.

The applicability of current CA methods for aircraft design and development of aerodynamic concepts is still being assessed. CA is currently used for a wide variety of commercial aircraft applications, as was shown for the 787 and A380, but significant improvements still need to be made in CA capabilities, as shown in Table 1.3. Improvements in the use of CA for prediction of high-lift flows and aircraft design efforts include a long list of future capabilities, such as improved methods for grid generation (see Section 1.8.2 and Chapter 7), more advanced turbulence models (see Chapter 8), and the ability to accurately predict boundary-layer transition (also in Chapter 8). The use of computational tools in the support of wind tunnel testing, a very important application of CA, also places demands on future CA development, including improved numerical methods and realistic boundary conditions. The process for integrating CA tools into estimating the aeroelastic issues that arise with modern aircraft also requires better coupling of fluid/structure interactions and the use of usable (affordable) nonlinear prediction methods. Finally, the use of CFD in stability and control predictions

Figure 1.14   A pitching F-16 with condensation making strake vortices and wing flow separation visible (U.S. Air Force photo by Josh Plueger).

for maneuvering aircraft is only now becoming feasible and places high-level requirements on grid fineness and time-accurate flow predictions.

You can see that many of these advances in CA have taken place within the very recent past, which shows that the capabilities are still evolving and improving. Another thing you will notice is that there are many areas within CA where we still have not reached full maturity. For example, within the area of predicting stability and control characteristics for maneuvering aircraft, the ability to accurately predict turbulent flow is critical to obtaining reasonable predictions. Very recent work has shown that modern CFD tools still need a great deal of improvement before we can confidently predict these types of flows.[36] In addition, codes combining optimization methods with other disciplines (called Multi-Disciplinary Optimization, or MDO) are important requirements for future improvements in aircraft design (see Chapter 10 for more details).

1.5.2 Military Aircraft Applications

Fighter aircraft have even more difficult requirements and critical flight conditions than commercial aircraft. These arise due to the sustained and instantaneous maneuver requirements across a range of Mach numbers that fighters may experience. For these types of maneuvers we expect large regions of massively separated flow, and the airplane flight path can vary rapidly, which results in a highly unsteady flowfield (as can be seen for the F-16 in Fig. 1.14). The high angle of attack characteristics and determination of acceptable control power are still challenging for all levels of computational aerodynamics codes.[37]

Figure 1.15   Example of computational aerodynamics predicting complex transonic flow over the F/A-18E/F; notice the highly separated flow on the back portion of the wings, which is caused by transonic shock waves (Ref. 38). (Courtesy of Jim Forsythe and the USAFA High Performance Computing Research Center; a full color version of this image is available on the website: www.cambridge.org/aerodynamics).

A recent example of a case where wind tunnel testing was not able to adequately define aerodynamic issues is with an F/A-18E/F Super Hornet that experienced "wing drop" or abrupt wing stall at transonic speeds. Only when computational fluid dynamics was applied to the appropriate flight condition was the cause of the wing stall more fully understood (see Fig. 1.15).[38] This problem turned out to be a complex, unsteady, shock/vortex interaction (visible in Fig. 1.15) that could cause asymmetric flow over the wings, leading to an abrupt change in rolling moment.

Computational Aerodynamics Concept Box

The Difficulty of Drag Prediction[39]

Whether in conceptual design or preliminary design, drag prediction is a key area of concern. This is true both for the managers "betting the company" on the design (because designing a commercial airliner requires so much monetary investment) and for the aerodynamicists making the predictions. Drag determines the size and number of engines on the aircraft, the range, and the maximum speed … accurate drag prediction is very important! To date, companies still rely on experienced aerodynamicists, wind tunnels, and computations for predicting the drag of an airplane. Although we can use our experience together with test and computations, there is still considerable uncertainty associated with the computational prediction of drag.

A large collaborative effort to assess the ability of CFD codes to predict drag has been ongoing. Five Drag Prediction Workshops have been organized by the American Institute of Aeronautics and Astronautics (AIAA), where participants used their best codes and practices to predict a set of cases, and the organizers collected and compared the results. A summary of the workshop results has appeared in several papers. The results show that more progress is needed before codes can accurately predict drag (especially in the area of grid generation), since the drag estimates are greatly dependent on the quality of the grid. However, students should be encouraged that the insights obtained from examining the predicted flowfields can be used to improve designs. The key area of concern identified in these workshops is an important component of CFD so far unmentioned here – the specification of a "grid" to use for the numerical solution of the governing equations. This will be an important part of the CFD discussion to follow. So, in addition to the selection of governing equations and definition of the geometry, the numerical grid (or mesh) is another key part of computational aerodynamics.

Although absolute values of drag are still considered difficult to compute using CFD, an AGARD Panel on CFD and Drag suggested that CFD-based drag prediction was very effective when "embedded in an increment/decrement procedure involving experimental results for complete configurations, and CFD results for simplified configurations."

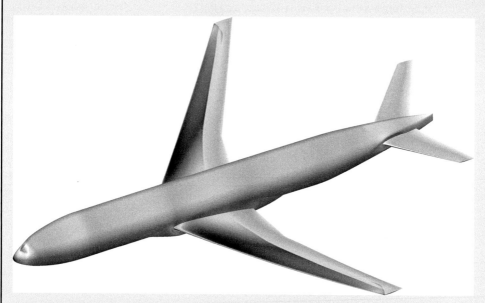

Common Research Model wing/body/tail drag prediction (courtesy of Dimitri Mavriplis of the University of Wyoming; a full color version of this image is available on the website: www.cambridge.org/aerodynamics)

1.6 Integration of CA and Experiments

Although widely used today, computational aerodynamics is still a developing capability and still depends on classical aerodynamic theory and experimentation with expensive models to give confidence in the resulting predictions. In fact, in order for engineers to validate their predictions for the full range of situations shown in Fig. 1.5, wind tunnel testing and flight testing will play an important role in aerodynamics for many years to come. Many advances are also being made in experimental technology at this time, which make the experimental results even more critical for aerodynamic understanding. Flight testing, wind tunnel testing, and computational aerodynamics are complementary capabilities, and aircraft development programs require the use of all three.[40] Hancock pointed out that advances in computational capability have actually led to increased demands on experimental aerodynamics in some cases.[41] For example, the evaluation of components and the overall design for the F-35, a new multipurpose fighter intended to replace the F-16 and F-18, involved extensive wind tunnel testing at a number of locations, including 3,500 hours at Arnold Engineering Development Center, as well as significant wind tunnel testing of the Short Take-Off and Vertical Landing (STOVL) variant at BAE Systems (formerly British Aerospace), NASA Ames, NASA Langley, and DNW, the German-Dutch wind tunnels.[42]

With current flight testing expenses on the order of $100,000 per hour and the cost of wind tunnel testing running as high as $10,000 per hour (or more), CA is an increasingly important tool in directing the testing to the most "valuable" areas where the most unknowns exist – not to mention that CA can help to reduce the risk to some very expensive test articles (such as wind tunnels or full-scale aircraft). As CA has progressed, and the ability to predict more and more details of the flowfield has increased, CA has shown the need for even more details from the experiments. Examples of the interplay and integration between computational and experimental work have been described by Neumann[43] and Cummings and Morton.[44] These include code validation at the very minimum (providing high-quality experimental data for showing how well CA can replicate results) but also can be much more than that. For example, CA can be used:

- before wind tunnel testing to decide where and how to measure (to know in advance where the flow is interesting), as well as to estimate loads to insure safety of the model design as it is used in the wind tunnel
- to determine how accurate and detailed the wind tunnel models need to be
- to check whether the experimental measurements make sense during the test
- to reduce the amount of time testing by involving CA in the design process

While it is critical to have good experimental data for validating CA predictions, it is equally important for CA practitioners to understand current

Figure 1.16   An engineer checks the model of the F-35 in the USAF Sixteen Foot Transonic Wind Tunnel prior to testing; notice the model support system and wind tunnel walls, all of which must be taken into account when making measurements (U.S. Air Force photo from Arnold Engineering Development Center).

experimental techniques, capabilities, and limitations. We believe this is important for the good practice of computational aerodynamics. Once you understand the intricacies of experimental data (and know what is required to obtain quality data), then you are better prepared to apply CA to useful problems. Understanding uncertainty for experimental data can be more complicated than most people think, since "typical wind tunnel data corrections applied to test data include corrections to model attitude (angle of attack), corrections to the model forces and moments, and corrections to the wind tunnel operating conditions."[45,46] These corrections are illustrated in Fig. 1.16, where the F-35 is being prepared for testing in the USAF sixteen-foot transonic wind tunnel. The model is supported by a sting which comes out the rear of the fuselage; the presence of the sting and model support system must be accounted for when measuring lift and drag on the vehicle. At very high angles of attack (as shown in this picture), the sting can bend slightly, as well as change the flow direction of the air in the vicinity of the model; both effects can change the results of the test. In addition, the wind tunnel walls impact the flow around the model and must also be taken into account.

For example, a number of researchers are using CA to attempt to resolve scale effects between wind tunnel testing and flight testing, so that wind tunnel data can be extrapolated correctly to full-scale conditions. This is especially important when considering dynamic motion in wind tunnel testing where, to get good comparison, nondimensional rotation rates have to be matched to flight testing rates.

Computational aerodynamics is heavily used in conjunction with wind tunnel tests once the design program moves into a demonstration/validation stage. Often the codes are used to design incremental modifications to the wind tunnel-tested configuration. Many examples of configuration refinements and modifications using computational aerodynamics have been documented at this stage. Perhaps the best example is the design of the nacelle-wing installation on Boeing transports, and especially on the re-engined Boeing 737.[47] In recent years CA has progressed from being simply a tool to obtain incremental aerodynamics (comparisons of one configuration to another) and has been shown to make predictions close enough to wind tunnel and flight test data to be useful for more advanced design applications. A good example of this is the excellent agreement between CA predictions for the F-16 and both wind tunnel and flight test data.[48] More details about these new capabilities will be presented in Chapter 10.

1.7 Design, Analysis, and Multidisciplinary Optimization

At the conceptual design level, decisions are made based on rapid evaluation of the performance *potential* of a variety of configurations, rather than the detailed study and development of a particular design (conceptual designers want to know what level of performance can be expected from a configuration after the aerodynamic design is done). Computational aerodynamics needs to be able to produce results for new configurations on a daily basis during the conceptual design phase. It takes much more time to make high-fidelity (very accurate) simulations for complex geometries and flight conditions. It may require weeks or even months to obtain reliable solutions over a completely new configuration. This would seem unacceptable in a conceptual design environment (see the Computational Aerodynamics Concept Box for an explanation of the differences between conceptual, preliminary, and detailed design).

So far we have been discussing computational aerodynamics in terms analogous to evaluating aerodynamic characteristics in a wind tunnel. If we want to develop new aerodynamic shapes computationally, we can change the shape and then recompute the flowfield. This can be a very effective application of CA and still plays a major role in aerodynamics. However, once we employ computational simulation appropriately, we can start to think about aerodynamic design in a fundamentally different way by using various numerical optimization schemes to determine a design. This is the most efficient use of computational aerodynamics in vehicle design.

Since the early days of theoretical aerodynamics research, the problem of finding the airfoil shape corresponding to a prescribed pressure distribution has been studied (this is known as the "inverse problem" since CA is typically used to find the pressure distribution for a prescribed geometry), and a variety of theoretical methods were developed to achieve this goal. They required significant computational effort and were early examples of computational aerodynamics.[16] An alternate approach to aerodynamic design was proposed once numerical optimization began to be developed. In this approach, an aerodynamics analysis code is coupled to an engineering optimization code, and the resulting program is used to optimize "something."[49] The "something" could be drag minimization or even the difference between a target and existing pressure distribution.

Once the optimization code is coupled to an aerodynamics code, other disciplinary codes can be included, such as a structural code or an aeroelasticity code. This is an area in computational design known as *multidisciplinary design optimization* (MDO). The work in inverse design and optimization methods has been successful but has also revealed a number of new issues that need to be resolved. Among other issues, the computational cost and time required to perform design using high-fidelity simulations is very high. Two current approaches address ways to achieve the goal of affordable CA-based design. The first is the use of surrogate models to represent disciplines in MDO.[50] The second is the extension of computational aerodynamics methods to include sensitivity information for use in optimization. See the survey by a pioneer in this area, Antony Jameson,[51] for a discussion of how this is done. As was mentioned previously, these types of simulations require state-of-the-art computer systems and methods in order to yield results in reasonable (and useful) amounts of time.

Computational Aerodynamics Concept Box

The Three Phases of Aircraft Design

This description of the three phases of the aircraft design process was largely written by a colleague of ours: Steve Brandt of the U.S. Air Force Academy.

The design process is usually described as having three phases. The initial phase is called *conceptual design*. The goal of conceptual design is to select a workable concept and optimize it as much as possible. The accompanying figure illustrates sketches made by Kelly Johnson in 1937 of nine possible configuration concepts for a twin-engined fighter that eventually became the Lockheed P-38 Lightning. Notice that the sketches, though crude, clearly communicate the fundamental configuration choices which Johnson's design team considered. A photo of a production P-38 is included for comparison.

Once a concept is selected, further iterations through the design cycle are needed to select the materials and work out the dimensions, structures, and functions of the design. Computer simulations are performed and physical models of the design are built and tested.

This phase is usually called *preliminary design*. Once the preliminary design phase is complete, *detail design* begins. In the detail design phase, the product is prepared for production. The design is described in complete detail, and the process by which it will be manufactured is also designed. A detailed set of drawings, a materials list, and a detailed cost estimate are prepared. In later iterations through the design cycle a prototype is often built and tested. The results of these tests are fed back into further design cycles to improve the performance and manufacturability of the design. Even after the design is in production, the design cycle continues. Information that is learned from continued testing of prototypes and initial operation of the early production models is typically used to further refine and improve later versions of the product. In some cases a design will continue to be improved for many years and even decades after the first versions have been produced and gone into service.

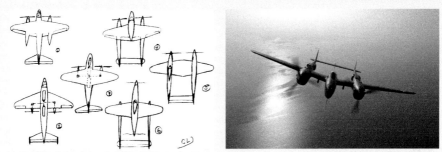

Configuration sketches used in the conceptual design phase of the Lockheed P-38 and the resulting P-38 (U.S. Air Force photo by Sgt. Ben Bloker)

1.8 The Computational Aerodynamics Process

Now that we have a fairly good background knowledge of what CA is and how it is currently applied, we need to explore how CA tools are used. This will require a good dose of engineering experience and judgment, since approaching computational aerodynamics simulations is similar to approaching any engineering process.

Given an aircraft (or some type of vehicle) to examine, we start with a physical problem and then represent the physical situation with a mathematical model. We then obtain a solution of the mathematical problem and use that solution to deduce something about the physical problem. Skill and experience are required to complete this process, but perhaps more critical are practice and remembering to be aware of the process as you go through it. Engineering judgment is very important in selecting the approach to be used, and as with all engineering projects, the allotted budget and time must also be considered. One example approach to the aerodynamic design process consists of the following steps (which are illustrated in Figure 1.17):[52]

- start with the real flow around the aircraft
- define a physical model of the flowfield, perhaps considering it to be appropriately modeled as an inviscid transonic flow, a boundary-layer flow, and a wake

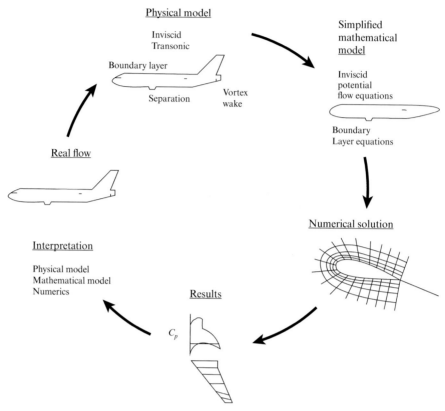

Figure 1.17 The relationship between a computation and a real flow (Ref. 52; Courtesy of Ed Tinoco).

- select the simplified mathematical model(s) to be solved
- define the geometry, and define the mathematical geometric model, including the mesh
- carry out the numerical solution
- examine the results
- interpret the sequence of physical model, mathematical model, and numerical solution, together with the computed results, to understand the physics of the solution

Notice here that the numerical solution of a computational problem is a small part of the total engineering process. Successful aerodynamicists must master the entire sequence of steps.

While the CA process has become fairly standardized over the years, intelligent users must never lose sight of the fact that CA provides a model of reality, as described in Fig. 1.17. The process of creating a geometric model, for example, involves simplifications to various geometric complexities (a model of a flap may or may not include an attachment mechanism, for example). Also, our equations of fluid motion do not always include all flow processes and phenomena (especially those relating to turbulence). We must never lose

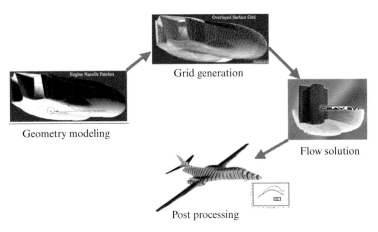

Figure 1.18   The Four Step CFD process.

sight of the difference between reality and our model of reality: the map of France is not France!

Once the overall engineering approach has been firmly established, a process for performing the computational analysis needs to be followed. Typically for CFD (a subset of CA as shown in Fig. 1.5), this is done in a four-step process, as shown in Fig. 1.18. The following sections describe the four steps in greater detail, but do not allow yourself to get lost in the forest looking at the trees; keep the overall approach in mind as you review these steps.

1.8.1 Geometry Modeling

The first step in the process is to have a definition of the geometry that is to be modeled. Simple geometries, especially during the conceptual design phase, may only involve knowing some basic geometric equations (circles, ellipses, straight lines, and other parametric geometry models); individuals can sometimes create their own geometry without using any specialty software. But as CA has been able to handle more and more complex flowfields, the geometries have also become more complex. Currently, aircraft geometries are typically defined using *computer-aided design* (CAD) programs (such as CATIA, ProEngineer, Inventor, SolidWorks, AutoCAD, etc.) using standardized formats, e.g., IGES (Initial Graphics Exchange Specification) files, or others. Figure 1.19 shows an aircraft design, including the integration of internal components, and the structure, using CAD. The CAD files are often obtained directly from the aircraft designers or manufacturers (a best case), but otherwise they may have to be re-created by the aerodynamicist if no numerical geometry definition is available (a worst case). A detailed, accurate geometry is critical when modeling the flowfield, and the end product of this process is a water-tight surface geometry over which a surface mesh may be readily generated. While this first

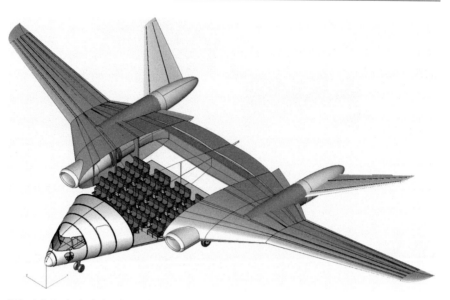

Figure 1.19 A blended wing/body airplane layout visualized with CAD software (Courtesy of David W. Hall; a full color version of this image is available on the website: www.cambridge.org/aerodynamics).

step in the process may be tedious and unexciting (and it often is), obtaining and modeling the geometry accurately is an essential and important component of the CA process.

1.8.2 Grid Generation

Once the geometry is defined in a computer-accessible format, the next step in the process involves creating a surface grid or mesh, which is known as *grid generation*. Since aerodynamicists are interested in the behavior of the flow of air around or through a geometry, the physical equations of motion must be solved in the "space" around or through the vehicle. This requires using one of several grid generation procedures to define a grid in the volume where the air will be flowing – the most common grid generation processes are those that create either structured or unstructured grids or meshes.

In general, the geometry model will not have the same elements that are required for modeling the flow field, so a new CA-friendly geometry often must be made. This can involve a variety of complicated features, such as projecting surfaces, smoothing elements, and discretizing the surface, but the end product is a *surface grid* or mesh. This surface mesh is most commonly composed of many triangular and quadrilateral elements that are distributed to approximate the surface geometry to an acceptable level of accuracy. The layout of the triangle or quadrilateral elements will have a profound impact on the ability of the model to compute an accurate flowfield. The smallest surface elements should usually be found in areas of high curvature (i.e.,

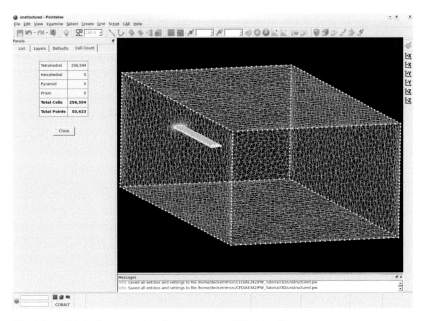

Wing being gridded with commercial software (courtesy of Pointwise, Inc.).

wing leading edge) or at junctions between vehicle components (i.e. wing-fuselage junction).

The grid that represents the flow of air is known as the *volume grid*. As in the case of the surface mesh, the density of volume elements has a profound effect on whether the simulation using the volume grid will be able to reproduce important flow features (i.e. shocks, vortices, boundary-layer separation, etc.).

For anything other than simple shapes, the grid will typically be generated by a commercial grid generation package. Some skill will be required to be able to use these codes effectively (remember, details of the grid are critical to the quality of the results!). Figure 1.20 shows a commercially available grid generation code setting up the initial stages for a grid around a wing for use on an aircraft.

1.8.3 Flow Solution

Once the grid is generated, a *flow solver* is selected. The flow solver is the heart of the CA process, and the choice of the particular solver will depend on several aspects of the problem and be closely connected to the physical model and specific mathematical flowfield representation selected in the process described earlier. The choice of the flow solver may also depend on what you have available at any given job location. However, it is important as a design engineer to resist the temptation to allow the deficiencies of the chosen flow solver to limit the design.

This leads to another important aspect of CA, which is the software used. This is an area where a major paradigm shift has taken place. In the past, aerodynamics software was traditionally "home grown" (meaning that everyone wrote their own codes). Today we see a commercialization of CA software, for both pre-/post-processing (grid generation and analysis of results) and for the flow solvers (the code that numerically predicts the aerodynamics). These commercial software suites are not inexpensive, and in addition, they are general-purpose codes that have been developed for a wide range of applications in addition to aerodynamics. Your ability to apply these codes to your specific applications may require a great deal of experience with the software prior to successfully using it.

Another consideration will be the availability of computer resources. For example, a high-end laptop may be sufficient for two-dimensional flow calculations. For three-dimensional calculations, some form of high-performance computer will probably be required, since you will be solving millions of simultaneous nonlinear algebraic equations. This could be a network of so-called workstations, a "cluster" of processors, or a state-of-the-art supercomputer. Frequently, the computer doing the actual calculations will be located at some remote geographical location. In the United States, the government has supercomputer centers associated with various agencies; for example, NSF, NASA, DOE, and DOD all maintain supercomputer centers at various locations. This requires high-speed communications with the computer site in order to transfer the pre- and post-processing data files. Being able to use the software and computers for modern CA programs can take a great deal of time, so do not underestimate how challenging this step will be.

1.8.4 Post Processing

As mentioned, the resulting "solution" is contained within a large file that can require significant computer storage space. In modern aircraft design studies, many solutions may need to be stored, and if the flow is unsteady, the entire time history of the flowfield may need to be stored. Once a solution is computed, the results require *post processing* using another commercial program, which is often called *flow visualization*. In some cases, you might only be interested in obtaining the force coefficients, but, if the force coefficient values are unexpected, the entire flowfield may need to be examined. In fact, one of the real powers of CA is often the ability to look at an entire flowfield and investigate the flow physics responsible for the system performance – whether it is good or bad; this is often called flow diagnostics. Figure 1.21 shows the F-18 fighter at a high angle of attack with the leading edge extension vortex and significant chaotic flow visible above the wing.

Figure 1.21   Flow visualization of the high angle of attack flowfield around the F-18 fighter. Notice the vortex above the leading edge extension and the highly chaotic flow over the wing (courtesy of the USAFA Modeling & Simulation Research Center; a full color version of this image is available on the website: www.cambridge.org/aerodynamics).

1.8.5 Code Validation

Code validation has become a field in its own right and will be discussed in greater detail in Chapter 2. Assessing the ability of a CA code to accurately predict aerodynamics, and to define the range of validity and accuracy of the code, is difficult and time consuming. However, the importance of this step cannot be overemphasized. It has become a continuing theme in computational aerodynamics development, and is part of the much broader field of computational science and engineering. For a history of CA code validation, read *Verification and Validation in Computational Science and Engineering*, written by an early pioneer in CFD.[53] The AIAA is addressing the issue of establishing standards for CFD, but this is proving to be a difficult task.[54] The American Society of Mechanical Engineers (ASME) is also working on validation issues.[55] This key aspect of CA will recur throughout this book, especially in Chapter 2; each engineer (or group of engineers) must test a code before using it to make design decisions.

Profiles in Computational Aerodynamics: David Vallespin

I was born in Cartagena in 1986, a small city in the southeast coast of Spain, not far from the birth place of renowned engineers like Juan De La Cierva and Isaac Peral. Due to my father's profession I grew up in different cities around Europe and became fluent in English and French at an early age. My parents tell me that when I was too young to remember I would dismantle my toys before I even started playing with them. It was not a desire for destruction but for understanding how things worked. It was at an early age that my dreams of becoming a football player or a fighter pilot faded to give way to my aching passion. Proof of this is my collection of model aircraft replicas still decorating my room. My father, an engineer at heart, made of my interest an aspiration in life. Mathematics and sciences became my favorite subjects in high school thanks to a group of inspiring teachers. I was always attracted to the more numerical subjects as opposed to the more descriptive ones. The fact that numbers can be used to predict phenomena that affect our day-to-day lives never ceases to amaze me.

In 2004 I began my studies in Aerospace Engineering at the University of Liverpool, from which I graduated in 2008 with a First Class Honours degree. During this time I was always more attracted to what I find to be the more fundamental fields of aeronautical engineering. This is why I chose the Flight Sciences stream, which entailed subjects in configurational aerodynamics, computational fluid dynamics, aeroelasticity and rotorcraft flight, as opposed to the more systems- or manufacturing-based routes. My masters thesis dealt with understanding and validating a potential flow code designed to analyze the performance of individual wind turbines. I developed a real passion for the use of numerical and computational methods for the prediction of the performance of real engineering machines, whether it was wind turbines or aircraft.

After my degree I had to choose between working for the large aircraft or helicopter manufacturers or to pursue further study doing a PhD. I decided I wanted to learn more about aerodynamics before moving into industry and so I decided to stay in Liverpool and study under the supervision of Prof. Kenneth Badcock. I carried out my research in the application of CFD tools for the prediction of flight dynamics models of various aircraft types. At the time I was using state of the art Navier-Stokes methods on structured grids for conventional and unconventional configurations. The work lead me to develop a process which generated an aerodynamic model based on CFD data and predicted aerodynamic loads for prescribed acrobatic maneuvers. The loads predicted using this method were then validated against time-accurate CFD results. Thanks to this project, I collaborated in a NATO-AVT Research group and worked with some of the most reputable CFD laboratories in the world. The feedback and friendships I gained from this experience were invaluable.

In 2012 I began working for an engineering consultancy in Toulouse in the Flight Dynamics department where I am still today. My role here is to give support to the Future Projects department at Airbus by developing their prediction tools, which help engineers iterate through future aircraft design concepts for the next generation of aircraft. This links very well with my previous work and gives me the opportunity to learn and gain experience in an industrial environment.

I usually occupy my spare time in social activities, team sports, traveling, and music involving my family and friends. I most enjoy playing football, whether it is with friends or colleagues, and I am also an avid fan of my home team. Thanks to the education my parents gave me, I find myself constantly looking for new projects and places to live and work.

(David was recently hired to work for Airbus in Madrid on aeroacoustics)

1.9 Computational Aerodynamics Users and Errors

The issues we have been describing so far, especially the process for creating a usable model, are essential to being an intelligent user of CA. But there is one more issue that must be discussed before we get into the details of how CA tools work, and that has to do with the user of the code … you. While it may seem obvious that you should not make errors while using CA tools, it is not always so easy to see your own mistakes. But, as we have all heard, "to err is human," and we are definitely human, so we will make errors.

The notion that we will make mistakes relates to our earlier discussion about the analogy between an engineer and a carpenter. If a carpenter builds a room that does not meet the specifications of the plans, whose fault is it? Could the plans be incorrect? Were the materials faulty? Were the tools being used manufactured with defects? Any of these items could be responsible for the building not meeting the requirements of the plans. But there is one more source of error that the carpenter (or engineer) might not be ready to admit – maybe they made a mistake!

Because of this, we should spend a little time talking about user error, which is a "rather embarrassing aspect of CFD, as it is one of the prime causes of uncertainty in the results of CFD simulations."[56] User error can be as obvious as making a typographical mistake while inputting values into a code, or as complicated to evaluate as having a broad bias in your trust in CA results (either trusting your results too much or too little). Because of this, you should follow good practices while using CA tools, most of which make common sense, but some of which you will only learn as you gain experience in working as an engineer within the CA community. These practices can include: having a good work approach and work habits, being able to clearly define the problem you are solving and what CA tools will best address your goals, having good code usage and handling processes (especially if you are modifying a code's programming[57]), having a rational approach to interpreting your results and analyzing their accuracy, making sure you understand as much as possible about the code you are using, and carefully documenting your results. Growing the attributes of meticulous attention to detail (and organization) and profuse note taking will help you to be a better CA user. We will discuss many of these issues as we proceed through this book, but start thinking about them now, since a code is only as good as the person using it: remember the GIGO Concept Box and the saying about computers: "garbage in, garbage out!" In this respect, wind tunnel testing procedures have been well established and can be used as a model for good CA processes.

1.10 Scope, Purpose, and Outline of the Book

The objective of this text is to provide an overview of computational aerodynamics as currently practiced. In addition, we want to provide an understanding of the basis for CA technology as well as the terminology of CA. We will emphasize the assumptions used in the various methods and illustrate the use of computational aerodynamics methods through numerous projects designed for students and professionals to use to develop their skills. Very few aerodynamicists today are developing the basic algorithms used in CA, but most aerodynamicists will be using the methods in already existing codes: we will spend most of our time addressing this issue. We will also include many examples showing what steps users must take to determine if the answers they are obtaining in their applications are reasonable. Engineering decisions are going to be made based on computational aerodynamics, thus the objective of any computational aerodynamics work must be to answer:

- How well have I modeled the physical reality of the flow?
- Assuming the modeling is good, what is computational aerodynamics revealing about the physics of the flowfield?

This book provides a systematic development of computational aerodynamics. We start in Chapter 2 with a discussion of the general engineering computing environment to set the stage for solving computational problems. This includes further discussion of code verification and validation and a brief review of computing related to aerodynamics.

In Chapter 3, we derive the governing equations to establish a basis for the terminology to be used in computational aerodynamics and then review and define the various levels of "fidelity" or approximation to the governing equations of fluid mechanics. All levels of fidelity have a role in computational aerodynamics. Today we have the capability to investigate just about any flowfield that could be of interest, and we need to emphasize the issues users should understand about the computational aerodynamics methods.

As a complement to Chapters 2 and 3, Chapter 4 is a review of some of the key concepts in aerodynamics. The effort and focus of the computational work might result in losing sight of the true objective of our computational analysis, so we need to keep aerodynamics foremost in our minds. Chapter 5 provides a description of the classical linear theory methods of computational aerodynamics. These are based on the potential flow approximation, and applications of these methods illustrate their value, especially in conceptual and preliminary design.

Chapter 6, which begins the CFD portion of the book, contains a description of the approaches for numerically solving partial differential equations. This is followed in Chapter 7 by the description of the main part of most computational aerodynamicists' day-to-day work: manipulation of geometry (design) and the construction of the computational grid required to obtain numerical solutions. Since the vast majority of flows of interest in computational aerodynamics are turbulent, one of the next critical components of the world of CFD is the use of turbulence models to predict flowfields, which is covered in Chapter 8. Flow visualization is described in Chapter 9, emphasizing the best ways to show results rather than the theoretical aspects of visualization techniques.

Chapter 10 covers the CFD process in much more detail than given previously and provides opportunities to examine the issues that a practitioner must master to obtain accurate solutions. This includes the use of CFD in both inviscid and viscous flow applications and emphases in subsonic/transonic flow computations and supersonic/hypersonic computations. We conclude Chapter 10 with some more elaborate applications, including projects for students and professionals to use CFD in much more realistic situations. We will provide a discussion of contemporary and emerging methods and the use of CA in the modern computational design environment.

Upon completion of the book, you should be able to assess a problem for analysis using computational aerodynamics, formulate the problem, select a method, and obtain a solution. Then you should be able to use engineering judgment to decide if you have a valid engineering answer. Once you are able to do that, you will be well on your way to becoming a competent computational aerodynamicist!

Summary of Best Practices

1. Becoming an intelligent user of computational aerodynamics requires that you begin learning about multiple important solution elements:
   a. flow physics
   b. numerical methods
   c. computer codes
   d. computer systems
   e. experimental and theoretical validation
2. A good grasp of the difference between a *real flow* and the *model* of a real flow is essential to understanding computational aerodynamics.
3. Computational aerodynamics (especially CFD) involves being proficient in a four-step process that includes:
   a. geometry modeling
   b. grid generation
   c. flow solution
   d. post processing
4. One of the biggest sources of error in computational aerodynamics is ... YOU! Always be sure that your interaction and use of CA tools is based on solid knowledge and experience, and never assume that, just because something is wrong with a prediction, the code must be the culprit.

Note: The best practices compiled at the end of each chapter in this book come from a variety of sources, including: the personal experiences of the authors and their colleagues, AIAA,[54] European Research Community on Flow Turbulence and Combustion (ERCOFTAC),[58] European Cooperation in Science and Technology (COST),[59] MARNETcfd,[56] QNET-CFD,[60] ECORA,[61] Machine Design,[62] and Nielsen Engineering and Research/NASA Langley.[63,64]

1.11 Project

Select a NACA 4- or 5-digit airfoil of your choice for which you can find geometry, the experimental pressure distributions, and force & moment data.

1. Plot the shape of the airfoil.
2. Plot the pressure distribution at one angle of attack.
3. Plot the force and moment data (lift, drag, and pitching moment) over a range of angles of attack. Make sure to include the drag polar (lift vs. drag).
4. Describe the airfoil and your data source (make sure to include the test conditions: Reynolds number, Mach number, and transition details, i.e., fixed or free transition. If fixed, where and how was it fixed?).

Note: You will compare these data with results from various computer programs and applications in several assignments: be sure you get a good, complete set of data. See Appendix A for additional information on airfoils, bodies, and wings, and Appendix B for sources of data.

1.12 References

1 Whitfield, D.L., "Perspective on Applied CFD," AIAA Paper 95–0349, January 1995.

2 Moore, J.A., Rutt, B.K., Karlik, S.J., Yin, K., and Ethier, C.R., "Computational Blood Flow Modeling Based on *In Vivo* Measurements," *Annals of Biomedical Engineering*, Vol. 27, No. 5, 1999, pp. 627–640.

3 Kalnay, E., *Atmospheric Modeling, Data Assimilation and Predictability*, New York: Cambridge University Press, 2006.

4 Katz, J., *Race Car Aerodynamics*, Cambridge: Bentley Publishers, 1995.

5 Lee, S. and Ditko, S., "*Amazing Fantasy #15*," New York: Marvel Entertainment, August 1962.

6 Blazek, J., *Computational Fluid Dynamics: Principles and Applications*, Oxford: Amsterdam, 2005.

7 Pletcher, R.H., Tannehill, J.C., and Anderson, D.A., *Computational Fluid Mechanics and Heat Transfer*, 3rd ed., Boca Raton: CRC Press, 2013.

8 Hoffmann, K.A. and Chiang, S.T., *Computational Fluid Dynamics*, Vols. 1 and 2, 4th ed., Wichita: Engineering Education System, 2000.

9 Löhner, R., *Applied CFD Techniques: An Introduction Based on Finite Element Methods*, West Sussex: John Wiley & Sons, 2001.

10 Hansell, S., "How Wall Street Lied to Its Computers," *New York Times*, September 18, 2008, p. C7.

11 Bertin, J.J. and Cummings, R.M., *Aerodynamics for Engineers*, 6th ed., Englewood Cliffs: Pearson, 2014.

12 Cosentino, G.B. and Holst, T.L., "Numerical Optimization Design of Advanced Transonic Wing Configurations," *Journal of Aircraft*, Vol. 23, No. 3, 1986, pp. 192–199.

13 Flores, J. and Chaderjian, N.M., "Zonal Navier-Stokes Methodology for Flow Simulation About a Complete Aircraft," *Journal of Aircraft*, Vol. 27, No. 7, 1990, pp. 583–590.

14 Weyer, R.M., "Predator Weaponization – An Application of Simulation Based Acquisition," AIAA Paper 2002–5058, August 2002.

15 Bradley, R.G. and Bhateley, I.C., "Computational Aerodynamic Design of Fighter Aircraft – Progress and Pitfalls," AIAA Paper 83–2063, August 1983.

16 Henne, P.A., ed., *Applied Computational Aerodynamics*, Washington, DC: AIAA, 1990.

17 Settles, G.S., "Schlieren and Shadowgraph Imaging in the Great Outdoors," Proceedings of PSFVIP-2, Paper PF302, Honolulu, HI, May 1999.

18 Mendenhall, M.R., Lesieutre, D.J., Whittaker, C.H., Curry, R.E., and Moulton, B. "Aerodynamic Analysis of Pegasus™ – Computations vs Reality," AIAA Paper 93–0520, January 1993.

19 Shevell, R.S., "Aerodynamic Anomalies: Can CFD Prevent or Correct Them?" *Journal of Aircraft*, Vol. 23, No. 8, 1986, pp. 641–649.

20 Tinoco, E.N., "The Impact of Computational Fluid Dynamics in Aircraft Design," *Canadian Aeronautics and Space Journal*, Vol. 44, No. 3, 1998, pp. 132–144.

21 Busch, R.J., "Computational Fluid Dynamics in the Design of the YF-23 ATF Prototype," AIAA Paper 91–1627, June 1991.

22 Wedderspoon, J.R., "The High Lift Development of the A320 Aircraft," ICAS Paper 86–2.3.2, 1986, pp. 343–351.

23 Nield, B.N., "An Overview of the Boeing 777 High Lift Aerodynamic Design," *Aeronautical Journal*, Vol. 99, Issue 989, 1995, pp. 361–371.

24 Reckzeh D., "Aerodynamic Design of the High-Lift-Wing for a Megaliner Aircraft," *Aerospace Science and Technology*, Vol. 7, Issue 2, 2003, pp. 107–119.

25 Büscher, A., Radespiel, R., and Streit, T., "Modelling and Design of Wing Tip Devices at Various Flight Conditions Using a Databased Aerodynamic

Prediction Tool," *Aerospace Science and Technology*, Vol. 10, Issue 8, 2006, pp. 668–678.

26 Tinoco, E.N., "Validation and Minimizing CFD Uncertainty for Commercial Aircraft Applications," AIAA Paper 2008–6902, August 2008.

27 Garrett, M., Testimony to the U.S. Senate Subcommittee on Technology, Innovation, and Competitiveness, July 2006.

28 Kingsley-Jones, M., "Playing Catch-Up: No Room for Delays of the Airbus A350," *Flight International*, June 12, 2007.

29 Rumsey, C.L. and Ying, S.X., "Prediction of High Lift: Review of Present CFD Capability," *Progress in Aerospace Sciences*, Vol. 38, Issue 2, 2002, pp. 145–180.

30 Van Dam, C.P., "The Aerodynamic Design of Multi-Element High-Lift Systems for Transport Airplanes," *Progress in Aerospace Sciences*, Vol. 38, Issue 2, 2002, pp. 101–144.

31 Fujii, K., "Progress and Future Prospects of CFD in Aerospace – Wind Tunnel and Beyond," *Progress in Aerospace Sciences*, Vol. 41, Issue 6, 2005, pp. 455–470.

32 Bosnyakov, S., Kursakov, I., Lysenkov, A., Matyash, S., Mikhailov, S., Vlasenko, V., and Quest, J., "Computational Tools for Supporting the Testing of Civil Aircraft Configurations in Wind Tunnels," *Progress in Aerospace Sciences*, Vol. 44, Issue 2, 2008, pp. 67–120.

33 Bushnell, D.M., Scaling: Wind Tunnel to Flight," *Annual Review of Fluid Mechanics*, Vol. 38, 2006, pp. 111–128.

34 Henshaw, M.J. de C., Badcock, K.J., Vio, G.A., Allen, C.B., Chamberlain, J., Kaynes, I., Dimitriadis, G., Cooper, J.E., Woodgate, M.A., Rampurawala, A.M., Jones, D., Fenwick, C., Gaitonde, A.L., Taylor, N.V., Amor, D.S., Eccles, T.A., and Denley, C.J., "Non-Linear Aeroelastic Prediction for Aircraft Applications," *Progress in Aerospace Sciences*, Vol. 43, Issues 4–6, 2007, pp. 65–137.

35 Cummings, R.M., Morton, S.A., and Siegel, S.G., "Numerical Prediction and Wind Tunnel Experiment for a Pitching Unmanned Combat Air Vehicle," *Aerospace Science and Technology*, Vol. 12, Issue 5, 2008, pp. 355–364.

36 Cummings, R.M. and Schütte, A. "Integrated Computational/Experimental Approach to UCAV Stability & Control Estimation," *Journal of Aircraft*, Vol. 49, No. 6, 2012, pp. 1542–1557.

37 Morrocco, J.D., "Lockheed ATF Team Cites Lessons Learned in Prototyping Effort," *Aviation Week*, November 5, 1990, p. 87.

38 Forsythe, J.R. and Woodsen, S.H., "Unsteady Computations of Abrupt Wing Stall Using Detached-Eddy Simulation," *Journal of Aircraft*, Vol. 42, No. 3, 2005, pp. 606–616.

39 Sloof, J.W., ed., "Technical Status Review on Drag Prediction and Analysis from Computational Fluid Dynamics: State of the Art," AGARD-AR-256, May 1989.

40 Skelley, M.L., Langham, T.F., Peters, W.L., and Frantz, B.G., "Lessons Learned During the Joint Strike Fighter Ground Testing and Evaluation at AEDC," AIAA Paper 2007–1635, February 2007.

41 Hancock, G.J., "Aerodynamics – The Role of the Computer," *Aeronautical Journal*, Vol. 89, No. 887, 1985, pp. 269–279.

42 Buchholz, M.D. 2002. Highlights of the JSF X-35 STOVL Jet Effects Test Effort," AIAA Paper 2002–5962, November 2002.

43 Neumann, R.D., "Requirements in the 1990s for High Enthalpy Ground Test Facilities for CFD Validation," AIAA Paper 90–1401, June 1990.

44 Cummings, R.M. and Morton, S.A., "Continuing Evolution of Aerodynamic Concept Development Using Collaborative Numerical and Experimental Evaluations," *Journal of Aerospace Engineering*, Vol. 220, No. 6, 2006, pp. 545–557.

45 Kmak, F., Hudgins, M., and Hergert, D., "Revalidation of the NASA Ames 11-by-11-Foot Transonic Wind Tunnel With a Commercial Airplane Model," AIAA Paper 2001–0454, January 2001.

46 Krynytzky, A.J., "Parametric Model Size Study of Wall Interference in Boeing Transonic Wind Tunnel Using TRANAIR," AIAA Paper 2004–2310, June 2004.

47 Baer, J.C. and Schuehle, A.L., "Re-Engining the 737," SAE Paper 821442, October 1982.

48 Dean, J.P., Clifton, J.D., Bodkin, D.J., Morton, S.A., and McDaniel, D.R., "Determining the Applicability and Effectiveness of Current CFD Methods in Store Certification Activities," AIAA Paper 2010–1231, January 2010.

49 Vanderplaats, G.N., Hicks, R.M., and Murman, E.M., "Application of Numerical Optimization Techniques to Airfoil Design," NASA Conference on Aerodynamic Analysis Using High Speed Computers, Langley Research Center, VA, NASA SP-347, Part II, 1975.

50 Mason, W.H., Knill, D.L., Giunta, A.A., Grossman, B., Watson, L.T., and Haftka, R.T., "Getting the Full Benefits of CFD in Conceptual Design," AIAA Paper 98–2513, June 1998.

51 Jameson, A., "Efficient Aerodynamic Shape Optimization," AIAA Paper 2004–4369, August 2004.

52 Rubbert, P.E. and Tinoco, E.N., "Impact of Computational Methods on Aircraft Design," AIAA Paper 83–2060, August 1983.

53 Roache, P.J., *Verification and Validation in Computational Science and Engineering*, Albuquerque, NM: Hermosa Publishers, 1998.

54 Cosner, R.R., Oberkampf, W.L., Rumsey, C.L., Rahaim, C.P., and Shih, T.I-P., "AIAA Committee on Standards for Computational Fluid Dynamics: Status and Plans," AIAA Paper 2006–0889, January 2006.

55 Schwer, L.E., "Guide for Verification and Validation in Computational Solid Mechanics," ASME VV 10, 2006.

56 Anonymous, "Best Practice Guidelines for Marine Applications of Computational Fluid Dynamics," MARNet CFD, Final Report, 2003.

57 Giles, M., "'Best Practices' in CFD Software Development," Oxford University Computing Laboratory, 1995.

58 Hutton, A.G. and Casey, M.V., "Quality and Trust in Industrial CFD – A European Initiative," AIAA Paper 2001–0656, January 2001.

59 Franke J., Hellsten, A., Schlünzen, H., Carissimo, B., eds., "Best Practice Guideline for the CFD Simulation of Flows in the Urban Environment," COST Action 732, May 2007.

60 Vos, J.B., Rizzi, A.W., and Darracq, D., "Overview of Application Challenges in the Aeronautical Industry," QNET-CFD Network Bulletin No. 1, July 2001.

61 Menter, F., "CFD Best Practice Guidelines for CFD Code Validation for Reactor-Safety Applications," EVOL–ECORA–D1, February 2002.

62 Dvorak, P., "Best Practices for CFD Simulations," *Machine Design*, Vol. 78, No. 11, 2006, pp. 104–110.

63 Mendenhall, M.R., Childs, R.E., and Morrison, J.H., "Best Practices for Reduction of Uncertainty in CFD Results," AIAA Paper 2003–0411, January 2003.

64 Stremel, P.M., Mendenhall, M.R., and Hegedus, M.C., "BPX – A Best Practices Expert System for CFD," AIAA Paper 2007–0974, January 2007.

2 Computers, Codes, and Engineering

For in much wisdom is much grief; and he who increases knowledge increases sorrow.
King Solomon; Ecclesiastes 1:18

(possibly looking forward to the problems that arise when using computers to solve aerodynamic problems.)

F-22 in maneuvering flight (U.S. Air Force photo by Senior Airman Vernon Young).

LEARNING OBJECTIVE QUESTIONS

After reading this chapter, you should know the answers to the following questions:

- What are the differences (in speed and accuracy) between engineering methods, potential flow methods, and CFD methods?
- What are empirical or semi-empirical methods and how do they work?
- How are potential flow methods and CFD methods used in analysis and design?
- What are the basic components of a computer and what are they for?
- What are computer codes and how do they "run" on a computer?
- What is Moore's law and how has it impacted computational aerodynamics?

- What is a parallel processor and why is it used? What is Amdahl's law and how does it relate to parallel computing?
- What are the differences between verification, validation, and certification of computer codes?
- What are the six numerical accuracy areas that should be addressed when reporting results from a CFD code?
- Can you name some of the elements that might be included in a computational aerodynamics checklist that would insure reasonable results from your code?

2.1 Introduction

The development of computer software has been referred to by F. P. Brooks as the "tar pits" of engineering.[1] Tar pits were pools of asphalt that seeped up through the ground and eventually became disguised with water and debris. When prehistoric animals walked by they didn't notice the pits, got stuck, and became completely bogged down in the primordial ooze. The trapped animals died, their bones becoming fossils, forever speaking to their negligence. Brooks is making the point that the details of writing (or even using) computer codes can be so absorbing that it is easy to forget the big picture and get lost in the details; we do not want you to get bogged down like those animals!

One of the reasons we wrote this chapter is to help you avoid getting bogged down. Rather than immediately pushing you into the "ooze" of computational aerodynamics, we will try to give you a few tools and concepts that will help you to navigate around the tar pits and be able to apply computational aerodynamics codes more successfully. Your ability to become an intelligent user of CA methods may depend more on the contents of this chapter than on the details presented in much of the remainder of the book!

Success in using computational aerodynamic methods depends on general engineering problem-solving skills as well as effective computer use to solve specific problems. An overview of the elements of computational aerodynamics (as discussed in Chapter 1) includes basic knowledge of fluid dynamics, computer systems, experimental validation, numerical methods, and computer codes and the ability to work with the solutions. In other words, there's a lot to know in order to successfully use CA!

This chapter presents various issues related to using computational resources to solve aerodynamics problems, along with some guidelines for effective use of computing systems. In the next few chapters we attempt to provide a background for dealing with each of these issues more intelligently.

2.2 From Engineering Methods to High-Performance Computing

With the computational resources available to engineers today, it is easy to "throw" the most complex tools at a particular problem. The thinking goes something like this: "if this code is robust enough to solve complex problems, it certainly should be able to solve my smaller problems." While this may or may not be true, it is often not the wise course of action to follow. The varieties of aerodynamics software include all types of problem-solving tools, from purely theory-based approaches to those that rely completely on interpolations of relevant experimental data (which are called empirical methods) to those that rely on a combination of the two (which are called semi-empirical methods or *engineering methods*).

The discussion in Chapter 1 described an analogy between engineering and carpentry; nowhere is that analogy more appropriate than when considering the relative tradeoffs between CFD methods and other "engineering" approaches to aerodynamics. There are a great variety of tools available for carpenters and engineers; CFD methods only represent a subset of those engineering tools. Table 2.1 shows the larger hierarchy of computational approaches (tools), with some pros and cons compared and examples of each type given. These CA approaches are often described as having either low fidelity or high fidelity, depending on the level of assumptions and simplifications that are made within the software. *Low-fidelity methods* (such as the engineering methods exemplified by *DATCOM* or *ACSYNT* or the linear potential flow methods such as *PANDA*, *XFOIL*, *VSAero*, or *LinAir*) make simplifications and assumptions and, because of that, are limited to specific application such as conceptual design. However, these methods are very fast on modern computers and can often be run easily on personal computers. For example, Fig. 2.1 shows the graphics user interface for *ACSYNT* (AirCraft

Table 2.1. Heirarchy of computational aerodynamics approaches

CA LEVEL	PURPOSE	ACCURACY (AVERAGE)	TIME TO USE (INCLUDING SET-UP)	EXAMPLES
Engineering Methods	Forces and moments for conceptual design or trade studies	±15%	Minutes on a PC	*DATCOM, ESDU, ACSYNT, RDS, AAA*, etc.
Potential Flow Methods (including viscous effects estimation)	Surface pressures, forces, and moments for analysis and design	±10%	10s of minutes on a PC	*PANDA, XFOIL, VORLAX, CEASIOM, LinAir, VSAero, PAN AIR*, etc.
CFD Methods	Detailed flow results (all flow variables throughout the flowfield) for analysis	±5% (or less)	Minutes to hours to days to weeks on a large computer	*FLUENT, OVERFLOW, Cobalt, Kestrel, CFL3D, CFD++, Tau, FLOWer, Edge*, etc.

Figure 2.1   Graphical User Interface for *ACSYNT*, an aircraft design code that also predicts aerodynamics (Ref. 2).

SYNThesis,[2] an aircraft design code) and Fig. 2.2 shows some sample input from *LinAir* (a potential flow vortex lattice code[3]). *High-fidelity methods,* such as computational fluid dynamics methods, make very few assumptions about the governing equations of fluid dynamics and attempt to solve them with minimal levels of modeling. While these methods supply a great deal of detail about the flow, they are very expensive to run on computers, in some cases requiring weeks or months to run on the fastest available computer systems. These methods are usually run directly on high-performance computers and often do not have a Graphical User Interface (GUI); rather, they require submission of lengthy job files containing input information for the program and create vast amounts of output information (which needs to be handled intelligently, as we will discuss in Chapter 9).

2.2.1 Semi-Empirical Methods

As mentioned previously, *empirical methods* are those that are solely based on experimental data. While there are numerous empirical methods in use today, most engineering methods, or *semi-empirical methods* (also sometimes called handbook methods), combine theory and experimental data. These

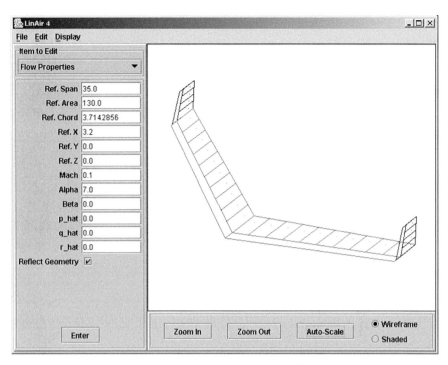

Figure 2.2 Sample aerodynamic input for *LinAir*, a linear potential flow-based code for wings (Ref. 3; Courtesy of Desktop Aeronautics, Inc.).

methods use experimental results to "fill in" shortcomings in theories and make predictive results more realistic. The resulting methods are often programmable and easily lend themselves to efficient computation (programs with thousands of lines of coding that took years to develop can run in seconds on a modern PC). Engineering methods are an important subset of the available computational aerodynamics techniques, especially when considering tools suitable for conceptual aircraft design.

Oftentimes, a variety of methods are combined to obtain full aircraft simulations. A good example of this is the design of high-lift systems for aircraft (including flaps and leading-edge devices), such as the high-lift system shown in Fig. 2.3 for the Boeing 747 as it lands. While the flow for high-lift systems contains many viscous effects, all levels of CA tools are used during the design process, including semi-empirical methods for conceptual design, inviscid methods for preliminary design, and viscous methods for final detailed design. This approach allows for a way to estimate aircraft aerodynamic coefficients in an increasingly more detailed manner as the design evolves and shows why semi-empirical methods are still valuable.

For example, if the pitching moment (the moment that rotates an airplane up or down) of a full aircraft was desired, you would normally simplify it in a series such as:

$$C_m = C_{m_o} + C_{m_\alpha}\alpha + C_{m_{\dot\alpha}}\dot\alpha + C_{m_q}q + \cdots \qquad (2.1)$$

Figure 2.3 Boeing 747 about to land with the various high-lift systems (flaps and slats) extended (Public domain photo by Adrian Pingstone).

where C_m is the pitching moment coefficient (see Chapter 4 for definitions of aerodynamic coefficients), C_{m_x} represents various stability coefficient derivatives, α is the angle of attack, $\dot{\alpha}$ is the plunging rate, and q is the pitching rate. Now that an overall modeling approach has been determined, methods would have to be found to approximate all of the stability derivatives (such as C_{m_α}) in Equation 2.1 using historical trends, experimental results, and theoretical development. It would be wrong to assume that this type of approach is primitive and useless, for a great deal of conceptual design is still accomplished with theoretical and semi-empirical methods. Once the individual building-block methods have been established (say for each of the terms in Equation 2.1 and many other equations like it), a comprehensive prediction tool can be developed, such as the USAF Stability & Control Data Compendium (*DATCOM*),[4] or the Engineering Sciences Data Unit (*ESDU*)[5] that provide methods for estimating all stability and control derivatives for aircraft based on theory, experiment, and semi-empirical relationships. A version of *DATCOM* (which is called *Missile DATCOM*[6]) has also been developed for missiles, and commercial software also exists that go beyond the capabilities of *Missile DATCOM* (e.g., *MISL3* or *AP09*).

So, how does a semi-empirical method actually work? As a simple example, consider the theoretical, incompressible lift-curve slope based on airfoil theory:[4]

$$C_{\ell_\alpha} = 6.28 + 4.7\left(\frac{t}{c}\right)[1 + 0.00375\phi_{TE}] \quad (1/\text{rad}) \qquad (2.2)$$

where (t/c) is the thickness-to-chord ratio of the airfoil, and ϕ_{TE} is the included angle of the trailing edge. One of the known deficiencies of Equation 2.2 is that boundary-layer effects cause the actual lift-curve slope to be less than the theoretical value. So, how can this theoretical equation be improved by including viscous effects and then also extended to include subsonic compressibility? First, remove the trailing edge angle function to obtain a baseline theory for the lift-curve slope of an airfoil:

$$\left(C_{\ell_\alpha}\right)_{theory} = 6.28 + 4.7\left(\frac{t}{c}\right) \qquad (2.3)$$

Next, add the Prandtl-Glauert compressibility correction (which is also theoretical) to take into account flight at subsonic, compressible speeds:

$$C_{\ell_\alpha} = \frac{1}{\sqrt{1 - M_\infty^2}}\left(C_{\ell_\alpha}\right)_{theory} \qquad (2.4)$$

where M_∞ is the freestream Mach number. Now, modify the compressibility correction in Equation 2.4 based on available experimental data, which shows that the lift-curve slope at compressible Mach numbers is slightly higher than the theory predicts. This is the first empirical "fix" to the theory, since we will add 5 percent to the lift-curve slope to better fit the experimental data:

$$C_{\ell_\alpha} = \frac{1.05}{\sqrt{1 - M_\infty^2}}\left(C_{\ell_\alpha}\right)_{theory}. \qquad (2.5)$$

Now apply another empirical correction to take the boundary-layer effects into account. This is much more complicated than the previous changes and is therefore represented by an empirical correction given by the term in the square brackets:

$$C_{\ell_\alpha} = \frac{1.05}{\sqrt{1 - M_\infty^2}}\left[\frac{C_{\ell_\alpha}}{\left(C_{\ell_\alpha}\right)_{theory}}\right]\left(C_{\ell_\alpha}\right)_{theory} \qquad (2.6)$$

where $\left(C_{\ell_\alpha}\right)_{theory}$ comes from Equation 2.3, and the ratio in the square brackets in Equation 2.6 is provided from experimental data as shown in Fig. 2.4. The relation in Equation 2.6 (along with Fig. 2.4) forms a single semi-empirical method.

Another example of a semi-empirical approach is the method often taught to students for estimating the lift-curve slope of a three-dimensional wing.

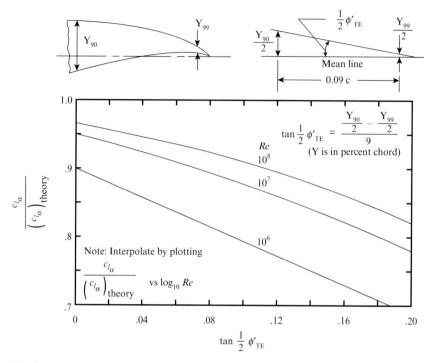

Figure 2.4   Ratio of experimental to theoretical lift-curve slopes for various airfoils (Refs. 7,8).

Prandtl's lifting-line theory gives the lift-curve slope for a wing with an elliptic lift distribution as:

$$C_{L_\alpha} = \frac{C_{l_\alpha}}{1 + \dfrac{C_{l_\alpha}}{\pi AR}} \quad (1/\text{rad}) \tag{2.7}$$

where C_{l_α} is the airfoil lift-curve slope, C_{L_α} is the wing lift-curve slope, and AR is the wing aspect ratio (see Chapter 4 for more details about lifting-line theory). A comparison of the lift-curve slope for wings with various aspect ratios is shown in Fig. 2.5. Notice that the slopes of the various lift curves are significantly different from each other, even if the wings used the same airfoils. Equation 2.7 is required to take into account the effect of aspect ratio on the lift-curve slope; otherwise the lifting capability of the wing would be grossly overestimated.

Re-deriving this expression for arbitrary lift distributions could be unnecessarily difficult and time consuming, so the usual approach for using lifting-line theory is to add a correction term to Equation 2.7, which is called the lift-curve slope parameter, τ:

$$C_{L_\alpha} = \frac{C'_{l_\alpha}}{1 + \dfrac{C_{l_\alpha}}{\pi AR}(1 + \tau)} \tag{2.8}$$

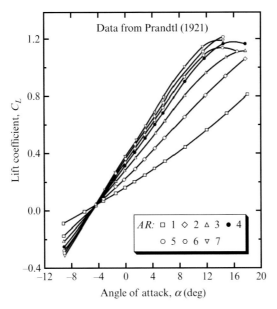

Data from Prandtl (1921)

Figure 2.5   Lift curves for wings with various aspect ratios.

where τ is usually found using wind tunnel data or from numerical approximations (see McCormick[9] for more details). Data compendiums such as *DATCOM* or *ESDU* contain literally thousands of similar methods that are combined together to give full aircraft (or missile) aerodynamic predictions quickly, oftentimes with very good results. Other programs, like *ACSYNT*, combine semi-empirical methods with other features in order to complete the full conceptual or preliminary design of an airplane.

Computational Aerodynamics Concept Box

How Well Does DATCOM Work?

As an example of how well a semi-empirical method like *DATCOM* can work, we will look at the basic aerodynamics for the T-38 aircraft as predicted by *DATCOM* and then compared with other CA methods and test data. The T-38 is a two-seat jet trainer (which is capable of speeds up to Mach 1.2) used by the U.S. Air Force.

The geometry of the airplane is input into *DATCOM* using basic measurements, such as wing span, fuselage length, control surface and flap locations, etc. The geometry inputs create a simplified configuration with some differences between the actual airplane and the *DATCOM* input (for example, the inlet is smoothed over with a solid surface). *DATCOM* then computes the various semi-empirical equations (such as Equation 2.6 and Equation 2.8) and predicts all of the aerodynamics characteristics of the airplane, including static and dynamic stability derivatives.

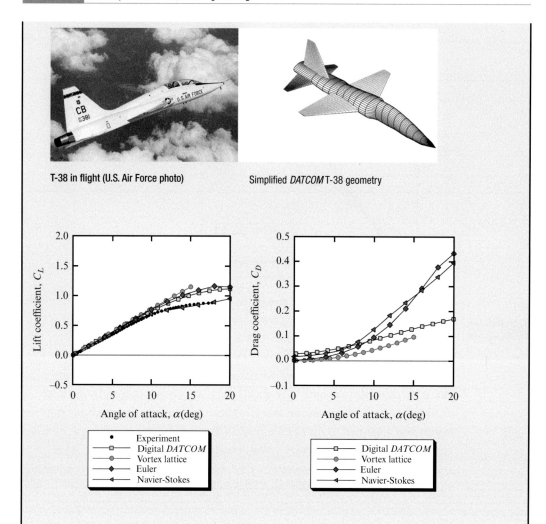

T-38 in flight (U.S. Air Force photo)   Simplified *DATCOM* T-38 geometry

    Notice that the lift coefficient, C_L, is well predicted by all methods at lower angles of attack (*DATCOM*, a potential flow approach called Vortex Lattice and CFD methods like Euler and Navier-Stokes codes). However, when the angle of attack reaches about 10 degrees, the lift curve starts to deviate from a linear relation and only the Navier-Stokes code is able to handle the viscous effects that are taking place. However, both *DATCOM* and the Euler code are giving some indication that there is a loss of lift. The drag values are also well predicted by *DATCOM* at low angles of attack but become increasingly poor as the angle of attack reaches 10 degrees. This makes *DATCOM* especially valuable for predicting cruise aerodynamics of aircraft, which takes place at relatively low angles of attack. *DATCOM* is also valuable as a "sanity check" for results being computed by other programs.

2.2.2 Linear Potential Flow Methods

The next level of CA tool shown in Table 2.1 is potential flow methods. These methods are based on the potential flow assumptions (including the inviscid flow assumption) and the computational approaches that result (such as panel methods and the vortex lattice method). Potential flow will be discussed in Chapter 4 in more detail, and the resulting methods will be developed in detail in Chapter 5. Linear potential flow methods are valuable for performing conceptual or preliminary design, since they are relatively fast on modern computers, and even run quite well on personal computers. Potential flow methods such as *PANDA*, *XFOIL*, *VORLAX*, and *PAN AIR* will also be discussed in Chapter 5.

2.2.3 CFD Methods

Finally, the high-fidelity methods, the "highest" level of simulation possible, is computational fluid dynamics (CFD). As we discussed, these methods solve the governing equations of fluid dynamics using various numerical methods, and often require some of the fastest, largest computer systems available (referred to as *high-performance computing*). Because of this, CFD methods historically have been reserved for determining details about a flowfield, rather than general trends, and are much more accurate than the other methods. CFD methods will be discussed in great detail in Chapters 6 through 10.

2.2.4 When Should You Use A Given Method?

Given the problem-solving choices shown in Table 2.1, it is not always easy to know which approach to use at any given time. The answer to that problem may not be simple, since this type of engineering judgment is usually based on experience, not just knowledge of the various approaches. One common sense approach to CA suggests: "don't use cannons to shoot flies and don't shoot spit wads at battleships." That is, use the method that is best suited for the results and purposes of your investigation. If you refer back to the Aircraft Design Box in Chapter 1, you will recall that conceptual design requires analyzing many aircraft configurations very quickly to determine which configurations are better than others. This type of analysis is strongly tied to the requirements for the airplane (how many passengers, how far will it fly, etc.) and rarely is driven only by aerodynamics. The key here is finding a fast, but reliable, method for estimating aerodynamics (as well as structure, propulsion, noise, emissions, etc.) for various designs. An example of this is shown in Fig. 2.6, where an airplane is being designed to

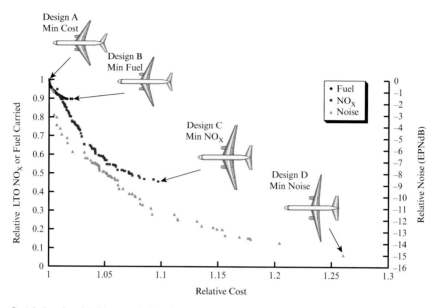

Figure 2.6 Best designs for minimizing cost, fuel carried, emissions, and noise as a function of operating cost. Notice that no single airplane design is optimum for all of these factors at once (Ref. 10).

minimize emissions, noise, fuel use, and cost.[10] In this case, literally hundreds of aircraft configurations were evaluated to determine viable candidates for the airplane design. For this type of study, a set of engineering methods will certainly give you results that will answer your questions (even if the results are not exactly accurate, you will probably find the better configuration). The individual aerodynamic simulations in this case can be completed rapidly and continuously on typical desktop computers.

If you want to know more details than just lift and drag, such as the pressures on an airplane during takeoff or landing while the airplane is close to the ground, panel methods will be able to fairly accurately give you what you want. An example is shown in Fig. 2.7, which shows the McDonnell-Douglas MD-11 near the surface of a runway during takeoff. This type of simulation can run in a matter of minutes on larger computer workstations and could even run in a few hours on a typical laptop.

And what should you do if you want to know about flow above a wing at a post-stall angle of attack, such as the F-15 during a spin, shown in Fig. 2.8? You definitely need CFD tools for that, and even then you may not be fully satisfied with your results! This level of CA requires significant computing power, often leading to the use of some of the largest and fastest computers in the world. In the case of the F-15, the simulation was performed on a CRAY XT3 computer using up to 4000 processors, and it required thousands of hours of computer time.

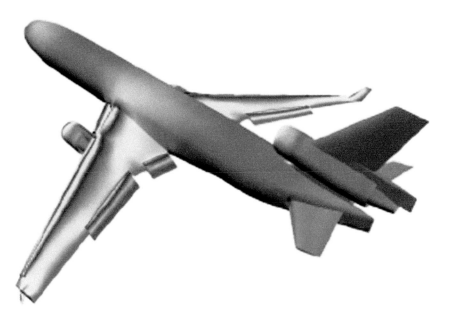

Figure 2.7   Surface pressures on an MD-11 near the runway during takeoff predicted with a potential method code, VSAERO. (Ref. 11; Courtesy of Analytical Methods, Inc.; a full color version of this image is available in the color insert pages of this text as well as on the website: www.cambridge.org/aerodynamics).

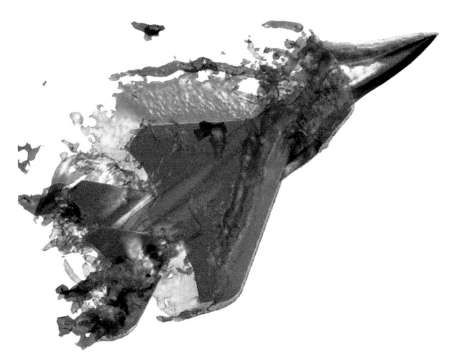

Figure 2.8   F-15 fighter during a spin simulation using a CFD code (courtesy of James Forsythe and the USAFA High Performance Computing Research Center; a full color version of this image is available in the color insert pages of this text as well as on the website: www.cambridge.org/aerodynamics).

Profiles in Computational Aerodynamics: Karen Gundy-Burlet

I was born in San Jose, California and have lived in the Bay Area all of my life. I became interested in aeronautics because of my grandmother, who told wonderful stories about her brother, Floyd Kelso. He was a barnstormer and he and his wing walker performed aerobatic stunts for audiences along the West Coast. My father was also a pilot and told funny stories about flying Ercoupes and Aeroncas in the San Jose area. Thus, when an opportunity to intern in aeronautics at NASA Ames Research Center was offered to me as a high school junior, I jumped at it.

As an intern, I worked in CFD on a full-potential airfoil analysis code with Dr. Terry Holst. I had always loved math and science, but developing algorithms, expressing them in computer code, visualizing them, and using the simulations to investigate physical processes was fascinating to me. To continue to work in the field, I earned a Bachelor of Science degree in Mechanical Engineering at UC Berkeley and returned to NASA Ames as a civil servant in computational aerodynamics. NASA then sent me to Stanford for a Master of Science degree and PhD in Aero/Astro Engineering. During this time, the state of the art in CFD was rapidly advancing, and I had the opportunity to work on increasingly higher fidelity applications, from inviscid wing simulations to viscous wing and then multiple-grid wing/body applications, then into the area of unsteady three-dimensional turbomachinery simulations with Dr. Man Mohan Rai. Verification and Validation techniques also evolved during this time, from simple analytical techniques and comparisons with available airfoil data, to purpose-built experimental rigs for specific validation of CFD data. Compilers and static analyzers improved in fidelity for catching common coding errors, and eventually "light" formal methods in the form of semantic analysis for verification of CFD code became available.

At this point in my career, NASA encouraged members of the CFD group to move into another area of aerospace. I started researching neural networks and learned control system theory to work on a damage-adaptive control system for aircraft. I became the program lead for the Intelligent Flight Control System. The system used online neural networks to augment control signals in case of damage or other degradation of aircraft performance. Pre-learned neural networks were used to supply stability derivatives (derived from experiment or CFD) to the feed-forward control algorithm, and the resulting commands were distributed across available control surfaces using a simplex algorithm. Validation of the algorithms was probably the most fun I've ever had! We devised a series of takeoff and landing scenarios with different types of damage, including failed tail surfaces, failed engines, and progressive hydraulic failure. We tested these scenarios with NASA, the U.S. Air Force, and commercial pilots in full-motion simulators. I served as co-pilot, test director, and recorded the pilot's evaluation of the aircraft's performance. The successful tests paved the way for piloted evaluations of the algorithms in an F-15 aircraft.

After the IFC program, I became a researcher in the Reliable Software Engineering group at Ames. My focus turned toward verification and validation algorithms for spaceflight simulations and software. The group has developed a range of techniques, including formal methods (rigorous mathematical proofs of correctness), model checkers, static analysis, and automated certification of software. I led the development of a code that combines advanced test case generation with machine learning algorithms to determine input parameter ranges that lead to success and failure. I am also the Flight Software Quality Engineer on the Lunar Atmosphere and Dust Explorer (LADEE) mission. For this mission, the flight software is modeled in a graphical environment and then is autocoded for use in the embedded environment. I helped develop the plans, processes, and testing framework that leverage the model-based environment to perform efficient and effective verification of requirements and validation of the flight software in the embedded environment.

I am busy at home as well, with a husband, kids, cats, dogs, horses, and chickens to enrich my life. I enjoy family vacations in our trailer, visiting historic sites and rockhounding. I like to ride horses and garden, and I am an avid reader in those rare quiet moments.

2.3 Computing Systems

You probably use computers a great deal in your engineering studies or work (the depiction in Fig. 2.9 may seem typical for today's students). Programs for word processing, spreadsheet calculations, making briefing slides, and even performing powerful mathematical calculations are common today. The modern engineering student must learn how to use these powerful and

Figure 2.9 Working with computers has become commonplace in today's world.

essential programs, but as we have progressed toward more advanced and user-friendly programs, the average student has become less and less knowledgeable about computers and how they work. A typical engineering student of twenty or thirty years ago would likely know a great deal about operating systems, programming languages, numerical methods, and data manipulation, but modern computer programs have left a great deal of that knowledge with the experts. In fact, our experience with students and new engineers is that they know much less about computing systems than students from the past, even though the computer power they carry with them is far greater.

Computational Aerodynamics Concept Box

How Do Computers Work?

Before we discuss the types of computer systems used to perform analysis and design for computational aerodynamics, we should first define a few computer terms. Computers are made up of various components, but the most basic part of a computer is the processor, which is often called the *central processing unit* (or CPU). This is the part of the computer where calculations are done and other functions are performed, as defined by a computer *program* (or code). The codes are written in a variety of computer languages, and for engineering purposes those languages are usually *high level programming languages* (such as FORTRAN or C++). Higher level languages were created for various user groups to enable them to program within their fields of interest without having to understand the most basic commands that the CPU uses. The CPU language is often referred to as a *low-level programming language*, and very few engineers and scientists utilize that level of programming.

In order to create commands that the computer understands (since it "speaks" the low-level language), the high-level language has to be "translated" into the low-level language by a *compiler*. The CPU and the other devices located with the computer (like the monitor, keyboard, etc.) are called the computer *hardware*, while the languages and operating systems that make the computer work are called the computer *software*. The total computer "system" is made up of the hardware and software, and different computers put all of these things together in different ways; how the computer is designed is often called the computer *architecture*.

The user interface with the computer hardware, and the system that runs programs, is called the *operating system*, which is the "manager" of the computer's resources. Modern computers use operating systems including DOS, MAC OS, and Unix, among others. There is a good chance that your computer uses a Windows environment, which is typically created to run on top of the computer's operating system, creating a user-friendly environment for you. Unlike when you use modern desktop or laptop computers, advanced CA users will often need to know how to use an operating system in order to run their programs. Finally, the user often needs to interact with the computer, which requires *input/output devices* (often called I/O devices). The I/O devices could be as basic as the keyboard, mouse, monitor, etc., or for more advanced computer systems these could include mass storage systems (among other things).

Most of you are familiar with many of the basic components of a computer system; if not, you can review the previous Concept Box to learn some basics. You might need to understand how to use the I/O devices (including the keyboard and monitor with which we communicate), the processors that run the software and perform the calculations, and the memory and storage capacity of your laptops and the communications devices (including wireless). The low-fidelity methods of CA will run on most of your computers with no problem. In contrast, the high-fidelity methods of CA may tax even the most sophisticated machines with the fastest processors and the largest memories.

2.3.1 Why CA Requires Large Computers

It is important to understand these limitations for the various levels of fidelity as we explore CA further. Not all CA capabilities will be accessible to all users – here's why. If you are solving the three-dimensional Reynolds-Averaged Navier-Stokes equations (which we will define in Chapter 3), you are calculating (in one way or another) the following flow parameters: density, velocity (three components), pressure, temperature, and viscosity. You will probably store these seven variables in your computer's memory as you calculate. In addition, if you represent the three-dimensional flowfield with grid points, you will require the (x,y,z) position of each grid point, which

Chimera domain decompostion of SSLV surface

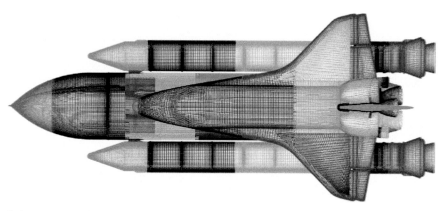

Figure 2.10   Surface grid on the Space Shuttle Orbiter and booster engines, showing the large number of locations where CFD calculations have to be performed (Ref. 12; a full color version of this image is available on the website: www.cambridge.org/aerodynamics).

requires three more variables. For example, Fig. 2.10 shows the surface grid on the Space Shuttle Orbiter and boosters.[12] Every intersection of lines on the surfaces represents a location where all the flow variables are computed. This "mesh" of lines and points extends into the space around the Space Shuttle to create millions of points where calculations need to be performed. So we now need to keep track of at least ten variables as we perform our calculation at each of these grid points. A three-dimensional grid around a full aircraft at flight Reynolds numbers could easily consist of 10 million grid points. So now we need to know ten values at 10 million grid points, or, in other words, at any one moment during the solution process we have to keep track of 100 million pieces of information. We will need to perform some sort of calculations on each of these pieces of information (a numerical *algorithm* will be used to find the solution), which will require performing thousands of operations on these 100 million pieces of information. How long will that take? Is it okay if it takes six months? Will you remember the question you were asking if you wait six months to find out the answer? Maybe you will, but maybe not!

These millions of calculations are the reason CA codes have traditionally been heavy users of state-of-the-art computers since the first CA codes were written. We often require the fastest, largest memory machines available to perform our work. The limited operating speed and memory capability of earlier computers reduced the level and complexity of potential CA simulations. Luckily for those of us who work in CA, computer processing speeds, memory, and storage have been increasing very rapidly. This has allowed us to attempt more and more complicated flow solutions, starting with simple two-dimensional inviscid flows decades ago to complex three-dimensional viscous flows today. This is why CA practitioners need to understand the systems and capabilities of modern high-speed computers. Not only has the development of computational aerodynamics been closely linked to the development of computers, but so have the computing systems and software being used (recall from Chapter 1, where we discussed that practitioners of CFD often need to know a great deal about computers in order to get their work done).

2.3.2 CA Historical Development

Long before there were useful computers, applied mathematicians developed a core of *numerical analysis* techniques. Numerical analysis is a field of mathematics that develops algorithms (or "recipes") for solving mathematical problems with approximate methods. These are especially useful when analytic techniques cannot solve a problem. Prior to the twentieth century, techniques such as Jacobi's method and the Gauss-Seidel method (both of which we will learn about in Chapter 6) had been developed to perform matrix calculations. Many of these techniques are still used today on computers. These techniques were often performed with human "computers," clerks who performed various mathematical operations using efficient hand calculating methods. Many early theoretical calculations in aerodynamics were done by rooms full of these clerks. With the emergence of the digital computer as a useful tool during the 1950s and 1960s, the amount of research into numerical analysis grew dramatically, and aerodynamicists eagerly awaited the new capabilities (or, in some cases, they even created the algorithms!).

These advances in computers and numerical algorithms led to some of the most important early computational aerodynamics work on "the blunt body problem."[13] At that time the prediction of the heat transfer and flight characteristics of ballistic missiles and manned space capsules entering the atmosphere were the "hot" items in fluid mechanics (the pun is intended!). The largest available computers being used for these calculations were slower and had much less memory storage than even the first personal computers. Trying to solve problems that tax the capability of the available computer is typical in computational aerodynamics. Despite rapid advances in computing technology, aerodynamicists always demand more computer speed and storage. For example, vortex lattice methods (a type of potential flow method) for aircraft applications were originally developed in the early 1960s.[14] Vortex lattice methods are still useful today but no longer require state-of-the-art computers. Panel methods (also potential flow methods) for full aircraft were developed at about the same time by Hess and Smith at Douglas Aircraft Company.[15] These programs required state-of-the-art computers when they were first written but are now easily run on PCs.

The introduction of the IBM System 360 in the mid-1960s revolutionized access to computers for nonspecialists. This was the first widely available, easily used computing system. However, it took hours or even days to get a simple job executed (often only to find out that there was a mistake in the input instructions, requiring an immediate resubmission). With the introduction of FORTRAN IV, the scientific computing community started using a language that would be stable for many years. Control Data Corporation (CDC) introduced the CDC 6000 series computers at about this time, and the CDC 6600 became the computer of choice for scientific computing (Seymour Cray was one of the key designers of that computer – he would later design many of the world's fastest supercomputers). The CDC 7600 was introduced later, and the machine at NASA Ames was a true workhorse for aerodynamics. Later, Cray formed his own company and provided the world with numerous innovative supercomputers, which will be shown in the following pages.

At a 1975 conference (Aerodynamic Analysis Requiring Advanced Computers) at NASA Langley, one speaker drew on the rapid advances in computer capability to present a chart (see Fig. 2.11) that could be used to project that computational aerodynamics would be fully developed by 1984.[16] This prediction was based on the rapid increase in computer capabilities coupled with the equally rapid development of new codes that could predict more and more complex flows. Unfortunately, this prediction alienated a large number of experimentalists (since the prediction implied that experimental research would no longer be needed), which resulted in a great deal of distrust between the computational and experimental approaches in aerodynamics. Hindsight has shown that aerodynamics requires using experimental, theoretical, and computational tools together in an integrated fashion, which is why aerodynamics can be so challenging.

Throughout the 1970s, aerodynamicists were routinely solving linearized inviscid three-dimensional incompressible flow problems, and two-dimensional boundary layer methods were available. The most important problem being tackled in 1970 was the computation of transonic flow. Richard Whitcomb had revolutionized airfoil design with his "supercritical" airfoils,[17] and that showed the need to be able to compute transonic flow. An important breakthrough for transonic flow solution methods was reported in 1970 by Murman and Cole,[18] and the first practical solution procedure was reported in 1971. The development of solutions for two-dimensional transonic flow dominated the first half of the 1970s. Three-dimensional transonic small disturbance theory solutions and some three-dimensional full potential solutions also began to appear in the mid- to late 1970s. These methods allowed designers to create "supercritical" airfoils and wings (see Fig. 2.12), which increased the maximum speed of commercial and military aircraft to new levels.

Most of these methods were based on the inviscid flow assumption with viscous effects included through the use of boundary layer calculations, if necessary. By the late 1980s and the early 1990s, researchers were able to

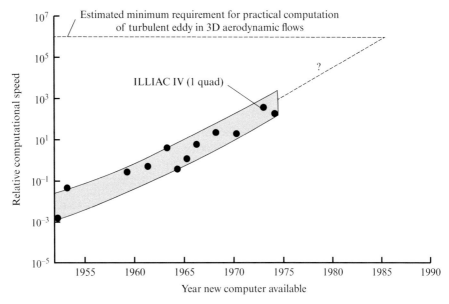

Figure 2.11 Estimated date for 3D aerodynamic viscous flow prediction from 1975 (Ref. 16).

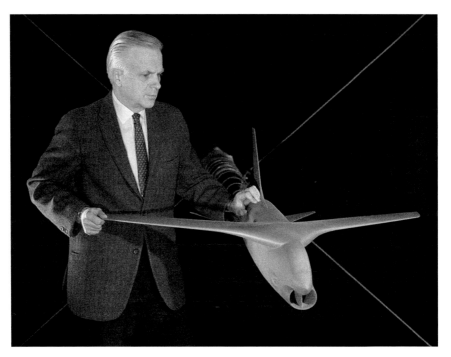

Figure 2.12 Richard Whitcomb, the designer of the supercritical wing concept, getting ready to test an aircraft with his design (Courtesy of NASA Langley Research Center).

perform viscous calculations on complete aircraft. All along, researchers were using every available computer tool at their disposal to solve increasingly more complex problems. This will continue to be true for many decades to come. In fact, Phillipe Spalart of The Boeing Company estimates that the growth in *processor* speed and memory capability will not allow direct numerical simulation of full aircraft at flight Reynolds numbers until the year 2080.[19]

2.3.3 Computer Measures of Merit

But how do we relate computers and computer speed when we are comparing one computer to another or one code to another? The basic speed unit for a computer is the floating point operations per second, or *flops*. A floating point operation is a basic arithmetic operation, such as addition, subtraction, or multiplication. CA codes are written with thousands of lines of commands using high-level programming languages, and each line requires various numbers of floating point operations. As the codes become more capable (and therefore longer), there are more and more operations to perform, which will take longer on the computer. This has led to a requirement to have faster and faster computer processors, whose speed is usually measured in three order-of-magnitude increments as:

Flops	Order	Defined as:
1 million	1×10^6	Megaflop or Mflop
1 billion	1×10^9	Gigaflop or Gflop
1 trillion	1×10^{12}	Teraflop or Tflop
1 thousand trillion	1×10^{15}	Petaflop or Pflop (current state of the art)
1 million trillion	1×10^{18}	Exoflop or Eflop

So how do we fit into the computer capabilities that exist, and how do we decide what we can do with the computer systems that we have? There has always been a very strong correlation between state-of-the-art computers and computational aerodynamics, as was mentioned previously. As new computer capabilities became available (such as the ILLIAC IV from the late 1960s shown in Fig 2.13), new applications were tried and improved.

Since the 1970s, a great deal of effort has been expended to make the fastest and largest computers available to CFD researchers, so supercomputer centers were established around the country, including the Numerical Aerodynamic Simulation (NAS; now called the NASA Advanced Supercomputing Division) facility at NASA Ames Research Center. These facilities obtained the latest *supercomputers*, which are very expensive computers available for specialized applications that require large numbers of calculations and memory storage. These facilities use computers that are at the cutting edge of *Moore's law* (which essentially states that processor performance doubles every 18 months), and researchers in CFD want these fast computers as soon as they become available. Some of these computers included the CRAY 1, the CRAY 2, the CRAY YMP, and the CRAY C90 (all shown in Fig. 2.14), which were among the first computers to take advantage of *vector processing* and *parallel processing* [vector processors

Figure 2.13   The ILLIAC IV computer, which was initially developed in the late 1960s (Courtesy of NASA).

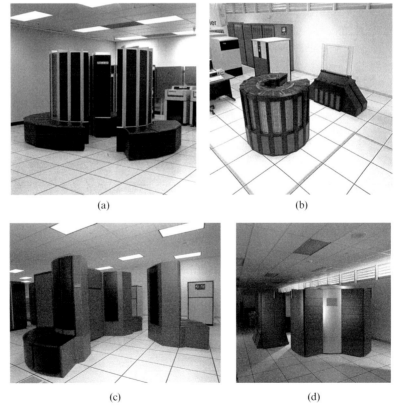

(a)   (b)

(c)   (d)

Figure 2.14   Various supercomputers previously used in NASA's NAS facility (Courtesy of NASA Ames Research Center). (a) CRAY 1S (c. 1976); (b) CRAY 2 (c. 1985); (c) CRAY YMP (c. 1988); (d) CRAY C-90 (c. 1991).

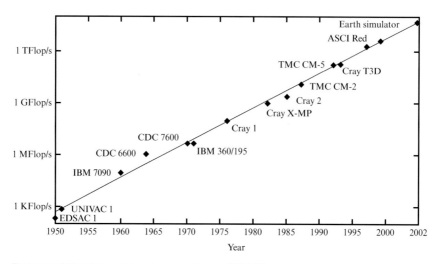

Figure 2.15   The impact of Moore's law on computer processing speed (Ref. 20).

used specialized processors which could simultaneously perform calculations on variables with several values (vectors), and parallel processors have multiple processors which are available for performing any calculations in parallel; these processors are in contrast to *serial processing*, which can only do one set of computations at a time].

These computers had state-of-the-art technology in processors and computer architecture and have shown that Moore's law is still alive and well, as seen in Fig. 2.15.[20] All evidence shows that the growth in computer speed has been maintained over the past forty years, and the future looks equally bright. The advent of computer "clusters" that take advantage of parallel processing (running concurrently on multiple processors) concepts has enabled Moore's law to extend far into the future. For example, the second "fastest" supercomputer in the world in 2004, Columbia (shown in Fig. 2.16), was a 10,240-processor supercomputer with a sustained performance of 42.7 teraflops (42.7 X 10^{12} floating point operations per second), sometimes called a *massively parallel computer*. By 2009, Columbia had been upgraded to a 14,336-processor machine with a theoretical peak performance of 88.8 teraflops. A follow-on computer to Columbia, NASA's Pleiades, garnered the sixth spot on the Top 500 list of the world's most powerful computers. The Pleiades supercomputer is an SGI Altix with 14,080 quad-core processors (56,320 cores, 110 racks) running at 544 trillion floating point operations per second (teraflops) on benchmark codes (the industry standard for measuring a system's floating point computing power). One of the most powerful general-purpose supercomputers ever built, Pleiades also features the world's largest interconnect network. The Air Force Research Laboratory's DoD Supercomputer Resource Center purchased a parallel machine containing more than 45,000 multi-processor computer cores capable of sustained petaflop performance. In addition, the Air Force Research Laboratory purchased about 1,760 PlayStation 3 computers

Figure 2.16   Columbia, NASA's massively parallel computer, c. 2004 (Courtesy of NASA Ames Research Center).

Figure 2.17   Condor, a U.S. Air Force Research Laboratory parallel computer using 1760 PlayStation 3 computers, c. 2010 (U.S. Air Force photo).

plus 168 graphics processing units, as well as other graphics-heavy computer components, to create a "homemade" supercomputer called Condor (shown in Fig. 2.17), which is capable of 500 teraflops for approximately fifteen times less power consumption than a typical supercomputer.

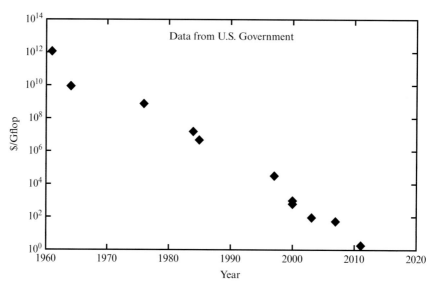

Cost of supercomputers per Gflop over time.

A paradigm shift has also taken place in that some of these parallel clusters have become very affordable (see Fig. 2.18 for a graph of computer cost per Gflop for the past fifty years), which means researchers do not have to be tied to major computing centers for computational work nearly as much as in the past. The Cray 2 cost approximately $15 million in the mid-1980s, and was only available at large computing centers with limited access and usage. Modern computer clusters (with much more memory and faster processors than the Cray 2) can cost a mere $100,000 and thus make high-end computing available to a large number of people in industry and academia. With this "maturing" of computational aerodynamics, the future looks bright for CA becoming an even more important part of aeronautical engineering, taking its place alongside wind tunnel testing and flight testing as pillars of aeronautical "experimentation."

Traditional *scalar processing* available on personal computers (and many standard mainframes) have been replaced by vector, parallel, and massively parallel machines. To use these machines effectively, the codes must exploit the specific advantages of the machine architectures.[21–24] A relatively new processor architecture is the *graphics processing unit* (GPU).[25] GPUs traditionally have been used for gaming acceleration on personal computers but are rapidly evolving for use in general-purpose scientific computations that could be termed *personal supercomputing* (PSC).

The key issue in advanced computing revolves around how to increase the computation speed. This includes the basic processor speed, the size of memory, and the speed of the data transfer through the machine. Although the "raw" computation speed can be misleading, it is nevertheless used

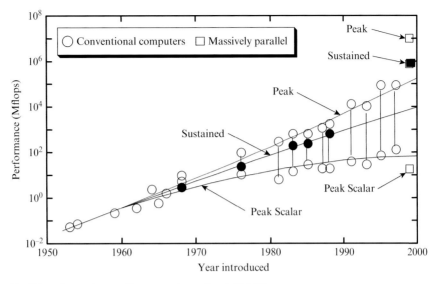

Figure 2.19 The history of computer speed increase and current goals (Ref. 26).

to quantify the speed of computers. With advanced computers there is a large difference between the peak speed and the maximum sustainable speed obtained in practice. A partial history of computer speed advances is shown in Fig. 2.19. The top line is the peak performance including advanced approaches (vector and parallel architectures). The bottom curve shows that the scalar, or serial, computing speeds may be starting to reach a limit.

2.3.4 Parallel Computer Scalability

In addition to the speed and cost of the computers, there are various ways to take advantage of parallel processors in computational aerodynamics. First, the machines can be used in either "coarse-" or "fine-" grain parallelization modes. The coarse-grained mode is of particular interest in aerodynamic and multidisciplinary design. Here, many solutions are required using the same program with different inputs. This is done to find the sensitivity of the design to various design variables, so the same code working on slightly different problems is run on different processors or nodes at the same time. This is one of the easiest ways to exploit the capability of parallel computing. Fine-grained parallel computing, in contrast, requires that the code be modified to make a single calculation using numerous nodes; achieving this has proven to be much more difficult. A scalability study was performed with the Navier-Stokes flow solver Cobalt (see Fig. 2.20) on a Cray XE6 computer for a grid with 49 million cells with up to 22,000 processors being used.[27] These results show the codes are well written, with minimal scalar computation, and can maintain linear speedup for a

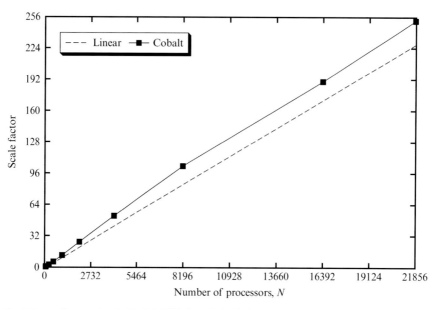

Figure 2.20 Parallel computing speed-up for the Cobalt Solutions Navier-Stokes solver (Ref. 27).

relatively large number of processors (linear speedup means a code taking one hour to run on 32 processors would take a half hour to run on 64 processors). These results were variable, however, depending on the number of cells per processor and the way that the computer communicates information. This type of speedup requires that the code programmers take full advantage of the particular computer architecture they are using, and it remains to be seen if these trends can be continued as massively parallel computer systems are used.

The issue with scalability is whether the speedup obtained using a small number of processors can be extrapolated to cases in which a large number of processors are used. Our experience shows that the performance achieved with a small numbers of processors, say twenty to thirty, does not scale up linearly when hundreds or thousands of processors are used. One standard computer science rule-of-thumb, *Ahmdahl's law*, says that the speedup decreases to a finite limit, which depends on the fraction of the code where serial (or scalar) computations are required (such as message passing). Figure 2.21 shows that if even small parts of the code require sequential computation, the speedup using parallel processing will not increase without limit. In the figure, R is the fraction of the code requiring serial computation and N is the number of processors used.[26] If none of the code requires serial computation ($R = 0$), then the linear trend is maintained; otherwise a slowdown is inevitable. Some computational scientists are currently trying to demonstrate that for CFD this "law" is not valid, and the trend can be shown to be approximately $R = 1/N$.

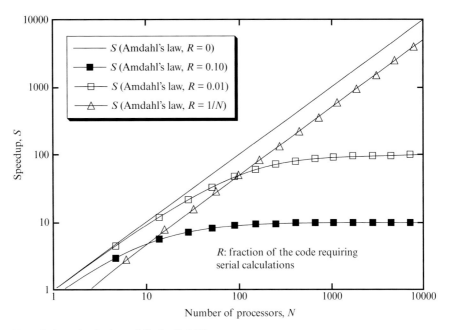

Speedup, S

- S (Amdahl's law, $R = 0$)
- S (Amdahl's law, $R = 0.10$)
- S (Amdahl's law, $R = 0.01$)
- S (Amdahl's law, $R = 1/N$)

R: fraction of the code requiring serial calculations

Number of processors, N

Figure 2.21 Theoretical speedup due to parallelization (Ref. 26).

Figure 2.22 depicts the trends in computational power from 1940 with extrapolation to 2020. The period from 1990 to today has followed a very predictable performance improvement path. In the 1990s, this performance was achieved by improved clock speed with modest core count increases (modern processors have multiple core components which are capable of independent work). In more recent years, this performance improvement is due to very large core count increases using parallel processing. This change in machine architecture design requires a significant change in the software being developed to run on the next generation of machines. If the trend continues, machines with one million cores are just around the corner. Unfortunately, the majority of older software packages typically run on 64 to 256 processors with dramatic performance penalties above 512 processors. The very best current CFD codes scale linearly to 5,000 cores (see Fig. 2.22), which is three orders of magnitude smaller than the million-core machine envisioned for the near future. In other words, our CFD codes are not written to adequately take advantage of our current and future computer systems. For this reason, future software programs should have a strong focus on algorithmic improvements for parallel scalability.

While the preceding section has described how computers work and how they have advanced over time, there is much more to CA than just the computers that we use. Computers perform calculations, but codes accomplish tasks, and CA is about intelligently running codes on computers to complete a task. Putting all of these aspects together is what makes CA challenging.

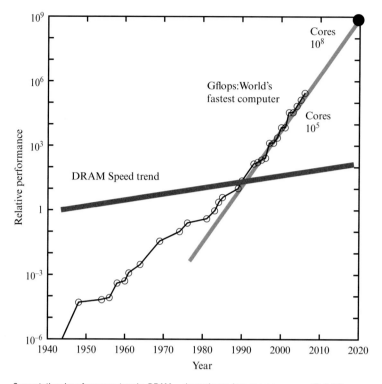

Figure 2.22 Computational performance trends; DRAM = dynamic random access memory (Ref. 28).

Flow Visualization Box

How to Obtain and Use Fieldview

One of the unique ways this book will approach computational aerodynamics is to use modern flow visualization software to help you explore the various concepts we are talking about. In order to do that, you will need access to flow visualization software – Intelligent Light has given us permission to offer you access to a demonstration version of FieldView®. Within the demonstration software is a dataset for the F-18 aircraft at 19 degrees angle of attack, as shown in the accompanying figure; we will also supply some solutions for flow over airfoils. We will look at these solutions throughout the following chapters, but first you need to get the demo version of FieldView.

How to access the software:

1. Go to the Intelligent Light web page for the Demo version: http://www.ilight.com/en/
2. Click on *Try Fieldview 14 (you will receive Fieldview 13 as the Demo version)*
3. Fill in information for *Demo Version*
4. You will receive an email with a link to the Demo version
5. Download and install Demo Version (Linux and Windows versions available)
6. Windows version is 72.3Mb
7. Once installed and started you will see the control window as shown:

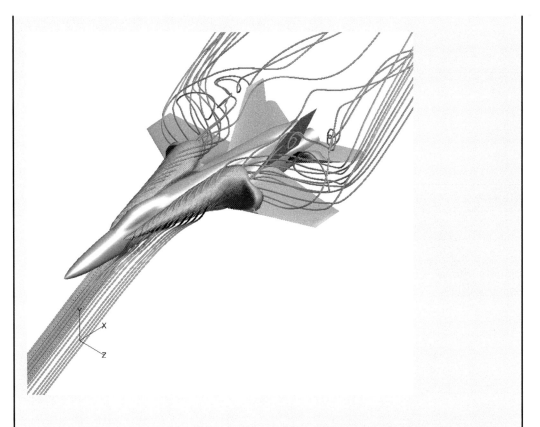

How to view the F-18 demo:

By pressing *Start/Restore* under the F-18 picture you will be able to see the picture shown above. There are multiple options for viewing the flow solution, each with a button below *Start/Restore*:

1. C_p Iso-surface
2. Streamline Displays
3. Show/Sweep Y Plane
4. Animated Cut-Away
5. C_p at Wing Tip

We will look at each of these to begin learning how to view a flow solution dataset. The default view is a combination of *Streamline Displays* and *Show/Sweep Y Plane*, and looks like the image shown in the figure:

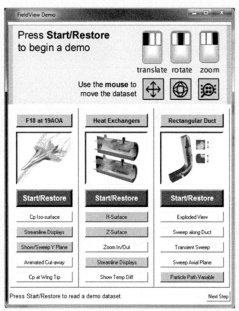

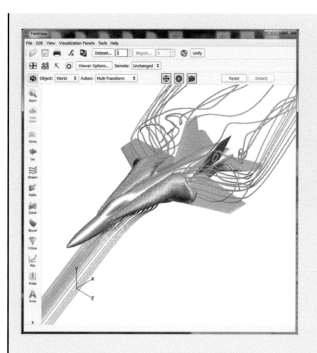

Now look at the *Cp Iso-Surface* result:

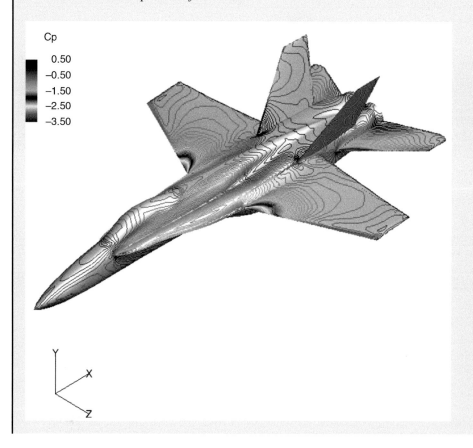

Do you know what *Cp* is? Could you define *Cp*? We will discuss pressure coefficient and other aerodynamic quantities used for flow visualization in Chapters 4 and 9.

The *Animated Cut-Away* is best viewed "live," so enjoy looking at that one yourself, and see if you can figure out how it was done. Finally, the *Cp at Wing Tip* looks like this:

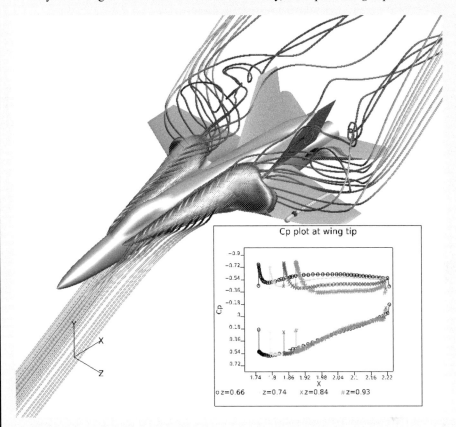

In this case both flow visualization and plot results are used together.

Notice that there are pull-down menus at the top of the window and various control buttons on the left – you can accomplish many tasks by using either of these approaches. As you explore these demos be sure to notice the following windows and start to become familiar with the various Visualization Panels, such as:

1. Computational Surface
2. Iso-Surface
3. Streamlines
4. Particle Paths
5. Coordinate Surface
6. Boundary Surface
7. Vortex Cores/Surface Flows
8. 2D Plots

Now that you have the flow visualization software you can experiment with it and see how it works. We will be using FieldView throughout the rest of the chapters to demonstrate various concepts, so it will be to your advantage to become adept at using the software.

2.4 Computer Codes: Verification, Validation, and Certification

Whenever a computer simulation models a physical process or system, it is critical to establish some level of trust in the simulation code. How do you know it works? When it does work, how well does it work? When doesn't it work? These are basic questions any engineer should ask (and answer) before using a simulation method. And while understanding the capabilities and limitations of a particular code is essential to good engineering practice, it is not always easy to quantify just how "accurate" a particular code is. When conducting experiments, it is possible to quantify precision errors (how repeatable the results are) and bias errors (how accurate the measurement is). Computer simulations may be very repeatable (you should usually get the same output for the same input), but it is quite difficult to quantify simulation accuracy.

The problem of defining accuracy in CA is so difficult, an American Institute of Aeronautics and Astronautics (AIAA) committee (the AIAA Committee on Standards for Computational Fluid Dynamics) has been working since the late 1980s trying to formalize the issues and concerns for computer simulation in engineering. The committee struggled with this task because "there were not any well-organized or accepted general procedures for building confidence in CFD ... there [also] was no widely accepted terminology for these activities within the CFD community or the wider communities of computational mechanics and heat transfer."[29] In spite of these difficulties, the AIAA committee, and other like-minded people in the American Society of Mechanical Engineers (ASME)[30] and at Los Alamos National Laboratory,[31] came up with two important concepts related to computer simulation: *verification* and *validation*. In addition to these two concepts, some researchers have also defined the requirements for the *certification* of a simulation approach. Establishing a scientific basis for certifying codes has also proven difficult in spite of the significant attention it has received.[32–34] Finally, in some cases, it is necessary to perform *validation for an intended purpose*. A good example of this is the AIAA Drag Prediction Workshop (DPW), where many experienced CFD practitioners predict the flow over complex fuselage-body configurations with codes that have passed all of the verification, validation, and certification requirements for transonic flow over simple configurations. Unfortunately, the results for the more detailed configurations used in the DPW are not nearly as good as you would hope.[35] The workshop practitioners have discovered that validation must include the entire CA process, from geometry definition to grid generation to flow solver to post processing. Ed Tinoco of Boeing says, "Two keys to effective CFD are consistency (or repeatability) of the 'lofts to plots' process, and [the process] being good enough for the problem of interest."[35] The human factor of these concepts should not be discounted either, since frustrations with the

Table 2.2 Verification assessment classifications and descriptions (Ref. 3, 30)

Classification	Focus	Focus	Responsibility	Methods
Code Verification	Software Quality Assurance	Reliability and robustness of the software	Code developer and Model developer	Configuration management, static & dynamic testing, etc.
	Numerical Algorithm Verification	Correctness of the numerical algorithms in the code	Model developer	Analytical solutions, benchmark problems, manufactured solutions, etc.
Calculation Verification	Numerical Error Estimation	Estimation of the numerical accuracy of a given solution to the governing equations	Model developer	Grid convergence, time convergence, etc.

acceptance of CA results have led some to lament, "Everyone believes experimental results except for the one doing the experiments; no one believes computational results except for the one doing the computations."

The semantics and definitions for verification, validation, and certification are still the subject of discussion, as can be seen in the papers by Roache[36] and Aeschliman et al.[37] In general, the basic terminology related to computational modeling codes may be defined this way:

1. verification: "the process of determining that a computational model accurately represents the underlying mathematical model and its solution,"[29] or in other words, that the mathematical model and solution algorithm are working correctly; this lies within the realm of mathematics
2. validation: "the process of determining the degree to which a model is an accurate representation of the real world from the perspective of the intended uses of the model,"[29] or in other words, that the discrete solution of the mathematical model is accurate; this lies within the realm of physics
3. certification: the process of establishing the range of applicability of a verified and validated computational model

Okay, that sounds good, but how do you do it? Again, that has proven to be a more difficult question to answer than to ask. Luckily, however, the AIAA and ASME committees have provided a guide to understanding these issues and have defined the verification process with the help of Table 2.2.[30,38] As you can see, the verification process includes code verification and calculation verification. This may begin with comparing the computational solution with well-known analytic solutions to the differential equations that describe fluid flow, such as laminar flow over a flat plate, flow between two plates, and other similar basic flow types. Other comparisons of the model can be made against theoretical results (such as inviscid vortex theory), benchmark problems, or manufactured solutions. Finally, the verification of the calculations

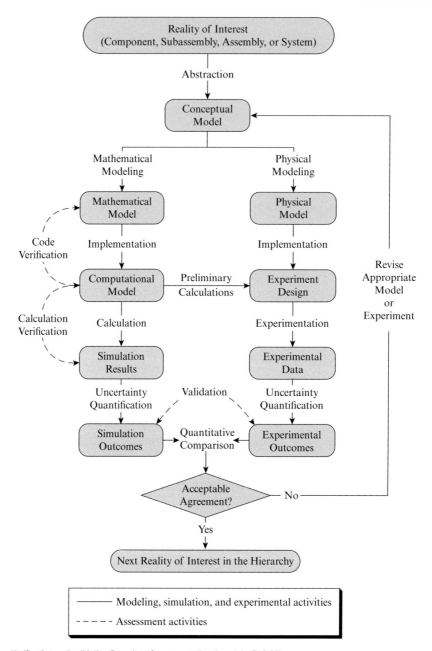

Figure 2.23 Verification and validation flow chart for computational models (Ref. 30).

must also be assured by performing grid and time-step convergence studies.[30] If the computer model is not working for these basic problems, then there probably is an error in the coding or implementation that must be corrected prior to full-scale use of the model.

If the computer code passes the verification tests, then it is ready for the validation process. This involves many more steps and is easier to see in context of the complete verification and validation process, as shown in Fig. 2.23.

This step probably is more difficult than the verification step and can take a great deal of time to perform correctly. The validation step requires comparing solutions from the computer code with real-world experimental data for basic flow problems and then having the model solve increasingly more complex problems for which there are experimental data. Some of these experimental cases are called "benchmark cases," since the experimental data have been used for many computer codes and offer a way to compare your results with the results of other codes or approaches. This step lends confidence that the model is valid and is producing results that provide insight into the flow physics of interest. In addition, going through a peer review process (including code "walk-throughs" and presenting results at a seminar) can be invaluable to the code developer.

In practice, users should develop a library of test cases for comparing with codes. Appendix B provides a list of basic experimental data that have been used for model comparisons. This allows for determining the model sensitivity to key solution control parameters, which should be well understood by the user. When looking at validation information provided by a code developer, you should always be wary if you only see solutions that compare well with experimental data. One common method of presenting validation results has been to use well-known validation data but to leave out some of the results in the available data. For example, one transonic wing test case includes data at a span station that shows a double-shock system. For years, code developers presented results for the wing without including this span station in their comparisons: they left out the most interesting, important, and difficult-to-predict data on the wing! Another misleading approach is to compare results with data but not with other theories. Modelers may present comparisons of results for Euler or Navier-Stokes solutions without including computations for those same cases from simpler theories. The impression they are providing is that they are presenting results for cases that could not previously be computed. The truth is that, very often, engineering methods, small disturbance, full potential, and full potential/boundary layer methods are able to demonstrate results that are as accurate as the solution of the more exact equations.[i] More discussion on this important topic will be presented in Chapter 10. There we will provide more details of the issues associated with code validation for computational aerodynamics. Other important information may be found in the book by Oberkampf and Roy.[39]

The AIAA committee (as of this writing) still has not agreed upon a complete set of requirements to achieve the goals of verification, validation, and

---

[i] A colleague we know who develops CFD codes objects to this observation. However, we believe that lower-order methods can often provide very good results for certain classes of flow and should be shown while performing validation of codes. Ed Tinoco of Boeing says, "Whatever the reason, we have routinely gotten as good or better agreement in predicting wing pressures with *TRANAIR* than with Navier-Stokes."

certification. However, they have agreed on six numerical accuracy areas that are important to understand when conducting a numerical simulation:[29]

1. Statement of *numerical methods*: authors of papers that use CA should "be clear and precise in the description of all important methods used in the investigation." In other words, don't just say you used *Code XYZ*, but state what numerical algorithms are employed, what orders of accuracy the numerical methods have (both for the flow interior and the boundary conditions), and what values of adjustable parameters were used.

2. Minimum *formal accuracy* of numerical methods: numerical methods should be formally at least second-order accurate for spatial simulations of continuous flows; non-continuous flows (such as shock waves and other singularities) should be at least first-order accurate; temporal accuracy for time-dependent solutions should be at least second-order accurate.

3. Statement of code verification activities: authors of papers that use CA codes should refer to papers that show how well the model works for a class of problems related to the work being reported.

4. *Spatial convergence* accuracy: a grid resolution study should be performed to show the level of grid dependence, or if possible, quantify the discretization error of the solution.

5. *Temporal convergence* accuracy: time-dependent solutions should quantify the error of the time integration method or show different results for different time steps.

6. *Iterative convergence* accuracy: authors must address the level of convergence of their solutions using a relative error comparison.

So, what does all of that mean? Right now, as you read through this book for the first time, many of the terms and concepts used in this section may seem difficult to understand. Our hope is that information provided in later chapters will make these concepts and ideas a little less confusing, but for now you should realize that there are standards for using aerodynamic codes, and you should always strive to achieve those standards. We will try to describe how these accuracy areas and verification and validation issues can be approached as we develop your knowledge of CA methods.

2.5 Some Comments on Programming

While most of you will never "code" a large CA program in your career, there is a chance you will modify one of these large "legacy" codes, or even write smaller programs to perform work with the results of these codes (gridding, graphing, etc.). You probably have heard college professors harping on the importance of good programming practices, and you also probably never fully understood why you needed to go to all of that trouble for your ten-line program. When you start writing programs that extend to hundreds or thousands of lines of code, those admonitions from your professors will begin to make sense.

A good friend of ours, Mike Giles at Oxford University,[40] took the time to write down what he considered the objectives of good programming, as well as some of the best practices that would help you write good codes. The first, most important thing you should do prior to writing a code is to keep some of these objectives in mind as you plan and execute your programming:

1. The code should execute correctly – no errors or "bugs."
2. The algorithm should be robust and accurate – best to compare with well-known test cases.
3. The program should be flexibly structured – you never know when you will adapt or modify the code in the future, so plan for that to happen while you are programming.
4. The program should be easy to use – obviously not all programs achieve this objective, but it is worth aiming for.
5. The program should be easy to maintain and update – if you make continuous changes to a code without keeping good track of what changes were made (and when they were made), you will get quickly lost in all of the versions of your program.
6. The program should be portable across platforms – will your program only run on your computer, or can you use it on many computers?

Once you have planned for these objectives in your initial phases of programming, you should also keep in mind some of the following "best practices" while you code.

Before starting to program:

1. Think about the objectives of your program and their relative importance
2. Review requirements for the program with future users
3. Review the relevant theory carefully
4. Plan complex software carefully
5. Think about how you will debug (correct) your code
6. Make the program modular (use self-contained subroutines and functions often)
7. Make the program readable – if no one else can tell what you've done, then no one else will be able to fully use your code
8. Put in extensive comments to help achieve the previous point
9. Show someone your code – just like you might have someone "proof read" a paper you have written, have someone look over your code

When you start debugging (fixing your code when it doesn't work):

1. Don't panic, and proceed logically in small steps
2. Find a concrete symptom and trace it back
3. Use the latest debugging tools
4. Remember that compilers aren't perfect, but neither are you (don't blame everything on your computer!)
5. Talk it through with someone
6. Suspect everything, including the diagnostic results
7. Beware of "cut and paste" errors

When your programming is complete, be sure to document:

1. Make extensive use of comments in the code (previously mentioned)
2. Show the results of validation test cases
3. Write a user's guide (for programs that will be used extensively by others)
4. Write down the theory used in the program
5. Maintain a "suite" of test cases and a guide for others to use

You should also plan to maintain your program after it is written:

1. Keep track of version numbers with dates of release
2. Document when and what changes have been made
3. Be sure to fully test all changes in the new version

By following these guidelines you will save yourself a great deal of pain and anguish when you program. Ignoring this advice can be seriously dangerous to your sanity!

Profile in Computational Aerodynamics: Christopher J. Roy

I was born in Mineral Wells, Texas in 1970. My father was in the military, so we moved around quite a bit during my childhood. Early on during high school in Fayetteville, North Carolina, I realized I had an aptitude for math, science, and computer programming. In 1988 I enrolled at Duke University as a mechanical engineer and also as a member of the varsity swim team. I really enjoyed my classes in thermodynamics, fluid mechanics, and heat transfer; however, it was a graduate course in aerodynamics that would provide the direction for my career. Between the exciting lectures of Professor Don Bliss and the excellent textbook *Fundamentals of Aerodynamics* by John Anderson, I was hooked on the subject of aerodynamics.

In fall 1992, I began my graduate work at Texas A&M in the aerospace engineering department. Quite early on, I realized that my strengths were in the computational aspects of aerodynamics rather than experiments. My master's degree research under Lee Carlson was in the area of computational fluid dynamics (CFD) simulation of radiation heating, an important phenomenon that occurs when a spacecraft uses a planet's atmosphere to slow down – a process called aerobraking. During such maneuvers, the high kinetic energy of the spacecraft causes the molecules in the atmosphere to dissociate and ionize, thus producing significant radiative heating.

I enrolled at North Carolina State University in the mechanical and aerospace engineering department in the fall of 1994, primarily because of its strength in CFD. That first summer I had the opportunity to teach two of my favorite courses from my undergraduate days: thermodynamics and fluid mechanics. I greatly enjoyed the teaching experience and knew at that point that I wanted to be a professor. My doctoral research was in CFD simulation of turbulent combusting flows spanning from low subsonic to supersonic speeds. We were focused on validating turbulence and turbulence/chemistry interaction models both for waste disposal applications and for scramjet propulsion. My doctoral advisor, Jack Edwards, was something of a CFD prodigy. Although he was a fairly new assistant professor when I started my doctoral work, he already excelled in many aspects of CFD, including coding, algorithm development, modeling, and applications.

When I graduated from North Carolina State University in 1998, I felt the need to first gain some industrial experience before embarking on a career in academia. Luckily, I was offered a job at Sandia National Laboratories in Albuquerque, New Mexico working in the Aerosciences and Compressible Fluid Mechanics Department. Sandia provided a good mix of real engineering problem solving and research. In addition, the Engineering Sciences Center where I worked encouraged – and rewarded – the publication of research papers and journal articles. It was at Sandia where I met Bill Oberkampf and Fred Blottner, two pillars in the developing field of Verification and Validation (V&V) for CFD. I quickly became involved in the estimation of grid-related numerical errors (verification) as well as the validation of different turbulence models in the presence of flow separation and shock waves.

In 2003, an opportunity arose for me to head back to academia, where I joined the aerospace engineering department at Auburn University as an assistant professor. The next year, Bill Oberkampf invited me to co-teach a professional short course on V&V. Over the next four years, Bill and I taught this course more than twenty times at AIAA conferences, NASA centers, DoD research labs, and private companies. My research work in V&V led to a Presidential Early Career Award for Scientists and Engineers in 2006 through the U.S. Department of Energy, including an exciting visit to the White House to meet President George W. Bush.

In 2007, I moved to Virginia Tech as an associate professor in the aerospace and ocean engineering department. Bill and I had often talked of writing a book on V&V, but never seemed to find the time to do it. Due to my reduced course load during my first year and Bill's recent retirement from Sandia, we finally had some time to work seriously on our book, which was completed in 2009. While the writing of the book was one of the biggest professional challenges I have ever undertaken, it was also one of the most rewarding, forcing me to learn in great depth about a wide variety of subjects.

My current research is focused on estimating grid-related (i.e., discretization) errors, mesh adaptation, and quantifying the total uncertainty in CFD predictions. I currently reside in Newport, Virginia (just northwest of Blacksburg) with my wife Rachel, five-year-old daughter Reagan, three-year-old son Cameron, four horses, and two dogs. My hobbies include swimming, reading, and watching Duke basketball.

2.6 Elements of a Solution

Before moving on to Chapter 3, it is important to note that computational aerodynamics codes are *always* changing. Every new problem seems to require new code developments and extensions, and this is especially true of aerodynamics codes. A new (and often small) modification will frequently be needed due to problem requirements or the innovation of a new solution algorithm or modeling approach; you may be the person who has to make this modification! Therefore, it is essential that newcomers into the field of computational aerodynamics make a consistent effort to stay abreast of the latest scientific computing developments, as well as understand how to use codes successfully.

Regardless of the type of tool or method you use for serious aerodynamic analysis effort, a significant investment in time will be necessary. Therefore, it is important to take time in advance to assess whether a particular approach will produce the needed results at the needed level of accuracy. In fact, experienced CA practitioners suggest that all code users should have a defined process in place to help guide their progress through the use of CA tools. Specifically, you might want to have "a clear checklist of issues that can arise which helps to ensure that all relevant problem areas have been dealt with."[41] Such a checklist could be very important for inexperienced CA users, but can even help experienced users stay on track (remember, airline pilots use checklists because they repeat similar processes every day of their working lives, and it would be easy to leave something out). A possible checklist for CA might include, among other things:[41]

1. What level of CA simulation method is really appropriate (semi-empirical, potential methods, CFD, etc.)?
2. Are the objectives of the simulation clearly defined?
3. What are the requirements on accuracy?
4. What local/global quantities are needed from the simulation?
5. What are the documentation/reporting requirements?
6. What are the important flow physics involved (steady, unsteady, laminar, turbulent, transitional, etc.)?
7. What is the area of primary interest (domain) for the flow calculation?

8. Is the geometry well defined?
9. What level of validation is necessary? (is this a routine application, where validation and calibration have already been carried out on similar flowfields, and where only relatively small changes can be expected from earlier similar simulations? Or is it a non-routine application, where little earlier validation work has been done?)
10. What level of computational resources is needed for the simulation (memory, disk space, CPU time) and are these available?
11. Solution strategy
12. Mathematical and physical models
13. Turbulence model
14. Available code/solver
15. Computational mesh
16. Boundary conditions
17. Code handling
18. Have the boundary conditions not only been properly defined but also properly applied?
19. Has the appropriate system of units been used?
20. Is the geometry correct?
21. Are the correct physical properties specified?
22. Have the intended physical and mathematical models been used?
23. Have default parameters been changed which may affect the solution?
24. Has the appropriate convergence criterion been defined and used?
25. Interpretation of results
26. Do not be seduced into believing that the solution is correct just because it has converged and produced high-quality color plots (or even seductive video presentations) of the CFD simulations. Make sure that an elementary interpretation of the flowfield explains the fluid behavior and that the trends of the flow analysis can be reconciled with a simple view of the flow.
27. Make sure that the mean values of engineering parameters derived from the simulation are computed consistently (e.g., mass-average values, area-average values, time-average values). Calculation of local and mean engineering parameters with external post-processing software may be inconsistent with the solution method of the code used (e.g., calculating shear stresses from the velocities). Check that any test data used for comparison with the simulations are also computed in the same way as the data from the simulation.
28. Consider whether the interpretation of the results and any decisions made are within the accuracy of your computation.
29. Documentation
30. Keep good records of the simulation with clear documentation of assumptions, approximations, simplifications, geometry, and data sources.

31. Organize the documentation of the calculations so that another CA user can follow what has been done.
32. Be aware that the level of documentation required depends strongly on the customer's requirements as defined in the problem definition.

Another important consideration is whether a particular method will produce the required results in time for it to impact future testing or important design decisions. Effective use of computational aerodynamics tools requires that you develop a sense for the required resources (both amount and cost) as well as the actual execution time to complete a calculation. This is much more difficult than might be expected. Simply referencing the documentation for a particular code or questioning the developer to determine the needed time and resources may not make much sense without qualifying the information. For example, a code developer may quote a computation time for a grid that is too crude for the intended application or without reaching a reliable level of solution convergence. What is really needed is a sort of "batting average" of the tool's performance over a number of applicable problems. Also, an advanced code may require many submissions on a single case to get all of the "issues" straightened out – execution of many advanced aerodynamic flow solvers is not a simple matter of "click here to run." In cases such as these, the required processing time to achieve the "final result" is not really meaningful; the required computational time (and calendar time) of all of the "attempts" leading up to the final computation should also be considered, as well as the time to analyze the solution and present your results in a coherent fashion.

Finally, before using a particular code – even one that is "off the shelf" – you will have to make a surprisingly large investment of time gaining enough confidence in the results to use it to make engineering decisions. It is *always* naïve to think deadlines do not exist; time is always a constraint in the analysis and design process.

Summary of Best Practices

1. There are many ways to predict aerodynamics using computational aerodynamics – you should understand the hierarchy of approaches and their strengths and weaknesses:
   a. engineering methods
   b. potential flow methods
   c. CFD methods
2. You should understand basic concepts about computing and computer architecture, including:
   a. computer speed
   b. Moore's law
   c. system architecture (serial, vector, parallel)
   d. code scalability
   e. Amdahl's law

3. Verification, validation, and certification are important concepts in computational aerodynamics – you should understand what these terms mean and how they apply to your work.
4. There are six areas of numerical accuracy that should be described when reporting on computational aerodynamics work:
   a. numerical methods used
   b. accuracy of the methods
   c. code verification activities
   d. spatial convergence accuracy
   e. temporal convergence accuracy
   f. iterative convergence accuracy
5. A CA solution process can greatly help in ensuring that you do not leave out important elements of obtaining good results, including:
   a. problem definition
   b. solution strategy
   c. code handling
   d. interpretation of results
   e. documentation
6. "A … system with checklists can help to support the inexperienced user to produce quality CFD simulations. It has been noted by Roache, however, that a CFD project can meet all formal [quality assurance] requirements and still be of low quality (or flatly erroneous). On the other hand high quality work can be done without a formal [quality assurance] system."[41]

2.7 Projects

1. Apply the engineering method for the lift-curve slope to the airfoil for which you found data in Chapter 1. Use Equation 2.6 (and values from Fig. 2.4) as well as the original theory of Equation 2.3. The included angle for the trailing edge is defined for your airfoil in Appendix A. Compare your lift-curve slope with the experimental data and discuss how well the semi-empirical methods were able to accurately model the data. Where do the methods work well? Where don't they work well?
2. Write a computer program to perform the analysis described in Project 1. Find a way to digitize the results of Fig. 2.4 so that the complete calculation process can be automated. Compare your results with the "hand" calculated results from Project 1.
3. Find estimates of computer speeds for new or future computer systems. Add your findings to Fig. 2.15 and comment on the accuracy of Moore's law relative to the data you found. How will future computer architecture (such as quantum computing) change the future growth rate of computer speed?

2.8 References

1 Brooks, F.P., Jr., *The Mythical Man-Month*, Reading: Addison-Wesley, 1975.

2 Jayaram, S., Myklebust, A., and Gelhausen, P, "ACSYNT – A Standards-Based System for Parametric Computer Aided Conceptual Design of Aircraft," AIAA Paper 92–1268, February 1992.

3 LinAir, Software Package, Ver. 4.0, Desktop Aeronautics, Inc., Palo Alto, CA, 2004.

4 Williams, J.E., and Vukelich, S.R., *The USAF Stability and Control Digital DATCOM*, AFFDL TR-79–3032, 1979.

5 Engineering Sciences Data Unit, Information Handling Services, Inc., London, England.

6 Blake, W.B. *Missile DATCOM*, AFRL VA-WP-TR-1998–3009, 1998.

7 Kinsey, D.W., and Bowers, D.L., "A Computerized Procedure to Obtain the Coordinates and Section Characteristics of NACA Designated Airfoils," AFFDL-TR-71–87, 1971.

8 Vukelich, S. R., "Missile DATCOM Volume 3 – Fin Alone Aerodynamic Methodology," March 1984.

9 McCormick, B.W., *Aerodynamics, Aeronautics, and Flight Mechanics*, 2nd Ed., Hoboken: John Wiley and Sons, 1995.

10 Antoine, N.E., and Kroo, I.M., "Framework for Aircraft Conceptual Design and Environmental Performance Studies,*" AIAA Journal*, Vol. 43, No. 10, 2005, pp. 2100–2109.

11 http://www.ami.aero/software-computing/amis-computational-fluid-dynamics-tools/vsaero

12 Pearce, D.G., Stanley, S.A., Martin, F.W., Gomez, R.J., LeBeau, G.J., Buning, P.G., Chan, W.M., Chan, I.T., Wulf, A., and Akdag, V., "Development of a Large Scale Chimera Grid System for the Space Shuttle Launch Vehicle," AIAA Paper 93–0533, January 1993.

13 Moretti, G., and Abbett, M., "A Time-Dependent Computational Model for Blunt Body Flows," *AIAA Journal*, Vol. 4, No. 12, 1966, pp. 2136–2141.

14 Rubbert, P.E., "Theoretical Characteristics of Arbitrary Wings by a Nonplanar Vortex Lattice Method," Boeing Report D6-9244, The Boeing Company, 1964.

15 Hess, J.L., and Smith, A.M.O., "Calculation of Nonlifting Potential Flow About Arbitrary Three-Dimensional Bodies," Douglas Report ES40622, Douglas Aircraft Company, 1962.

16 Chapman, D.R., "Aerodynamic Analysis Requiring Advanced Computers," NASA SP-347, 1975, pp. 4–7.

17 Whitcomb, R., "Review of NASA Supercritical Airfoils," ICAS Paper 74-10, 1974.

18 Murman, E.M., and Cole, J.D., "Calculation of Plane Steady Transonic Flows," *AIAA Journal*, Vol. 9, No. 1, 1971, pp. 114–121.

19 Spalart, P.R., "Strategies for Turbulence Modeling and Simulations," *International Journal of Heat & Fluid Flow*, Vol. 21, 2000, pp. 252–263.

20 Dongarra, J., ed., *Sourcebook of Parallel Computing*, San Francisco: Morgan Kaufmann Publishers, 2003.

21 Rizzi, A., and Engquist, B., "Selected Topics in the Theory and Practice of Computational Fluid Dynamics," *Journal of Computational Physics*, Vol. 72, No. 1, 1987, pp. 1–69.

22 Neves, K.W., "Hardware Architecture" in *Computational Fluid Dynamics: Algorithms and Supercomputers*, AGARD-AG-311, 1988.

23 Deane, A., Brenner, G., Ecer, A., Emerson, D.R., McDonough, J., Periaux, J., Satofuka, N., and Tromeur-Dervout, D., eds., *Parallel Computational Fluid Dynamics: Theory and Applications*, Oxford: Elsevier, 2006.

24 Rajasekaran, S., and Reif, J., eds., *Handbook of Parallel Computing: Models, Algorithms and Applications*, Boca Raton: Chapman & Hall, 2008.

25 Asanovic, K., Bodik, R., Catanzaro, B.C., Gebis, J.J., Husbands, P., Keutzer, K., Patterson, D.A., Plishker, W.L., Shalf, J., Williams, S.W., Yelick, K.A., "The Landscape of Parallel Computing Research: A View from Berkeley," Technical Report No. UCB/EECS-2006-183, 2006.

26 Holst, T.L., Salas, M.D., and Claus, R.W., "The NASA Computational Aerosciences Program – Toward Teraflops Computing," AIAA Paper 92–0558, January 1992.

27 Tomoro, R.F., Strang, W.Z., and Wurtzler, K.E., "Can Legacy Codes Scale on Tens of Thousands of PEs or Do We Need to Reinvent the Wheel?," DoD HPC User Group Conference, June 2012.

28 Morton, S.A., McDaniel, D.R., Sears, D.R., Tillman, B., and Tuckey, T.R., "Kestrel v2.0 – 6DoF and Control Surface Additions to a CREATE Simulation Tool," AIAA Paper 2010-0511, January 2010.

29 Cosner, R.R., Oberkampf, W.L., Rumsey, C.L., Rahaim, C.P., and Shih, T.I-P., "AIAA Committee on Standards for Computational Fluid Dynamics: Status and Plans," AIAA Paper 2006-0889, January 2006.

30 Schwer, L.E., "Guide for Verification and Validation in Computational Solid Mechanics," ASME VV 10, 2006.

31 Thacker, B.H., Doebling, S.W., Hemez, F.M., Anderson, M.C., Pepin, J.E., and Rodriguez, E.A., "Concepts of Model Verification and Validation," Los Alamos National Laboratory, LA-14167-MS, October 2004.

32 Sacher, P.W., "Technical Evaluation Report on the Fluid Dynamics Panel Symposium on Validation of Computational Fluid Dynamics," AGARD Advisory Report No. 257, May 1989.

33 Melnik, R.E., Siclari, M.J., Barber, T., and Verhoff, A., "A Process for Industry Certification of Physical Simulation Codes," AIAA Paper 94–2235, June 1994.

34 Melnik, R.E., Siclari, M.J., Marconi, F., Barber, T., and Verhoff, A., "An Overview of a Recent Industry Effort at CFD Code Certification," AIAA Paper 95–2229, June 1995.

35 Tinoco, E.N., "Validation and Minimizing CFD Uncertainty for Commercial Aircraft Applications," AIAA Paper 2008–6902, August 2008.

36 Roache, P.J., "Verification of Codes and Calculations," *AIAA Journal*, Vol. 36, No. 5, 1998, pp. 696–702.

37 Aeschliman, D.F., Oberkampf, W.L., and Blottner, F.G., "A Proposed Methodology for Computational Fluid Dynamics Code Verification, Calibration, and Validation," International Congress on Instrumentation in Aerospace Simulation Facilities (ICIASF), July 18–21, 1995, Wright-Patterson AFB, OH.

38 "Guide for the Verification and Validation of Computational Fluid Dynamics Simulations," AIAA G-077-1998, 1998.

39 Oberkampf, W.L., and Roy, C.J., *Verification and Validation in Scientific Computing*, New York: Cambridge University Press, 2010.

40 Giles, M., "'Best Practices' in CFD Software Development," Oxford University Computing Laboratory, 1995.

41 Anonymous, "Best Practice Guidelines for Marine Applications of Computational Fluid Dynamics," MARNet CFD, Final Report, 2003.

3 Getting Ready for Computational Aerodynamics: Fluid Mechanics Foundations

Don't be too ambitious for you will not understand these things

George G. Stokes
written to someone who was trying to understand some of
Stokes's mathematical formulas[1]

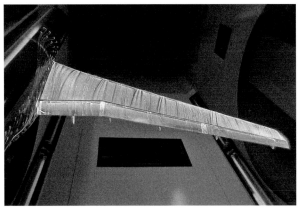

Oil flow visualization on a wing with flaps extended (Courtesy of NASA).

LEARNING OBJECTIVE QUESTIONS

After reading this chapter, you should know the answers to the following questions:

- What are governing equations and what are they based on (what basic physical concepts or principles)?
- What is the difference between the Eulerian and Lagrangian reference frames? How can you change between these two frames?
- What is conservation of mass? How is the physical observation turned into an equation?
- What is conservation of momentum? How is the physical observation turned into an equation?
- What is conservation of energy? How is the physical observation turned into an equation?
- What is the Conservation (or Divergence) form of the governing equations? Why is this form useful in CFD?
- What are the inviscid simplifications to the equations of motion? What are the assumptions and limitations for these simplified equations?

- What are the viscous simplifications to the equations of motion? What are the assumptions and limitations for these simplified equations?
- What are the implications of the equation "type" on their solution strategy and boundary conditions?

3.1 Introduction

If you are going to use computers to solve aerodynamic problems, you really need to know which equations to use and what they mean. In the same way that computational aerodynamics requires you to have a basic understanding of computers and their uses, you also need to understand fluid dynamics. Since we will be obtaining numerical solutions to various forms of the *governing equations* of fluid dynamics, we need to review those equations before examining the methods of computational aerodynamics in detail. However, as George Stokes said (see opening quote), truly understanding the equations that describe fluid motion can be quite a challenge, so do not be too discouraged if everything takes a little while to figure out. To make matters even more challenging, the traditional format for the equations of motion that are used in most textbooks is not the same as the format commonly used in computational codes, since developments in computational methods have resulted in a slightly different approach to the fundamental conservation law statements. Additionally, we will have to establish the nomenclature (the symbols and words that we use to describe the governing equations) which we will then use. In case you need additional background information on the developments in this chapter, numerous excellent discussions of the foundations of fluid mechanics for aerodynamic applications are available: Karamcheti[2] does a good job with aerodynamic theory, and the books by Bertin and Cummings,[3] Anderson,[4] and Moran[5] all contain good discussions of the material.

3.2 Governing Equations of Fluid Dynamics

When developing any kind of theory or model of physical reality, we have to have a starting point, or a place to begin our thinking. For classical fluid dynamics, our most fundamental premise is that the fluid is assumed to be a *continuum*. This means that we consider the medium to be the aggregate of the individual molecular motion (a macroscopic viewpoint) and ignore the motion of individual molecules (a microscopic viewpoint). For nearly all aerodynamics work, this is a valid assumption. The situation where this is not true is called rarefied gas dynamics, where the flow has such a low density that the actual molecular motion must be taken into account. While low-density flows can be important for atmospheric re-entry, even in very high altitude flight our continuum assumption will usually serve us well.

We will also assume that the fluid is defined by an *equation of state* and the relationships between the thermodynamic and transport properties (i.e., the ratio of specific heats, γ, viscosity, μ, and the coefficient of heat conduction, k). The governing equations of fluid dynamics, as well as the appropriate *boundary conditions*, will control the motion of the fluid, so we need to have a good understanding of all of these aspects in order to obtain valid solutions.

We said earlier that we needed a place to start when developing our equations, and, as in many fields of physics, the starting place is with the "laws" of physics (see the next Concept Box for a discussion about physical laws). In fluid dynamics we also start with basic physical laws and develop our governing equations based on physical observations which are known as *conservation laws*. The conservation laws of physics are well known, even to non-scientists or engineers. A good example is the conservation of mass, which as an observation is very simple to state: mass can neither be created nor destroyed. This observation is then applied to fluid dynamics in equation form, which makes the observation useful for predictions and measurements. The three conservation laws commonly used in fluid dynamics, as well as their commonly used names, are:

- mass   continuity
- momentum   Newton's 2nd Law, $\sum \vec{F} = m\vec{a}$
- energy   1st Law of Thermodynamics

We will show how to take these physical laws and derive useful equations for them, such as the Navier-Stokes equations and their simplifications. Since these equations will be partial differential equations, we will also discuss how to classify the equations and then provide a general approach to solving them (both theoretically and numerically). All of these concepts are important to computational aerodynamics, so while the details may seem daunting (remember the quote from Stokes at the beginning of the chapter!), the importance of going through the process will hopefully be evident.

Computational Aerodynamics Concept Box

What are the Laws of Physics and where did they come from?

The so-called Laws of Physics are based in science and therefore are a result of an application of the scientific method. In the seventeenth century, as modern science developed, a basic approach to scientific discovery also developed. The approach had four basic steps:

- Characterizations (observations, definitions, and measurements)
- Hypotheses (theoretical, hypothetical explanations of the observations and measurements of the subject)

- Predictions (results from the hypothesis or theory)
- Experiments (tests of all of the above)

The cycle is then repeated to refine and improve the hypothesis and resulting theories. Each aspect of the scientific method is subject to peer review for possible mistakes, which is why scientists need to carefully record their observations and measurements so that other people can try to re-create the cycle and either refute or verify the results. When a hypothesis is supported by a large body of experimental data and helps to explain a particular set of observations, it may begin to be referred to as a theory. If a theory can be written down in a concise form (sometimes in a mathematical equation) and be shown to be generally applicable, it may become known as a physical law.

There are a large number of physical laws that we use often in science and engineering, such as gravity, electricity, thermodynamics, and the conservation of mass, momentum, and energy. You probably studied Isaac Newton's laws in your physics classes, such as the conservation of linear and angular momentum (which are known as Newton's Second Law).

In fluid dynamics, most of the important equations and concepts are based on the Conservation Laws. Since mass, momentum, and energy are conserved (which is an observation), we can write down those observations as equations for various applications. Knowing how to derive those expressions, and understanding their basis in the physical laws, is an important part of aerodynamics.

Isaac Newton (1642–1727)

The choice of coordinate systems is also very important in aerodynamics. While the general equations of fluid motion are independent of the coordinate system, simplifying assumptions frequently introduce a *directional bias* into approximate forms of the equations and require that the equations be used with a specific coordinate system orientation relative to the flowfield. In addition, in this book we assume that the body is stationary and the air is moving over the body, which is what takes place in a wind tunnel test. This can be shown to be equivalent to the case of an airplane flying at a constant speed through still air as occurs in flight; the equivalence between the two frames of reference is known as *Galilean invariance*. However, there are times when we need to fix our coordinate system to a body that is not moving at a constant speed relative to the air. When this occurs, additional terms are required in the equations of motion that, for the sake of simplicity, we will neglect in this text. One example of

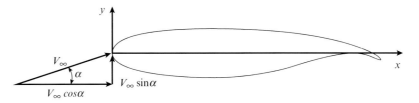

Figure 3.1 Standard coordinate system for two-dimensional flow.

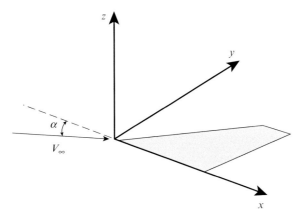

Figure 3.2 A typical coordinate system for three-dimensional aircraft flowfields.

this situation occurs when we need to attach our coordinate system to a rotating helicopter rotor blade; Owczarek[6] includes the details required for treating these special cases (you may also want to see the comments about these special cases made by White[7]).

Cartesian coordinates are normally used to describe the geometry of an aircraft, so we will work entirely in the Cartesian coordinate system in this chapter. Other coordinate systems could be used appropriately, such as spherical and cylindrical coordinate systems, or a more general body-fitted coordinate system. It is frequently desirable to make calculations in non-Cartesian coordinate systems that are distorted to fit a particular shape. General non-orthogonal curvilinear coordinates are discussed briefly in this chapter and in Appendix E. Even when using Cartesian coordinates, the x, y, and z coordinates are oriented differently depending on whether the flow is two- or three-dimensional. Figure 3.1 shows the usual two-dimensional coordinate system, where V_∞ denotes the freestream velocity (the velocity far from the aircraft) and α is the angle of attack. A standard aerodynamic coordinate system in three dimensions is illustrated in Fig. 3.2, but other coordinate systems are certainly used in various computer codes (make sure you understand the coordinate system being used before using any code!).

In general Cartesian coordinates, the independent variables are x, y, z, and t. We want to know the corresponding fluid velocities, u, v, w, and the fluid field properties; p, ρ, T.[i] These are the dependent variables of the flow represented by six unknowns and therefore requiring six equations. The six equations are provided by the following conservation laws and relations:

continuity	1	equation(s)
momentum	3	"
energy	1	"
equation of state	1	"

We also need to specify the fluid properties, and this leads to additional equations or relations. These additional relations include transport properties such as an equation for the viscosity as a function of temperature, as well as relationships for the various thermodynamic properties.

Simplifications for specific flowfields frequently reduce the number of equations required. Examples of these simplifications include incompressible, inviscid, and irrotational flow, which can be described by a single equation, as will be shown later in this chapter. Prior to the 1980s, almost all aerodynamics work used a single partial differential equation, possibly coupled with another algebraic equation. Another approach is the calculation of potential flow for the inviscid portion of the flowfield, and use of the boundary-layer equations to compute the flowfield when an estimate of the viscous effects is required.

3.3 Derivation of Governing Equations

We want to develop a mathematical model of fluid motion suitable for use in our numerical calculations. In order to know the aerodynamics of the vehicle, we want to find the flowfield velocity, density, pressure, and temperature distributions. As stated previously, the mathematical model required to achieve this is based on conservation laws and fluid properties. Two reference frames can be used to obtain the mathematical description defining the governing equations.

- **Lagrangian frame**: In this reference frame each fluid particle is followed as it moves around the body. Even in steady flow, the forces encountered by the particle will be a function of its time history as it moves relative to a coordinate system fixed to the body, as defined in Figures 3.1 and 3.2. This method corresponds to the conventional concept of Newton's Second Law.

---

[i] We will assume in this book that the fluid is a perfect (or ideal) gas. For very high speed flows, the flow may be chemically reacting, which makes the situation much more complicated, and leads to the need for additional chemical "species" equations. (For more information, see Anderson, *Hypersonic and High Temperature Gas Dynamics*, 2nd Ed., Reston: AIAA, 2006.)

- **Eulerian frame**: In this method we look at the entire space around the body as a field and determine flow properties at various points in the field while the fluid particles stream past. Once this viewpoint is adopted, we consider the distribution of velocity and pressure throughout the field and ignore the motion of individual fluid particles.

Virtually all computational aerodynamics methods use the Eulerian approach, at least partially because it relates well to how we measure fluid flows in wind tunnels. The use of this approach requires careful attention to the application of the conservation concepts (especially for Newton's Second Law). Since these two approaches describe the same physical phenomena, they can be mathematically related if needed.

Newton's Second Law governs the motion of a specific fluid particle. Here we consider that particle as it moves through our control volume, where, to establish a viable method for computation, aerodynamicists employ the Eulerian approach and define a control volume that maintains a fixed location relative to the coordinate system (see the description of a control volume in the next Concept Box). The connection between the rate of change of the properties of our fluid particle (velocity, density, pressure, *etc.*) and the rate of change of fluid properties flowing through our fixed control volume requires special consideration. The *substantial derivative* is used to define the rate of change of the fluid particle properties as the particle moves through the flowfield relative to the fixed coordinate system. To connect the motion of a "clump" of fluid obeying Newton's Laws to the description of the flow in the coordinate system fixed to geometry, we can use the *Reynolds Transport theorem*. A derivation is contained in Owczarek,[6] as well as other sources.

Once we have chosen a coordinate system and a frame of reference, the conservation equations can be expressed from either a differential or integral viewpoint. The differential form is more frequently used in fluid dynamics analysis; however, many numerical methods use the integral form. One of the reasons for this is that, when performing numerical calculations, integrals are easier to compute accurately when compared with derivatives. The integral form of the equations also handles discontinuities (such as shocks) better than the differential form, which assumes that the fluid properties are continuous. We will use aspects of each approach in our derivations.

Computational Aerodynamics Concept Box

The Concept of a Control Volume

"The statement is sometimes made, usually in the abstract, that engineers think about their problems differently from scientists." This is a quote from Walter G. Vincenti, discussing one of the ways that engineers do things differently than scientists: the control volume.[8] The concept of a "control volume" arose as an engineering requirement for a means to formulate a physical description that allowed practical calculations to be made. "Engineering texts speak frequently of a so-called control volume, an imagined spatial volume having certain characteristics and introduced for purposes of analysis. These texts are replete with diagrams showing, conventionally by dashed lines, the imaginary 'control surface' enclosing such a volume. Textbooks on thermodynamics for physicists, on the other hand, are notable for the complete absence of such ideas and diagrams."

So, how do engineers use control volumes? The convention typically adopted requires that the control volume is shown with dashed lines to make the point that the boundaries are fictitious, and fluid can flow across the line. An historical example is show here, from Ludwig Prandtl for flow in a sudden pipe enlargement.[8]

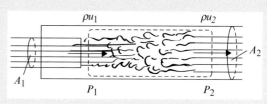

Control volume for flow in a sudden pipe enlargement from Ludwig Prandtl

Once the arbitrary control volume is chosen (and the choice can make solving a problem easier or more difficult), a set of points is identified around which the integral form of the governing equations is typically applied. The integration direction typically follows the mathematical convention to integrate with the interior on the left (counterclockwise). One of the corners of the control volume is chosen as a starting point, and then the integration is conducted around the entire volume until the starting point is reached again. If a line or surface is aligned with a solid surface (such as in the example given), then no flow can cross that boundary.

In addition to allowing practical results to be obtained, another important advantage of using control volumes was described by Prandtl: "The undoubted value of theorems (of control-volume analysis) lies in the fact that their application enables one to obtain results in physical problems from just a knowledge of the boundary conditions. There is no need to be told anything about the interior of the fluid or about the mechanism of the motion."[8] In other words, the complexity of the flow on the interior of the problem (as shown in the figure) does not need to be known in order to find out the impact of that region of complex flow. Of course, some detail is also lost in this approach, but very practical and applicable results are obtained by using the control volume approach.

3.3.1 Conservation of Mass: The Continuity Equation

In this section we derive the conservation of mass equation (sometimes called the Continuity Equation) from a control volume viewpoint (but only in two dimensions), and then we look at the equivalent integral statement and the use of the Gauss Divergence Theorem to establish the connection between the two viewpoints.

The statement of the conservation of mass is based on a physical observation (as discussed in the Laws of Physics Concept Box) and in words is simply stated as:

| *net outflow of mass through the surface surrounding the volume* | = | *decrease of mass within the volume* |

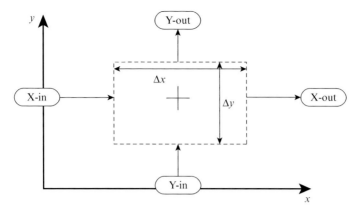

Figure 3.3 Control volume for conservation of mass.

To translate this statement into a mathematical form, consider the control volume given in Fig. 3.3. Here, u is the velocity in the x-direction, v is the velocity in the y-direction, and ρ is the fluid density.

The net mass flow rate, or mass flux, is a physical quantity that flows across the boundary of the control volume – typically we account for mass, momentum, and energy fluxes. The net mass flux *out* of the volume is:

$$[\text{X-out}] - [\text{X-in}] + [\text{Y-out}] - [\text{Y-in}] = \text{change of mass (decrease)}$$

$$= -\frac{\partial \rho}{\partial t} \Delta x \Delta y. \tag{3.1}$$

We now use a Taylor's series expansion of the mass fluxes into the volume written around the origin of the volume. We could use many terms in the series (we could include the next terms in the series), but they will vanish when we take the limit of Δ going to 0. The flux per unit length through the surface is multiplied by the length of the surface to obtain:

$$[\text{X-out}] = \left[\rho u + \frac{\partial \rho u}{\partial x} \cdot \frac{\Delta x}{2} \right] \Delta y$$

$$[\text{X-in}] = \left[\rho u - \frac{\partial \rho u}{\partial x} \cdot \frac{\Delta x}{2} \right] \Delta y$$

$$[\text{Y-out}] = \left[\rho v + \frac{\partial \rho v}{\partial x} \cdot \frac{\Delta y}{2} \right] \Delta x \tag{3.2}$$

$$[\text{Y-in}] = \left[\rho v - \frac{\partial \rho v}{\partial x} \cdot \frac{\Delta y}{2} \right] \Delta x$$

Adding these terms we obtain:

$$\left[\rho u + \frac{\partial \rho u}{\partial x} \cdot \frac{\Delta x}{2} \right] \Delta y - \left[\rho u - \frac{\partial \rho u}{\partial x} \cdot \frac{\Delta x}{2} \right] \Delta y$$

$$+ \left[\rho v + \frac{\partial \rho v}{\partial y} \cdot \frac{\Delta y}{2} \right] \Delta x - \left[\rho v - \frac{\partial \rho v}{\partial y} \cdot \frac{\Delta y}{2} \right] \Delta x = -\frac{\partial \rho}{\partial t} \Delta x \Delta y \tag{3.3}$$

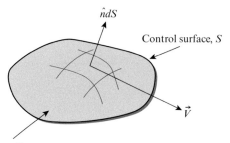

$\hat{n}dS$

Control surface, S

$\vec{V}$

Control volume, V

Figure 3.4   Arbitrary fluid control volume.

Summing up and canceling $\Delta x \Delta y$ (which verifies that our choice of control volume was arbitrary) we get:

$$\frac{\partial \rho u}{\partial x} + \frac{\partial \rho v}{\partial y} = -\frac{\partial \rho}{\partial t} \qquad (3.4)$$

or in three dimensions:

$$\frac{\partial \rho}{\partial t} + \frac{\partial \rho u}{\partial x} + \frac{\partial \rho v}{\partial y} + \frac{\partial \rho w}{\partial z} = 0. \qquad (3.5)$$

This is the differential form of the continuity equation. The more general vector form of the equation is:

$$\frac{\partial \rho}{\partial t} + \vec{\nabla} \cdot \left(\rho \vec{V} \right) = 0. \qquad (3.6)$$

Another way to derive the conservation of mass uses the arbitrary control volume shown in Fig. 3.4. Using the control volume and accounting for the flux of mass across the control surface, the conservation of mass can then be written in an integral form: the surface integral of the flow out of the volume simply equals the change of mass in the volume,

$$\iint_S \rho \vec{V} \cdot \hat{n} dS = -\frac{\partial}{\partial t} \iiint_V \rho dV. \qquad (3.7)$$

This observation is true without making any assumption requiring continuous variables and differentiability. Equation 3.7 is valid for all flows, unsteady or steady, viscous or inviscid, compressible or incompressible (the only assumption made in the development is that the fluid is a continuum).

To relate the integral expression (Equation 3.7) to the differential form (Equation 3.5), we make use of the Gauss Divergence Theorem,[2] which assumes continuous partial derivatives:

$$\iint_S \vec{A} \cdot \hat{n} dS = \iiint_V \vec{\nabla} \cdot \vec{A} dV \qquad (3.8)$$

and the equivalent statement for a scalar is:

$$\iint_S \phi \hat{n} dS = \iiint_V \vec{\nabla} \phi dV. \tag{3.9}$$

Using this theorem, the differential and integral forms of the continuity equation can be shown to be the same. First, rewrite the surface integral in the conservation of mass equation, Equation 3.7, using the divergence theorem, Equation 3.8, as:

$$\iint_S \rho \vec{V} \cdot \hat{n} dS = \iiint_V \vec{\nabla} \cdot \left(\rho \vec{V}\right) dV. \tag{3.10}$$

Now substitute Equation 3.10 into Equation 3.7, and the integral form of the continuity equation becomes:

$$\iiint_V \vec{\nabla} \cdot \left(\rho \vec{V}\right) dV = -\frac{\partial}{\partial t} \iiint_V \rho dV, \tag{3.11}$$

and since V refers to a fixed volume, we can move $\partial / \partial t$ inside the integral to obtain

$$\iiint_V \left[\frac{\partial \rho}{\partial t} + \vec{\nabla} \cdot \left(\rho \vec{V}\right)\right] dV = 0. \tag{3.12}$$

Since the volume we chose in Fig. 3.4 was arbitrary (any volume of any shape would have led us to this same result), the only way for the integral to always be zero for any limits of integration is for the integrand to be identically zero, which is the conservation of mass in partial differential equation form, given in Equations 3.5 or 3.6. We have now shown that the differential and integral forms of the conservation of mass are equivalent.

The conservation of mass equation is a scalar equation with four unknowns: (ρ, u, v, w) and is valid for:

- compressible or incompressible flow
- steady or unsteady flow
- one-, two-, or three-dimensional flow

The equation can be simplified in a number of ways, including:

- assume steady flow, $\frac{\partial \rho}{\partial t} = 0$, and the equation becomes

$$\vec{\nabla} \cdot \rho \vec{V} = \frac{\partial \rho u}{\partial x} + \frac{\partial \rho v}{\partial y} + \frac{\partial \rho w}{\partial z} = 0 \tag{3.13}$$

which still has four unknowns (ρ, u, v, w);

- assume incompressible flow, $\partial \rho / \partial t = 0$, $\vec{\nabla} \cdot \rho \vec{V} = \rho \left(\vec{\nabla} \cdot \vec{V} \right)$, and the equation becomes

$$\vec{\nabla} \cdot \vec{V} = \frac{\partial u}{\partial x} + \frac{\partial v}{\partial y} + \frac{\partial w}{\partial z} = 0 \qquad (3.14)$$

and the equation now has only three unknowns (u, v, w).

Equation 3.7 forms the basis of *finite volume* formulations in computational fluid dynamics, which relates the net mass flux across the boundary of a "cell" to the time rate of change of mass within the "cell" (a cell is a simple geometric shape, like a pyramid or a hexahedra, used to apply the integral equations). In a similar fashion, Equation 3.5 forms the basis for *finite difference* formulations, which relate the time rate of change of density to the convection of density by approximating partial derivatives on a "grid" of points. We will investigate both of these approaches in Chapter 6.

3.3.2 Conservation of Momentum and the Substantial Derivative

In this section we derive the general equations for the Conservation of Momentum. This is a statement of Newton's Second Law: the time rate of change of momentum of a body equals the net force exerted on it. For a fixed mass this is the famous equation:

$$\sum \vec{F} = m\vec{a} = m\frac{D\vec{V}}{Dt}. \qquad (3.15)$$

3.3.2.1 SUBSTANTIAL DERIVATIVE

We need to apply Newton's Second Law to a moving fluid element (the "body" in the Second Law statement given earlier) using our fixed coordinate system, which introduces some extra complications. From our fixed coordinate system, we can begin to understand what the substantial derivative, D/Dt, means by considering Fig. 3.5.[2] First, we can consider any fluid property as a function of space and time, $Q(\vec{r}, t)$.

The change in position of the particle between the position r at t, and $r + \Delta r$ at $t + \Delta t$, is:

$$\Delta Q = Q\left(\vec{r} + \Delta s, t + \Delta t\right) - Q\left(\vec{r}, t\right) \qquad (3.16)$$

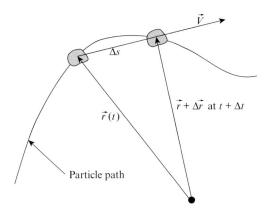

Figure 3.5   Moving particle viewed from a fixed coordinate system (Ref. 2).

where the spatial change Δs is simply equal to $\vec{V}\Delta t$. So we can now write:

$$\Delta Q = Q\left(\vec{r}+\vec{V}\Delta t, t+\Delta t\right)-Q\left(\vec{r},t\right), \tag{3.17}$$

which is in a form that can be used to find the rate of change of Q:

$$\frac{DQ}{Dt}=\lim_{\Delta t\to 0}\frac{\Delta Q}{\Delta t}=\lim_{\Delta t\to 0}\frac{Q\left(\vec{r}+\vec{V}\Delta t, t+\Delta t\right)-Q\left(\vec{r},t\right)}{\Delta t}. \tag{3.18}$$

Notice that the rate of change is in two parts: one for a change in time and one for a change in space. Therefore, we write the change of Q as a function of both time and space using the following Taylor's series expansion:

$$Q\left(\vec{r}+\vec{V}\Delta t, t+\Delta t\right)=Q\left(\vec{r},t\right)+\left.\frac{\partial Q}{\partial t}\right|_{\vec{r},t}\Delta t+\cdots+\left.\frac{\partial Q}{\partial s}\right|_{\vec{r},t}V\Delta t+\cdots \tag{3.19}$$

where the direction of s is shown in Fig. 3.5. If we substitute Equation 3.19 into Equation 3.18 and take the limit, we obtain:

$$\lim_{\Delta t\to 0}\frac{\Delta Q}{\Delta t}=\underbrace{\underbrace{\frac{\partial Q}{\partial t}}_{\substack{\text{local time}\\\text{derivative, or}\\\text{local derivative}}}+\underbrace{\frac{\partial Q}{\partial s}}_{\substack{\text{variation with}\\\text{change of position,}\\\text{convective derivative}}}\vec{V}}_{\text{substantial derivative}}. \tag{3.20}$$

This is the important consequence of applying Newton's Second Law for a moving particle to a point fixed in a stationary coordinate system. The second term in Equation 3.20 has the unknown velocity $\vec{V}$ multiplying a term containing the unknown fluid property Q. This is important because *the convective derivative introduces a fundamental nonlinearity* into the governing equations.

Computational Aerodynamics Concept Box

What is "nonlinear"?

You often hear engineers and mathematicians talk about equations being "linear" or "nonlinear," but you may not clearly understand what that means and why it is important. A simple example will help to explain the difference. Consider a simple two-dimensional function, $f(x) = x$, which when graphed gives a straight (or linear) curve. By making a small change to the function, say $f(x) = x^2$, the resulting graph is no longer straight. In mathematics, the first function is linear and the second is nonlinear.

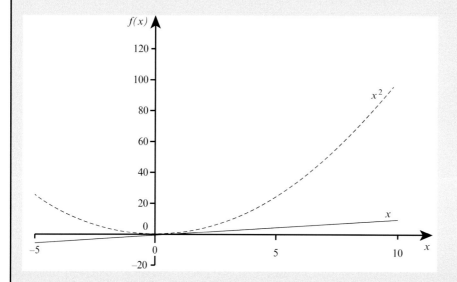

The importance of this is evident if you try to use either of these functions, such as by adding together two separate values of the function. For example, for $f(x) = x$, you can evaluate the function at two locations and add them together: $f(2) + f(5) = 2 + 5 = 7$. Because the function is linear, this is the same result you would get for $f(2 + 5) = 7$. But look what happens for the nonlinear function, $f(x) = x^2$: $f(2) + f(5) = 4 + 25 = 29$, and $f(2 + 5) = 49$. In other words, you get different results depending on how you use the function.

When the nonlinear concept is extended to differential equations, one of the most important implications is that the nonlinear equations are much more difficult to solve than the linear equations. So a differential equation like $dy/dt + y + 5 = 0$ is linear, while $dy/dt + y^2 + 5 = 0$ is nonlinear. We will see the difference as we derive and solve our fluid dynamic equations.

We now put this result into a specific coordinate system:

$$\frac{\partial Q}{\partial s} = \hat{e}_V \cdot \vec{\nabla} Q \tag{3.21}$$

where $\hat{e}_V$ denotes the unit vector in the direction of $\vec{V}$. Thus, $\vec{V} = V\hat{e}_V$ and:

$$\frac{\partial Q}{\partial s} V = \vec{V} \cdot \vec{\nabla} Q. \tag{3.22}$$

Now we can write the substantial derivative, Equation 3.18, using Equations 3.20 and 3.22 as:

$$\frac{D}{Dt} = \frac{\partial}{\partial t} + \left(\vec{V} \cdot \vec{\nabla}\right) \tag{3.23}$$

which can be applied to a scalar quantity as:

$$\frac{DQ}{Dt} = \frac{\partial Q}{\partial t} + \left(\vec{V} \cdot \vec{\nabla}\right)Q \tag{3.24}$$

or to a vector quantity as:

$$\frac{D\vec{V}}{Dt} = \frac{\partial \vec{V}}{\partial t} + \left(\vec{V} \cdot \vec{\nabla}\right)\vec{V}. \tag{3.25}$$

$\partial \vec{V}/\partial t$ is typically called the unsteady (or acceleration) term and $\left(\vec{V} \cdot \vec{\nabla}\right)\vec{V}$ is called the convection term. In Cartesian coordinates, $\vec{V} = u\hat{i} + v\hat{j} + w\hat{k}$ and the substantial derivative becomes:

$$\frac{Du}{Dt} = \frac{\partial u}{\partial t} + u\frac{\partial u}{\partial x} + v\frac{\partial u}{\partial y} + w\frac{\partial u}{\partial z}$$

$$\frac{Dv}{Dt} = \frac{\partial v}{\partial t} + u\frac{\partial v}{\partial x} + v\frac{\partial v}{\partial y} + w\frac{\partial v}{\partial z} \tag{3.26}$$

$$\frac{Dw}{Dt} = \frac{\partial w}{\partial t} + u\frac{\partial w}{\partial x} + v\frac{\partial w}{\partial y} + w\frac{\partial w}{\partial z}.$$

To solve equations containing these nonlinear terms analytically is quite difficult (if not impossible), which is why we typically use numerical methods.

3.3.2.2 FORCES

Now that we have defined the acceleration terms, we need to find the net forces acting on the system. There are only three general categories of forces to consider:

- body forces (acting on the interior of the volume)
- pressure forces (acting perpendicular to the surface of the volume)
- shear forces (acting parallel to the surface of the volume)

Each of these forces applies to the control volume shown in Fig. 3.6, where τ will be used as a general symbol for stresses. The general notation for the stress term uses the first subscript to indicate the direction normal to the surface, and the second subscript defines the direction in which the force acts. For example, τ_{xy} is the stress on the face normal to the x-axis from forces

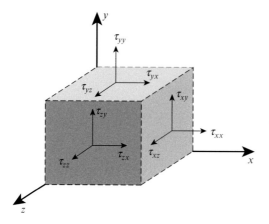

Figure 3.6 Control volume with surface forces shown.

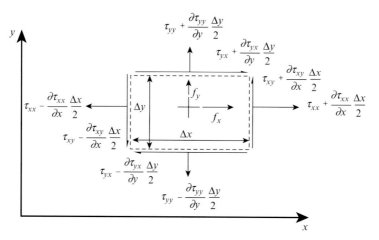

Figure 3.7 Details of forces acting on a two-dimensional control volume.

in the y-direction. Fluids of interest in aerodynamics are *isotropic*, which means that the fluid properties are the same in any direction:

$$\tau_{xy} = \tau_{yx}, \quad \tau_{yz} = \tau_{zy}, \quad \tau_{zx} = \tau_{xz}. \tag{3.27}$$

The connection between pressure and stress is defined more specifically when the properties of a specific fluid are prescribed. Figure 3.7 shows the details of the forces, expanded about the center of the control volume using a Taylor's series. The force f is defined to be the body force per unit mass.

Considering the x-direction as an example and using the Taylor's series expansion shown in Fig. 3.7, the net forces are found in a manner exactly analogous to the approach used in the derivation of the continuity equation. Thus, the net force in the x-direction is found to be:

$$\rho \Delta x \Delta y f_x + \frac{\partial}{\partial x}\left(\tau_{xx}\right)\Delta x \Delta y + \frac{\partial}{\partial y}\left(\tau_{yx}\right)\Delta x \Delta y \tag{3.28}$$

where the first term represents the body force acting on the fluid within the volume.

Now we can combine the forces, including the z-direction terms (as shown in Fig. 3.6). Substituting for the forces into the original statement of $\sum \vec{F} = m\vec{a}$ (Equation 3.15), and using the substantial derivative and the definition of the mass, $m = \rho \Delta x \Delta y \Delta z$, the x-momentum equation becomes (writing Equation 3.15 as $m\vec{a} = \sum \vec{F}$, the usual fluid mechanics convention, and considering the x component, $ma_x = F_x$),

$$\rho \Delta x \Delta y \Delta z \frac{Du}{Dt} = \rho \Delta x \Delta y \Delta z \, f_x + \frac{\partial}{\partial x}\left(\tau_{xx}\right) \Delta x \Delta y \Delta z$$
$$+ \frac{\partial}{\partial y}\left(\tau_{yx}\right) \Delta y \Delta x \Delta z + \frac{\partial}{\partial z}\left(\tau_{zx}\right) \Delta y \Delta x \Delta z \tag{3.29}$$

The $\Delta x \Delta y \Delta z$'s cancel and can be dropped, which again shows that the choice of control volume did not affect our results. The final equations can now be written by completing the system with the y- and z-equations to obtain

$$\rho \frac{Du}{Dt} = \rho f_x + \frac{\partial \tau_{xx}}{\partial x} + \frac{\partial \tau_{yx}}{\partial y} + \frac{\partial \tau_{zx}}{\partial z}$$

$$\rho \frac{Dv}{Dt} = \rho f_y + \frac{\partial \tau_{xy}}{\partial x} + \frac{\partial \tau_{yy}}{\partial y} + \frac{\partial \tau_{zy}}{\partial z} \tag{3.30}$$

$$\rho \frac{Dw}{Dt} = \rho f_z + \frac{\partial \tau_{xz}}{\partial x} + \frac{\partial \tau_{yz}}{\partial y} + \frac{\partial \tau_{zz}}{\partial z}.$$

These are general conservation of momentum relations, valid for anything!

To make Equation 3.30 more specific, we relate the stresses to the motion of the fluid. For gases and water, stress is a linear function of the rate of strain: such a fluid is called a *Newtonian fluid*, i.e.:

$$\tau = \mu \frac{\partial u}{\partial y} \tag{3.31}$$

where μ is the coefficient of viscosity. In our work we consider μ to be a function of temperature only. Note that in air the viscosity increases with increasing temperature, and in water the viscosity decreases with an increase in temperature.[ii] We will only consider air in this text, since our applications will be in the field of aerodynamics.

To complete the specification of the connection between stress and rate of strain, we need to define precisely the relation between the stresses and the

---

[ii] This led the famous aerodynamicist Edward van Driest to suggest heating the surface of a submarine to reduce viscosity and, hence, drag, leading to his supersonic submarine concept.

Table 3.1 Constants for Sutherland's Law

Unit System	C_1	C_2
meter-kilogram-second	1.458×10^{-6} $N\text{-}s/(m^2\text{-}K^{1/2})$	$110.4K$
Foot-slug-sec	2.275×10^{-8} $lb\text{-}s/(ft^2\text{-}R^{1/2})$	$198.72R$

motion of the fluid. This can become quite complicated. In general the fluid description requires two coefficients of viscosity, μ and λ. The coefficient of viscosity arising from the shear stress, μ, is well defined. The second coefficient of viscosity, λ, is not. This coefficient depends on the normal stress, and is only important in computing the detailed structure of shock waves. Various assumptions relating the coefficients of viscosity have been made, but the set of assumptions which leads to the equations known as the Navier-Stokes equations are:

- The stress-rate-of-strain relations must be independent of coordinate system (isotropic).
- When the fluid is at rest and the velocity gradients are zero (the strain rates are zero), the stress reduces to the hydrostatic pressure.
- Stoke's Hypothesis, $\lambda + 2\mu/3 = 0$, provides a connection between λ and μ, and is used to eliminate the issue of mean pressure versus thermodynamic pressure (which is true for monatomic gases but is typically assumed to be true for air).

Details of the theory associated with these requirements can be found in Schlichting.[9]

Since the viscous terms in Equation (3.30) contain an additional unknown, the viscosity, μ, we need another equation relating μ to the flow variables. For air, the viscosity is generally assumed to satisfy *Sutherland's law*,[9] given by:

$$\mu = C_1 \frac{T^{3/2}}{T + C_2} \tag{3.32}$$

where C_1 and C_2 are constants that are defined in Table 3.1.

Using these conditions leads to the following relations connecting the stresses to the fluid motion:

$$\tau_{xx} = -p - \frac{2}{3}\mu \vec{\nabla} \cdot \vec{V} + 2\mu \frac{\partial u}{\partial x}$$

$$\tau_{yy} = -p - \frac{2}{3}\mu \vec{\nabla} \cdot \vec{V} + 2\mu \frac{\partial v}{\partial y} \tag{3.33}$$

$$\tau_{zz} = -p - \frac{2}{3}\mu\vec{\nabla}\cdot\vec{V} + 2\mu\frac{\partial w}{\partial z}$$

and

$$\tau_{xy} = \tau_{yx} = \mu\left(\frac{\partial u}{\partial y} + \frac{\partial v}{\partial x}\right)$$

$$\tau_{xz} = \tau_{zx} = \mu\left(\frac{\partial u}{\partial z} + \frac{\partial w}{\partial x}\right)$$ (3.34)

$$\tau_{yz} = \tau_{zy} = \mu\left(\frac{\partial v}{\partial z} + \frac{\partial w}{\partial y}\right).$$

Substituting Equations 3.33 and 3.34 into Equation 3.30 and neglecting the body force (a standard assumption in aerodynamics since air is so light), we obtain:

$$\rho\frac{Du}{Dt} = -\frac{\partial p}{\partial x} + \frac{\partial}{\partial x}\left(2\mu\frac{\partial u}{\partial x} - \frac{2}{3}\mu\vec{\nabla}\cdot\vec{V}\right) + \frac{\partial}{\partial y}\left[\mu\left(\frac{\partial u}{\partial y} + \frac{\partial v}{\partial x}\right)\right] + \frac{\partial}{\partial z}\left[\mu\left(\frac{\partial w}{\partial x} + \frac{\partial u}{\partial z}\right)\right]$$

$$\rho\frac{Dv}{Dt} = -\frac{\partial p}{\partial y} + \frac{\partial}{\partial x}\left[\mu\left(\frac{\partial u}{\partial y} + \frac{\partial v}{\partial x}\right)\right] + \frac{\partial}{\partial y}\left(2\mu\frac{\partial v}{\partial y} - \frac{2}{3}\mu\vec{\nabla}\cdot\vec{V}\right) + \frac{\partial}{\partial z}\left[\mu\left(\frac{\partial w}{\partial y} + \frac{\partial v}{\partial z}\right)\right]$$

(3.35)

$$\rho\frac{Dw}{Dt} = -\frac{\partial p}{\partial z} + \frac{\partial}{\partial x}\left[\mu\left(\frac{\partial w}{\partial x} + \frac{\partial u}{\partial z}\right)\right] + \frac{\partial}{\partial y}\left[\mu\left(\frac{\partial v}{\partial z} + \frac{\partial w}{\partial y}\right)\right] + \frac{\partial}{\partial z}\left(2\mu\frac{\partial w}{\partial z} - \frac{2}{3}\mu\vec{\nabla}\cdot\vec{V}\right)$$

These are the classic *Navier-Stokes equations* (written in the standard aerodynamic form that neglects the body force). They have multiple characteristics, including that they are:

• partial differential equations
• nonlinear (remember the definition of D/Dt includes nonlinearities, so that superposition of solutions is not allowed)
• second order (notice the second derivatives in the viscous terms)
• highly coupled (the variables appear in all of the equations), and
• long!

As written in the preceding discussion, it is easy to identify $\vec{F} = m\vec{a}$, written in the fluid mechanics form $m\vec{a} = \vec{F}$.

Computational Aerodynamics Concept Box

Euler and His Equation

Leonhard Euler was an important Swiss mathematician and physicist. During his life he contributed to a variety of mathematic fields, including infinitesimal calculus and graph theory, and he introduced a great deal of the modern mathematical terminology and notation that we use today. In addition, he did work in astronomy, fluid dynamics, mechanics, and optics.

Within fluid dynamics, he derived the set of mass and momentum equations that are applicable when the viscous terms are negligible; in this case, the flow is called *inviscid*. If you start with the Navier-Stokes equations (Equation 3.35) and neglect the viscous terms, you have (in the *x*-direction):

$$\rho \frac{Du}{Dt} = -\frac{\partial p}{\partial x}.$$

Applying this assumption to all three directions, the resulting equation set is known as the *Euler equations*. Euler derived his equations prior to Navier and Stokes (rather than simplifying from the Navier-Stokes equations like we just did), which would have been quite a challenge in 1755! These equations form the basis of inviscid flow theory, and can be shown to lead directly to Bernoulli's equation.

Leonhard Euler (1707–1783)

There are also alternate integral formulations of the Navier-Stokes equations. Consider the momentum flux through an arbitrary control volume in a manner similar to the integral statement of the continuity equation pictured in Fig. 3.4 and given in Equation 3.7. Here, the time rate of change of momentum, $\rho \vec{V}^2$, is proportional to the force, and the integral statement is:

$$\iint_S \rho \vec{V} \left(\vec{V} \cdot \hat{n} \right) dS + \frac{\partial}{\partial t} \iiint_V \rho \vec{V} dV = \vec{F} = \vec{F}_{volume} + \vec{F}_{surface}. \tag{3.36}$$

This statement can also be converted to the differential form using the Gauss Divergence Theorem. Notice that we use the derivative notation $\partial / \partial t$ to denote the change in the fixed control volume that has fluid moving across the boundaries.

The derivation of the Navier-Stokes equations was made for general unsteady fluid motion. These equations are valid for all flow simulations for a Newtonian fluid, but, due to limitations in our computational capability (a limitation that will be around for some time to come), we cannot solve these equations for the full spectrum of turbulent flows (which will be discussed in greater detail in Chapter 8). Therefore, currently we can only solve these equations directly for laminar flow or for flow with very low Reynolds numbers. When the flow is turbulent, the usual approach is to Reynolds-average (or time average) the equations, with the result being that additional Reynolds stress terms appear. Clearly, the addition of new unknowns requires additional equations, which is known as the closure problem. The closure problem is treated by the use of *turbulence models* and is discussed in greater detail in Chapter 8.

3.3.3 The Energy Equation

The equation for the conservation of energy is required to complete our system of equations. This is a statement of the First Law of Thermodynamics: The sum of the work and heat added to a system will equal the increase of energy. Following the derivation given by White:[7]

$$\underbrace{dE_t}_{\substack{\text{change of total energy} \\ \text{of the system}}} = \underbrace{\delta Q}_{\substack{\text{change of heat added}}} + \underbrace{\delta W}_{\substack{\text{change of work done} \\ \text{on the system}}} \tag{3.37}$$

where δ represents a change rather than a differential, since heat and work are not fluid properties. For our fixed control volume coordinate system, the rate of change is:

$$\frac{DE_t}{Dt} = \dot{Q} + \dot{W} \tag{3.38}$$

where

$$E_t = \rho\left(e + V^2/2 - \vec{g}\cdot\vec{r}\right) \tag{3.39}$$

and e is the internal energy per unit mass. The second term in the brackets is the kinetic energy, and the last term is the potential energy, i.e., the body force (which is usually neglected in aerodynamics). E_t can also be written in terms of specific energy as:

$$E_t = \rho e_0, \tag{3.40}$$

where

$$e_0 = e + V^2/2. \tag{3.41}$$

To obtain the energy equation we need to write the right-hand side of Equation 3.38 in terms of flow properties. First, we will consider the heat

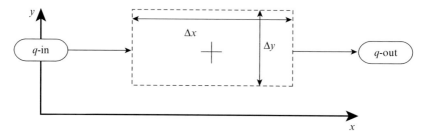

Figure 3.8   *x*-component of heat flux into and out of the control volume.

added to the system (neglecting heat addition due to radiation). The heat flow into the control volume is found in the identical manner to the mass flow. Using Fig. 3.8 for reference, we will obtain the expression for the net heat flow.

The heat fluxes across the control volume are:

$$
\begin{aligned}
q_{x_{in}} &= \left(q_x - \frac{\partial q}{\partial x}\frac{\Delta x}{2} \right)\Delta y \\
q_{x_{out}} &= \left(q_x + \frac{\partial q}{\partial x}\frac{\Delta x}{2} \right)\Delta y
\end{aligned}
\tag{3.42}
$$

and the net heat flow into the control volume in the *x*-direction is $q_{x_{in}} - q_{x_{out}}$, or:

$$
-\frac{\partial q_x}{\partial x}\Delta x \Delta y.
$$

Similarly, using the same analysis in the *y*- and *z*-directions, we obtain the net heat flux into the control volume (realizing that the $\Delta x \Delta y \Delta z$ terms will cancel):

$$
\dot{Q} = -\left(\frac{\partial q_x}{\partial x} + \frac{\partial q_y}{\partial y} + \frac{\partial q_z}{\partial z} \right) = -\vec{\nabla}\cdot\vec{q}.
\tag{3.43}
$$

Now we relate the heat flow to the temperature field using *Fourier's law:*[3]

$$
\vec{q} = -k\vec{\nabla}T
\tag{3.44}
$$

where k is the coefficient of thermal conductivity, which is also a function of temperature (see White[7] for more details). Equation 3.44 is then put into Equation 3.43 to get the heat conduction in terms of the temperature gradient:

$$
\dot{Q} = -\vec{\nabla}\cdot\vec{q} = +\vec{\nabla}\cdot\left(k\vec{\nabla}T \right).
\tag{3.45}
$$

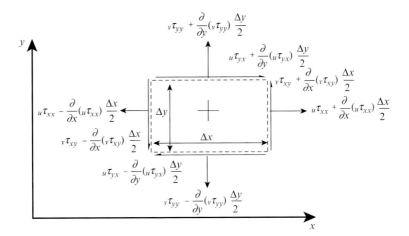

Figure 3.9   Work done on a control volume.

Next we need to find the work done on the system. Using the definition of *work = force x distance*, the *rate* of work is *force x distance / time*, or:

$$\dot{W} = force \times velocity. \tag{3.46}$$

Using the control volume again we find the work, which is equal to the velocity times the stress. The work associated with the x-face of the control volume (for two-dimensional flow) is:

$$w_x = u\tau_{xx} + v\tau_{xy}. \tag{3.47}$$

The complete description of the work on the control volume is shown in Fig. 3.9.

Using the x-component of net work as an example again, the work done on the system is $w_{xin} - w_{xout}$, or:

$$\left(w_x - \frac{\partial w_x}{\partial x}\frac{\Delta x}{2}\right)\Delta y - \left(w_x + \frac{\partial w_x}{\partial x}\frac{\Delta x}{2}\right)\Delta y = -\frac{\partial w_x}{\partial x}\Delta x \Delta y. \tag{3.48}$$

Including the other directions (and dropping the $\Delta x \Delta y \Delta z$ terms, which cancel out) we obtain:

$$\dot{W} = -\vec{\nabla}\cdot\vec{w} = \frac{\partial}{\partial x}\left(u\tau_{xx} + v\tau_{xy} + w\tau_{xz}\right) + \frac{\partial}{\partial y}\left(u\tau_{yx} + v\tau_{yy} + w\tau_{yz}\right) + \frac{\partial}{\partial z}\left(u\tau_{zx} + v\tau_{zy} + w\tau_{zz}\right).$$

$$\tag{3.49}$$

Substituting Eqns. 3.40 and 3.41 into Equation 3.38 for E_t, Equation 3.45 for the heat, and Equation 3.49 for the work, we obtain:

$$\frac{D\rho(e + V^2/2)}{Dt} = \vec{\nabla}\cdot\left(k\vec{\nabla}T\right) - \vec{\nabla}\cdot\vec{w}. \tag{3.50}$$

Many equivalent forms of the energy equation are found in the literature. Often the equation is thought of as an equation for the temperature. We now describe how to obtain one other specific form. Substituting in the relations for the τ's in terms of μ and the velocity gradients, Eqns. 3.31 and 3.33, we obtain the following lengthy expression (see Bertin and Cummings[3]). Making use of the momentum and continuity equations to expand the equation, and introducing the definition of enthalpy, $h = e + p/\rho$, we obtain a form that is also frequently used. This classical form of the energy equation is:

$$\rho \frac{Dh}{Dt} - \frac{Dp}{Dt} = \underbrace{\vec{\nabla} \cdot \left(k \vec{\nabla} T \right)}_{\text{heat conduction}} + \underbrace{\Phi}_{\substack{\text{viscous dissipation} \\ \text{(always positive)}}} \qquad (3.51)$$

where

$$\Phi = \mu \left\{ \begin{array}{l} 2\left[\left(\dfrac{\partial u}{\partial x}\right)^2 + \left(\dfrac{\partial v}{\partial y}\right)^2 + \left(\dfrac{\partial w}{\partial z}\right)^2 \right] + \left(\dfrac{\partial v}{\partial x} + \dfrac{\partial u}{\partial y}\right)^2 + \left(\dfrac{\partial w}{\partial y} + \dfrac{\partial v}{\partial z}\right)^2 \\[4mm] + \left(\dfrac{\partial u}{\partial z} + \dfrac{\partial w}{\partial x}\right)^2 - \dfrac{2}{3}\left(\dfrac{\partial u}{\partial x} + \dfrac{\partial v}{\partial y} + \dfrac{\partial w}{\partial z}\right)^2 \end{array} \right\}. \qquad (3.52)$$

Also, the energy equation is frequently written in terms of the total enthalpy, $H = h + V^2/2$, to good advantage in inviscid and boundary-layer flows. A good discussion of the energy equation is given by White.[7]

There is also an integral form of this equation, just as there was for the mass and momentum equations:

$$\iint_S \rho\left(e + V^2/2\right)\left(\vec{V} \cdot \hat{n}\right) dS + \frac{\partial}{\partial t} \iiint_V \rho\left(e + V/2\right) d\mkern-1mu\mathchar'26\mkern-9mu V = \dot{Q} + \dot{W}. \qquad (3.53)$$

Here again note that we use the derivative notation $\partial / \partial t$ to denote the change in the fixed control volume that has fluid moving across the boundaries.

3.4 Solution of the Set of Governing Equations

As we just found out, when we combine our three conservation laws, the equation of state and relations for viscosity and thermal conductivity as a function of temperature, we end up with eight equations in eight unknowns. If we were to mathematically characterize these equations, we would give them the following attributes:

- they are a set of five partial differential equations and three algebraic relations in eight unknowns (density, three components of velocity, pressure, temperature, viscosity, and thermal conductivity)

- the equations are coupled (the same unknown variables appear in multiple equations)
- some of the equations are nonlinear equations (due to products of unknowns, or dependent variables appearing in equations)
- some of the equations are second-order equations (the second derivatives appear in the viscous terms)

Unfortunately, there is no known general, closed-form analytic solution to this set of equations. Therefore, we have only three possibilities for finding solutions to the equations:

- make simplifying assumptions (like steady flow, inviscid flow, or two-dimensional flow; or that the flow is only moving at a slightly different speed than the freestream value) and drop terms until we obtain a set of equations that we can solve – this is done for potential flow or supersonic flow theory[10,3]
- apply the full set of equations to simple flows (like laminar flow over a flat plate) and get analytic solutions, or possibly ordinary differential equations, as in the case of similarity solutions in boundary-layer theory; there are only a handful of useful solutions from this approach[9]
- solve the equation set numerically

The approach in this book is to find ways to use computers to find solutions to the preceding equations using either the first or third of these methods. When we use the first method (simplifying assumptions), we end up with what are known as panel methods or vortex-lattice methods. When we use the third method (solving the equations numerically), we are working in the area known as Computational Fluid Dynamics (CFD). While both approaches are valuable, as computers have increased in speed and memory capability, CFD has become the more dominant of the two methods, although panel methods that use a computational approach to solve large systems of equations continue to have an important role in computational aerodynamics and aircraft design.

3.5 Standard Forms and Terminology of Governing Equations

To understand the literature in computational aerodynamics, several other aspects of the terminology for the governing equations must be discussed. This section provides details about several of these considerations. For example, there are other forms of the governing equations, but we will not be able to show them all here.

3.5.1 Nondimensionalization

The governing equations used for CFD are often nondimensionalized. In fluid mechanics theory, nondimensionalization is used to reveal important similarity parameters for the flow. Similarity parameters are usually nondimensional terms that help describe the category or type of flow (e.g., Mach number tells us about whether the flow is greater than or less than the speed of sound). In practice, many different nondimensionalizations could be used.

Sometimes the dimensional quantities are defined by ()*'s or (ˆ)'s. In other schemes the nondimensionalized variables are designated by other special symbols. In the example given here, the nondimensionalized values are denoted by ()*. In our derivation, once the quantities are defined, the *'s will be dropped, and the nondimensionalization is understood.

As we said, many different approaches can be used for the nondimensionalization process. We will give one example here that uses the freestream velocity (V_∞), together with a reference length (L) as follows:

$$x^* = \frac{x}{L} \quad y^* = \frac{y}{L} \quad z^* = \frac{z}{L} \quad t^* = \frac{tV_\infty}{L}$$

$$u^* = \frac{u}{V_\infty} \quad v^* = \frac{v}{V_\infty} \quad w^* = \frac{w}{V_\infty} \quad p^* = \frac{p}{\rho_\infty V_\infty^2} \tag{3.54}$$

$$T^* = \frac{T}{T_\infty} \quad \rho^* = \frac{\rho}{\rho_\infty} \quad \mu^* = \frac{\mu}{\mu_\infty} \quad e_o^* = \frac{e_o}{V_\infty^2}.$$

Each computer code will have a set of reference nondimensionalizations similar to these; a specific example is given in Section 3.5.3. For example, it is quite common to use the speed of sound as the reference velocity (to understand why, take a look at the various velocity terms in Equation 3.54 and see what happens for $V_\infty = 0$). Making sure that you understand the nondimensionalization is an important part of using computational codes for aerodynamic applications properly.

3.5.2 Use of Divergence Form

The classical forms of the governing equations normally given in textbooks usually are not used for computations (as we gave them earlier). Instead, the *divergence form*, or *conservation form*,[iii] is used. This form is found to

---

[iii] Be careful here since the terminology can be confusing. The continuity, momentum, and energy equations are all conservation equations. Conservation *form* refers to the situation where all of the variables are inside the derivatives. That's why we prefer the use of divergence form to describe this mathematical arrangement. Conservation form is the more widely used terminology. They are both the same.

be required for reliable numerical calculations, and the papers by Viviand[11] and Vinokur[12] are often cited as the modern origin of the conservation equations used in CFD computations.[iv] If discontinuities in the flowfield exist, this form must be used to correctly account for discontinuities (for example, across a shock wave the density and velocity both jump in value; however, the product of these quantities, the mass flux, is a constant). Therefore, we can easily see why it is better numerically to work with the product of these variables rather than the individual variables. In this section we show how the divergence forms are obtained from the standard classical form.

We can use the inviscid two-dimensional steady x-momentum equation (a simplified version of the full equation given as Equation 3.35) as an example:

$$\rho u \frac{\partial u}{\partial x} + \rho v \frac{\partial u}{\partial y} = -\frac{\partial p}{\partial x}. \tag{3.55}$$

This equation can be rewritten using the following identities: for the first term in Equation 3.55, write

$$\frac{\partial \rho u u}{\partial x} = \rho u \frac{\partial u}{\partial x} + u \frac{\partial \rho u}{\partial x} \tag{3.56}$$

or

$$\rho u \frac{\partial u}{\partial x} = \frac{\partial \left(\rho u^2\right)}{\partial x} - u \frac{\partial \rho u}{\partial x}, \tag{3.57}$$

and similarly for the second term in Equation 3.55, write

$$\frac{\partial \rho v u}{\partial y} = \rho v \frac{\partial u}{\partial y} + u \frac{\partial \rho v}{\partial y} \tag{3.58}$$

or

$$\rho v \frac{\partial u}{\partial y} = \frac{\partial \rho v u}{\partial y} - u \frac{\partial \rho v}{\partial y}. \tag{3.59}$$

Substituting Equations 3.57 and 3.59 into Equation 3.55 we obtain

$$\frac{\partial \rho u^2}{\partial x} - u \frac{\partial \rho u}{\partial x} + \frac{\partial \rho v u}{\partial y} - u \frac{\partial \rho v}{\partial y} + \frac{\partial p}{\partial x} = 0 \tag{3.60}$$

[iv] The use of conservation form actually dates to 1960, as shown in Richtmyer and Morton, *Difference Methods for Initial Value Problems*, 2nd Ed., New York: Interscience Publishers, 1967, pp. 300–306.

which can be written as

$$\frac{\partial \rho u^2}{\partial x} + \frac{\partial \rho v u}{\partial y} - u\underbrace{\left(\frac{\partial \rho u}{\partial x} + \frac{\partial \rho v}{\partial y}\right)}_{=0 \text{ from continuity}} + \frac{\partial p}{\partial x} = 0. \qquad (3.61)$$

Finally, the x-momentum equation written in divergence form for inviscid two-dimensional steady flow is:

$$\frac{\partial \left(\rho u^2 + p\right)}{\partial x} + \frac{\partial \left(\rho v u\right)}{\partial y} = 0. \qquad (3.62)$$

All of the governing equations must be written in divergence form to be valid when shock waves are present, which will be illustrated in the next section.

3.5.3 The "Standard" or "Vector" Form of the Equations

Even after writing the governing equations in divergence form, the equations that you see in the literature will not look like the ones we have been showing. A standard form for the equations called the *vector form* is used in the literature for numerical solutions of the fluid dynamic equations of motion. In this section, we will provide one representative form for these equations which comes from two NASA codes, CFL3D[13] and ARC3D[14] (see Pletcher et al.[15] for additional derivation details about the vector form).

The Navier-Stokes equations (and the other equations required in the system) are written in the so-called vector divergence form as follows:

$$\frac{\partial Q}{\partial t} + \frac{\partial \left(F - F_v\right)}{\partial x} + \frac{\partial \left(G - G_v\right)}{\partial y} + \frac{\partial \left(H - H_v\right)}{\partial z} = 0 \qquad (3.63)$$

where the conserved variables are:

$$Q = \left\{ \begin{array}{c} \rho \\ \rho u \\ \rho v \\ \rho w \\ E_t \end{array} \right\} = \left\{ \begin{array}{l} \text{density} \\ \text{x-momentum} \\ \text{y-momentum} \\ \text{z-momentum} \\ \text{total energy per unit volume} \end{array} \right\}. \qquad (3.64)$$

The flux vectors in the x-direction are:

$$
\begin{array}{cc}
\text{Inviscid terms} & \text{Viscous terms}
\end{array}
$$

$$
F = \begin{bmatrix} \rho u \\ \rho u^2 + p \\ \rho uv \\ \rho uw \\ (E_t + p)u \end{bmatrix} \quad
F_v = \begin{bmatrix} 0 \\ \tau_{xx} \\ \tau_{xy} \\ \tau_{xz} \\ u\tau_{xx} + v\tau_{xy} + w\tau_{xz} + \dot{q}_x \end{bmatrix}. \tag{3.65}
$$

Notice that the pressure term has now been taken out of the stress term, so the various τ terms only represent shear stress.

Similar expressions can be written down for the y- and z-direction fluxes, with the y-direction given as:

$$
\begin{array}{cc}
\text{Inviscid terms} & \text{Viscous terms}
\end{array}
$$

$$
G = \begin{bmatrix} \rho v \\ \rho vu \\ \rho v^2 + p \\ \rho vw \\ (E_t + p)v \end{bmatrix} \quad
G_v = \begin{bmatrix} 0 \\ \tau_{yx} \\ \tau_{yy} \\ \tau_{yz} \\ u\tau_{yx} + v\tau_{yy} + w\tau_{yz} + \dot{q}_y \end{bmatrix} \tag{3.66}
$$

and the z-direction as:

$$
\begin{array}{cc}
\text{Inviscid terms} & \text{Viscous terms}
\end{array}
$$

$$
H = \begin{bmatrix} \rho w \\ \rho wu \\ \rho wv \\ \rho w^2 + p \\ (E_t + p)w \end{bmatrix} \quad
H_v = \begin{bmatrix} 0 \\ \tau_{zx} \\ \tau_{zy} \\ \tau_{zz} \\ u\tau_{zx} + v\tau_{zy} + w\tau_{zz} + \dot{q}_z \end{bmatrix}. \tag{3.67}
$$

The equation of state (for a perfect, or ideal, gas) is written in this formulation as:

$$
p = (\gamma - 1)\left[E_t - \rho(u^2 + v^2 + w^2)/2 \right]. \tag{3.68}
$$

To complete the flow equations, we need to define the shear stress and heat transfer nomenclature as well as the nondimensionalization we will use. The shear stress and heat transfer terms are written in indicial (or index[v]) notation as:

$$
\tau_{x_i x_j} = \frac{M_\infty}{Re_L}\left[\mu\left(\frac{\partial u_i}{\partial x_j} + \frac{\partial u_j}{\partial x_i} \right) + \lambda \frac{\partial u_k}{\partial x_k}\delta_{ij} \right] \tag{3.69}
$$

---

[v] Index notation is a shorthand notation. x_i denotes x, y, z for $i = 1,2,3$. δ_{ij} is the Kronecker delta, equal to 1 if $i = j$, and 0 otherwise. If an index is repeated in a single term a summation takes place over all three values, i.e., $u_i u_i = u_1^2 + u_2^2 + u_3^2$.

and

$$\dot{q}_{x_i} = -\left[\frac{M_\infty \mu}{Re_L \, Pr(\gamma-1)}\right]\frac{\partial(a^2)}{\partial x_i} = -\left[\frac{M_\infty \mu}{Re_L \, Pr(\gamma-1)}\right]\frac{\partial T}{\partial x_i}. \tag{3.70}$$

The molecular viscosity is found using Sutherland's law, which was defined in Equation 3.32. The other quantities are defined as: Reynolds number, $Re_L = \rho_\infty V_\infty L / \mu_\infty$; Mach number, $M_\infty = V_\infty / a_\infty$; and Prandtl number, $Pr = c_p \mu / k$. Stokes's hypothesis for bulk viscosity is also used in this case, which states that $\lambda + 2\mu/3 = 0$, and the freestream velocity magnitude is $V_\infty = \left[u_\infty^2 + v_\infty^2 + w_\infty^2\right]^{1/2}$.

The velocity components are given by (with α and β being the angles of attack and sideslip, respectively):

$$\begin{aligned} u &= \tilde{u} / \tilde{a}_\infty & u_\infty &= M_\infty \cos\alpha \cos\beta \\ v &= \tilde{v} / \tilde{a}_\infty & v_\infty &= -M_\infty \sin\beta \\ w &= \tilde{w} / \tilde{a}_\infty & w_\infty &= M_\infty \sin\alpha \cos\beta \end{aligned} \tag{3.71}$$

and the thermodynamic variables are given by:

$$\begin{aligned} \rho &= \tilde{\rho} / \tilde{\rho}_\infty & \rho_\infty &= 1 \\ p &= \tilde{p} / \tilde{\rho}\tilde{a}_\infty^2 & p_\infty &= 1/\gamma \\ T &= \tilde{T} / \tilde{T}_\infty = \gamma p / \rho = a^2 & T_\infty &= 1 \end{aligned} \tag{3.72}$$

and

$$E_t = \tilde{E}_t / \tilde{\rho}_\infty \tilde{a}_\infty^2 \qquad E_{t_\infty} = 1/\left[\gamma(\gamma-1)\right] + M_\infty^2 / 2. \tag{3.73}$$

This completes the nomenclature for one typical example of the application of the Navier-Stokes equations in current computer codes. Notice that these equations are for a Cartesian coordinate system; other coordinate systems could have been used. Although these equations look like they could not become any more complicated, they certainly can! For example, we will discuss the necessary extension of these equations to general coordinate systems in Appendix E, and we will also discuss the modifications required to include turbulence models in Chapter 8.

Profiles in Computational Aerodynamics: Kozo Fujii

I was born in 1951 and grew up in Kofu, Japan, where I lived until I was eighteen years old and finished high school. I moved to Tokyo for my undergraduate studies at the University of Tokyo in 1970. After two years of liberal arts study, I decided to study in the Department of Aeronautics. There were no strong reasons for the choice except that I liked airplanes and was interested in "manufacturing," which is why I was not a motivated student for my aeronautical studies. After I finished my undergraduate work, I entered graduate school in the same department in 1974 and physically moved to the Institute of Space and Aeronautical Science (ISAS). The ISAS, a part of the University of Tokyo at that time, was (and is still now) a unique institute that had developed the first Japanese rocket in 1955 and the first Japanese satellite in 1970.

I first became familiar with Computational Fluid Dynamics (CFD) when I was at ISAS. The early to middle 1970s was the period when computational techniques in aeronautics made tremendous progress. I read many articles by the famous researchers, but I did not even imagine that they would become good friends of mine, as I was only a graduate student far away in Asia. After receiving a PhD in 1980, I had a hard time finding a job, but an opportunity to give a presentation at the well-known International Conference on Numerical Methods in Fluid Dynamics in 1980 opened the door of my career. It gave me an opportunity to work at NASA Ames Research Center in California as a National Research Council (NRC) research associate in 1981. At that time, NASA Ames was the place where top CFD scientists worked with advanced computer systems. Dr. Paul Kutler, my NRC research advisor, suggested that I work on the research topic, "Navier-Stokes simulation of vortex flows over a delta wing." This was totally a different topic than I had proposed, and nobody had tried to simulate such flowfields in the past. Fortunately, the first general-purpose supercomputer, the CRAY-1S, had just become available, and although the memory size was 32MB (which only allowed 10,000 grid points), a leading-edge separation vortex was successfully simulated for the first time in the world.

Two years later, I came back to Japan, and started working for the National Aerospace Laboratory (NAL) in 1984. The NAL was just about to introduce the first Japanese supercomputer. Japanese heavy industries were interested in supercomputing for wing design and flow analysis due to their negotiations with The Boeing Company for a new aircraft development (the Boeing 7J7). Two-hour computations using the Fujitsu Supercomputer VP400 (which had 1 GFPLOPS and 256MB memory) enabled us to conduct transonic flow simulations over a newly developed practical wing geometry. The AIAA magazine *Aerospace America* chose this effort as one of the important aerospace topics for the year 1986.

In 1986, I returned to NASA Ames as a senior NRC research associate, now working with Lew Schiff and Terry Holst. This time, my research topic was simulation of a more practical delta wing having a strake. With the timely arrival of the CRAY-2 supercomputer, my simulation captured two types of vortex breakdown of leading-edge flow separation over the strake-delta wing. I felt very honored by the fact that NASA Ames held a press conference to highlight this achievement as a pioneering work for the future of supersonic transport and fighter aircraft design. It was also an honor that the result, together with my photograph, was displayed in the main Numerical Aerodynamic Simulation (NAS) building at Ames for several years as an important contribution from NAS. In 1988, I moved back to the ISAS and worked there until the present time. The ISAS became one division of the Japan Aerospace Exploration Agency (JAXA) by the merger of three aerospace organizations in 2003.

In 2005, I was assigned to become a first director of JAXA's Engineering Digital Innovation center (JEDI, which I intentionally named), which is responsible for computational mechanics, eventually making CFD a feasible tool for space development. I am now responsible for all space science activity in Japan as a deputy director general of the ISAS/JAXA. However, I am still a professor and continue my CFD research with my colleagues and students. Several years ago, I found one article written as an ISAS report in 1930. This article includes boundary-layer theory, stability analysis for the transition, linear subsonic and supersonic theory, transonic hodograph methods, and others. I felt that our research only focuses on a very small fraction of what the author of this article had foreseen. The author's name was Professor Ludwig Prandtl. Since then, I have focused on the research topics that may introduce engineering innovation in fluid dynamic design, which will only be possible with the use of leading-edge supercomputers.

Fortunately, I started my CFD research at the right time, at the right place, and with the right people. I was always given opportunities to use leading-edge supercomputers and was able to work with excellent researchers. CFD still has room to make additional progress in the future, although it experienced remarkable progress in the 1980s with the appearance of supercomputers. I believe CFD not only will replace certain aspects of wind tunnel experiments, but will also do much more than that as a third pillar of science.

3.6 Boundary Conditions, Initial Conditions, and the Mathematical Classification of Partial Differential Equations (PDEs)

Now that we have developed the governing equations, we need to understand how to specify boundary conditions in order to solve the governing equations. If all flowfields are governed by the same equations, what makes one flowfield different from another? Application-specific boundary conditions are the means through which the solution of the governing equations produces differing results for different situations. In computational aerodynamics, the specification of boundary conditions constitutes an important part of any effort. Presuming that the flowfield algorithm selected for a particular problem is already developed and tested, the application of the method requires the user to specify problem-specific boundary conditions to obtain a solution. Also, to understand what constitutes a properly posed set of initial and boundary conditions requires us to understand the mathematical classification or "type" of partial differential equations. The mathematical type of a problem also plays a key role in understanding how we can simplify the equations in specific cases.

A key property of any system of PDEs is the "type" of the equations, since in mathematics an equation "type" has a very precise meaning and importance. Essentially, the *type* of the equation determines the domain on which boundary or initial conditions must be specified. The mathematical theory has been developed over a number of years for PDEs, and the details may be found in various books, including the books by Sneddon,[16] Chester,[17] and Carrier and Pearson.[18] Discussions from the computational fluid dynamics viewpoint are also available in Pletcher, Tannehill, and Anderson,[15] Fletcher,[19] and Hoffman and Chiang.[20]

To successfully obtain the numerical solution of a PDE, you must satisfy the "spirit" of the theory for that type of PDE. For PDEs describing physical systems, the type will be related to the following categorization:

- *Equilibrium problems.* Examples include steady-state temperature distributions and steady incompressible flow. These are similar to boundary value problems for ordinary differential equations.
- *Marching or propagation problems.* These are transient or transient-like problems. Examples include transient heat conduction and steady supersonic flow. These are similar to initial value problems for ODEs.

The three PDE *types* are *elliptic*, *parabolic*, and *hyperbolic*. A linear equation will have a single type, but nonlinear equations of fluid flow can change type locally depending on the local values of the variables in the equation. This "mixed-type" feature has a profound influence on the development of numerical methods for computational aerodynamics (especially during the early development of transonic airfoil codes). A mismatch between the type of the PDE and the prescribed boundary conditions dooms any attempt at numerical solution to failure!

The standard illustration for the mathematical equation type uses a single second-order PDE in two dimensions given by:

$$a\phi_{xx} + b\phi_{xy} + c\phi_{yy} + d\phi_x + e\phi_y + f\phi + g = 0 \tag{3.74}$$

where a, b, c, d, e, f, and g can be constants or functions of x and y. Depending on the values of the coefficients of the second-order partial derivatives (a, b, and c), the PDE will take on one of the three types. The specific type of the PDE depends on the characteristic curves of the PDE. One of the important properties of characteristic curves is that the second derivative of the dependent variables is allowed to "jump" (have a discontinuity) on a characteristic curve, although there can be no discontinuity in the first derivative. The slopes of the characteristic curves are found from a, b, and c depending on the sign of the determinant:

		Characteristics	PDE Type
$(b^2 - 4ac)$	>0	real	hyperbolic
	=0	real, equal	parabolic
	<0	imaginary	elliptic

$$\tag{3.75}$$

Now we can look at each type and see what distinguishes them from one another.

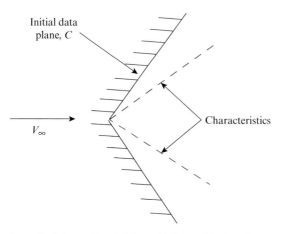

Figure 3.10 Connection between characteristics and initial condition data planes.

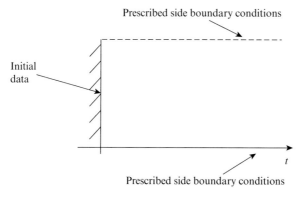

Figure 3.11 Initial data curve for a parabolic PDE.

3.6.1 Hyperbolic Type

The basic property of a hyperbolic PDE is a limited domain of dependence. Initial data are required on a curve C, which does not coincide with a characteristic curve (Fig. 3.10 illustrates this requirement).

Classical linearized supersonic aerodynamic theory is an example of a hyperbolic system. We will use a one-dimensional unsteady wave equation as our model for a hyperbolic system.

3.6.2 Parabolic Type

The parabolic PDE type is associated with a diffusion process. Data must be specified at an initial plane and "marched" forward in a time or a time-like direction. There is no limited zone of influence equivalent to the hyperbolic case. Data are required on the entire initial data surface and along the side boundaries (Fig. 3.11 illustrates the requirement).

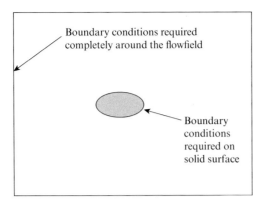

Boundary conditions required completely around the flowfield

Boundary conditions required on solid surface

Figure 3.12 Boundary conditions required for elliptic PDEs.

In aerodynamics, the boundary-layer equations have a parabolic type. We will use a heat conduction equation as our model for a parabolic system.

3.6.3 Elliptic Type

These are equilibrium problems. They require boundary conditions everywhere around the solution space, as shown in Fig. 3.12. Incompressible potential flow is an example of a governing equation of elliptic type. Our model equation for an elliptic PDE is Laplace's equation.

3.6.4 Equations of Mixed Type

We will also want to consider examples of equations that are of mixed type. One good example is the Prandtl-Glauert equation, which we will derive later in the chapter:

$$(1 - M_\infty^2)\phi_{xx} + \phi_{yy} = 0. \tag{3.76}$$

Depending on the coefficient of the leading term, the equation can have different types:

$$\begin{aligned} M_\infty &< 1 \quad \text{type is elliptic} \\ &> 1 \quad \text{type is hyperbolic} \end{aligned} \tag{3.77}$$

so that although the equation is the same for both subsonic and supersonic flow, the physical behaviors of subsonic and supersonic flows are entirely different, as are the numerical approaches used to obtain a solution.

If we use a transonic flow model that includes a nonlinear term (such as in the transonic small disturbance equation shown in Equation 3.78), then the equation can change type locally in the flowfield. The equation models the real physics, and the solution can actually include a shock wave. Looking at

Equation 3.78 we can see how the coefficient of the ϕ_{xx} term determines the type based on Equations 3.74 and 3.75:

$$\underbrace{\left[\left(1 - M_\infty^2\right) - \left(\gamma + 1\right) M_\infty^2 \frac{\phi_x}{V_\infty} \right]}_{\substack{\text{sign depends on the solution:} \\ \text{-locally subsonic flow is elliptic} \\ \text{-locally supersonic flow is hyperbolic}}} \phi_{xx} + \phi_{yy} = 0. \tag{3.78}$$

This is an equation of mixed type, which is required to treat transonic flows, where both subsonic and supersonic flows exist in the flowfield. The added complexity of the equation is required to treat the nonlinearities of transonic flow.

We have seen that *type* plays a key role in computational approaches and, at times, can be used to our advantage. As an example, consider the numerical solution of the Euler equations. Experience has shown us that the steady-state Euler equations are quite hard to solve. It has become standard procedure instead to solve the unsteady equation, which is hyperbolic in time, and obtain the steady-state solution by marching in time until the solution is constant (steady).

3.6.5 Elaboration on Characteristics

It is often difficult to understand why the various PDE types give such different results, but a closer look at the characteristics of each type can be informative. Here we provide additional details that offer some insight into the reason that the determinant of the coefficients of the second derivative terms defines the type of the equation.[15] Consider Equation 3.74 again:

$$a\phi_{xx} + b\phi_{xy} + c\phi_{yy} + d\phi_x + e\phi_y + f\phi + g = 0. \tag{3.74}$$

- Assume ϕ is a solution describing a curve in space
- These curves "patch" various solutions, known as characteristic curves
- Discontinuity of the second derivative of the dependent variable is allowed, but no discontinuity of the first derivative. The differentials of ϕ_x and ϕ_y that represent changes from x, y to $x + dx, y + dy$ along characteristics are:

$$d\phi_x = \frac{\partial \phi_x}{\partial x} dx + \frac{\partial \phi_x}{\partial y} dy = \phi_{xx} dx + \phi_{xy} dy \tag{3.79}$$

$$d\phi_y = \frac{\partial \phi_y}{\partial x} dx + \frac{\partial \phi_y}{\partial y} dy = \phi_{yx} dx + \phi_{yy} dy \tag{3.80}$$

Express Equation 3.74 as

$$a\phi_{xx} + b\phi_{xy} + c\phi_{yy} = h \tag{3.81}$$

with

$$h = -(d\phi_x + e\phi_y + f\phi + g). \tag{3.82}$$

We will assume Equation 3.81 is linear and solve Equation 3.81 with Equation 3.79 and Equation 3.80 for the second derivatives of ϕ:

$$\begin{aligned} a\phi_{xx} &+ b\phi_{xy} &+ c\phi_{yy} &= h \\ dx\phi_{xx} &+ dy\phi_{xy} & &= d\phi_x \\ &dx\phi_{xy} &+ dy\phi_{yy} &= d\phi_y \end{aligned} \tag{3.83}$$

or

$$\begin{bmatrix} a & b & c \\ dx & dy & 0 \\ 0 & dx & dy \end{bmatrix} \begin{bmatrix} \phi_{xx} \\ \phi_{xy} \\ \phi_{yy} \end{bmatrix} = \begin{bmatrix} h \\ d\phi_x \\ d\phi_y \end{bmatrix}. \tag{3.84}$$

Now solve for ϕ_{xx}, ϕ_{xy}, ϕ_{yy}. Since second derivatives can be discontinuous on the characteristics, the derivatives are indeterminate and the coefficient matrix would be singular:

$$\begin{vmatrix} a & b & c \\ dx & dy & 0 \\ 0 & dx & dy \end{vmatrix} = 0. \tag{3.85}$$

Expanding this yields

$$a(dy)^2 - b\,dx\,dy + c(dx)^2 = 0. \tag{3.86}$$

The slopes of the characteristics curves are found by dividing by $(dx)^2$:

$$a\left(\frac{dy}{dx}\right)^2 - b\left(\frac{dy}{dx}\right) + c = 0. \tag{3.87}$$

Solve for dy/dx to obtain:

$$\frac{dy}{dx} = \frac{b \pm \sqrt{b^2 - 4ac}}{2a} \tag{3.88}$$

and hence the requirement for $\sqrt{b^2 - 4ac}$ to define the type of the PDE as related to the characteristics of the equation. Depending on whether the term under the radical is positive, zero, or negative, the PDE will be hyperbolic, parabolic, or elliptic. See the PDE references cited previously for more details on this derivation.

3.7 Hyperbolic PDEs

The first PDE type, the hyperbolic PDE, occurs when $b^2 - 4ac > 0$ in Equation 3.88. The two characteristic curves represented by Equation 3.88 are real functions of x and y. When we prescribe the initial data, known as prescribing Cauchy data, we cannot choose the characteristic curves.

An example of a Hyperbolic PDE is the wave equation. The wave equation for an infinite region with arbitrarily defined initial conditions is given by

$$u_{tt} = c^2 u_{xx}$$

$$u(x,0) = h(x) \tag{3.89}$$

$$u_t(x,0) = p(x)$$

where $-\infty < x < \infty$ and $t > 0$. For c as a real constant this is a very interesting partial differential equation. The final form of the solution to the wave equation is known as D'Alembert's solution (see Carrier and Pearson[18] for more details of the solution) and is given by:

$$u(x,t) = \frac{1}{2}\{h(x+ct) + h(x-ct)\} + \frac{1}{2c}\int_{x-ct}^{x+ct} p(\tau)d\tau. \tag{3.90}$$

The solution presented in Equation 3.90 is extraordinary in that it is only dependent on the initial conditions, $h(x)$ and $p(x)$. Once the initial conditions are established and the problem is "begun," no further alterations to the solution are possible; it is completely predetermined. Also notice the simplicity of the solution for the special case where $u_t(x,0) = p(x) = 0$:

$$u(x,t) = \frac{1}{2}\{h(x+ct) + h(x-ct)\} \tag{3.91}$$

Equation 3.91 is a simple functional relationship consisting of two nearly identical solutions, differing only in their sense of time. If you think of c as the wave speed, then these two solutions represent two identical unchanging waves, each traveling in opposite directions. The wave represented by $h(x+ct)$ travels along the x-axis with velocity $-c$; this is a wave pattern moving to the left with constant speed and non-changing form. The wave represented by $h(x-ct)$ travels along the x-axis with velocity $+c$; this is a wave pattern moving to the right with constant speed and non-changing form. The total solution is the average of the individual waves – superposition holds since the

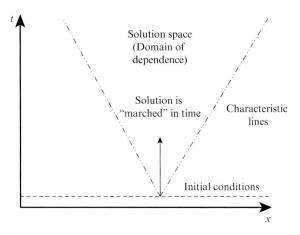

Figure 3.13 Solution space for a hyperbolic PDE.

equation is linear. Also, note that the initial conditions have the capability to affect the solution in both directions, restricted only by the magnitude of dx/dt, as shown in Fig. 3.13.

The fluid dynamic analogy for a hyperbolic PDE is supersonic flow, where pressures can only propagate within a limited space downstream from the point of interest (the space is defined by two lines, known in mathematics as the characteristic lines and known in fluid dynamics as the Mach lines). Because of this, when we solve this problem numerically, we should use "upwind" differencing (backward differencing) so we do not use any information from the future and make sure that the analytic domain of influence lies within the numeric domain of influence (which will be described in Chapter 6). We will also be required to supply initial conditions in order to get a solution.

3.8 Parabolic PDEs

The next PDE type we will explore is the parabolic PDE, which takes place when $b^2 - 4ac = 0$. For such an equation to be nontrivial (to remain a second-order partial differential equation), $a, c \neq 0$. If we follow the same procedure as for the hyperbolic case, the characteristic relation is

$$\frac{dy}{dx} = \frac{b}{2a}. \tag{3.92}$$

Equation 3.92 represents a solution with only one "characteristic." Unlike the hyperbolic case, the solution propagates from the initial conditions in only one direction, restricted by the magnitude of dy/dx. An example of a parabolic PDE is the diffusion (or heat) equation:

$$u_t = a^2 u_{xx}. \tag{3.93}$$

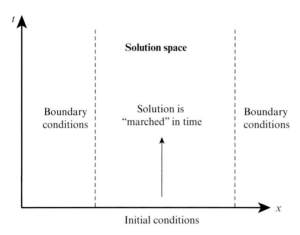

Figure 3.14 Solution space for a parabolic PDE.

The diffusion equation can also be solved, although the parabolic nature of this PDE makes the solutions more complicated than those for hyperbolic PDEs. Fourier series analysis is required to solve this equation, and, for the initial and boundary conditions $u(x,0) = f(x)$ and $u(0,t) = u(1,t) = 0$, the solution is given by:

$$u(x,t) = \sum_{n=1}^{\infty} A_n e^{-\alpha k^2 t} \sin(kx) \tag{3.94}$$

where the Fourier coefficients are found from:

$$A_n = 2\int_0^1 f(x)\sin(kx)dx \quad k = n\pi. \tag{3.95}$$

A parabolic PDE requires initial conditions along a data curve, with boundary conditions that form "sides" to the solution space. The solution is then obtained by "marching" (or integrating) in time through the space, as shown in Fig. 3.14.

The aerodynamic analogy for a parabolic PDE is a boundary-layer calculation, where the boundary layer develops moving downstream along the surface. To compute the boundary layer we need initial conditions, a starting velocity profile, and then we need wall boundary conditions and the velocity at the outer edge of the boundary layer to be prescribed.

3.9 Elliptic PDEs

The third and final type of PDE is the elliptic PDE, which arises when $b^2 - 4ac < 0$. In this case the characteristic equation, Equation 3.88, has no real roots. Analytic solutions of elliptic PDEs are usually found using infinite

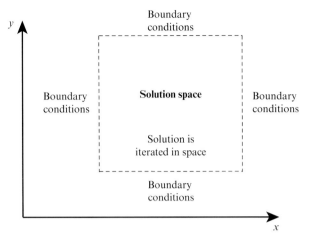

Figure 3.15 Solution space for an elliptic PDE.

series of complex functions, which is quite complicated. An example of an elliptic PDE is Laplace's equation:

$$\Phi_{xx} + \Phi_{yy} = 0 \tag{3.96}$$

where, using the coefficient definitions of Equation 3.74, $a = c = 1$ and all other coefficients are zero. This yields $b^2 - 4ac = -4 < 0$ and hence the equation is classified as elliptic. There is no analytic, closed-form solution of Laplace's equation for general boundary conditions, showing the complexity of the elliptic PDE type. An elliptic PDE requires a boundary that completely encloses the solution space with continuous data on the boundary, as shown in Fig. 3.15.

The fluid analogy for an elliptic PDE is subsonic flow, where pressures propagate in all directions (everything affects everything else). When we use numerical methods, we will use central difference finite difference formulas so that the information all around the point of interest is used in finding the solution. We will also be required to supply boundary conditions all around the solution region in order to get a solution.

3.10 Boundary Conditions

We have now seen where boundary and initial conditions must be applied to solve the equations depending on the *type* of the flowfield being solved. In general, the aerodynamicist must specify the boundary conditions for a number of different situations; we will only describe some of the details and issues that must be addressed in order to obtain a solution. Perhaps the easiest boundary condition (and the physically most obvious) is the condition on a solid surface. The statement of the boundary conditions is tightly connected to the flowfield model in use. For an inviscid flow over a solid surface, the statement of the boundary condition is:

$$\vec{V}_R \cdot \hat{n} = 0 \tag{3.97}$$

which is the *"slip" condition* that says the difference between the velocity of the component of flow normal to the surface and the surface normal velocity (the relative velocity, $\vec{V}_R$) is zero. This simply means that the flow must be parallel to the surface. If $\vec{V}$ is the fluid velocity and $\vec{V}_S$ is the surface velocity, then this becomes,

$$\left(\vec{V} - \vec{V}_S\right) \cdot \hat{n} = 0. \tag{3.98}$$

Finally, if the surface is fixed,

$$\vec{V} \cdot \hat{n} = 0. \tag{3.99}$$

If the flow is viscous, the statement becomes even simpler: $\vec{V} = 0$, which is the *"no-slip" condition*. If the surface is porous and there is mass flow, the values of the surface velocity must be specified as part of the problem definition. To provide more specifics, recall that the unit normal for the body is defined (in two dimensions) in the form $F(x,y) = 0$, the traditional analytic geometry nomenclature. In terms of the usual two-dimensional notation, the body shape is given by $y = f(x)$, which is then written as:

$$F(x,y) = 0 = y - f(x) \tag{3.100}$$

and

$$\hat{n} = \frac{\vec{\nabla} F}{\left|\vec{\nabla} F\right|}. \tag{3.101}$$

The combination of Equations 3.98 or 3.99 and 3.101 is one of the reasons why obtaining solutions to flow over "realistic" shapes is so difficult. The details of the geometry dictate the surface normal in Equation 3.101, and the computational code must force Equations 3.98 or 3.99 to be true. To capture all of the fine details of the geometry, we must specify hundreds of thousands (or even millions) of locations where Equations 3.98 or 3.99 are true.

Numerical solutions of the Euler and Navier-Stokes solutions require that other boundary conditions be specified as well, since there are other unknowns. In particular, conditions for pressure and temperature are required.[vi] Specifications of the boundary conditions have a significant

---

[vi] Traditionally either the temperature, T, or the heat transfer, $\partial T / \partial n$, is specified. If $\partial T / \partial n = 0$ there is no heat transfer, and this is called an *adiabatic* wall.

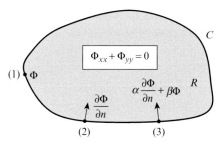

Figure 3.16 Boundary condition types (Ref. 15).

impact on the accuracy and robustness of the simulation.

Next we look at the different types of boundary conditions and specify conditions that are commonly used in fluid dynamics. We already described a specific example of applying boundary conditions. Now we describe the possibilities in a little more generality.

An important aspect of specifying boundary conditions in aerodynamics is to realize that we will be specifying boundary conditions both on the surface of the body and also at a farfield boundary, since when we make a grid to use in the computation, we will not be going all the way out to infinity (and beyond, as Buzz Lightyear would say).

Given a region, R, with boundary, C, as shown in Fig. 3.16, we can specify three types of boundary conditions (see Pletcher et al.[15] for details):

- Dirichlet – the function, Φ, is prescribed on the boundary.
- Neumann – the normal gradient of the function, $\partial\Phi / \partial n$, is prescribed on the boundary.
- Third (or "Robin") Condition[21] – a combination of the first two types, where $\alpha\Phi_n + \beta\Phi$ is prescribed on the boundary.

In practice there are generally multiple equations and unknowns requiring combinations of the three types of boundary conditions at the same time and location on the boundary. It is also important to note that the location of a boundary can significantly affect the solution. The following Concept Box lists commonly used boundary conditions for aerodynamic problems.

Considering the conditions that must be specified away from the body, this commonly means that, at large distances from the body, the flowfield must approach the freestream conditions. In numerical computations, the question of the farfield boundary condition can become troublesome. How far away is infinity? Exactly how should you specify the farfield boundary condition numerically? While it is possible to prescribe boundary conditions at infinity using coordinate transformations,[22] to avoid having to go extremely far away from the body, researchers are continuing to develop accurate, efficient boundary condition approaches.[23,24] We will look into these issues in greater detail in Chapter 7.

As shown in the Boundary Conditions Concept Box, *Riemann invariants* are often used to apply boundary conditions at the farfield boundary. Formally, Riemann invariants are the integrated characteristic equation for one-dimensional unsteady inviscid flow, assuming a perfect gas,[25] and can be used as the basis for implementing numerical schemes to be described later.

Another important use of boundary conditions arises as a means of modeling physics that would be neglected otherwise. When an approximate flow-field model is used, the boundary conditions frequently provide a means of including key elements of the physics in the problem without having to include the physics explicitly. The most famous example of this is the Kutta condition, where the viscous effects at the trailing edge can be accounted for in an inviscid calculation without treating the trailing edge problem explicitly. Karamcheti[2] discusses aerodynamic boundary conditions in more detail, and Pletcher et al.[15] discuss the farfield boundary for numerical solutions.

Computational Aerodynamics Concept Box

Commonly Used Boundary Conditions in Fluid Dynamics

While there are many boundary conditions possible for modeling fluid flow, only a few are used frequently. Typically, a computational domain is made up of a farfield (inflow and/or outflow) and solid surfaces (inviscid and viscous). Many of these common boundary conditions are described here for your convenience.

Surface conditions

Invisicid
$\dfrac{\partial p}{\partial n} = \dfrac{\rho v_t^2}{R}$, v_t is the tangential velocity, R is the surface curvature ($R = \infty$ for a flat surface, yielding $dp/dn = 0$)

$\vec{V} \cdot \hat{n}_{wall} = 0$ (no flow through the surface)

Kutta condition enforced for potential flows

Viscous
$\dfrac{\partial p}{\partial n} = \dfrac{\rho v_t^2}{R}$ v_t is the tangential velocity, R is the surface curvature ($R = \infty$ for a flat surface, yielding $dp/dn = 0$)

$\vec{V}_{wall} = 0$ (no flow at the surface)

$\vec{V} = \vec{V}_{wall}$ (for a moving ground plane)

$\dot{q}_{wall}$ is prescribed (= 0 for an adiabatic wall) or T_{wall} is prescribed

Farfield

Inflow
$V_{in} = V_\infty$ (overprescribing)

Riemann invariants

Outflow
$\dfrac{\partial V_{out}}{\partial x} = 0$ (extrapolation)

Riemann invariants

Symmetry plane
$\dfrac{\partial V}{\partial n} = 0$ $\vec{V} \cdot \hat{n} = 0$

Periodic
$V_1 = V_{max+1}$

Flow Visualization Box

NACA 0012 Airfoil Boundary Conditions

The boundary conditions for the NACA 0012 airfoil flow solutions should be apparent as we look at the flow solution in FieldView. The 16 degree angle of attack case was chosen to make the freestream vectors more apparent, as shown in the first figure which shows the upper left corner of the flow volume. The mesh has been included so you can see that the flow vectors start at the center of the mesh (this is from a cell-centered flow solver). Notice that the flow enters from the left at 16 degrees and is able to flow out of the top at the same angle (these are modified Riemann invariant boundary conditions). The same is true for the outflow condition, which is shown in the second figure.

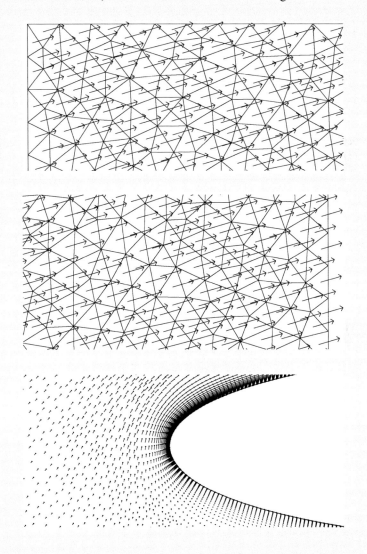

These are velocity vectors in the vicinity of the leading edge of the airfoil. Notice how the vectors are aligned with the surface, but also notice that a boundary layer is developing as the flow travels over the upper surface of the airfoil. This is a viscous (no-slip) solid wall boundary condition.

3.11 Using and Simplifying these Equations: High- to Low-Fidelity Flowfield Models

Historically, since it was not possible to solve the Navier-Stokes equations analytically, the only option available to researchers and engineers was to simplify the equations until an approximate solution could be obtained. This results in a hierarchy of equations developed for specific applications, ranging from *high-fidelity models* (such as solutions to the Navier-Stokes equations) to *low-fidelity models* (such as panel methods based on potential flow methods). As computer capability increased, the simplifications became less severe (as described in Chapter 2). A large body of both aerodynamic understanding and methodology is associated with the various levels of approximation that have been used over the years.

The following sections describe some of the different levels of equations that have been developed. While some of these equations were derived to find analytic solutions, they are also useful for finding numerical solutions. As we discuss the various approximations and when they can be used, it will become apparent why we had to understand the various *types* of partial differential equations. An overview of the equations that result from various types of simplifications is presented in Fig. 3.17. The equations are shown in two categories: viscous simplifications and inviscid simplifications. We emphasize the assumptions required to obtain the various equations and the limitations of the equations in predicting various types of aerodynamic phenomena. We start with inviscid flows in Section 3.12 and then describe the viscous flow models in Section 3.13.

3.12 Inviscid Flow Models

We noted that if viscous effects are small in a particular flow, useful information can be found by neglecting viscosity. This means that the Navier-Stokes equations without the viscous terms can be used; the resulting equations are known as the Euler equations. It is actually surprising how often this assumption is useful for aerodynamic flow solutions when we are interested in flows over streamlined shapes that are designed to minimize the effects of viscosity (such as an aircraft in cruise conditions). However, the Euler equations are still nonlinear coupled PDEs. There are few, if any, analytical solutions of these equations for aerodynamic flows of interest, so further assumptions are made to obtain flow models that can be solved more easily. Often, these models were used in early computational methods development, and they are still useful. They form the left-hand column in Fig. 3.17.

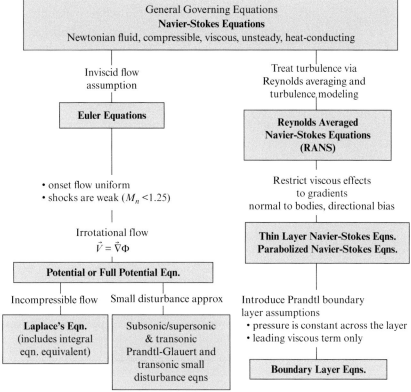

Notes:
1. **The only linear PDEs here are Laplace's Eqn. and the P-G (Prandtl – Glauert) equation.**
2. **All of these formulations except the PNS equations can be unsteady.**
3. **Only Laplace's equation is incompressible.**

Figure 3.17   Connection between various approximations to the governing equations.

3.12.1 Potential Flow

Additional simplifying assumptions result in the potential flow models. For aerodynamic shapes, we can obtain good estimates of the pressure distribution making the assumption that outside of the thin attached viscous boundary layer, the flow can be assumed to be irrotational (see Chapter 4 for details). This requires that the onset flow be uniform (and hence irrotational) and that there cannot be any shock waves. However, we often continue to assume that the flow can be represented approximately as irrotational when the Mach number normal to any shock wave is close to one (say $M_n <$ 1.2). Recall that the irrotational flow assumption is stated mathematically as $\vec{\nabla} \times \vec{V} = 0$, and, when this is true, $\vec{V}$ can be defined as the gradient of a scalar quantity, $\vec{V} = \vec{\nabla}\Phi$. Using the common subscript notation to represent partial derivatives, the velocity components are $u = \Phi_x$, $v = \Phi_y$, and $w = \Phi_z$. Using this assumption, all three velocity components can be found as the gradient of a single scalar function, $\vec{V} = \vec{\nabla}\Phi$, the velocity potential. In addition, for steady flow, since the total enthalpy is a constant, the energy equation can be reduced to an algebraic equation. This means that the computational problem requires the solution of a single partial differential equation. The general potential equation can be found in two ways that are equivalent. This equation is still nonlinear and is generally known as the "full potential equation" (or Velocity Potential equation) to avoid confusion with Laplace's equation (which is often referred to somewhat erroneously in aerodynamics as "the potential equation").

The "full potential equation" is given most simply by writing the continuity equation with the velocity given in terms of the potential function. Starting with Equation 3.5 and substituting for the velocity components as defined above gives:

$$\frac{\partial \rho}{\partial t} + \frac{\partial}{\partial x}(\rho\Phi_x) + \frac{\partial}{\partial y}(\rho\Phi_y) + \frac{\partial}{\partial z}(\rho\Phi_z) = 0 \tag{3.102}$$

and density is found from a general form of the Bernoulli equation:[26]

$$\rho = [1 + (\tfrac{\gamma-1}{2})(M_\infty^2 - 2\Phi_t - \Phi_x^2 - \Phi_y^2 - \Phi_z^2)]^{\frac{1}{\gamma-1}}. \tag{3.103}$$

where ρ is nondimensionalized by ρ_∞, and the velocity is nondimensionalized by a_∞. The choice of the speed of sound as a nondimensionalization is interesting. In unsteady flow, the speed at any point may vary with time, so that V_∞ is hard to understand as a constant. Clearly, if the density can be assumed to be constant, Equation 3.102 reduces to Laplace's equation. The importance of Laplace's equation is that it is a linear PDE, so that many well-developed numerical methods can be used to obtain solutions (as will be shown in Chapters 4 and 5).

The more traditional form of the full potential equation is derived through a more circuitous route (which will be left to the reader as an exercise at the end of this chapter). We write it here including the unsteady flow terms:

$$\Phi_{tt} + (\Phi_x^2 - a^2)\Phi_{xx} + (\Phi_y^2 - a^2)\Phi_{yy} + (\Phi_z^2 - a^2)\Phi_{zz}$$
$$+ 2\left(\Phi_x\Phi_{xt} + \Phi_y\Phi_{yt} + \Phi_z\Phi_{zt}\right)$$
$$+ 2\left(\Phi_x\Phi_y\Phi_{xy} + \Phi_y\Phi_z\Phi_{yz} + \Phi_z\Phi_x\Phi_{zx}\right) = 0. \tag{3.104}$$

This is the classic form of the full potential equation that has been used for many years to obtain physical insight into a wide variety of flows. It is not in divergence (or conservation) form, whereas Equation 3.102 is in divergence form. In this formulation, Bernoulli's equation is written in terms of the speed of sound, a, and is:

$$\frac{a^2}{a_\infty^2} = 1 - (\gamma - 1)\left[\frac{\Phi_t}{a_\infty^2} + \frac{1}{2}\left(\frac{u^2 + v^2 + w^2}{a_\infty^2} - M_\infty^2\right)\right] \qquad (3.105)$$

where we have left the equation in dimensional form for clarity (see Ashley and Landahl[27] for details of the derivation). For steady flow, Equation 3.105 is generally thought of as a statement of the energy equation, because, in that case, the inviscid energy equation is simply a statement that the total enthalpy is a constant, and Equation 3.105 is an algebraic equation. In that case, it is equivalent to writing the isentropic relation for total temperature in terms of the speed of sound squared.

Probably the most important reason to use Equation 3.104 is that it is easy to see that the coefficient of the second derivative terms change sign depending on whether the local flow is greater or less than the speed of sound, changing the PDE type from elliptic to hyperbolic. A more complete discussion and example applications are available from Holst.[28]

3.12.2 Small Disturbance Expansion of the Full Potential and Energy Equation

In this section we show how further simplifying approximations to the full potential equation can be obtained in a rational manner. These simplified equations are entirely adequate for many of the problems of computational aerodynamics and, until recently, were used nearly exclusively for airfoils and wings. The ability to obtain simpler relations that provide explicit physical insight into the flowfield process has played an important role in the development of aerodynamic concepts. One key idea is the notion of small disturbance equations. The assumption is that the flowfield is only slightly disturbed by the body. This assumption is expressed mathematically using small perturbation or asymptotic expansion methods, and is elegantly described in the book by Van Dyke.[29] We use this approach to show how to obtain small disturbance versions of the full potential equation and the related energy equation.

For simplicity, we illustrate this approach assuming two-dimensional steady flow (not required). We start with the steady algebraic version of the energy equation, Equation 3.105, and show how we can express the square of the speed of sound (or equivalently the temperature) approximately as a linear function of the velocity. Using the steady energy equation, Equation 3.106, written in two dimensions and using the stagnation speed of sound:

$$a^2 = a_0^2 - \left(\frac{\gamma - 1}{2}\right)(u^2 + v^2). \qquad (3.106)$$

Now express a_0^2 in terms of freestream values:

$$a_0^2 = const = a^2 + \left(\frac{\gamma-1}{2}\right)(u^2 + v^2) = a_\infty^2 + \frac{\gamma-1}{2}V_\infty^2. \tag{3.107}$$

Letting $u = V_\infty + u'$, $v = v'$, and substituting a_0^2 from Equation 3.107 into Equation 3.106:

$$a^2 = a_\infty^2 + \frac{\gamma-1}{2}V_\infty^2 - \left(\frac{\gamma-1}{2}\right)[V_\infty^2 + 2V_\infty u' + u'^2 + v' \tag{3.108}$$

and combining terms gives:

$$a^2 = a_\infty^2 - \left(\frac{\gamma-1}{2}\right)[2V_\infty u' + u'^2 + v'^2]. \tag{3.109}$$

At this point the relation is still exact, but it is now written so that it can easily be simplified. The basic idea will be to take advantage of the assumption:

$$u' \ll V_\infty \quad v' \ll V_\infty \tag{3.110}$$

and thus,

$$\frac{u'}{V_\infty} \ll 1 \Rightarrow \left(\frac{u'}{V_\infty}\right)^2 \approx 0, \tag{3.111}$$

so Equation 3.109 becomes:

$$a^2 = a_\infty^2 - \left(\frac{\gamma-1}{2}\right)\left[2V_\infty u' + \underbrace{u'^2 + v'^2}_{\substack{\text{neglect as small} \\ \text{henceforth}}}\right]. \tag{3.112}$$

After the last two terms are dropped (because they are products of small values), this is a linear relation between the disturbance velocity and the speed of sound. It is a heuristic example of the procedures used in a more formal approach known as perturbation theory.

Now we examine the full potential equation. We will rewrite the full potential equation given earlier in the steady two-dimensional form for simplicity:

$$(\Phi_x^2 - a^2)\Phi_{xx} + 2\Phi_x\Phi_y\Phi_{xy} + (\Phi_y^2 - a^2)\Phi_{yy} = 0 \tag{3.113}$$

and write the velocity as a difference from the freestream velocity. We also introduce a disturbance potential ϕ, defined by:

$$\begin{aligned}
\Phi &= V_\infty x + \phi(x, y) \\
\Phi_x &= u = V_\infty + \phi_x \\
\Phi_y &= v = \phi_y
\end{aligned} \tag{3.114}$$

where we have introduced a directional bias: the x-direction is the direction of the freestream velocity. We will assume that ϕ_x and ϕ_y are small compared to V_∞. Using the idea of a small disturbance to the freestream, simplified (and even linear) forms of a small disturbance potential equation and an energy equation can be derived.

As an example of the expansion process, consider the first term in Equation 3.104. Use the definition of the disturbance potential and the simplified energy equation, Equation 3.112, to obtain:

$$\begin{aligned}
\left(\Phi_x^2 - a^2\right) &\cong \left(V_\infty + \phi_x\right)^2 - \left\{a_\infty^2 - \left(\frac{\gamma-1}{2}\right)\left[2V_\infty u'\right]\right\} \\
&\cong V_\infty^2 + 2V_\infty \phi_x + \phi_x^2 - a_\infty^2 + \frac{\gamma-1}{2} 2V_\infty \underbrace{u'}_{=\phi_x}.
\end{aligned} \tag{3.115}$$

Regroup and drop the square of the disturbance velocity as small:

$$\begin{aligned}
\left(\Phi_x^2 - a^2\right) &\cong V_\infty^2 - a_\infty^2 + 2V_\infty \phi_x + \left(\gamma-1\right)V_\infty \phi_x \\
&\cong V_\infty^2 - a_\infty^2 + \underbrace{\left[2 + \left(\gamma-1\right)\right]}_{\gamma+1}V_\infty \phi_x \\
&\cong V_\infty^2 - a_\infty^2 + \left(\gamma+1\right)V_\infty \phi_x.
\end{aligned} \tag{3.116}$$

Dividing by a_∞^2,

$$\begin{aligned}
\left(\frac{\Phi_x^2}{a_\infty^2} - \frac{a^2}{a_\infty^2}\right) &\cong \frac{V_\infty^2}{a_\infty^2} - 1 + \left(\gamma+1\right)\frac{V_\infty}{a_\infty}\frac{\phi_x}{a_\infty} \\
&\cong M_\infty^2 - 1 + \left(\gamma+1\right)M_\infty \underbrace{\frac{V_\infty}{V_\infty}\frac{\phi_x}{a_\infty}}_{\frac{V_\infty}{a_\infty}\frac{\phi_x}{V_\infty}} \\
&\cong \left(M_\infty^2 - 1\right) + \left(\gamma+1\right)M_\infty^2 \left(\frac{\phi_x}{V_\infty}\right).
\end{aligned} \tag{3.117}$$

Rewrite the potential equation, Equation 3.113, dividing by a_∞^2 and replace the coefficient of the first term using Equation 3.117 to obtain:

$$\underbrace{\left(\frac{\Phi_x^2}{a_\infty^2} - \frac{a^2}{a_\infty^2}\right)}_{\left(M_\infty^2 - 1\right) + \left(\gamma + 1\right) M_\infty^2 \left(\frac{\phi_x}{V_\infty}\right)} \Phi_{xx} + 2\frac{\Phi_x}{a_\infty}\frac{\Phi_y}{a_\infty}\Phi_{xy} + \left(\frac{\Phi_y^2}{a_\infty^2} - \frac{a^2}{a_\infty^2}\right)\Phi_{yy} = 0. \quad (3.118)$$

Now, by definition

$$\Phi_{xx} = \phi_{xx}, \quad \Phi_{yy} = \phi_{yy}, \quad \Phi_{xy} = \phi_{xy}, \quad (3.119)$$

while

$$\frac{\Phi_x}{a_\infty} = M_\infty\left(1 + \frac{\phi_x}{V_\infty}\right) \quad \frac{\Phi_y}{a_\infty} = M_\infty\left(1 + \frac{\phi_y}{V_\infty}\right) \quad (3.120)$$

and using the same previously demonstrated approach we can write:

$$\left(\frac{\Phi_y^2}{a_\infty^2} = \frac{a^2}{a_\infty^2}\right) \cong -1 + \left(\gamma - 1\right) M_\infty^2\left(\frac{\phi_y}{V_\infty}\right). \quad (3.121)$$

Putting these relations all into the potential equation, Equation 3.118, we obtain:

$$\left[M_\infty^2 - 1 + \left(\gamma + 1\right)M_\infty^2\frac{\phi_x}{V_\infty}\right]\phi_{xx} + 2M_\infty^2\left(1 + \frac{\phi_x}{V_\infty}\right)\frac{\phi_y}{V_\infty}\phi_{xy}$$
$$+ \left[-1 + \left(\gamma - 1\right)M_\infty^2\frac{\phi_y}{V_\infty}\right]\phi_{yy} = 0 \quad (3.122)$$

where the ϕ_x^2, ϕ_y^2 terms are neglected in the coefficients. This equation is still nonlinear, but is in a form ready for the further simplifications described next.

3.12.3 Transonic Small Disturbance Equation

Transonic flows contain regions with both subsonic and supersonic velocities. Any equation describing this flow must simulate the correct physics in the two different flow regimes. As we will show, this makes the problem difficult to solve numerically. Indeed, the numerical solution of transonic flows was one of the primary thrusts of research in CFD throughout the 1970s and 1980s. We can derive a small disturbance equation that captures the essential nonlinearity of transonic flow, which is the rapid streamwise variation of flow disturbances in the x-direction, primarily through normal shock waves. Therefore, in transonic flows:

$$\frac{\partial}{\partial x} \gg \frac{\partial}{\partial y}.$$

The transonic small disturbance equation retains the key term in the convective derivative, $u(\partial u/\partial x)$, which allows the shock to occur in the solution. Retaining this key nonlinear term from Equation 3.122, the small disturbance equation given becomes:[30]

$$\left[\left(1 - M_\infty^2\right) - \left(\gamma + 1\right)M_\infty^2 \frac{\phi_x}{V_\infty}\right]\phi_{xx} + \phi_{yy} = 0. \tag{3.123}$$

Note that, using the definition of the potential from Equation 3.114, we can identify the nonlinear term $u(\partial u/\partial x)$ that appears as the product of the second term in the bracket, $u = \phi_x$, and the ϕ_{xx} term, which is $\partial u / \partial x$.

This is one version of the transonic small disturbance equation. It is still nonlinear and can change mathematical *type*. This means that the sign of the coefficient of ϕ_{xx} can change in the flowfield, depending on the value of the nonlinear term. It is valid for transonic flow, and, as written, it is not in a divergence form. Typically, transonic flows occur for Mach numbers from 0.6 to 1.2, depending on the degree of flow disturbance. They also can occur under other circumstances: at high-lift conditions, the flow around the leading edge of an airfoil may become locally supersonic at freestream Mach numbers as low as 0.20 or 0.25. Transonic flow also occurs on rotor blades and propellers at high rotation speeds. At hypersonic speeds the flow between the bow shock and the body will be locally subsonic, rapidly re-accelerating to hypersonic speed and is therefore transonic.

The unsteady version of Equation 3.123 is:

$$\left[\left(1 - M_\infty^2\right) - \left(\gamma + 1\right)M_\infty^2 \frac{\phi_x}{V_\infty}\right]\phi_{xx} + \phi_{yy} - \frac{2M_\infty^2}{V_\infty}\phi_{xt} - \frac{M_\infty^2}{V_\infty^2}\phi_{tt} = 0. \tag{3.124}$$

If the ϕ_{tt} term is neglected, the resulting equation is the so-called low-frequency transonic small disturbance equation. The low-frequency equation can be used when the reduced frequency is less than 0.2[vii] (for more details see Nixon[30]).

3.12.4 Prandtl-Glauert Equation

When the flowfield is entirely subsonic or supersonic, all terms involving products of small quantities can be neglected in the small disturbance equation. When this is done, we obtain the Prandtl-Glauert equation (shown in three dimensions here):

$$\left(1 - M_\infty^2\right)\phi_{xx} + \phi_{yy} + \phi_{zz} = 0. \tag{3.125}$$

[vii] The reduced frequency is associated with motions with an identifiable frequency and is defined as $k = \omega L / V_\infty$

This is a linear equation valid for small disturbance flows that are either entirely supersonic or subsonic. For subsonic flows this equation can be transformed to Laplace's equation, while at supersonic speeds this equation takes the form of a wave equation. The difference is important, since at subsonic speed the PDE is elliptic, while at supersonic speed the PDE is hyperbolic. This equation requires that the onset flow be in the x-direction (which is another example of the importance of coordinate systems).

Although we have a PDE for the potential function, generally we want to know the pressure. In aerodynamics this is given as a pressure coefficient, C_p. The definition of the pressure coefficient is:

$$C_p = \frac{p - p_\infty}{q_\infty} \qquad (3.126)$$

where $q_\infty = \rho_\infty V_\infty^2 / 2 = \gamma p_\infty M_\infty^2 / 2$. With the assumption of small disturbances (which will be derived in detail in Section 5.3.1), a linear relation exists between the velocity perturbation and the pressure coefficient:

$$C_p = -2\phi_x. \qquad (3.127)$$

The extension to three-dimensional unsteady flow is:[30]

$$\left(1 - M_\infty^2\right)\phi_{xx} + \phi_{yy} + \phi_{zz} - \frac{2M_\infty^2}{V_\infty}\phi_{xt} - \frac{M_\infty^2}{V_\infty^2}\phi_{tt} = 0 \qquad (3.128)$$

and the calculation of pressure also requires an unsteady term. Using the consistent perturbation relation for the pressure coefficient, extending Equation 3.127 leads to:

$$C_p = -2\phi_x - \frac{2}{V_\infty}\phi_t \qquad (3.129)$$

3.12.5 Incompressible Irrotational Flow: Laplace's Equation

Assuming that the flow is incompressible, the density, ρ, is a constant and can be removed from the modified continuity equation, Equation 3.102. Alternately, divide the full potential equation, Equation 3.104, by the square of the speed of sound, a, and take the limit as a goes to infinity, which yields a Mach number of $M = 0$ (incompressible flow). Either way, the following equation is obtained:

$$\Phi_{xx} + \Phi_{yy} + \Phi_{zz} = \nabla^2\Phi = 0. \qquad (3.130)$$

This is Laplace's equation, which is also frequently called the potential equation. For that reason the complete potential equation given by Equation 3.104 is known as the full potential equation. Do not confuse the true potential flow equation with Laplace's equation, which requires the assumption of incompressible flow. When the flow is incompressible, this equation is exact when using the inviscid, irrotational flow model and does not require the assumption of small disturbances.

3.13 Viscous Flow Models

Depending on whether or not viscous effects are important in a particular application, the aerodynamicist needs to decide whether equations incorporating viscous effects should be used. Viscous effects are critical in computing aerodynamic heating in hypersonic flows, and much of the early computational work was directed toward prediction of aerodynamic heating.

Subsequently, Whitcomb reinvigorated transonic aerodynamics with the invention of the supercritical airfoil.[31] It became possible to solve the inviscid transonic flow equations using the small disturbance or full potential equations, and it was discovered that viscous effects had to be included in the calculations to obtain computational results that agreed with experiments for supercritical airfoils. Initially this was done using boundary-layer equations coupled with the inviscid equations and what are known as viscous-inviscid interactions. Murman and Cole presented the first transonic small disturbance results in 1970,[32] Jameson published his first full potential equation results in 1974,[33] and Bavitz's program that included viscous-inviscid interactions became available in 1975.[34] A good overview of the viscous-inviscid equations and solution approaches is presented by Rom.[35]

Profiles in Computational Aerodynamics: Earll Murman

My engagement with CFD and aerospace engineering was shaped by serendipitous events, as has my entire career. I entered Princeton in the fall of 1959 intending to major in chemical engineering. By spring I knew I needed a different destiny. With the launch of Sputnik in October 1957, aerospace engineering seemed really interesting and Princeton faculty were energetic and exciting. So I enrolled in aeronautical engineering and soon found I had a great affinity for all aspects of fluid mechanics: aerodynamics, gas dynamics, aerothermodynamics, theory, and experiments. I was inspired by the faculty and decided an academic career

would be my goal. Eight years later, completing a thesis in experimental hypersonics and a PhD, I decided I needed a change. An academic career would have to wait.

I was fortunate to join the Flight Sciences Laboratory of the Boeing Scientific Research Laboratory in Seattle, Washington. When I told the lab director, Dr. Arnold Goldberg, that I wanted to do something different, he suggested looking into numerically solving the Navier-Stokes equations. He asked Princeton Professor Sin-I Cheng to mentor me through this career transition, and off we went. I bought books on numerical methods and FORTRAN, and soon was solving the NS equations for shockwave/boundary layer interactions using the Lax-Wendroff (LW) method. Since our IBM 360/44 could only handle a few hundred grid points, we were limited to Reynolds numbers around 10, a totally useless regime. Needless to say, I was disappointed – but my life was about to dramatically change.

Dr. Goldberg invited Professor Julian Cole of Cal Tech to spend a year at the Lab, and asked that I work with Julian to make some progress on solving the transonic flow equations. One thing led to another and by midyear we had come up with a new numerical algorithm for a mixed elliptic-hyperbolic equation that included embedded shock waves. We used a relaxation method, but switched from central to upwind differencing for supersonic flow. The method was an order of magnitude more efficient than the LW method, and our calculations of transonic airfoil flows opened many people's eyes to the applicability of CFD to aerodynamic analysis and design. As I was trying to understand a vexing issue about the shock capturing fidelity, others quickly adopted the method for wings, bodies, and complete configurations. Eventually Antony Jameson and others came up with far superior methods and the original work is now only of historical interest.

I would have happily stayed at Boeing, but in 1971 the company was in a nosedive. I was fortunate to land at NASA's Ames Research Center, where I continued my transonic work (eventually figuring out the shock capturing issue) and collaborated with many of the wonderful CFD researchers there at the time. For personal reasons, we returned to the Pacific Northwest in 1974, to join Flow Research in Kent, Washington. In 1980 I was so lucky to be offered a tenured professor position in MIT's Aeronautics and Astronautics Department as part of their effort to bring CFD into the mainstream of their research and education. My goal of an academic career was now to be realized.

For my first 10 years at MIT I continued CFD work, moving on to solving the Euler and sometimes Navier-Stokes equations for compressible flow, particularly vortex-dominated ones. With other MIT colleagues, we established a CFD curriculum and attracted a number of highly talented graduate students to our CFD lab. The rapid advances in computing technology took us to parallel computing and flow visualization. As I felt the pounding feet of younger, more talented people taking over, two events happened which brought my CFD career more or less to a conclusion in the early 1990s. The first event was that I was selected as Department Head as the Berlin Wall was coming down and a new future course needed charting for our Department. That eventually carried me into system engineering. The second was a hard disk crash. It turned out the automatic backup system was not properly installed and all my FORTRAN and LaTeX files were gone! Once again, it was time to move on to something different.

My Department leadership opened my eyes and interest to how large aerospace systems (aircraft, air transportation, space) can be more efficiently designed, developed, and deployed. Through more serendipity I ended up leading a major research and implementation effort to bring the management principles of Toyota – popularly called Lean Thinking – into all aspects of aerospace, including engineering. It is a totally different field with no differential equations or computations, but desperately needed by aerospace, whose programs constantly overrun cost and schedule goals.

Previously, aerodynamicists had used skin friction formulas with form factors to estimate drag due to viscous effects (as will be discussed in Chapter 4) and relied on wind tunnel data to estimate maximum lift. There was an urgent need to be able to compute viscous effects, and the simplifications to the Navier-Stokes codes described here were closely connected to the computing power available at the time. The question is whether the full Navier-Stokes equations are necessary to solve a problem, or whether a simpler version of the viscous flow equations can be used (the main reason to use simpler versions of the Navier-Stokes equations is to save computation time). If computation time is not an issue, then the complete equations are certainly the best to use in any simulation, avoiding the mis-application of simplifying assumptions.

3.13.1 Thin-Layer Navier-Stokes Equations

Historically the boundary-layer equations (see Section 3.13.3) were the only equations that could be solved on computers. However, the boundary-layer equations, solved in the traditional fashion, cannot compute separated flows. They do, however, demonstrate that the key viscous effects are included in the terms describing the rapid variation in the velocity normal to the wall. This is much more important than the viscous diffusion terms parallel to the flow. Thus, these terms are neglected by comparison. The other major simplification in boundary-layer theory is the assumption that the pressure across the boundary layer is constant. This is the assumption that prevents the boundary-layer equations from computing separated flows when used in the classical manner.[viii] The boundary-layer equations are of parabolic type and typically start at a stagnation point, and the calculations proceed downstream until flow separation is predicted. This is very convenient but does not accommodate the situation for separated flow, where there is reversed flow, and downstream information is required. Nevertheless, the boundary-layer concept is one of the most important concepts in fluid mechanics.

In the late 1970s, at NASA Ames Research Center,[36-38] it was decided to investigate the possibility of solving a simplified form of the Navier-Stokes

---

[viii] If inverse boundary-layer solutions are used, where the wall skin friction is specified, and the pressure is found, the boundary-layer equations *can* be solved in regions of flow separation.

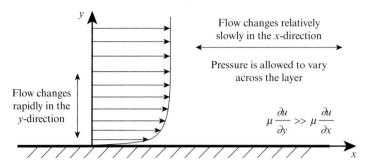

Fig. 3.18   Illustration of the Thin Layer Navier-Stokes idea.

equations that maintained the spirit of the boundary-layer concept but that would also allow the computation of separated flows. Thus, the pressure was allowed to vary across the boundary layer. Another reason for making this simplification was that the grid resolution required to resolve the terms in the equations in both the direction normal to the surface and parallel to the surface would result in a larger grid than the computers of the day could handle. The resulting equations are known as the *Thin Layer Approximation* to the Navier Stokes equations (TLNS); the basic idea is illustrated in Fig. 3.18. They were intended to be applied to situations where the boundary-layer concept was valid: high Reynolds numbers where the viscous effects were confined to a thin region near the surface. However, one equation set could be used to solve the entire flowfield, including some modest amount of separated flow. The equations proved to be very successful and were applied to many situations well beyond the original intention.

While this simplification has been done in a variety of ways (depending on the application), the simplest approximation is to assume that all viscous terms along (longitudinal to) any solid surface are small when compared with viscous terms normal to or around (circumferential to) a body. What does this accomplish? Look at Equation 3.34, for example, where $\tau_{xy} = \tau_{yx} = \mu(\partial u/\partial y + \partial v/\partial x)$. If we assume that the flow is largely aligned with the x-axis (say, for example, we are calculating flow over a flat plate, where y is the direction normal to the surface of the plate), then the thin-layer assumption would allow all viscous terms with x-derivatives to be neglected, so the term would become $\tau_{xy} = \tau_{yx} = \mu \partial u/\partial y$. The reduction in the number of viscous terms also decreases the computational time for analyzing these terms, which is especially important because the viscous terms are the most time-consuming calculations in the Navier-Stokes equations. If you look at Equations 3.33–3.35, you will notice that each shear stress term (τ_{xy} for example) also has a derivative taken, yielding a fairly complicated second-derivative term. These two-step derivatives are what make the viscous terms computationally time consuming (and in most applications the grids are highly stretched near solid surfaces in the longitudinal direction, so the resolution for computing longitudinal derivatives would probably yield poor results). The thin-layer equations have been used with success for a variety of aerodynamic applications, especially for applications to full aircraft flying at flight Reynolds numbers.[39] Although the equations are simpler, they retain the PDE type of the full Navier-Stokes equations and are solved as hyperbolic PDEs by marching forward in time. The viscous terms for the Navier-Stokes equations presented in Equation 3.35 have been revised to take into account the thin-layer assumption and are presented here (see Ref. 38 for details), with the dropped terms identified,

$$
\rho \frac{Du}{Dt} = -\frac{\partial p}{\partial x} + \overbrace{\frac{\partial}{\partial x}\left(2\mu\frac{\partial u}{\partial x} - \frac{2}{3}\mu\vec{\nabla}\cdot\vec{V}\right)}^{\text{neglect}} + \frac{\partial}{\partial y}\left[\mu\left(\frac{\partial u}{\partial y} + \frac{\partial v}{\partial x}\right)\right] + \overbrace{\frac{\partial}{\partial z}\left[\mu\left(\frac{\partial w}{\partial x} + \frac{\partial u}{\partial z}\right)\right]}^{\text{neglect}}
$$

$$
\rho \frac{Dv}{Dt} = -\frac{\partial p}{\partial y} + \overbrace{\frac{\partial}{\partial x}\left[\mu\left(\frac{\partial u}{\partial y} + \frac{\partial v}{\partial x}\right)\right]}^{\text{neglect}} + \frac{\partial}{\partial y}\left(2\mu\frac{\partial v}{\partial y} - \frac{2}{3}\mu\vec{\nabla}\cdot\vec{V}\right) + \overbrace{\frac{\partial}{\partial z}\left[\mu\left(\frac{\partial w}{\partial y} + \frac{\partial v}{\partial z}\right)\right]}^{\text{neglect}}
$$

$$
(3.35)
$$

$$
\rho \frac{Dw}{Dt} = -\frac{\partial p}{\partial z} + \overbrace{\frac{\partial}{\partial x}\left(\mu\left(\frac{\partial w}{\partial x} + \frac{\partial u}{\partial z}\right)\right)}^{\text{neglect}} + \frac{\partial}{\partial y}\left[\mu\left(\frac{\partial v}{\partial z} + \frac{\partial w}{\partial y}\right)\right] + \overbrace{\frac{\partial}{\partial z}\left(2\mu\frac{\partial w}{\partial z} - \frac{2}{3}\mu\vec{\nabla}\cdot\vec{V}\right)}^{\text{neglect}}.
$$

With the identified terms removed, we obtain one version of the thin layer equations as given in Equation 3.131. There is also a corresponding energy equation based on the thin-layer assumption (see Ref. 38 for details).

$$
\rho \frac{Du}{Dt} = -\frac{\partial p}{\partial x} + \frac{\partial}{\partial y}\left[\mu\left(\frac{\partial u}{\partial y}\right)\right]
$$

$$
\rho \frac{Dv}{Dt} = -\frac{\partial p}{\partial y} + \frac{\partial}{\partial y}\left(\frac{4}{3}\mu\frac{\partial v}{\partial y}\right)
$$

$$
(3.131)
$$

$$
\rho \frac{Dw}{Dt} = -\frac{\partial p}{\partial z} + \frac{\partial}{\partial y}\left(\mu\frac{\partial w}{\partial y}\right)
$$

Because these equations apply in the direction normal and parallel to the surface, we need to introduce one more aspect of using the governing equations in practice. The Cartesian coordinate system needs to be replaced by a coordinate system that fits the surface, so that the body surface corresponds to a constant coordinate line. An easy shape to think about might be a cone attached to a circular cylinder. This means that the governing equations need to be transformed to the new coordinate system (as will be discussed in Chapter 7 and shown in detail in Appendix E). Unfortunately for CFD, most aerodynamic shapes are not simple Cartesian shapes! Figure 3.19 illustrates

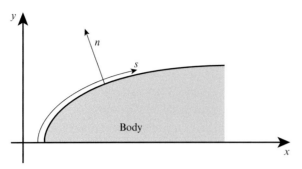

Fig. 3.19
Coordinate system for application of the thin layer Navier-Stokes equations.

this type of s,n coordinate system. Here the equations should be transformed to the s,n coordinate system in order to identify the appropriate coordinate parallel to the surface and the direction normal to the surface.

3.13.2 Parabolized Navier-Stokes Equations

For supersonic and hypersonic flow, a useful approximation to the Navier-Stokes equations for finding numerical solutions is the parabolized form of the equations (PNS). The word parabolic refers to the *type* of the partial differential equation, as we discussed in Section 3.6. The reason that we are interested in this approximation is that typically we find the steady-state solution to the Navier-Stokes equations by solving an unsteady problem until the steady state is reached, and the unsteady Navier-Stokes solution may take thousands of time steps to obtain a final steady-state solution; this is called *time marching*. The entire flowfield is computed at each step. A major time savings can be achieved if we can solve the flowfield just once. For supersonic and hypersonic flow, this is sometimes possible by exploiting the physical characteristics of these flows. As in the case of the boundary-layer equations (which are parabolic), we would like to start at the nose of the body and march downstream. This may be possible as long as there is no streamwise flow separation and the streamwise flow everywhere outside the boundary layer is supersonic. For this to be the case, the PNS equations are only applicable to steady flows. This procedure is referred to as *space marching* the solution as opposed to time marching (the Euler equations can also be "space marched" in many supersonic flow cases).

The parabolized Navier-Stokes equations have been obtained in various forms by various researchers. They are very similar to the thin-layer Navier-Stokes equations, and, in fact, several TLNS codes that use upwind differencing contain "space marching" as an option. The reason that these equations are termed "parabolized" is that they are not exactly parabolic. Even though the vast majority of the flow is supersonic, and hence there is no upstream influence, the no-slip condition results in a thin portion of the boundary layer adjacent to the surface that is subsonic. This small subsonic region allows the boundary layer to provide a path for upstream influence of the solution

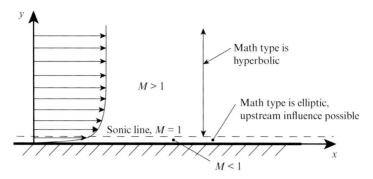

Fig. 3.20 Parabolized Navier-Stokes equation concept, illustrating that they contain a region allowing upstream influence.

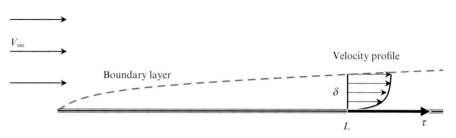

Figure 3.21 Boundary-layer development over a flat plate.

from downstream, preventing the flow from being precisely of parabolic type (Fig. 3.20 illustrates the situation). Significant research has been done to determine how to handle this situation. Pletcher et al.[15] provide a detailed explanation of the issues and several approaches to handle the effect. The upstream influence can be suppressed by neglecting the streamwise pressure gradient term in the x-momentum equation in the subsonic portion of the viscous layer. However, this approach can lead to errors in the results. A way to include the streamwise pressure gradient was proposed by Vigneron et al.,[40] and a refinement to that procedure was demonstrated by Morrison and Korte.[41] PNS continues to be valuable when doing aerodynamic optimization, where thousands of solutions may be required.[42] Anderson[43] and Degani and Schiff[44] also provide examples of the results that can be obtained using the PNS equations.

3.13.3 Boundary-Layer Equations

Historically, the boundary-layer equations formed the starting point for analytic solutions that include viscous effects. The boundary-layer equations begin directly with the Navier-Stokes equations and make some key simplifying assumptions to obtain a "manageable" equation set. Although they can be compressible, three-dimensional, and unsteady, assume for this discussion that the flow is incompressible, two-dimensional, and steady, and that we are trying to find the solution for a "thin" boundary layer that has developed over a flat plate, as shown in Fig. 3.21. Since the boundary layer is thin, we will require that $\delta \ll L$.

Based on the flow assumptions outlined, the conservation of mass equation, Equation 3.5, is reduced to:

$$\frac{\partial u}{\partial x} + \frac{\partial v}{\partial y} = 0 \tag{3.132}$$

and the nondimensionalized x- and y-momentum equations (see Equation 3.54) become:

$$u\frac{\partial u}{\partial x} + v\frac{\partial u}{\partial y} = -\frac{\partial p}{\partial x} + \frac{1}{Re}\left(\frac{\partial^2 u}{\partial x^2} + \frac{\partial^2 u}{\partial y^2}\right) \tag{3.133}$$

$$u\frac{\partial v}{\partial x} + v\frac{\partial v}{\partial y} = -\frac{\partial p}{\partial y} + \frac{1}{Re}\left(\frac{\partial^2 v}{\partial x^2} + \frac{\partial^2 v}{\partial y^2}\right) \tag{3.134}$$

where the *Reynolds number* is defined as $Re \equiv \rho V L / \mu$. Now we will perform an order-of-magnitude analysis based on the term $\varepsilon = \delta / L$ and the requirement that $\delta \ll L$ (which means that $\varepsilon \ll 1$). This results in terms like u and x being of $O(1)$ (which is a mathematical concept meaning that these terms are on the "order" of magnitude of 1), while v and y terms are of $O(\varepsilon)$. Now take a close look at the continuity equation, Equation 3.132. Since u and x are $O(1)$, the derivative $\partial u/\partial x$ is also $O(1)$. Likewise, since v and y are $O(\varepsilon)$, the derivative $\partial v/\partial y$ is $O(1)$ [since the ratio of two $O(\varepsilon)$ terms gives a result that is $O(1)$]. Since both terms in the equation are of the same order [namely $O(1)$], neither of the terms can be dropped; we must use Equation 3.132 as it is. Now look again at the x-momentum equation, Equation 3.133.

$$u\frac{\partial u}{\partial x} \quad + \quad v\frac{\partial u}{\partial y} = \quad -\frac{\partial p}{\partial x} \quad + \quad \frac{1}{Re}\left(\frac{\partial^2 u}{\partial x^2} \quad + \quad \frac{\partial^2 u}{\partial y^2}\right) \tag{3.133}$$

$$O(1 \times 1) \quad O\left(\varepsilon \times \frac{1}{\varepsilon}\right) \quad O(1) \quad O(\varepsilon^2)\left(O(1) + O\left(\frac{1}{\varepsilon^2}\right)\right)$$

where $1/Re$ is $O(\varepsilon^2)$ in order for the largest viscous terms to be of the same order as the convective terms,[7] namely

$$O\left(\frac{\partial^2 u}{\partial y^2}\right) = O\left(\frac{\partial u}{\partial x}\right) \tag{3.135}$$

and the pressure gradient is the same order as the convection term at the edge of the boundary layer where there is no viscosity:

$$O\left(\frac{\partial p}{\partial x}\right) = O\left(\frac{\partial u}{\partial x}\right). \tag{3.136}$$

Dropping terms in Equation 3.133 that are less than $O(1)$ results in a simplified x-momentum equation:

$$u\frac{\partial u}{\partial x} + v\frac{\partial u}{\partial y} = -\frac{\partial p}{\partial x} + \frac{1}{Re}\frac{\partial^2 u}{\partial y^2}. \qquad (3.137)$$

The y-momentum equation may also be analyzed in a similar fashion:

$$u\frac{\partial v}{\partial x} \;+\; v\frac{\partial v}{\partial y} \;=\; -\frac{\partial p}{\partial y} \;+\; \frac{1}{Re}\left(\frac{\partial^2 v}{\partial x^2} \;+\; \frac{\partial^2 v}{\partial y^2}\right) \qquad (3.134)$$

$$O(1\times\varepsilon)\; O(\varepsilon\times1) \quad O(1) \quad O(\varepsilon^2)\left(O(\varepsilon)+O\!\left(\frac{\varepsilon}{\varepsilon}\right)\right).$$

Thus, in a similar fashion, the y-momentum equation reduces to:

$$\frac{\partial p}{\partial y} = 0, \qquad (3.138)$$

which is the basis for the notion that pressure does not vary normal to the wall within the boundary layer. These equations are greatly simplified and provide the starting point for a variety of analytic and numerical solutions that include viscous effects. For more information on analytic and numerical solutions to the boundary-layer equations, see the books by Schlichting et al.,[9] Cebeci and Cousteix,[45] and Cebeci and Smith.[46]

The boundary-layer equations were traditionally used to compute skin friction, displacement, and momentum thicknesses and to predict the separation location. Together with the energy equation, they are used to compute heat transfer or surface temperature. To make the calculation, wall boundary conditions (usually no-slip and either a prescribed temperature or that the flow is adiabatic) are required. In addition, the edge velocity or pressure is prescribed, which comes from an inviscid calculation. Finally, a starting velocity profile must be provided. Typically this is a similarity solution (a solution that is valid at many locations using transformed variables), either for a flat plate or at a stagnation point.

3.14 Examples of Zones of Application

The appropriate version of the governing equation depends on the type of flowfield being investigated. For high Reynolds number attached flow, the pressure can be obtained very accurately without considering viscosity. Recall that the use of a Kutta condition provides a simple way of enforcing key physics associated with viscosity by specifying this feature as a boundary condition on an otherwise inviscid solution. If the onset flow is uniform, and any shocks are weak, $M_n < 1.25$, then the potential flow approximation is valid. If a slight flow separation exists, a special approach using the boundary-layer equations can be used interactively with the inviscid solution to obtain an overall solution. As speed increases, shocks begin to get strong and are curved. Under these circumstances, the solution of the complete Euler equations is required.

When significant separation occurs, or you cannot figure out the preferred direction to apply a boundary-layer approach, the Navier-Stokes equations are used. As previously described, many different "levels" of the Navier-Stokes equations have been used over the years.

One other important consideration not explicitly described so far is that we have been describing the viscous flow models in terms of the Navier-Stokes equations. For most airplane aerodynamics, the flow is turbulent, and this must be included. To do this requires the use of turbulence models, which will be discussed in Chapter 8. Generally, this means that the governing equations must be Reynolds averaged, and the resulting equations are called the Reynolds Averaged Navier-Stokes (RANS) equations. This process introduces additional terms to represent the effects of turbulence, and development of turbulence models has been a major research activity since the days of Prandtl. Often, users of commercial programs will be given a menu of models to choose from, so it is important to have some knowledge of the models. Furthermore, most computational codes do not predict the transition location from laminar to turbulent flow. The choice of the appropriate turbulence model and transition location is perhaps an unfair burden on the user and is an example of why the material in this text is important to understand before undertaking computational aerodynamics analysis and design.

3.15 Requirements for a Complete Problem Formulation

When formulating a mathematical representation of a fluid flow problem, you have to consider carefully both the flowfield model equations and the boundary conditions. An evaluation of the mathematical type of the PDEs that are being solved plays a key role in this. Boundary conditions must be properly specified. Either over- or under-specifying boundary conditions will doom your calculation before you start. A proper formulation requires:

- governing equations
- boundary conditions
- coordinate system specification

All of these are necessary before computing the first number! If this is done, then the mathematical problem being solved is considered to be *well posed*.

Summary of Best Practices

1. Knowing the equations of motion is essential to being an intelligent user of CA!
2. Understanding the various levels of equations that are used is also a very important aspect of dealing with simulations – which equation will work and why? What are the computer requirements for performing your simulation with one equation set relative to another?
3. Codes use a variety of nondimensionalizations and other modifications to the equations of motion – you should know which equations your code is using and be familiar with how the input and output impact your usage.
4. Knowing boundary conditions and their implementation in codes is essential to obtaining good results.
5. Coordinate systems are different for different codes, and you should understand how your code deals with geometry and coordinates.
6. Understand the three PDE types and the impact of the type on the solution of the equations (boundary conditions, initial conditions, solution region, etc.).

3.16 Exercises

1. Convert the unsteady 3D Euler equations from classical nonconservative form to divergence form.
2. Equation 3.73 is an unusual form of the equation of state. It is from viewgraphs defining the equations used in *CFL3D*. Turn in your derivation of this equation. Is there a typo?
3. Show how Equation 3.78 can be obtained.
4. Why is Equation 3.81 not in divergence form?
5. Show that point source and point vortex singularities are solutions of Laplace's equation in two dimensions.
   Recall that a point source can be expressed as:

$$\phi(x, y) = \frac{q}{4\pi} \ln(x^2 + y^2)$$

and a point vortex is:

$$\phi(x, y) = \frac{\Gamma}{2\pi} \tan^{-1}\left(\frac{y}{x}\right)$$

6. Consider the point source of Problem 5. What is the behavior of the velocity as the distance from the source becomes large? What is the potential function for a point source? How does it behave as the distance from the source becomes large? Comment from the standpoint of having to satisfy the "infinity" boundary condition in a program for a potential flow solution.

7. Find the classification type of the following equations:
   a. Laplace: $\phi_{xx} + \phi_{yy} = 0$
   b. Heat equation: $\phi_y = \sigma \phi_{xx}$, σ real
   c. Wave equation: $\phi_{xx} = c^2 \phi_{yy}$, c real
8. Classify the following partial differential equations:
   a. $3\phi_{xx} + \phi_{xy} + 2\phi_{yy} = 0$
   b. $\phi_t + \beta \phi_x + \alpha \phi_{xx} = 0$
   c. $x\phi_{xx} + \phi_{xy} + y\phi_{yy} = 0$
   d. $\left(1 - M_\infty^2\right)\phi_{xx} + \phi_{yy} = 0$

3.17 References

1 *Memoir and Scientific Correspondence of the Late Sir George Gabriel Stokes*, Joseph Larmor, ed., Vol. 1, Cambridge: Cambridge University Press, 1907.

2 Karamcheti, K., *Principles of Ideal-Fluid Aerodynamics*, 2nd Ed., Melbourne: Krieger Publishing Co., 1980.

3 Bertin, J.J., and Cummings, R.M., *Aerodynamics for Engineers*, 6th Ed., Upper Saddle River: Pearson, 2014.

4 Anderson, J.D., *Fundamentals of Aerodynamics*, 5th Ed., New York: McGraw-Hill, 2011.

5 Moran, J., *An Introduction to Theoretical and Computational Aerodynamics*, New York: John Wiley and Sons, 1984.

6 Owczarek, J.A., *Fundamentals of Gas Dynamics*, Scranton: International Textbook Co., 1964.

7 White, F.M., *Viscous Fluid Flow*, 3rd Ed., New York: McGraw-Hill, 2005.

8 Vincenti, W.G., "Control-Volume Analysis: A Difference in Thinking Between Engineering and Physics," *Technology and Culture*, Vol. 23, No. 2, 1982, pp. 145–174.

9 Schlichting, H., Gersten, K., Krause, E., and Oertel, H., *Boundary-Layer Theory*, 8th Ed., New York: Springer, 2004.

10 Katz, J., and Plotkin, A., *Low-Speed Aerodynamics*, 2nd Ed., Cambridge: Cambridge University Press, 2001.

11 Viviand, H., "Conservation Forms of Gas Dynamics Equations," *La Recherche Aerospatiale*, No. 1974-1, 1974, pp. 65–66.

12 Vinokur, M., "Conservation Equations of Gas-Dynamics in Curvilinear Coordinate Systems," *Journal of Computational Physics*, Vol. 14, Issue 2, 1974, pp. 105–125.

13 Thomas, J.L., and Walters, R.W., "Upwind Relaxation Algorithms for the Navier Stokes Equations," AIAA Paper 85–1501, July 1985.

14 Pulliam, T. H., "Euler and Thin Layer Navier-Stokes Codes: ARC2D, ARC3D," Computational Fluid Dynamics User's Workshop, University of Tennessee Space Institute, Tullahoma, TN, March 1984.

15 Pletcher, R.H., Tannehill, J.C., and Anderson, D.A., *Computational Fluid Mechanics and Heat Transfer*, 3rd Ed., Boca Raton: CRC Press, 2013.

16 Sneddon, I.N., *Elements of Partial Differential Equations*, New York: McGraw-Hill, 1957.

17 Chester, C.R., *Techniques in Partial Differential Equations*, New York: McGraw-Hill, 1971.

18 Carrier, G.F., and Pearson, C.E., *Partial Differential Equations*, New York: Academic Press, 1976.

19 Fletcher, C.A.J., *Computational Techniques for Fluid Dynamics*, Vol. 1, 2nd Ed., Berlin: Springer-Verlag, 1991.

20 Hoffman, K.A., and Chiang, S.T., *Computational Fluid Dynamics*, Vols. 1 and 2, 4th Ed., Wichita: Engineering Education System, 2000.

21 Greenberg, M.D., *Advanced Engineering Mathematics*, 2nd Ed., Upper Saddle River: Prentice Hall, 1998.

22 Jafroudi, H., and Yang, H.T., "Steady Laminar Forced Convection from a Circular Cylinder," *Journal of Computational Physics*, Vol. 65, No. 1, 1986, pp. 46–56.

23 Allmaras, S.R., Venkatakrishnan, V., and Johnson, F.T., "Farfield Boundary Conditions for 2-D Airfoils," AIAA Paper 2005–4711, June 2005.

24 Colonius, Tim, "Modeling Artificial Boundary Conditions for Compressible Flow," *Annual Review of Fluid Mechanics*, Vol. 36, 2004, pp. 315–345.

25 Garabedian, P.R., *Partial Differential Equations*, New York: John Wiley & Sons, 1964.

26 Steger, J.L., and Van Dalsem, W.R., "Basic Numerical Methods," in *Unsteady Transonic Aerodynamics*, Progress in Astro. and Aero., Vol. 120, Nixon, D., ed., Reston: AIAA, 1989.

27 Ashley, H., and Landahl, M., *Aerodynamics of Wings and Bodies*, Reading: Addison-Wesley, 1965.

28 Holst, T.L., "Transonic Flow Computational Using Nonlinear Potential Methods," *Progress in Aerospace Sciences*, Vol. 36, No. 1, 2000, pp. 1–61.

29 Van Dyke, M., *Perturbation Methods in Fluid Mechanics*, Stanford: Parabolic Press, 1975.

30 Nixon, D., "Basic Equations for Unsteady Transonic Flow," in *Unsteady Transonic Aerodynamics*, Progress in Astro. and Aero., Vol. 120, Nixon, D., ed., Reston: AIAA, 1989.

31 Whitcomb, R.T., "Review of NASA Supercritical Airfoils," ICAS Paper No. 74-10, Aug. 1974.

32 Murman, E.M., and Cole, J.D., "Calculation of Plane Steady Transonic Flows," *AIAA Journal*, Vol. 9, No. 1, 1971, pp. 114–121.

33 Jameson, A., "Iterative Solution of Transonic Flow over Airfoils and Wings, including Flows at Mach 1," *Comm. Pure Appl. Math.*, Vol. 27, 1974, pp. 283–309.

34 Bavitz, P.C., "An Analysis Method for Two-Dimensional Transonic Viscous Flow," NASA TN D-7718, 1974.

35 Rom, J., "Flow with Strong Interaction Between the Viscous and Inviscid Regions," *SIAM Journal of Applied Math.*, Vol. 29, No. 2, 1975, pp. 309–328.

36 Steger, J.L., "Implicit Finite-Difference Simulation of Flow about Arbitrary Two-Dimensional Geometries," *AIAA Journal*, Vol. 16, No. 7, 1978, pp. 679–686.

37 Pulliam, T.H., and Steger, J.L., "On Implicit Finite-Difference Simulations of Three Dimensional Flow," AIAA Paper 78–0010, January 1978.

38 Baldwin, B.S., and Lomax, H., "Thin Layer Approximation and Algebraic Model for Separated Turbulent Flows," AIAA Paper 78–0257, January 1978.

39 Cummings, R.M., Forsythe, J.R., Morton, S.A., and Squires, K.D., "Computational Challenges in High Angle of Attack Flow Prediction," *Progress in Aerospace Sciences*, Vol. 39, No. 5, 2003, pp. 369–384.

40 Vigneron, Y.C., Rakich, J.V., and Tannehill, J.C., "Calculation of Supersonic Flow Over Delta Wings with Sharp Subsonic Leading Edges," AIAA Paper 78–1137, July 1978.

41 Morrison, J.H., and Korte, J.J., "Implementation of Vigneron's Streamwise Pressure Gradient Approximation in the PNS Equations," AIAA Paper 92–0189, January 1992.

42 Knill, D.L., Giunta, A.A., Baker, C.A., Grossman, B., Mason, W.H., Haftka, R.T. and Watson, L.T., "Response Surface Models Combining Linear and Euler Aerodynamics for Supersonic Transport Design," *Journal of Aircraft*, Vol. 36, No. 1, 1999, pp. 75–86.

43 Anderson, J.D., *Hypersonic and High Temperature Gas Dynamics*, 2nd Ed., Reston: AIAA, 2006.

44 Degani D., and Schiff L.B., "Computation of Turbulent Flows Around Pointed Bodies Having Crossflow Separation," *Journal of Computational Physics*, Vol. 66, No. 1, 1986, pp. 173–196.

45 Cebeci, T., and Cousteix, J., *Modeling and Computation of Boundary Layer Flows*, Long Beach: Horizons Publishing, 1999.

46 Cebeci, T., and Smith, A.M.O., *Analysis of Turbulent Boundary Layers*, Orlando: Academic Press, 1974.

4 Getting Ready for Computational Aerodynamics: Aerodynamic Concepts

I'm still trying to learn about aerodynamics.

A.M.O. Smith, Chief Aerodynamics Engineer for Research,
Douglas Aircraft Company[1]

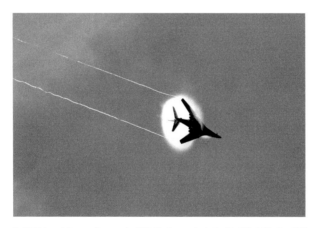

B-1B flying at transonic speeds (U.S. Air Force photo by Staff Sgt. Shelley Gill).

LEARNING OBJECTIVE QUESTIONS

After reading this chapter you should be able to answer the following questions:

- What are the basic assumptions of potential flow theory? How do these assumptions impact the applicability of the results?
- Why does the potential flow solution for flow over a cylinder result in no drag?
- What is the Kutta condition and why is it important to the development of airfoil theory?
- What are the differences between laminar and turbulent boundary layers and what impact do the differences have on airfoils?
- How do camber and thickness impact airfoil aerodynamics?
- How do wing-tip vortices affect wing aerodynamics?
- What are the characteristics of a good wing planform for subsonic flight?

- What is the definition of "transonic"?
- How does a supercritical airfoil work and why?
- Why do high-speed aircraft have swept wings?
- What is the difference between subsonic and supersonic airfoils?
- Do supersonic wings experience induced drag?
- What are the defining characteristics of hypersonic flow?

4.1 Introduction

Determining the aerodynamics of a modern aircraft is a complex job, one made more difficult because there is so much about aerodynamics that we still do not fully understand. As A.M.O. Smith stated in the 1970s, learning about aerodynamics, even for someone who knew as much about the subject as he did, is a never-ending job. As an example of just how complex aerodynamics can be, look at the C-17 shown in Fig. 4.1, where the wing-tip vortices are visualized using smoke. This airplane is taking off and has various control surfaces deflected, as well as high-lift devices in use. The airplane is near the ground, which impacts the aerodynamics of the vehicle as well. The engines are being used to augment the lift and they greatly influence the flow near them. In fact, the C-17 uses the exhaust from the engine to blow over the flaps to augment lift. Being able to computationally model all of these

Figure 4.1 Boeing C-17 taking off with wing-tip vortices marked with smoke (U.S. Air Force photo).

effects is quite a challenge, but in this chapter we will discover some basic concepts that will aid in our understanding as we continue in our study of computational aerodynamics.

Prior to performing these simulations, however, we will spend some time trying to understand the basic aerodynamic concepts that are essential to performing good computations. We will look at how airfoils and wings work, what causes the forces acting on an airplane, and why viscous effects are important as well. This knowledge will help to make us intelligent users of CA and will also help us to be able to validate our methods with experimental data or theory, an important consideration in CA (see Fig. 1.3). By no means is this chapter meant to be a comprehensive review of aerodynamics or a replacement for a basic course in fluid dynamics. Interested readers can find good sources of aerodynamic information in the books by Anderson,[2] Bertin and Cummings,[3] and McCormick.[4]

4.2 Review of Potential Flow Theory

The potential flow model for air moving over a highly streamlined shape such as an airplane in cruising flight can be used to provide a framework for understanding aerodynamics. Perhaps the oldest and longest used computer method for simulating aerodynamic flows is the panel method, which is a direct outgrowth of potential flow concepts. Panel methods have allowed designers to make great strides in predicting the aerodynamics of vehicles at cruise conditions, which is a very important capability. While the assumptions of potential flow will not allow for panel methods to be useful when a great deal of flow separation is present, these methods are extremely useful for evaluating large numbers of configurations for streamlined flow cases and for making comparisons between them.

So, before we delve into the details about panel methods in Chapter 5, we will spend a little time understanding potential flow. The assumptions we make and the results we obtain from potential flow will have a direct bearing on how we use panel methods, so the background of potential flow theory is important for becoming an intelligent user of panel method programs.

Assuming that the fluid flow is steady, inviscid, and incompressible[i] with no body forces, we have from the conservation of mass and momentum (Equation 3.14 and Equations 3.25 and 3.30 modified using a vector identity)[2,3]

$$\vec{\nabla} \cdot \vec{V} = 0 \qquad (4.1)$$

---

[i] Compressible flow can also be modeled as a potential flow, as was described in Chapter 3.

$$\rho\left(\vec{V}\cdot\nabla\vec{V}\right)\vec{V} = \rho\vec{\nabla}\left(\frac{V^2}{2}\right) - \rho\vec{V}\times\vec{\nabla}\times\vec{V} = -\nabla p. \qquad (4.2)$$

These equations form the basis for the potential flow concept.

4.2.1 Vorticity

Now that we have our basic equations for potential flow, we can begin to try to understand some flow concepts. We will take advantage of the mathematical concept of the curl of the velocity vector and relate it to the fluid dynamic concept of *vorticity*, given by:

$$\vec{\omega} = \vec{\nabla}\times\vec{V} = \begin{vmatrix} \hat{i} & \hat{j} & \hat{k} \\ \dfrac{\partial}{\partial x} & \dfrac{\partial}{\partial y} & \dfrac{\partial}{\partial z} \\ u & v & w \end{vmatrix} = \left(\frac{\partial w}{\partial y} - \frac{\partial v}{\partial z}\right)\hat{i} - \left(\frac{\partial w}{\partial x} - \frac{\partial u}{\partial z}\right)\hat{j} + \left(\frac{\partial v}{\partial x} - \frac{\partial u}{\partial y}\right)\hat{k}$$

$$= \omega_x\hat{i} + \omega_y\hat{j} + \omega_z\hat{k} \qquad (4.3)$$

where $\vec{\omega}$ is the vorticity vector. But what is vorticity? If we look at the partial derivative terms in the vorticity equation (Equation 4.3), we will begin to understand the physical meaning of vorticity. Look at a fluid element in the x-y plane with arbitrary velocities at each edge, as shown in Fig. 4.2.

The vorticity terms that apply in this situation are $\partial v/\partial x$ and $\partial u/\partial y$, which represent the change in velocity acting on the edges of the fluid element. If there are net changes in velocity in both the x- and y-directions, then the fluid element will rotate (physically), which is represented by the non-zero term in the vorticity equation (mathematically)

$$\vec{\nabla}\times\vec{V} = \left(\frac{\partial v}{\partial x} - \frac{\partial u}{\partial y}\right) \neq 0. \qquad (4.4)$$

In other words, the vorticity is directly related to the fluid element rotational velocity; in fact, the vorticity turns out to be twice the *rotational velocity*, $\vec{\xi}$, of the fluid element

$$\vec{\omega} = 2\vec{\xi}. \qquad (4.5)$$

If the flowfield does not have any vorticity (which means that the fluid elements are not rotating), then the flow is called *irrotational* (see Fig. 4.3). If the flowfield does have vorticity, then the flow is called *rotational*. It is

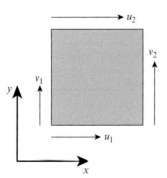

Figure 4.2   Vorticity of a fluid element.

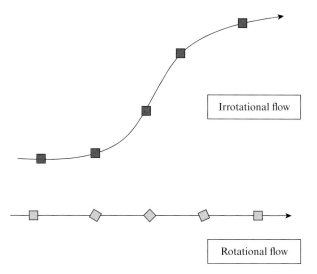

Irrotational flow

Rotational flow

Figure 4.3 Illustration of irrotational and rotational flow.

very important at this point not to confuse rotation of a fluid element with curvature (or turning) of the flow; these are two separate (although related) concepts. It is theoretically possible for a turning flow to be irrotational, just as it is possible for a straight flow to be rotational.

If we assume that the flowfield is irrotational ($\vec{\nabla} \times \vec{V} = 0$) in addition to steady, incompressible, and inviscid, then the conservation of mass and momentum equations (Equations 4.1 and 4.2) reduce to:

$$\vec{\nabla} \cdot \vec{V} = 0 \tag{4.6}$$

$$\rho \vec{\nabla} \left(\frac{V^2}{2} \right) = -\vec{\nabla} p. \tag{4.7}$$

In addition, since the flow is irrotational, we can represent the velocity vector as the gradient of a scalar function, $\vec{V} = \vec{\nabla} \Phi$, where Φ is called the velocity potential (see Chapter 3 for more details, especially the Control Volume Box).

Flow Visualization Box

Vorticity in a Flow Field

You just learned how to define vorticity, but you probably don't have a good feel for where fluid particle rotation would take place in a flowfield. Now that you know how to access the F-18 flow solution in FieldView, we will use that solution to "see" where vorticity takes place so that you can start to develop a physical feeling for vortical flow. Start the FieldView Demo version and open the F-18 solution that you first saw in Chapter 3.

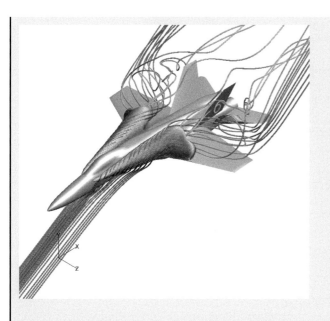

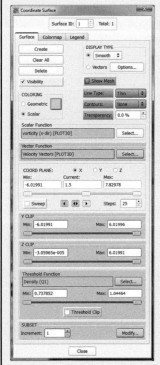

How to look at vorticity:

First turn off the streamlines and iso-surfaces that are contained in the flowfield visualization shown in the figure – all you should see now is the surface of the aircraft. Now use *Coordinate Surface* to define some planes of vorticity. Click *Create*, choose *Scalar* coloring and *Smooth* display type, choose *Vorticity* and the Scalar Function, and define an x Coordinate Plane.

You will need to choose the *Colormap* tab and choose settings that allow for the appropriate viewing of vorticity in this

plane, and after some experimentation you might be able to see a single plane of vorticity that looks like the figure shown here (this view uses the *Transparency* capability as well).

Notice how the regions with no vorticity show up in this cutting plane as light blue – it would be nice to crop those values to only see where high values of vorticity take place. Once you see how to do one plane of vorticity, you can add several planes to begin to develop a picture of the flowfield, and change from *Smooth* to *Contours* so you can see each plane. Where is vorticity coming from on the airplane? Where are there regions of high vorticity? You can clearly see the vortex forming over the leading-edge extension and along the leading edge of the wing, and how these flow features convect downstream and interact with each other. Now that you know how to do this, you should try out other ways to visualize this flowfield.

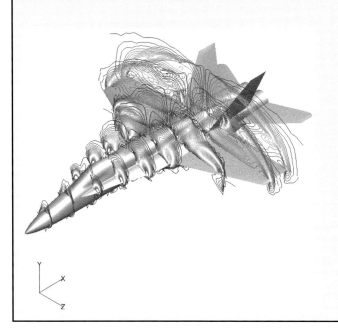

4.2.2 Simplified Equations of Motion

Equation 4.7, which is also known as Euler's equation (see Chapter 3 for details), can be integrated along a streamline in order to obtain *Bernoulli's equation*, resulting in a simplified set of equations for steady, two-dimensional, incompressible, inviscid, irrotational flow of a perfect fluid with no body forces:

Continuity equation

$$\frac{\partial u}{\partial x} + \frac{\partial v}{\partial y} = 0 \tag{4.8}$$

Bernoulli's equation

$$p + \frac{\rho V^2}{2} = p_o \tag{4.9}$$

Irrotational condition

$$\frac{\partial v}{\partial x} - \frac{\partial u}{\partial y} = 0 \tag{4.10}$$

where p_o is the total pressure in the flow. These linear equations (Equations 4.8 and 4.10) and Bernoulli's equation (Equation 4.9), coupled with the Perfect Gas Law

$$p = \rho R T \tag{4.11}$$

give us a set of equations that allows for an analytic solution, which is what we have been trying to find. From a historical perspective, these are the equations that could be analytically solved prior to the development of computers and numerical methods. While these equations would later be solved on computers, they form a framework for a classic analytic solution method known as potential flow theory. The derivation of potential flow theory is presented in Appendix C. Readers who have not been through the derivation of potential flow theory (or who do not readily remember it well) are encouraged to review the derivation.

4.3 Potential Flow Applications

The basic concept of linear potential flow is to find flow functions that satisfy Equations 4.8 and 4.10, rather than attempting to solve the equations directly. The traditional way to accomplish this is to use the basic flow types defined and developed in Appendix C: uniform flow, source/sink flow, vortex flow, and doublet flow. In terms of the velocity potential, Φ, or the stream function, ψ, Equations 4.8 and 4.10 can be combined and rewritten as Laplace's equation:

$$\frac{\partial^2 \Phi}{\partial x^2} + \frac{\partial^2 \Phi}{\partial y^2} = \Phi_{xx} + \Phi_{yy} = \nabla^2 \Phi = 0$$
$$\frac{\partial^2 \psi}{\partial x^2} + \frac{\partial^2 \psi}{\partial y^2} = \psi_{xx} + \psi_{yy} = \nabla^2 \psi = 0 \tag{4.12}$$

Since Laplace's equation is a linear partial differential equation, the individual solutions can be added together to create more complicated flowfields (the superposition concept). The definitions of the four potential functions

Table 4.1 Basic potential flow functions (Ref. 3 and Appendix C)

Type of Flow	Velocity	Φ (Velocity Potential)	ψ (Stream Function)
Uniform flow	$u = V_\infty$	$V_\infty x$	$V_\infty y$
Source/Sink	$u_r = \Lambda/2\pi r$	$\dfrac{\Lambda}{2\pi}\ln r$	$\dfrac{\Lambda}{2\pi}\theta$
Vortex	$u_\theta = -\Gamma/2\pi r$	$-\dfrac{\Gamma}{2\pi}\theta$	$\dfrac{\Gamma}{2\pi}\ln r$
Doublet	$u_r = -m\cos\theta/2\pi r^2$ $u_\theta = -m\sin\theta/2\pi r^2$	$m\cos\theta/2\pi r$	$-m\sin\theta/2\pi r$

Note: Λ, Γ, and m are the strengths of the source/sink, vortex, and doublet, respectively.

that are used to create arbitrary flowfields are summarized in Table 4.1. These building-block functions form the basis for creating more complex flowfields through the superposition concept.

4.3.1 Flow Over a Circular Cylinder

We continue our journey toward predicting more practical aerodynamic flows by finding the flow over a simple geometry, namely a circle (which simulates a circular cylinder). This flow is created by adding together uniform flow and a doublet, yielding a stream function in cylindrical coordinates given by

$$\psi = V_\infty r\sin\theta - \frac{m}{2\pi r}\sin\theta. \tag{4.13}$$

Since both terms have some variables in common, the function may be rewritten in terms of R, which represents the radius of the circular cylinder, as:

$$\psi = V_\infty r\sin\theta\left(1 - \frac{R^2}{r^2}\right) \tag{4.14}$$

where $R^2 = m/(2\pi V_\infty)$. Now that we have the stream function, we can easily obtain the velocity field for the flow using the cylindrical coordinate velocity relationships (see Ref. 3 for details):

$$u_r = \frac{1}{r}\frac{\partial\psi}{\partial\theta} = V_\infty\cos\theta\left(1 - \frac{R^2}{r^2}\right) \quad u_\theta = -\frac{\partial\psi}{\partial r} = -V_\infty\sin\theta\left(1 + \frac{R^2}{r^2}\right) \tag{4.15}$$

The velocity field can be used to find the stagnation points of the flow, since stagnation points take place where $u_r = u_\theta = 0$. Setting the two velocities equal to each other and noting that they also must equal zero yields

$$V_\infty\cos\theta\left(1 - \frac{R^2}{r^2}\right) = -V_\infty\sin\theta\left(1 + \frac{R^2}{r^2}\right) = 0 \tag{4.16}$$

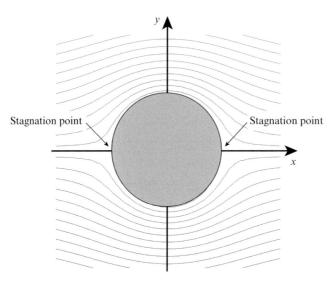

Uniform flow over a circular cylinder.

which is satisfied for $r = R$ and $\theta = 0, \pi$. In other words, there are two stagnation points, both of which are on the surface of the cylinder, one at the front of the cylinder (relative to the flow direction) and one at the rear, as shown in Fig. 4.4.

This picture of the flowfield is showing us more than just the streamlines, it is also telling us about various properties of the flow. For example, since we are modeling incompressible flow where pressure disturbances propagate in all directions equally (and instantly), the flow "sees" the cylinder long before it reaches the solid body. This is why the streamlines curve prior to reaching the cylinder. In addition, we see that the streamlines move closer together as the flow goes around the cylinder, which means that the flow must have higher velocity in the smaller area between the streamlines (since the mass flow rate between any two streamlines is constant). In fact, if we calculate the velocity at the top and bottom of the cylinder we find that $u = 2V_\infty$; the flow has doubled its speed as it goes up and over the cylinder. The streamlines farther from the cylinder have lower velocities, since the doublet velocity decreases with distance from the origin.

Now we have a prediction for the flowfield over a cylinder, but how realistic is it? Figure 4.5 shows the flow over a circular cylinder in a water tunnel. The flow over the front half of the cylinder looks very similar to the flow in Fig. 4.4, but separation is clearly evident at the top and bottom of the cylinder in Fig. 4.5. In addition, the wake is showing evidence of vortex shedding, a highly unsteady flow phenomenon (even though the cylinder is stationary, the flow behind it is not!). Our potential flow theory cannot predict flow separation, since we assumed the flow was inviscid during the derivation. Thus, potential flow theory will not predict the wake flow shown in Fig. 4.5. It is this type of limitation that anyone using potential flow theory or panel methods must understand!Fig. 4.6

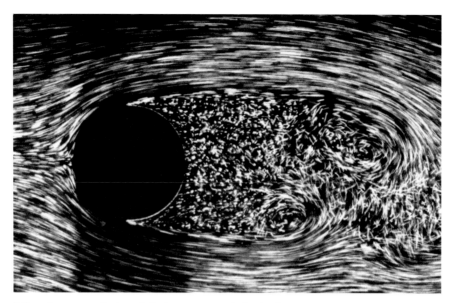

Uniform flow over a circular cylinder in a water tunnel showing flow separation and wake vortices (Ref. 5; Photograph by Henri Werlé © ONERA, the French Aerospace Lab).

We can define a pressure coefficient for potential flow by using Bernoulli's equation (Equation 4.9) and applying it to two points in the flow (namely, a point in the freestream and a point on the surface of the cylinder):

$$p_\infty + \frac{\rho V_\infty^2}{2} = p + \frac{\rho V^2}{2} \qquad (4.17)$$

or, rearranging we have

$$\left(p - p_\infty\right) = \frac{\rho}{2}\left(V_\infty^2 - V^2\right). \qquad (4.18)$$

Now define a *pressure coefficient* as

$$C_p \equiv \frac{p - p_\infty}{\frac{1}{2}\rho V_\infty^2} \qquad (4.19)$$

which, for incompressible flow, can be simplified using Equation 4.18 to give

$$C_p = \frac{p - p_\infty}{\frac{1}{2}\rho V_\infty^2} = \frac{\frac{\rho}{2}\left(V_\infty^2 - V^2\right)}{\frac{1}{2}\rho V_\infty^2} = 1 - \left(\frac{V}{V_\infty}\right)^2. \qquad (4.20)$$

The pressure coefficient is a nondimensional measure of the local static pressure relative to the freestream static pressure. When $C_p = 0$, the local static pressure is the same as the freestream static pressure. When $C_p > 0$, the local

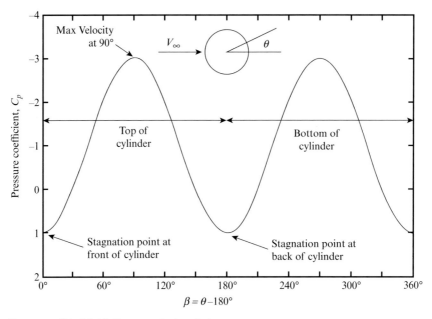

Figure 4.6 Pressure coefficient distribution over a circular cylinder.

static pressure is greater than freestream static pressure. Finally, when $C_p < 0$ the local static pressure is less than the freestream static pressure.

The velocities on the surface of the cylinder ($r = R$) are given by

$$u_r = V_\infty \cos\theta \left(1 - \frac{R^2}{r^2}\right) = 0 \quad u_\theta = -V_\infty \sin\theta \left(1 + \frac{R^2}{r^2}\right) = -2V_\infty \sin\theta \quad (4.21)$$

which yields a pressure coefficient of

$$C_p = 1 - \left(\frac{V}{V_\infty}\right)^2 = 1 - \left(\frac{-2V_\infty \sin\theta}{V_\infty}\right)^2 = 1 - 4\sin^2\theta. \quad (4.22)$$

The plot of the pressure coefficient over the surface of the cylinder is shown in .

Notice that the flow is symmetric top and bottom, as well as right and left, meaning that the pressures over the top of the cylinder are the same as the pressures over the bottom, and the pressures over the left half of the cylinder are the same as the pressures over the right half. To find the forces acting on the cylinder, we will have to integrate these forces around the cylinder, taking components in the x and y directions to find drag and lift, respectively. The lift and drag per unit span will be[3]:

$$l = -\int_0^{2\pi} p\sin\theta R\, d\theta \quad d = -\int_0^{2\pi} p\cos\theta R\, d\theta \quad (4.23)$$

and if we nondimensionalize the equations by the freestream dynamic pressure, $q_\infty = \rho V_\infty^2 / 2$, and the cylinder diameter, $2R$, we obtain

$$C_l \equiv \frac{l}{q_\infty 2R} = -\frac{1}{2}\int_0^{2\pi} C_p \sin\theta d\theta$$

$$C_d \equiv \frac{d}{q_\infty 2R} = -\frac{1}{2}\int_0^{2\pi} C_p \cos\theta d\theta \qquad (4.24)$$

Now substitute the pressure coefficient for the flow over the cylinder (Equation 4.22) into the lift coefficient equation (Equation 4.24a) to obtain

$$C_l = -\frac{1}{2}\int_0^{2\pi} C_p \sin\theta d\theta = -\frac{1}{2}\int_0^{2\pi}\left(1 - 4\sin^2\theta\right)\sin\theta d\theta = 0 \qquad (4.25)$$

which is not surprising since the flow over the top of the cylinder is the same as the flow over the bottom. Now repeat the process for the drag coefficient equation (Equation 4.24b):

$$C_d = -\frac{1}{2}\int_0^{2\pi} C_p \cos\theta d\theta = -\frac{1}{2}\int_0^{2\pi}\left(1 - 4\sin^2\theta\right)\cos\theta d\theta = 0 \qquad (4.26)$$

which is considerably more disturbing – the cylinder has no drag! This famous result is known as *D'Alembert's paradox*, since D'Alembert reached this conclusion and was quite annoyed that his elegant theoretical derivation arrived at the wrong answer, since he knew that flow over a cylinder produced drag. But why did this derivation produce no drag? Look again at the assumptions we have made for potential flow: incompressible, irrotational, inviscid, and steady flow. Since we assumed that the flow was inviscid, there will be no skin friction acting on the body. In addition, boundary-layer separation is a viscous effect, so there will also be no pressure drag if we assume inviscid flow. So, while D'Alembert despaired, we do not – the results make perfect sense considering the assumptions we have made. By the way, D'Alembert performed this derivation before the concept of the boundary layer was developed, so he did not know he was making the inviscid flow assumption, which further increased his paradox! A real flow, such as the flow shown in Fig. 4.5, will have pressure drag (since a great deal of flow separation is present) as well as skin friction drag.

4.3.2 Flow Over a Circular Cylinder with Circulation

While the previous flow is interesting, it would be more informative to find a flow that creates a force on the cylinder. In order to accomplish this, we will add a vortex to the flow over the cylinder as given in Equation 4.14:

$$\psi = V_\infty r \sin\theta\left(1 - \frac{R^2}{r^2}\right) + \frac{\Gamma}{2\pi}\ln r + C. \qquad (4.27)$$

where Γ is the circulation (or strength) of the vortex (see Appendix C for details).

Any stream function can have an arbitrary constant added – we have used one here in order to simplify the mathematics. Specifically, we will define the constant to be

$$C = -\frac{\Gamma}{2\pi} \ln R \qquad (4.28)$$

and obtain the following stream function for our flow

$$\psi = V_\infty r \sin\theta \left(1 - \frac{R^2}{r^2}\right) + \frac{\Gamma}{2\pi} \ln\left(\frac{r}{R}\right). \qquad (4.29)$$

We now have a stream function for uniform flow over a circular cylinder with circulation, which is a model for a rotating cylinder where Γ is proportional to the rate of rotation of the cylinder. The velocity components for this flow are:

$$
\begin{aligned}
u_r &= \frac{1}{r}\frac{\partial \psi}{\partial \theta} = V_\infty \cos\theta \left(1 - \frac{R^2}{r^2}\right) \\
u_\theta &= -\frac{\partial \psi}{\partial r} = -V_\infty \sin\theta \left(1 + \frac{R^2}{r^2}\right) - \frac{\Gamma}{2\pi r}
\end{aligned}
\qquad (4.30)
$$

Notice that the radial component of velocity is the same as Equation 4.21, since the vortex we added does not have a radial velocity component. Also, the tangential velocity is the same as before, but with the velocity for a vortex added. This is an important result, since we see that potential flow theory, where superposition applies for the flow functions, also allows superposition of the velocities. If we add two or three functions together we' obtain a velocity field that is the sum of the parts.

Now let's look at the stagnation points for the flowfield. As before, we will require the two velocity components to be zero at a stagnation point, which results in:

$$\theta = \sin^{-1}\left(\frac{-\Gamma}{4\pi V_\infty R}\right). \qquad (4.31)$$

For increasing values of Γ, the two stagnation points move down the sides of the cylinder, as shown in Fig. 4.7. Eventually, if the circulation is increased to high enough levels, the stagnation points will both be at the bottom of the cylinder. If the circulation is increased to still higher levels, the stagnation point will move off the cylinder, as shown in Fig. 4.8.

Now find the pressures and forces acting on the cylinder. The pressure coefficient equation (Equation 4.20) can be used with the velocity component for

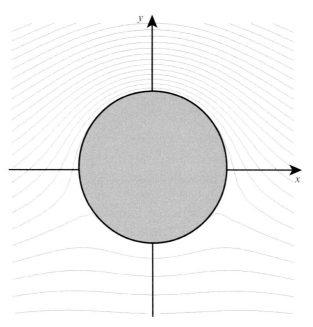

Figure 4.7   Flow over a circular cylinder with circulation.

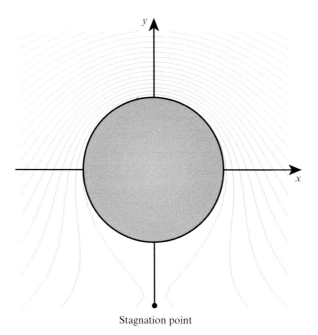

Stagnation point

Figure 4.8   Stagnation point of the cylinder when circulation is very high.

the vortex (Equation 4.30) to obtain the pressure coefficient on the surface of the cylinder as:

$$
\begin{aligned}
C_p &= 1 - \left(\frac{V}{V_\infty}\right)^2 = 1 - \left(\frac{-2V_\infty \sin\theta - \dfrac{\Gamma}{2\pi R}}{V_\infty}\right)^2 \\
&= 1 - \left[4\sin^2\theta + \frac{2\Gamma}{\pi R V_\infty}\sin\theta + \left(\frac{\Gamma}{2\pi R V_\infty}\right)^2\right].
\end{aligned}
\tag{4.32}
$$

We can easily see that when there is no circulation we obtain the same result as we did in Equation 4.22 for flow over the cylinder with no circulation. We can now place the pressure coefficient into the lift and drag coefficient equations (Equation 4.24) and obtain

$$
\begin{aligned}
C_l &= -\frac{1}{2}\int_0^{2\pi} C_p \sin\theta\, d\theta = \frac{\Gamma}{V_\infty R} \\
C_d &= -\frac{1}{2}\int_0^{2\pi} C_p \cos\theta\, d\theta = 0.
\end{aligned}
\tag{4.33}
$$

If we use the definition for the lift coefficient from Equation 4.24, we can find the lift per unit span acting on the cylinder as

$$
l = \rho V_\infty \Gamma,
\tag{4.34}
$$

which is known as the *Kutta-Joukowski theorem* (note that the drag is still zero). This is a very important result – the lift on the cylinder is directly proportional to the circulation of the flow! Also recall that Γ relates to the rotation speed of the cylinder, so the circulation is directly related to how fast the flow is going around the cylinder. While the Kutta-Joukowski theorem is interesting for the flow over a cylinder, the theorem goes far beyond this result, since it also states that the lift for uniform flow over any arbitrary body (not just a circular cylinder) will also be given by Equation 4.34.[6]

What are the important results from this derivation? First, the drag on the cylinder is always zero, whether there is circulation or not (due to our inviscid flow assumption). Second, if there is no circulation there is no lift. Finally, if there is circulation, the lift per unit span is $l = \rho V_\infty \Gamma$. This means that circulation is the key to producing lift, which should not be totally surprising. If you recall from your studies of fluid dynamics (see Appendix C), circulation is the turning of the flow. This result states that if the flow is turning, it is creating lift – that is the same result you would obtain from the conservation of linear momentum – if the flow is turned there must be a force, and that force is lift. We will see that this is a fundamental aerodynamic concept that must be understood in order to fully comprehend many aspects of lift on airfoils and wings.

4.4 Applications to Airfoils

At this point we have a number of ways to proceed with potential flow theory in order to obtain flow over an airfoil. These approaches include: conformal mapping, panel methods, and thin airfoil theory. We will briefly describe conformal mapping, leave panel methods to Chapter 5, and then develop thin airfoil theory later in this chapter.

4.4.1 Conformal Mapping

Traditionally, potential flow theory was used to obtain the flow over airfoil shapes using *conformal transformations* from complex variable theory.[7] In this approach, we take the already completed solutions for flow over the cylinder (with or without circulation) and transform that flow into the flow over an airfoil. Although the potential flow around a cylinder is not an accurate model, when the cylinder is transformed into a streamlined shape like an airfoil it works very well. One of the more common transformations used is the Joukowski transformation (shown in Fig. 4.9), which produces a family of airfoils with interesting and informative results about the effects of thickness, camber, and angle of attack on airfoil aerodynamics.[6]

Once the airfoil shape is mapped using the transformation function, the flowfield for the cylinder can be mapped to the airfoil flow. In addition, the

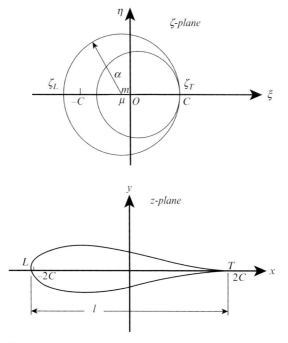

Figure 4.9   Conformal transformation of a circle into a Joukowski airfoil (Ref. 6).

pressures acting on the cylinder can be mapped to the airfoil, leading to a fairly straightforward method for obtaining the lift on the airfoil. Interested readers should refer to Ref. 6 for more details about conformal transformations, since they lend themselves to computational solutions that are easily accomplished on personal computers.

4.4.2 Singularity Distribution Approaches

The other two methods commonly used to find the flow over airfoils are the panel method and thin airfoil theory. Both of these methods require the use of the potential flow singularities described in Section 4.3 to create airfoils and wings by distributing sources, vortices, and doublets to create appropriate flowfields. Panel methods are a computational method (described in Chapter 5) that simulate the airfoil surface with potential flow functions and then determine the velocity and pressure fields due to those functions. Panel methods are very useful since they are able to predict the flow over a variety of shapes, including full aircraft configurations. Thin airfoil theory accomplishes the simulation analytically using vortices to represent the mean camber line of the airfoil, and therefore the results are limited to thin airfoils.

4.4.3 Kutta Condition

If we obtain potential flow solutions over any airfoil shape there will always be an infinite number of solutions possible, one solution for each value of circulation chosen. In order to find the one solution that is physically realistic, we need an additional boundary condition, the *Kutta condition*. The Kutta condition is a physical (viscous) boundary condition based on observation of the flow in the vicinity of the trailing edge of the airfoil. Researchers noticed that the viscous flow near the trailing edge never went around the tip of the trailing edge, but rather left the trailing edge smoothly, as shown in Fig. 4.10. This led to the observation that the conditions at the trailing edge must be such that the flow reaching the trailing edge from the upper surface and the flow reaching the trailing edge from the lower surface must have the same velocity and pressure when they reach the trailing edge. This is a viscous flow boundary condition that will help to make our inviscid theory work. If the trailing edge has a finite angle, then the only way for the velocities to be the same is for the trailing edge to be a stagnation point. Another way of saying this is that the local circulation at the trailing edge is zero, $\gamma(TE) = 0$. If the trailing edge were to have an infinitely thin thickness, then it would not to be have to be a stagnation point, but the velocity at the trailing edge would have to be the same regardless of which side of the airfoil the fluid came from (top or bottom).

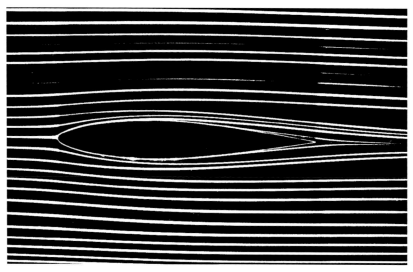

Figure 4.10   Kutta condition at airfoil trailing edge as visualized in a water tunnel (Ref. 5; Photograph by Henri Werlé © ONERA, the French Aerospace Lab).

4.5 Boundary Layers and Viscous Effects

An important concept to understand in order to predict aerodynamics is the boundary layer. As with other concepts in this chapter, we assume that the reader has some familiarity with fluid dynamics, and we will just review concepts here that will be required later in the book.

4.5.1 Boundary Layer Concepts

As we mentioned earlier, if there are no viscous effects in a flowfield, there is no friction or flow separation. If there is no friction and no separation, there is no drag, which is why D'Alembert had a paradoxical moment. He knew that bodies moving through a fluid had drag, but he did not understand the concept of viscosity or its effects. In fact, it would not be until the early 1900s that someone would provide an explanation for drag, and that was Ludwig Prandtl. Prandtl hypothesized that there was a thin layer of fluid near the surface of a body where viscous effects took place, and he called this the *boundary layer*.[8]

Figure 4.11 shows the development of a boundary layer over a flat plate. As the boundary layer initially begins to develop, it is in a laminar state. Laminar boundary layers are layered, steady, and thinner than turbulent boundary layers. At some point on the plate, the boundary layer will transition to a turbulent boundary layer. Transition takes place over a finite distance on the plate and is a little-understood fluid dynamic process. Transition can be influenced by surface roughness or imperfections, vibration, or excessive

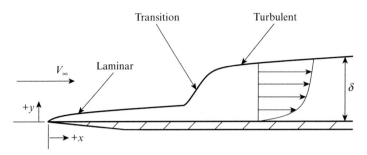

Figure 4.11   Development of a boundary layer over a flat plate.

turbulence in the freestream flow. Primarily, however, transition is a function of the *Reynolds number* along the plate, given by

$$Re_x = \frac{\rho_\infty V_\infty x}{\mu_\infty}. \tag{4.35}$$

Low speed flows transition at a Reynolds number of $Re_x \approx 500,000$, for typical freestream turbulence levels, and as long as other disturbances are not present. In fact, transition does not happen all at once, nor does it happen at a single location, but for engineering purposes the 500,000 value gives reasonable results.

4.5.1.1 LAMINAR BOUNDARY LAYERS

As stated previously, laminar boundary layers are thin, layered, and steady. For this reason a great deal of theoretical analysis of these boundary layers has been developed over the years. These theories have been shown to be very accurate in describing the basic physical features of laminar boundary layers, including their thickness, velocity profile, and surface shear stress for a zero pressure gradient flow. According to the Blasius solution, the boundary layer thickness, δ, is given by[3]

$$\delta = \frac{5.0x}{\sqrt{Re_x}} \tag{4.36}$$

and the velocity profile is approximately given by

$$\frac{u}{V_\infty} = \frac{3}{2}\left(\frac{y}{\delta}\right) - \frac{1}{2}\left(\frac{y}{\delta}\right)^3 \tag{4.37}$$

The shear stress at the wall is calculated using the *Newtonian fluid* relationship:

$$\tau = \left(\mu \frac{\partial u}{\partial y}\right)_{y=0}. \tag{4.38}$$

Finding the skin friction on a flat plate requires knowledge about this local shear stress acting at each point along the length of the plate. As with other aerodynamic properties, we will define a nondimensional version of the shear stress, known as the local skin friction coefficient, C_f, as

$$C_f \equiv \frac{\tau}{q_\infty} \qquad (4.39)$$

where the freestream dynamic pressure is defined as $q_\infty \equiv \rho_\infty V_\infty^2 / 2$. The local skin friction coefficient for laminar flow is given by

$$C_f = \frac{0.664}{\sqrt{Re_x}} \qquad (4.40)$$

which is known as the *Blasius formula*. In this formulation, x is the distance from the front of the plate to the point where the shear stress is being estimated, and we have assumed that the velocity at the edge of the boundary layer is essentially the same as the freestream velocity, which is a good assumption for flow over a flat plate. This relationship is very accurate and will do a good job of estimating laminar skin friction.

4.5.1.2 TURBULENT BOUNDARY LAYERS

Since turbulent boundary layers are highly unsteady, relationships for features such as thickness, velocity profile, and shear stress are necessarily time-averaged values. While this averaging may seem highly inaccurate, reasonable estimates are possible. The boundary-layer thickness, δ, is given by[3]

$$\delta = \frac{0.3747x}{\left(Re_x\right)^{0.2}}. \qquad (4.41)$$

Comparing the boundary-layer thickness for a turbulent boundary layer (Equation 4.41) with that for a laminar boundary layer (Equation 4.36) you can see that a turbulent boundary layer develops much faster and is much thicker than a laminar boundary layer.

The 1/7th power law velocity profile for a turbulent boundary layer is given by

$$\frac{u}{V_\infty} = \left(\frac{y}{\delta}\right)^{1/7}. \qquad (4.42)$$

However, the actual velocity profile in a turbulent boundary layer is much more complicated than Equation 4.42 shows. In fact, if you look at Fig. 4.12 you see that the turbulent boundary layer has several regions, each with a different velocity profile.

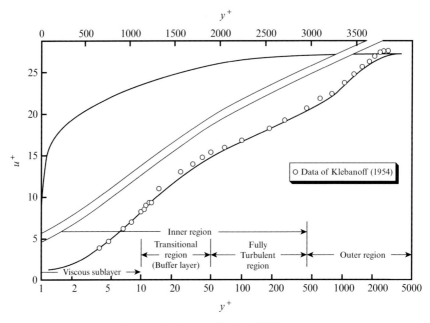

Figure 4.12 Velocity distribution for a turbulent boundary layer on a flat plate (Ref. 9).

Note: The velocity profile is shown twice: using a linear scale on the top half and a logarithmic scale on the bottom half.

Figure 4.12 uses "wall units" for velocity and distance above the flat plate, u^+ and y^+, which are defined in Chapter 8. y^+ is the nondimensional distance normal to the wall (and can be thought of as a local Reynolds number) and u^+ is the streamwise velocity nondimensionalized by the turbulent velocity and length scales, u_τ, known as the *friction velocity*. Since we are interested primarily in the shear stress at the surface, it is a little easier to estimate the velocity profile in the viscous sublayer, which is also based on a derivation by Blasius[3]

$$C_f = \frac{\tau}{q_\infty} = \frac{0.0583}{\left(Re_x\right)^{0.2}}. \tag{4.43}$$

This will provide us with usable estimates of the skin friction at the surface of the wall.Fig. 4.13

4.5.1.3 RELATIVE FEATURES OF BOUNDARY LAYERS

While it may seem obvious that a laminar boundary layer might be better from an aerodynamics perspective, that is not necessarily the case. You must understand two features of boundary layers that are essential to aerodynamics: skin friction and flow separation.

From a skin friction perspective, the important physical feature is the velocity gradient at the surface (as defined by Equation 4.38). The velocity profiles for a laminar and turbulent boundary layer are shown in . You can

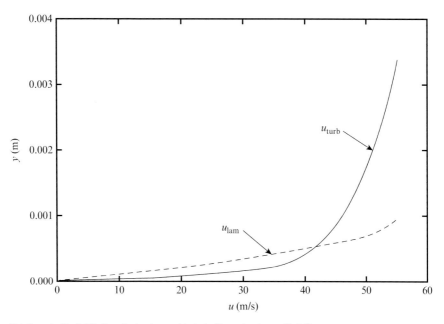

Figure 4.13 Relative velocity distributions for laminar and turbulent boundary layers (Ref. 3).

see that the laminar boundary layer actually has less skin friction than the turbulent boundary layer

$$\left(\frac{\partial u}{\partial y}\right)_{lam} < \left(\frac{\partial u}{\partial y}\right)_{turb} \qquad \tau_{lam} < \tau_{turb}. \qquad (4.44)$$

Therefore, from the perspective of skin friction, a laminar boundary layer is definitely better than a turbulent boundary layer. But that is not the whole story – what about flow separation?

Look again at Fig. 4.13, but now notice the fullness of the two velocity profiles. The turbulent boundary layer has higher velocities at closer distances to the surface. When a boundary layer encounters an adverse pressure gradient ($dp/dx > 0$, which is an increasing pressure), the momentum within the boundary layer is what overcomes the increasingly high pressures as the flow moves along the flat plate. For this reason, laminar boundary layers are more likely to separate away from the surface than turbulent boundary layers, and flow separation leads to an increase in pressure drag. A good example of laminar flow separation is the *laminar separation bubble* that forms near the leading edge of a low-speed airfoil (Fig. 4.14 shows a laminar separation bubble at the front of a blunt body). This type of separation is very detrimental to the lifting properties of an airfoil, and is to be avoided if possible. For this reason, turbulent boundary layers are actually better at overcoming adverse pressure gradients, and therefore have an advantage in keeping pressure drag low.

Figure 4.14   Laminar separation bubble on a blunt body (Ref. 5; Photograph by Henri Werlé © ONERA, the French Aerospace Lab).

So, which type of boundary layer should an airfoil have? It depends! The flow conditions, design goals, and overall requirements for the airfoil should determine whether a laminar or turbulent boundary layer is desired. There are usually no easy answers in aerodynamics. Airfoil design is a tricky balance between creating high levels of lift without allowing flow separation that produces high levels of drag.

4.5.2 Skin Friction Estimation

A convenient formulation for determining all of the skin friction on a flat plate is found by integrating the local skin friction coefficient, C_f, found in Equation 4.39, to obtain a "total" or "average" skin friction drag coefficient over one side of the flat plate.[10] The use of the total skin friction drag coefficient avoids performing the same integration numerous times with different flat plate lengths accounting for different results. The total skin friction coefficient is defined as

$$C_F \equiv \frac{D_f}{q_\infty S_{wet}} \tag{4.45}$$

where D_f is the friction drag on the plate and S_{wet} is the wetted area of the plate (the wetted area is the area of the plate in contact with the fluid; for one side of the plate, $S_{wet} = Lb$). The total skin friction coefficient for laminar boundary layers is obtained by taking the local skin friction relation in Equation 4.40 and integrating

$$C_F = \frac{b}{q_\infty S_{wet}} \int_0^L \tau dx = \frac{b}{q_\infty Lb} \int_0^L C_f(x) q_\infty dx = \frac{1}{L} \int_0^L \frac{0.664}{\sqrt{Re_x}} dx = 2C_f(L) \quad (4.46)$$

which is just twice the value of the local skin friction coefficient evaluated at $x = L$. The total skin friction coefficient for laminar flow simply becomes

$$C_F = \frac{1.328}{\sqrt{Re_L}} \quad (4.47)$$

where Re_L is the Reynolds number evaluated at $x = L$, the end of the flat plate.

Since drag coefficients are normally nondimensionalized by a reference area rather than a wetted area, the drag coefficient due to skin friction is obtained from Equation 4.47 as

$$C_D \equiv \frac{D_f}{q_\infty S_{ref}} = C_F \frac{S_{wet}}{S_{ref}}. \quad (4.48)$$

It can be tempting to add together the total skin friction coefficients for various flat plates to obtain a total skin friction drag – this must never be done! Since each total skin friction coefficient is defined with a different wetted area, doing this would result in an incorrect result. In other words:

$$C_{F_{total}} \neq \sum_{i=1}^N C_{F_i}. \quad (4.49)$$

Always convert total skin friction coefficients into drag coefficients (based on a single reference area) and then add the drag coefficients to obtain a total skin friction drag coefficient:

$$C_D = \sum_{i=1}^N C_{D_i}. \quad (4.50)$$

A total skin friction coefficient can be found for turbulent flow by integrating Equation 4.43 over the length of a flat plate:

$$C_F = \frac{1}{L} \int_0^L C_f(x) dx = \frac{1}{L} \int_0^L \frac{0.0583}{(Re_x)^{0.2}} dx$$

$$C_F = \frac{0.074}{(Re_L)^{0.2}}. \quad (4.51)$$

Table 4.2 Empirical constants for transition
correction in Eqn. 4.53 (Ref. 11)

$Re_{x,tr}$	A
300,000	1050
500,000	1700
1,000,000	3300
3,000,000	8700

This formula, known as the *Prandtl formula*, is a theoretical representation of the turbulent skin friction drag. However, when compared with experimental data, it is found to be only ±25% accurate. A semi-empirical turbulent skin friction coefficient relation, which is more accurate than the Prandtl formula, is given by the Prandtl-Schlichting formula – this formula is ±3% accurate:[10]

$$C_F \equiv \frac{0.455}{\left(\log_{10} Re_L\right)^{2.58}} \tag{4.52}$$

A straightforward approach to model transitional flow is to use an empirical correction to the Prandtl-Schlichting turbulent skin friction relation, Equation 4.52:[11]

$$C_F \equiv \frac{0.455}{\left(\log_{10} Re_L\right)^{2.58}} - \frac{A}{Re_L} \tag{4.53}$$

where the correction term reduces the skin friction since laminar boundary layers produce less skin friction than turbulent boundary layers. The experimentally determined constant, A, varies depending on the transition Reynolds number, as shown in Table 4.2. The value $A = 1700$ represents the laminar correction for a transition Reynolds number of $Re_{x,tr} = 500,000$. You can see from this formulation that, if the Reynolds number at the end of the plate is very high, then the laminar correction term plays a fairly insignificant role in the total skin friction drag on the plate. A good rule of thumb is to assume that, if transition takes place at less than 10 percent of the length of the plate, then the laminar correction can usually be ignored, since it is relatively small.

Figure 4.15 shows how the total skin friction coefficient varies from the laminar value, Equation 4.47, through transition, and finally to the fully turbulent value, Equation 4.52.[13]

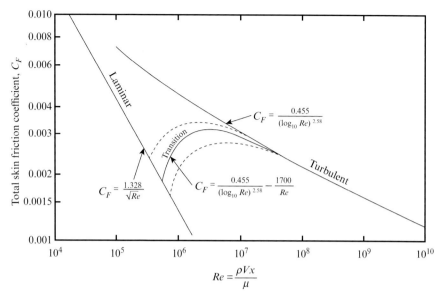

Figure 4.15 Variation of total skin friction coefficient with Reynolds number for a smooth, flat plate (note that both scales are logarithmic) (Ref. 12).

Profiles in Computational Aerodynamics: Kerstin Huber

I was born in Tübingen, a small town located in the southern part of Germany. As a child I spent my time riding bikes and exploring nature with friends. During my childhood I was always exposed to engineering sciences, since my dad was a hardware engineer and my mum was an X-ray technician; I also liked traveling and learning about other cultures and wanted to learn a second language fluently. When I was in eleventh grade I decided to spend a year at college in England, so in 2002 I started studying at Lancing College in West Sussex. After I had spent a year at college it was time for me to return to Germany, but I really wanted to stay in England and finish off my A-Levels (England's university entrance exams), and I am grateful to my parents for making this wish possible for me.

When it came to deciding what I wanted to study at university, I recalled motivational memories from my childhood, especially of Selma Lagerlöf's story about a small boy, Nils Holgersson, who discovered the world by riding on a goose, which captivated me to flying. In 2004, I started a Bachelors Degree in Aerospace Engineering at the University of Surrey in England. During the first two years of my study I was working as a volunteer at the Brooklands Museum, where I spent most of my spare weekends helping to restore an old British Airways Concorde. This activity not only taught me a lot about the theory behind building aircraft, but it also enhanced my practical skills and taught me about the history of flight. In my third year I carried out a 14-month internship at AIRBUS Industries in Toulouse, France, working at Customer Service and at the A380 final assembly line. After returning to Surrey, I started writing my bachelor thesis on turbulent drag reduction using a zero-mass-flux flow control device. This work led me in the direction of aerodynamics, fluid mechanics, and wind tunnel testing, and I finished my Bachelor of Engineering degree in 2008.

As I wanted to continue to my studies by pursuing a master's degree, but at the same time discover Europe a little more, I applied to the Royal Institute of Technology (KTH) in Stockholm, Sweden and started my graduate studies in fall 2008. Sweden is a beautiful country, although after battling through two very cold and dark winters, I longed for a return back home to Germany.

Under the supervision of my tutor, Dipl.-Ing. Andreas Schuette, I started writing my master's thesis at the Institute of Aerodynamics and Flow Technology at the German Aerospace Center (DLR) in Braunschweig. This work dealt with the investigation of the aerodynamic properties of a flying wing configuration with a highly swept leading edge and varying leading edge radius. During this time I not only had the opportunity to use the numerical and computational tools, such as the DLR CFD code, TAU, in conjunction with the high-performance computing cluster at $C^2A^2S^2E$, but also had the chance to participate in various wind tunnel experiments in low- and high-speed facilities. These six months really taught me first-hand how important it is to couple CFD solutions with experimental findings. I was offered a job at DLR immediately after I had finished my master's degree, and am currently continuing to research the aerodynamic behavior of flying wing configurations. The flowfield around such configurations is dominated by complex vortex structures and strong vortex-to-vortex interactions. In order to verify the CFD results, traditional experimental methods, such as force and moment measurements, are used. However, more complex experimental methods, such as Particle Image Velocimetry (PIV) and Pressure Sensitive Paint (PSP) studies, are also applied to gather more detailed information about the flowfield. As part of this research I work closely with researchers from around the world.

From the start of my studies up till now I discovered that there are still only a few women working in engineering sciences. During my future work, I hope that I will be able to encourage and support other young women to choose a career in engineering.

4.6 Airfoil Aerodynamics

The most important aerodynamic feature on an airplane is the wing, and wings are made up primarily of airfoil sections. Understanding these aerodynamic devices is essential to gaining knowledge about how airplanes fly. Also of great importance is having an understanding of how various control surfaces and high-lift devices work, which will lead to a basic understanding of viscous effects and the complexities of aerodynamics.

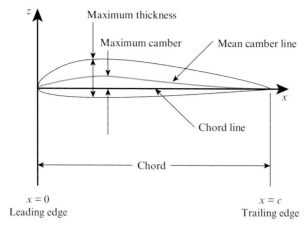

Figure 4.16 Airfoil terminology (Ref. 13).

4.6.1 Airfoil Terminology

Airfoil geometry definition uses a standard terminology to describe the shape, largely based on what we know about how airfoils work, as well as which geometric parameters are important to airfoil aerodynamics. Obviously, an airfoil could take an infinite number of shapes, so no terminology system is perfect, but we shall use the common definitions. A typical airfoil is shown in Fig. 4.16, and the geometric definitions used to describe the airfoil are:

- Leading edge (LE): the most forward point of the airfoil ($x=0$, $x/c=0$)
- Trailing edge (TE): the most aft point of the airfoil ($x=c$, $x/c=1$)
- Chordline: typically a straight line connecting leading edge and trailing edge
- Chord, c: length of chordline from leading edge to trailing edge (not all airfoil designers use this definition, however, so care should be taken!)
- Mean camber line: line connecting points halfway between upper and lower airfoil surfaces (same as chordline for a symmetric airfoil)
- Maximum camber, h: maximum $\perp$ distance between chordline and mean camber line (usually represented as h/c)
- Maximum thickness, t: maximum $\perp$ distance between upper and lower airfoil surfaces (usually represented as t/c)

4.6.2 Forces and Moments on an Airfoil

Another aspect of airfoils that is important to know is how they behave in flight. In order to understand this, you first need to understand the concept of angle of attack. As shown in Fig. 4.17, an airfoil is shown as it travels through a body of air. In order to deal with this, we usually assume that an airfoil traveling to the left can be represented by a stationary airfoil with air

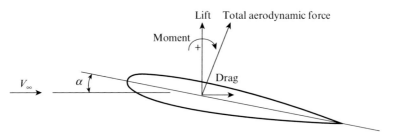

Definition of angle of attack and aerodynamic forces and moment on an airfoil (Ref. 13).

flowing over it from the left. The air is traveling at the speed, V_∞, which represents the *freestream* velocity (the air flow far from the airfoil and out of its influence). The angle between the freestream velocity and the chordline of the airfoil is called the angle of attack, α. It is often confusing when first seeing the concept of angle of attack, since either the chordline or the velocity is usually shown as a horizontal line, as it is in Fig. 4.3. This gives the impression that the angle of attack is relative to horizontal, when in fact the angle of attack has no connection to horizontal, but rather is relative to the flight path angle. An airfoil moving vertically could be at a small angle of attack, just as an airfoil traveling horizontally could be at a large angle of attack. The essential feature is the relation between the airfoil and the freestream flow direction, not between the airfoil and the horizontal direction.

As the air flows over the airfoil, it creates a combination of normal stresses (*pressure*) and tangential stresses (*friction*) on the surface of the airfoil. If these stresses are integrated over the surface of the airfoil they create a total aerodynamic force (as shown in Fig. 4.17), which acts at some location on the airfoil (called the *center of pressure*). There are several choices for an axis system to resolve this total force, the two most common being an axis relative to the chordline (called a body axis) or an axis relative to V_∞ (called a wind axis). Airplane aerodynamics has traditionally used a wind axis system to resolve the total aerodynamic force, leading to two forces acting on the airfoil: *lift* and *drag*.

Lift is the force perpendicular to V_∞, while drag is the force parallel to V_∞. A moment could also exist if the total aerodynamic force is resolved about any location on the airfoil other than the center of pressure; the moment is known as the *pitching moment*. The sign convention for the pitching moment is positive for a moment that would rotate the leading edge up (or clockwise, as shown in Fig. 4.17). If the total aerodynamic force is assumed to act at the center of pressure, the pitching moment is zero. Since the pitching moment could be resolved about any point on the airfoil, the value of the moment depends on the location where it is resolved, an important consideration for the stability of an airplane.

Another important point on the airfoil is the *aerodynamic center*. This is a fixed point (does not vary with angle of attack) where the total aerodynamic force could be assumed to act such that the pitching moment does not vary with angle of attack. For most low-speed airfoils the aerodynamic center is at the quarter chord of the airfoil, $c/4$. For a symmetric airfoil, the center of pressure and the aerodynamic center are both located at $c/4$. For airfoils with camber (the mean camber line is not a straight line), the center of pressure is not the same as the aerodynamic center, and the center of pressure varies with angle of attack.

Flow Visualization Box

NACA 0012 Airfoil Pressures

Using FieldView, look at the flow solutions for the NACA 0012 airfoil at $\alpha = 4$, 10, and 16 degrees. Specifically, look at the pressure coefficient and be sure to include a legend that defines the scale and colors being used. Also be sure to use the same scale for all three solutions so that you can compare the results with each other.

What do you notice about the three solutions? In general, they all have lower pressures acting over the upper surface of the airfoil and higher pressures acting over the lower surface, which shows the lift on the airfoil. However, as you progress from 4 to 10 degrees you see that the low-pressure region (the green on the graph) is much larger at the high angle of attack. When you reach 16 degrees there is a significantly lower pressure coefficient near the leading edge, but a smaller region of lower pressure behind that location: why? We will look at that a little later in the chapter.

$\alpha = 4$ degrees

$\alpha = 10$ degrees

$\alpha = 16$ degrees

4.6.3 Airfoil Aerodynamic Coefficients

Since the forces and moments on an airfoil depend on a number of factors (such as the airfoil chord, angle of attack, density of the air, viscosity, and freestream velocity), it would be difficult to compare results between airfoils, as they would all have been tested under different conditions. To overcome this, aerodynamicists use nondimensional aerodynamic coefficients. These coefficients are defined for each of the forces and the moment shown in Fig. 4.17: lift, drag, and pitching moment. Since an airfoil is a two-dimensional shape, the forces we are talking about are actually forces per unit length (or span) of the wing, so airfoil *lift coefficient, drag coefficient,* and *pitching moment coefficient* are defined as:

$$C_l \equiv \frac{l}{q_\infty c} \quad C_d \equiv \frac{d}{q_\infty c} \quad C_m \equiv \frac{m}{q_\infty c^2}. \tag{4.54}$$

These definitions allow the aerodynamic coefficients to be functions of angle of attack, Reynolds number, and Mach number, but not directly a function of the size of the airfoil or the altitude or velocity of the airfoil. The Reynolds number and *Mach number* are nondimensional numbers that are important in understanding how airfoils work, and are defined as:

$$Re \equiv \frac{\rho V l}{\mu} \quad M \equiv \frac{V}{a} \tag{4.55}$$

where *l* is a characteristic length for the flow (taken as the chord for airfoils and wings) and *a* is the speed of sound of the air. Two flows with the same Mach and Reynolds numbers are considered to be dynamically similar.

4.6.4 Airfoil Lift and Drag Variations

As an airfoil flies at different angles to the relative wind (the angle of attack), the lift varies in a fairly predictable fashion, as shown in Fig. 4.18. As the angle of attack increases, the flow moves faster over the top of the airfoil, causing even lower pressure on the top relative to the bottom. This creates more lift; therefore, as α increases, lift increases, as long as viscous effects do not dominate the flowfield. This portion of the lift curve is known as the linear portion, since there is a straight-line relation between lift and angle of attack. Since the relationship is linear, it can be properly defined by an intercept and a slope, labeled in Fig. 4.18. These factors are defined as:

1. *Zero lift angle of attack,* $\alpha_{l=0}$: the angle of attack where the airfoil produces no lift ($C_l - 0$)
2. *Lift curve slope,* a_0: the change in lift coefficient as angle of attack increases

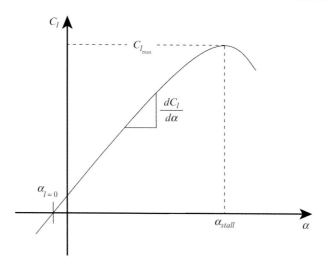

Figure 4.18 Lift coefficient variation with angle of attack for a typical airfoil (Ref. 13).

$$a_0 = C_{l_\alpha} = \frac{dC_l}{d\alpha} \approx 2\pi \; 1/\text{rad} = 0.106 \; 1/\text{deg for a two-dimensional airfoil.}$$

Eventually, as the angle of attack continues to increase, viscous effects begin to play an important role on the airfoil. The adverse pressure gradient over the upper surface of the airfoil increases to such a level that the boundary layer separates. When separation takes place over an appreciable portion of the airfoil, the airfoil is said to have undergone *stall*. Two more important factors about the lift curve now are defined:

3. *Maximum lift coefficient*, $C_{l_{max}}$: the highest lift coefficient the airfoil can produce prior to full stall
4. *Stall angle of attack*, α_{stall}: the angle of attack corresponding to $C_{l_{max}}$, the highest angle of attack the airfoil can sustain prior to full stall

These factors are used to relate the performance of one airfoil to another. While it might seem that the easiest way to increase lift would be to increase the lift-curve slope, in fact lift-curve slope is very insensitive to geometric changes on the airfoil. While changing thickness does alter the lift-curve slope slightly, it has more of an impact on $C_{l_{max}}$. Changes in camber have a large impact on $\alpha_{l=0}$, where positive camber causes $\alpha_{l=0}$ to shift to the left, which creates higher lift on the airfoil for any angle of attack in the linear range. Finally, leading-edge radius has a substantial impact on the type of stall experienced by the airfoil, with large radii (very round) leading edges producing mild stall characteristics and low radii (sharper) leading edges having more abrupt stall.

A typical drag variation for a two-dimensional airfoil is shown in Fig. 4.19. The drag variation is not linear but forms what is known as a *drag polar*. This

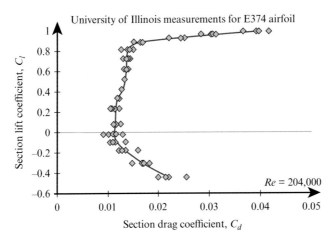

Figure 4.19 Drag polar for the Eppler 374 airfoil at a Reynolds number of 204,000 (Ref. 14).

nonlinear variation of drag with lift is caused by separation of the boundary layer from the airfoil upper surface, and, as the lift is increased, the drag increases approximately as the square of the lift. This is one of the features of aerodynamics that make design of optimum aircraft challenging. The minimum value of drag is given the symbol C_{d_0}. This level of drag is primarily due to friction acting on the upper and lower surfaces of the airfoil, since there is very little flow separation at lower angles of attack. When the airfoil is symmetric, C_{d_0} takes place at $C_l = 0$, and, when the airfoil has positive camber, C_{d_0} takes place at $C_l > 0$.

Computational Aerodynamics Concept Box

How Airfoils Work

Airfoils work by creating a pressure difference (or a velocity difference, ΔV) between the upper and lower surfaces. The lift of the airfoil is proportional to the integral of that pressure difference:

$$C_l \approx \int_0^c \Delta V dx$$

This is typically accomplished by having a round leading edge followed by a surface with variable curvature ending in a sharp trailing edge. The shape causes the flow to turn, aided by the fact that air (and other fluids) tend to stay attached to a solid surface, which is known as the Coanda effect. The following two figures show the velocity difference between the upper and lower surfaces, so that the gray shaded area is proportional to the lift. Since the airfoil has camber, it produces a small amount of lift even at zero degrees angle of attack (left figure). As the angle of attack increases to 6 degrees (right figure), the velocity difference (and thus the pressure difference) also increases, producing more lift.

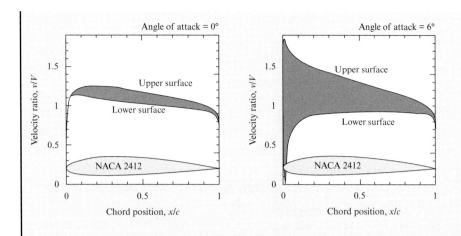

But how high an angle of attack can still produce increasing lift? The limiting factor is usually boundary-layer separation, which is strongly influenced by adverse pressure gradients. Regions where the pressure gradient is increasing (an adverse pressure gradient) are also regions where the velocity is decreasing. On the right-hand figure, you can see regions of large adverse pressure gradient immediately after the velocity peak (also called the "suction" peak) and near the trailing edge (because the trailing edge is a stagnation point). Eventually, the boundary layer will not be able to overcome these adverse pressure gradients and will separate, leading to stall and loss of lift. So airfoil design is a delicate balance between creating high velocities (in order to create lift) without creating flow separation (which leads to the loss of lift). (The information contained in this Concept Box is based on the excellent book by Richard Eppler, "Airfoil Design and Data," Berlin: Springer, 1990.)

4.6.5 NACA Airfoil Families

The National Advisory Committee for Aeronautics (N.A.C.A., but we will abbreviate it as NACA from now on), which became NASA in 1958, developed and tested several families of airfoil shapes from the 1930s to the 1950s.[2] One of the first sets of airfoils tested was known as the four-digit series family, which are represented by a four-digit code that described the airfoil shapes. Each airfoil had a name like NACA $XYZZ$, where the digits represented:

- X – maximum camber in percent chord (maximum h/c)
- Y – location of maximum camber along chordline (from leading edge) in tenths of chord
- ZZ – maximum thickness in percent chord (maximum t/c)

See Appendix A for more details about airfoil family definitions. The other airfoil families, such as the five-digit series and the six-digit series, have more complicated geometric definitions, as the airfoils became more advanced over time. Many general aviation aircraft use six-digit series airfoils, and many of these more advanced airfoils can be found in the very useful book by Abbott and von Doenhoff.[15]

4.6.6 How to Use NACA Airfoil Data

NACA airfoil data are presented in a uniform style that can be quite confusing to understand at first. The data charts assume that the user is familiar with airfoil nomenclature and the presentation of airfoil data, so there is very little extra information to help explain what you are seeing. Most of the airfoils were tested at various Reynolds numbers (usually $Re_c = 3.0 \times 10^6$, 6.0×10^6, and 9.0×10^6), and some of the airfoils have results for flap deflections as well. The results are mostly for airfoils with aerodynamically smooth surfaces, but often one set of data is presented for the airfoil with standard surface roughness to show the impact of roughness on the results. Figure 4.20 shows the standard presentation format for airfoils that is used in Ref. 15. There are two sets of graphs for each airfoil, and each graph contains two sets of data. The legend for the graphs is at the bottom of the right-hand graph, but it pertains to the left-hand graph as well. The left-hand graph contains variations with angle of attack for lift coefficient, C_l, and pitching moment coefficient about the quarter chord, $C_{m_{c/4}}$. The right-hand graph

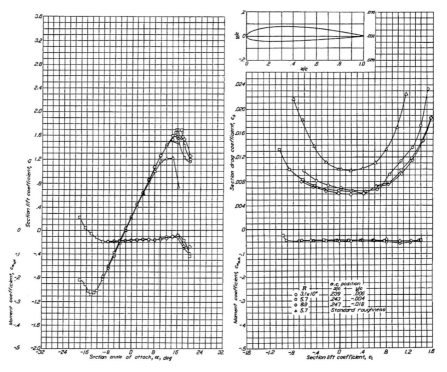

Figure 4.20   Airfoil data for the NACA 2412 airfoil (Ref. 15).

shows variations with the lift coefficient for drag coefficient, C_d, and pitching moment coefficient about the aerodynamic center, $C_{m_{ac}}$. The airfoil shape is also shown at the top of the right-hand graph. Care should be taken when using the graphs, since each coefficient has its own vertical scale, and, in the case of the left-hand graph, these scales overlap.

Computational Aerodynamics Concept Box

How to Use Airfoil Data

Here is how to find the lift, drag, and pitching moment about the aerodynamic center for a smooth NACA 2412 airfoil with a chord of $4\,ft$ flying at an angle of attack of $\alpha = 12°$. Assume the airfoil is flying at $V_\infty = 122\,ft/s$ at sea level standard day conditions.

First calculate the Reynolds number. For standard day sea level conditions:

$$\rho_\infty = 0.002377\,slug/ft^3 \quad \mu_\infty = 3.737 \times 10^{-7}\,slug/ft-s$$

and the Reynolds number is:

$$Re_c = \frac{\rho_\infty V_\infty c}{\mu_\infty} = \frac{(.002377\,slug/ft^3)(122\,ft/s)(4\,ft)}{3.737 \times 10^{-7}\,slug/ft-s} = 3.10 \times 10^6.$$

Now check the legend to find the symbols that correspond to a smooth airfoil at $Re_c = 3.1 \times 10^6$, which are the circles. At $\alpha = 12°$, the lift coefficient for $Re_c = 3.1 \times 10^6$ is approximately $C_l = 1.4$. Now use the right-hand graph to find the drag coefficient and the pitching moment for $C_l = 1.4$, which is approximately $C_d = 0.0173$ and $C_{m_{ac}} = -0.05$. Now use the definitions of the coefficients from Equations 4.54 to find the lift, drag, and moment per unit span of the airfoil.

$$l = C_l q_\infty c \quad d = C_d q_\infty c \quad m_{ac} = C_m q_\infty c^2.$$

All of these forces and moments require knowing the dynamic pressure, which for these conditions is:

$$q_\infty = \rho_\infty V_\infty^2 / 2 = (0.02377\,slug/ft^3)(122\,ft/s)^2 / 2 = 17.69\,lb/ft^2$$

and the forces and moments are:

$$l = C_l q_\infty c = (1.4)(17.69\,lb/ft^2)(4\,ft) = 99.1\ lb\,\text{per foot of span}$$

$$d = C_d q_\infty c = (0.0173)(17.69\,lb/ft^2)(4\,ft) = 1.22\ lb\,\text{per foot of span}$$

$$m_{ac} = C_m q_\infty c^2 = (-0.05)(17.69\,lb/ft^2)(4\,ft)^2 = -14.15\ ft\text{-}lb\,\text{per foot of span}$$

4.6.7 Factors That Affect Airfoil Aerodynamics

A variety of factors can change the airfoil aerodynamics. Some of these factors are directly visible in the experimental data and others will become clearer. The factors that we will discuss are Reynolds number, camber, and thickness effects, as they are the most important.

4.6.7.1 REYNOLDS NUMBER

Higher Reynolds numbers produce thinner boundary layers, which are better able to overcome adverse pressure gradients, and therefore delay separation. Figure 4.21 shows the impact of changing Reynolds number on the lift and drag coefficients (which can also be seen in the airfoil data in Fig. 4.20). The linear portion of the lift curve is not changed by Reynolds number (since the flow is largely inviscid in nature), but the stall regions for the two Reynolds numbers are very different. The delay in separation at the higher Reynolds number leads to higher values of the maximum lift coefficient and the stall angle of attack. Both of these effects are beneficial, since the airfoil will be able to produce more lift and fly slower than at the lower Reynolds number. The drag coefficient is also changed as the Reynolds number increases, since the delayed separation reduces pressure drag. The zero-lift drag coefficient stays approximately the same, and the drag curve flattens out at both positive and negative values of lift coefficient.

4.6.7.2 CAMBER

Camber has an entirely different impact on the lift coefficient than does Reynolds number. Rather than extending the lift curve to higher angles of attack, positive camber shifts the lift curve up and to the left (see Fig. 4.22). This yields a lower angle of attack for zero lift, a higher maximum lift

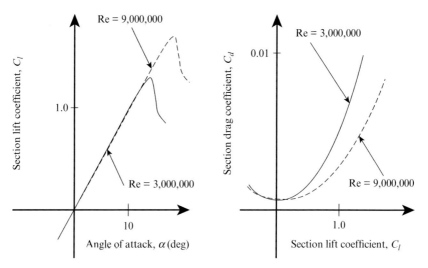

Figure 4.21   Impact of Reynolds number on lift and drag coefficients (Ref. 13; reprinted by permission of the American Institute of Aeronautics and Astronautics, Inc.).

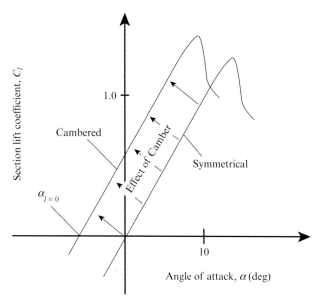

Figure 4.22 Impact of camber on lift coefficient (Ref. 13; reprinted by permission of the American Institute of Aeronautics and Astronautics, Inc.).

coefficient, a lower stall angle of attack, and the same lift-curve slope. The camber produces higher velocities over the upper surface of the airfoil when compared with the lower surface and causes the airfoil to produce lift at $\alpha = 0°$. This is why the lift curve is essentially just shifted. One of the characteristics of camber is the decrease in the stall angle of attack, since the cambered airfoil could stall at significantly lower angles of attack, although at a higher lift coefficient. This could be a real safety issue for an airplane.

4.6.7.3 THICKNESS

While thickness has a small impact on the lift-curve slope of airfoils, the primary impact of thickness is on the maximum lift coefficient. Figure 4.23 shows the variation of maximum lift coefficient with varying thickness for various NACA four-digit airfoils. As thickness is increased to about 12%, the maximum lift coefficient increases. After a thickness ratio of 12%, there is a dramatic reduction in maximum lift coefficient, since the thicker airfoil is unable to maintain attached flow at the same angles of attack as thinner airfoils. Notice that the impact is most dramatic at the highest Reynolds numbers, where increasing thickness past 12 percent can reduce the maximum lift coefficient by approximately 10 percent.

4.6.8 How Airfoils Work

Airfoils are very complicated lift-producing devices that require a great deal of experience and knowledge to design (see the previous Concept Box on airfoils). Since the goal of an airfoil is to produce lift, we will again look at how

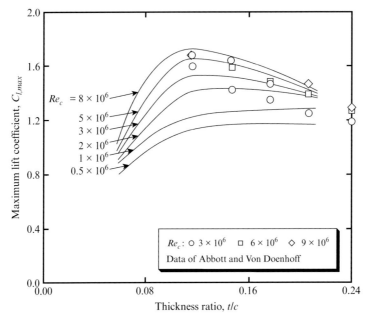

Figure 4.23 Impact of thickness on maximum lift coefficient (Ref. 3; Data courtesy of NASA.).

lift relates to the velocity difference between the upper and lower surfaces of the airfoil, ΔV:

$$C_l \approx \int_0^c \Delta V dx \qquad (4.56)$$

The purpose of the shape of an airfoil is to increase the velocity over the upper surface compared with the lower surface. There are three basic ways to do this: increase the angle of attack, add camber to the airfoil, or add thickness to the airfoil. Each of these methods will yield higher velocities over the upper surface of the airfoil, especially near the leading edge. However, adding thickness alone does not aid in producing lift without increasing the angle of attack or adding camber. Since the upper surface has higher velocities, Bernoulli's equation will yield lower pressures on the upper surface when compared with the lower surface, and, when the pressures are integrated over the two surfaces, there will be a net force upward, which is lift. Most airfoils produce a very high velocity peak near the leading edge, which corresponds to a very low pressure, which is often called the *suction peak*.[16]

Unfortunately, increasing the velocity over the upper surface is not the whole story of how an airfoil works. Since boundary layers are strongly affected by the pressure gradient, dp/dx, increasing the angle of attack too much could cause the flow to separate, and thus decrease the lift. Due to Bernoulli's equation, the pressure gradient is inversely related to the velocity gradient, which gives two important types of pressure gradients:

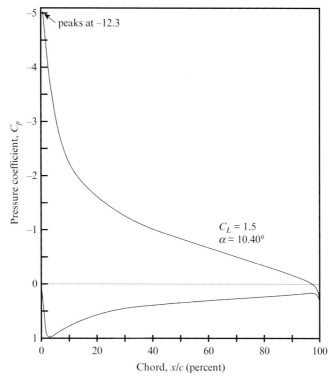

peaks at –12.3

$C_L = 1.5$
$\alpha = 10.40°$

Pressure coefficient, C_p

Chord, x/c (percent)

Figure 4.24   Pressure coefficient variation for the Clark-Y-8 airfoil (Ref. 17).

- favorable pressure gradient, where $dV/dx > 0$, $dp/dx < 0$
- adverse pressure gradient, where $dV/dx < 0$, $dp/dx > 0$

Since a strong adverse pressure gradient usually exists near the trailing edge of the airfoil, the trailing edge is often the location where separation begins. There is also a mild adverse pressure gradient over the upper surface of the airfoil and a mild favorable pressure gradient usually over the lower surface of the airfoil (see Fig. 4.24). Thicker airfoils have higher maximum velocities and stronger adverse pressure gradients, while very blunt leading-edge airfoils have lower differential velocities. Note that a thicker airfoil has a larger leading-edge radius which reduces the severity of the suction peak, helping to reduce the possibility of flow separation near the leading edge. The differential velocity is nearly independent of thickness except near the leading edge (the "flat plate distribution"), and camber shifts the location of maximum velocity and the local lift toward the trailing edge.[3] All of these effects directly affect how well any airfoil works, since the airfoil not only has to create high velocities over the upper surface, the airfoil also cannot create too much lift, since that creates higher adverse pressure gradients and leads to separation and stall.

Airfoil data are usually plotted in terms of the pressure coefficient, which was defined in Equation 4.19. If a velocity graph were converted into a

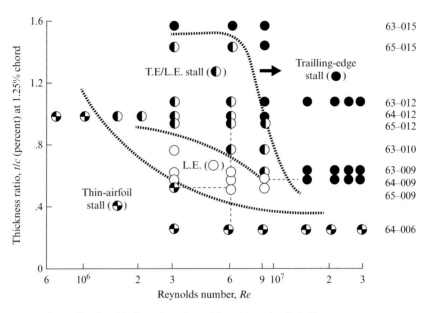

Stall boundaries as a function of leading-edge radius and Reynolds number (Ref. 18).

pressure coefficient graph, it would look the same but would be inverted (due to the inverse relation between velocity and pressure for incompressible flow). If this were then graphed, the upper surface pressures would be on the bottom of the plot and the lower surface pressures would be on the top of the plot. Over the years, aerodynamicists have decided that the confusion caused by this situation is best alleviated by plotting $-C_p$ on the top of the vertical axis (upper surface pressures on the top) and $+C_p$ on the bottom of the vertical axis (lower surface pressures on the bottom). This is illustrated in Fig. 4.24. Regions with strong adverse pressure gradients are easy to visualize, since they are portions of the curve with negative slopes. The more negative the slope, the more adverse the pressure gradient. Here we can see why the velocity peak near the leading edge, while good for producing lift, is also a disadvantage, since it creates a strong adverse pressure gradient. There is also a strong adverse pressure gradient near the trailing edge since the trailing edge is a stagnation point (the Kutta condition). This shows the challenge of designing airfoils: airfoils are designed to produce lift, but if they produce too much lift too quickly, then separation will take place, which leads to a loss of lift and increased drag.

A comprehensive description of stall boundaries for NACA symmetric airfoils is shown in Fig. 4.25.[18] Very thin airfoils with almost no leading-edge radius (like a flat plate) exhibit thin-airfoil stall, which takes place at the leading edge at very low angles of attack. Airfoils with larger leading-edge radii (but still relatively small amounts of camber) will exhibit leading-edge stall, where the flow suddenly separates from the leading edge and stall is abrupt. Thicker airfoils with large leading-edge radius tend to stall at the trailing edge and stall is often benign. Airfoils with "in between" leading-edge radii (depending on the Reynolds number) can also exhibit both leading- and trailing-edge stall at the same time (see combined region in Fig. 4.25). Needless to say, there is a great deal to understand about airfoil stall; readers wanting to know more about airfoil design are encouraged to read the informative book by Eppler[16] or the high lift paper by A.M.O. Smith.[1]

Flow Visualization Box

NACA 0012 Airfoil Stall

Using the *Coordinate Surface* window in FieldView, create velocity vectors on the upper surface for the NACA 0012 airfoil at α = 4, 10, and 16 degrees. You will have to scale the vectors (use *Vectors, Options*) as shown here. Maintain the scalar coloring using *Cp* for comparison to our previous results.

Be sure to take advantage of FieldView's ability to save a Restart File that can recreate a picture you have developed. Once you have the results you like, you can load the other solutions by using *File, Data Input* and then load the *Restart File* using *Open Restart, Complete No Data Read.*

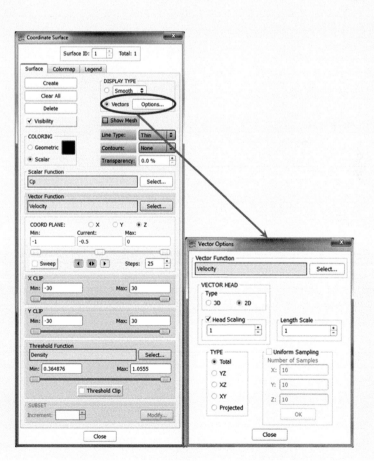

When you finish, your three results should look similar to the pictures shown here.

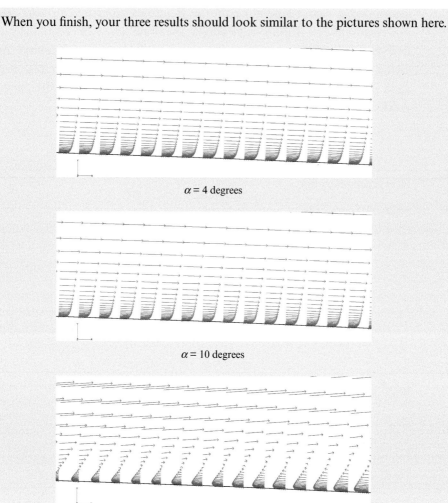

$\alpha = 4$ degrees

$\alpha = 10$ degrees

$\alpha = 16$ degrees

What do you notice about the three solutions? The solutions at 4 and 10 degrees show attached turbulent boundary layers along the upper surface of the airfoil (you can scan toward the trailing edge to see if attached flow is maintained everywhere on the upper surface). When you reach 16 degrees, there is reversed flow near the surface (vectors are pointing upstream) which is a sign that the flow has separated and the airfoil is stalled. Compare these results with the pressure plots in the previous Flow Visualization Box and notice that the pressures didn't necessarily make it obvious that the flow had separated. Also compare these results with the typical lift curve for an airfoil and see why the lift stops increasing once you reach the stall region.

4.6.9 Thin Airfoil Theory

Now that we have a basic understanding about how airfoils work and what the aerodynamics for an airfoil are, we will take a look at a basic theoretical development of the aerodynamics for "thin" airfoils. Most subsonic

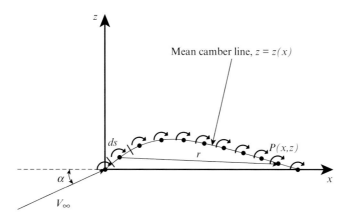

Figure 4.26   Vortex sheet simulating an airfoil.

airfoils used today are less than 15 percent thick, and many are 12 percent thick or less, so the assumption that the airfoil is thin is not very restrictive.

We will simulate a thin airfoil by a continuous sheet of vortices placed along the camber line, as shown in Fig. 4.26. In order to properly model the flow, we need to use an infinite number of vortex filaments side by side where the strength of each filament is infinitesimally small. Once we have established the position of each vortex, we have to find the strength as well. This is done by applying the surface boundary condition for inviscid flow across the airfoil: no flow can cross the mean camber line. This means that the total velocity induced at any point on the airfoil (due to the other vortices and the freestream component) must be tangential to the surface all along the length of the airfoil surface.

Each vortex located along a differential arc length, ds, induces a velocity at some other point on the airfoil, $P(x,z)$ given in Table 4.1.

$$dV = -\frac{\gamma(s)ds}{2\pi r} \tag{4.57}$$

The total circulation for the airfoil can be found by integrating the vortex strength distribution along the chord of the airfoil (assuming that the airfoil is thin and that all angles are small):

$$\Gamma = \int_0^c \gamma(s)ds. \tag{4.58}$$

The local vortex strength can be shown to be equal to the difference in velocity above and below the airfoil,[2,3]

$$\gamma(s) = V_{upper} - V_{lower}, \tag{4.59}$$

which is an important relation for understanding lift, since circulation is the turning of the flow, and lift is directly related to the circulation. The Kutta condition also requires that the vortex strength at the trailing edge yield a stagnation point, or

$$\gamma(T.E.) = 0. \tag{4.60}$$

The fundamental equation of thin airfoil theory is obtained when the boundary condition for the airfoil is enforced, namely that the mean camber line must be a streamline. This requires that there is no flow across the mean camber line, which is the inviscid slip boundary condition. When the influence of all of the vortices is taken into account at $P(x,z)$, the velocity normal to the mean camber line must be equal and opposite to the normal component of the freestream velocity component at the same point. The resulting equation is

$$\frac{1}{2\pi}\int_0^c \frac{\gamma(\xi)d\xi}{x-\xi} = V_\infty\left(\alpha - \frac{dz}{dx}\right). \tag{4.61}$$

At this point, it is important to remember the assumptions that went into the development of thin airfoil theory, namely, inviscid, incompressible, and irrotational flow of thin airfoils at low angles of attack. The last two restrictions are the result of assuming that the airfoil could be represented by the mean camber line and of making small angle assumptions throughout the derivation.

If the airfoil is symmetric, then $dz/dx = 0$, and Equation 4.61 becomes an integral of an unknown function (the vortex strength distribution), that has to equal a known constant, $V_\infty\alpha$ (where α is in radians). The solution of this integral is a vortex strength distribution given by

$$\gamma(\theta) = 2\alpha V_\infty \frac{1+\cos\theta}{\sin\theta} \tag{4.62}$$

where a change of variables was performed in order to simplify the integration such that

$$\xi = \frac{c}{2}(1-\cos\theta) \tag{4.63}$$

where $\theta = 0$ corresponds to the leading edge and $\theta = \pi$ corresponds to the trailing edge. The vortex strength distribution given in Equation 4.61 yields the following very important results:

$$C_l = 2\pi\alpha \quad C_{m_{L.E.}} = -\frac{C_l}{4} \quad C_{m_{c/4}} = 0 \quad x_{c.p.} = x_{a.c.} = \frac{c}{4}. \tag{4.64}$$

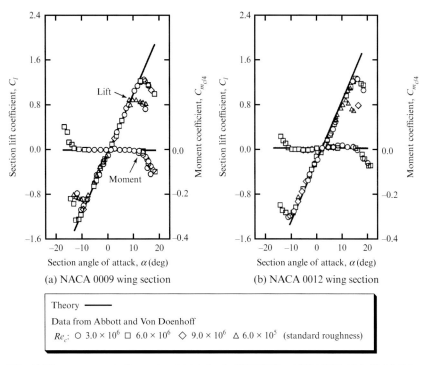

(a) NACA 0009 wing section

(b) NACA 0012 wing section

Theory ——————

Data from Abbott and Von Doenhoff

Re_c: ○ 3.0×10^6 □ 6.0×10^6 ◇ 9.0×10^6 △ 6.0×10^5 (standard roughness)

Figure 4.27 Thin airfoil theory compared with experimental data for the NACA 0009 and NACA 0012 airfoils (Ref. 3; Data courtesy of NASA.).

Application of these relations to symmetric airfoil experimental data for the NACA 0009 airfoil is shown in Fig. 4.27. Notice that the results are excellent in the linear range of angle of attack (for this airfoil the linear range extends to $\alpha \approx 12°$). Stall characteristics are not predicted by the theory because the theory assumes inviscid flow, and stall is a viscous phenomenon. These results are remarkably accurate and form the basis for theoretical airfoil results. Every student of aerodynamics should know these results!

When the airfoil has camber, $dz/dx \neq 0$ and the integration of Equation 4.61 becomes much more complicated, since the integral of the unknown function must now equal a known function. This integration requires the use of Fourier analysis and results in the following relations:[2,3]

$$C_l = \pi(2A_0 + A_1) \quad C_{m_{L.E.}} = -\left[\frac{C_l}{4} + \frac{\pi}{4}(A_1 - A_2)\right] \quad C_{m_{c/4}} = \frac{\pi}{4}(A_2 - A_1) \quad (4.65)$$

$$\alpha_{l=0} = -\frac{1}{\pi}\int_0^\pi \frac{dz}{dx}(\cos\theta - 1)d\theta \quad x_{c.p.} = \frac{c}{4}\left[1 + \frac{\pi}{C_l}(A_1 - A_2)\right] \quad x_{a.c.} = \frac{c}{4}$$

where the A's are the Fourier coefficients given by

$$A_0 = \alpha - \frac{1}{\pi}\int_0^\pi \frac{dz}{dx}d\theta \quad A_n = \frac{2}{\pi}\int_0^\pi \frac{dz}{dx}\cos n\theta d\theta. \quad (4.66)$$

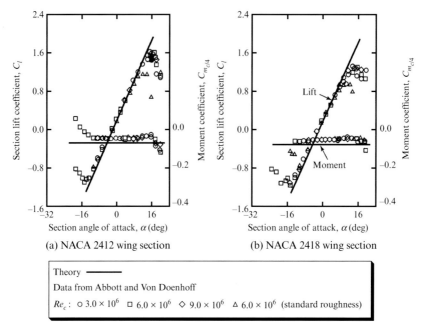

Thin airfoil theory compared with experimental data for the NACA 2412 and NACA 2418 airfoils (Ref. 3). Data courtesy of NASA.

A comparison of these results with experimental data for the NACA 2418 airfoil is shown in Fig. 4.28, and as with the symmetric airfoil results, the comparison is excellent in the linear range of angle of attack.

Notice that the NACA 2418 prediction is nearly as good as the prediction for the NACA 2412, even though an 18 percent thick airfoil probably could not be considered "thin." These results show the power of thin airfoil theory to accurately model inviscid airfoil aerodynamics, even though the theory does not predict drag.

4.7 Wing Aerodynamics

Although it would be nice if three-dimensional wings had the same aerodynamics as two-dimensional airfoils stacked side by side, real life is rarely so convenient. In fact, there can be quite large differences that take place when a wing of finite span is created out of an airfoil section, and none of the differences are beneficial. Since the wing produces lift, there is lower relative pressure on the upper surface of the wing compared with the lower surface, and in the vicinity of the wing tip, this pressure difference causes the air to flow around the wing tip and create a *wing-tip vortex* or *trailing vortex* (see Fig. 4.29). These vortices alter the air over the entire span of the wing by creating a downward velocity for every airfoil section. This downward velocity is known as *downwash*. The impact of the downwash is to decrease the lift and increase the drag of the wing, an undesirable condition. A great deal of study

Figure 4.29 Wing-tip vortices on a Boeing 727 (Courtesy of NASA Dryden Flight Research Center).

has taken place to alleviate the impact of these wing-tip vortices, including the use of various wing-tip devices such as winglet or raked wing tips, but the differences afforded by such devices are usually quite small. Luckily, the direct impact on a commercial transport aircraft in reducing drag even 1 percent or 2 percent is a direct increase in the range, or the ability to add payload.

4.7.1 Wing Terminology

As was true for airfoils, wing geometry definitions use a standard terminology to describe the shape. A wing could also take an infinite number of shapes, so no terminology system is perfect, but we shall use the common definitions. A trapezoidal wing is shown in Fig. 4.30. The geometric definitions used to describe the airfoil are:

- Wing planform area, S: projected area of the wing (including the part of the wing covered by the fuselage)
- Root chord, c_r: chord at the centerline of the aircraft
- Tip chord, c_t: chord at the tip of the wing
- Wing span, b: straight-line distance from one wing tip to the other
- Mean aerodynamic chord, $\bar{c}$: the geometric average chord value for a wing
- Wing sweep, Λ: measured aft
- Taper ratio, λ: the ratio of the tip chord to the root chord, c_t/c_r
- Aspect ratio, AR: a measure of the slenderness of the wing, b^2/S

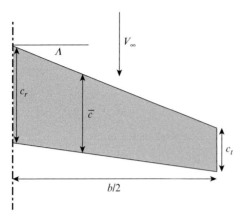

Figure 4.30 Geometry of a wing with a trapezoidal planform.

The wing planform area for a trapezoidal wing is given by

$$S = \frac{b}{2}(c_r + c_t).$$ (4.67)

The mean aerodynamic chord (m.a.c.) can be calculated using the following formulation, which is valid for a trapezoidal wing:

$$\bar{c} = \frac{2}{3}\left(c_r + c_t - \frac{c_r c_t}{c_r + c_t}\right) = \frac{2}{3}c_r\left(\frac{\lambda^2 + \lambda + 1}{\lambda + 1}\right).$$ (4.68)

The mean aerodynamic chord can then be used to define a "mean" Reynolds number for the wing:

$$Re_{\bar{c}} = \frac{\rho_\infty U_\infty \bar{c}}{\mu_\infty}$$ (4.69)

4.7.2 Wing Aerodynamic Coefficients

Although we defined aerodynamic coefficients for airfoils in Section 4.6.3, we need to define a set of coefficients for wings, since the nondimensionalization process will be slightly different. While the basic definitions for lift, drag, and moment do not change, we are now dealing with forces and moment, rather than forces and moments per unit span, which is usually handled using the concept of the wing planform area, as shown in Fig. 4.30. In order to keep the coefficients separate from each other, we use uppercase subscripts for these coefficients. Since a wing is a three-dimensional shape, the forces and moment are nondimensionalized by the wing planform area, the dynamic pressure, and the wing chord (usually the mean aerodynamic chord) as:

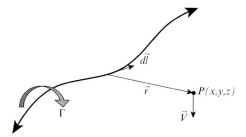

Figure 4.31 The velocity induced by a vortex filament.

$$C_L \equiv \frac{L}{q_\infty S} \quad C_D \equiv \frac{D}{q_\infty S} \quad C_M \equiv \frac{M}{q_\infty S \bar{c}} \tag{4.70}$$

Compare these coefficients to the ones defined in Equations 4.54 to see the differences between airfoil (section, 2-D) coefficients and wing (3-D) coefficients. Also, the moment has to be taken about a specified moment reference center, which is often taken to be the airplane's center of gravity.

4.7.3 The Vortex Filament

Another important potential flow concept that we will need to understand is that of the *vortex filament*. While somewhat complicated to derive mathematically (see Ref. 6 for details), a convenient way to think of a vortex filament is as a three-dimensional line around which the flow rotates as an irrotational element. Figure 4.31 shows an arbitrarily shaped vortex filament. In order to use vortex filaments, we need to understand some basic limitations of the filament, known as the *Helmholtz Vortex Theorems*, which state:

1. A vortex filament cannot end in the fluid (it can end at a solid boundary, close on itself, or go to infinity).
2. A vortex filament must have constant strength along its length.
3. An initially irrotational, inviscid flow will remain irrotational.

Related to these theorems we state an important result: a sheet of vortices can support a jump in tangential velocity [i.e., a force], while the normal velocity is continuous. This means you can use a vortex sheet to represent a lifting surface. These "rules" will help us as we find useful applications for vortex filaments.

Another important concept, borrowed from electro-magnetic theory, is the *Biot-Savart Law*, which states that the velocity induced by a segment of a vortex filament is given by:

$$d\vec{V} = \frac{\Gamma}{4\pi} \frac{d\vec{l} \times \vec{r}}{|\vec{r}|^3}. \tag{4.71}$$

Computational Aerodynamics Concept Box

The Meaning of the Cross Product

What does $\vec{a} \times \vec{b}$ mean? Consider the sketch shown here.

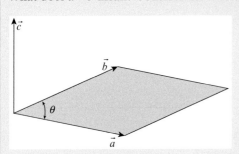

The vectors $\vec{a}$ and $\vec{b}$ form a plane, and the result of the cross product operation is a vector $\vec{c}$, where $\vec{c}$ is perpendicular to the plane defined by $\vec{a}$ and $\vec{b}$ (the direction is determined by the right-hand rule). The value of the cross product is given by:

$$\vec{c} = \vec{a} \times \vec{b} = |\vec{a}||\vec{b}|\sin\theta \, \hat{e}$$

and $\hat{e}$ is perpendicular to the plane of $\vec{a}$ and $\vec{b}$.

One consequence of this is that if $\vec{a}$ and $\vec{b}$ are parallel, then $\vec{a} \times \vec{b} = 0$.

Also: $|\vec{a} \times \vec{b}|$ = area of the parallelogram

which is the shaded area of the figure. Finally, the cross product is evaluated as:

$$\vec{a} \times \vec{b} = \begin{vmatrix} \hat{i} & \hat{j} & \hat{k} \\ a_x & a_y & a_z \\ b_x & b_y & b_z \end{vmatrix} = \left(a_y b_z - a_z b_y\right)\hat{i} - \left(a_x b_z - a_z b_x\right)\hat{j} + \left(a_x b_y - a_y b_x\right)\hat{k}.$$

When this concept is applied to an infinitely long straight-line vortex filament as shown in Fig. 4.32, the resulting velocity at point $P(x,y,z)$ is:

$$V = \frac{\Gamma}{2\pi h} \tag{4.72}$$

which is the same magnitude we found for a point vortex! If we only consider the influence of half of the infinite vortex, then the velocity induced at a perpendicular distance h from the "end" of the line is:

$$V = \frac{\Gamma}{4\pi h} \tag{4.73}$$

which seems perfectly reasonable, since it is one-half of the value in Equation 4.72.

Finally, the velocity induced by a segment of a straight-line vortex filament, as shown in Fig. 4.33, is:

$$V = \frac{\Gamma}{4\pi h}\left(\cos\theta_1 - \cos\theta_2\right). \tag{4.74}$$

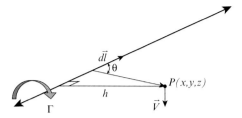

Figure 4.32 The velocity induced by an infinite straight vortex filament.

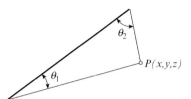

Figure 4.33 The velocity induced by a segment of an infinite straight vortex filament.

These basic formulas turn out to be the only results we will require from the Biot-Savart Law, since these formulas form the basis for modeling a wing-tip vortex.

4.7.4 Prandtl's Lifting Line Theory

When Ludwig Prandtl developed theories, he always began by trying to understand the physical features he was trying to model. If you look at Fig. 4.34, you see the vortex system formed at the tips of a wing. How can you create a mathematical model of this? If you remember the concept of the vortex filament and the restrictions of the Helmholtz Vortex Theorems, the concept of the *lifting line* is not that hard to understand.[19]

First, it is fairly obvious that you could model either of the wing-tip vortices (known as the *trailing vortices*) with a segment of a straight vortex filament. Because Helmholtz's theorems require that a vortex filament cannot end in the fluid, it would be convenient to connect the two trailing vortex filaments with a straight-line segment along the wing. When you think about it, however, the wing itself is producing lift (or there wouldn't be any trailing vortices), so the vortex filament along the span of the wing is also essential to the model (it is modeling the wing itself), and is known as the *bound vortex*. We now have a complete model of our wing, which is known as a *horseshoe vortex*, since it is shaped something like a horseshoe, as shown in Fig. 4.35.

Interested readers are referred to Refs. 3 or 8 for further details about the theory behind lifting-line theory, but the main result of the theory is the key element here. What is the impact of the trailing vortices

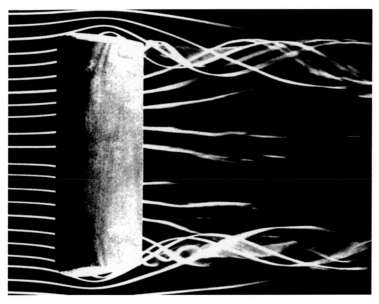

Figure 4.34   The trailing vortices from a rectangular wing (Ref. 5; Photograph by H. Werlé, © ONERA, the French Aerospace Lab.).

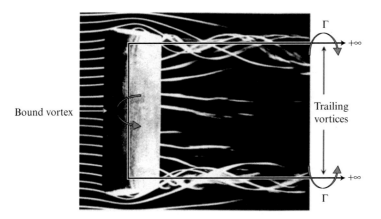

Figure 4.35   The horseshoe vortex model for a wing. (Ref. 5; Photograph by H. Werlé, © ONERA, the French Aerospace Lab.).

on the wing? Each trailing vortex filament induces a velocity given by Equation 4.73:

$$V = \frac{\Gamma}{4\pi h} \qquad (4.75)$$

which means all along the wing there is a variation of velocity induced by the vortices, known as the *downwash velocity*. The downwash velocity reduces the angle of attack of the wing and thereby reduces the lift on the wing. In addition, the wing-tip vortices create a component of drag known as the *induced drag* (or *vortex drag*). Both the decrease in lift and increase in drag can be computed using lifting-line theory as:

$$C_{L_\alpha} = \frac{C_{\ell_\alpha}}{1 + \frac{C_{\ell_\alpha}}{\pi AR}(1 + \tau)} \qquad C_{Di} = \frac{C_L^2}{\pi e AR} \qquad (4.76)$$

which is the explanation for why wings do not behave exactly as the airfoil sections from which they are made. This leads to a total drag coefficient for the wing, which could be stated as:

$$C_D = C_{d_0} + C_{D_i}. \qquad (4.77)$$

The value, e, in Equation 4.76 is the span efficiency factor, where a wing with an elliptic distribution of lift will have $e = 1$, and most other wings will have values of $e < 1$. An approximation for the span efficiency factor is given by[20]

$$e = \frac{2}{2 - AR + \sqrt{4 + AR^2 \left(1 + \tan^2 \Lambda_{t_{max}}\right)}} \qquad (4.78)$$

where $\Lambda_{t_{max}}$ is the sweep angle of the line connecting the maximum thickness locations of the wing. Aerodynamic designers often use wing twist and airfoil camber changes to increase the value of e.

While we could certainly spend a great deal of time trying to understand lifting-line theory, our main purpose in presenting it is so we can use the lifting-line concept in advanced panel methods. Figure 4.36 shows the impact of the lift-curve slope reduction and the additional induced drag from Equation 4.76. As the aspect ratio of the wing is reduced from infinity (2-D airfoil), the lift-curve slope decreases, meaning that a wing composed of an airfoil section (that achieves one level of lift coefficient) will actually experience a smaller lift coefficient at the same angle of attack. The smaller the aspect ratio of the wing, the greater is the reduction in lift-curve slope and the greater the induced drag.

For this reason, low-speed wings should have the maximum possible aspect ratio, within the constraints of the wing structure and weight of the aircraft, since the higher aspect ratio results in less induced downwash and a reduced drag coefficient (wing span, however, is what reduces the induced drag force). High-performance gliders, for example, may have aspect ratios higher than $AR = 20$, such as the TG-15A at the U.S. Air Force Academy shown in Fig. 4.37. Typical general aviation aircraft have aspect ratios of about 8 to 10.

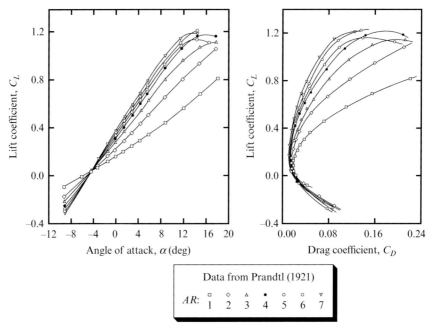

Figure 4.36   Impact of downwash on wing lift curve and drag polar for several wings with aspect ratios ranging from 1 to 7 (Ref. 21).

Figure 4.37   TG-15A high-performance glider at the U.S. Air Force Academy (U.S. Air Force photo).

Unfortunately, the induced drag predicted by Equation 4.76 increases as the square of the lift coefficient, which means that this drag component grows quite rapidly as the angle of attack is increased (as the aircraft flies slower at a constant altitude).

Flow Visualization Box

Wing-tip Vortex

Seeing a wing-tip vortex in a flow solution is not as easy as you might think, but one way to find the extent of the vortex flowfield is to use the streamline capability in FieldView. First, open the Streamlines window as shown below. There are many ways to create the "seeds" that make a streamline, but one way is to place the "seeds" on the grid itself. By choosing Auxiliary Seed Plane, you can visualize "planes" of the grid and add seeds to the grid itself. It may take a little while to find the right place to put the seeds, but after some trial and error you may find a good place to start the streamlines.

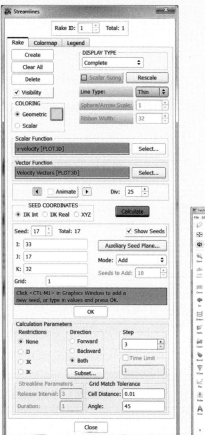

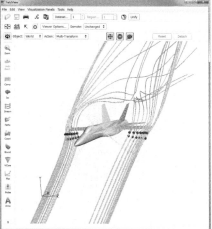

Once you've found streamlines you like, you can extend them forward and backward in space by choosing the *Direction* you like.

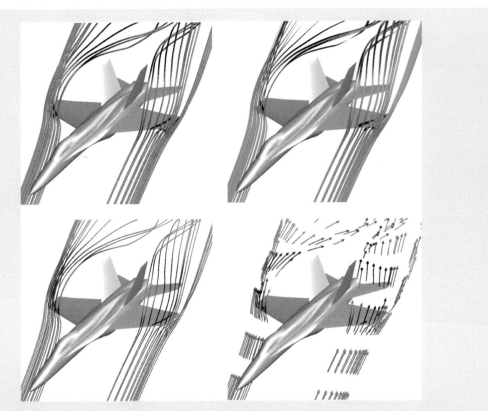

Now you can color the streamline with various scalar properties, such as vorticity or pressure or velocity. You can also change the Display Type of the streamline to show lines, ribbons, or even filaments with arrows (which can be animated to see the flow in an even more useful way).

4.7.5 Subsonic Compressibility Effects

While the incompressible results shown previously are quite valuable, they are only valid for lower speeds where compressibility can be assumed to be negligible (say, $M_\infty \leq 0.3$). Compressibility changes the aerodynamics of a vehicle, and, depending on the Mach number, the change can be significant. You can begin to appreciate the impact of these changes at various Mach numbers by viewing Fig. 4.38. In Fig. 4.38a, the point disturbance is not moving and the pressure waves travel at the speed of sound, a, in all directions. Once the disturbance starts moving, however, the situation starts to change (as can be seen in Fig. 4.38b). Here the disturbance is moving to the left at a Mach number less than the speed of sound, $M < a$. In this case, pressure information is not traveling in all directions equally, since the relative motion between the disturbance and the pressure waves is different depending on the direction. Waves start to "pile up" in front of the disturbance, since the speed of the disturbance is "catching up" to

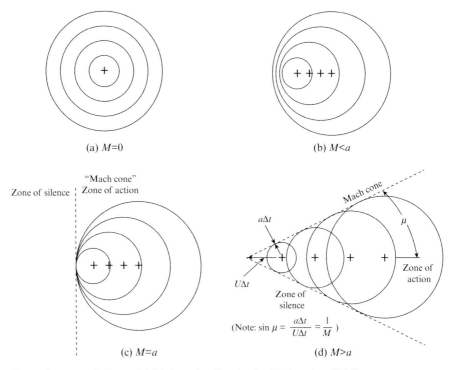

(a) $M=0$

(b) $M<a$

(c) $M=a$

(d) $M>a$

(Note: $\sin\mu = \dfrac{a\Delta t}{U\Delta t} = \dfrac{1}{M}$)

Figure 4.38   Wave patterns generated by a point disturbance traveling at various Mach numbers (Ref. 3).

the speed of the wave (the speed of sound). Finally, the speed of the disturbance increases until it is the same as the speed of sound, $M = a$, as shown in Fig. 4.38c. Now information can no longer propagate in front of the disturbance since the pressure waves are traveling at the same speed as the disturbance. This creates a "zone of silence" in front of the disturbance (no information can be propagated into this region) and a "zone of action" behind the disturbance (information can only be propagated into this region). The creation of a zone of action shows why flowfields in supersonic flight are significantly different from flows in subsonic flight and why we will need very different theories to determine the aerodynamics for each speed regime. Finally, as the speed of the disturbance is increased even more, the zone of action falls within a cone, as shown in Fig. 4.38d. Since the pressure waves travel at the speed of sound, they can only travel a distance $a\Delta t$ over any period of time. For the same time interval, the point disturbance travels a distance $U\Delta t$, which creates a zone of action that takes the shape of a cone. This cone is known as the Mach cone and is found to have an angle, μ, which is the Mach angle. The Mach angle can be found from the geometry in Fig. 4.38d as $\sin\mu = 1/M$.

When the speed of an aircraft increases above the incompressible range, Laplace's equation, Equation 4.12, is no longer valid. Researchers in the 1930s and 1940s used compressibility corrections on their incompressible data in order to determine what would happen as the airplane flew faster. These corrections were based on a theoretical development for subsonic compressible flow, which yields the Prandtl-Glauert equation (Equation 3.125):

$$\left(1 - M_\infty^2\right)\phi_{xx} + \phi_{yy} = 0. \tag{4.79}$$

This equation looks very similar to Laplace's equation, and, in fact, can be shown to mathematically transform into Laplace's equation. Researchers were able to equate incompressible results to subsonic compressible results through the use of the *Prandtl-Glauert rule*:

$$C_p = \frac{C_{p_{inc}}}{\sqrt{1 - M_\infty^2}}. \tag{4.80}$$

Since pressures are corrected according to Equation 4.80, forces and moments due to pressure, such as lift, lift-curve slope, and pitching moment are corrected in the same way:

$$C_l = \frac{C_{l_{inc}}}{\sqrt{1 - M_\infty^2}} \qquad C_m = \frac{C_{m_{inc}}}{\sqrt{1 - M_\infty^2}}. \tag{4.81}$$

Since drag is primarily a viscous-related force, it does not vary in this way and is usually assumed to be constant throughout the subsonic compressible range. More advanced compressibility corrections can be found in Ref. 2.

Equation 4.79 can also be applied at supersonic speeds (as will be shown later in this chapter). When $M > 1$, the mathematical "type" of the equation changes (see Chapter 3 for a discussion of type) and the mathematical solution method changes to reflect the different physics, as shown in Fig. 4.38.

Profiles in Computational Aerodynamics: Bryan Richards

I grew up in Hornchurch, east of London, UK in the 1940s and 1950s. Maybe it was the dog-fights and V1 flying bombs overhead during the war, but certainly the varied aircraft designed, built, and flown in the fifties attracted me to take a degree course in aeronautical engineering at Queen Mary College (chaired by Professor Alec Young) at the end of that decade and into a career of research and development in aircraft and their simulation. In those days computations were generally done using slide rules and mathematical tables. When I joined the Bristol Aircraft Company in 1960, we were doing sums using Monroe and Facit calculating machines (we had no digital computers available then) to convince airlines, governments, and safety regulators to introduce supersonic transport aircraft (finally resulting in Concorde). I then joined Imperial College, London working for Professor John Stollery) where my focus was

on experimental research in hypersonic aerodynamics (other research students at the time were Hans Hornung – now Caltech; Mike Holden – CUBRC; and Chul Park – NASA Ames, all having been prominent in the progress of high-speed aerodynamics). We used analog computation techniques using resistance-capacitance networks there to enable us to extract heat transfer rates from short-duration surface temperature measurements on ceramic walls of the models.

CFD took a more important role when I joined the faculty of the von Karman Institute in Belgium in 1967 to set up a High Enthalpy Laboratory, which included a heavy piston gun tunnel called Longshot that was capable of simulating the re-entry of the Space Shuttle at Mach 15–20, and another piston tunnel, called CT2, for testing the hot turbine flows needed for the new high-bypass ratio fan engines. We had an IBM 1130 computer (64Kbytes of core memory, 512 Kbyte disc drive, 3.6 microsecond memory cycle time, and a card reader) to serve the whole institute of twelve faculty and fifty postgraduates. Exclusive time on it needed to be booked by the hour, and with such a modest resource (by present standards) heroic calculations were made. For example, Kurt Enkenhus with my assistance carried out accurate simulation on this computer of the run cycle of Longshot toward extending its performance with real gas effects implemented by creating subroutines by clever curve-fitting to Mollier charts. Even with today's tools this would be a significant task. Through this exercise I became interested at VKI in numerical methods, and devised and taught a postgraduate course in collaboration with some U.S. postdocs who were visiting at the time. An exercise in analyzing the latest papers in CFD techniques was included in the course, as well as attendance at the now well-known annual VKI Lecture Series in CFD organized by Professor Jean Smolderen.

In 1980 I moved as professor to the University of Glasgow in Scotland (past faculty members included Watt, Kelvin, and Rankine!) where my interest in CFD to study external aerodynamics was pursued more exclusively. After some early work in solving the Navier-Stokes equations for practical incompressible/super-/hyper-sonic flows (in particular with Ning Qin – now University of Sheffield, UK; ZJ Wang – Iowa State University; and Chang Shu – University of Singapore, important practitioners of applied CFD) our main focus turned to unsteady trans/supersonic flows with the then ultimate aim of tackling the very demanding task of simulating aircraft flutter. The resultant developments were carried out by a team (always important in good developments) in the 1990s including Ken Badcock, Bill McMillan – now Platform Computing, and more latterly George Barakos, among many others. While we settled on parallel implicit upwind approaches to the problem (leading to the Parallel Multi-block CFD solver – PMB Solver), the challenges were not only in solver development, but also concerned with accessing increasingly large computing resources necessary to carry out useful tasks at low cost. In the early days we were favored with shared access to the university mainframe computers for which we had frequent systems upgrades, but this situation was shattered when arts and social science researchers in the university demanded desktop capabilities, with the demise of the mainframe. We were forced to search for other solutions, and these included a 32-transputer platform, which was soon overtaken in power with just one IBM RS6000 workstation (RISC/UNIX based). We then ran parallel codes on spare capacity available from other users on university networks of such worksta-tions – we ran a demonstration project using LSF for our researches for British universities. The ultimate solution came with using a PC cluster, another UK demonstration, by wiring up multiple low-cost PCs running Linux and obtained the necessary high speed and distrib-uted memory that PMB required. We no longer needed to share our resource and with the same modest budget were able to upgrade by an order of magnitude in cpu and memory every three years while increasing our capability to simulating full aircraft flows. In industry, design and development requires not only the ability to tackle large geometrically and physically complex problems but to do it cheaply. Our success in doing this contributed to industry's ability to do this.

Presently as an emeritus professor at Glasgow I keep up-to-date with aerospace engineering, made easier by editing the journal *Progress in Aerospace Sciences*. During undergraduate studies in London, I developed an interest in sailing which has become a lifetime passion. It was inevitable that, with my wife Margaret, also a sailor, this became our main pastime with a growing family; I built sailing dinghies for the children when we moved to Scotland. It was with some astonishment and pride that our daughter, Emma, from these beginnings sailed around the world single-handed in an Open 60 (ft) boat in the Around Alone race (youngest and only female) in 2002/3, and our two sons have became top sailors in tricky-to-sail eighteen-foot skiff boats while employed in demanding engineering jobs. So time is also spent following the family, now expanded to include nine grandchildren.

4.8 Transonic Aerodynamics

Transonic aerodynamics is one of the most difficult areas in aerodynamics for deriving theoretical models, mainly due to the nonlinear nature of the flow and the resulting nonlinear nature of the equations that model the flow. Rather than attempt to discuss the theoretical aspects of transonic flow, we will undertake a description of the flowfields that are common to this flight regime. The interested reader may find more detailed discussion of transonic aerodynamics in Refs. 2 and 3.

Perhaps the most important feature of transonic flow from a design perspective is the rapid increase in drag that takes place at transonic speeds. Figure 4.39 shows the drag coefficient of a simple projectile as a function of Mach number. Notice the sharp increase in drag that takes place just prior to $M = 1$. In this case the drag coefficient more than doubles after reaching transonic speeds. In order to understand how and why this increase takes place, we will need to learn more about shock waves and their formation.

First of all, shock waves can exist as either normal shocks (the shock is perpendicular to the flow direction) or oblique shocks (the shock is at an angle to the flow). Shocks create nearly discontinuous changes to the flow properties. These changes must be understood in order to gain a feeling for transonic and supersonic flow. While the details of how these flow properties change requires a good background in gas dynamics,[23] we will summarize the essential aspects of the flow changes here in Table 4.3. Also included in Table 4.3 are the changes in flow properties through an expansion wave.

Notice that a shock compresses the flow (all static properties increase) as the flow decelerates. Also of great importance is that a shock decreases total pressure (the flow is irreversible) while maintaining total temperature (the flow is adiabatic). Therefore, shocks are not isentropic! Expansion waves, however,

Table 4.3 Flow property changes across normal shocks and expansion waves

Flow Property	Normal Shock	Expansion Wave
M	$M_2 < 1 < M_1$	$M_2 > M_1$
ρ	$\rho_2 > \rho_1$	$\rho_2 < \rho_1$
p	$p_2 > p_1$	$p_2 < p_1$
T	$T_2 > T_1$	$T_2 < T_1$
p_0	$p_{0,2} < p_{0,1}$	$p_{0,2} = p_{0,1}$
T_0	$T_{0,2} = T_{0,1}$	$T_{0,2} = T_{0,1}$

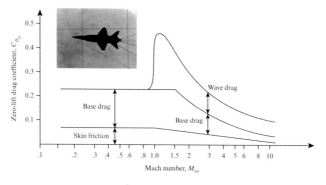

Figure 4.39   Zero-lift drag coefficient for the X-15 as a function of Mach number (graph based on Ref. 22; photo courtesy of NASA).

speed up the flow while decreasing the static properties. Expansion waves also maintain total pressure and total temperature and are therefore isentropic.

So now that we have a basic understanding of supersonic waves, we will try to understand transonic flow. *Transonic flow* is defined as a flowfield where both subsonic and supersonic flows are present. There is no precise Mach number definition for the transonic flight regime, since the preceding definition will result in different Mach number boundaries depending on the airplane geometry and attitude of the airplane while flying. A good description of the types of flow that occur can be seen in Fig. 4.40.[24]

As the aircraft accelerates from lower subsonic Mach number (say $M = 0.50$, as shown in the figure), the first sign that something significant is about to change takes place when the *critical Mach number* (M_{cr}) is reached. Since flow accelerates around the upper surface of a lifting airfoil, the velocities over the airfoil can be significantly higher than the freestream velocity. Remember that the flow over a circular cylinder, discussed in Section 4.3.1, had velocities twice that of the freestream on the top of the cylinder. Because of this, a freestream Mach number will eventually be reached where sonic flow (flow at $M = 1$) first takes place over the airfoil or wing (see Fig. 4.40 for $M_\infty = 0.72$).

While this sounds important (and it is), nothing significant actually happens at the critical Mach number, but this Mach number is the gateway to

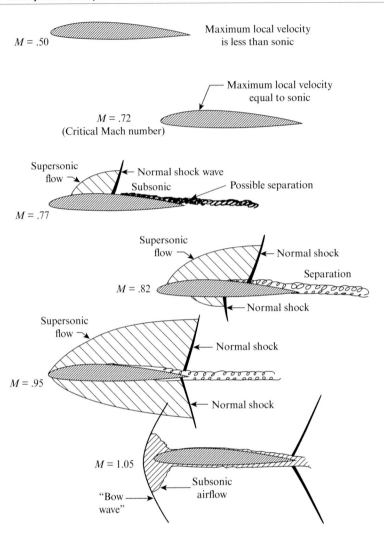

Figure 4.40 Flow features in transonic flight (Ref. 24).

the transonic flight regime. Immediately after the critical Mach number is passed, a region of supersonic flow develops over the upper surface of the wing. While flow may accelerate to supersonic speeds without penalty, a locally supersonic flow within a subsonic freestream must eventually decelerate to subsonic speeds, which takes place by the formation of a nearly normal shock wave on the wing (see Fig. 4.40 for $M_\infty = 0.77$). Normal shocks decelerate supersonic flow to subsonic speeds (see Table 4.3 for details), and the static properties (pressure, temperature, and density) all rise across the shock. These increases in static properties take place within the thickness of the shock, which is approximately 10^{-7} ft at these speeds. This represents a very large adverse pressure gradient on the wing, which often results in boundary-layer separation behind the shock. The existence of the shock and the boundary-layer separation leads to a sharp increase in the drag of the wing, called *wave drag*. The lambda shock pattern for an airfoil in

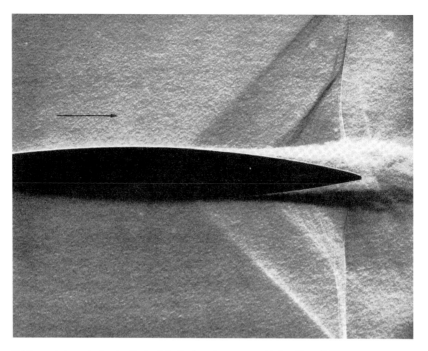

Figure 4.41   Schlieren photograph showing the lambda shock pattern over an airfoil at $M_\infty = 0.895$ (Ref. 25; reprinted by permission of the American Institute of Aeronautics and Astronautics, Inc.).

viscous flow is shown in Fig. 4.41, which is a complex region of interaction between the normal shock, the boundary layer, and oblique shocks that form upstream of the normal shock. As you can imagine, computationally predicting such a flow pattern challenges even the most advanced computer simulation approaches. Aerodynamic designers shape the wing upper surface to reduce the strength of shocks.

As the freestream Mach number continues to increase, the supersonic region above the airfoil continues to grow in size, which leads to a larger and stronger shock (see Fig. 4.40 for $M_\infty = 0.82$). Of course, this means that the drag of the wing has increased even more. At these high subsonic speeds, shocks can even begin to form on the lower surface of the airfoil, mimicking the flow at lower Mach numbers that took place on the upper surface. As the freestream Mach number has increased, the normal shock has moved aft on the wing's surface and continues to move aft until it eventually reaches the trailing edge of the wing. Eventually, at very high subsonic Mach numbers, both the upper and lower surfaces of the wing are engulfed in supersonic flow, with twin normal shocks located at the trailing edge (see Fig. 4.40 for $M_\infty = 0.95$ and Fig. 4.41).

Finally, when the freestream Mach number passes into the supersonic regime, a bow shock forms at the front of the wing (see Fig. 4.40 for $M_\infty = 1.05$). Since the bow shock is nearly normal to the flow directly in front of the wing, there is a region of subsonic flow directly in front of the leading-edge region – this is also classified as transonic flow since both subsonic and supersonic flow are still present in the flowfield.

Flow Visualization Box

How to Visualize Shock Waves

Once the Mach number of the freestream flow reaches high subsonic values (say above M = 0.8), shock waves start to form on the airfoil. This is because the airfoil accelerates the flow around the leading edge, possibly to supersonic values. When this happens, the supersonic flow region is ended with a nearly normal shock wave and possible boundary-layer separation. How can you visualize this supersonic region and the shock wave? One way is to look at a *coordinate surface* colored by a scalar (like Mach number) as shown in the accompanying figure. The Mach number scale was chosen so that when the flow reaches M = 1.0 the color is red, making it quite easy to see the supersonic region, both on the upper and lower surfaces of the airfoil. Since this airfoil is at 4 degrees angle of attack, the flow accelerates to higher values over the upper surface and even higher supersonic Mach numbers are reached. This creates a stronger shock wave that also leads to boundary-layer separation.

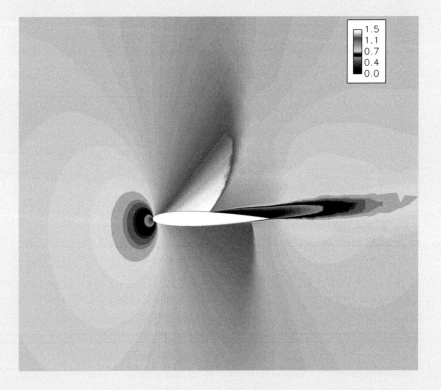

The boundary-layer separation is not as easy to see in the Mach contours, so find another way to visualize the flowfield. One possibility is to use the Streamline capability and release particles along the inflow, as shown here. The recirculation region created by the shock-induced separation is clearly visible, and now your view of the flowfield is expanded.

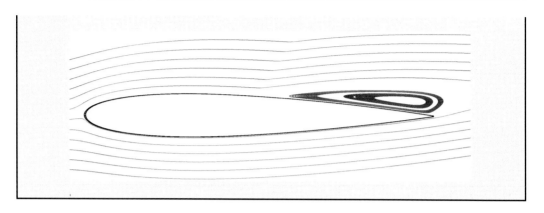

4.8.1 Transonic Theories

While theoretical formulations exist for transonic flow, as was mentioned earlier, they are quite complicated and beyond the scope of this book. Such theoretical approaches include the derivation of the full potential equation or the transonic small disturbance equation as shown in Chapter 3. The steady, two-dimensional full potential equation is given by (see Equation 3.102):

$$\left(\rho\Phi_x\right)_x + \left(\rho\Phi_y\right)_y = 0 \tag{4.82}$$

where the most important feature of the equation is the nonlinear nature of the terms, which is caused by the density being an unknown variable. Any term with a product of flow variables will be nonlinear, which makes solving the equation much more difficult than a linear equation. Theoretical solutions for transonic flow can be obtained using Tricomi's equation, performing a hodograph transformation, or by finding solutions using the shock polar approach.[26] A more useful approach in recent decades has been to solve the various levels of fluid dynamic governing equations computationally, a subject that was discussed in greater detail in Section 3.12.

4.8.2 Supercritical Airfoils

The shock formation and resulting boundary-layer separation that takes place at high subsonic Mach numbers makes most low-speed airfoils unsuitable for transonic flight. This led Richard Whitcomb, a researcher at NASA Langley Research Center, to develop supercritical airfoils in the 1960s.[27] The supercritical airfoil is significantly different from conventional low-speed airfoils, as shown in Fig. 4.42. The transonic flow patterns for a conventional airfoil at $M_\infty = 0.72$ are shown in Fig. 4.42a. The supersonic flow region is significant (the cross-hatched region above the airfoil), and the normal shock is quite strong, leading to flow separation. The pressure coefficient data show that the airfoil has accelerated the flow to significant levels above $C_{p,sonic}$,

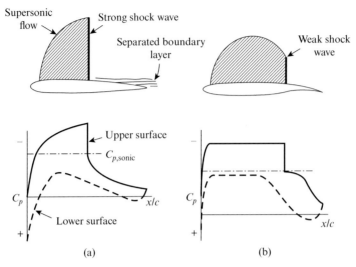

Figure 4.42   Comparison of transonic flow over a NACA 64A series airfoil with that over a "supercritical" airfoil section: (a) NACA 64A series, $M = 0.72$ (b) supercritical airfoil, $M = 0.80$ (Ref. 28).

which is the pressure coefficient for $M = 1$ flow. This causes the normal shock to form further upstream and for the shock to be stronger (as shown by the large pressure increase in the pressure coefficient). Since lift is directly related to the area between the upper and lower surface pressure coefficient curves, the shock has caused the back half of the airfoil to lose most of its lift, which also impacts the pitching moment of the airfoil.

Supercritical airfoils attempt to solve these problems through novel shaping of the upper and lower surfaces of the airfoil. First the upper surface is made relatively flat when compared with the conventional airfoil (see Fig. 4.42b). This allows for low pressures on the upper surface but maintains control of these pressures by keeping them constant over a large percentage of the airfoil chord. When the normal shock finally occurs, it is further aft and weaker than when compared with the conventional airfoil. Since the back one-third of the airfoil chord still has loss of lift due to the presence of the shock, the lower surface is cambered in this region to increase the high pressures acting on the bottom of the airfoil, thus maintaining lift in this region. The supercritical airfoil not only has less drag than the conventional airfoil (due to the weaker shock which is further aft), the lift is also maintained. Note also that the supercritical airfoil in Fig. 4.42 is flying at $M_\infty = 0.80$, a considerably higher speed than the conventional airfoil with which it is being compared.

The supercritical airfoil concept was flight tested on a modified F-8 in the early 1970s, as shown in Fig. 4.43. Results of the NASA flight research for the supercritical wing (SCW) "demonstrated that aircraft using the supercritical wing concept would have increased cruising speed, improved fuel efficiency, and greater flight range than those using conventional wings. As a

Figure 4.43 F-8 modified with a supercritical wing undergoing flight testing (Courtesy of NASA Dryden Flight Research Center).

result, supercritical wings are now commonplace on virtually every modern subsonic commercial transport. Results of the NASA project showed the SCW had increased the transonic efficiency of the F-8 as much as 15 percent and proved that passenger transports with supercritical wings, versus conventional wings, could save $78 million (in 1974 dollars) per year for a fleet of 280 200-passenger airliners."[29]

4.8.3 Korn Airfoil Equation

Attempts have been made to estimate the capability of transonic airfoils for the purposes of design studies without performing wind tunnel or detailed computational design work. This is important in the initial stages of aircraft design, where airfoil performance needs to be estimated before the actual airfoil design has been done. Here we provide an approximate method for estimating the transonic performance of airfoils based on an empirical relation developed by Korn in the early 1970s. Based on his experience, it appeared that airfoils could be designed for a variety of Mach numbers, thickness to chord ratios, and design lift coefficients, but in all cases there seemed to be a limit to the combination. In particular, the Korn relation is:

$$M_{DD} + \frac{C_l}{10} + \frac{t}{c} = \kappa_A \qquad (4.83)$$

where κ_A is an airfoil technology factor (the airfoil technology factor has a value of 0.87 for an NACA 6-series airfoil section, and a value of 0.95 for a supercritical section). M_{DD} is the drag divergence Mach number (a Mach number slightly above the critical Mach number where the drag starts to rise appreciably), C_l is the lift coefficient, and t/c is the airfoil thickness to chord ratio. This relation provides a simple means of estimating the possible combination of Mach, lift, and thickness that can be obtained using modern airfoil design methods (note that the Korn equation is very sensitive to the value of the technology factor). Figure 4.44 compares the prediction from the Korn equation with other estimates. In Fig. 4.44a, the estimates of both older airfoils and modern supercritical airfoil performance presented by Shevell are compared,[30] and the agreement is good, with the exception of being overly pessimistic regarding older conventional airfoils at lower thickness ratios. Figure 4.44b compares the Korn equation with NASA projections[31] for supercritical airfoils based on a wealth of data and experience. In this case the Korn equation is extremely good at lift coefficients of 0.4 and 0.7 but overly optimistic at higher lift coefficients.

4.8.4 Wing Sweep

Another important design feature for transonic aircraft is wing sweep. In the late 1930s, Adolf Busemann and Albert Betz, two aerodynamicists who had been taught by Ludwig Prandtl, discovered that drag at transonic and supersonic speeds could be reduced by sweeping back the wings. Betz was the first person to draw attention to the significant reduction in transonic drag which comes when the wing is swept back enough to avoid the formation of shock waves that occur when the flow over the wing is locally supersonic. R.T. Jones was working on similar wing sweep concepts at the same time in the United States.

The basic principle is that the component of the main flow parallel to the wing leading edge is not perturbed by the wing, so the critical conditions are reached only when the component of the freestream velocity normal to the leading edge has been locally accelerated, at some point on the wing, to the local sonic speed. This simple principle is only true on an infinite-span wing of constant section. Nevertheless, Betz's initial suggestion led to wind tunnel tests which substantiated the essence of the theory, as shown in Fig. 4.45. These data show that the effect of the shock waves that occur on the wing at high subsonic speeds is delayed to higher Mach numbers by wing sweepback.

One way to achieve these results at various Mach numbers is to use a variable-geometry (or swing-wing) design to obtain a suitable combination of low-speed and high-speed characteristics. In the highly swept, low-aspect-ratio configuration, the variable-geometry wing provides low wave drag. At the opposite end of the sweep range, efficient subsonic cruise, loiter, and good maneuverability at the lower speeds are obtained. Negative factors in

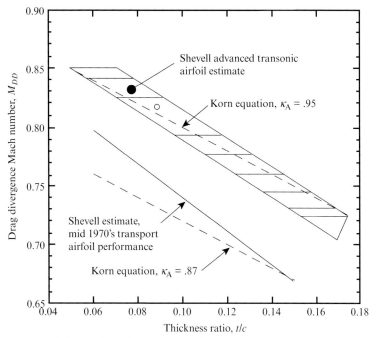

(a) Comparison of the Korn equation with Shevell's estimates (Ref. 30).

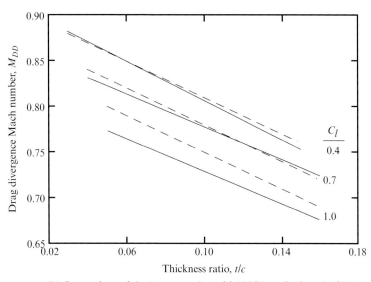

(b) Comparison of the Korn equation with NASA projections (Ref. 31).

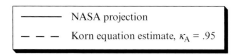

Figure 4.44   Validation of the Korn Equation for Airfoil Performance Projections. (a) Comparison of the Korn equation with Shevell's estimates (Ref. 30). (b) Comparison of the Korn equation with NASA projections (Ref. 31).

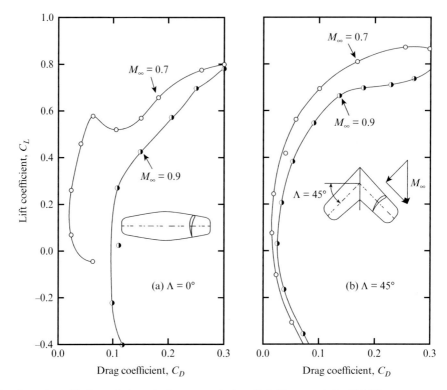

Comparison of the transonic drag polar for an unswept wing with that for a swept wing (Ref. 32).

The USAF B-1B bomber with variable-sweep wings (U.S. Air Force photo).

a swing-wing design are complexity, a loss of internal fuel capacity, and the considerable weight of the hinge/pivot structure. A variable-geometry design is used on the USAF B-1B, as shown in Fig. 4.46.

Most current aircraft, however, do not use variable sweep to attain improved aerodynamic efficiency at transonic speeds. Commercial transports such as the Boeing 787 and the Airbus A380 have swept wings with supercritical

Figure 4.47   The USAF C-17 transport aircraft with swept wings using supercritical airfoils and wing-tip devices (U.S. Air Force photo).

airfoils that are designed in conjunction with one another to obtain an optimal design for cruising at high subsonic speeds. A military transport that also incorporates these design features is the USAF C-17, which also uses supercritical airfoils in a swept-wing design that includes winglets, another invention of Richard Whitcomb (see Fig. 4.47).

These wings are generally fairly high aspect ratio swept tapered wings that clearly have an airfoil embedded in them. Normally, the aerodynamic designer is given the planform and maximum thickness and asked to design the twist and camber, as well as shifting the thickness envelope slightly. The goal is to obtain "good" isobars on the wing (pressure isobars that are as constant in the spanwise direction as possible). The natural tendency is for the flow to unsweep at the root and tip, so the designer tries to reduce this tendency to obtain an effective aerodynamic sweep as large as the geometric sweep. Possibly the best tutorial paper on the problem of isobar unsweep is by Haines,[33] who is actually considering the thickness effects at zero lift.

We can illustrate the problem of isobar unsweep with an example taken from work at Grumman for the initial G-III wing design studies. In Fig. 4.48 we see the isobar pattern at transonic speed on the planform in the upper left-hand side of the figure. This wing was designed with subsonic methods, which were essentially all that was available at that time. Note that the isobars are tending to unsweep at the tip of the wing. On the upper right-hand side of the figure, we see that at subsonic speed the pressure distributions at the 30 percent and 70 percent span stations lie on top of each other, so the

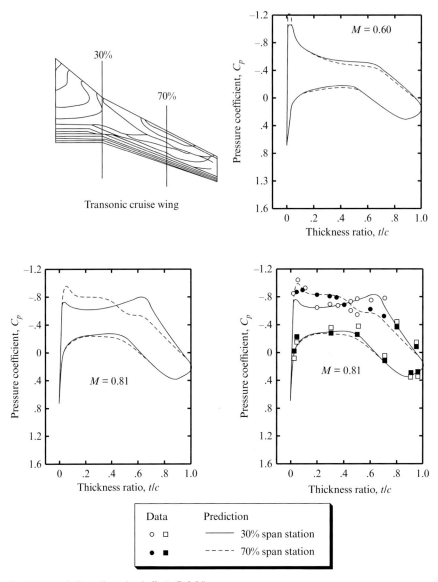

Figure 4.48 Explicit transonic three-dimensional effects (Ref. 34).

isobars are good. The lower left-hand side of the figure shows the predicted pressure distributions at these same two stations when the Mach number is increased to the transonic cruise Mach. There has been a large change in the distributions at the two stations, with the outboard shock ahead of the inboard shock. Finally, on the lower right-hand side we see the same predictions, but with wind tunnel data included. Clearly, the prediction and the test results agree and show that extra effort is required to design the wing when the flow is transonic. Although for a given span the aerodynamicist generally prefers an elliptic spanload, it might be better for the design if the load is shifted inboard slightly, reducing the root bending moment and hence wing structural weight.[35] The optimum spanload with a winglet present (as is true for the C-17) is not elliptic either.

Essentially, the twist distribution is found to generate the design spanload. Note that spanloads are predicted fairly well using linear theory codes. It is primarily the chord load that reflects the nonlinearity of transonic flow; that prediction would require nonlinear full potential codes or Euler codes. Once the basic twist is found, root and tip modifications are developed to maintain the isobar pattern. Without special effort, the chord load is drawn aft at the root and shifts forward at the tip. Changes in camber and thickness are introduced to counter these effects. The planform may deviate from pure trapezoidal due to trailing-edge variations (called a Yehudi) and a leading-edge glove inboard. This allows the t/c to be lower for the same t, and increasing the chord lowers the section C_l required to obtain the spanload required, as well as helping to maintain isobar sweep. Breaks in the planform chord distribution produce rapid variations in the section lift distribution because the spanload will tend to remain smooth. The section lift distribution may be smoothed out by using several smaller spanwise breaks. This has been done on modern Boeing and Airbus designs. The designer also has to consider buffet margins. This means the wing C_L has to be capable of a 1.3g turn at the highest cruise Mach number without predicting any significant flow separation. Other important details include nacelle/pylon interference and the resulting detailed shaping and manufacturing constraints. This means considering the limits to curvature and the manufacturing department's desire for straight-line wrap or ruled surfaces. Once the design starts to get close to the desired properties, local inverse methods can be applied to achieve the target pressure distributions. A review of the transonic wing design process by Jameson is worth reading to get a better idea of the complexity of the design process and future possibilities.[36]

4.8.5 Korn Wing Equation

As described previously, the Korn equation (Equation 4.83) can be used to estimate the drag divergence Mach number for airfoils. This equation has been extended to include sweep using simple sweep theory;[37,38] the result is given by:

$$M_{DD} = \frac{\kappa_A}{\cos \Lambda} - \frac{(t/c)}{\cos^2 \Lambda} - \frac{C_l}{10 \cos^3 \Lambda}. \tag{4.84}$$

This model estimates the drag divergence Mach number as a function of an airfoil technology factor (κ_A), the thickness-to-chord ratio (t/c), the airfoil lift coefficient (C_l), and the sweep angle (Λ) (recall that the airfoil technology factor has a value of 0.87 for a NACA 6 series airfoil section and a value of 0.95 for a supercritical section). With this approximation for the drag divergence Mach number, we can now calculate the critical Mach number. The definition of the drag divergence Mach number is taken to be where the drag rise attains a slope of 10 percent:

$$\frac{\partial C_D}{\partial M} = 0.1. \tag{4.85}$$

Next, we can make use of Lock's empirically derived shape of the drag rise[39]

$$C_D = 20\left(M - M_{crit}\right)^4. \tag{4.86}$$

The definition of the drag divergence Mach number is equated to the derivative of the drag rise formula given in Equation 4.85 to produce the following equation:

$$\frac{\partial C_D}{\partial M} = 0.1 = 80\left(M - M_{crit}\right)^3. \tag{4.87}$$

We can then solve this equation for the critical Mach number:

$$M_{crit} = M_{DD} - \left(\frac{0.1}{80}\right)^{1/3} \tag{4.88}$$

where the drag divergence Mach number is given by the extended Korn equation, Equation 4.84.

Grasmeyer then developed a method to compute the wave drag coefficient for use in multidisciplinary design optimization studies of a transonic strut-braced wing concept using the following relation:[40]

$$C_{d_{wave}} = 20\left(M - M_{crit}\right)^4 \frac{S_{strip}}{S_{ref}} \qquad \text{for } M > M_{crit} \tag{4.89}$$

where the local t/c, C_l, and half-chord sweep angle are specified for a number of spanwise strips along the wing, and the drag of each strip is combined to form the total wave drag. In the preceding equation, the wave drag for each strip is multiplied by the ratio of the strip area (S_{strip}) to the reference area (S_{ref}). The method has been validated with the Boeing 747–100 using eight spanwise strips, as shown in Fig. 4.49. The solid lines represent the current model predictions, and the discrete data points represent the Boeing 747 flight test data. The predictions show good agreement with the data over a wide range of Mach numbers and lift coefficients (a value for the technology factor of 0.89 was used for the Boeing 747 results in Fig. 4.49). Based on an analysis of the Boeing 777, a value of 0.955 was used to simulate that aircraft's wave drag characteristics.

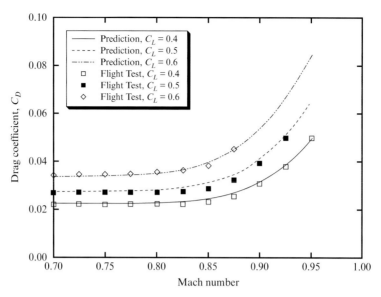

Figure 4.49   Comparison of approximate drag rise methodology with Boeing 747–100 flight test data from Mair and Birdsall (Ref. 41).

4.9 Supersonic Aerodynamics

There actually have been only a few truly supersonic airplanes, that is, airplanes that could *cruise* supersonically. Before the F-22, classic "supersonic" fighters had to use brute force (in the form of afterburners) to get to supersonic speeds, and once they did they had extremely limited duration. As an example, consider one of the supersonic missions for the F-14A, the combat air patrol mission: (a) 150 miles subsonic cruise to station, (b) loiter, (c) accelerate from $M = 0.7$ to 1.35, then dash out 25 miles (which would take 4 1/2 minutes and a total of 50 miles, and (d) head home or find a tanker! This supersonic capability is typical of the so-called supersonic fighters, and obviously the supersonic performance is limited. The small number of actual "supersonic cruisers" prior to the F-22 are shown in Table 4.4.

Table 4.4. True Supersonic Cruise Aircraft Prior to the F-22

Aircraft	L/D	First Flight(s)
Concorde	7.4	March 1969; Mach 2 in November 1970
SR-71 (A-12)	6.6	April 1962
B-58	4.5	November 1956
XB-70	7.2	September 1964; Mach 3 in October 1965

Figure 4.50   Convair B-58A in flight (U.S. Air Force photo).

Figure 4.51   North American XB-70A in flight with wing-tips drooped (U.S. Air Force photo).

Note the low L/D values associated with supersonic flight (compared with values from 10 to 20 or higher for subsonic flight); nevertheless, these airplanes were all remarkable. In case you are not familiar with the B-58[42] and XB-70,[43] they are shown in Figures 4.50 and 4.51. The B-58 weighed around 160,000 pounds, of which more than 100,000 pounds was fuel, and the structural weight fraction was an amazing 14 percent. To achieve the required aerodynamic performance, the wings were thin (4.08 percent at the tip and 3.46 percent at the root). The XB-70 deflected its wing tip down in supersonic flight, which can also be seen in Fig. 4.51. The tip deflection helped control the aerodynamic center shift problem that takes place on supersonic aircraft.

Flow Visualization Box

Supersonic Airfoil Visualization

What does the flow around our NACA 0012 airfoil look like in supersonic flow? Would it be a viable option for use on a supersonic aircraft? If you use the *Coordinate Surface* capability colored by pressure coefficient (see Equation 4.19 for the definition) you will find that the flowfield is significantly different than it was for transonic flow. Notice that the legend defines red to represent high pressure, and the region with the highest pressure is in the vicinity of the nose. That is because the shock in front of the nose, which is a detached bow shock, is nearly normal to the airfoil nose. Normal shocks reduce the Mach number of the flow from supersonic to subsonic speeds and greatly increase the static pressure, creating a great deal of drag on the airfoil. That is why the airfoils described in this section have sharp leading edges, so that this region of high pressure doesn't form on the front of the airfoil.

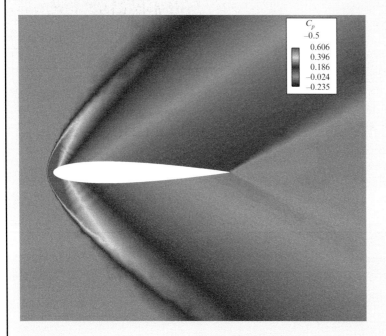

Now look at the flow using Mach number contours. You see that not only has the flow changed from supersonic to subsonic in front of the nose, but the flow then accelerates (expands) back to supersonic Mach numbers over the curved surface of the airfoil. This causes a trailing-edge shock to form, eventually bringing the flow back to freestream conditions.

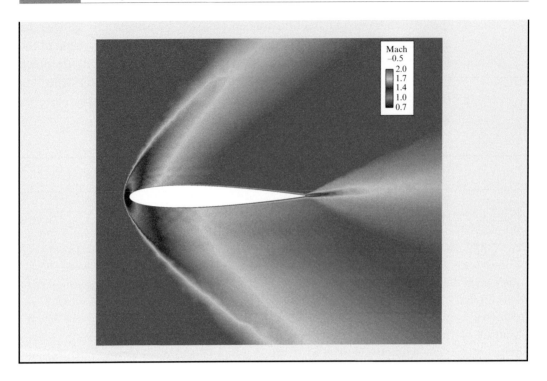

4.9.1 Supersonic Linear Theory and Airfoil Aerodynamics

Once a vehicle is flying fully in the supersonic flow regime, there are no longer any regions of subsonic flow (neglecting the boundary layer), and normal and oblique shock theories, coupled with the Prandtl-Meyer shock expansion approach, do a good job of predicting the aerodynamics of airfoils and wings (see Ref. 3 for details). These theories are relatively straightforward to apply to simple shapes, but the calculations can be somewhat cumbersome for more complex three-dimensional shapes. A simplified, linear, version of these theories can be obtained by solving the Prandtl-Glauert equation for supersonic flow. This linear approach to predicting supersonic flow can be quite useful in understanding the essential features of flight at these Mach numbers. Specifically, the Prandtl-Glauert equation used to develop the subsonic compressibility corrections can also be used to develop a linear supersonic flow theory. This derivation is accomplished by looking at Equation 4.79 for two-dimensional flow:

$$\left(1 - M_\infty^2\right)\phi_{xx} + \phi_{yy} = 0$$

which can be rewritten for supersonic flow as:

$$\left(M_\infty^2 - 1\right)\phi_{xx} - \phi_{yy} = 0. \tag{4.90}$$

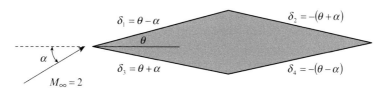

$$\delta_1 = \theta - \alpha \qquad \delta_2 = -(\theta + \alpha)$$
$$\theta$$
$$\alpha$$
$$\delta_3 = \theta + \alpha \qquad \delta_4 = -(\theta - \alpha)$$
$$M_\infty = 2$$

Figure 4.52   Symmetric diamond wedge airfoil with local surface slopes defined.

This type of partial differential equation has wave solutions that are quite useful in determining where information propagates in high-speed flight (see Ref. 3 for details of the derivation). The resulting solution for supersonic aerodynamics is quite useful: the local pressures at any point in a supersonic flow may be estimated by:

$$C_p = \frac{2\delta}{\sqrt{M_\infty^2 - 1}} \tag{4.91}$$

where δ is the angle between the freestream velocity and the local surface on a vehicle. Notice that the linear theory is only a function of two factors: the freestream Mach number (the only upstream information in the equation) and the local surface slope. The use of supersonic linear theory is quite straightforward, as we can see by applying the theory to a symmetric diamond wedge airfoil, as shown in Fig. 4.52.

The angles used to determine the pressure coefficient are not defined relative to the horizontal axis, as is usually the case; rather they represent an angle relative to the freestream. In this case, positive angles represent surfaces where the flow will "impact" the surface, and negative angles represent surfaces that are hidden from the flow ("shadow" surfaces). Therefore the pressure coefficients for the four surfaces at an angle of attack of $\alpha = 5°$ for a wedge angle of $\theta = 5°$ will be:

$$C_{p_1} = \frac{2((5-5)\pi/180)}{\sqrt{2^2-1}} = 0$$

$$C_{p_2} = \frac{2(-(5+5)\pi/180)}{\sqrt{2^2-1}} = -0.202$$

$$C_{p_3} = \frac{2((5+5)\pi/180)}{\sqrt{2^2-1}} = 0.202$$

$$C_{p_4} = \frac{2(-(5-5)\pi/180)}{\sqrt{2^2-1}} = 0.$$

A positive pressure coefficient represents a pressure that is higher than freestream, which will be true for the forward-facing surfaces, since a shock would exist in front of these surfaces and shocks compress the flow. A negative pressure coefficient represents a pressure that is lower than freestream, which will be true for the aft-facing surfaces, since an expansion fan would exist in front of these surfaces: expansion fans expand the flow. A pressure coefficient of zero means that freestream pressure acts on the surface, which is what linear theory gives for Regions 1 and 3.

The lift and drag of the airfoil at an angle of attack can be found taking into account the linear nature of the solution method. Specifically, a generally shaped airfoil can be broken down into three relatively simple shapes representing angle of attack, camber, and thickness effects (as shown in Fig. 4.53).

The application of Equation 4.91 to these shapes results in the following lift, drag, and pitching moment coefficients for the supersonic airfoil (the derivation of these results is made assuming that all angles are small, so the results are only valid for thin airfoils with small amounts of camber at low angles of attack; since we need to keep the drag low this is not a severe restriction):

$$C_l = \frac{4\alpha}{\sqrt{M_\infty^2 - 1}} \tag{4.92}$$

$$C_d = \frac{4}{\sqrt{M_\infty^2 - 1}}\left(\alpha^2 + K_1 + K_2\right) \tag{4.93}$$

$$C_{m_{LE}} = \frac{4}{\sqrt{M_\infty^2 - 1}}\left(-\frac{\alpha}{2} + K_3 + 0\right) \tag{4.94}$$

where

$$K_1 = \frac{1}{c}\int_{LE}^{TE}(dz_c/dx)^2\,dx, \; K_2 = \frac{1}{c}\int_{LE}^{TE}(dz_t/dx)^2\,dx, \; K_3 = \frac{1}{c^2}\int_{LE}^{TE}x(dz_c/dx)^2\,dx,$$

dz_c/dx is the local slope of the mean camber line of the airfoil, and dz_t/dx is the local angle of the thickness distribution. These results are quite different from the results for subsonic airfoils, if for no other reason than it is possible to obtain results for the supersonic airfoil by only knowing the geometry of the mean camber line and the thickness distribution. The lift of the supersonic airfoil (to first order) is only a function of angle of attack: camber and thickness do not aid in the production of lift! Subsonic airfoils need camber and thickness to produce the desired low pressures on the upper surface, and therefore lift. Supersonic airfoils do not work in the same way. They obtain very high pressures on the lower surface near the leading edge

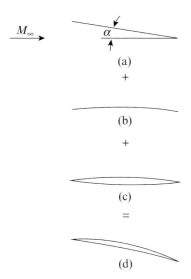

Figure 4.53 Effects of angle of attack, camber, and thickness are additive in linear theory: (a) angle of attack; (b) camber distribution; (c) thickness distribution; (d) resultant wing.

due to the shocks that form ahead of that region, which provide a large amount of the lift for the airfoil. This method of producing lift is not very efficient, however, and the lift-to-drag ratio of a supersonic airfoil is much lower than its subsonic counterpart. The lift-curve slope for the airfoil is also quite easy to estimate, since $C_{l_\alpha} = \dfrac{4}{\sqrt{M_\infty^2 - 1}}$. At a Mach number of $M_\infty = 2$,

for example, the lift-curve slope is $C_{l_\alpha} = 2.309\,/\,rad$, whereas a subsonic airfoil has a theoretical lift-curve slope of $C_{l_\alpha} = 2\pi\,/\,rad = 6.283\,/\,rad$ (nearly three times the supersonic value).

Other features of interest in the supersonic linear theory results are that the drag is a function of angle of attack, camber, and thickness. Likewise, the pitching moment is a function of angle of attack and camber (but not thickness). This means that thickness only produces negative effects on a supersonic airfoil, and camber is only beneficial for controlling pitching moment but also adds drag. This leads designers of supersonic airfoils to use airfoils similar to the symmetric diamond wedge shown in Fig. 4.52, since the sharp, flat surfaces minimize the drag for given thickness.

Finally, the center of pressure for supersonic airfoils will be near the 50 percent chord location of the airfoil, which is also quite different than subsonic airfoils that have their center of pressure near the 25 percent chord location. This represents a significant shift in center of pressure for an airfoil as it accelerates from subsonic to supersonic flight, and a factor that must be taken into account in the design of the configuration and the aircraft control system.

Computational Aerodynamics Concept Box

How to Use Supersonic Linear Theory

Find the lift coefficient, wave drag coefficient, lift-to-drag ratio, pitching moment coefficient, and center of pressure for a symmetric diamond wedge airfoil at $\alpha = 10°$ and $M_\infty = 2$. The airfoil has a wedge angle of $\delta_w = 10°$ (problem and figure from Ref. 3).

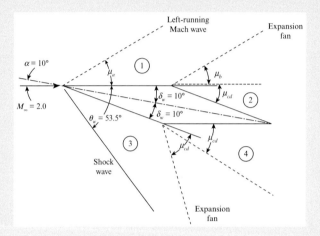

The lift coefficient for linear theory is given by Equation 4.92 and is not a function of camber or thickness, only angle of attack:

$$C_l = \frac{4\alpha}{\sqrt{M_\infty^2 - 1}} = \frac{4(10\pi/180)}{\sqrt{2^2 - 1}} = 0.4031.$$

According to Equation 4.93, the drag coefficient is a function of angle of attack, camber, and thickness, as given by:

$$C_d = \frac{4}{\sqrt{M_\infty^2 - 1}}\left(\alpha^2 + K_1 + K_2\right).$$

Since the airfoil has no camber, $K_1 = 0$. Since the local thickness variation is constant for the airfoil, $K_2 = \delta_w^2$, which yields:

$$C_d = \frac{4}{\sqrt{2^2 - 1}}\left((10\pi/180)^2 + (10\pi/180)^2\right) = 0.1407.$$

The lift-to-drag ratio (neglecting skin friction) is:

$$\frac{l}{d} = \frac{C_l}{C_d} = \frac{0.4031}{0.1407} = 2.865.$$

The pitching moment coefficient is given by Equation 4.94 as:

$$C_{m_{LE}} = \frac{4}{\sqrt{M_\infty^2 - 1}}\left(-\frac{\alpha}{2} + K_3 + 0\right).$$

But since there is no camber, $K_3 = 0$ and

$$C_{m_{LE}} = -\frac{2\alpha}{\sqrt{M_\infty^2 - 1}} = -\frac{2(10\pi/180)}{\sqrt{2^2 - 1}} = -0.2015,$$

and the center of pressure is given by:

$$x_{cp} = -\frac{C_{m_{LE}}}{C_l}c = -\frac{-0.2015}{0.4031}c = 0.50c.$$

In other the words, the center of pressure is at the mid-chord of the airfoil.

4.9.2 VOLUMETRIC WAVE DRAG

The key idea to supersonic flight is the area rule, which basically states that we want a smooth area distribution along the length of the vehicle. Anything that leads to discontinuities in the area distribution will lead to increased wave drag, and anything that minimizes discontinuities will decrease wave drag. Moreover, there are specific shapes that produce minimum drag for axisymmetric bodies, which we will discuss. The derivation of the wave drag integral is given in Ashley and Landahl,[44] and results in what is known as slender body theory:

$$D_{wave} = -\frac{\rho_\infty V_\infty^2}{4\pi} \int_0^l \int_0^l S''(x_1) S''(x_2) \ln|x_1 - x_2| dx_1 dx_2 \qquad (4.95)$$

where $S(x)$ is the cross-sectional area at each station along the body, and x_1 and x_2 are variables of integration representing different positions along the length of the body. This equation is used to analyze different cross-sectional area distributions and also to find shapes with minimum wave drag. In particular, note that the Mach number does not appear as a function, and this integral requires that the ends be closed, that is $S'(l) = 0$.

The integral shows that it is actually the second derivative of the area distribution that is required for the computation, which is why we want to make changes in the cross-sectional area small. Also, unless care is taken, the numerical method may result in values that are too high because of artificial noise in the interpolation procedures and the quality of the input data for the area distribution, S. This difficulty was substantially reduced with a rather ingenious scheme due to Eminton,[45] and adopted for use in the now-standard wave drag program written by Boeing for Roy Harris at NASA Langley (known as the Harris Wave Drag program).[46] Eminton's approach to finding the value of the integral was to find the interpolating curve passing through the specified input points that minimized the value of the wave drag. Thus she solved an optimization problem to eliminate issues arising artificially from the interpolation procedure.

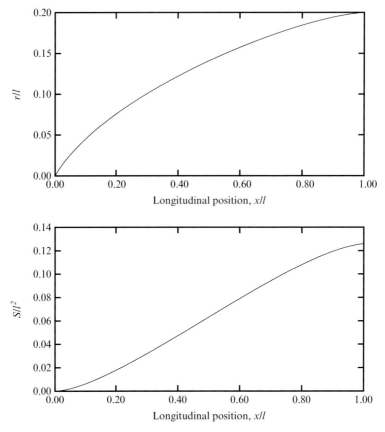

Figure 4.54 Radius and area distribution of a von Kármán ogive.

Minimum wave drag shapes have been found using the wave drag integral: for an open base the minimum wave drag is the von Kármán ogive (see Appendix A for the equation of the shape). Given the base area and length, the drag is:

$$D_{wave} = \frac{2\rho_\infty V_\infty^2}{\pi} \frac{S(l)^2}{l^2}.$$
(4.96)

Or, if the reference area is taken as the base area:

$$C_{D_{wave}} = \frac{4}{\pi} \frac{S(l)}{l^2}.$$
(4.97)

The radius and area distribution for the von Kármán ogive are given in Fig. 4.54.

If the body is closed at both ends, then for a given length, l, and volume, $\mathcal{V}$, the wave drag is:

$$D_{wave} = \frac{\rho_\infty V_\infty^2}{\pi} \frac{64\mathcal{V}^2}{l^4},$$
(4.98)

which yields a drag coefficient based on maximum cross-sectional area of:

$$C_{D_{wave}} = \frac{24\mathcal{V}}{l^3}. \tag{4.99}$$

Note: this is *not* the minimum drag for a given maximum cross-sectional area, which may often be a more relevant constraint. However, using the connection between volume and maximum body radius:

$$\mathcal{V} = \pi r_{\max}^2 \frac{3\pi l}{16}. \tag{4.100}$$

we get a form that shows the connection between the drag coefficient and fineness ratio, l/d:

$$C_{D_{wave}} = \frac{9\pi^2}{8} \frac{1}{\left(l/d_{\max}\right)^2}. \tag{4.101}$$

The radius and area distribution for this shape, known as the Sears-Haack body, are given in Fig. 4.55. Once we get past saying "smooth area," the next step is to make the area distribution for the entire airplane close to the axisymmetric minimum wave drag shapes described earlier. A summary of minimum wave drag axisymmetric body shapes subject to a variety of constraints is contained in the paper by Adams.[47]

4.9.3 Wing Aerodynamics

Three-dimensional wings in supersonic flow have very different wing-tip effects than do subsonic wings. While wing-tip vortices still exist on supersonic wings, the fact that information cannot propagate upstream in supersonic flow greatly limits the impact of the vortices. Figure 4.56 shows how this is applied to a rectangular wing traveling at supersonic speeds. The assumption is that disturbances in the flow are only felt with the region of the Mach cone, which is defined by the Mach angle, $\mu = \sin^{-1}(1/M_\infty)$. Therefore, the majority of the wing in this case actually experiences two-dimensional aerodynamics, which would be defined by the airfoil results. A detailed discussion of how to evaluate supersonic wings is presented in Ref. 3.

In addition, supersonic wings are designed based on the wing sweep angle relative to the Mach angle, as shown in Fig. 4.57. When the wing sweep angle is less than the Mach angle, as shown in Fig. 4.57a, the component of the freestream flow normal to the leading edge will be supersonic. This wing will require supersonic airfoil characteristics, such as a thin airfoil with sharp leading and trailing edges and little or no camber. This will create a wing with good supersonic aerodynamic characteristics, but also with poor subsonic performance. When the wing sweep angle is greater than the Mach

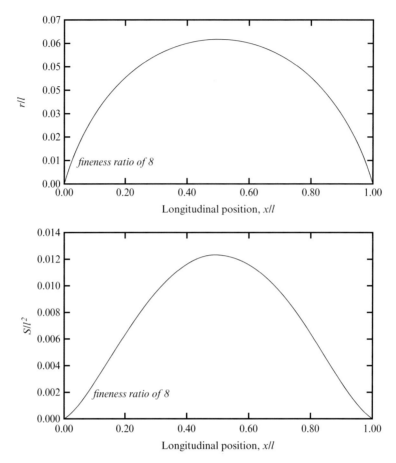

Radius and area distribution of a Sears-Haack body.

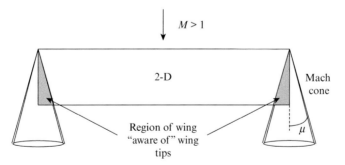

Rectangular wing in supersonic flow showing regions of two- and three-dimensional flow (Ref. 3).

angle, as shown in Fig. 4.57b, the component of the freestream flow normal to the leading edge will be subsonic. This wing will require subsonic airfoil characteristics, such as a thicker airfoil with rounded leading edges and fairly significant camber, but also with poorer supersonic characteristics. Typically, subsonic leading edges have been more commonly used since they are able to produce significant amounts of leading-edge suction.

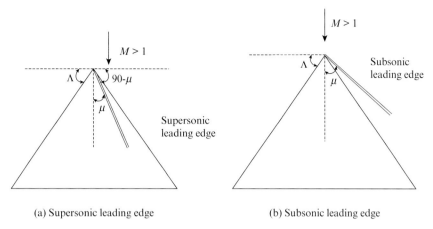

(a) Supersonic leading edge   (b) Subsonic leading edge

Figure 4.57   Wings with supersonic and subsonic leading edges. (a) Supersonic leading edge; (b) Subsonic leading edge.

Figure 4.58   USAF F-15E Super Eagle with NACA airfoils and a 45 degree swept leading edge wing (U.S. Air Force photo).

In general, aircraft that fly both at subsonic and supersonic speeds use airfoils with a combination of both characteristics, such as on the F-15 Eagle shown in Fig. 4.58. The F-15 has thin subsonic airfoils (the wing root uses a NACA 64A(.055)5.9 airfoil, 5.9% thickness ratio, and the tip uses a NACA 64A203 airfoil, 3% thickness ratio) with a wing sweep of 45 degrees.

4.10 Hypersonic Aerodynamics

Vehicles have been flying at hypersonic speeds since Pete Knight flew the North American X-15 at 7274 km/h (4520 miles per hour) at 58,400 m (102,000 ft). This remarkable accomplishment of flight at a Mach number of 6.7 occurred more than forty years ago. But what is hypersonic flow and

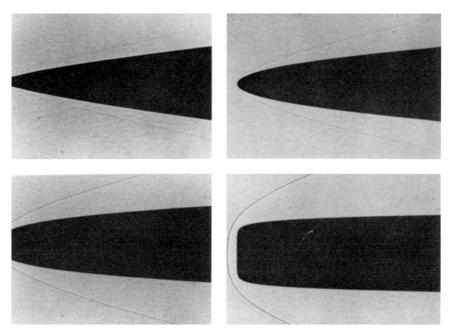

Figure 4.59 Hypersonic flow past various power-law bodies (Ref. 5).

why is it different than supersonic flow? Using the definition for the Mach number, the basic assumption for all *hypersonic flow* theories is:

$$M_\infty \equiv \frac{V_\infty}{a_\infty} \gg 1. \qquad (4.102)$$

This definition is usually assumed to be true for Mach numbers greater than about $M_\infty \approx 5$. The effects that distinguish hypersonic flow from supersonic flow are:[2,3]

- thin shock layers
- viscous-inviscid interactions that cannot be neglected
- entropy layer
- high-temperature effects and extreme heat transfer
- low-density flows (typically due to re-entry or an attempt to achieve orbit or high altitude)

4.10.1 Importance of Temperature in Hypersonic Flow

Thus, for hypersonic flows, the internal thermodynamic energy of the freestream fluid particles is small when compared with the kinetic energy of the freestream. For flight applications, M_∞ is very large because the freestream velocity is very high while the freestream thermodynamic state remains fixed. The flow slows down as it crosses the shock wave that envelops the vehicle, producing extremely high temperatures in the shock layer. The kinetic energy of the air particles in the flowfield associated with a vehicle in hypersonic flight is converted into increasing the temperature of the air and into endothermic reactions, such as dissociation and ionization of the air near the vehicle's surface. The mechanisms for this conversion include adiabatic compression and viscous energy dissipation. Heat is transferred to the surface from the high-temperature air in the shock layer. The rate at which heat is transferred to the surface depends upon many factors, including the freestream conditions, the configuration of the vehicle and its orientation to the flow, the difference between the temperature of the air in the shock layer and the temperature of the vehicle's surface, and the surface catalycity. In order to generate solutions for the flowfield, the designer must simultaneously solve the continuity equation, the momentum equation, and the energy equation. Some of these flow features can be seen in Fig. 4.59, including the very thin shock layers for the pointed body configurations.

The importance of temperatures in hypersonic flow is easily illustrated using the relation between stagnation temperature and static temperature:

$$\frac{T_0}{T} = 1 + \frac{\gamma - 1}{2} M_\infty^2 \tag{4.103}$$

or we could also look at the wall temperature:

$$T_{\substack{\text{adiabatic} \\ \text{wall}}} = \left(1 + r \frac{\gamma - 1}{2} M_\infty^2\right) T_e \tag{4.104}$$

where r is the recovery factor. With these relations, you would find that the limit for an aluminum structure is around Mach 2, which was the Concorde's cruise Mach number. The SR-71 is made of titanium, and temperature limits the speed to slightly over Mach 3.

Thus hypersonic aerodynamic configuration design means that you must always deal with heating effects. In general, at sustained high speeds, surfaces must be cooled, and since heating is a critical concern, this means that viscous effects are crucial immediately. Also, unlike low speed flight, hypersonic Reynolds numbers may be so low at high altitudes that the boundary layer is laminar, so laminar flows are often of interest. This is important because the heat transfer is much lower when the flow is laminar. In fact, being able to estimate the transition location with certainty is a critical requirement in hypersonic vehicle design, and is the subject of current research.[48]

The X-15, shown in Fig. 4.60, is the only real manned hypersonic airplane flown to date. It was rocket powered, and started flight by being dropped

Figure 4.60 The X-15 experimental hypersonic aircraft (Courtesy of NASA Dryden Flight Research Center).

from a B-52, so it was purely a research airplane. The first flight was by Scott Crossfield in June 1959. The X-15 reached 314,750 feet in July 1962; an improved version reached a Mach number of 6.7 and an altitude of 102,100 feet in October 1967 with Pete Knight at the controls. The X-15 program flew 199 flights, with the last one occurring in October 1968.

4.10.2 Newtonian and Modified Newtonian Flow Theory

For the windward surface of relatively simple shapes, we can assume that the speed and direction of the gas particles in the freestream remain unchanged until they strike the solid surface exposed to the flow. For this flow model, which is termed *Newtonian flow theory* (since it is similar in character to the one described by Newton in the seventeenth century), the normal component of momentum of an impinging fluid particle is wiped out, while the tangential component is conserved. Using the nomenclature of Fig. 4.61, the equation for the local surface pressure coefficient is:

$$C_p = \frac{p_s - p_\infty}{\rho_\infty U_\infty^2 / 2} = 2 \sin^2 \theta_b = 2 \cos^2 \phi. \qquad (4.105)$$

This result is even simpler than the linear theory we looked at for supersonic flow. The local pressure coefficient in this case is only a function of the local flow angle (the result is independent of Mach number). The pressure

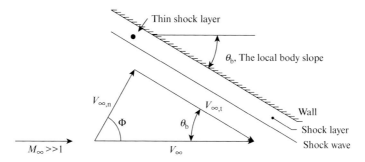

Figure 4.61 Notation for Newtonian flow theory (Ref. 49).

coefficient presented in Equation 4.105 is specifically for "impact" surfaces (surfaces that the freestream flow directly impacts). For surfaces that are not impacted by the flow (known as "shadow" surfaces), the theory assumes

$$C_p = 0. \tag{4.106}$$

Or, in other words, shadow surfaces are assumed to have freestream pressure acting on them.

However, since measurements and computations of the stagnation-point value of the pressure coefficient are always less than two, Lees[50] proposed an alternative representation of the pressure coefficient for hypersonic flow, replacing the factor "2" by the appropriate value of the stagnation-point pressure at the flow conditions of interest, which is known as *Modified Newtonian flow theory*:

$$C_p = C_{p,t2} \sin^2 \theta_b \tag{4.107}$$

where $C_{p,t2}$ could be found experimentally or by using the Rayleigh-pitot formula[51] which calculates the stagnation pressure behind a normal shock as:

$$C_{p,t2} = \frac{2}{\gamma M_\infty^2} \left\{ \frac{p_{t2}}{p_\infty} - 1 \right\}, \quad \frac{p_{t2}}{p_\infty} = \left[\frac{(\gamma+1)}{2} M_\infty^2 \right]^{\frac{\gamma}{\gamma-1}} \left[\frac{\gamma+1}{2\gamma M_\infty^2 - (\gamma-1)} \right]^{\frac{1}{\gamma-1}} \tag{4.108}$$

Despite the dramatic improvements in computational fluid dynamics (CFD) in the fifty years since the introduction of Modified Newtonian flow theory, the theory continues to be used often to generate engineering approximations for the pressures acting on configurations in hypersonic flows.

4.10.3 Aerodynamic Heating

Finally, one of the most important issues for hypersonic vehicles is the maximum heating rates which often take place in the vicinity of the stagnation

Figure 4.62   X-38 experimental crew rescue vehicle (Courtesy of NASA Dryden Flight Research Center).

region at the nose. The flowfield in the vicinity of the stagnation point often exhibits many of the characteristics listed at the beginning of this section, including: thin shock layers, viscous-inviscid interactions, entropy layers, high-temperature effects, and extreme heat transfer. Calculations for this region are necessarily quite challenging, and the classic treatise on the subject was developed by Fay and Riddell.[52] A more useful correlation for the stagnation point heat transfer was developed by Detra et al.,[53] and is given by:

$$\dot{q}_t = \frac{17,600}{(R_N)^{0.5}} \left(\frac{\rho_\infty}{\rho_{SL}}\right)^{0.5} \left(\frac{U_\infty}{U_{co}}\right)^{3.15} \tag{4.109}$$

where R_N is the node radius in feet, ρ_{SL} is the sea level density, U_{co} is the velocity for a circular orbit, and $\dot{q}_t$ is the stagnation point heat transfer in $Btu / ft^2 - s$. Notice the importance of the nose radius on the heat transfer, with a sharp nose creating very high heat transfer and blunter noses creating much lower values. This is why most re-entry vehicles have blunt noses, which not only have lower heat transfer but also have more mass in the nose (due to the nose radius) for dealing with the high heat loads.

Since hypersonic flow is often so severe, vehicles that are designed to fly hypersonically have unusual characteristics, such as the X-38 experimental crew rescue vehicle shown in Fig. 4.62. Notice the blunt nose and rounded bottom of the vehicle, which helps reduce heating rates on re-entry. Also,

X-43A hypersonic vehicle (Courtesy of NASA Dryden Flight Research Center).

all of the control surfaces are located on the leeside of the vehicle so that they are not directly in the high-velocity, high-temperature regions of the flowfield.

4.10.4 Engine/Airframe Integration

Any air-breathing hypersonic vehicle will have a highly integrated engine and airframe; it is hard to tell exactly what counts as the vehicle and what counts as the propulsion system. This is exemplified in the concepts developed for recent aerospace planes. In these concepts the propulsion is at least in part provided by a scramjet engine, which obtains thrust with a combustion chamber in which the flow is supersonic. Figure 4.63 shows that the entire forebody of the vehicle underside is used as an external inlet to provide flow at just the right conditions to the engine. The entire underside afterbody is the exhaust nozzle.

The first U.S. vehicle to use a scramjet engine was the Hyper-X, now known as the X-43. The initial flight attempt was made in June 2001, but failed before having a chance to demonstrate the operation of the scramjet. After a long investigation and a revised procedure, it flew successfully on March 27, 2004; the X-43A is shown in Fig. 4.63. The November-December 2001 issue of the *Journal of Spacecraft and Rockets* had a special section devoted to the Hyper-X.

Computational Aerodynamics Concept Box

How to Use Hypersonic Theory on a Cone

Determine the drag coefficient for a 15 degree half-angle sharp cone at zero degrees angle of attack for $1.5 \leq M_\infty \leq 7$ using supersonic cone theory, Newtonian flow theory, and Modified Newtonian flow theory.

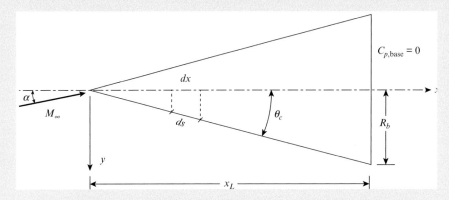

The incremental area of the cone is given by:

$$dA = 2\pi r ds = 2\pi r \frac{dx}{\cos\theta}$$

From the figure

$$\tan\theta = \frac{r}{x} \qquad x = \frac{r}{\tan\theta} \qquad dr = dx\tan\theta$$

The drag on the cone is:

$$D = \int_0^{r_b} p_c \sin\theta \, dA - p_b \pi r_b^2 = \int_0^{r_b} p_c \sin\theta\left(2\pi r \frac{dx}{\cos\theta}\right) - p_b \pi r_b^2$$

$$= \int_0^{r_b} 2\pi r p_c \, dr - p_b \pi r_b^2 = \pi r_b^2 \left(p_c - p_b\right)$$

The drag coefficient for the cone is defined to be $C_D \equiv D / (q_\infty S_b)$ which yields:

$$C_D = \frac{p_c - p_b}{q_\infty}$$

If the base pressure is assumed to be freestream pressure:

$$C_D = \frac{p_c - p_\infty}{q_\infty} = C_p.$$

For a cone with a 15 degree half angle, supersonic cone theory values can be found in Ref. 3. From Newtonian flow theory:

$$C_D = C_p = 2\sin^2\theta = 0.134.$$

From Modified Newtonian flow theory:

$$C_D = C_p = C_{p,t2} \sin^2 \theta$$

which will be a function of Mach number based on using Equation 4.92. The comparison between the three methods is shown in the table.

M_∞	C_D (cone theory)	C_D (Newtonian)	C_D(ModNewtonian)
1.5	0.240	0.134	0.103
2.0	0.202	0.134	0.111
3.0	0.173	0.134	0.118
4.0	0.161	0.134	0.120
5.0	0.154	0.134	0.121
6.0	0.150	0.134	0.122
7.0	0.148	0.134	0.122

A comparison of Newtonian theory predictions with experimental data for the lift and drag coefficients of various cones at $M_\infty = 6.8$ was made by Penland.[54] The comparison shown here is for a cone angle of 5 degrees and shows how good the results can be for realistic conical configurations at hypersonic Mach numbers.

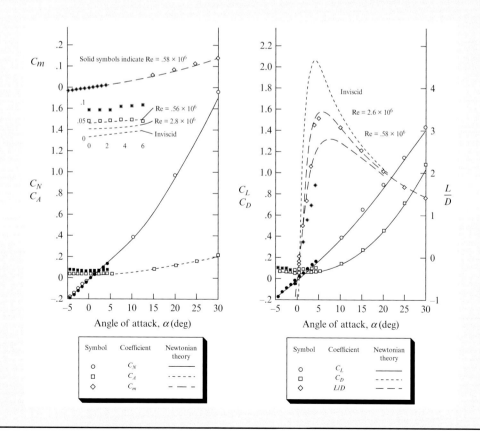

Summary of Best Practices

1. As was mentioned in Chapter 1, understanding flow physics is a key element in computational aerodynamics. This includes having a working knowledge of concepts such as:
   a. lift, drag, and pitching moment
   b. viscous flow concepts such as laminar and turbulent boundary layers and their effects, transition, and separation
   c. pressure distributions and their impact
   d. how airfoils work and which geometric parameters affect their aerodynamics
   e. how wings work and which geometric parameters affect their aerodynamics
2. You need to understand the assumptions and limitations of various aerodynamic theories which you may use to compare with CA predictions.
3. You also need to understand the specific characteristics of the flow regimes you are interested in computing:
   a. subsonic flow (incompressible and compressible)
   b. transonic flow
   c. supersonic flow
   d. hypersonic flow
4. Understand different airplanes and their uses in order to have a better understanding of their aerodynamics – all of this can come in handy when trying to predict and understand aerodynamics using CA tools.
5. It would be good to adopt the humble attitude of A.M.O. Smith that was stated at the beginning of the chapter: "I'm *still* trying to learn about aerodynamics!"

4.11 Exercises

1. Find the lift, drag, and pitching moment about the quarter chord for the NACA 4- or 5-digit airfoil you obtained data for in Chapter 1. Assume the airfoil has a chord of $2\,ft$ flying at an angle of attack of $\alpha = 12°$ at $V_\infty = 244\,ft\,/\,s$ at sea level standard day conditions.

2. Find the lift coefficient, wave drag coefficient, lift-to-drag ratio, pitching moment coefficient, and center of pressure for a symmetric diamond wedge airfoil for $\alpha = 0° - 10°$ and $M_\infty = 3$. The airfoil has a wedge angle of $\delta_w = 5°$.

3. Determine the drag coefficient for a 5 degree half-angle sharp cone at zero degrees angle of attack for $1.5 \leq M_\infty \leq 7$ using supersonic cone theory, Newtonian flow theory, and Modified Newtonian flow theory. Assume that the base pressure equals freestream pressure.

4. Aerodynamic results are often found in terms of axial and normal force coefficients (body axis), not lift and drag coefficients (wind axis). Derive the relation between the wind axis and the body axis.

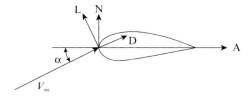

5. Derive an expression for the aerodynamic center location. The final answer (knowing the value of the moment at several alphas about x_1/c) is approximately:

$$\frac{x_{ac}}{c} = \frac{x_1}{c} - \frac{dC_{m_{x_1/c}}}{dC_L}$$

6. Derive the expression relating the center of pressure to the aerodynamic center (approximately) as:

$$\frac{x_{cp}}{c} = \frac{x_{ac}}{c} - \frac{C_{m_{ac}}}{C_L}$$

7. CFD codes should always have an input to define the moment reference center; however you do not have to rerun a case to find the pitching moment about another reference location! Show that the general relation for the pitching moment coefficient at a point x_2/c relative to the pitching moment coefficient about a point at x_1/c, where x is measured positive aft from the leading edge, is:

$$C_{m_{x_2/c}} = C_{m_{x_1/c}} + \left(C_L \cos\alpha + C_D \sin\alpha\right)\left(\frac{x_1}{c} - \frac{x_2}{c}\right)$$

8. Using the parabolic drag polar, Equation 4.76, derive the expression for L/D_{max} and the related value of C_{Lmax}. At L/D_{max} how are the drag due to lift and the zero lift drag related?

4.12 Projects

1. Consider the NACA 4- or 5-digit airfoil you obtained data for in Chapter 1. Using the mean camber line coordinate definition found in Appendix A, write a thin airfoil theory program (following the development in Section 4.6.9). Use the program to predict the following aerodynamic information: C_l vs. α, $\alpha_{l=0}$, C_{l_α}, $C_{m_{c/4}}$ vs. α, x_{cp} vs. α. Compare your results with the experimental data of your chosen airfoil and the semi-empirical prediction you made for the airfoil in Chapter 2. Comment on the ability

of airfoil theory to predict airfoil aerodynamics. Where does the theory work well? Where does the theory not work well? Why or why not?

2. Show how the lift-curve slope for the airfoil you chose in Chapter 1 would vary at subsonic compressible Mach numbers using the Prandtl-Glauert rule (see Section 4.7.5). Also, find the minimum pressure coefficient for your airfoil at two or three relatively low angles of attack ($\alpha \leq 10°$) and estimate the critical Mach number by graphing the compressibility correction results on the same graph as the sonic pressure coefficient given by:

$$C_{p,sonic} = \frac{2}{\gamma M_\infty^2} \left(\left[\frac{1 + \frac{\gamma - 1}{2} M_\infty^2}{1 + \frac{\gamma - 1}{2}} \right]^{\gamma/\gamma - 1} - 1 \right)$$

where γ is the ratio of specific heats (for air we can assume that $\gamma = 1.4$). Where the two curves cross the pressure coefficient at the minimum pressure location is the same as the sonic pressure coefficient, meaning that the velocity at that point is sonic, which is the definition of the critical Mach number. Graph M_{cr} versus α and comment on your results. How do you think the results would change if you made the airfoil thicker rather than increased the angle of attack?

3. Create a rectangular wing which uses your chosen airfoil from Chapter 1. The wing should have an aspect ratio of $AR = 8$. Using the results of lifting-line theory from Section 4.7.4, find the lift-curve slope of your wing assuming the airfoil sections are the same as your chosen airfoil and are untwisted. Also find the induced drag coefficient for your wing. Finally, using the methods of Section 4.5.2, estimate the drag coefficient of your airfoil due to skin friction assuming the Reynolds number is $Re = 6.0 \times 10^6$. Add the skin friction drag coefficient (which we will assume to be the zero-lift drag coefficient for your airfoil) to the induced drag coefficient and graph the drag polar for your wing (C_L vs. C_D).

4. Write a program to find the pressure coefficients on all four faces of a symmetric diamond wedge airfoil at supersonic speeds assuming the pressures are given by: (1) supersonic linear theory, (2) Newtonian flow theory, and (3) Modified Newtonian flow theory. Also find the lift, drag, and pitching moment coefficients as a function of Mach number from $M_\infty = 2$ to $M_\infty = 6$ for $\delta_w = 10°$.

4.13 References

1 Smith, A.M.O., "High-Lift Aerodynamics," *Journal of Aircraft*, Vol. 12, No. 6, 1975, pp. 501–530.

2 Anderson, J.D., *Fundamentals of Aerodynamics*, 5th Ed., Boston: McGraw-Hill, 2010.

3 Bertin, J.J., and Cummings, R.M., *Aerodynamics for Engineers*, 6th Ed., Upper Saddle River: Pearson, 2014.

4 McCormick, B.W., *Aerodynamics, Aeronautics, and Flight Mechanics*, 2nd Ed., Hoboken: John Wiley and Sons, 1995.

5 Van Dyke, M., Ed., *An Album of Fluid Motion*, Stanford: Parabolic Press, 1982.

6 Karamcheti, K., *Principles of Ideal-Fluid Aerodynamics*, Malabar: Krieger Publishing Co., 1980.

7 Churchill, R.V., and Brown, J.W., *Complex Variables and Applications*, New York: McGraw-Hill, 1984.

8 Prandtl, L., and Tietjens, O.G., *Applied Hydro- and Aeromechanics*, New York: Dover Publications, 1957.

9 Cebeci, T., and Smith, A.M.O., *Analysis of Turbulent Boundary Layers*, Orlando: Academic Press, 1974.

10 White, F.M., *Viscous Fluid Flow*, 3rd Ed., New York: McGraw-Hill, 2006.

11 Schlichting, H., Gersten, K., Krause, E., and Oertel, H., *Boundary-Layer Theory*, 8th Ed., New York: Springer, 2004.

12 Dommasch, D.O., Sherby, S.S., and Connolly, T.F., *Airplane Aerodynamics*, 4th Ed., New York: Pitman Publishing Corp., 1967.

13 Brandt, S.A., Stiles, R.J., Bertin, J.J., and Whitford, R., *Introduction to Aeronautics: A Design Perspective*, 2nd Ed., Reston: AIAA, 2004.

14 Lyon, C.A., Broeren, A.P., Giguere, P., Gopalarathnam, A., and Selig, M.S., *Summary of Low-Speed Airfoil Data – Volume 3*, Virginia Beach: SoarTech Publications, 1997.

15 Abbott, I.H., and von Doenhoff, A.E., *Theory of Wing Sections*, New York: Dover, 1959.

16 Eppler, R. *Airfoil Design and Data*, Berlin: Springer-Verlag, 1990.

17 Theodorson, T., and Naimen, I., "Pressure Distributions for Representative Airfoils and Related Profiles," NACA TN 1016, February 1946.

18 Polhamus, E.C., "A Survey of Reynolds Number and Wing Geometry Effects on Lift Characteristics in the Low Speed Stall Region," NASA CR 4745, June 1996.

19 Prandtl, L., and Tietjens, O.G., *Fundamentals of Hydro- and Aeromechanics*, New York: Dover Publications, 1957.

20 Williams, J.E., and Vukelich, S.R., *The USAF Stability and Control Digital DATCOM*, AFFDL TR-79–3032, 1979.

21 Prandtl, L., "Applications of Modern Hydrodynamics to Aeronautics," NACA Report 116, January 1923.

22 Hoerner, S.F., *Fluid-Dynamic Drag*, Brick Town: Hoerner Fluid Dynamics, 1965.

23 Liepmann, H.W., and Roshko, A., *Elements of Gasdynamics*, New York: Dover Publications, 2001.

24 H.H. Hurt, Jr., *Aerodynamics for Naval Aviators*, Rev. ed., Renton: Aviation Supplies and Academics, January 1965.

25 Liepmann, H.W., "The Interaction Between Boundary Layer and Shock Waves in Transonic Flow," *Journal of the Aeronautical Sciences*, Vol. 13, No. 12, 1946, pp. 623–637.

26 Ferrari, C., and Tricomi, F., *Transonic Aerodynamics*, New York: Academic Press, 1968.

27 Whitcomb, R.T., "Review of NASA Supercritical Airfoils," ICAS Paper 74-10, 1974 and "Advanced Transonic Aerodynamic Technology," *Advances in Engineering Science*, Vol. 4, 1976, pp. 1521–1537.

28 Ayers, T.G., "Supercritical Aerodynamics: Worthwhile over a Range of Speeds," *Astronautics and Aeronautics*, Vol. 10, No. 8, 1972, pp. 32–36.

29 NASA Dryden Flight Research Center, "F-8 Supercritical Wing (SCW) Project Description."

30 Shevell, R.S., *Fundamentals of Flight*, 2nd Ed., Englewood Cliffs: Prentice-Hall, 1989.

31 Harris, C.D., "NASA Supercritical Airfoils," NASA TP 2969, March 1990.

32 Schlichting, H., "Some Developments in Boundary Layer Research in the Past Thirty Years," *Journal of the Royal Aeronautical Society*, Vol. 64, 1960, pp. 64–79.

33 Haines, A.B., "Wing Section Design for Swept-Back Wings at Transonic Speed," *Journal of the Royal Aeronautical Society*, Vol. 61, 1957, pp. 238–244.

34 Mason, W.H., MacKenzie, D.A., Stern, M.A., and Johnson, J.K., "A Numerical Three-Dimensional Viscous Transonic Wing-Body Analysis and Design Tool," AIAA Paper 78–0101, January 1978.

35 Iglesias, S. and Mason, W.H., "Optimum Spanloads Including Wing Structural Weight," AIAA Paper 2001–5234, October 2001.

36 Jameson, A., "Re-Engineering the Design Process Through Computation," AIAA Paper 97–0641, January 1997.

37 Mason, W.H., "Analytic Models for Technology Integration in Aircraft Design," AIAA Paper 90–3262, September 1990.

38 Malone, B., and Mason, W.H., "Multidisciplinary Optimization in Aircraft Design Using Analytic Technology Models," *Journal of Aircraft*, Vol. 32, No. 2, 1995, pp. 431–438.

39 Hilton, W.F., *High Speed Aerodynamics*, London: Longmans, Green & Co., 1952.

40 Grasmeyer, J.M., Naghshineh, A., Tetrault, P.-A., Grossman, B., Haftka, R.T., Kapania, R.K., Mason, W.H., Schetz, J.A., "Multidisciplinary Design Optimization of a Strut-Braced Wing Aircraft with Tip-Mounted Engines," Virginia Tech MAD Center Report MAD 98-01-01, January 1998.

41 Mair, W.A., and Birdsall, D.L., *Aircraft Performance*, Cambridge: Cambridge University Press, 1992.

42 Erickson, B.A., "Flight Characteristics of the B-58 Mach 2 Bomber," *Journal of the Royal Aeronautical Society*, Vol. 66, No. 623, 1962, pp. 665–671.

43 Pike, I., "B-70: The State of the Art Improver," *Flight International*, June 25, 1964 and July 2, 1964.

44 Ashley, H., and Landahl. M., *Aerodynamics of Wings and Bodies*, Reading: Addison-Wesley, 1965.

45 Eminton, E., and Lord, W.T., "Note on the Numerical Evaluation of the Wave Drag of Smooth Bodies Using Optimum Area Distributions for Minimum Wave Drag," *Journal of the Royal Aeronautical Society*, January 1956, pp. 61–63.

46 Harris, R.V., "An Analysis and Correlation of Aircraft Wave Drag," NASA TM X-947, 1964.

47 Adams, M.C., "Determination of Shapes of Boattail Bodies of Revolution for Minimum Wave Drag," NACA TN 2550, November 1951.

48 Bertin, J.J., and Cummings, R.M., "Fifty Years of Hypersonics: Where We've Been, Where We're Going," *Progress in Aerospace Sciences*, Vol. 39, Issue 6–7, 2003, pp. 511–536.

49 Bertin, J.J., *Hypersonic Aerothermodynamics*, Reston: AIAA, 1994.

50 Lees, L., "Hypersonic Flow," Proc. of the 5th International Aeronautical Conference, Los Angeles CA, 1955, pp. 241–275.

51 Ames Research Staff, "Equations, Tables, and Charts for Compressible Flow," NACA Report 1135, 1953.

52 Fay, J.A., and Riddell, F.R., "Theory of Stagnation Point Heat Transfer in Dissociated Air," *Journal of the Aeronautical Sciences*, Vol. 25, 1958, pp. 73–85, 121.

53 Detra, R.W., Kemp, N.H., and Riddell, F.R., " Addendum to Heat Transfer to Satellite Vehicles Reentering the Atmosphere." *Jet Propulsion*, Vol. 27, 1957, pp. 1256–1257.

54 Penland, J.A., "Aerodynamic Force Characteristics of a Series of Lifting Cone and Cone-Cylinder Configurations at a Mach Number of 6.83 and Angles of Attack up to 130 deg," NASA TN-D-840, Jan. 1961.

5 Classical Linear Theory Computational Aerodynamics

I am interested in almost any numerical method that aids engineering analysis and design

Forrester T. Johnson,
Senior Technical Fellow of The Boeing Company,
talking about, among other things, panel methods

VSAERO Panel Method Simulation of MD-11 (Courtesy of Analytical Methods, Inc.; a full color version of this image is available in the color insert pages of this text as well as on the website: www.cambridge.org/aerodynamics).

LEARNING OBJECTIVE QUESTIONS

After reading this chapter, you should know the answers to the following questions:

- What are the basic assumptions of linear theory methods?
- What are the limitations of the methods due to these assumptions?
- What are panel methods and how are they useful?
- What is the vortex lattice method and how is it useful?
- What type of designs are linear theory methods applied to?
- What are the true benefits of using linear theory methods?

5.1 Introduction

Forrester Johnson is a highly experienced methodology developer for aircraft designers at Boeing who knows that analyzing and designing an airplane requires a variety of tools and approaches, as is evident from his comment. It may not be obvious now, but one of the things that he was referring to is the ability to perform a large number of aerodynamic predictions in a very short period of time, something important during the initial phases of any aircraft design project. But what methods can give these "quick and dirty" predictions? That is what this chapter is all about: the linear theory aerodynamic methods that form the basis of conceptual aircraft design.

Computational methods based on the linear potential flow theory (described in Chapter 4) have been used, and still are used, to perform a great deal of good aerodynamic design.[1] While some people believe that these methods are outdated, we believe they still have a valuable place in computational aerodynamics. In fact, these methods can be invaluable in conducting conceptual design trade studies to begin the process of identifying feasible aircraft configurations (see the Aircraft Design Box in Chapter 1 for a discussion about the three phases of aircraft design). The usefulness of the linear methods makes it important to understand their development, assumptions, and limitation for use as a CA tool.

In order to understand how linear methods are used in aircraft design, we will discuss two related linear theory approaches that are based on the potential flow equations and play an important role in aerodynamics: panel methods and vortex lattice methods. For the potential flow assumption to be useful for aerodynamics calculations, the essential requirement is that viscous effects are negligible, meaning that the flow is largely attached. This does not mean that there are no viscous effects, rather that the viscous effects are small. Typically, this means that the airplane is flying in a cruise configuration, but linear methods are also used for predicting flap and control surface effectiveness early in the design process. Another important limitation to linear methods is that the flow must be either completely subsonic or completely supersonic. If the flow contains regions of both subsonic and supersonic flow, then it is called transonic flow and would require a higher fidelity (nonlinear) flowfield model. Supersonic velocities can occur locally at surprisingly low freestream Mach numbers, so you should be careful to ensure that the flow is not locally supersonic when using these methods. For example, airfoils at very high lift coefficients have peak velocities around the leading edge that can become locally supersonic at freestream Mach numbers as low as $0.20 \sim 0.25$. Flows in the vicinity of extended flap systems can also yield supersonic velocities at relatively low freestream conditions. In general, however, the flowfield is at a low speed everywhere, and we can assume that the flow is incompressible (which is usually true for $M \leq 0.3$), which means that Laplace's equation (Equation 3.130) is essentially an exact representation of the inviscid flow.

For higher subsonic Mach numbers with small disturbances to the freestream flow (which is true for relatively thin airfoils at low angles of attack), the Prandtl-Glauert equation (Equation 3.125) can be used. The Prandtl-Glauert equation can be converted into Laplace's equation by a simple transformation, which means that incompressible results can be transformed into subsonic compressible results quite easily (as discussed in Chapters 3 and 4).[2] This provides the basis for estimating the initial effects of compressibility on the flowfield, i.e., "linearized" subsonic flow.

The Prandtl-Glauert equation can also be used to describe supersonic flows (in that case the mathematical type of the equation is hyperbolic, as described in Chapter 3). Recall the important distinction between the two cases:

subsonic flow: elliptic PDE where each point in the flowfield influences every other point

supersonic flow: hyperbolic PDE where discontinuities can exist and there is a strong "zone of influence" solution dependency

Although there are supersonic as well as subsonic panel methods,[3] vortex lattice methods are only subsonic (although they have been extended to high subsonic Mach numbers). We will discuss the similarities and differences between these methods in this chapter, but we will only consider incompressible flow in our derivation of panel method theory.

Computational Aerodynamics Concept Box

How Llinear Methods Are Used for Conceptual Design

Many people consider linear methods to be old-fashioned and not important to modern aircraft analysis, but nothing could be further from the truth. Because linear methods simplify the governing equations, they can run on readily accessible computer systems, including laptops. They often take only a few minutes for aircraft geometry definition, and even less time to run. The results for one configuration can then be compared to those of other configurations, making it relatively easy to determine the relative merits of many design choices.

As an example, an aircraft conceptual design phase might include looking at a variety of configuration concepts for achieving a given set of objectives. In the case shown in the figure, the designers were trying to find an optimal aircraft for a solar-powered aircraft. This is the creative side of aircraft design, when a group of designers brainstorm a relatively large number of candidate configurations that might work. Once the candidate configurations are imagined, however, there needs to be an objective way to tell which one(s) work the best, and that is where linear aerodynamic methods (such as panel methods and vortex lattice methods) come in.

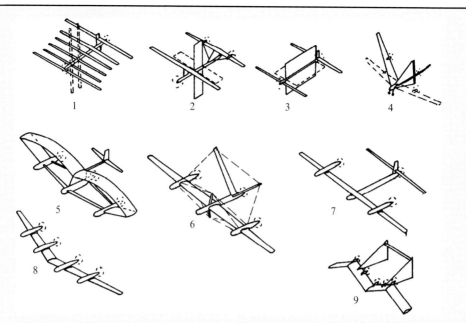

Configuration study from "A Preliminary Study of Solar Powered Aircraft and Associated Power Trains," NASA
CR 3699, 1983 by D.W. Hall, C.D. Fortenbach, E.V. Dimiceli, and R.W. Parks.

Once the configurations have been analyzed using the linear methods, comparisons can
be made of aerodynamic and performance parameters such as lift, drag, lift-to-drag ratio,
range, maximum speed, etc. These comparisons will lead the designers to determine that
certain configurations work better than others at achieving the objectives for the design,
which then allows them to concentrate their efforts on the best configurations for further
development. Typically, these methods allow for comparisons to be made quickly.

5.2 Panel Methods

Although we saw in Chapter 4 how basic shapes like circular cylinders can be
modeled with combinations of singularities that are solutions of Laplace's
equation, completely arbitrary shapes cannot be modeled exactly with singularities placed on the axis, nor can the streamlines of an arbitrary body be
determined by specifying the strengths of the singularities. Instead, we need
to use another approach for distributing singularities to model these more
complicated flowfields.

One of the key features of Laplace's equation is the property that allows
the equation governing the flowfield to be converted from a three-dimensional problem throughout the field (a PDE) to a two-dimensional problem for finding the potential variation on the surface (an integral equation).
The flowfield solution is then found using this property by distributing
"singularities" of unknown strength over discretized portions of the surface: *panels*.[4] The flowfield solution is found by representing the surface by

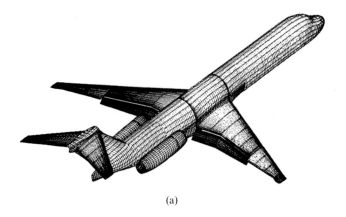

(a)

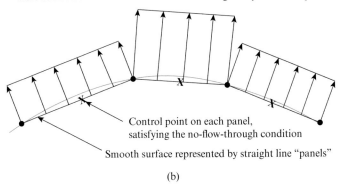

Arrows indicate constant distribution of the singularity across the panel

Control point on each panel,
satisfying the no-flow-through condition

Smooth surface represented by straight line "panels"

(b)

Figure 5.1 Representation of an airplane by a panel model. (a) a 7000 panel representation of a transport aircraft with flaps deflected (Ref. 5; reprinted by permission of the American Institute of Aeronautics and Astronautics, Inc.). (b) local surface cut showing distribution of singularities on a panel.

a number of panels and solving a linear set of algebraic equations to determine the unknown strengths of the singularities.[i] Figure 5.1 illustrates the idea, with Fig. 5.1a showing a panel representation of a commercial transport aircraft, and Fig. 5.1b showing how singularities may be distributed over each panel. The flexibility and relative economy of the panel methods is so important in practice that the methods continue to be widely used despite the availability of more exact, but more difficult to use, computational methods. More detailed discussion of the theory is available from Hess,[4,5] and, if you ever want to write your own panel methods code, you should study the works by Erickson[3] and/or Katz and Plotkin.[6]

The overall approach to the derivation of panel method theory will be to first derive the integral equation for the potential solution of Laplace's equation (Section 5.2.1), followed by the details for one specific approach

---

[i] The singularities are distributed across the panel, they are not specified at a point. However, the boundary conditions are usually satisfied at a specific location.

to solving the integral equation (Section 5.2.2). For clarity and simplicity of the algebra, the analysis will use the two-dimensional case to illustrate the methods (note that Moran[7] and Cebeci[8] contain computer programs based on linear potential theory that can be run on many computer systems). This approach will result in two ironic aspects of panel methods:

- The algebraic forms of the singularities are different between two dimensions and three dimensions (due to three-dimensional relief).
- The power of panel methods arises in three-dimensional applications.

The majority of this chapter will be devoted to examples of applications to aerodynamic analysis so you can gain a feel for how well panel methods work in different situations, but the basic concept development will also be included. Then you can decide if these linear methods are useful for your particular application.

5.2.1 The Integral Equation for the Potential

Potential theory is a well-developed (meaning it is fairly old) and elegant mathematical theory, devoted to the solution of Laplace's equation (as discussed in Sections 3.12.5 and 4.3 and Appendix C):

$$\Phi_{xx} + \Phi_{yy} + \Phi_{zz} = \nabla^2 \Phi = 0. \tag{5.1}$$

There are several ways to view the solution of this equation. The one most familiar to aerodynamicists uses the notion of "singularities." These are algebraic functions that satisfy Laplace's equation and can be combined to construct flowfields. Since Laplace's equation is linear, superposition of solutions can be used (that means any two solutions can be added together to create another solution). The most familiar singularities used in aerodynamics are the point source (or sink), doublet, and vortex. In classical examples the singularities are located inside the body; unfortunately, as noted, an arbitrary body shape cannot be created using singularities placed inside the body. A more sophisticated approach has to be used to determine the potential flow over arbitrary shapes, and mathematicians have spent a great deal of time developing this theory. We will draw on a few selected results to help understand the development of panel methods.

Before we develop the theory, however, we will look at the specification of the boundary conditions for this approach, as illustrated in Fig. 5.2. The flow pattern is uniquely determined by giving one of two types of boundary conditions (as discussed in Sec. 3.10):

$$\Phi \text{ on } \Sigma + \kappa \ \{\text{Dirichlet problem : used for design}\} \tag{5.2}$$

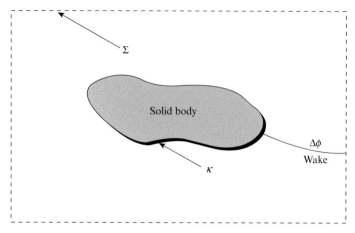

Figure 5.2   Boundaries for flowfield analysis.

or

$$\partial\Phi\,/\,\partial n \text{ on } \Sigma + \kappa \left\{ \text{Neumann problem : used for analysis} \right\} \qquad (5.3)$$

Potential flow theory states that you cannot specify both conditions arbitrarily but you can have a mixed boundary condition, $a\Phi + b\partial\Phi\,/\,\partial n$ on $\Sigma + \kappa$ (which is known as a Robin or "third" condition, as discussed in Ch. 3). The Neumann problem is identified as "analysis" because it naturally corresponds to the problem where the flow across a surface is specified (usually zero to serve as an inviscid boundary condition). The Dirichlet problem is identified as "design" because it tends to correspond to the aerodynamic case where a surface pressure distribution is specified and the body shape corresponding to that pressure distribution is sought. Because of the wide range of problem formulations available in linear theory, some analysis procedures appear to be Dirichlet problems, but Equation 5.3 must still be used. Figure 5.2 also shows a wake behind the body, across which the value of the potential is allowed to "jump." This allows for different velocities across the wake (but not different pressures) and is required to allow the flowfield to produce a value for the lift on the body.

Some other key properties of potential flow theory are:

- If either Φ or $\partial\Phi\,/\,\partial n$ is zero everywhere on $\Sigma + \kappa$, then $\Phi = 0$ at all interior points
- Φ cannot have a maximum or minimum at any interior point; the maximum value can only occur on the surface boundary, and therefore the minimum pressure (and maximum velocity) occurs on the surface

Now that the boundary conditions are described, we need to obtain the equation for the potential in a form suitable for use in panel method

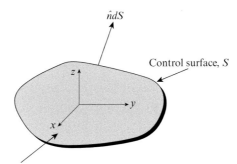

Figure 5.3   Nomenclature for integral equation derivation.

calculations (we will follow the basic approach of Karamcheti,[9] Katz and Plotkin,[6] and Moran[7]). The objective is to obtain an expression for the potential anywhere in the flowfield in terms of values on the surface bounding the flowfield. Starting with the Gauss Divergence Theorem, which relates a volume integral and a surface integral (given previously in Ch. 3 as Equation 3.8),

$$\iint_S \vec{A} \cdot \hat{n} dS = \iiint_V \vec{\nabla} \cdot \vec{A} dV;$$ (5.4)

the classical derivation considers the interior problem as shown in Fig. 5.3.

To start the derivation, we introduce the vector function of two scalars:

$$\vec{A} = \omega \vec{\nabla} \chi - \chi \vec{\nabla} \omega$$ (5.5)

If we substitute this function into the Gauss Divergence Theorem, Equation 5.4, we can find:

$$\iiint_V \vec{\nabla} \cdot \left(\omega \vec{\nabla} \chi - \chi \vec{\nabla} \omega\right) dV = \iint_S \left(\omega \vec{\nabla} \chi - \chi \vec{\nabla} \omega\right) \cdot \hat{n} dS.$$ (5.6)

Now we can use the vector identity $\vec{\nabla} \cdot \sigma \vec{F} = \sigma \vec{\nabla} \cdot \vec{F} + \vec{F} \cdot \vec{\nabla} \sigma$ to simplify the left-hand side of Equation 5.6 as:

$$\begin{aligned}
\vec{\nabla} \cdot \left(\omega \vec{\nabla} \chi - \chi \vec{\nabla} \omega\right) &= \vec{\nabla} \cdot \left(\omega \vec{\nabla} \chi\right) - \vec{\nabla} \cdot \left(\chi \vec{\nabla} \omega\right) \\
&= \omega \vec{\nabla} \cdot \vec{\nabla} \chi + \vec{\nabla} \chi \cdot \vec{\nabla} \omega - \chi \vec{\nabla} \cdot \vec{\nabla} \omega - \vec{\nabla} \omega \cdot \vec{\nabla} \chi \\
&= \omega \vec{\nabla}^2 \chi - \chi \vec{\nabla}^2 \omega.
\end{aligned}$$ (5.7)

Substituting the result of Equation 5.7 for the integrand in the left-hand side of Equation 5.6, we find that:

$$\iiint_V \left(\omega \vec{\nabla}^2 \chi - \chi \vec{\nabla}^2 \omega\right) dV = \iint_S \left(\omega \vec{\nabla} \chi - \chi \vec{\nabla} \omega\right) \cdot \hat{n} dS$$ (5.8)

or equivalently (recalling that $\vec{\nabla}\chi \cdot \hat{n} = \partial\chi / \partial n$),

$$\iiint_{\mathcal{V}}\left(\omega\vec{\nabla}^2\chi - \chi\vec{\nabla}^2\omega\right)d\mathcal{V} = \iint_{S}\left(\omega\frac{\partial\chi}{\partial n} - \chi\frac{\partial\omega}{\partial n}\right)dS. \tag{5.9}$$

Either statement (Equation 5.8 or 5.9) is known as Green's Theorem of the second form.

Now we can define $\omega = 1/r$ and $\chi = \Phi$, where Φ is a harmonic function (a function that satisfies Laplace's equation). The $1/r$ term is a source singularity in three dimensions, which makes our analysis three-dimensional. In two dimensions, the source singularity takes the form $\ln r$, and a two-dimensional analysis starts by defining $\omega = \ln r$. Now rewrite Equation 5.8 using the definitions of ω and χ given earlier and switching sides to obtain

$$\iint_{S}\left[\frac{1}{r}\vec{\nabla}\Phi - \Phi\vec{\nabla}\left(\frac{1}{r}\right)\right]\cdot\hat{n}dS = \iiint_{\mathcal{V}}\left[\frac{1}{r}\vec{\nabla}^2\Phi - \Phi\vec{\nabla}^2\left(\frac{1}{r}\right)\right]d\mathcal{V}. \tag{5.10}$$

Recall from Fig. 5.3 that $\mathcal{V}$ is the region enclosed by the surface S. Recognize that on the right-hand side of the first term, $\vec{\nabla}^2\Phi = 0$ from Equation 5.1, so that Equation 5.10 becomes:

$$\iint_{S}\left[\frac{1}{r}\vec{\nabla}\Phi - \Phi\vec{\nabla}\left(\frac{1}{r}\right)\right]\cdot\hat{n}dS = -\iiint_{\mathcal{V}}\Phi\vec{\nabla}^2\left(\frac{1}{r}\right)d\mathcal{V}. \tag{5.11}$$

If any point P is external to S, then $\vec{\nabla}^2(1/r) = 0$ everywhere since $1/r$ is a source and thus has to satisfy Laplace's equation. This leaves the right-hand side of Equation 5.11 equal to zero, with the following result:

$$\iint_{S}\left[\frac{1}{r}\vec{\nabla}\Phi - \Phi\vec{\nabla}\left(\frac{1}{r}\right)\right]\cdot\hat{n}dS = 0. \tag{5.12}$$

However, we have included the origin in our region S as defined. If P is inside S, then $\vec{\nabla}^2(1/r) \to \infty$ at $r = 0$. Therefore, we exclude this point by defining a new region which excludes the origin by drawing a sphere of radius ε around $r = 0$ and applying Equation 5.12 to the region between ε and S:

$$\underbrace{\iint_{S}\left[\frac{1}{r}\vec{\nabla}\Phi - \Phi\vec{\nabla}\left(\frac{1}{r}\right)\right]\cdot\hat{n}dS}_{\text{arbitrary region}} - \underbrace{\iint_{\varepsilon}\left(\frac{1}{r}\frac{\partial\Phi}{\partial r} + \frac{\Phi}{r^2}\right)dS}_{\text{sphere}} = 0 \tag{5.13}$$

or:

$$\iint_{\varepsilon}\left(\frac{1}{r}\frac{\partial\phi}{\partial r} + \frac{\phi}{r^2}\right)dS = \iint_{S}\left[\frac{1}{r}\vec{\nabla}\phi - \phi\vec{\nabla}\left(\frac{1}{r}\right)\right]\cdot\hat{n}dS. \tag{5.14}$$

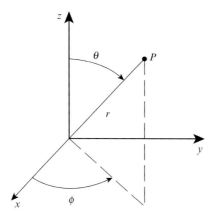

Figure 5.4   Spherical coordinate system nomenclature.

Consider the first integral on the left-hand side of Equation 5.14. First we need to let $\varepsilon \to 0$ and assume that $\Phi \approx$ constant ($\partial\Phi / \partial r = 0$, assuming that Φ is well behaved). Then we need to evaluate:

$$\iint_\varepsilon \frac{dS}{r^2}$$

over the surface of the sphere where $\varepsilon = r$. Recall that for a sphere the elemental area is:[10]

$$dS = r^2 \sin\theta \, d\theta d\phi \qquad (5.15)$$

where we defined the angles being used in Fig. 5.4 (do not confuse the classical notation for the spherical coordinate angles with the potential function; the spherical coordinate ϕ will disappear as soon as we evaluate the integral).

Substituting for dS in the integral above, we obtain:

$$\iint_\varepsilon \sin\theta \, d\theta d\phi$$

Integrating for $\theta = 0$ to π, and $\phi = 0$ to 2π, we get:

$$\int_{\phi=0}^{\phi=2\pi} \int_{\theta=0}^{\theta=\pi} \sin\theta \, d\theta d\phi = 4\pi. \qquad (5.16)$$

The final result for the first integral in Equation 5.14 is:

$$\iint_\varepsilon \left(\frac{1}{r} \frac{\partial\Phi}{\partial r} + \frac{\Phi}{r^2} \right) dS = 4\pi\Phi. \qquad (5.17)$$

Replacing the result of this integral into Equation 5.14, we can write the expression for the potential at any point P as (where the origin can be placed anywhere inside S):

$$\Phi(P) = \frac{1}{4\pi} \iint_S \left[\frac{1}{r} \vec{\nabla}\Phi - \Phi\vec{\nabla}\left(\frac{1}{r}\right) \right] \cdot \hat{n} \, dS \tag{5.18}$$

and the value of Φ at any point P is now known as a function of Φ and $\partial\Phi / \partial n$ on the boundary.

We used the interior region to allow the origin to be written at point P. This equation can be extended to the solution for Φ for the region exterior to V. Apply the results to the region between the surface S_B of the body and an arbitrary surface Σ enclosing S_B, and then let Σ go to infinity. The integrals over Σ go to Φ_∞ as Σ goes to infinity. Thus potential flow theory is used to obtain the important result that the potential at any point P' in the flowfield outside the body can be expressed as:

$$\Phi(P') = \Phi_\infty - \frac{1}{4\pi} \iint_{S_B} \left[\frac{1}{r} \vec{\nabla}\Phi - \Phi\vec{\nabla}\left(\frac{1}{r}\right) \right] \cdot \hat{n} \, dS. \tag{5.19}$$

Here the unit normal $\hat{n}$ is now considered to be pointing outward and the area can include not only solid surfaces but also wakes. Equation 5.19 can also be written using the dot product of the unit normal and the gradient as:

$$\Phi(P') = \Phi_\infty - \frac{1}{4\pi} \iint_{S_B} \left[\frac{1}{r} \frac{\partial\Phi}{\partial n} - \Phi\frac{\partial}{\partial n}\left(\frac{1}{r}\right) \right] dS. \tag{5.20}$$

The $1/r$ term in Equation 5.19 can be interpreted as a source of strength $\partial\Phi / \partial n$, and the $\nabla(1/r)$ term in Equation 5.19 as a doublet of strength Φ. Therefore, we can find the potential as a function of a distribution of sources and doublets over the surface. The integral in Equation 5.20 is normally broken up into body and wake pieces, where the wake is generally considered to be infinitely thin. Therefore, only doublets are used to represent the wakes.

Now consider the potential to be given by the superposition of two different known functions, the first and second terms in the integral, Equation 5.20. These can be taken to be the distribution of the source and doublet strengths, σ and μ, respectively (refer to Chapter 4 and Appendix C for discussions about sources and doublets). Now Equation 5.20 can be written in the form usually seen in the literature,

$$\Phi(P') = \Phi_\infty - \frac{1}{4\pi} \iint_{S_B} \left[\sigma\frac{1}{r} - \mu\frac{\partial}{\partial n}\left(\frac{1}{r}\right) \right] dS. \tag{5.21}$$

The problem is to find the values of the unknown source and doublet strengths, σ and μ, for a specific geometry and given freestream, Φ_∞.

What just happened? We replaced the requirement to find the solution over the entire flowfield (a three-dimensional problem) with the problem of finding the solution for the singularity distribution over a surface (a two-dimensional problem). In addition, we now have an integral equation to solve for the unknown surface singularity distributions instead of a partial differential equation. The problem is linear, which allows us to use the super-position principle to construct solutions. We also have the freedom to pick whether to represent the solution as a distribution of sources or doublets distributed over the surface. In practice it has been found best to use a com-bination of sources and doublets. The theory can also be extended to include other singularities if required.

At one time, the change from a three-dimensional to a two-dimensional problem was considered significant, although the total information content is the same computationally since the three-dimensional matrix is large but "sparse," and the two-dimensional matrix is small but "dense." As compu-tational methods for sparse matrix solutions evolved, the problems became nearly equivalent. The real advantage in using the panel methods arises because there is no need to define a grid throughout the flowfield since we only have to create panels on the solid surface.

This is the theory that justifies panel methods, i.e., that we can represent the surface by panels with distributions of singularities placed on them. Special precautions must be taken when applying the theory described here. Care should be used to ensure that the region S_B is in fact completely closed, and you must ensure that the outward normal is properly defined.

Furthermore, in general, the interior problem cannot be ignored. Surface distributions of sources and doublets affect the interior region as well as the exterior. In some methods the interior problem is implicitly satisfied. In other methods the interior problem requires explicit attention. The need to consider this subtlety arose when advanced panel methods were first devel-oped. The problem is not well posed unless the interior problem is consid-ered, and numerical solutions failed when this aspect of the problem was not addressed (Refs. 3 and 6 provide further discussion). When the exterior and interior problems are formulated properly, the boundary value problem is properly posed.[11,12]

We implement these ideas by:

- approximating the surface by a series of line segments (which creates panels)
- placing distributions of sources and vortices or doublets on each panel

There are many ways to tackle the solution of the panel method problem (and many resulting computer codes are available). As we will discuss,

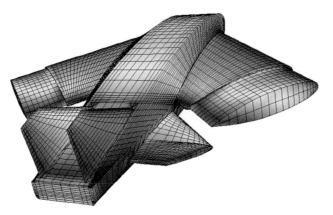

Panel model representation of a "morphing" airplane where surface shades depict pressure levels (wakes not shown) (Ref. 13; a full color version of this image is available on the website: www.cambridge.org/aerodynamics).

possible differences in approaches to the implementation include the use of:

- various types of singularities
- various distributions of the singularity strength over each panel
- panel geometry (panels do not have to be flat)

Recall that superposition allows us to construct the solution by adding separate contributions. Thus we write the potential as the sum of several contributions. Figure 5.5 provides an example of a panel representation of an airplane being used to develop the aerodynamic characteristics of a morphing airplane for a flight simulation. The surface is colored to represent the pressure distribution on the plane predicted by the panel model. The wakes are not shown, and a more precise illustration of a panel method representation will be given in Section 5.2.4.

An example of the implementation of a panel method is carried out in Section 5.2.2 in two dimensions. To do this, we write down the two-dimensional version of Equation 5.21. In addition, we use a vortex singularity in place of the doublet singularity (Refs. 6 and 7 provide details on this change). The resulting expression for the potential is:

$$\Phi = \underbrace{\Phi_\infty}_{\substack{\text{uniform onset flow} \\ =V_\infty x\cos\alpha+V_\infty y\sin\alpha}} + \int_S \left[\underbrace{\frac{q(s)}{2\pi}\ln r}_{\substack{\lambda \text{ is the 2D} \\ \text{source strength}}} - \underbrace{\frac{\gamma(s)}{2\pi}\theta}_{\substack{\text{vortex singularity} \\ \text{of strength } \gamma(s)}} \right] dS \qquad (5.22)$$

and $\theta = \tan^{-1}(y/x)$. Equation 5.22 shows contributions from various components of the flowfield, but the relation is still exact; no small disturbance assumption has been made. Note that $q(s)$ is the source strength, and $\gamma(s)$ is the vortex strength.

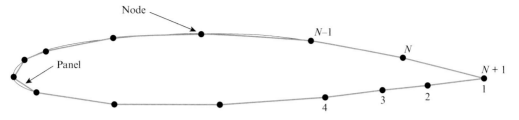

Figure 5.6 Representation of a smooth airfoil with straight-line segments.

5.2.2 An Example of a Panel Code: The Classic Hess and Smith Method

A.M.O. Smith at Douglas Aircraft directed an incredibly productive aerodynamics development group from the late 1950s through the early 1970s. In this section we describe the implementation of the theory that originated in his group.[14] The approach is to: (a) break up the surface into straight line segments, (b) assume the source strength distribution is constant over each line segment (panel) but has a different value for each panel, q_i, and (c) distribute a vortex singularity distribution over each panel, but with the vortex strength constant and equal over each panel, γ. You can think of the constant vortices as adding up to the circulation to satisfy the Kutta condition, while the sources are required to satisfy flow tangency on the surface (thickness).

Figure 5.6 illustrates the representation of the smooth airfoil surface by a series of line segments. The numbering system starts at the lower surface trailing edge and proceeds forward, around the leading surface and aft to the upper surface trailing edge. $N+1$ points define N panels (note that other implementations may use other numbering schemes).

The potential relation given in Equation 5.22 can then be evaluated by breaking up the integral into segments along each panel:

$$\Phi = V_\infty \left(x \cos\alpha + y \sin\alpha \right) + \sum_{j=1}^{N} \int_{\text{panel } j} \left[\frac{q(s)}{2\pi} \ln r - \frac{\gamma}{2\pi} \theta \right] dS \qquad (5.23)$$

where $\lambda(s)$ is taken to be constant on each panel, allowing us to write $\lambda(s) = \lambda_i$, $i = 1, \ldots N$. Here we need to find N values of λ_i and one value of γ.

We will use Fig. 5.7 to define the nomenclature on each panel. Let the ith panel be the one between the ith and $i+1^{\text{th}}$ nodes, and let the ith panel's inclination to the x axis be θ (see Fig. 5.7a). Under these assumptions, the sine and cosine of θ are given by (see Fig. 5.7b):

$$\sin\theta_i = \frac{y_{i+1} - y_i}{l_i}, \qquad \cos\theta_i = \frac{x_{i+1} - x_i}{l_i} \qquad (5.24)$$

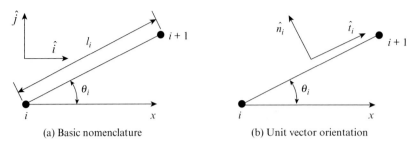

(a) Basic nomenclature   (b) Unit vector orientation

Figure 5.7   Nomenclature for local coordinate systems.

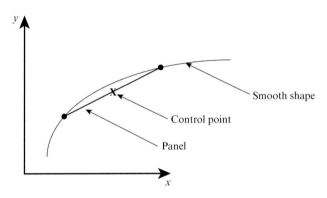

Smooth shape

Control point

Panel

Figure 5.8   Local panel nomenclature.

and the normal and tangential unit vectors are:

$$\hat{n}_i = -\sin\theta_i \hat{i} + \cos\theta_i \hat{j}$$
$$\hat{t}_i = \cos\theta_i \hat{i} + \sin\theta_i \hat{j}.$$

(5.25)

We will find the unknowns by satisfying the flow tangency condition on each panel at one specific control point (also known as a collocation point) and requiring the overall solution to satisfy the Kutta condition (see Section 4.4.3). The control point will be picked to be at the midpoint of each panel, as shown in Fig. 5.8.

Thus the coordinates of the midpoint of the control point are given by:

$$\bar{x}_i = \frac{x_i + x_{i+1}}{2}, \qquad \bar{y}_i = \frac{y_i + y_{i+1}}{2}$$

(5.26)

and the velocity components at the control point $\bar{x}_i, \bar{y}_i$ are $u_i = u(\bar{x}_i, \bar{y}_i)$, $v_i = v(\bar{x}_i, \bar{y}_i)$.

The flow tangency boundary condition is given by $\vec{V} \cdot \hat{n} = 0$ and is written using the relations given here as (in the original coordinate system in Fig. 5.7b):

$$\left(u_i \hat{i} + v_i \hat{j}\right) \cdot \left(-\sin\theta_i \hat{i} + \cos\theta_i \hat{j}\right) = 0$$

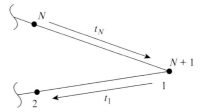

Figure 5.9 Trailing-edge panel nomenclature.

or

$$-u_i \sin \theta_i + v_i \cos \theta_i = 0, \quad \text{for each } i,\, i = 1, \ldots, N. \tag{5.27}$$

The remaining relation is found from the Kutta condition. This condition states that the flow must leave the trailing edge smoothly. Many different numerical approaches have been adopted to satisfy this condition, but, in practice, this implies that at the trailing edge the pressures on the upper and lower surfaces are equal. Here the Kutta condition is satisfied approximately by equating velocity components tangential to the panels adjacent to the trailing edge on the upper and lower surfaces. Because of the importance of the Kutta condition in determining the flow, the solution is extremely sensitive to the flow details at the trailing edge. Since the assumption is made that the velocities are equal on the top and bottom panels at the trailing edge, we need to make sure that the last panels on the top and bottom are small and of equal length, otherwise we have an inconsistent approximation (accuracy will deteriorate rapidly if the trailing-edge panels are not the same length). The specific numerical formula is developed using the nomenclature for the trailing edge shown in Fig. 5.9. In two dimensions, and especially for a single airfoil, the Kutta condition is sufficient to handle the wake, and we do not have to address the wake explicitly in the formulation (this is not the case in three dimensions, where a full wake representation is required).

Equating the magnitude of the tangential velocities on the upper and lower surface:

$$u_{t_1} = u_{t_N} \tag{5.28}$$

and taking the difference in direction of the tangential unit vectors into account, this is written as:

$$\vec{V} \cdot \hat{t}\big|_1 = -\vec{V} \cdot \hat{t}\big|_N . \tag{5.29}$$

Carrying out the operation in the original coordinate system we get the relation:

$$\left(u_1\hat{i} + v_1\hat{j}\right) \cdot \left(\cos\theta_1\hat{i} + \sin\theta_1\hat{j}\right) = -\left(u_N\hat{i} + v_N\hat{j}\right) \cdot \left(\cos\theta_N\hat{i} + \sin\theta_N\hat{j}\right)$$

which is expanded to obtain the final relation:

$$u_1 \cos\theta_1 + v_1 \sin\theta_1 = -u_N \cos\theta_N + v_N \sin\theta_N. \qquad (5.30)$$

The expression for the potential in terms of the singularities on each panel and the boundary conditions derived earlier for the flow tangency and Kutta conditions are used to construct a system of linear algebraic equations for the strengths of the sources and the vortex. The steps required to obtain a solution are summarized as:

1. Find the algebraic equations defining the "influence" coefficients. These are the relations connecting the velocities induced by the singularity distribution of unit strength over a panel at a control point. Each control point will have an influence coefficient for each of the panels on the surface and is a function of the geometry.
2. Write down the velocities, u_i, v_i, in terms of contributions from all the singularities. This includes q_i, γ from each panel and the influence coefficients.

To generate the system of algebraic equations:

3. Write down flow tangency conditions in terms of the velocities (N equations, $N+1$ unknowns).
4. Write down the Kutta condition equation to get the $N+1$ equation.
5. Solve the resulting linear algebraic system of equations for the unknown q_i and γ.
6. Given q_i and γ, write down the equations for u_{ti}, the tangential velocity at each panel control point.
7. Determine the pressure distribution from Bernoulli's equation using the tangential velocity on each panel.

While the details for this process are easily carried out, the algebra can get quite tedious. You can see why panel methods are best done on a computer, since solving large matrices requires a great deal of computational power.

Profiles in Computational Aerodynamics: A.M.O. Smith

The "Smith" in the Hess-Smith panel code was Apollo Milton Olin Smith (1911–1997). A.M.O., often called Amo by his friends and co-workers, was chief aerodynamics engineer for research at Douglas Aircraft Company in Long Beach, California. Prior to working at Douglas, he studied mechanical and aeronautical engineering at the California Institute of Technology (Caltech), receiving master of science degrees in both fields. He was interested in aeronautics because of the time he spent in high school building gliders with his friends. After graduation from Caltech in 1938, he joined Douglas as assistant chief aerodynamicist.

During this period, he worked on aerodynamic and preliminary design problems of the DC-5 and the SBD dive bomber, as well as the A-20, DB-7, and B-26 attack bombers (he had prime responsibility for detailed aerodynamic design of the B-26).

Because of earlier work with rockets at Caltech, he was asked by General H. H. "Hap" Arnold to organize and head the engineering department at Aerojet as their first chief engineer (on leave of absence from Douglas, from 1942 to 1944). After expanding the department from 6 to more than 400 people, and overseeing Aerojet into production on Jet Assisted Take Off (JATO) units, he returned to Douglas and aerodynamics. There he handled aerodynamics for the D-558-1 Skystreak and the F4D-1 Skyray, both of which held world speed records. In 1948, he moved into the research aspect of aerodynamics, where he developed powerful methods of calculating potential and boundary-layer flows, culminating in a book co-authored with Tuncer Cebeci entitled *Analysis of Turbulent Boundary Layers*. In this period, he oversaw development of practical methods of analyzing laminar and turbulent boundary-layer flow, new and improved static pressure probes, the hydrogen bubble technique of flow visualization, potential flow analysis, analysis of stability and transition of boundary layers, and the e^n method of predicting boundary-layer transition. Of his work on panel methods he said that his boss at Douglas had asked him to look over the available methods, which he did, and he "discovered an entirely new method that was far more powerful than any of the existing ones. I got an OK to try and work out the details of a computer programming method and got it to work quite successfully. The method is now known as the Panel Method and is universally used for low speed flow analysis" (from *Legacy of a Gentle Genius: The Life of A.M.O. Smith*, Ed. T. Cebeci, Long Beach: Horizons Publishing, 1999). In June 1975 he retired from what was then McDonnell Douglas and was appointed adjunct professor at UCLA, a position he held until 1980.

For his work at Aerojet, he received the Robert H. Goddard Award from the American Rocket Society (which would later become AIAA). For his early rocket work at Caltech, he is commemorated in bronze at the NASA Jet Propulsion Laboratory. In 1970, he received the Casey Baldwin Award of the Canadian Aeronautics and Space Institute (CASI). He was also named a member of the National Academy of Engineering and a Fellow of the AIAA and was awarded an honorary D.Sc. degree from the University of Colorado.

(This is the only profile in the book of someone who is no longer living. Much of this sketch was written by A.M.O. Smith in his article, "High-Lift Aerodynamics," *Journal of Aircraft*, Vol. 12, No. 6, 1975, pp. 501–530; additional details were provided by Professor Tuncer Cebeci.)

5.2.3 Program *PANEL*

In this section we illustrate the results of the procedure outlined above. Program *PANEL* is an exact implementation of the analysis just described and is essentially the program given by Moran.[7] Other panel method programs are available in the textbooks by Cebeci,[8] Houghton and Carpenter,[15] and Kuethe and Chow.[16] Two other similar programs are available: a MATLAB program, *PABLO*, written at KTH in Sweden that is available on the web,[17] as well as the program by Mark Drela at MIT, *XFOIL*.[18] Moran's and Drela's programs include routines to generate the ordinates for NACA 4-digit and 5-digit airfoils (see Appendix A for a description of these airfoil sections and Appendix D for descriptions and/or links to some of the programs). The main drawback of several of these approaches is the requirement for a trailing-edge thickness that is exactly zero. To accommodate this restriction, the ordinates generated are often altered slightly from the "official" ordinates (the extension of the program to handle arbitrary airfoils is an exercise at the end of the chapter). The freestream velocity in *PANEL* is assumed to be unity, since the inviscid solution in coefficient form is independent of scale.

PANEL's node points are distributed employing the widely used cosine spacing function. The equation for this spacing is given by defining the points on the thickness distribution to be placed at:

$$\frac{x_i}{c} = \frac{1}{2}\left[1 - \cos\left\{\frac{(i-1)\pi}{(N-1)}\right\}\right] \qquad i = 1, ..., N. \qquad (5.31)$$

These locations are then altered when camber is added (see Equations A.1 and A.2 in Appendix A). This approach is used to provide a smoothly varying distribution of panel node points that concentrate points around the leading and trailing edges.

An example of the accuracy of program *PANEL* is illustrated in Fig. 5.10, where the results from *PANEL* for the NACA 4412 airfoil pressure distribution

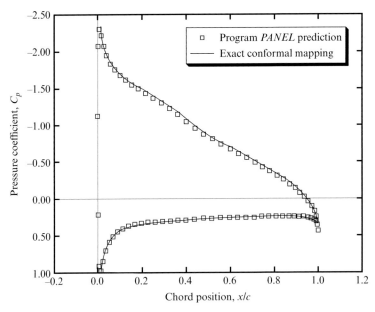

Figure 5.10   Comparison of results from program *PANEL* with an exact mapping solution for the NACA 4412 airfoil at 6° angle of attack.

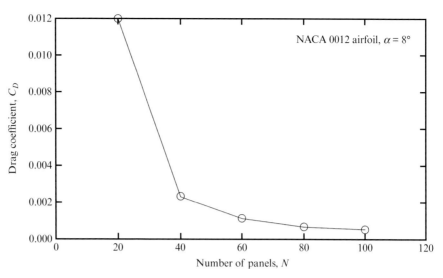

Figure 5.11 Change of drag with number of panels.

are compared with results obtained from an exact conformal mapping of the airfoil (conformal mapping methods were described in Chapter 4; conformal transformations can also be used to generate meshes of points for use in CFD methods). The agreement between the panel method and the conformal transformation approach is nearly perfect.

Whenever starting the prediction for a geometry, you need to conduct numerical studies to determine how many panels are required to obtain accurate results. Both forces and moments and pressure distributions should be examined. Since you select the number of panels used to represent the surface, how many panels should you use? Most computational programs provide the user with freedom to decide how detailed (expensive – in money or time) the calculations should be. One of the first things the user should do is evaluate how detailed the calculation should be to obtain the level of accuracy desired. In *PANEL*, as in most codes, the control is obtained through the number of panels used.

We can check the sensitivity of the solution to the number of panels by comparing force and moment results and pressure distributions as a function of the numbers of panels. Figures 5.11 and 5.12 present the change of drag and lift, respectively, by varying the number of panels. For *PANEL*, which uses an inviscid incompressible flowfield model, the drag should be exactly zero, and the drag coefficient found by integrating the pressures over the airfoil is an indication of the error in the numerical scheme. The drag obtained using a surface (or "nearfield") pressure integration is a numerically sensitive calculation and is a strict test of the method. The figures show the drag going to zero and the lift becoming constant as the number of panels increases. In this style of presentation, it is hard to see exactly how quickly the solution is converging to a fixed value.

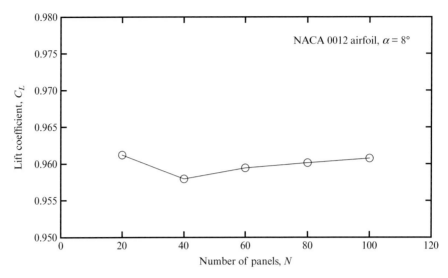

Change of lift with number of panels.

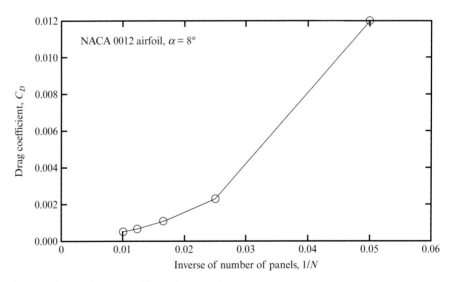

Change of drag with the inverse of the number of panels.

The results given in Figures 5.11 and 5.12 indicate that 60–80 panels (30 upper panels and 30 lower panels, for example) should be enough. Note that the lift coefficient is presented in an extremely expanded scale, and the drag coefficient presented in Fig. 5.11 also uses an expanded scale. Because drag is typically a small number, it is frequently described in drag counts, where 1 drag count is a C_D of 0.0001.

To estimate the limit for an infinitely large number of panels, the results can be plotted as a function of the reciprocal of the number of panels. Therefore, the limit result occurs as $1/n$ goes to zero. Figures 5.13, 5.14, and 5.15 present

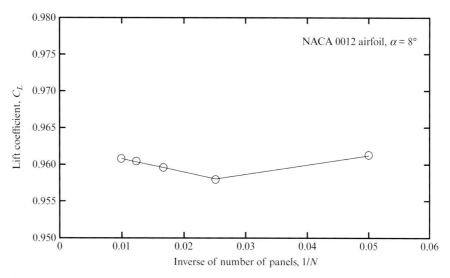

Figure 5.14 Change of lift with the inverse of the number of panels.

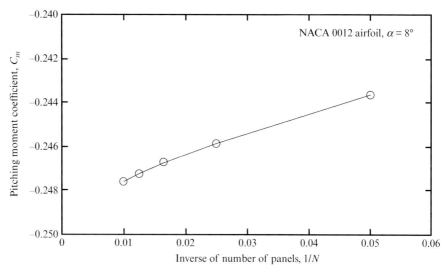

Figure 5.15 Change of pitching moment with the inverse of the number of panels.

the results in this manner for the case given previously and with the pitching moment included for examination in the analysis.

The results given in Fig. 5.13 through 5.15 show that the program *PANEL* produces results that are relatively insensitive to the number of panels once fifty or sixty panels are used, and by extrapolating to $1/n = 0$, an estimate of the limiting value can be obtained.

In addition to forces and moments, the sensitivity of the pressure distributions to changes in panel density must also be investigated: pressure distributions are shown in Figures 5.16 and 5.17. The twenty- and sixty-panel results are given in Fig. 5.16. In this case, it appears that the pressure distribution

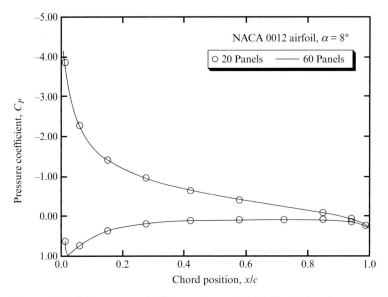

Figure 5.16 Pressure distribution from program *PANEL*, comparing results using 20 and 60 panels.

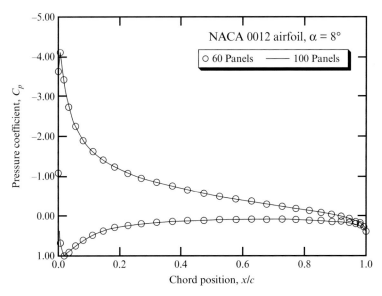

Figure 5.17 Pressure distribution from program *PANEL*, comparing results using 60 and 100 panels.

is well defined with sixty panels. This is confirmed in Fig. 5.17, which demonstrates that it is almost impossible to identify the differences between the sixty- and one hundred-panel cases. This type of study should (in fact *must*) be conducted when using computational aerodynamics methods. The results of this study show that sixty panels are sufficient for obtaining good estimates of the forces and moments on the airfoil, since there is no added benefit to using larger numbers of panels.

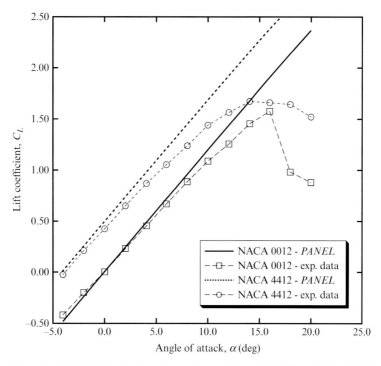

Figure 5.18 Comparison of *PANEL* lift predictions with experimental data for NACA 0012 and 4412 airfoils (Ref. 19).

Having examined the convergence of the mathematical solution, we can now investigate the agreement of our results with experimental data. Figure 5.18 compares the lift coefficients from the inviscid solutions obtained from *PANEL* with experimental data from Abbott and von Doenhoff.[19] Agreement is good at low angles of attack, where the flow is fully attached. The agreement deteriorates as the angle of attack increases, and viscous effects start to show up as a reduction in lift with increasing angle of attack, until, finally, the airfoil stalls. The inviscid solutions from *PANEL* cannot capture this part of the flow physics. The different stall character between the two airfoils arises due to different flow separation locations on the airfoils. The cambered airfoil separates at the trailing edge first. Stall occurs gradually as the separation point moves forward on the airfoil with increasing incidence. The uncambered airfoil stalls due to a sudden separation at the leading edge. An examination of the difference in pressure distributions could be made to see why this might be the case.

The pitching moment characteristics are also important. Figure 5.19 provides a comparison of the *PANEL* pitching moment predictions (taken about the quarter chord point) with experimental data. In this case, the calculations indicate that the computed location of the aerodynamic center, $dC_m/dC_L = 0$, is not exactly at the quarter chord, although the experimental data are very close to this value. The uncambered NACA 0012 data show nearly zero pitching moment until flow separation starts to occur. The

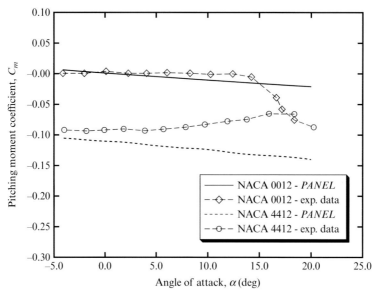

Figure 5.19   Comparison of *PANEL* moment predictions with experimental data for NACA 0012 and 4412 airfoils (Ref. 19).

cambered airfoil shows a significant nose-down pitching moment, C_{m0}, and a trend with angle of attack due to viscous effects that is exactly opposite the inviscid prediction. This occurs because the separation location is moving forward from the trailing edge of the airfoil and the load over the aft portion of the airfoil does not increase as fast as the forward loading. This leads to a nose-up pitching moment until eventually the separation causes the airfoil to stall, resulting in a nose-down pitching moment (taken about the quarter chord).

We will not compare the drag prediction from *PANEL* with experimental data. For two-dimensional incompressible inviscid flow, the drag is theoretically zero. In the actual case, drag arises from skin friction effects, further additional form drag due to the small change of pressure on the body due to the boundary layer (which primarily prevents full pressure recovery at the trailing edge), and from drag due to increased viscous effects at higher angles of attack. A well-designed airfoil will have a drag value very nearly equal to the skin friction and nearly invariant with incidence until the maximum lift coefficient is approached.

In addition to the force and moment comparisons, we need to compare the pressure distributions predicted with *PANEL* to experimental data (Fig. 5.20 provides one example). The NACA 4412 experimental pressure distribution is compared with *PANEL* predictions; in general the agreement is very good. The primary area of disagreement is at the trailing edge. Here viscous effects act to prevent the recovery of the experimental pressure to the levels predicted by the inviscid solution. The disagreement on the lower surface is a little surprising and suggests that the angle of attack from the experiment may not be as precise as would be desirable.

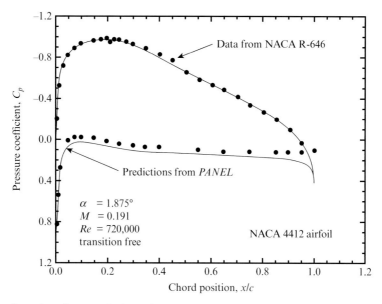

Figure 5.20 Comparison of pressure distribution from *PANEL* with experimental data for a NACA 0012 airfoil (Ref. 20).

Panel methods often have trouble with accuracy at the trailing edge of airfoils with cusped trailing edges, when the included angle at the trailing edge is zero. Figure 5.21 shows the predictions of program *PANEL* compared with an exact mapping solution (a *FLO36*[21] run at low Mach number) for two cases. Figure 5.21a is for a case with a small trailing-edge angle: the NACA 65_1-012, while Fig. 5.21b is for the more standard 6A version of the airfoil. The corresponding airfoil shapes are shown in Fig. 5.22. The "loop" in the pressure distribution in Fig. 5.21a is an indication of a problem with the panel method.

This case demonstrates a situation where this particular panel method is not accurate, which can have practical implications for the user. The 6-series airfoils were theoretically derived by specifying a pressure distribution and determining the required shape. The small trailing-edge angles (less than half those of the 4-digit series), cusped shape, and the unobtainable zero thickness specified at the trailing edge resulted in objections from the aircraft industry. These airfoils were very difficult to manufacture and use on operational aircraft. Subsequently, the 6A-series airfoils were introduced to remedy the problem. These airfoils had larger trailing-edge angles (approximately the same as the 4-digit series), and were made up of nearly straight (or flat) surfaces over the last 20 percent of the airfoil. Most applications of 6-series airfoils today actually use the modified 6A-series thickness distribution. This is definitely an area where the user should check the performance of a particular panel method with a variety of airfoils prior to using the code to make a general prediction.

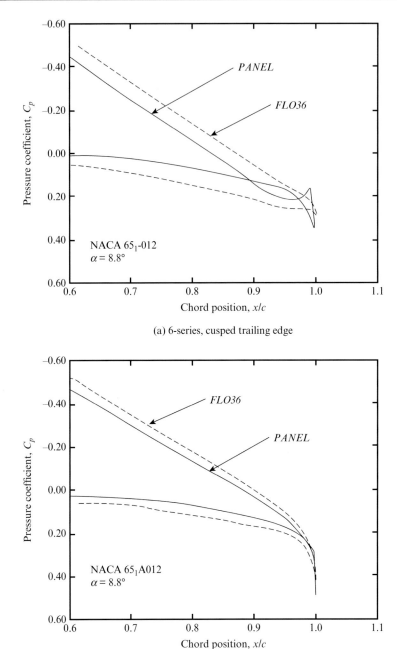

(a) 6-series, cusped trailing edge

(b) 6A-series, finite trailing edge angle

Figure 5.21   *PANEL* performance near the airfoil trailing edge compared with results from *FLO36,* a full potential CFD code.

5.2.4 Geometry and Design

Up to this point, we have been discussing aerodynamic prediction using panel methods from an analysis point of view. Oftentimes, however, designers want to determine a geometry that will give them the aerodynamic characteristics they desire, which can be accomplished in multiple ways: making

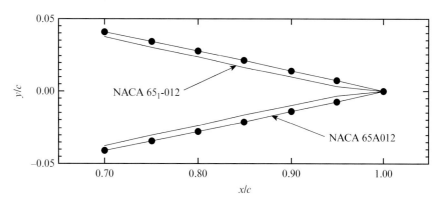

Figure 5.22 Comparison at the trailing edge of 6- and 6A-series airfoil geometries.

modifications to airfoil geometry and finding the resulting change in the pressure distribution and finding the shape that will yield a desired pressure distribution. We will describe both of these approaches in the following sections.

5.2.4.1 EFFECTS OF SHAPE CHANGES ON PRESSURE DISTRIBUTIONS
To develop an understanding of the typical effects of adding local modifications to the airfoil surface, you should try taking a typical panel method code and changing the shape at various locations on the airfoil. For example, see Project 5 at the end of the chapter, which provides a framework for the reader to carry out an investigation to help understand what happens when changes are made to the airfoil shape. It is also worthwhile to investigate the very powerful effects that small deflections of the trailing edge can produce. This reveals the power of the Kutta condition and alerts the aerodynamicist to the basis for the importance of viscous effects at the trailing edge.

   Making ad hoc changes to an airfoil shape is extremely educational when implemented in an interactive computer program, where the aerodynamicist can easily make shape changes and see the effect on the pressure distribution immediately. An outstanding code that does this has been created by Ilan Kroo and is known as *PANDA*.[22] Strictly speaking, *PANDA* is not a panel method, but it is an accurate subsonic airfoil prediction method.

5.2.4.2 SHAPE FOR A SPECIFIED PRESSURE DISTRIBUTION
There is another way that aerodynamicists view the design problem often referred to as the "inverse" problem. Although the local modification approach previously described is useful to make minor changes in airfoil pressure distributions, often the aerodynamic designer wants to find the geometric shape corresponding to a prescribed pressure distribution. This problem is more difficult than the analysis problem since it is possible to prescribe a pressure distribution for which no geometry exists. Even if the geometry exists, it may not be acceptable from a structural standpoint. For

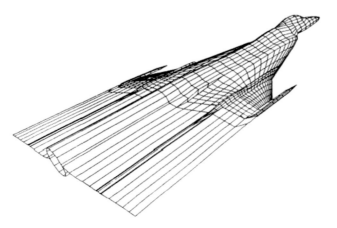

Illustration of the panel model of an F-16XL, including the wakes usually not shown in figures of panel models, but critical to the model (Ref. 3).

two-dimensional incompressible flow it is possible to obtain conditions on the surface velocity distribution that ensure that a closed airfoil shape exists. Excellent discussions of this problem have been given by Volpe[23] and Labrujere and Sloof.[24] A two-dimensional inverse panel method has been developed by Bristow,[25] and *XFOIL* also has an inverse design option.[18] Numerical optimization can also be used to find the shape corresponding to a prescribed pressure distribution.[26]

5.2.5 Issues in the Problem Formulation for 3D Potential Flow Over Aircraft

The extension of panel methods to three dimensions leads to fundamental questions regarding the proper specification of the potential flow problem for flow over an aircraft. The main problem is how to model the wake coming from the fuselage aft of the wing and wing tips, specifically, how to specify the wake behind surfaces without sharp edges. The Kutta condition applies to distinct edges and is not applicable if there are not well-defined trailing edges.

In some methods wakes are handled automatically. In other methods the wakes must be specified by the user. This provides complete control over the simulation but means that the user must understand precisely what the problem statement should be. The details of the wake specification often cause users difficulties in making panel models. Figure 5.23, from Erickson,[3] shows an example of a panel model including the details of the wakes. For high-lift cases and for cases where wakes from one surface pass near another, wake deflections must be computed as part of the solution. Figure 5.24 comes from a one week "short" course that was given to prospective users of an advanced panel method, *PAN AIR*.[27] Each surface has to have a wake, and the wakes need to be connected, as illustrated in Fig. 5.24; the modeling and execution of an aircraft can be quite complicated, especially since the

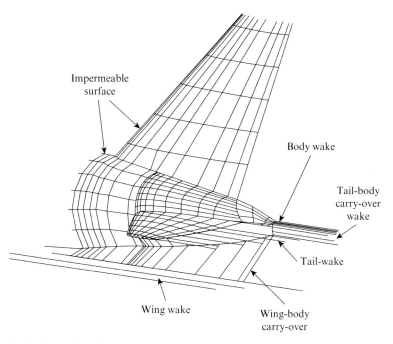

Impermeable surface

Body wake

Tail-body carry-over wake

Tail-wake

Wing wake

Wing-body carry-over

Figure 5.24   Details of a panel model showing the wake model details and that the wakes are connected (from a viewgraph presented at a *PAN AIR* user's short course, Ref. 27).

wake layout must be defined by the user. To ensure that the problem is properly specified and to examine the entire flowfield in detail, a complete graphics capability is required. Hess[28] provides an excellent discussion of these problems and outlines many different approaches to deal with them, while Carmichael and Erickson[29] also provide good insight into the requirements for a proper panel method formulation.

As illustrated, a practical aspect of using panel methods is the need to pay attention to details (which is actually true for all engineering work). This includes making sure that the outward surface normal is oriented in the proper direction and that all surfaces are properly enclosed. Aerodynamic panel methods generally use quadrilateral panels to define the surface. Since three points determine a plane, the quadrilateral may not necessarily define a consistent flat surface. In practice, the methods actually subdivide panels into triangular elements to determine an estimate of the outward normal. It is also important that edges fit so that there is no leakage in the panel model representation of the surface. Nathman has extended a panel method to have panels that include "warp."[30]

There is one other significant difference between two-dimensional and three-dimensional panel methods. Induced drag occurs even in inviscid, irrotational flow, and this component of drag can be computed by a panel model. However, its calculation by integration of pressures over the surface requires extreme accuracy, as we saw for the two-dimensional example. The use of a farfield momentum approach is much more accurate. For drag, this is known as a Trefftz plane analysis (see Katz and Plotkin[6] for details of this analysis).

5.2.6 Example Applications of Panel Methods

Many examples of panel methods have been presented in the literature. Figure 5.25 shows a classic example of the use of a panel model to evaluate the effect of the Space Shuttle Orbiter on the NASA Boeing 747. Other uses include the simulation of wind tunnel walls, support interference, and ground effects. Of special recent interest is the use of panel methods to simulate flapping wing aerodynamics for low-speed vehicles.[31] Panel methods are also used in ocean engineering; America's Cup designs have been dependent on panel methods for hull and keel design for many years, including the effects of the ocean free surface.

We have chosen the Boeing 737 as one example to present in extended detail.[32] It is an excellent illustration of how a panel method is used in design and provides a realistic example of the typical agreement that can be expected between a panel method and experimental data in a demanding real-world application. Figure 5.26 shows the geometry alterations required to modify a

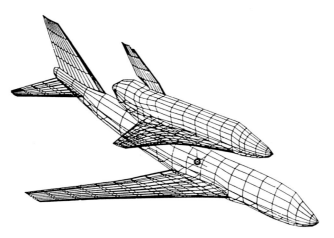

Figure 5.25   The Space Shuttle Orbiter mounted on the NASA Boeing 747 (Courtesy of Ed Tinoco).

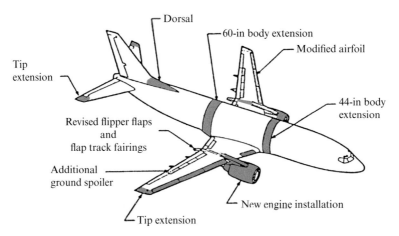

Figure 5.26   The Boeing 737–300 relative to the model 737–200 (Ref. 32; reprinted by permission of the American Institute of Aeronautics and Astronautics, Inc.).

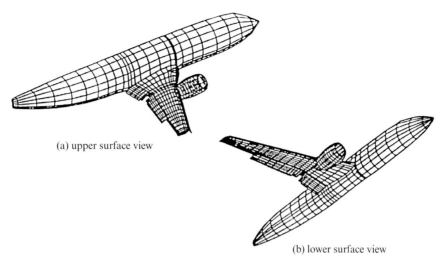

(a) upper surface view

(b) lower surface view

Figure 5.27   The panel representation of the 737–300 with 15° flap deflection (Ref. 32; reprinted by permission of the American Institute of Aeronautics and Astronautics, Inc.).

Boeing 737–200 into the 737–300 configuration. *PAN AIR*, which is a Boeing-developed advanced panel method,[33] was used to investigate the design of the new high-lift system, among other things. Figure 5.27 shows the panel method representation of the airplane, including flap deflections.

An understanding of the wing flowfield for two different takeoff flap settings was desired: the "flaps 1" case, which is the high-altitude, hot day setting, and the "flaps 15" case, which is the normal takeoff setting. The work was conducted in concert with the flight test program to give insight into the flight test results by providing complete flowfield details not available from the flight test. The computational models used 1,750 panels for "flaps 1" and 2,900 panels for "flaps 15." The modeling used to simulate this flowfield illustrates typical idealizations employed when applying panel methods to actual aircraft. Although typical, it is one of the most geometrically complicated examples ever published.

Figure 5.28 shows the 737–300 wing leading edge and nacelle. The inboard Krueger flap was actually modeled as a doublet of zero thickness. The position was adjusted slightly to allow the doublet sheet to provide a simple matching of the trailing edge of the Krueger and the leading edge of the wing. These types of slight adjustments to keep panel schemes relatively simple are commonly used. The outboard leading- and trailing-edge flap geometries were also modified for use in this inviscid simulation. Figure 5.29a shows the actual and computational "flaps 1" geometry; in this case, the airfoil was modeled as a single element airfoil. The "flaps 15" trailing-edge comparison between the actual and computational geometry is shown in Fig. 5.29b; the triple slotted flap was modeled as a single-element flap. At this setting, the gap between the forward vane and main flap is closed, and the gap between the main and aft flap is very small.

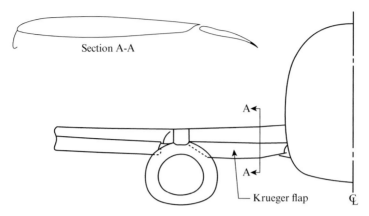

Figure 5.28 Inboard wing leading edge and nacelle details (Ref. 32; reprinted by permission of the American Institute of Aeronautics and Astronautics, Inc.).

Several three-dimensional modeling considerations also required attention. In the "flaps 1" case shown in Fig. 5.30, spanwise discontinuities included the end of the outboard leading-edge slat and trailing-edge discontinuities at the back of the nacelle installation (called the thrust gate) between the inboard and outboard flaps. At the outboard leading edge, the edges of the slat and wing were paneled to prevent leakage (a 0.1 inch gap was left between these surfaces). At the trailing-edge discontinuity, a wake was included to model a continuous trailing edge from which a trailing vortex sheet could be shed. Similar considerations are required for the "flaps 15" case. Special care was taken to ensure that the configuration was closed and contained no holes in the surface at the ends of the flap segments.

Another important consideration is the nacelle model. This required the specification of the inlet flow at the engine face, a model of the strut wake, and both the outer bypass air plume and the primary wake from the inner hot gas jet (Fig. 5.31 shows the details of this model). Complete details of the model are contained in Ref. 32.

With the model complete, the *PAN AIR* code was run and solutions were obtained. The spanwise distribution of airfoil section lift coefficients is presented in Fig. 5.32; Fig. 5.32a shows the results for the "flaps 1" case, and Fig. 5.32b presents the "flaps 15" case. In both cases, the jig shape and flight shape including aeroelastic deformation are included for evaluation purposes (modeling these affects is another important consideration in making a proper aerodynamic simulation). In both cases, the shape including the deformation under load shows much better agreement with flight test and wind tunnel data. Notice the loss of lift on the wing at the nacelle station and the decrease in lift outboard of the trailing-edge flap location.

Figure 5.33 presents the change in section lift coefficient with angle of attack at several span stations. The agreement between *PAN AIR* and flight test data is better for the "flaps 1" case; viscous effects are becoming important for the "flaps 15" case and the predictions are not as good.

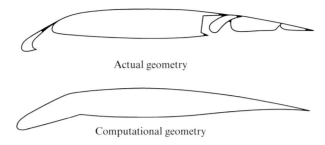

(a) Comparison of actual and computational wing geometry for the flaps 1 case

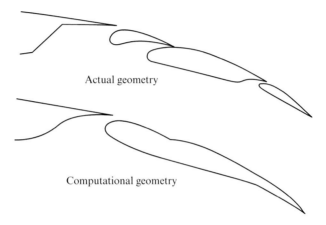

(b) Actual and computational trailing edge geometry for the flaps 15 case

Figure 5.29   Examples of computational modeling for a real application. (a) Comparison of actual and computational wing geometry for the flaps 1 case. (b) Actual and computational trailing-edge geometry for the flaps 15 case (Ref. 32; reprinted by permission of the American Institute of Aeronautics and Astronautics, Inc.).

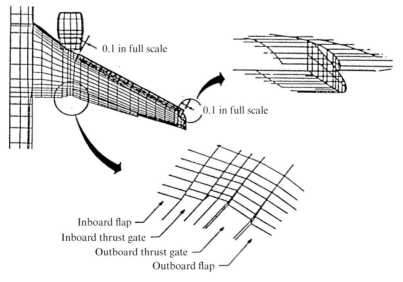

Figure 5.30   Spanwise discontinuity details requiring modeling for flaps 1 case (Ref. 32; reprinted by permission of the American Institute of Aeronautics and Astronautics, Inc.).

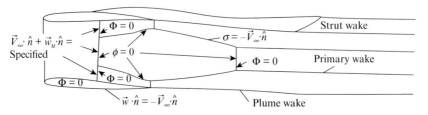

Figure 5.31 Nacelle model illustrating the application of boundary conditions (Ref. 32; reprinted by permission of the American Institute of Aeronautics and Astronautics, Inc.).

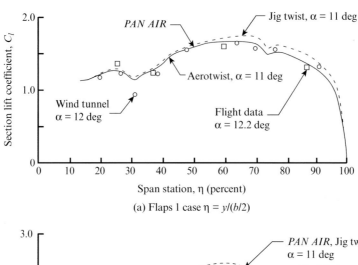

(a) Flaps 1 case η = y/(b/2)

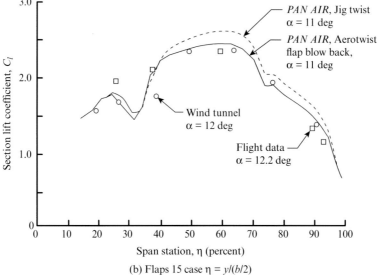

(b) Flaps 15 case η = y/(b/2)

Figure 5.32 Spanwise distribution of lift coefficient on the Boeing 737–300 (Ref. 32; reprinted by permission of the American Institute of Aeronautics and Astronautics, Inc.). (a) flaps 1 case. (b) flaps 15 case.

Figure 5.34 completes this example by presenting the comparison of pressure distributions for the two cases at four spanwise stations. The "flaps 1" case agreement is generally quite good. Calculations are presented for both the actual angle of attack and the angle of attack that matches the lift coefficient (matching lift coefficient instead of angle of attack is a common

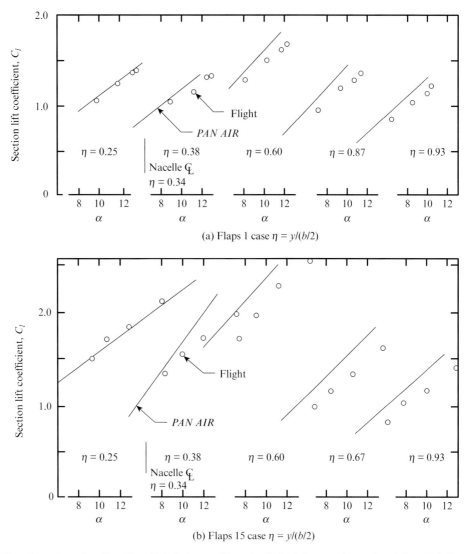

(a) Flaps 1 case $\eta = y/(b/2)$

(b) Flaps 15 case $\eta = y/(b/2)$

Figure 5.33 Comparison of section lift coefficient change with angle of attack for several spanwise stations (Ref. 32; reprinted by permission of the American Institute of Aeronautics and Astronautics, Inc.). (a) flaps 1 case. (b) flaps 15 case.

practice in computational aerodynamics). Considering the simplifications made to the geometry and the absence of the simulation of viscous effects, the agreement is very good. The "flaps 15" case starts to show the problems that arise from the simplifications that were made. This is a good example of the use of a panel method that illustrates almost all of the considerations that must be addressed in actual applications.

5.2.7 Using Panel Methods

The previous example of the Boeing 737–300 panel method prediction study shows that there is a big difference between the concept of the panel method

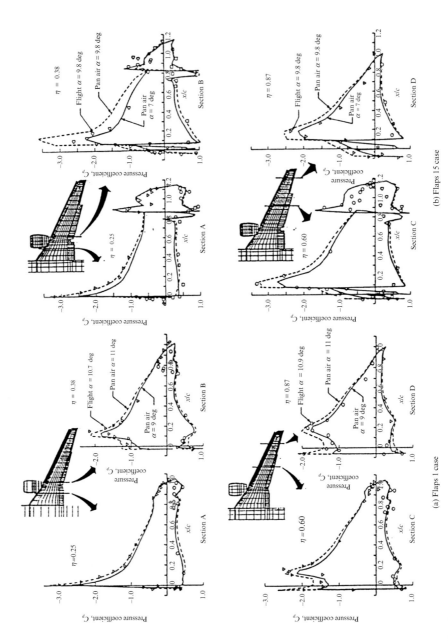

Figure 5.34 Comparison of pressure distributions between flight and computations for the *737–300*: solid line is *PAN AIR* at flight lift, dashed line is *PAN AIR* at flight angle of attack (Ref. 32; reprinted by permission of the American Institute of Aeronautics and Astronautics, Inc.). (a) flaps 1 case (b) flaps 15 case.

and the applicability of the method. Therefore, we will provide some basic rules for helping to successfully use the panel method.

5.2.7.1 COMMONSENSE RULES FOR PANELS
Panels should have the following characteristics:

- Vary the size of panels smoothly
- Concentrate panels where the flowfield and/or geometry is changing rapidly
- Do not spend more money and time (i.e., numbers of panels) than required

Panel placement and variation of panel size affect the quality of the solution. However, extreme sensitivity of the solution to the panel layout is an indication of an improperly posed problem. If this happens, the user should investigate the problem thoroughly.

Panel methods are an aid to the aerodynamicist. You must use the results as a guide to help you develop your own judgment. Remember that the panel method solution is an approximation of the real-life problem and an idealized representation of the flowfield (recall Fig. 1.17). An understanding of aerodynamics that provides an intuitive expectation of the types of results that may be obtained and an appreciation of how to relate your idealization to the real flow is required to get the most from the methods. This insight requires experience and study.

5.2.7.2 WHAT A PANEL METHOD CAN AND CANNOT DO
While panel methods are quite useful, they cannot do everything; for example:

1. Panel methods are inviscid solutions. You will not capture viscous effects except via user "modeling" by changing the geometry.
2. Solutions are invalid as soon as the flow develops local supersonic zones, [i.e., $C_p < C_{p_{crit}}$]. For two-dimensional isentropic flow, the exact value of C_p for critical flow is:

$$C_{p_{crit}} = -\frac{2}{\gamma M_\infty^2}\left[1-\left\{\frac{1+\dfrac{\gamma-1}{2}M_\infty^2}{\left(\dfrac{\gamma+1}{2}\right)}\right\}^{\frac{\gamma}{\gamma-1}}\right] \qquad (5.32)$$

5.2.8 Advanced Panel Methods: What Is a "Higher Order" Panel Method?

Higher-order panel methods use singularity distributions that are not constant over the panel and may also use panels that are non-planar. Higher-order methods were actually found to be crucial in obtaining accurate solutions for the Prandtl-Glauert equation (Equation 3.125) at supersonic speeds. At supersonic speeds, the Prandtl-Glauert equation is actually a wave equation (hyperbolic) and requires much more accurate numerical solution than the subsonic case to avoid "noisy" pressure distributions in the solution.[33] However, subsonic higher-order panel methods, although not as important as for the supersonic flow case, have been studied in great detail. In theory, good results can be obtained using far fewer panels with higher-order methods. In practice, the need to resolve geometric details often leads to the need to use small panels anyway, and all the advantages of higher-order paneling methods are not necessarily obtained. Nevertheless, since a higher-order panel method may also be a relatively new program taking advantage of many years of experience, the higher-order code may still be a good candidate for use.

5.2.9 Current Standard Panel Method Programs: A Brief Survey

Panel methods are widely used in the aircraft industry, and have been for a long time. Comparisons between various panel codes have been made,[34] and, in general, all the new professionally-developed codes can be made to work well (recall the attention to detail illustrated earlier). The selection of a specific code will likely be based on non-technical considerations, such as cost or availability. One of the relatively newer panel codes is known as *PMARC*,[35] for Panel Method Ames Research Center, and has received a great deal of development effort. We will now provide a brief description of the codes a new aerodynamicist will most likely encounter, either at work or by reading the literature.

5.2.9.1 *PAN AIR* [29,33]

This is a Boeing-developed code, funded by a variety of government agencies, including NASA. This code provides total flexibility, i.e., it is really an integral equation solver and not an aerodynamicist's tool per se. It is a higher-order panel method and can handle both subsonic and supersonic flow. It is relatively expensive and difficult to run (a *PAN AIR* user would take months to train, and it would probably become his or her primary job). To effectively use the code, good pre- and post-processing systems are available. Although Boeing has these systems in place, they were internally developed and are not available outside the company.

5.2.9.2 *VSAERO*

The code was originally a low-order method but has been extended to include variations of the singularity strength over a panel and has now been described as "multi-order,"[30] but is for subsonic flow only. It handles general geometries and includes options to treat viscous effects and vortex flows.

Table 5.1 Comparison of Some Major Panel Method Programs: Early Codes

Originator and Method Name	Year	Panel Geometry	Source Type	Doublet Type	Boundary Conditions	Restrictions	Comments
Hess and Smith[39] (Douglas)	1962	flat	constant	none	specification of normal flow	non-lifting wings and bodies only	
Rubbert[40] (vortex lattice)	1964	flat	none	constant	normal flow	planar wings only	
Woodward[41] (Woodward I)	1967	flat	constant	linear	normal flow	wings must be planar	
Rubbert and Saaris[42] (Boeing A-230)	1968	flat	constant	constant	normal flow	nearly constant panel density	
Hess I[43]	1972	flat	constant	linear	normal flow	wings and bodies only	
USSAERO[44] (Woodward II)	1973	flat					subsonic and supersonic, analysis only
W12SC3[37] (Grumman)	1983	flat			mixed design and analysis		combines Woodward I & II features

Table 5.2 Comparison of Some Major Panel Method Programs: Advanced Methods

Originator and Method Name	Year	Panel Geometry	Source Type	Doublet Type	Boundary Conditions	Restrictions	Comments
Roberts and Rundle[45]	1973	paraboloidal	quadratic	quadratic	normal flow		numerical integrations, very expensive
Mercer, Weber and Lesford[46]	1973	flat	none	smooth, cubic, quadratic	normal flow in least squares sense	planar wings	subsonic/ supersonic, cubic spanwise, quadratic chordwise
Morino and Kuo[47] (SOUSSA)	1974	continuous, hyperboloidal	constant	constant	potential	no thin configurations	unsteady
Johnson and Rubbert[48]	1975	paraboloidal	linear	quadratic	normal flow		
Ehlers and Rubbert[49] (Mach line paneling)	1976	flat	linear	continuous quadratic	normal flow	planar wings, special paneling	supersonic flow
Ehlers et al.[50] (PAN AIR pilot code)	1977	continuous piecewise flat	linear	continuous quadratic	arbitrary in $\Phi, \Delta\Phi$		subsonic and supersonic

Originally developed for NASA, the code has been much further developed by AMI as a commercial product (there is also a plotting package and other supporting software available). Typical development enhancements of the code are described by Nathman.[36] The public domain version of this code

Table 5.3 Comparison of Major Panel Method Programs: Production Codes

Originator and Method Name	Year	Panel Geometry	Source Type	Doublet Type	Boundary Conditions	Restrictions	Comments
MCAIR[51] (McDonnell)	1980	flat	constant	quadratic			design option
PAN AIR[33] (Boeing)	1980	continuous piecewise flat	continuous linear	continuous quadratic	arbitrary in Φ, $\Delta\Phi$		subsonic and supersonic
Hess II[52] (Douglas)	1981	parabolic	linear	quadratic	normal flow		
VSAERO[53] (AMI)	1981	flat	constant	constant	exterior and interior normal flow		subsonic
QUADPAN[54] (Lockheed)	1981	flat	constant	constant			
PMARC[35] (NASA Ames)	1988	flat	constant	constant			Unsteady, wake rollup

was obtained by several groups that worked on the design of the America's Cup yacht competition in the mid-1980s; the code was used for hull and keel design. One of the modifications that was made for this application was the addition of the free surface representing the air-water interface (recall that the free surface problem means that the surface displacement is unknown, and the boundary condition is that a constant pressure exists at the interface).

5.2.9.3 WOODWARD CODE

An old panel method that is sometimes encountered is the code known as the "Woodward" or "Woodward-Carmichael" code. Woodward's first methods were developed while he was at Boeing and were supported by NASA Ames Research Center, primarily for the U.S. Supersonic Transport program (which was an important national effort in the 1960s). Subsequently, Woodward went into business and continued to develop codes. *USSAERO* treats both supersonic and subsonic flow, and a version that combines a number of features of each code, as well as additional design options, is also available, and is known as "Woodward 12" or *W12SC3*.[37]

5.2.9.4 *PMARC*[35]

This is one of the newest panel method codes, and was developed at NASA Ames Research Center to provide an extremely flexible method to simulate a wide range of very general geometries. An example is the simulation of high-lift systems and jet exhausts for VSTOL aircraft. The code is a lower-order panel method and can simulate unsteady as well as steady flow. The wake position can be obtained as part of the solution. It has been used for underwater applications as well as for aircraft.

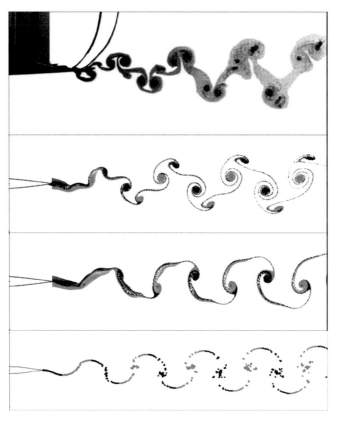

Figure 5.35 Comparison of flapping wing wake from four sources: (a) experiment, (b) Navier-Stokes codes (second and third frames), and (c) unsteady panel method (Ref. 31; Courtesy of Max Platzer of the Naval Postgraduate School).

Panel methods can also treat unsteady flows, with a recent example being the developments made by van Zyl,[38] and Fig. 5.35 shows the comparison of an unsteady panel method with Navier-Stokes predictions and experimental data for a flapping wing case.[31]

References for these codes are cited in Tables 5.1–5.3. Table 5.1 summarizes some of the key early panel methods that have been developed.[ii] Table 5.2 reviews the extremely active era of the development of advanced panel methods. Finally, Table 5.3 provides details on the current production codes likely to be used on current aerodynamic design and analysis projects.

5.3 Vortex Lattice Methods

Vortex Lattice Methods (VLM) are similar to panel methods, but they are easier to use and capable of providing remarkable insight into wing aerodynamics and component interaction. The VLM approach was among the

[ii] *W12SC3* is included because it was a valuable combination of two early codes,[37] providing significant new design capability, particularly at supersonic speeds; despite the title of the report, *W12SC3* can be run at subsonic speeds.

earliest methods utilizing computers to actually assist aerodynamicists in estimating aircraft aerodynamics. Vortex lattice methods are based on solutions to Laplace's equation, and are subject to the same basic theoretical restrictions that apply to panel methods. As a comparison, most vortex lattice methods are:

Similar to Panel methods:

- singularities are placed on a surface
- the no flow-through condition is satisfied at a number of control points
- a system of linear algebraic equations is solved to determine singularity strengths

Different from Panel methods:

- Oriented toward lifting effects, and classical formulations ignore thickness
- Boundary conditions are applied on a mean surface, not the actual surface (not an exact solution of Laplace's equation over a body, but embodies some additional approximations, i.e., together with the first item, we find ΔC_p, not $C_{p_{upper}}$ and $C_{p_{lower}}$)
- Singularities are not distributed over the entire surface
- Oriented toward combinations of thin lifting surfaces (recall that panel methods had no limitations on thickness).

Vortex lattice methods were first formulated in the late 1930s, and the method includes wing-tip vortex effects through the use of a trailing vortex filament system. It was first called a "Vortex Lattice" in 1943 by Faulkner.[55] The concept is extremely simple, but, because of its purely numerical approach (i.e., no answers are possible without finding the numerical solution of a matrix too large for routine hand calculation), practical applications awaited sufficient development of computers – the early 1960s saw widespread adoption of the method, and the use was so pervasive that a workshop was devoted to these methods at NASA in the mid-1970s.[56] A nearly universal standard for vortex lattice prediction capabilities was established by a code developed at NASA Langley Research Center. The authors include Margason and Lamar,[57] Lamar and Gloss,[58] and Lamar and Herbert,[59,60] and the code is generally known as the "Lamar Vortex Lattice Method" (the "final" development in this series is designated *VLM4.997*). The original codes could handle two lifting surfaces, while *VLM4.997* could handle up to four lifting surfaces. Many other people have written vortex lattice method codes since that time – the current methods that are widely used are: the code by Drela and Youngren, *AVL*;[61] *Tornado*, a MATLAB code developed by Tomas Melin at KTH in Sweden (now incorporated into a comprehensive aerodynamic design code called *CEASIOM*);[62] and *VORLAX*,[63] as part of the *HASC* code,[64] is also widely used. The authors of the *VSAERO* panel method described previously have also released a vortex lattice code, *VLAERO*.[65] The VLM method is simple enough that many students have written their own codes, which are easily executed on current personal computers (see Ref. 2 for details).

Noteworthy variations on the basic VLM approach have been developed by Lan[66] (Quasi-Vortex Lattice Method) and Mook[67] and co-workers. Mook developed vortex lattice class methods that treat flowfields that contain leading-edge vortex-type separation and also handle general unsteady motions. The book by Katz and Plotkin[6] contains another variation, using vortex rings, and includes a computer code. Kay also wrote a code,[68] *JKayVLM*, using the method of Katz and Plotkin to estimate stability derivatives, and that code is also readily available.

To understand the VLM method, a number of basic concepts must first be reviewed. Then we will describe one implementation of the VLM method and use it to obtain insights into wing and wing-canard aerodynamics. Naturally, the method is based on the idea of a vortex singularity as the solution of Laplace's equation. A good description of the basic theory for vortices in inviscid flow and thin wing analysis is contained in Karamcheti,[9] and the theory was reviewed in Chapter 4. A good general description of the vortex lattice method is also given by Bertin and Cummings.[2] Following some illustrations of the results from VLM methods, an example of a vortex lattice method used in a design mode is presented, where the camber line required to produce a specified loading is found.

5.3.1 Boundary Conditions on the Mean Surface and the Pressure Relation

An important difference between vortex lattice methods and panel methods is the way in which the boundary conditions are handled. Typically, the vortex lattice method uses an approximate boundary condition treatment; this boundary condition can also be used in other circumstances to good advantage. This is a good "trick" applied aerodynamicists should know and understand, and that is why it is covered in detail here. In general, this approach results in the so-called thin airfoil boundary condition and arises by linearizing and transferring the boundary condition from the actual surface to a flat mean "reference" surface that is typically a constant coordinate surface. Consistent with the boundary condition simplification, a simplified relation between the pressure and velocity is also possible. The simplification in the boundary condition and pressure-velocity relation provides a basis for treating the problem as a superposition of the lift and thickness contributions to the aerodynamic results. Karamcheti[9] provides an excellent discussion of this approach.

To understand the thin airfoil theory boundary condition treatment, we provide an example in two dimensions. Recall from Equation 3.99 that the exact surface boundary condition for steady inviscid flow is:

$$\vec{V} \cdot \hat{n} = 0 \tag{5.33}$$

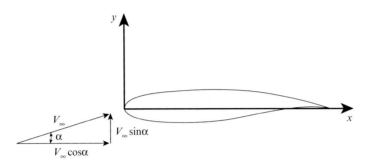

Figure 5.36 Basic coordinate system for boundary condition analysis.

on $F(x, y) = 0 = y - f(x)$. The unit normal vector is $\hat{n} = \vec{\nabla} F(x, y) / \left| \vec{\nabla} F(x, y) \right|$, and the velocity field is defined using the notation defined in Fig. 5.36. We will define the velocity components of $\vec{V}$ as:

$$\vec{V} = \vec{V}_\infty + \underbrace{\vec{q}(x, y)}_{\text{disturbance velocity}} \tag{5.34}$$

where $\vec{q}$ is a disturbance velocity with components u and v. The total velocity $\vec{V}$ then becomes (in terms of velocity components):

$$\begin{aligned} u_{\text{TOT}} &= V_\infty \cos \alpha + u(x, y) \\ v_{\text{TOT}} &= V_\infty \sin \alpha + v(x, y) \end{aligned} \tag{5.35}$$

and we can write out the boundary condition as:

$$\vec{V} \cdot \hat{n} = \left(u_{\text{TOT}} \hat{i} + v_{\text{TOT}} \hat{j} \right) \cdot \left(\frac{\partial F}{\partial x} \hat{i} + \frac{\partial F}{\partial y} \hat{j} \right) = 0 \tag{5.36}$$

or

$$\left[V_\infty \cos \alpha + u(x, y) \right] \frac{\partial F}{\partial x} + \left[V_\infty \sin \alpha + v(x, y) \right] \frac{\partial F}{\partial y} = 0 \tag{5.37}$$

on $F(x, y) = 0$. Now recall the relationship between F and f given below Equation 5.33:

$$\begin{aligned} \frac{\partial F}{\partial x} &= \frac{\partial}{\partial x} \{ y - f(x) \} = -\frac{df(x)}{dx} \\ \frac{\partial F}{\partial y} &= \frac{\partial}{\partial y} \{ y - f(x) \} = 1 \end{aligned} \tag{5.38}$$

Substituting for F in Equation 5.37, we have:

$$\left(V_\infty \cos \alpha + u\right)\left(-\frac{df}{dx}\right) + \left(V_\infty \sin \alpha + v\right) = 0 \tag{5.39}$$

which, solving for v, is:

$$v = \left(V_\infty \cos \alpha + u\right)\frac{df}{dx} - V_\infty \sin \alpha \tag{5.40}$$

on $y = f(x)$. Note that v is defined in terms of the unknown u. Thus Equation 5.40 is a nonlinear boundary condition and further analysis is needed to obtain a useful relation.[iii]

5.3.1.1 LINEARIZED FORM OF THE BOUNDARY CONDITION

The relation given by Equation 5.40 is exact. It has been derived as the starting point for the derivation of useful relations when the body (which is assumed to be a thin surface at a small angle of attack) induces disturbances to the freestream velocity that are small in comparison to the freestream velocity. Thus we assume: $u \ll V_\infty$, $v \ll V_\infty$, and $\partial f/\partial x < \partial f/\partial y$. Note that this introduces a bias in the coordinate system to simplify the analysis, a typical consequence of introducing simplifying assumptions. Consistent with this assumption, the components of the freestream velocity are (remembering that α is in radians and small):

$$V_\infty \cos \alpha \approx V_\infty$$
$$V_\infty \sin \alpha \approx V_\infty \alpha \tag{5.41}$$

and the expression for v in Equation 5.40 becomes:

$$v = \left(V_\infty + u\right)\frac{df}{dx} - V_\infty \alpha. \tag{5.42}$$

Dividing by V_∞,

$$\frac{v}{V_\infty} = \left(1 + \frac{u}{V_\infty}\right)\frac{df}{dx} - \alpha \tag{5.43}$$

and the linearized boundary condition is obtained by neglecting u/V_∞ compared with unity (consistent with the previous approximations). With this assumption, the linearized boundary condition becomes:

[iii] Notice that even when the flow field model is defined by a linear partial differential equation, an assumption which we have not yet made, the boundary condition can make the problem nonlinear.

$$\frac{v}{V_\infty} = \frac{df}{dx} - \alpha \quad \text{on } y = f(x). \tag{5.44}$$

This form of the boundary condition is not valid if the flow disturbance is large compared to the freestream velocity (for aerodynamically stream-lined shapes, this is usually valid everywhere except at the leading edge of the airfoil, where a stagnation point exists ($u = -V_\infty$), and the slope is infinite ($df/dx = \infty$)). In practice, a local violation of this assumption leads to a local error. So, if the details of the flow at the leading edge are not important to the analysis, which surprisingly is often the case at low angles of attack, the linearized boundary condition can be used.

5.3.1.2 TRANSFER OF THE BOUNDARY CONDITIONS

Although Equation 5.44 is linear, it is hard to utilize because it is not applied on a coordinate line.[iv] We now use a further approximation of this relation to get the useful form of the linearized boundary condition. Using a Taylor's series expansion of the v component of velocity about the coordinate axis, we obtain the v velocity on the surface:

$$v\{x, y = f(x)\} = v(x,0) + f(x)\frac{\partial v}{\partial y}\bigg|_{y=0} + \dots \tag{5.45}$$

For the thin surfaces under consideration, $f(x)$ is small, and because the disturbances are assumed small, $\partial v/\partial y$ is also small. For example, assume that v and $\partial v/\partial y$ are the same size, equal to 0.1, and df/dy is also about 0.1. The relation between v on the airfoil surface and the axis is:

$$v\{x, y = f(x)\} = (0.1) + (0.1)(0.1) = 0.1 + \underbrace{0.01}_{\text{neglect}}. \tag{5.46}$$

Neglecting the second term, we assume:

$$v\{x, f(x)\} \approx v(x,0). \tag{5.47}$$

We now apply both the upper and lower surface boundary conditions on the axis $y = 0$, and distinguish between the upper and lower surface shapes by using:

$$\begin{aligned} f &= f_u \quad \text{on the upper surface} \\ f &= f_l \quad \text{on the lower surface.} \end{aligned} \tag{5.48}$$

[iv] Note the simplification introduced by applying boundary conditions on a constant coordinate surface. In this case we introduced some approximations to be able to do this; often there is only a minor loss in accuracy. However, to satisfy the boundary conditions exactly on arbitrary shapes we often use rather elaborate coordinate trans-formations, which will be discussed in Chapter 7.

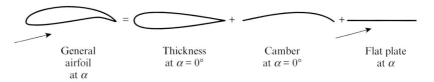

General
airfoil
at α

Thickness
at $\alpha = 0°$

Camber
at $\alpha = 0°$

Flat plate
at α

Figure 5.37   Decomposition of a general shape at incidence.

Using Equation 5.48, we write the upper and lower surface boundary conditions as:

$$\left.\frac{v(x,0^+)}{V_\infty}\right|_{up} = \frac{df_u}{dx} - \alpha, \qquad \left.\frac{v(x,0^-)}{V_\infty}\right|_{low} = \frac{df_l}{dx} - \alpha. \tag{5.49}$$

These are the linearized and transferred boundary conditions. Frequently, these boundary conditions result in a surprisingly good approximation to the flowfield, even in transonic and supersonic flow.

5.3.1.3 DECOMPOSITION OF BOUNDARY CONDITIONS INTO CAMBER/THICKNESS/ALPHA

Further simplification and insight can be gained by considering the airfoils in terms of the combination of thickness and camber, a natural point of view. We could write the upper and lower surface shapes in terms of camber, f_c, and thickness, f_t, as:

$$\begin{aligned} f_u &= f_c + f_t \\ f_l &= f_c - f_t \end{aligned} \tag{5.50}$$

and the general problem is then divided into the sum of three parts as shown in Fig. 5.37.

The decomposition of the problem is somewhat arbitrary, since linearity allows for any decomposition to be used. Camber could also be considered to include angle of attack effects using the boundary condition relations given previously. To proceed further, we make use of the basic vortex lattice method assumption: *the flowfield is governed by a linear partial differential equation (Laplace's equation), allowing for superposition of solutions.* Superposition allows us to solve the problem in pieces and add up the contributions from the various parts of the problem. This results in the final form of the thin airfoil theory boundary conditions, substituting Equation 5.50 into Equation 5.49:

$$\begin{aligned} \left.\frac{v(x,0^+)}{V_\infty}\right|_{up} &= \frac{df_c}{dx} + \frac{df_t}{dx} - \alpha \\ \left.\frac{v(x,0^-)}{V_\infty}\right|_{low} &= \frac{df_c}{dx} - \frac{df_t}{dx} - \alpha \end{aligned} \tag{5.51}$$

The problem can be solved for the various contributions, and the contributions are added together to obtain the complete solution. If thickness is neglected, the boundary conditions are the same for the upper and lower surfaces.

5.3.1.4 THIN AIRFOIL THEORY PRESSURE RELATION

Consistent with the linearization of the boundary conditions, we can also obtain a useful relation between the pressure and velocity. For incompressible flows, the exact relation between the pressure coefficient and velocity is:

$$C_p = 1 - \left(\frac{V}{V_\infty}\right)^2. \tag{5.52}$$

Assuming two-dimensional flow for this example, we express the velocity considering disturbances to the freestream velocity using the approximations discussed earlier:

$$V^2 = \left(V_\infty \cos\alpha + u\right)^2 + \left(V_\infty \sin\alpha + v\right)^2 \approx \left(V_\infty + u\right)^2 + \left(V_\infty \alpha + v\right)^2. \tag{5.53}$$

Expanding and dividing by V_∞^2 we get:

$$\frac{V^2}{V_\infty^2} = 1 + 2\frac{u}{V_\infty} + \frac{u^2}{V_\infty^2} + \alpha^2 + 2\alpha\frac{v}{V_\infty} + \frac{v^2}{V_\infty^2}. \tag{5.54}$$

Substituting into the C_p relation, Equation 5.52, we get:

$$\begin{aligned} C_P &= 1 - \left(1 + 2\frac{u}{V_\infty} + \frac{u^2}{V_\infty^2} + \alpha^2 + 2\alpha\frac{v}{V_\infty} + \frac{v^2}{V_\infty^2}\right) \\ &= 1 - 1 - 2\frac{u}{V_\infty} - \frac{u^2}{V_\infty^2} - \alpha^2 - 2\alpha\frac{v}{V_\infty} - \frac{v^2}{V_\infty^2} \end{aligned} \tag{5.55}$$

and if α, u/V_∞ and v/V_∞ are $<< 1$, then the last four terms can be neglected in comparison with the third term. The final result is:

$$C_p = -2\frac{u}{V_\infty}. \tag{5.56}$$

This is the linearized, or thin airfoil theory, pressure coefficient formula. From experience gained comparing various computational results, we have found that this formula is a slightly more severe restriction on the accuracy of the solution than the linearized boundary condition. Equation 5.56

shows that, under the small disturbance approximation, the pressure is a linear function of u, and we can add the C_p contribution from thickness, camber, and angle of attack by superposition. A similar derivation can be used to show that Equation 5.56 is also valid for compressible flow up to moderate supersonic speeds.

5.3.1.5 ΔC_p DUE TO CAMBER/ALPHA (THICKNESS EFFECTS CANCEL!)

Next, we use the result in Equation 5.56 to obtain a formula for the load distribution on the wing, which is the difference between the pressure on the top and bottom surfaces:

$$\Delta C_p = C_{p_l} - C_{p_u}. \tag{5.57}$$

Using superposition, the pressures can be obtained as the contributions from wing thickness, camber, and angle of attack effects:

$$\begin{aligned}
C_{p_l} &= C_{p_{\text{Thickness}}} + C_{p_{\text{Camber}}} + C_{p_{\text{Angle of attack}}} \\
C_{p_u} &= C_{p_{\text{Thickness}}} - C_{p_{\text{Camber}}} - C_{p_{\text{Angle of attack}}}
\end{aligned} \tag{5.58}$$

so that:

$$\begin{aligned}
\Delta C_p &= \left(C_{p_{\text{Thickness}}} + C_{p_{\text{Camber}}} + C_{p_{\text{Angle of attack}}} \right) \\
&\quad - \left(C_{p_{\text{Thickness}}} - C_{p_{\text{Camber}}} - C_{p_{\text{Angle of attack}}} \right) \\
&= 2\left(C_{p_{\text{Camber}}} + C_{p_{\text{Angle of attack}}} \right).
\end{aligned} \tag{5.59}$$

Equation 5.59 demonstrates that, for cases where the linearized pressure coefficient relation is valid, thickness does not contribute to lift (to first order) in the velocity disturbance.

The importance of this analysis is that we have shown:

1. how the lifting effects can be obtained without considering thickness, and
2. that the cambered surface boundary conditions can be applied on a flat coordinate surface, resulting in an easy-to-apply boundary condition.

The principles demonstrated here for transfer and linearization of boundary conditions can be applied in a variety of situations other than the application to vortex lattice methods. Often this idea can be used to handle complicated geometries that cannot easily be treated exactly.

The analysis shown here produces an entirely consistent problem formulation. This includes the linearization of the boundary condition, the transfer

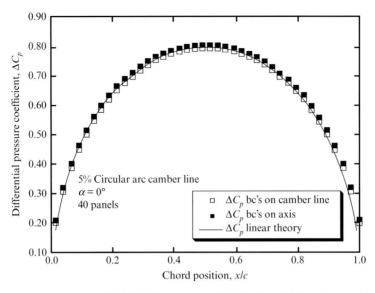

5% Circular arc camber line
$\alpha = 0°$
40 panels

□ ΔC_p bc's on camber line
■ ΔC_p bc's on axis
——— ΔC_p linear theory

Figure 5.38   Comparison in 2D of the 1/4–3/4 rule for vortex-control point locations with linear theory, and including a comparison between placing the vortex and control point on the camber line or on the axis.

of the boundary condition, and the approximation between velocity and pressure. All of the approximations are consistent with one another, and improving one of these approximations without improving all of them in a consistent manner may actually lead to worse results. Sometimes you can make agreement with data better, sometimes it may get worse; you have to be careful when trying to improve a theory on an ad hoc basis.

To examine the use of these ideas, we present a two-dimensional example. For the numerical calculation the airfoil is divided into a number of equal length panels. A vortex is placed at the quarter-chord point of each panel and the no-flow-through boundary condition is satisfied at the three-quarter-chord point of the individual panel. This can be shown to produce exact thin airfoil theory results for a small number of panels, and is known as the "1/4–3/4 rule." In Fig. 5.38, we compare the results obtained for a 5 percent circular arc camber airfoil[v] in order to illustrate the accuracy of the method. The curve labeled "linear theory" uses classical analytical thin airfoil theory with results obtained satisfying the boundary condition on the mean surface. These results are compared with numerical results for the case where the boundary condition is applied exactly on the camber line and the result obtained applying the boundary condition on the mean surface using the approximate method described. The difference between placing the vortex on the actual camber surface and satisfying the boundary condition on the actual surface and the more approximate traditional approach of locating the vortex and control point on the mean surface is extremely small.

---

[v]   A relatively large camber for a practical airfoil, the NACA 4412 example we used earlier illustrating panels methods was an extreme case with 4 percent camber.

Flow Visualization Box

CEASIOM Aerodynamic Prediction

We have been looking at various fluid dynamic properties within flowfields using *FieldView* in previous chapters. Panel methods, however, do not concentrate on the details of the flowfield as much as the pressures and forces/moments that act on the aircraft. The code *CEASIOM*[vi] can be used to perform *DATCOM* (see Chapter 2) and vortex lattice predictions, where the first step is setting up the geometry, as shown here.

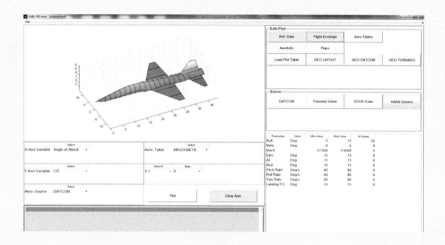

Notice that the geometry (in this case a T-38) has been simplified for this situation, with faired-over inlets and a simplified fuselage. Next, control surfaces can be defined, as shown to the right.

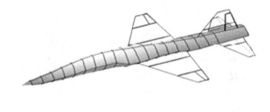

A *DATCOM* geometric model can also be defined in order to obtain a semi-empirical method prediction.

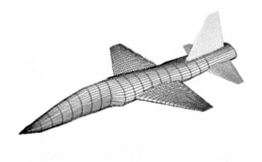

[vi] www.ceasiom.com

After performing a panel sensitivity study (as discussed earlier in Chapter 5), a variety of flight conditions can be run very quickly by defining the various angles of attack and sideslip, as well as the Mach numbers, where predictions are desired. Comparisons can then be made between *DATCOM*, Vortex Lattice, and experimental data to see how well the predictions match reality (another important part of the verification and validation process discussed in Chapter 2).

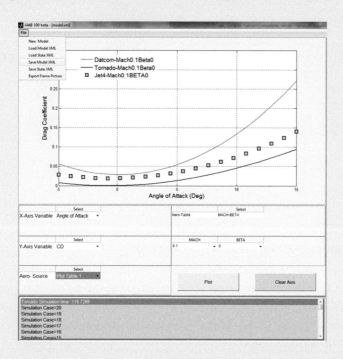

In addition to the forces and moments, the Vortex Lattice code can show the differential pressure distribution on the lifting surfaces for design purposes.

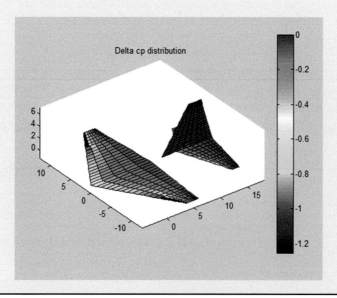

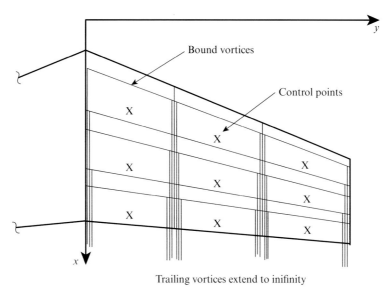

Figure 5.39 The horseshoe vortex layout for the classical vortex lattice method.

5.3.2 The classical Vortex Lattice Method

There are many different vortex lattice schemes; in this section we will describe the "classical" implementation, in which a horseshoe vortex is placed on each panel.[vii] Knowing that, from our airfoil analysis, vortices can represent lift, this approach places a vortex filament at the 1/4 chord location of a "panel" and then the boundary condition is satisfied at the 3/4 chord point on the panel (the so-called "1/4–3/4 rule"). This was the approach used to obtain the results shown in Fig. 5.38. This approach is illustrated in Fig. 5.39, where the "horseshoe vortex" is shown on each panel. Each horseshoe vortex consists of the "bound vortex" attached to the panel and two trailing vortex legs that attach to the end of the bound vortex and extend to downstream infinity (since a vortex cannot end in a fluid). Using this approach, we proceed as follows:

1. Divide the planform into a lattice of quadrilateral panels and put a horseshoe vortex on each panel.
2. Place the bound vortex of the horseshoe vortex on the 1/4 chord element line of each panel.
3. Place the control point on the 3/4 chord point of each panel at the midpoint in the spanwise direction (sometimes the lateral panel centroid location is used).
4. Assume a flat wake in the usual classical method.

[vii] Alternatively, methods have been developed that employ vortex rings around each panel. The panel at the trailing edge still has to use a vortex that extends to downstream infinity to model the wake. A curious property of vortex rings is that they are equivalent to a constant strength doublet panel, as pointed out by Paul Rubbert in a 1978 AIAA short course on applied computational aerodynamics.

5. Determine the strengths of each Γ_n required to satisfy the boundary conditions by solving a system of linear equations. This system comes from writing down the sum of all the horseshoe-vortex-induced velocity contributions at each control point and then using that velocity in the equation required to satisfy the no-flow-through condition. This results in a system for the unknown values of the Γ_n's.

Note that the lift is on the bound vortices. To understand why, consider the vector statement of the Kutta-Joukowski theorem, $\vec{F} = \rho \vec{V} \times \vec{\Gamma}$ (this is the vector form of Equation 4.34). Assuming that the freestream velocity is the primary contributor to the velocity, the trailing vortices are parallel to the velocity vector and hence the force on the trailing vortices is zero (recall that, if vectors are parallel, the cross product is zero). More accurate methods find the wake deformation required to eliminate the force in the presence of the complete induced-flowfield.

As in the case of panels methods, the algebra required to obtain the influence coefficients connecting the velocities induced at a specific location by each horseshoe vortex is lengthy. The velocity induced by a vortex filament is given by the Biot-Savart Law (see Equation 4.71). Each horseshoe vortex makes a contribution with a single value of Γ (Bertin and Cummings[2] contains a detailed derivation). To illustrate the essence of the method, we consider the simple planar surface case, without dihedral. Using the basic boundary condition relation given in Equation 5.51 and realizing that in three dimensions the downwash velocity is w, the boundary condition is:

$$\frac{w}{V_\infty} = \frac{df_c}{dx} - a. \tag{5.60}$$

Now, the downwash w at a point m, due to a horseshoe vortex, n, can be written as:

$$w_m = C_{m,n} \Gamma_n \tag{5.61}$$

where $C_{m,n}$ is an influence coefficient based on the Biot-Savart Law and the geometric distance between point m and the horseshoe vortex n. The next step is to sum the induced velocities from all the horseshoe vortices at a point m and use that velocity in the boundary condition, Equation 5.60,

$$w_m = \sum_{n=1}^{2N} C_{m,n_k} \Gamma_n = V_\infty \left(\frac{df_c}{dx} - \alpha \right)\bigg|_m , \tag{5.62}$$

where there are N horseshoe vortices on each side of the wing, resulting in a total of $2N$ horseshoe vortices representing the wing. Thus we have the following equation that satisfies the boundary conditions and can be used to relate the circulation distribution and the wing camber and angle of attack:

$$\sum_{n=1}^{2N} C_{m,n_k} \left(\frac{\Gamma_n}{V_\infty}\right) = \left(\frac{df_c}{dx} - \alpha\right)_m \qquad m = 1,...,2N. \qquad (5.63)$$

Equation 5.63 includes two possible applications:

1. *The Analysis Problem.* Given camber slopes and α, solve for the circulation strengths, (Γ/V_∞) [a system of $2N$ simultaneous linear equations].

or

2. *The Design Problem.* Given (Γ_n/V_∞), which corresponds to a specified surface loading, find the camber and α required to generate this loading (only requires simple algebra, no system of equations must be solved).

Notice that the way df_c/dx and α are combined illustrates that the division between camber, angle of attack, and wing twist is arbitrary (twist can be considered a separate part of the camber distribution and is useful for wing design). However, care must be taken to keep the bookkeeping straight.

A reduction in the size of the problem is possible in many cases. If the geometry is symmetrical and the camber and twist are also symmetrical, then Γ_n is the same on each side of the planform (but not the influence coefficient). Therefore, we only need to solve for N Γ's, not $2N$. This is also true if ground effects are desired, where the image of a vortex placed symmetrically below the ground plane can be included in the influence coefficient matrix without adding additional unknowns, as shown in Katz and Plotkin.[6] The system of equations for the case of planform symmetry becomes:

$$\sum_{n=1}^{N} \left[C_{m,n_{k\text{left}}} + C_{m,n_{k\text{right}}} \right] \left(\frac{\Gamma_n}{V_\infty}\right) = \left(\frac{df_c}{dx} - \alpha\right)_m \qquad m = 1,...,N. \qquad (5.64)$$

Doing this seems easy! Why not just program it for yourself? You can, but most of the work is taken up by:

1. Automatic layout of panels for arbitrary geometry. As an example, when considering multiple lifting surfaces, the horseshoe vortices on each surface must "line up." The downstream leg of a horseshoe vortex cannot pass through the control point of another panel.

and

2. Converting Γ_n to the aerodynamics values of interest, C_L, C_m, etc., and the spanload is tedious for arbitrary configurations.

Nevertheless, many people, including many students, have written VLM codes. The method is widely used in industry and government for aerodynamic estimates for conceptual and preliminary design predictions. It provides good insight into the aerodynamics of wings, including interactions between lifting surfaces. Typical analysis uses (in a design environment) include:

- Predicting the configuration neutral point during initial configuration layout and studying the effects of wing placement and canard and/or tail size and location.
- Finding the induced drag, C_{D_i}, from the spanload in conjunction with far-field methods.
- With care, estimating control and device deflection effectiveness. (Estimates where viscous effects may be important require calibration. Some examples are shown in the next section. For example, take 60 percent of the inviscid value to account for viscous losses, and also realize that a deflection of $\delta_f = 20$–$25°$ is about the maximum useful device deflection in practice.)
- Investigating the aerodynamics of interacting surfaces.
- Finding the lift-curve slope, C_{L_α}, approach angle of attack, etc.

Typical design applications include:

- Initial estimates of twist to obtain a desired spanload, or root bending moment.
- Starting point for finding a camber distribution in purely subsonic cases.

Before examining how well the method works, two special cases require comments. The first case arises when a control point is in line with the projection of one of the finite length vortex segments. This problem occurs when the projection of a swept bound vortex segment from one side of the wing intersects with a control point on the other side of the wing; this happens surprisingly frequently. The velocity induced by this vortex is zero, but the equation as usually written degenerates into a singular form, with the denominator going to zero; thus, a special form of the equation should be used. In practice, when this happens, the contribution can be set to zero without invoking the special form. Figure 5.40 shows how this happens for the Warren 12 planform and 36 vortices on each side of the wing. We see that the projection of the line of bound vortices on the last row of the left-hand side of the planform intersects one of the control points on the right-hand side.

A model problem illustrating this can be constructed for a simple finite-length vortex segment. The velocity induced by this vortex is shown in Fig. 5.41. When the vortex is approached directly, $x/l = 0.5$, the velocity is singular for $h = 0$. However, as soon as you approach the axis ($h = 0$) off the end of the segment ($x/l > 1.0$), the induced velocity is zero. This illustrates why you can set the induced velocity to zero when this happens.

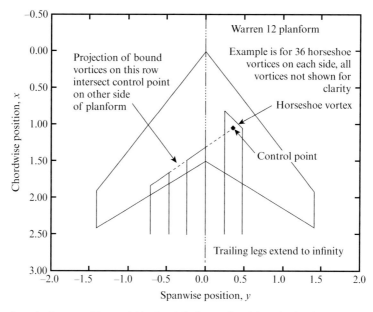

Figure 5.40   Example of case requiring special treatment, the intersection of the projection of a vortex with a control point.

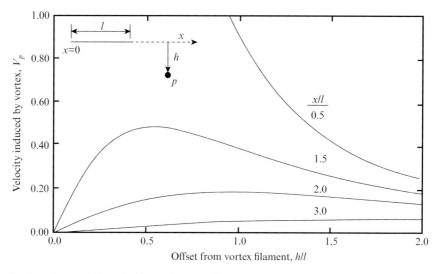

Figure 5.41   Velocity induced by finite straight-line section of a vortex.

The second special case that needs to be discussed arises when two or more planforms are used with this method, which is one of the most powerful applications of the vortex lattice method. However, you must take care to make sure that the trailing vortices from the first surface do not intersect the control points on the second surface. In this case, the induced velocity is in fact infinite, and the method breaks down. Usually, this problem is solved by using the same spanwise distribution of horseshoe vortices on each surface. This aligns the vortex legs, and the control points are well removed from the trailing vortices of the forward surfaces.

Profiles in Computational Aerodynamics: Zach Hoisington

"When I was around six years old, I learned that a few folds can turn a sheet of paper into a flying aircraft. After a few too many "experiments" at home and school, I was encouraged to consider aerospace engineering as a future career. My interest in aviation continued with remote control airplanes. One memorable experience involved rescuing a severely damaged model from a garbage bin and bringing it back to life with help from popsicle sticks and glue. After one heroic flight, the plane quickly returned to the state in which I found it.

"Flying model aircraft got me excited about actually getting myself airborne. Jumping my bicycle was not always successful, but it helped me appreciate how aircraft deal with crosswind landings. My father was a private pilot, giving me opportunities to take the controls at a young age. When I was in my early teens, we learned how to paraglide and hang glide.

"My favorite subjects in school were science and math, although I was more interested in the application of the principles than the textbooks. I spent much of my high school years on flying trips and paragliding competitions, with a third-place finish at the U.S. nationals when I was a senior.

"I attended Cal Poly, San Luis Obispo, to study aeronautical engineering. It was a bit discouraging in the beginning when I discovered how much work would be required to get a degree and begin a career. My interest in school increased after an internship with a paragliding manufacturer in Austria, where I was able to see results of new ideas within days. After the main test pilot was injured, I filled in for the flying duties. Although it was educational, I was happy to return to the United States alive. The following summer, I worked for Boeing in the mechanisms group for the International Space Station. Although aeronautics is my main passion, this was a great introduction to opportunities available within large companies.

"After graduating from college, I went to Huntington Beach, California to work in the conceptual design group for Boeing Research & Technology. I work as a conceptual designer in the aircraft configuration group, where we generate new aircraft designs and evaluate their performance. We have an energetic team of designers working closely with experts in aircraft stability, structures, propulsion, aerodynamics, and other key disciplines. I've worked on subsonic, supersonic, underwater, and lighter-than-air vehicles during my ten-year career. Recent work has concentrated on airliner concepts with significant reductions in fuel burn and environmental impact for the 2035 and beyond timeframe. Big advancements may be possible if hybrid electric propulsion, laminar flow, riblets, and new structural concepts are all incorporated into one vehicle.

"I teach an aircraft design lab at USC each spring. This helps me stay current with the fundamentals, and keeps me up to date with new analysis programs and ideas. I still fly model aircraft and paragliders, and I also make paper airplanes on occasion."

Zach Hoisington won the AIAA National Student Paper competition and the B.F. Goodrich Collegiate Inventors competition in 1998 for his design of a variable aspect ratio paraglider. He uses panel methods and vortex lattice methods on a regular basis for his aircraft design work at Boeing!

5.3.3 Examples of the Use and Accuracy of the Vortex Lattice Method

How well does the vortex lattice method work? In this section, we describe how the method is normally applied and present several example results obtained using the method. The vortex lattice layout is clear for most wings and wing-tail or wing-canard configurations. The method can be used for wing-body cases by simply specifying the projected planform of the entire configuration as a flat lifting surface made up of a number of straight-line segments. The exact origin of this somewhat surprising approach is unknown, but its success is illustrated in examples given later in this chapter.

To get good, consistent, and reliable results using Lamar's program, some simple rules for panel layout should be followed (likely equally valid for any VLM method). This requires that a few common rules of thumb be used in selecting the planform break points: (a) the number of line segments should be minimized; (b) breakpoints should line up streamwise on front and rear portions of each planform and should line up between planforms; (c) streamwise tips should be used; (d) small spanwise distances should be avoided by making edges streamwise if they are actually highly swept, and (e) trailing vortices from forward surfaces cannot hit the control point of an aft surface. Figure 5.42 illustrates these requirements.

5.3.3.1 THE WARREN 12 TEST CASE

One reference wing case that is a standard used to check the accuracy of vortex lattice codes is the Warren 12 planform. It provides a check case for

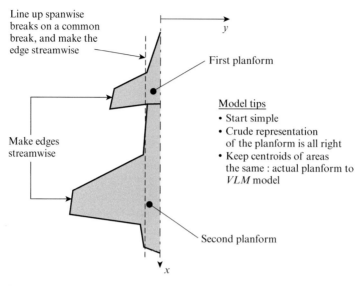

Figure 5.42 Example of a VLM model of an aircraft configuration. Note that one side of a symmetrical planform is shown.

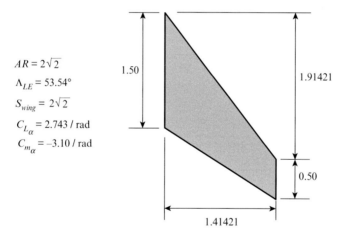

$AR = 2\sqrt{2}$

$\Lambda_{LE} = 53.54°$

$S_{wing} = 2\sqrt{2}$

$C_{L_\alpha} = 2.743\ /\ \text{rad}$

$C_{m_\alpha} = -3.10\ /\ \text{rad}$

Figure 5.43 Definition and reference results for the Warren 12 wing.

the evaluation of any new or modified VLM code, as well as a check on the panel scheme layout, and is defined, together with the "official" characteristics from previous calculations, in Fig. 5.43. For the Warren 12 results, the reference chord used in the moment calculation is the average chord (which is slightly nonstandard; normally the reference chord used is the mean aerodynamic chord) and the moment reference point is located at the wing apex (which is also nonstandard).

As with all computational aerodynamics methods, VLM should be checked to assess the convergence characteristics. Figure 5.44 is taken from the original NASA report describing the Lamar program,[57] and shows that the basic force results can be obtained with a relatively small number of

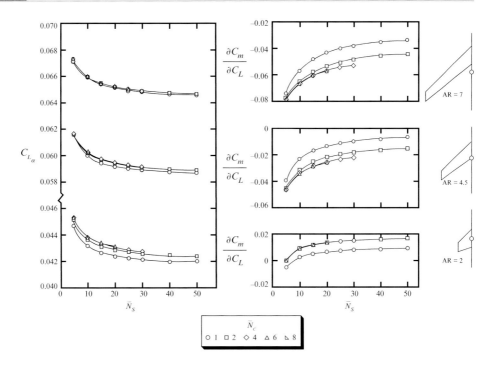

Figure 5.44 Convergence of the lift-curve slope and stability with increasing numbers of panels, $\bar{N}_s$ is the number of spanwise panels, $\bar{N}_c$ is the number of chordwise panels (from Ref. 57).

chordwise panels, but requires a fairly large number of spanwise rows of panels. The original program distributed by NASA allowed for a maximum of 200 horseshoe vortices, which is about what is needed to obtain results that are converged for the number of vortices.

5.3.3.2 ISOLATED SWEPT WING

Next we show how a VLM calculation compares with experimental data for an isolated wing case.[69] Here we selected an aspect ratio 10 wing with a quarter chord sweep of 35° and a taper ratio of 0.5. The airfoil was a 12% thick NACA 65A012 section. The Mach number is low, 0.14, and the Reynolds number is 6 million for the lift comparison and 10 million for the pitching moment comparison (the results are given in Fig. 5.45). Figure 5.45a shows the results for the lift coefficient, where we see very good agreement at low angles of attack where the flow is attached. Figure 5.45b compares the pitching moment variation with the lift coefficient. Predicting pitching moment is one of the most important considerations in airplane aerodynamics, and the agreement is excellent until the flow starts to separate at the wing tip and the wing "pitches up," a problem that occurs with swept wings.

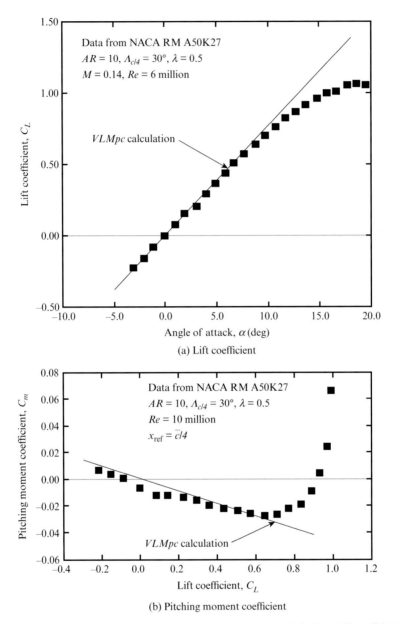

Figure 5.45 Comparison of *VLMpc* with wind tunnel data for an isolated wing (Ref. 69). (a) lift coefficient. (b) pitching moment.

5.3.3.3 WING-BODY-TAIL

The comparison of VLM results with wind tunnel data for a wing-body-tail configuration[70] is given in Fig. 5.46. Figure 5.46a shows the configuration, and Figures 5.46b and 5.46c compare the results with the wind tunnel data for the lift and pitching moment coefficients. The effect of the horizontal tail is included in each figure. Notice that the agreement with the lift coefficient is very good, both for the "tails on" and "tails off" cases. The pitching moment

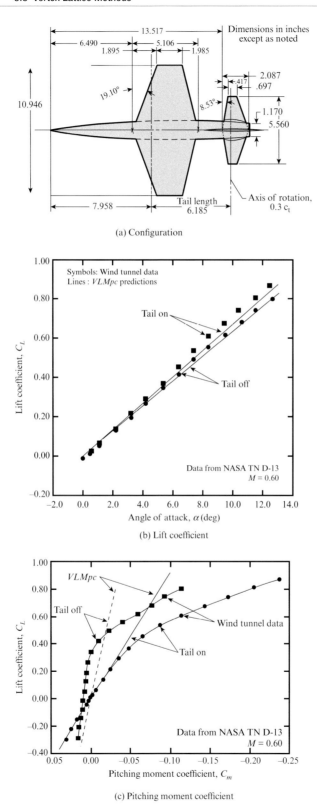

(a) Configuration

(b) Lift coefficient

(c) Pitching moment coefficient

Figure 5.46   Wing body tail case (Ref. 70). (a) configuration. (b) lift. (c) comparison of pitching moment estimate from *VLMpc* with wind tunnel data.

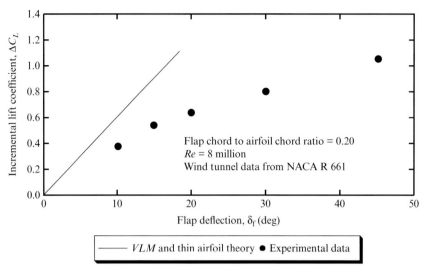

Figure 5.47 Control surface deflection effectiveness comparison between a VLM calculation and wind tunnel data (data taken from Ref. 71).

comparisons are fairly good at low angles of attack (low lift coefficients), but the VLM method underpredicts the pitching moment at the high incidence angles due to viscous effects.

5.3.3.4 CONTROL SURFACE DEFLECTION

The VLM method can also be used to estimate control surface deflection effects (an example is shown in Fig. 5.47). Because the method is inviscid and the control device is at the back end of the surface and subjected to significant viscous effects, the method generally overpredicts the effectiveness. The case shown is for an NACA 23012 airfoil with a 20% chord plane flap at a Reynolds number of 8 million.[71] Notice that the prediction at low flap deflection angles (up to 10 degrees) is actually fairly good when compared to an extrapolation of the experimental data to low-flap deflection angles.

5.3.3.5 PITCH AND ROLL DAMPING ESTIMATION

The VLM method can also be used to predict the value of the pitch and roll damping coefficients, C_{m_q} and C_{l_p}. These characteristics are important in establishing the stability characteristics of airplanes, and the method described in Etkin[72] is easily implemented. Figure 5.48 shows a comparison of the VLM pitching and roll damping results using another VLM method, $JKayVLM$,[68] for an F-18-type airplane compared to estimates from $DATCOM$ (a semi-empirical program that we discussed in Chapter 2). The case shows reasonably good, but not perfect, agreement.[68]

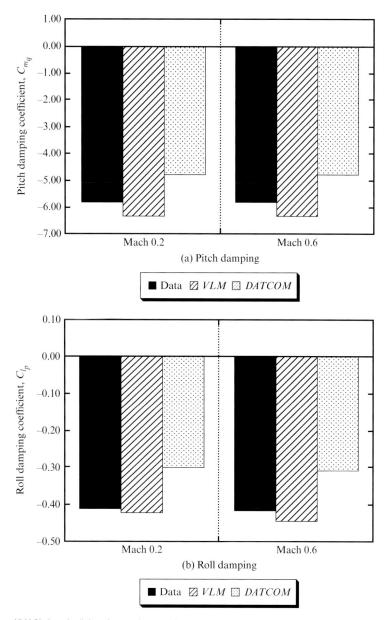

(a) Pitch damping

■ Data ▨ *VLM* ⊞ *DATCOM*

(b) Roll damping

■ Data ▨ *VLM* ⊞ *DATCOM*

Figure 5.48   VLM Pitch and roll damping result comparison with data and *DATCOM*. (Ref. 68).

5.3.3.6 SLENDER LIFTING BODY RESULTS[73]

To illustrate the capability of the vortex lattice method for bodies that are more fuselage-like than wing-like, we present the lifting body comparison of the experimental and VLM results published by Pittman of NASA Langley Research Center.[73] Figure 5.49 shows the configuration used, and Fig. 5.50 provides the results of the vortex lattice method compared with experimental data. In this case, the camber shape was modeled by specifying camber slopes on the mean surface (the model used 138 vortex panels). For highly swept wings, leading-edge vortex flow effects are included. Lamar's program

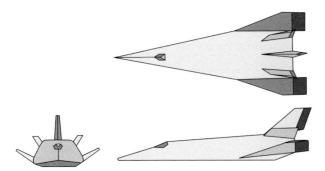

Figure 5.49   Highly swept lifting body type hypersonic concept (Ref. 73; reprinted by permission of the American Institute of Aeronautics and Astronautics, Inc.).

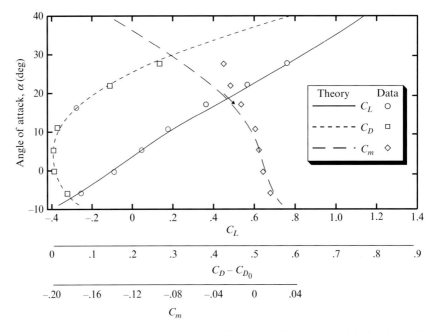

Figure 5.50   Comparison of *CL*, *Cm*, and *CD* predictions with data (Ref. 73; reprinted by permission of the American Institute of Aeronautics and Astronautics, Inc.).

contains the option of using the leading-edge suction analogy[74] to model these effects.[viii] Remarkably good agreement with the force and moment data is demonstrated in Fig. 5.50. The nonlinear variation of lift and moment with angle of attack arises due to the inclusion of the vortex lift effects. The agreement between data and computation breaks down at higher angles of attack because the details of the distribution of vortex flow separation are not provided by the leading-edge suction analogy. The drag prediction is also

---

[viii] The leading edge suction analogy is the empirical observation by Eddie Polhamus, after examining experimental data, that the magnitude of the vortex lift is approximately equal to the value of the leading-edge suction for attached flow, but rotated 90°.

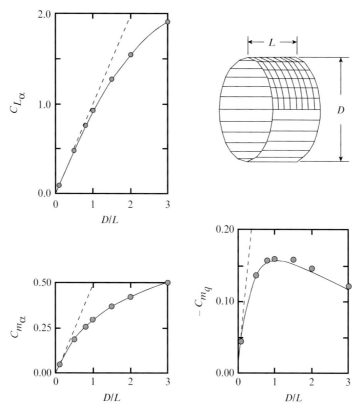

Figure 5.51 Predictions by Kalma n, Rodden, and Giesing (Ref. 75) using VLM (symbols) of the aerodynamic characteristics of a ring wing, compared with linear theory (solid lines) and slender wing theory (dashed lines), both from Belotserkovskii (Ref. 76).

very good; keep in mind that the experimental drag is adjusted by removing the zero lift drag, which contains the drag due to friction and separation. The resulting drag due to lift is compared with the VLM estimates. The comparisons are good, primarily because this planform is achieving, essentially, no leading-edge suction (see Katz and Plotkin,[6] to review the leading-edge suction concept) and, hence, the drag is simply $C_D = C_L \tan\alpha$, which is easy to predict, but aerodynamically inefficient.

5.3.3.7 NON-PLANAR RESULTS[75]

All of the examples presented so far considered essentially planar lifting surface cases. The vortex lattice method can also be used for highly non-planar analysis, and the example cases used at Douglas Aircraft Company in a classic paper[75] have been selected to illustrate the capability. Several of the cases were re-computed using the *JKayVLM* code[68] and provide an interesting comparison with the original results from Douglas.

Figure 5.51 provides an example of the results obtained for an extreme non-planar case: the ring, or annular, wing. In this case, the estimates are compared with other theories and are seen to be very good. The figure also

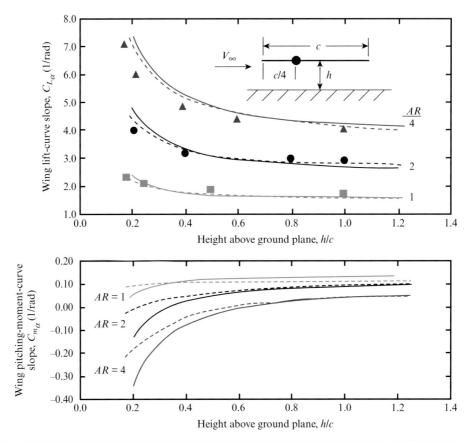

Figure 5.52 Example of ground effects for a simple rectangular wing. Dashed lines from Kalman, Rodden, and Giesing (Ref. 75); solid lines are predictions by *JKayVLM* (Ref. 68) Symbols are experimental results by Saunders (Ref. 77). Pitch moment reference taken about the quarter chord.

includes the estimate of C_{m_q}. As illustrated earlier, C_{m_q} and C_{l_p} can be computed using VLM methods, and this capability is included in Lamar's vortex lattice method, *VLMpc*.[ix]

5.3.3.8 GROUND EFFECTS AND DIHEDRAL EFFECTS

Figure 5.52 provides an example of the effects of the presence of the ground on the aerodynamics of simple unswept rectangular wings. The lift and pitching moment slopes are presented for calculations made using *JKayVLM* and compared with the results published by Kalman, Rodden, and Giesing[75] and experimental data. A mirror image of the configuration is used below the ground plane to represent the ground effect. The agreement between the data and calculations is excellent for the lift curve slope. The $AR = 1$ wing shows the smallest effects of ground proximity because of the three-dimensional relief provided around the wing tips. As the aspect ratio increases,

---

[ix] *JKayVLM* and *VLMpc* are available at www.cambridge.org/aerodynamics along with links to a few other VLM codes.

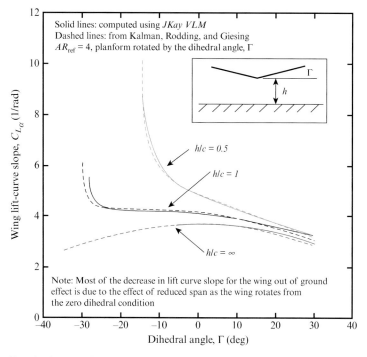

Figure 5.53 Example of ground effects for a wing with dihedral (a case from Kalman, Rodden, and Giesing, Ref. 75), including a comparison with results from *JKayVLM* (Ref. 68).

the magnitude of the ground effects increases. The lift-curve slope starts to increase rapidly as the ground is approached.

The wings also experience a significant change in the pitching moment slope (aerodynamic center shift).[x] Note that the predictions start to differ as the ground is approached. *JKayVLM* actually rotates the entire surface to obtain another solution to use in estimating the lift curve slope. The standard procedure used by most methods is to simply change the slope condition at the mean line, as discussed previously in this chapter. Because of the proximity to the ground, this might be a case where the transfer of the boundary condition may not be accurate.

Figure 5.53 presents similar information for the effect of dihedral angle on a wing. In this case, the effects of anhedral, where the wing tips approach the ground, are extremely large. The results of dihedral changes for a wing out of ground effect are also shown for comparison. The methods agree well with each other, with differences appearing only as the wing tips approach the ground. Here, again, *JKayVLM* actually rotates the entire geometry, apparently resulting in an increase in the effects as the tips nearly contact the ground. It also prevents calculations from being obtained as close to

---

[x] This shift is the reason that ground effects machines cannot just "take off" and why WIG vehicles cannot fly out of ground effect.

the ground as the published results. In making these calculations, it was discovered that the wing panel was rotated and not sheared, so that the projected span decreases as the dihedral increases, and this produces much more pronounced changes in the lift-curve slope due to the reduction in projected span.

5.3.4 Inverse Design Methods and Program *DesCam*

Although most of the examples discussed previously correspond to the analysis problem, the design problem can also be treated. In this section, we will provide one example: the determination of the camber line shape required to obtain a specific chord load in the two-dimensional case. We take the opportunity to illustrate a method due to Lan[66] that uses a mathematically based selection of vortex and control point placements instead of the 1/4–3/4 rule used previously.

Recall that a line of vortex singularities induces a vertical velocity on the singularity line given by (see Chapter 4)

$$w(x) = -\frac{1}{2\pi}\int_0^c \frac{\gamma(x')}{x - x'}\,dx'. \tag{5.65}$$

In thin wing theory, the vertical velocity, w, can be related to the camber line slope as shown in Equation 5.51. Vortex strengths can be related to the streamwise velocity by $\gamma = u_l/V_\infty - u_u/V_\infty$, where the induced velocities are anti-symmetric above and below the camber line. This, in turn, can be used to relate the vorticity to the change in pressure, ΔC_p, through Equation 5.56:

$$\Delta C_p = C_{p_l} - C_{p_u} = \left(-2\frac{u_l}{V_\infty}\right) - \left(-2\frac{u_u}{V_\infty}\right) = 2\left(\frac{u_u}{V_\infty} - \frac{u_l}{V_\infty}\right) \tag{5.66}$$

which leads to:

$$\frac{\Delta C_p(x)}{2} = \gamma(x) \tag{5.67}$$

resulting in the expression for camber line slope ($dz/dx = w$) in terms of the design chord load:

$$\frac{dz}{dx} = -\frac{1}{4\pi}\int_0^c \frac{\Delta C_p}{x - x'}\,dx'. \tag{5.68}$$

Here dz/dx includes the slope due to the angle of attack. Note that the integral contains a singularity, and this singularity introduces the extra complications that require special analysis for numerical integration. The original Lan theory[66] was used to find ΔC_p (in a slightly different form), but it can also be used to obtain dz/dx directly from ΔC_p. To do this, Lan derived a summation formula to obtain the slope. Once the slope is known, it is integrated to obtain the camber line.

Lan showed that the integral in Equation 5.68 can be found very accurately from the summation:

$$\frac{dz}{dx}\bigg|_i = -\frac{1}{N}\sum_{k=1}^{N}\frac{\Delta C_p}{4}\frac{\sqrt{x_k(1-x_k)}}{x_i-x_k} \tag{5.69}$$

where

$$x_k = \frac{1}{2}\left[1-\cos\left\{\frac{(2k-1)\pi}{2N}\right\}\right] \qquad k=1,2,...,N \tag{5.70}$$

and

$$x_i = \frac{1}{2}\left[1-\cos\left\{\frac{i\pi}{N}\right\}\right] \qquad i=0,1,2,...,N. \tag{5.71}$$

Here, $N+1$ is the number of stations on the camber line at which the slopes are obtained.

Given dz/dx, the camber line is then computed by integration using the trapezoidal rule (marching forward starting at the trailing edge):

$$z_{i+1} = z_i - \left[\frac{x_{i+1}-x_i}{2}\right]\left[\frac{dz}{dx}\bigg|_i + \frac{dz}{dx}\bigg|_{i+1}\right]. \tag{5.72}$$

The design angle of attack is then:

$$\alpha_{DES} = \tan^{-1}z_0. \tag{5.73}$$

The camber line can then be redefined in standard nomenclature, i.e., $z(x=0) = z(x=1) = 0.0$:

$$\bar{z}_i = z_i - (1-x_i)\tan\alpha_{DES}. \tag{5.74}$$

How well does this work? Here we compare the results from *DesCam* with the analytic formula given in Appendix A for the NACA 6 Series mean line with $a = 0.4$ (the results are shown in Fig. 5.54). The camber scale is greatly enlarged to demonstrate the excellent comparison. Even though the chord

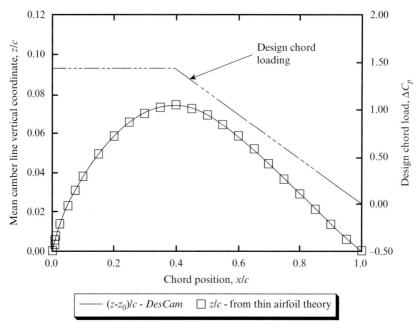

Figure 5.54   Example and verification of camber design using *DesCam*.

load is constructed by two straight-line segments, the resulting required camber line is highly curved over the forward portion of the airfoil. Note that thin airfoil theory allows only two possible values for the pressure differential at the leading edge, zero or infinity. A close examination of the camber line shape required to produce a finite load reveals a singularity, since the slope is infinite. This feature is much easier to study using the analytic solution, as given in Appendix A. The inverse problem approach shown here can easily be extended to three dimensions, and notice that the design problem is direct, in that it does not require the solution of a system of equations. This makes VLM a very powerful conceptual design tool.

5.3.5 Alternate and Advanced VLM Methods

Many variations of the vortex lattice method have been proposed over the years. They address both the improvement in accuracy for the traditional case with a planar wake and extensions to include wake position and rollup as part of the solution. The traditional vortex lattice approaches assume that the wing wake remains flat and aligned with the freestream. This assumption is acceptable for most cases, although not technically correct. The effect of the wake on the wing that generates it is small unless the wing is highly loaded. However, the interaction between the wake from an upstream surface and a trailing lifting surface can be influenced by the rollup and position.

In the basic case where the wake is assumed to be flat and at a specified location, the primary extensions of the method have been directed toward improving the accuracy using a smaller number of panels. Hough[78] demonstrated that improvement in accuracy could be achieved by using a lattice that was slightly smaller than the true planform area. Basically, he proposed a 1/4 panel width inset from the wing tips.

Perhaps the most important revision of the vortex lattice method was proposed by Lan,[66] which is called the "quasi vortex lattice method." In this method, Lan used mathematical methods, rather than the more heuristic arguments described earlier, to find an approximation for the thin airfoil integral in the streamwise direction. The result was, in effect, a method in which the vortex and control point locations were established from the theory of Chebychev polynomials to obtain an accurate value of the integrals with a small number of panels. The mathematically based approach also led to an ability to compute leading-edge suction very accurately.

The wake rollup and position problem has been addressed by Mook,[67] among others, and his work should be consulted for details. A method similar to Mook's has been presented in the book by Katz and Plotkin.[6] They propose a vortex ring method, which has advantages when vortices are placed on the true surface of a highly cambered shape.

5.3.6 Unsteady Flow Extension

Extensions to include unsteady flow have also been made. For the case of harmonically oscillating surfaces and an assumed flat wake, the extension was given by Albano and Rodden.[79] In this case, the vortex is augmented with an oscillating doublet, and the so-called doublet-lattice method is obtained. The doublet-lattice method is widely used for subsonic flutter calculations. Kalman, Giesing, and Rodden[75] provide additional details and examples (they also demonstrated the highly nonplanar capability of the VLM method as given earlier). A refinement of this method and an investigation of panel density requirements was made by Rodden et al.[80,81]

A general unsteady flow calculation method, including wake rollup and incorporation of leading-edge vortices, has been developed by Mook and his co-workers.[67] Katz and Plotkin[6] also describe VLM methods for unsteady motion in some detail, and biologically inspired interest in flapping flight has also led to the development of general unsteady methods.[82] The resulting codes have the potential to be used to model time-accurate aerodynamics of vehicles in arbitrary maneuvering flight, including the high angle of attack cases of interest in fighter aerodynamics. These methods are also currently being used in studies where the aircraft aerodynamics is coupled with advanced control systems as well as structural dynamics. In this case, active control is incorporated and adverse aeroelastic behavior can be suppressed.

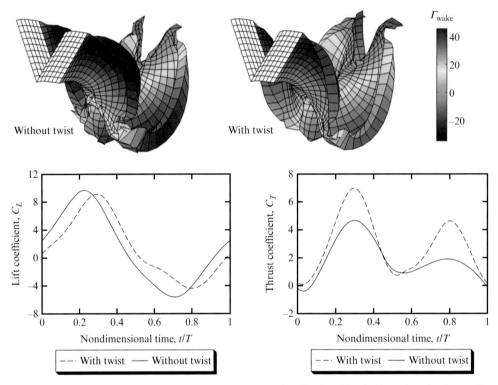

Figure 5.55 Comparison of unsteady vortex lattice methods predictions for a flapping with and without wing twist; the resulting lift and thrust coefficients are also shown (Ref. 84; courtesy of Phil Beran of the Air Force Research Laboratory; a full color version of this image is available on the website: www.cambridge.org/aerodynamics).

An unsteady compressible vortex lattice method has also been developed by Soviero and Hernandes.[83] An example of this is an unsteady VLM code which includes a novel wake method applied to a biologically inspired case shown in Fig. 5.55.[84] Notice the ability of the wake to "follow" the flapping wing and create a realistic unsteady wake, which impacts the aerodynamics a great deal. Also included in the analysis is the effect of wing twist on the lift and thrust coefficients, which makes this approach very valuable in understanding low-speed flyers.[85]

5.3.7 Vortex Lattice Method Summary

Classical vortex lattice methods, per se, are not currently being developed, although extensions of lifting-line theory approaches have been made for compressible flow,[86,87] and unsteady methods continue to be developed.[88] The steady flow methods are continuously used in a variety of applications for conceptual design trade studies and in advanced simulation methods where several disciplines are being studied simultaneously and an affordable model of the aerodynamics is required.

F-16 fighter in formation flight with a computational aerodynamic simulation showing strake vortices, surface pressures, and the exhaust of the engine (courtesy of Stefan Görtz and the USAFA High Performance Computing Research Center).

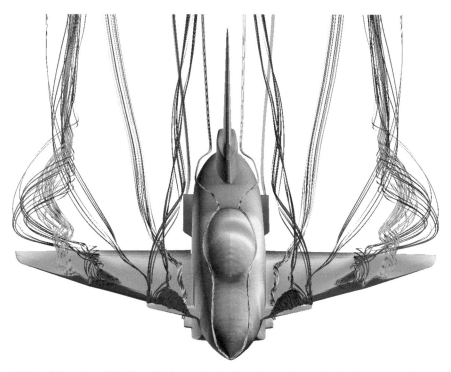

X-31 simulation at $\alpha = 16°$ with particle traces showing the various vortices on the aircraft (courtesy of Andreas Schütte of DLR).

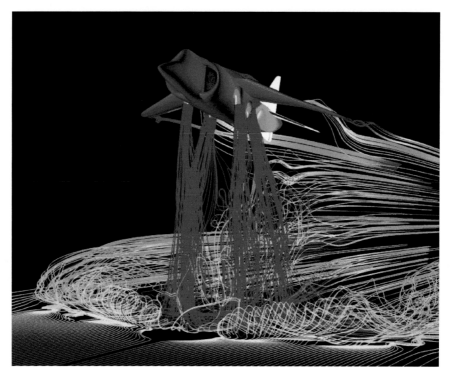

**Plate 3
(Figure 1.3)** AV-8B Harrier hover simulation showing interaction of various jets with the ground plane.

**Plate 4
(Figure 1.4)** F-16 fighter with surface colored to show the pressures plus surfaces of constant vorticity showing various vortex flow structures (courtesy of Stefan Görtz and the USAFA High Performance Computing Research Center).

CFD simulation of a modern commercial transport (courtesy of Airbus).

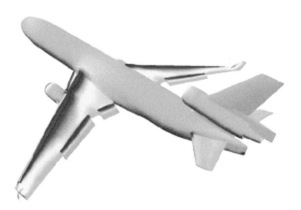

Surface pressures on an MD-11 near the runway during takeoff predicted with a potential method code, VSAERO. (Ref. 11) Notice the regions of low pressure (colored in red) over the upper surface of the wing and near the deflected flaps (Courtesy of Analytical Graphics, Inc.).

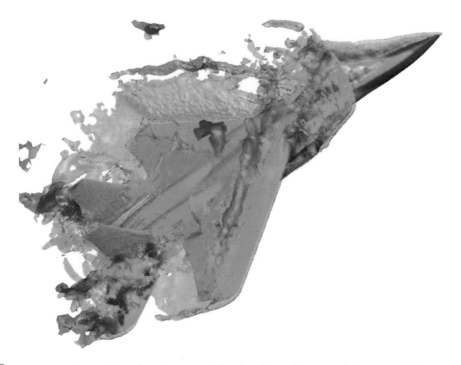

F-15 fighter during a spin simulation using a CFD code (courtesy of James Forsythe and the USAFA High Performance Computing Research Center).

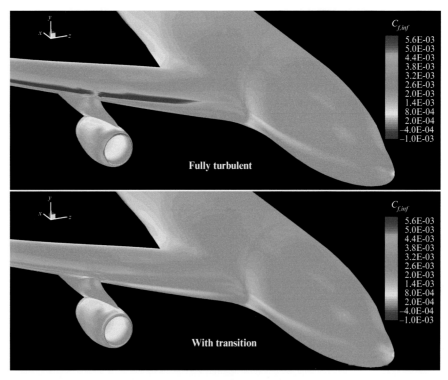

Computed surface friction distribution of a wing-body-pylon-nacelle configuration, $M = 0.8$, $Re = 9.97 \cdot 10^6$, $\alpha = 2.2°$, k-ω model (fully turbulent or with fixed transition) (Ref. 22; Courtesy of Andreas Krumbein of DLR).

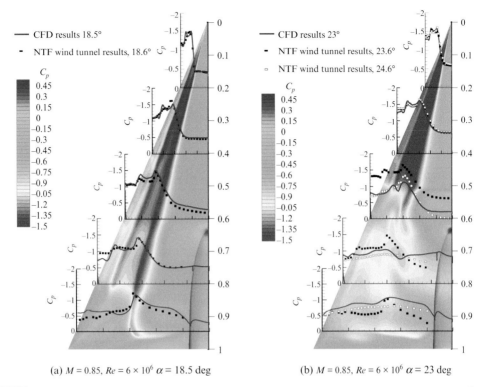

(a) $M = 0.85$, $Re = 6 \times 10^6$ $\alpha = 18.5$ deg

(b) $M = 0.85$, $Re = 6 \times 10^6$ $\alpha = 23$ deg

**Plate 9
(Figures 8.23
and 9.13)**

Comparisons of predictions of surface pressure for a sharp leading-edge delta wing at $M = 0.85$ and $Re_c = 6 \times 10^6$ using the Wilcox k-ω model (Ref. 54).

(a) SST

(b) SA

(c) SA-DES

**Plate 10
(Figure. 8.36)**

RANS and DES simulations of flow over an F/A-18C at $\alpha = 30$ deg, $Re_c = 13 \times 10^6$, leading-edge flaps set to -33 deg, trailing-edge flaps set to 0 deg, with no diverter slot present (Ref. 70).

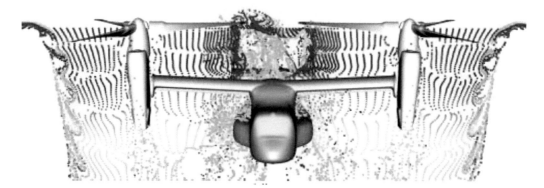

Plate 11 (Figure. 9.1) Time-dependent particle traces for the V-22 in hover, colored by particle release time: blue − earliest, red − latest (Ref. 2; Courtesy of Mark Potsdam of the U.S. Army Aeroflightdynamics Center).

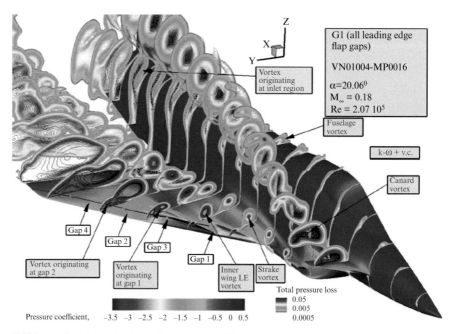

Plate 12 (Figure 9.12) Multiple crossflow planes showing total pressure loss with pressure coefficients on the surface of the X-31 (Ref. 16; Courtesy of Okko Boelens of NLR).

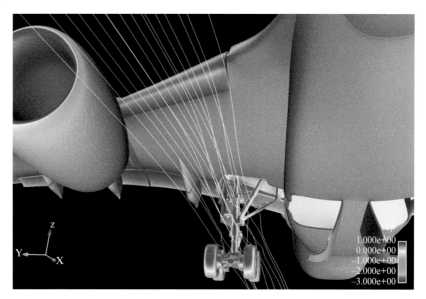

Plate 13 (Figure 9.16) Streamlines in the vicinity of the landing gear on a generic commercial transport (Courtesy of Airbus).

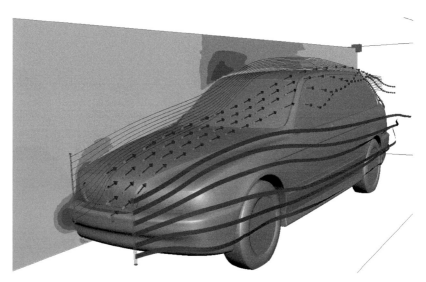

Plate 14 (Figure 9.17) Streaklines, stream ribbons, and glyphs, used in conjunction with a cutting plane, to show the flow around an automobile (Ref. 27; Courtesy of Wolf Bartelheimer of BMW).

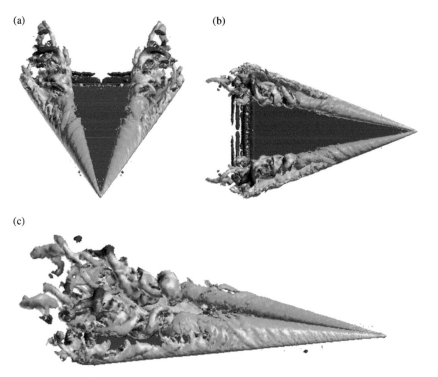

**Plate 15
(Figure. 9.19)** Instantaneous isosurfaces of vorticity magnitude colored by the spanwise component of vorticity for three views of a 70 deg delta wing; $\alpha = 27$ deg, $M = 0.069$, and $Re = 1.56 \times 10^6$ (Ref. 28).

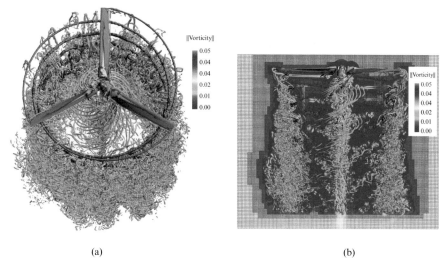

**Plate 16
(Figure. 9.23)** Isosurfaces of vorticity magnitude in the wake of a V-22 Osprey rotor. The complex flowfield includes the rotor tip vortex as well as the shear layers formed by the rotor (Ref. 30; Courtesy of Neal Chaderjian of NASA Ames Research Center).

Vortices in a dataset with turbulent vortex structures, visualized using isosurfaces and ellipsoids (Ref. 36; Courtesy of Frits Post and the TU Delft Visualization Group).

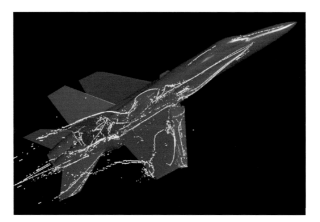

Vortices over the F-18 aircraft, showing the leading-edge extension vortex, as well as vortices over the wing (Ref. 37; Courtesy of NASA Ames Research Center).

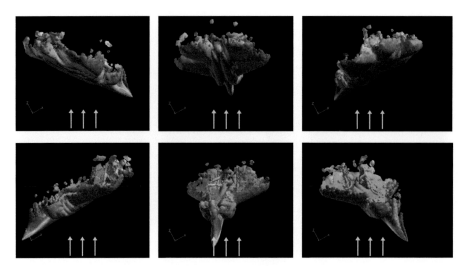

Flow over the F-15 aircraft during a prescribed spin, with individual snapshots of the solution serving to give the impression of motion (arrows from the bottom show the direction of the flow) (Ref. 39; Courtesy of James Forsythe of the U.S. Air Force Academy).

CFD Simulation of F/A-18 at high angle of attack.

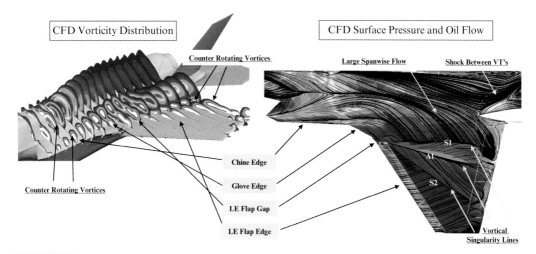

CFD Vorticity Distribution

Counter Rotating Vortices

Counter Rotating Vortices

Chine Edge

Glove Edge

LE Flap Gap

LE Flap Edge

CFD Surface Pressure and Oil Flow

Large Spanwise Flow

Shock Between VT's

S1

A1

S2

Vortical Singularity Lines

Plate 21 (Figure 10.16) F-35 upper surface on- and off-body flowfield visualization (Ref. 18; Courtesy of Brian Smith of Lockheed Martin).

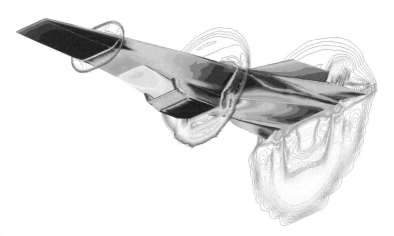

Plate 22 (Figure 10.30) CFD prediction of Hyper-X flowfield at $M = 7$ with engine operating (Courtesy of NASA Dryden Flight Research Center).

Differences between C-130 wind tunnel model and original CAD definition: color scale: −2 to +2 mm (Ref. 78).

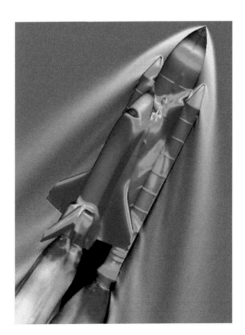

CFD simulation of Space Shuttle during ascent (Courtesy of NASA).

Summary of Best Practices

1. Linear theory aerodynamics prediction is an important part of computational aerodynamics – you should understand how and when to use these methods in aerodynamic analysis and design.
2. You should understand the basic assumptions of panel methods and the vortex lattice method and gain a feeling for when they work well and when they do not work well; always keep the assumptions in mind when using these methods!
3. Panels should have the following characteristics:
   a. Vary the size of panels smoothly
   b. Concentrate panels where the flowfield and/or geometry is changing rapidly
   c. Do not spend more money and time (i.e., numbers of panels) than required
4. You need to understand the assumptions and limitations of various aerodynamic theories that you may use to compare with CA predictions.
5. Always perform a panel sensitivity study before applying the results of these methods to analysis or design.
6. Understand which aerodynamic theories you should use along with panel methods to give you confidence in your results.
7. Realize that panel methods work well on "inverse" problems, and are therefore very useful for design as well as analysis.

5.4 Projects

1. *XFOIL* is a 2D panel code developed at MIT. You are going to use the code and determine how accurate it is by comparing predictions from *XFOIL* with experimental data (which you collected in Chapter 1) and your thin airfoil theory predictions (which you performed in Chapter 4). Go to the *XFOIL* website at: http://web.mit.edu/drela/Public/web/xfoil/ and download the appropriate zip file and the *XFOIL* User Guide. Run *XFOIL*, which is an executable program that is prompt driven (no GUI). At any time you can type "?" to see what commands are currently available. *XFOIL* tutorials are available online which you should follow and complete as practice. Run the airfoil at a Reynolds number of 6 million and a Mach number of 0.1. Use Excel to make the following plots: C_l vs α, C_d vs α, C_m vs α, and C_l vs C_d. Plot a comparison of your *XFOIL* prediction and thin airfoil theory results with your experimental data and plot the percentage difference between *XFOIL*, theory, and experiment as a function of angle of attack. Comment on how well the numerical prediction and theoretical prediction compare with the data, and relate your results to the assumptions of the theory where appropriate.

2. Use *CEASIOM* (www.ceasiom.com) to solve the aerodynamics of the T-38 aircraft at a variety of conditions. You should become familiar with the program and its use prior to working on the project.

The basic geometric parameters for the T-38 are:

Wing area, *S*:	170 ft²
Aspect ratio, *AR*:	3.75
Taper ratio, *λ*:	0.30

Other geometric dimensions for the T-38 can be found in *Introduction to Aeronautics*, Appendix B.[89] You will need this information to properly predict the aerodynamic coefficients.

The first thing you should do is perform a "panel sensitivity study" on the configuration. Run *CEASIOM* at $\alpha = 10°$ and M = 0.4 with 100 vortex panels and find the resulting C_L and C_D. Now rerun with 200 panels, 300 panels, etc., until you find the number of panels that give you "converged" results. You may need to continue this up to 1,500 or more panels! Graph your results (C_L and C_D vs. number of panels) and comment on the results, as appropriate.

Now run the program for the following combination of conditions: M = 0.0, 0.4, 0.8, and $\alpha = 0°$, $2°$, $4°$, $6°$, $8°$, $10°$, $12°$, $14°$, $16°$, $18°$. Graph the results for C_L and C_D as a function of angle of attack, with one graph for each Mach number (you can plot both coefficients on one graph by using two vertical scales if you want). Since the vortex lattice program only predicts drag due to lift, you should add the experimental value of zero-lift drag coefficient to your results. For the C_L and C_D results at M = 0.4, compare with the actual *aircraft data*. Also for M = 0.4 compare your results with the *theoretical values* you obtain using the theoretical lift-curve slope for a wing:[90]

$$C_{L_\alpha} = \frac{2\pi AR}{2 + \sqrt{4 + AR^2\beta^2(1 + \frac{\tan^2 \Lambda}{\beta^2})}}.$$

Where the 2π term in the numerator is the classical 2D thin airfoil theory lift-curve slope (per radian), AR is the aspect ratio (b^2/S_{ref}), $\beta = \sqrt{1 - M_\infty^2}$, and Λ is the sweep of the maximum thickness line. You should correct the theoretical lift-curve slope for compressibility using the Prandtl-Glauert rule:[2]

$$C_{L_\alpha} = \frac{C_{L_\alpha}(M_\infty = 0)}{\sqrt{1 - M_\infty^2}}.$$

Data for the T-38 are given from Brandt et al (reprinted courtesy of the American Institute of Aeronautics and Astronautics, Inc.).[89]

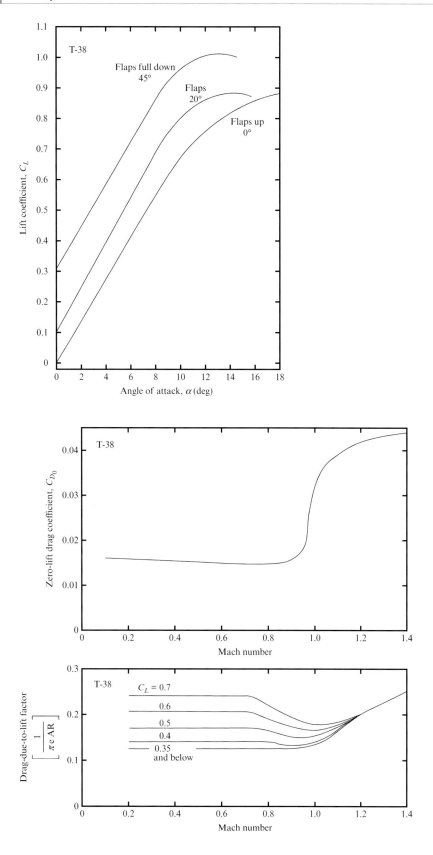

Comment on how well the vortex lattice program was able to predict the aerodynamics of the vehicle. What did it do well? What didn't it do well? What didn't it do at all? How well did the results of the program compare with both the aircraft data and the theoretical values? Comment on the shortcomings relative to the assumptions for a potential flow method.

The following projects use a panel method and can be done as a whole or in parts. Links to *XFOIL* or *PABLO* can be obtained from Appendix D or the text website: www.cambridge.edu/aerodynamics:

3. Conduct a panel resolution study with *XFOIL* or *PABLO*.
    (a) Run your panel code with variable numbers of panels on the upper and lower surfaces (start with 20 panels for each and increase until your predictions stop changing. How many panels do you need to get results independent of the number of panels? What happens to the computer time as the number of panels increases?
    (b) If possible, check the coordinates generated by the program vs. exact (consider using the NACA 0012, see Appendix A for geometry definition), including examination of the coordinates at the trailing edge. This is best done by making a table of exact and computed values at selected values of *x/c*. What did you find out?
    (c) Locate the source strengths and sum the source strengths x panel lengths to get the total source strength. Does it sum to zero? Should it?
    (d) Where is the moment reference center in this code?

4. You will perform an airfoil design using *XFOIL*. Start with your reference airfoil:
    (a) add thickness on the bottom (mid chord) and describe what happens
    (b) shave some thickness off the bottom (mid chord) and describe what happens
    (c) add thickness on the top (mid chord) and describe what happens
    (d) deflect the trailing edge down a couple of degrees (how sensitive is the airfoil to changes at the trailing edge?)
5. In order to determine the accuracy of panel methods, compare the results with thin airfoil theory. Compare the thin airfoil ΔC_p for a flat plate with program *XFOIL* or *PABLO*.
    Recall thin airfoil theory for an uncambered flat plate:

$$\Delta C_p = 4\alpha \sqrt{\frac{(1 - x/c)}{x/c}}.$$

    (a) pick an NACA 0012 airfoil at $\alpha = 2°$ and $12°$ and run your panel program.

    (b) plot $\Delta C_p/\alpha$ as a function of x/c.

    (c) how many panels do you need to get a converged solution?

    (d) what conclusions do you reach?

6. Get a copy of *VLMpc* from the text website, or an equivalent 3D VLM method such as *Tornado* contained within *CEASIOM*. Install the program on your computer and repeat the sample case, checking that your output is the same as the sample output files. Study the output to familiarize yourself with the variety of information generated.

7. Determine how well thin airfoil theory compares to vortex lattice results. Compare the thin airfoil theory ΔC_p for a 2D flat plate airfoil with program *VLMpc*.

Compare your results with flat plate thin airfoil theory:

$$\Delta C_p = 4\alpha\sqrt{\frac{(1 - x/c)}{x/c}}$$

    (a) Pick an aspect ratio 10 unswept wing at $\alpha = 3°$ and $12°$ and run *VLMpc*.

    (b) Plot $(\Delta C_p)/\alpha$ as a function of x/c at the wing root.

    (c) How many panels do you need to get a converged solution from VLM?

    (d) What conclusions do you reach?

8. You will compare the validity of an aerodynamic strip theory using *VLMpc* or *Tornado*. Consider an uncambered, untwisted wing, $AR = 4$, $\lambda = .4$, $\Lambda_{le} = 50°$, at a lift coefficient of 1. Plot the spanload and the ΔC_p distribution at approximately the center section, the midspan station, and the 85% semispan station. Compare your results with a spanload constructed assuming that the wing flow is approximated as 2D at the angle of attack required to obtain the specified lift. Also compare the chord loads, ΔC_p, at the three span stations. How many panels do you need to obtain converged results? Do you consider this aerodynamic strip theory valid based on this investigation?

    (a) Compare the wing aerodynamic center location relative to the quarter chord of the *mean aerodynamic chord* for the wing, as well as for similar wings. Consider one wing with zero sweep on the quarter chord and a forward swept wing with a leading-edge sweep of $-50°$. Compare the spanloads. Document and analyze these results. What did you learn from this comparison?

(b) Compare the section lift coefficients. Where would each one stall first? Which wing appears to be able to reach the highest lift coefficient before the section stalls?

(c) Add twist to each wing to obtain near elliptic spanloads. Compare the twist distributions required in each case.

9. Choose a NASA or NACA report describing wind tunnel results for a simple one- or two-lifting surface configuration at subsonic speeds. Compare the lift-curve slope and stability level predicted by *VLMpc* with wind tunnel data. Submit a report describing your work and assessing the results.

10. Add a canard to the aft and forward swept wings analyzed in Project 8.

(a) Plot the sum of the spanloads. How does the canard affect the wing spanload?

(b) How does lift change with canard deflection? Add an equivalent tail. Compare the effect of tail or canard deflection on total lift and moment. Did you learn anything? What?

5.5 References

1 De Resende, O.C., "The Evolution of the Aerodynamic Design Tools and Transport Aircraft Wings at Embraer," *Journal of the Brazilian Society of Mechanical Science and Engineering*, Vol. 26, No. 4, 2004, pp. 379–390.

2 Bertin, J.J., and Cummings, R.M., *Aerodynamics for Engineers*, 6th Ed., Upper Saddle River: Pearson, 2014.

3 Erickson, L.L., "Panel Methods – An Introduction," NASA TP-2995, December 1990.

4 Hess, J.L., "Panel Methods in Computational Fluid Dynamics," *Annual Review of Fluid Mechanics*, Vol. 22, 1990, pp. 255–274.

5 Hess, J.L., "Linear Potential Schemes," in *Applied Computational Aerodynamics*, Ed. Henne, P.A., Reston: AIAA, 1990.

6 Katz, J., and Plotkin, A., *Low-Speed Aerodynamics*, 2nd Ed., Cambridge: Cambridge University Press, 2001.

7 Moran, J., *An Introduction to Theoretical and Computational Aerodynamics*, New York: Dover Publications, 2010.

8 Cebeci, T., *An Engineering Approach to the Calculation of Aerodynamic Flows*, Long Beach: Horizon Publishing, 1999.

9 Karamcheti, K., *Principles of Ideal-Fluid Aerodynamics*, 2nd Rev. Ed., Melbourne: Krieger Publishing, 1980.

10 Hildebrand, F.B., *Advanced Calculus for Applications*, 2nd Ed., Englewood Cliffs: Prentice-Hall, 1976.

11 Ashley, H., and Landahl, M., *Aerodynamics of Wings and Bodies*, Reading: Addison-Wesley, 1965.

12 Curle, N., and Davis, H.J., *Modern Fluid Dynamics*, Volume 1: Incompressible Flow, London: Van Nostrand, 1968.

13 Neal, D., PhD candidate, Virginia Tech, 2007, private communication.

14 Smith, A.M.O., "The Panel Method: Its Original Development," in *Applied Computational Aerodynamics*, ed. Henne, P.A., Reston: AIAA, 1990.

15 Houghton, E.L., and Carpenter, P.W., *Aerodynamics for Engineering Students*, 4th Ed., New York: Halsted Press, 1993.

16 Kuethe, A.M., and Chow, C-Y., *Foundations of Aerodynamics*, 5th Ed., New York: John Wiley, 1998, pp. 156–164.

17 Wauquiez, C., http://www.nada.kth.se/~chris/pablo/pablo.html, *Pablo* theory described as part of his Licentiate's Thesis, "Shape Optimization of Low Speed Airfoils using MATLAB and Automatic Differentiation," KTH, Stockholm, 2000.

18 Drela, M., "XFOIL: An Analysis and Design System for Low Reynolds Number Airfoils," in *Low Reynolds Number Aerodynamics*, Ed. Mueller, T.J., Lecture Notes in Engineering No. 54, Berlin: Springer Verlag, 1989.

19 Abbott, I.H., and von Doenhoff, A.E., *Theory of Wing Sections*, New York: Dover Publications, 1959.

20 Stack, J., Lindsey, W.F., and Littell, R.E., "The Compressibility Burble and the Effect of Compressibility on Pressures and Forces Acting on an Airfoil," NACA R-646, 1938.

21 Jameson, A., "Acceleration of Transonic Potential Flow Calculations on Arbitrary Meshes by the Multiple Grid Method," AIAA Paper 79–1458, July 1979.

22 Kroo, I., "Aerodynamic Analyses for Design and Education," AIAA Paper 92–2664, June 1992.

23 Volpe, G., "Inverse Airfoil Design: A Classical Approach Updated for Transonic Applications," in *Applied Computational Aerodynamics*, Ed. Henne, P.A., Reston: AIAA, New York, 1990.

24 Labrujere, T.E., and Sloof, J.W., "Computational Methods for the Aerodynamic Design of Aircraft Components," *Annual Review of Fluid Mechanics*, Vol. 25, 1993, pp. 183–214.

25 Bristow, D.R., "A New Surface Singularity Method for Multi-Element Airfoil Analysis and Design," AIAA Paper 76–0020, January 1976.

26 Aidala, P.V., Davis, W.H., and Mason, W.H., "Smart Aerodynamic Optimization," AIAA Paper 83–1863, July 1983.

27 PAN AIR User's Class Short Course Presentation Material, NASA Ames Research Center 1981.

28 Hess, J.L., "The Problem of Three-Dimensional Lifting Potential Flow and Its Solution by Means of Surface Singularity Distributions," *Computer Methods in Applied Mechanics and Engineering*, Vol. 4, 1974, pp. 283–319.

29 Carmichael, R.L., and Erickson, L.L., "PAN AIR – A Higher Order Panel Method for Predicting Subsonic or Supersonic Linear Potential Flows About Arbitrary Configurations," AIAA Paper 81–1255, June 1981.

30 Nathman, J.K., "Improvement of a Panel Method by Including Panel Warp," AIAA Paper 2004–0721, January 2004.

31 Platzer, M., Jones, K., Young, J., and Lai, J., "Flapping Wing Aerodynamics: Progress and Challenges," *AIAA Journal*, Vol. 46, No. 9, 2008, pp. 2136–2149.

32 Tinoco, E.N., Ball, D.N., and Rice, F.A., "PAN AIR Analysis of a Transport High-Lift Configuration," *Journal of Aircraft*, Vol. 24, No. 3, 1987, pp. 181–188.

33 Magnus, A.E., and Epton, M.A., "PAN AIR – A Computer Program for Predicting Subsonic or Supersonic Linear Potential Flows About Arbitrary Configurations Using a Higher Order Panel Method," Volume I – Theory Document (Version 1.0), NASA CR-3251, April 1980.

34 Margason, R.J., Kjelgaard, S.O., Sellers, W.L., Morris, C.E.K., Walkley, K.B., and Shields, E.W., "Subsonic Panel Methods – A Comparison of Several Production Codes," AIAA Paper 85–0280, January 1985.

35 Ashby, D.L., Dudley, M.R., Iguchi, S.K., Browne, L., and Katz, J., "Potential Flow Theory and Operation Guide for the Panel Code PMARC," NASA TM 102851, January 1991.

36 Nathman, J.K., "Subsonic Panel Methods – Second (Order) Thoughts," AIAA Paper 98–5563. September 1998.

37 Mason, W.H., and Rosen, B.S., "The COREL and W12SC3 Computer Programs for Supersonic Wing Design and Analysis," NASA CR-3676, December 1983.

38 Van Zyl, L.H., "Unsteady Panel Method for Complex Configurations Including Wake Modeling," *Journal of Aircraft*, Vol. 45, No. 1, 2008, pp. 276–285.

39 Hess, J.L., and Smith, A.M.O., "Calculation of Nonlifting Potential Flow About Arbitrary Three-Dimensional Bodies," Douglas Report ES40622, Douglas Aircraft Company, 1962.

40 Rubbert, P.E., "Theoretical Characteristics of Arbitrary Wings by a Nonplanar Vortex Lattice Method," Boeing Report D6-9244, The Boeing Company, 1964.

41 Woodward, F.A., Tinoco, E.N., and Larsen, J.W., "Analysis and Design of Supersonic Wing-Body Combinations, Including Flow Properties in the Near Field," Part I – Theory and Application, NASA CR-73106, 1967.

42 Rubbert, P.E., and Saaris, G.R., "A General Three-Dimensional Potential Flow Method Applied to V/STOL Aerodynamics," SAE Paper 680304, 1968.

43 Hess, J.L., "Calculation of Potential Flow About Arbitrary 3-D Lifting Bodies," Douglas Report MDC-J5679-01, Douglas Aircraft Company October 1972.

44 Woodward, F.A., "An Improved Method for the Aerodynamic Analysis of Wing-Body-Tail Configurations in Subsonic and Supersonic Flow," NASA CR-2228, Parts I and II, 1973.

45 Roberts, A., and Rundle, K., "Computation of First Order Compressible Flow About Wing-Body Configurations," AERO MA No. 20, British Aircraft Corporation, February 1973.

46 Mercer, J.E., Weber, J.A., and Lesfor, E.P., "Aerodynamic Influence Coefficient Method Using Singularity Splines," NASA CR-2423, May 1974.

47 Morino, L., and Kuo, C.-C., "Subsonic Potential Aerodynamics for Complex Configurations: A General Theory," *AIAA Journal*, Vol. 12, No. 2, 1974, pp. 191–197.

48 Johnson, F.T., and Rubbert, P.E., "Advanced Panel-Type Influence Coefficient Methods Applied to Subsonic Flow," AIAA Paper 75–0050, January 1975.

49 Ehlers, F.E., and Rubbert, P.E., "A Mach Line Panel Method for Computing the Linearized Supersonic Flow," NASA CR-152126, 1979.

50 Ehlers, F.E., Epton, M.A., Johnson, F.T., Magnus, A.E., and Rubbert, P.E., "A Higher Order Panel Method for Linearized Flow," NASA CR-3062, 1979.

51 Bristow, D.R., "Development of Panel Methods for Subsonic Analysis and Design," NASA CR-3234, 1980.

52 Hess, J.L., and Friedman, D.M., "An Improved Higher Order Panel Method for Three-Dimensional Lifting Flow," Douglas Report No. NADC-79277-60, Douglas Aircraft Company, 1981.

53 Maskew, B., "Prediction of Subsonic Aerodynamic Characteristics: A Case for Lower Order Panel Methods," *Journal of Aircraft*, Vol. 19, No. 2, 1982, pp. 157–163.

54 Coopersmith, R.M., Youngren, H.H., and Bouchard, E.E., "Quadrilateral Element Panel Method (QUADPAN)," Lockheed-California Report LR 29671, 1981.

55 Falkner, V.M., "The Calculations of Aerodynamic Loading on Surfaces of Any Shape," British A.R.C. R&M 1910, 1943.

56 *Vortex Lattice Utilization*, NASA SP-405, May 1976.

57 Margason, R.J., and Lamar, J.E., "Vortex-Lattice FORTRAN Program for Estimating Subsonic Aerodynamic Characteristics of Complex Planforms," NASA TN D-6142, February 1971.

58 Lamar, J.E., and Gloss, B.B., "Subsonic Aerodynamic Characteristics of Interacting Lifting Surfaces With Separated Flow Around Sharp Edges Predicted by a Vortex-Lattice Method," NASA TN D-7921, 1975.

59 Lamar, J.E., and Herbert, H.E., "Production Version of the Extended NASA-Langley Vortex Lattice FORTRAN Computer Program," Vol. I, User's Guide, NASA TM-83303, 1982.

60 Herbert, H.E., and Lamar, J.E., "Production Version of the Extended NASA-Langley Vortex Lattice FORTRAN Computer Program," Vol. II, Source Code, NASA TM-83304, 1982.

61 Drela, M., and Youngren, H., "AVL, An Extended Vortex Lattice Method, http://web.mit.edu/drela/Public/web/avl.

62 Melin, T., *Tornado*, http://www.redhammer.se/tornado/.

63 Miranda, L.R., Eliott, R.D., and Baker, W.M., "A Generalized Vortex Lattice Method for Subsonic and Supersonic Flow Applications," NASA CR-2865, December 1977.

64 Albright, A.E., Dixon, C.J., and Hegedus, M.C., "Modification and Validation of Conceptual Design Aerodynamic Prediction Method HASC95 With VTXCHN," NASA CR-4712, March 1996.

65 Nathman, J.K., and McComas, A., "Comparison of Stability and Control Calculations from Vortex Lattice and Panel Methods," AIAA Paper 2008–0314, January 2008.

66 Lan, C.E., "A Quasi-Vortex-Lattice Method in Thin Wing Theory," *Journal of Aircraft*, Vol. 11, No. 9, 1974, pp. 518–527.

67 Mook, D.T., and Nayfeh, A.H., "Application of the Vortex-Lattice Method to High-Angle-of-Attack Subsonic Aerodynamics," SAE Paper 851817, October 1985.

68 Jacob, K., Mason, W.H., Durham, W., Lutze, F., and Benoliel, A., "Control Power Issues in Conceptual Design: Critical Conditions, Estimation Methodology, Spreadsheet Assessment, Trim and Bibliography," VPI-Aero-200, November 1993. (Software and report available at www.cambridge.org/aerodynamics.)

69 Tinling, B.E., and Kolk, W.R., "The Effects of Mach Number and Reynolds Number on the Aerodynamic Characteristics of Several 12-Percent-Thick Wings Having 35° of Sweepback and Various Amounts of Camber," NACA RM-A50K27, February 1951.

70 Stivers, L.S., "Effectiveness of an All-Movable Horizontal Tail on an Unswept-Wing and Body Combination for Mach Numbers from 0.60 to 1.40," NASA TN D-13, August 1959.

71 Abbott, I.H., and Greenberg, H., "Test in the Variable-Density Wind Tunnel of the NACA 23012 Airfoil with Plain and Split Flaps," NACA Report 661, 1938.

72 Etkin, B., *Dynamics of Flight – Stability and Control*, 2nd Ed., New York: John Wiley & Sons, 1982.

73 Pittman, J.L., and Dillon, J.L., "Vortex Lattice Prediction of Subsonic Aerodynamics of Hypersonic Vehicle Concepts," *Journal of Aircraft*, Vol. 14, No. 10, 1977, pp. 1017–1018.

74 Polhamus, E.C., "Prediction of Vortex Lift Characteristics by a Leading-edge Suction Analogy," *Journal of Aircraft*, Vol. 8, No. 4, 1971, pp. 193–199.

75 Kalman, T.P., Rodden, W.P., and Giesing, J., "Application of the Doublet-Lattice Method to Nonplanar Configurations in Subsonic Flow," *Journal of Aircraft*, Vol. 8, No. 6, 1971, pp. 406–415.

76 Belotserkovskii, S.M., *The Theory of Thin Wings in Subsonic Flow*, New York: Plenum Press, 1967.

77 Saunders, G.H., "Aerodynamic Characteristics of Wings in Ground Proximity," *Canadian Aeronautics and Space Journal*, Vol. 11, 1965, pp. 185–192.

78 Hough, G.R., "Remarks on Vortex-Lattice Methods," *Journal of Aircraft*, Vol. 10, No. 5, 1973, pp. 314–317.

79 Albano, E., and Rodden, W.P., "A Doublet-Lattice Method for Calculating Lift Distributions on Oscillating Surfaces in Subsonic Flows," *AIAA Journal*, Vol. 7, No. 2, 1969, pp. 279–285; errata *AIAA Journal*, Vol. 7, No. 11, 1969, p. 2192.

80 Rodden, W.P., Taylor, P.F., and McIntosh, S.C., Jr., "Further Refinement of the Subsonic Doublet-Lattice Method," *Journal of Aircraft*, Vol. 35, No. 5, 1998, pp. 720–727.

81 Rodden, W.P., Taylor, P.F., McIntosh, S.C., Jr., and Baker, M.L., "Further Convergence Studies of the Enhanced Doublet-Lattice Method," *Journal of Aircraft*, Vol. 36, No. 4, 1999, pp. 682–688.

82 Fritz, T.E., and Long, L.N., "Object-Oriented Unsteady Vortex Lattice Method for Flapping Flight," *Journal of Aircraft*, Vol. 41, No. 6, 2004, pp. 1275–1289.

83 Soviero, P.A., and Hernandes, F., "Compressible Unsteady Vortex Lattice Method for Arbitrary Two-Dimensional Motion of Thin Profiles," *Journal of Aircraft*, Vol. 44, No. 5, 2007, pp. 1494–1498.

84 Stanford, B.K., and Beran, P.S., "Analytical Sensitivity Analysis of an Unsteady Vortex Lattice Method for Flapping Wing Optimization," AIAA Paper 2009–2614, May 2009.

85 Shyy, W., Lian, Y., Tang, J., Viiery, D., and Liu, H., *Aerodynamics of Low Reynolds Number Flyers*, Cambridge: Cambridge University Press, 2008.

86 Hernandes, F., and Soviero, P.A.O., "Unsteady Aerodynamic Coefficients Obtained by a Compressible Vortex Lattice Method," *Journal of Aircraft*, Vol. 46, No. 4, 2009, pp. 1291–1301.

87 Jacobs, R.B., Ran, H., Kirby, M.R., and Mavris, D.N., "Extension of a Modern Lifting-Line Method to Transonic Speeds and Application to Multiple-Lifting-Surface Configurations," AIAA Paper 2012–2889, June 2012.

88 Murua, J., Palacios, R., and Graham, J.M.R., "Applications of the Unsteady Vortex-Lattice Method in Aircraft Aeroelasticity and Flight Dynamics," *Progress in Aerospace Sciences*, Vol. 55, 2012, pp. 46–72.

89 Brandt, S.A., Stiles, R.J., Bertin, J.J., and Whitford, R., *Introduction to Aeronautics: A Design Perspective*, 2nd Ed., Reston: AIAA, 2004.

90 Nicolai, L.M., *Fundamentals of Aircraft Design*, San Jose: METS, Inc., 1975.

6 Introduction to Computational Fluid Dynamics

The only way to learn CFD is to do it!

Robert W. MacCormack,
numerical algorithm developer

Candidate supersonic transport geometry (Courtesy of NASA; a full color version of this image is available on the website: www.cambridge.org/aerodynamics).

LEARNING OBJECTIVE QUESTIONS

After reading this chapter you should know the answers to the following questions:

- What is an algorithm and why do we need to use it?
- What are the options for numerically solving the governing equations of fluid dynamics?
- Can you briefly describe how finite difference methods work? Can you briefly describe how finite volume methods work? Which form of the governing equations do each of these approaches use?
- What is truncation error and how can it be controlled with the numerical method being used?
- What is the difference between a partial differential equation and a finite difference equation?
- What are consistency, stability, and convergence? How do they relate to one another?
- How do dissipative and dispersive errors show up in numerical predictions?
- What is the difference between explicit and implicit numerical methods? What are the advantages and disadvantages of each?
- What are some of the most common numerical methods used to solve the governing equations of fluid dynamics? What are the advantages and disadvantages of these approaches?

6.1 Introduction

We have been using the idea of distributions of singularities on surfaces to study the aerodynamics of airfoils and wings (see Chapter 5). This approach is useful and provides us with methods that can be used easily on personal computers to solve real problems. Considerable insight into aerodynamics can be obtained using these methods. However, the class of flow effects that can be examined with these methods is somewhat restricted; in particular, practical methods for computing fundamentally nonlinear flow effects are excluded. This includes both inviscid transonic and boundary-layer flows.

In this chapter we examine the basic ideas behind the direct numerical solution of partial differential equations. This approach leads to methods that can handle nonlinear equations. The simplest methods to understand are developed using numerical approximations to the derivative terms in the partial differential equation (PDE) form of the governing equations (Equations 3.63–3.73). Direct numerical solutions of the partial differential equations of fluid mechanics constitute the field of computational fluid dynamics (CFD). Although the field is still developing, a number of books have been written about CFD.[1-5] In particular, the book by Pletcher et al.[1] covers many of the technical aspects of CFD theory used in current codes. Fundamental concepts for solving partial differential equations in general using numerical methods are presented in a number of basic texts, including books by Morton and Mayers,[6] Smith,[7] and Ames.[8]

The basic idea is to model the derivatives in the governing equations by some numerical approximation, such as the *finite difference* approach. When this approach is used, the entire flowfield must be discretized, with the field around the vehicle defined in terms of a mesh of grid points or cells. We need to find the flowfield values at every mesh (or grid) point by writing down the discretized form of the governing equation at each mesh point. Discretizing the equations leads to a system of simultaneous algebraic equations. A large number of mesh points is usually required to accurately obtain the details of the flowfield, and this leads to a very large system of equations. Especially in three dimensions, this generates demanding requirements for computational resources since millions of grid points are required to obtain the solution over a complete three-dimensional aerodynamic configuration!

In contrast to the finite difference idea, approximations to the integral form of the governing equations result in the *finite volume* approach. Several books have been written devoted to this approach,[9-11] and we will cover the approach briefly here. Other methods currently used include the *finite element* method and the *pseudo spectral* approach, which will also be briefly discussed in Section 6.2.3.

CFD is usually associated with computers with large memories and high processing speeds. In addition, massive data storage systems must be available to store computed results, and ways to transmit and examine the massive amounts of data associated with a computed result must be available. Before the computation of the solution is started, the mesh of grid points must be established. Thus the broad area of CFD leads to many different closely related, but nevertheless specialized, technology sub areas. These include (see Figures 1.18 and 1.21):

- Grid/mesh generation
- flowfield discretization algorithms
- efficient solution of large systems of equations
- massive data storage and transmission technology methods
- computational flow visualization

Originally, CFD was only associated with the second and third items listed, but the flows being computed were fairly simple at that time. Then, the problem with establishing a suitable mesh for complex, arbitrary geometries became apparent, and the specialization of grid generation emerged. Finally, the availability of large computers and remote processing led to the need for work in the last two items listed, since it became more and more difficult to manage the results as the size and complexity of the flowfield output increased. A current limiting factor in the further improvement in CFD capability is the continued development of good turbulence models, which will be discussed in Chapter 8. In addition, there are researchers constantly trying to improve the algorithms and solution methods used in CFD.

This chapter provides an introduction to the concepts required for developing discretized forms of the governing equations and a discussion of the solution of the resulting algebraic equations. Although the basic idea of CFD appears straightforward, we will find that a successful numerical method depends on considerable analysis to formulate an accurate, robust, and efficient solution method. We will see that the classification of the mathematical type of the governing equations (Section 3.6) plays an important role in the development of the numerical methods. Although we adopt finite difference/finite volume methods to solve nonlinear equations, to establish the basic ideas we will consider linear equations. Application to nonlinear equations will be addressed more fully later in this chapter where additional concepts are introduced and applied to the solution of nonlinear equations. When we allude to numerical solutions of the governing equations, we usually mean more specifically the Navier-Stokes equations or RANS equations; we will begin with an overview of the most common approaches used to solve those equations.

Computational Aerodynamics Concept Box

Example of an Algorithm

Algorithms are a set of specific rules that allow you to calculate or estimate something. For example, you were probably taught in your calculus course that finding the area under a curve could be done with integration, that is, the area under the curve can be found by:

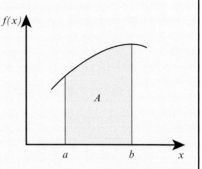

$$A = \int_a^b f(x)dx.$$

When computers first started being used for performing calculations, the users had to find some way to approximate analytic mathematics, such as integration, with an algorithm. Many of the basic concepts for algorithms already existed in mathematics, which can also be shown for the preceding integration example. Since integration is really an infinite summation, you could approximate the integration by first approximating the area under the curve as:

$$A \approx (b-a)f\left(\frac{a+b}{2}\right)$$

which is known as the mid-point rule, since it approximates the curve with a constant value using a point midway between a and b. Of course, you could improve the accuracy of the result by using the trapezoidal rule, which evaluates the function at the end points:

$$A \approx (b-a)\frac{f(a)+f(b)}{2}.$$

You can make the result even more accurate by breaking the interval into sub-intervals and applying the trapezoidal rule, which would yield:

$$A \approx \frac{(b-a)}{n}\left(\frac{f(a)+f(b)}{2} + \sum_{k=1}^{n-1} f\left(a+k\frac{(b-a)}{n}\right)\right)$$

where each interval has a width kh, where $h = (b-a)/n$. This last equation then becomes the cornerstone for an algorithm, since it can be coded into a computer and used to approximate integration.

6.2 Options for Numerically Solving the Navier-Stokes Equations

There are many ways of numerically solving the Navier-Stokes equations. The most commonly used are finite difference, finite volume, finite element, and spectral methods. Each of these methods has positive aspects, and historically each of them has experienced various degrees of popularity. This section will introduce each of these methods,[12] and the methods will be described in greater detail later in the chapter. We will concentrate our efforts on learning the finite difference approach, since it lends itself to understanding computational methods at an introductory level.

6.2.1 Finite Difference Methods

The finite difference method was developed by applying the Navier-Stokes governing equations to a point in the flowfield (i,j,k), as shown in Fig. 6.1. A *grid* of lines is created in three-dimensional space, and a *grid point* is placed at the intersection of the lines. The point of interest is given an index reference of (i,j,k), and all points around that point are referred to relative to that point, such as $(i+1,j,k)$ or $(i,j+1,k-1)$, etc.[i]

The vector form of the governing equations, Equation 3.63, can be used to show how finite differencing is done on this grid of points. Assuming the flow is inviscid, then Equation 3.63 becomes

$$\frac{\partial Q}{\partial t} + \frac{\partial F}{\partial x} + \frac{\partial G}{\partial y} + \frac{\partial H}{\partial z} = 0. \tag{6.1}$$

These equations could be *finite differenced* by taking differences in the x, y, z, and t dimensions to approximate the partial derivatives as

$$\frac{Q_{i,j,k}^{n+1} - Q_{i,j,k}^{n}}{\Delta t} + \frac{F_{i+1,j,k}^{n} - F_{i-1,j,k}^{n}}{2\Delta x} + \frac{G_{i,j+1,k}^{n} - G_{i,j-1,k}^{n}}{2\Delta y} + \frac{H_{i,j,k+1}^{n} - H_{i,j,k-1}^{n}}{2\Delta z} = 0 \tag{6.2}$$

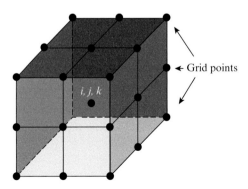

← Grid points

i, j, k

Figure 6.1   Finite difference grid scheme.

---

[i] Figure 6.1 shows an assembly of grid points in a systematically organized arrangement. This is called a *structured grid*. In Chapter 7 we will show an alternate approach to this grid arrangement, called an *unstructured grid*.

where the superscript (n) represents the current time level and $(n+1)$ represents the time level Δt in the future.

We will learn how these differences were taken in a later section of this chapter, but for now look at what this differencing has allowed us to accomplish. We can perform time integration of the equation by solving for the unknown flow variable, Q^{n+1} in Equation 6.2 at the next step in time by using the known values in the equation (Q^n, etc.) at time t:

$$Q_{i,j,k}^{n+1} = Q_{i,j,k}^{n} - \frac{\Delta t}{2}\left(\frac{F_{i+1,j,k}^{n} - F_{i-1,j,k}^{n}}{\Delta x} + \frac{G_{i,j+1,k}^{n} - G_{i,j-1,k}^{n}}{\Delta y} + \frac{H_{i,j,k+1}^{n} - H_{i,j,k-1}^{n}}{\Delta z} \right) \quad (6.3)$$

There are also more complicated time integration methods, as we will learn. Since Equation 6.3 is one of many possible finite difference representation of Equation 6.1, it may not be successful or accurate, as we will also discover in later sections. Many CFD solutions seek to find a steady-state solution, and the use of time is a convenient means of arriving at the steady result numerically. After finding Q^{n+1} at all grid points in the flowfield (which is called an *iteration*), the process is repeated until a *converged* (or steady) solution is obtained.

6.2.2 Finite Volume Methods

The finite volume method is developed by applying the integral form of the Navier-Stokes equations to a control volume in the domain of interest. A *mesh* of volumes is used to define the flowfield, usually made up of tetrahedra, pyramids, hexahedra, prisms, etc. The integral form of the conservation equations, such as the conservation of mass integral form given in Equation 6.4, is solved over an arbitrary region, as shown in Fig. 6.2.[ii] The advantage of this approach is that it lends itself well to irregularly shaped domains and complex geometry, and the integral form of the equations handles discontinuities (normally shocks) much better. An excellent advanced summary of this approach has been given by Mavriplis.[13]

$$\frac{\partial}{\partial t}\iiint_{\Psi} Q d\Psi + \iint_{S} Q \vec{V} \cdot \hat{n} dS = 0. \quad (6.4)$$

The solution is obtained by computing the second term in Equation 6.4

$$\iint_{S} Q \vec{V} \cdot \hat{n} dS \quad (6.5)$$

[ii] You should look at Equation 3.7 for the *primitive variable* form of the equation; primitive variables are the basic variables of flow such as density or velocity.

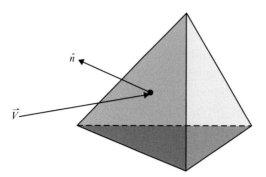

Figure 6.2   Finite volume mesh scheme.

on the surface of the geometry and then integrating in time to obtain Q^{n+1} by the following method:

$$\int_{t_1}^{t_2} \frac{\partial}{\partial t}\left[\iiint_{\mathbb{V}} Q d\mathbb{V} \right] = -\int_{t_1}^{t_2}\left[\iint_{S} Q\vec{V}\cdot\hat{n}dS \right]. \tag{6.6}$$

This integration is then performed for all cells in the mesh and the process is repeated until a converged solution is obtained. To evaluate the integral shown in Equation 6.5, approximations to integration are used over the surface of the cell, which can be done using either fluid properties at the cell vertices or an average value of the properties at the cell center.[6]

6.2.3 Finite Element/Pseudo Spectral Methods

Finite element[14–16] and *pseudo spectral*[17] methods are significantly different than finite difference or finite volume methods. Finite difference methods replace partial derivatives with numerical differences in the governing equations and then solve a set of algebraic relations. Finite element and pseudo-spectral methods (both approaches are subsets of the spectral method[18,19]) assume a form of the solution with unknown coefficients, substitute the assumed solution into the governing equations, apply boundary conditions, and then solve a set of algebraic equations to determine the set of unknown coefficients. A simple example will help to make this clear – look at the convection/diffusion equation:

$$\frac{\partial f}{\partial t} + \frac{\partial f}{\partial x} = v\frac{\partial^2 f}{\partial x^2} \quad \text{for } 0 \le x \le L. \tag{6.7}$$

We can assume a form of the solution as

$$f(x,t) = \sum_{i=0}^{N} a_i(t)F_i(x) \tag{6.8}$$

where the a_i coefficients are the unknowns and the $F_i(x)$ are the assumed functions. Finite element simulations typically use linear functions or polynomials for $F_i(x)$ and pseudo-spectral methods typically use trigonometric functions (like *sine*, *cosine*, and *exp*), and both approaches take advantage of fast Fourier transforms during the solution process. Once the unknown function has been defined, substitute Equation 6.8 into Equation 6.7 and obtain a set of algebraic equations for the various a_i coefficients. These equations are then solved to determine the values of the a_i coefficients, and a solution is obtained using Equation 6.8.

Boundary conditions in finite element methods are usually handled by choosing a function, $F_i(x)$, that automatically satisfies the boundary conditions of the physical problem. Pseudo-spectral methods often have a more difficult time with boundary conditions, since most boundary conditions do not typically lend themselves to being modeled with trigonometric functions. A great deal of research has been done to determine the best $F_i(x)$ functions to use and the best way to handle boundary conditions using these methods.[17]

Profiles in Computational Aerodynamics: Ken Badcock

"I grew up in Glasgow, Scotland and took a while to settle on CFD as a career. Maths became interesting when we starting studying calculus at school, and I followed this interest at the University of Strathclyde, studying for a maths degree. I liked all forms of maths, but the idea of applications particularly appealed, and I liked the idea of really understanding something deeply. This combination led me to continue my studies for a doctoral thesis in numerical methods for flow simulation, this time at the Computing Laboratory at Oxford University. Apart from spending three years in a stimulating and beautiful place, I was lucky to have the sympathetic supervision of Dr. Ian Sobey, who allowed me to follow my interests in different directions. During this time I really had the time to learn about discretization schemes for compressible flows, and I developed a liking for writing computer codes to solve flow problems. Whilst being hard work, and frustrating at times, there is a precision in understanding what is required to write a successful CFD code – the computer after all will only do as it is told. This is something that appeals to me.

"At the time I submitted my thesis I had the great good fortune that there was a research job in roughly my research specialty at the University of Glasgow, working with Professor Bryan Richards. This allowed me to continue an interest in implicit flow solvers applied to unsteady flow problems. A couple of years later I was appointed to the academic staff and started teaching and supervising research students. At this time, the basis for all my future work was established through an effort to write a flow code that was general enough to cope with all of our envisaged uses. This code has proved quite resilient and is still being used by many researchers fifteen years later. The code has proved useful in a number of ways. As it has been used for a wide range of aerodynamic studies, a large body of experience and confidence has been established. I strongly believe that the best asset a code can have is a wide user base. Second, the intimate knowledge of the workings of the code has allowed research in numerical methods to continue alongside application studies in aerodynamics. This combination is powerful for driving forward the state of knowledge in both fields, since advanced aerodynamics is often right at the limit of the capabilities of the best numerical methods.

"My CFD career has followed several strands ever since: developing and testing CFD methods for flows relevant to aircraft; finding ways of using CFD predicted aerodynamics for predicting the interaction with aircraft structures; predicting aircraft motion based on CFD predicted aerodynamics. This combination indicates the maturity of CFD in that some of the research action is in making it work for aircraft applications in cooperation with other disciplines. I feel very privileged to have had the chance to do this work, as it has provided an education in many fields I would have understood a fraction as well from just reading a book. The focus on applications has meant that almost from the start I have had to make use of the best computers available, meaning large calculations on distributed memory parallel computers.

"Without doubt the high point of my career so far has been the opportunity to work with many doctoral students and research assistants who have then gone on to develop their own careers, mostly in the aerospace industry. I hope that there will be many more in the years to come."

Ken Badcock is Professor of Computational Aerodynamics and Executive Pro-Vice Chancellor for Science and Engineering at the University of Liverpool. He was honored in 2008 as the Goldstein Lecturer for the Royal Aeronautical Society.

6.3 Approximations to Derivatives

There are many ways to obtain finite difference representations of derivatives; Fig. 6.3 illustrates the approach intuitively. Suppose we use the values of f at a point x_0 and a point a distance Δx away. Then we can approximate the slope at x_0 by taking the slope between these points. The sketch illustrates the difference between this simple slope approximation and the actual slope at the point x_0. Clearly, accurate slope estimation depends on the method used to estimate the slope and the use of suitably small values of Δx. Instead of taking a point Δx to the right of x_0, we could have estimated the slope using a point Δx to the left. Would that have given a better approximation to the true slope? At times, yes, and at other times, no, depending on the function being used – in the example of Fig. 6.3, using a point to the left probably would have resulted in a worse approximation to the derivative. What if we had used points Δx to the left and Δx to the right (yielding an interrogation size of $2\Delta x$)? Again, we would have to investigate the accuracy of such an approach, which would depend on the function. All of these approaches,

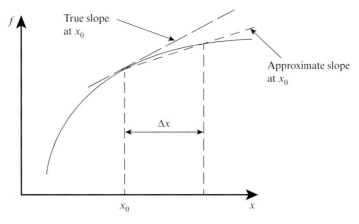

Figure 6.3 Example of slope approximation using two values of the function.

however, would give better and better results as the size of Δx was decreased, so we will have to determine which methods are the best, which we will do more formally in forthcoming sections of this chapter. Other methods for approximating derivatives exist, some of which are based on the mathematical definition of the derivative, and all of which have different levels of accuracy.

6.4 Finite Difference Methods

To solve a partial differential equation (or a system of partial differential equations), we can "discretize" the equations, i.e., we replace the fluid continuum on which the equations apply with a discrete set of points and assume that the variables of the problem only exist at those discrete points, as shown in Fig. 6.4 for a two-dimensional field of fluid.

In the case of interest here, namely fluid dynamics, the variables could be pressure, density, velocity, and/or temperature. As an example, the u-velocity field for a two-dimensional field could be represented by

$$
\begin{aligned}
u_{i,j} &= u(x_o, y_o) \\
u_{i+1,j} &= u(x_o + \Delta x, y_o) \\
u_{i,j+1} &= u(x_o, y_o + \Delta y). \\
&\vdots
\end{aligned}
\tag{6.9}
$$

We need to determine an easy way to represent derivatives in our equations of motion based on discrete locations in the flowfield. Many possibilities are available, but we will first use the mathematical definition of the derivative

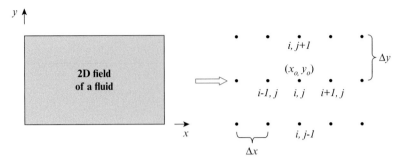

Figure 6.4 Representation of a fluid field with a grid system.

$$\frac{\partial u}{\partial x} = \lim_{\Delta x \to 0} \frac{u(x_o + \Delta x, y_o) - u(x_o, y_o)}{\Delta x}. \tag{6.10}$$

We will not be able to have $\Delta x \to 0$, since that would require an infinite number of grid points, but we can make Δx small enough to represent the derivative to within a certain level of accuracy. What we need is to be able to formally determine how accurate any finite difference representation of a derivative might be.

We can estimate the value of a function at some value $x_o + \Delta x$ if we know the value at x_o by using a Taylor's series expansion

$$u(x_o + \Delta x) = u(x_o) + \Delta x \frac{\partial u(x_o)}{\partial x} + \frac{\Delta x^2}{2!} \frac{\partial^2 u(x_o)}{\partial x^2} + \cdots + \frac{\Delta x^n}{n!} \frac{\partial^n u(x_o)}{\partial x^n} + \cdots \tag{6.11}$$

If we want to represent $\partial u / \partial x$, we merely solve Equation 6.11 for that derivative:

$$\frac{\partial u(x_o)}{\partial x} = \frac{1}{\Delta x}\left(u(x_o + \Delta x) - u(x_o)\right) - \frac{\Delta x}{2!} \frac{\partial^2 u(x_o)}{\partial x^2} - \cdots \tag{6.12}$$

If we use our discrete notation, the derivative in Equation 6.12 may be written as

$$\frac{\partial u}{\partial x} = \frac{u_{i+1,j} - u_{i,j}}{\Delta x} + O(\Delta x) \tag{6.13}$$

where the *truncation error* represented by $O(\Delta x)$ is the order of the lowest-order term which was "dropped" from the infinite series. We call Equation 6.13 a *first-order accurate* representation of a first derivative, since we have truncated terms of $O(\Delta x)$. One important feature of this representation is that it mimics the definition of a derivative, namely, as $\Delta x \to 0$, the error tends toward zero.[iii] The form of the truncation error term is frequently important in developing numerical methods. Specifically, this representation is referred to as a *forward difference*, since the derivative is evaluated using the discrete points at or ahead of the point of interest.

We can now proceed with creating whatever differencing schemes we want, to any desired level of accuracy. We could obtain a *backward difference* by writing the Taylor's series expansion as

$$u(x_o - \Delta x) = u(x_o) - \Delta x \frac{\partial u(x_o)}{\partial x} + \frac{\Delta x^2}{2!} \frac{\partial^2 u(x_o)}{\partial x^2} - \cdots \qquad (6.14)$$

and solving for the first derivative as before

$$\frac{\partial u}{\partial x} = \frac{u_{i,j} - u_{i-1,j}}{\Delta x} + O(\Delta x) \qquad (6.15)$$

which is called the backward difference formula, since the derivative is evaluated using the points at or behind the point of interest. Both the forward difference (Equation 6.13) and the backward difference (Equation 6.15), however, are first-order accurate approximations.

We can find a "higher order" approximation for the derivative if we subtract Equation 6.14 from Equation 6.11 (that is the "backward" Taylor's series from the "forward" Taylor's series). The resulting equation is:

$$u(x_o + \Delta x) - u(x_o - \Delta x) = u(x_o) - u(x_o)$$

$$+ \Delta x \frac{\partial u(x_o)}{\partial x} + \Delta x \frac{\partial u(x_o)}{\partial x} + \frac{\Delta x}{2!} \frac{\partial^2 u(x_o)}{\partial x^2} - \frac{\Delta x}{2!} \frac{\partial^2 u(x_o)}{\partial x^2} + \cdots \qquad (6.16)$$

As you can see, all of the even-power terms will cancel and the odd-power terms will add. We can rewrite Equation 6.16 using our discrete notation as

$$u_{i+1,j} - u_{i-1,j} = 2\Delta x \frac{\partial u}{\partial x} + \cdots. \qquad (6.17)$$

And, solving for the derivative, we obtain a second-order accurate finite difference representation for the first derivative:

[iii] This assumes that the numerical results are exactly accurate. There is a lower limit to the size of the difference step in Δx due to the use of finite length arithmetic. Below that step size, roundoff error becomes important. In most cases, the step size used for practical finite difference calculations is larger than the limit imposed by roundoff errors. We can't afford to compute using grids so finely spaced that roundoff becomes a problem.

$$\frac{\partial u}{\partial x} = \frac{u_{i+1,j} - u_{i-1,j}}{2\Delta x} + O(\Delta x^2) \tag{6.18}$$

which is known as a *central difference*, since it uses points on both sides of the point of interest (this is the formula that was used to develop Equation 6.2). This formula is second-order accurate, which is an improvement over the previous two first-derivative formulas, since the representation is more accurate at no increase in computational work.

If we add the two Taylor's series (Eqns. 6.11 and 6.14), the terms containing the first derivatives cancel, and we find our first representation for the second derivative

$$\frac{\partial^2 u}{\partial x^2} = \frac{u_{i+1,j} - 2u_{i,j} + u_{i-1,j}}{\Delta x^2} + O(\Delta x^2). \tag{6.19}$$

In this case the truncation error is second order, $O(\Delta x^2)$, so it is called a second-order accurate central difference.

These formulas are the most frequently used approximations to the derivatives using finite difference representations. Other methods can be used to develop finite difference approximations, but, in most cases, we want to use no more than two or three function values to approximate derivatives (higher-order methods will be an exception to this as we will discuss in Section 6.16.2).

Forward and backward finite difference approximations for the second derivative can also be derived. Note that these expressions are only first-order accurate. They are:

- a forward difference expression:

$$\left.\frac{\partial^2 u}{\partial x^2}\right|_{x_0} = \frac{u(x_0) - 2u(x_0 + \Delta x) + u(x_0 + 2\Delta x)}{(\Delta x)^2} + O(\Delta x) \tag{6.20}$$

- a backward difference expression:

$$\left.\frac{\partial^2 u}{\partial x^2}\right|_{x_0} = \frac{u(x_0) - 2u(x_0 - \Delta x) + u(x_0 - 2\Delta x)}{(\Delta x)^2} + O(\Delta x). \tag{6.21}$$

In addition, expressions can be derived for cases where the points are not evenly distributed. In general, the formal truncation error for unevenly spaced points is not as high as for evenly spaced point distributions. In practice, for reasonable variations in grid spacing, this may not be a serious problem. We present the derivation of these expressions here; however, a better way of handling non-uniform grid points is presented in Chapter 7.

The one-sided first-derivative expressions Equation 6.13 and Equation 6.15 are already suitable for use in unevenly spaced situations. We need to obtain a central difference formula for the first derivative and an expression for the second derivative. First, consider the Taylor's series expansion as given in Eqns. 6.11 and 6.14. However, the spacing will be different in the two directions, so we will use Δx^+ and Δx^- to distinguish between the grid spacing in the two directions. Equations 6.11 and 6.14 can then be rewritten as:

$$u(x_0 + \Delta x^+) = u(x_0) + \Delta x^+ \left.\frac{\partial u}{\partial x}\right|_{x_0} + \frac{\left(\Delta x^+\right)^2}{2} \left.\frac{\partial^2 u}{\partial x^2}\right|_{x_0} + \frac{\left(\Delta x^+\right)^3}{6} \left.\frac{\partial^3 u}{\partial x^3}\right|_{x_0} + \ldots$$

$$(6.22)$$

$$u(x_0 - \Delta x^-) = u(x_0) - \Delta x^- \left.\frac{\partial u}{\partial x}\right|_{x_0} + \frac{\left(\Delta x^-\right)^2}{2} \left.\frac{\partial^2 u}{\partial x^2}\right|_{x_0} - \frac{\left(\Delta x^-\right)^3}{6} \left.\frac{\partial^3 u}{\partial x^3}\right|_{x_0} + \ldots$$

$$(6.23)$$

Now define $\Delta x^+ = \alpha \, \Delta x^-$. To obtain the forms suitable for derivation of the desired expressions, replace Δx^+ in Equation 6.22 with $\alpha \, \Delta x^-$, and multiply Equation 6.23 by α. The resulting expressions are:

$$u(x_0 + \Delta x^+) = u(x_0) + \alpha \Delta x^- \left.\frac{\partial u}{\partial x}\right|_{x_0} + \frac{\left(\alpha\Delta x^-\right)^2}{2} \left.\frac{\partial^2 u}{\partial x^2}\right|_{x_0} + \frac{\left(\alpha\Delta x^-\right)^3}{6} \left.\frac{\partial^3 u}{\partial x^3}\right|_{x_0} + \ldots$$

$$(6.24)$$

$$\alpha u(x_0 - \Delta x^-) = \alpha u(x_0) - \alpha\Delta x^- \left.\frac{\partial u}{\partial x}\right|_{x_0} + \alpha\frac{\left(\Delta x^-\right)^2}{2} \left.\frac{\partial^2 u}{\partial x^2}\right|_{x_0} - \alpha\frac{\left(\Delta x^-\right)^3}{6} \left.\frac{\partial^3 u}{\partial x^3}\right|_{x_0} + \ldots$$

$$(6.25)$$

To obtain the expression for the first derivative, subtract Equation 6.25 from Equation 6.24,

$$u(x_0 + \Delta x^+) - \alpha u(x_0 - \Delta x^-) = u(x_0) - \alpha u(x_0) + 2\alpha\Delta x^- \left.\frac{\partial u}{\partial x}\right|_{x_0}$$
$$+ \left[\frac{\left(\alpha\Delta x^-\right)^2}{2} - \alpha\frac{\left(\Delta x^-\right)^2}{2}\right] \left.\frac{\partial^2 u}{\partial x^2}\right|_{x_0} + \ldots$$

$$(6.26)$$

and rearrange to obtain a central difference formula for the first derivative $\partial u / \partial x$ where the points are unequally spaced:

$$\left.\frac{\partial u}{\partial x}\right|_{x_0} = \frac{u(x_0 + \Delta x^+) + (\alpha - 1)u(x_0) - \alpha u(x_0 - \Delta x^-)}{2\alpha \Delta x^-} + O(\Delta x^-). \quad (6.27)$$

To obtain a central difference expression for the second derivative, add Eqns. 6.24 and 6.25 and then solve for $\partial^2 u / \partial x^2$:

$$u(x_0 + \Delta x^+) + \alpha u(x_0 - \Delta x^-) = u(x_0) + \alpha u(x_0) +$$

$$\left[\frac{(\alpha \Delta x^-)^2}{2} + \alpha \frac{(\Delta x^-)^2}{2}\right] \left.\frac{\partial^2 u}{\partial x^2}\right|_{x_0} + O(\Delta x)^3 \dots$$

$$\left.\frac{\partial^2 u}{\partial x^2}\right|_{x_0} = \frac{u(x_0 + \Delta x^+) - (1 + \alpha)u(x_0) + \alpha u(x_0 - \Delta x^-)}{\frac{\alpha}{2}(1 + \alpha)(\Delta x^-)^2} + O(\Delta x^-). \quad (6.28)$$

Note that both Equations 6.27 and 6.28 reduce to the forms given in Equation 6.18 and Equation 6.19 when the grid spacing is uniform ($\alpha = 1$). Finally, note that a slightly more sophisticated analysis (see Ref. 1) will lead to a second-order central difference expression for the first derivative on unevenly spaced points:

$$\left.\frac{\partial u}{\partial x}\right|_{x_0} = \frac{u(x_0 + \Delta x^+) + (\alpha^2 - 1)u(x_0) - \alpha^2 u(x_0 - \Delta x^-)}{\alpha(\alpha + 1)\Delta x^-} + O(\Delta x^-)^2. \quad (6.29)$$

So, with a minimum amount of work, we have developed many finite-difference formulas for first-derivatives and second-derivatives (as well as formulas for non-uniform grids), which we will be able to use in representing our partial differential equations as finite difference equations. We can develop almost any required formula with a little ingenuity and some often tedious work. A list of frequently used finite difference formulas is shown in Table 6.1. Reference 1 gives additional details and a collection of difference approximations using more than three points and difference approximations for mixed partial derivatives. Numerous other methods of obtaining approximations for the derivatives are also possible. The most natural one is the use of a polynomial fit through the points; polynomials are frequently used to obtain derivative expressions on non-uniformly spaced grid points.

Table 6.1 List of Commonly Used Finite Difference Formulas

First Derivatives

Forward
$$\frac{\partial f}{\partial x} = \frac{f_{i+1} - f_i}{\Delta x} + O(\Delta x) \qquad \frac{\partial f}{\partial x} = \frac{-3f_i + 4f_{i+1} - f_{i+2}}{2\Delta x} + O(\Delta x^2)$$

Central
$$\frac{\partial f}{\partial x} = \frac{f_{i+1} - f_{i-1}}{2\Delta x} + O(\Delta x^2)$$

$$\frac{\partial f}{\partial x} = \frac{-f_{i+2} + 8f_{i+1} - 8f_{i-1} + f_{i-2}}{12\Delta x} + O(\Delta x^4)$$

Backward
$$\frac{\partial f}{\partial x} = \frac{f_i - f_{i-1}}{\Delta x} + O(\Delta x) \qquad \frac{\partial f}{\partial x} = \frac{3f_i - 4f_{i-1} + f_{i-2}}{2\Delta x} + O(\Delta x^2)$$

Second Derivatives

Forward
$$\frac{\partial^2 f}{\partial x^2} = \frac{f_i - 2f_{i+1} + f_{i+2}}{\Delta x^2} + O(\Delta x)$$

$$\frac{\partial^2 f}{\partial x^2} = \frac{-f_{i+3} + 4f_{i+2} - 5f_{i+1} + 2f_i}{\Delta x^2} + O(\Delta x^2)$$

Central
$$\frac{\partial^2 f}{\partial x^2} = \frac{f_{i+1} - 2f_i + f_{i-1}}{\Delta x^2} + O(\Delta x^2)$$

$$\frac{\partial^2 f}{\partial x^2} = \frac{-f_{i+2} + 16f_{i+1} - 30f_i + 16f_{i-1} - f_{i-2}}{12\Delta x^2} + O(\Delta x^4)$$

Backward
$$\frac{\partial^2 f}{\partial x^2} = \frac{f_i - 2f_{i-1} + f_{i-2}}{\Delta x^2} + O(\Delta x)$$

$$\frac{\partial^2 f}{\partial x^2} = \frac{2f_i + 5f_{i-1} + 4f_{i-2} - f_{i-3}}{\Delta x^2} + O(\Delta x^2)$$

Computational Aerodynamics Concept Box

How to Develop a Finite Difference Algorithm

Here is an example of how to develop an expression for $\partial f/\partial x$ with order Δx^2 that uses one-sided differencing. This approach could be used to derive a wide variety of difference schemes, given some patience and some algebraic agility.

Start with two Taylor's series:

$$f(x_0 + \Delta x) = f(x_o) + \Delta x \frac{\partial f(x_o)}{\partial x} + \frac{\Delta x^2}{2!} \frac{\partial^2 f(x_o)}{\partial x^2} + \frac{\Delta x^3}{3!} \frac{\partial^3 f(x_o)}{\partial x^3} + O(\Delta x^4)$$

$$f(x_o + 2\Delta x) = f(x_o) + 2\Delta x \frac{\partial f(x_o)}{\partial x} + \frac{(2\Delta x)^2}{2!} \frac{\partial^2 f(x_o)}{\partial x^2} + \frac{(2\Delta x)^3}{3!} \frac{\partial^3 f(x_o)}{\partial x^3} + O(\Delta x^4)$$

Combine these two equations in such a way as to eliminate the Δx^2 term. Do this by multiplying the first equation by a constant value a, and the second equation by a constant value b, then add the two equations together.

$$af(x_o + \Delta x) + bf(x_o + 2\Delta x) = (a+b)f(x_o) + (a+2b)\Delta x \frac{\partial f(x_o)}{\partial x}$$

$$+ (a/2 + 2b)\Delta x^2 \frac{\partial^2 f(x_o)}{\partial x^2} + (a/6 + 4b/3)\Delta x^3 \frac{\partial^3 f(x_o)}{\partial x^3} + O(\Delta x^4)$$

Now, solve for $\partial f / \partial x$.

$$\frac{\partial f}{\partial x} = \frac{-(a+b)f(x_o) + af(x_o + \Delta x) + bf(x_o + 2\Delta x)}{(a+2b)\Delta x} - \frac{(a/2 + 2b)\Delta x^2}{(a+2b)\Delta x} \frac{\partial^2 f}{\partial x^2} + O(\Delta x^2)$$

We need $\dfrac{a}{2} + 2b = 0$ in order to eliminate the second-derivative term, which gives $a = -4b$.

Substituting this value of a gives

$$\frac{\partial f}{\partial x} = \frac{3bf(x_o) - 4bf(x_o + \Delta x) + bf(x_o + 2\Delta x)}{-2b\Delta x} + O(\Delta x^2)$$

or, if we set $b = 1$ (since a and b are arbitrary)

$$\frac{\partial f}{\partial x} = \frac{-3f(x_o) + 4f(x_o + \Delta x) - f(x_o + 2\Delta x)}{2\Delta x} + O(\Delta x^2)$$

which gives us the second-order accurate one-sided difference for the first derivative we were looking for (which can be found in Table 6.1).

While we have obtained several finite difference formulas, what is the impact of the leading-order truncation term? We can gain some understanding if we look at the leading-order term in the previous example problem (assuming we had kept more terms than we did):

$$\frac{\partial f}{\partial x} = \frac{-3f(x_o) + 4f(x_o + \Delta x) - f(x_o + 2\Delta x)}{2\Delta x} - \frac{2}{3}\Delta x^2 \frac{\partial^3 f}{\partial x^3} + O(\Delta x^4) \qquad (6.30)$$

Note that we have assumed that the leading-order truncation term, $-\frac{2}{3}\Delta x^2 (\partial^3 f / \partial x^3)$, is small, which can happen in two different ways. If $(\partial^3 f / \partial x^3) \to 0$, then Δx can be "large," and the truncated term will still be small. Also, as $\Delta x \to 0$, then $(\partial^3 f / \partial x^3)$ can be "large," and the truncated term will also be small. This implies that, if flow gradients are small (see Fig. 4.5 as an example of a flow with both small and large flow gradients), we can have large grid spacing (or what we call a *coarse grid*) without increasing the error, and if the flow gradients are large, we will need small grid spacing (or what we call a *dense grid*) to keep error low. This is a fundamental concept in CFD and you should always keep it in mind as you perform calculations.

Steps and requirements to obtain a valid numerical solution

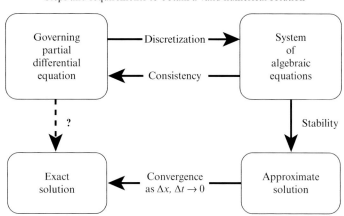

Figure 6.5 Overall approach used to develop a CFD solution procedure.

6.5 Representing Partial Differential Equations

We can use the approximations to the derivatives obtained to replace the individual terms in partial differential equations. Figure 6.5 provides a schematic of the steps required, and some of the key terms used to ensure that the results obtained are in fact the solution of the original partial differential equation. We will define each of the new terms used in Fig. 6.5.

Successful numerical methods for partial differential equations demand that the physical features of the PDE be reflected in the numerical approach. The selection of a particular finite difference approximation depends on the physics of the problem being studied. In large part the *type* of the PDE is crucial, and thus a determination of the type, i.e., elliptic, hyperbolic, or parabolic, is extremely important (see Section 3.6 for details). The mathematical type of the PDE must be used to construct the numerical scheme for approximating partial derivatives (some advanced methods obscure the relationship, but it still exists). Consider the example given in Fig. 6.6, illustrating how information in a grid must be used. In an elliptic PDE system (shown in Fig. 6.6a), the results at any point must have a continuous dependence on the boundary conditions in all directions. In this case, therefore, central differences would be used in both directions to allow information to propagate throughout the region freely. A hyperbolic PDE system (shown in Fig. 6.6b) has a zone of dependence, so only those points within that zone should be used to find results at the point of interest. Flows with mixed elliptic/hyperbolic systems (such as is common in transonic flows), would require a differencing system that could switch between the two approaches (see the article by Murman and Cole for a classic example of dealing with a mixed system[20]).

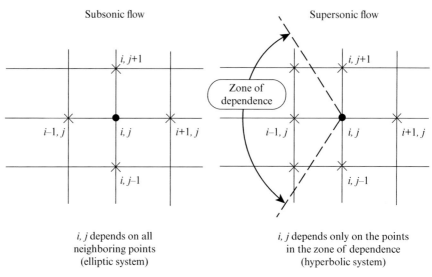

Subsonic flow   Supersonic flow

i, j depends on all
neighboring points
(elliptic system)

i, j depends only on the points
in the zone of dependence
(hyperbolic system)

Figure 6.6   Connection between grid points used in numerical method and equation type.

Any scheme that fails to represents the physics correctly will fail when you attempt to obtain a solution. Furthermore, in this case we are looking at a uniformly spaced Cartesian grid. In "real life" applications we have to consider much more complicated non-uniform grids in non-Cartesian coordinate systems. In this section, we will use simple uniform Cartesian grid systems to illustrate the basic ideas; the necessary extensions of the methods illustrated in this chapter are outlined in Chapter 7 and Appendix E.

In Fig. 6.5, we introduced several important terms requiring definition and discussion:

- discretization
- consistency
- stability
- convergence

Before defining these terms, we provide an example using the heat equation (Equation 3.93):

$$\frac{\partial u}{\partial t} = \alpha \frac{\partial^2 u}{\partial x^2}. \tag{6.31}$$

We discretize the equation using a forward difference in time and a central difference in space following the notation shown in Fig. 6.7. This results in replacing the PDE with a finite difference equation (FDE) and neglecting the truncation error (TE).

The heat equation can now be written as:

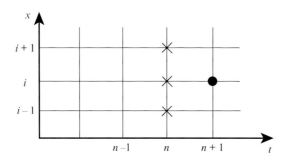

Figure 6.7   Grid nomenclature for discretization of heat equation.

$$\underbrace{\frac{\partial u}{\partial t} - \alpha \frac{\partial^2 u}{\partial x^2}}_{\text{PDE}} = \underbrace{\frac{u_i^{n+1} - u_i^n}{\Delta t} - \frac{\alpha}{(\Delta x)^2}\left(u_{i+1}^n - 2u_i^n + u_{i-1}^n\right) +}_{\text{FDE}}$$

$$\underbrace{\left[-\left.\frac{\partial^2 u}{\partial t^2}\right|_i^n \frac{\Delta t}{2} + \alpha \left.\frac{\partial^4 u}{\partial x^4}\right|_i^n \frac{(\Delta x)^2}{12} + \cdots \right]}_{\text{TE}} = 0$$

(6.32)

where we use the superscripts to denote time and the subscripts to denote spatial location. In Equation 6.32, the PDE has been converted to the related FDE. The TE is $O(\Delta t) + O(\Delta x)^2$ or $O\left[\Delta t, (\Delta x)^2\right]$. As discussed, an understanding of the truncation error for a particular scheme is very important. Using the model equation given in Equation 6.32, we can define the terms in Fig. 6.5, including discretization, consistency, stability, and convergence.

6.5.1 Discretization

This is the process of replacing derivatives by finite difference approximations: we will replace continuous derivatives with an approximation at a discrete set of points (the grid or mesh). This introduces an error due to the *truncation error* arising from the finite difference approximation and any errors due to treatment of boundary conditions. A re-examination of the Taylor's series representation is worthwhile in thinking about the possible error arising from the discretization process:

$$\frac{\partial u}{\partial x} = \frac{u_{i+1,j} - u_{i-1,j}}{2\Delta x} + \underbrace{\frac{\Delta x^2}{6} \frac{\partial^3 u}{\partial x^3}}_{\substack{\text{formally valid for } \Delta x \to 0, \\ \text{but when } \Delta x = \text{finite}, \partial^3 u / \partial x^3 \\ \text{can be big for rapidly changing} \\ \text{solutions (such as shock waves)}}} . \qquad (6.18)$$

Thus we see that the size of the truncation error will depend locally on the solution.

6.5.2 Consistency

A finite difference representation of a PDE is *consistent* if the difference between the PDE and its difference representation vanishes as the mesh is refined, i.e.,

$$\lim_{mesh \to 0} (PDE - FDE) = \lim_{mesh \to 0} (TE) = 0. \tag{6.33}$$

When might this be a problem? Consider a case where the truncation error is $O(\Delta t/\Delta x)$. In this case we must let the mesh go to zero such that:

$$\lim_{\Delta t, \Delta x \to 0} \left(\frac{\Delta t}{\Delta x} \right) = 0. \tag{6.34}$$

This would require meeting some very difficult requirements as the mesh was refined, something that would normally be difficult to do. Some finite difference representations have been tried that were not consistent; an example is the DuFort-Frankel differencing of the wave equation.[1]

6.5.3 Stability

A stable numerical scheme is one for which errors from any source (round-off[iv], truncation, etc.) are not permitted to grow in the sequence of numerical procedures as the calculation proceeds from one marching step, or iteration, to the next; thus:

$$\text{errors grow} \to \text{unstable}$$
$$\text{errors decay} \to \text{stable}$$

and

- Stability is normally thought of as being associated with marching problems
- Stability requirements often dictate allowable step sizes
- In many cases a stability analysis can be performed to define the stability requirements

One example of a stability analysis is given in Section 6.6.

6.5.4 Convergence

The solution of the FDEs should approach the solution of the PDE as the mesh is refined, as was described previously. In the case of a linear equation,

---

[iv] Roundoff error is the difference between a computer approximation for a number and the actual number—computers store numbers with a fixed set of digits and the last digit is typically rounded.

there is a theorem which proves that the numerical solution to the FDE is in fact the solution of the original PDE.

Lax Equivalence Theorem[21] (linear, initial value problem): For a properly posed problem, with a consistent finite difference representation, stability is the necessary and sufficient condition for convergence. Or, in other words, for a well-posed problem:

Consistency + Stability = Convergence.

In practice, numerical experiments must be conducted to determine if the solution appears to be converged with respect to mesh size (spatial convergence).[v] Machine capability and computing budget (time as well as money) dictate limits to the mesh size. Many, many results presented in the literature are not completely converged with respect to the mesh.

So far we have represented the PDE by an FDE at the point (i,n). The PDE is now a set of algebraic equations written at each mesh point. If the grid is (in three dimensions) defined by a grid with $IMAX$, $JMAX$, and $KMAX$ mesh points in each direction, then we have a grid with $IMAX \times JMAX \times KMAX$ grid points; this can be a very large number of points. As an example, the flow over an entire F-18 was simulated with an unstructured mesh using 10 million cells[22] and a calculation for an X-31 with a block structured grid used 28 million cells.[23] Thus, the ability to carry out aerodynamic analysis depends on our ability to solve large systems of algebraic (and often nonlinear) equations efficiently.

We have found ways to represent the derivatives, and now we need to obtain the solution for the values at each grid point; we will now consider how this is actually accomplished. Since the computer requirements and approach are influenced by the mathematical type of the equation being solved, we illustrate the basic types of approaches to the solution with examples based on the different mathematical types of equations. But first we will see how a stability analysis is performed and obtain a stability restriction for the heat equation.

6.6 Stability Analysis

The preceding analysis makes the solution of the fluid governing equations appear deceptively simple; however, in many cases, it proves impossible to obtain solutions. Frequently the reason was the choice of an inherently unstable numerical algorithm. In this section, we present one of the classical approaches to the determination of stability criteria for use in CFD. This

---

[v] This is convergence with respect to grid. Another convergence requirement is associated with the satisfaction of the solution of a system of equations by iterative methods on a fixed grid (iterative convergence).

type of analysis provides insight into grid and step size requirements (the term step size tends to denote time steps, whereas a grid size is thought of as spatial size). In addition, this analysis is directly applicable to a linear equation; applications in nonlinear problems are not as fully developed at this time.

6.6.1 Fourier or Von Neumann Stability Analysis

Consider the heat equation used previously (Equation 3.93)

$$\frac{\partial u}{\partial t} = \alpha \frac{\partial^2 u}{\partial x^2} \tag{6.35}$$

and examine the stability of the explicit representation of this equation (which will be discussed in more detail in Section 6.9.1). Assume that at $t = 0$, an error, possibly due to finite length arithmetic used in the computation, is introduced in the form:

$$\underbrace{u(x,t)}_{\substack{\text{"error" is} \\ \text{introduced}}} = \psi(t) \underbrace{e^{j\beta x}}_{\substack{\text{actually could be a series;} \\ \text{take one term here}}} \tag{6.36}$$

where

$$\beta = \text{a real constant.}$$
$$j = \sqrt{-1}$$

Here we will use an explicit finite difference representation for the heat equation:

$$\frac{u(x,t+\Delta t) - u(x,t)}{\Delta t} = \alpha \frac{u(x+\Delta x,t) - 2u(x,t) + u(x-\Delta x,t)}{(\Delta x)^2}. \tag{6.37}$$

Substitute Equation 6.36 into Equation 6.37 in order to determine the impact of the errors on this method, and then solve for $\psi(t + \Delta t)$. Start with

$$\frac{\psi(t+\Delta t)e^{j\beta x} - \psi(t)e^{j\beta x}}{\Delta t} = \alpha \frac{\psi(t)}{(\Delta x)^2}\left\{e^{j\beta(x+\Delta x)} - 2e^{j\beta x} + e^{j\beta(x-\Delta x)}\right\} \tag{6.38}$$

and collect terms to obtain:

$$\psi(t+\Delta t)e^{j\beta x} = \psi(t)e^{j\beta x} + \alpha \frac{\Delta t}{(\Delta x)^2}\psi(t)e^{j\beta x}\left\{\underbrace{e^{j\beta\Delta x} - 2 + e^{-j\beta\Delta x}}_{\substack{-2+e^{j\beta\Delta x}+e^{-j\beta\Delta x} \\ 2\cos\beta\Delta x}}\right\}. \tag{6.39}$$

Note that the $e^{j\beta x}$ term cancels, and Equation 6.39 can be rewritten as:

$$\psi(t+\Delta t) = \psi(t)\left[1 + \alpha \frac{\Delta t}{(\Delta x)^2}(-2 + 2\cos\beta\Delta x)\right]$$

$$= \psi(t)\left[1 - 2\alpha\frac{\Delta t}{(\Delta x)^2}\left(1 - \underbrace{\cos\beta\Delta x}_{\substack{\text{double angle formula}\\ =1-2\sin^2\frac{\beta\Delta x}{2}}}\right)\right] \tag{6.40}$$

which reduces to:

$$\psi(t+\Delta t) = \psi(t)\left[1 - 2\alpha\frac{\Delta t}{(\Delta x)^2}\left(1 - 1 + 2\sin^2\beta\frac{\Delta x}{2}\right)\right]$$

$$= \psi(t)\left[1 - 4\alpha\frac{\Delta t}{(\Delta x)^2}\sin^2\beta\frac{\Delta x}{2}\right]. \tag{6.41}$$

Now, look at the ratio of $\psi(t+\Delta t)$ to $\psi(t)$, which is defined as the *amplification factor*, G:

$$G = \frac{\psi(t+\Delta t)}{\psi(t)} = \left[1 - 4\alpha\frac{\Delta t}{(\Delta x)^2}\sin^2\beta\frac{\Delta x}{2}\right]. \tag{6.42}$$

For stability, the requirement is clearly

$$|G| < 1, \tag{6.43}$$

which means that any error introduced during the process will decay. For arbitrary β, what does this condition mean? Observe that the maximum value of the sine term is 1. Thus, the condition for stability is reduced to:

$$\left|1 - 4\alpha\underbrace{\frac{\Delta t}{(\Delta x)^2}}_{\lambda}\right| < 1 \tag{6.44}$$

and the limit of Equation 6.44 will be:

$$|1 - 4\lambda| = 1. \tag{6.45}$$

The largest λ that can satisfy this requirement is:

$$1 - 4\lambda = -1$$

or

$$-4\lambda = -2 \tag{6.46}$$

and

$$\lambda = \frac{1}{2}.$$

Thus, the largest λ for $|G| < 1$ means

$$\lambda = \alpha \frac{\Delta t}{\left(\Delta x\right)^2} < \frac{1}{2} \tag{6.47}$$

or:

$$\alpha \frac{\Delta t}{\left(\Delta x\right)^2} < \frac{1}{2}. \tag{6.48}$$

This sets the condition on Δt and Δx for stability of the model equation: this is a real restriction. It can be applied locally for nonlinear equations by assuming constant coefficients. An analysis of the implicit formulation for the heat equation that will be derived in Section 6.9.1 demonstrates that the implicit formulation is unconditionally stable (it will converge for any combination of values of Δx and Δt). This is a very important result: in general, explicit methods have stability restrictions and implicit methods do not, which means that implicit methods may take much larger time steps than explicit methods, primarily being limited by their accuracy level.

6.6.2 Examples of Stability and Instability

If a mathematical derivation states that there is a restriction on stability, what does that really mean? Do the restrictions on Δt and Δx mean anything when we try to solve the equations using a computer? Richtmyer and Morton[24] provide a dramatic example that demonstrates the stability criteria; their numerical experiments show the importance of the stability condition. Figure 6.8 repeats the analysis of Richtmyer and Morton as they solved the heat equation for the following initial and boundary conditions:

$$u(x,0) = \phi(x) \quad \text{for } 0 \le x \le \pi$$
$$u(0,t) = u(\pi,t) = 0 \quad \text{for } t > 0.$$

Figure 6.8a presents the development of the solution and shows the particular choice of initial value shape, ϕ, using a value of $\lambda < \frac{1}{2}$, namely 5/11. Figures 6.8b–d provide the results for a value of $\lambda > \frac{1}{2}$, namely 5/9. Theoretically, this step size will lead to an unstable numerical method, and

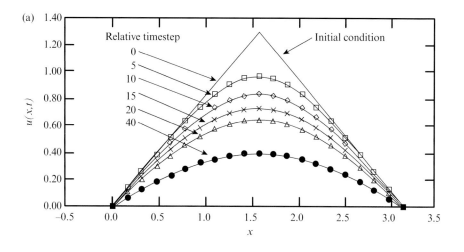

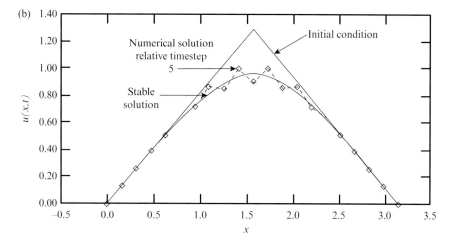

Figure 6.8a-b Demonstration of the step size stability criteria on numerical solutions (Ref. 17). (a) numerical solution using a theoretically stable step size, $\lambda = 5/11$. (b) numerical solution using a theoretically unstable step size, $\lambda = 5/9$, at $\alpha \Delta t / (\Delta x) = 5$.

the figure demonstrates that this is, in fact, starting to happen in Fig. 6.8b. Figures 6.8c and 6.8d shows what happens as the solution continues to march forward in time.

Our model problem for this example was the heat equation, which is parabolic. Another famous example considers a hyperbolic equation, namely the wave equation, where c is the wave speed (see Equation 3.89):

$$\frac{\partial^2 u}{\partial t^2} - c^2 \frac{\partial^2 u}{\partial x^2} = 0. \tag{6.49}$$

This equation represents one-dimensional acoustic disturbances. The two-dimensional small disturbance equation for the potential flow can also be written in this form for supersonic flow. Recall,

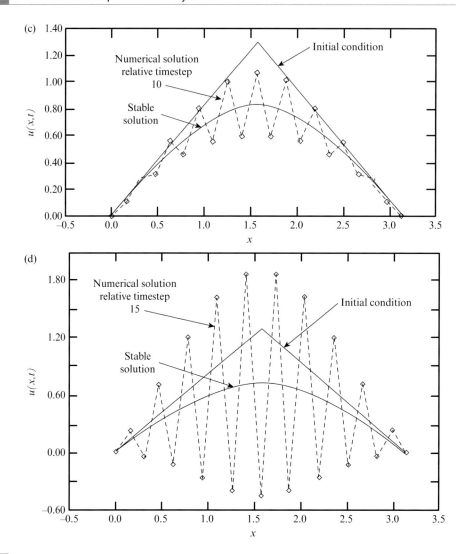

Figure 6.8c-d   Demonstration of the step size stability criteria (concluded). (c) numerical solution using a theoretically unstable step size, $\lambda = 5/9$, at $\alpha \Delta t / (\Delta x)^2 = 10$. (d) numerical solution using a theoretically unstable step size, $\lambda = 5/9$, at $\alpha \Delta t / (\Delta x)^2 = 15$.

$$\left(1 - M_\infty^2\right)\phi_{xx} + \phi_{yy} = 0. \tag{3.125}$$

Or, when the flow is supersonic:

$$\phi_{xx} - \frac{1}{\left(M_\infty^2 - 1\right)}\phi_{yy} = 0 \tag{6.50}$$

and we see here that x is the time-like variable for supersonic flow.

Performing a stability analysis similar to the one shown in Section 6.6.1, the stability requirement for the wave equation is found to result in a specific parameter for stability, namely

$$v = c\frac{\Delta t}{\Delta x} \tag{6.51}$$

which is called the *Courant number*. For many explicit schemes for hyperbolic equations, the stability requirement is found to be

$$|v| \le 1. \tag{6.52}$$

This requirement is known as the *CFL condition*, which is named after its discoverers, Courant, Friedrichs, and Lewy.[25] It has a physical interpretation: the analytic domain of influence must lie within the numerical domain of influence. Since the evolution of the solution for an elliptic system had a definite time-like quality, a stability analysis for elliptic problems can also be carried out.

6.7 The Wave Equation

Now that we have seen the importance of stability on our algorithms, we will return to the wave equation and develop some more meaningful solution methods. The wave equation is given by (see Equation 3.89):

$$u_{tt} = c^2 u_{xx} \tag{6.53}$$

which is a second-order PDE with coefficients $a = c^2$, $b = 0$, $c = -1$, so $b^2 - 4ac = 0 - 4(c^2)(-1) = 4c^2 > 0$ (see Chapter 3 for details on classifying a PDE). Therefore, the wave equation is hyperbolic. The wave equation has a well-known solution (D'Alembert's solution):

$$u(x,t) = \frac{1}{2}\{h(x+ct) + h(x-ct)\} + \frac{1}{2c}\int_{x-ct}^{x+ct} p(\tau)d\tau \tag{6.54}$$

where $u(x,0) = h(x)$ and $u_t(x,0) = p(x)$ are the initial conditions for the problem being solved. This solution shows the wave-like nature of the equation, since, if $p(x) = 0$, the solution is made up of two waves traveling in opposite directions.

We will model the wave equation with what is known as the linearized wave equation. It includes first derivatives instead of the second derivatives in Equation 6.53, but has similar characteristics:

$$u_t + cu_x = 0. \tag{6.55}$$

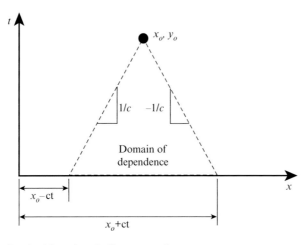

Figure 6.9   Domain of dependence for the wave equation.

You can verify that these equations are equivalent by taking the time derivative of Equation 6.55:

$$u_{tt} + cu_{xt} = 0 \tag{6.56}$$

and a spatial derivative of Equation 6.55:

$$u_{xt} + cu_{xx} = 0. \tag{6.57}$$

Now combine Equations 6.56 and 6.57 to obtain:

$$u_{tt} + c(-cu_{xx}) = 0 \ \text{ or } \ u_{tt} = c^2 u_{xx} \tag{6.58}$$

which is Equation 6.53! Therefore, Equation 6.55 can be used as a model for the original wave equation.

Hyperbolic equations have the fundamental property that the solution at a point can only be dependent on the rest of the solution within a certain "domain of dependence" and must have continuous dependence on the initial conditions. For the wave equation, the solution $u(x,t)$ at the point (x_o, y_o) only depends on the initial data contained in the interval $x_o - ct \le x \le x_o + ct$, as shown in Fig. 6.9.

Information from outside the domain of dependence cannot affect, or *influence*, the solution at (x_o, y_o). We will construct solutions starting with an initial "data curve" of the function and its derivatives. What options do we have to finite difference the linearized wave equation? We will start with some straightforward approaches and then see why some more complicated approaches are required in order to obtain accurate solutions.

6.7.1 Forward Difference in x

The first approach to solve the linearized wave equation is to use a one-sided difference for the time derivative (in order to integrate forward in time) as well as a one-sided difference for the spatial derivative, which is called a forward difference:

$$\frac{u_i^{n+1} - u_i^n}{\Delta t} + c\frac{u_{i+1}^n - u_i^n}{\Delta x} = 0. \tag{6.59}$$

This algorithm has an accuracy of $O(\Delta t, \Delta x)$ but is unconditionally unstable when a von Neumann stability analysis (described in Section 6.6) is performed. This method is unstable and will not work!

6.7.2 Central Difference in x

For our next attempt at finding a working algorithm, we will still use the forward difference in time but replace the forward spatial difference with a central spatial difference:

$$\frac{u_i^{n+1} - u_i^n}{\Delta t} + c\frac{u_{i+1}^n - u_{i-1}^n}{2\Delta x} = 0. \tag{6.60}$$

This approach has the same accuracy in time as the forward difference but has a higher spatial accuracy given by $O(\Delta t, \Delta x^2)$. However, this method is also unconditionally unstable when a von Neumann stability analysis is performed. This method also will not work!

6.7.3 Backward Difference in x

There is only one basic differencing direction we can use for the spatial derivative, namely, to use a backward difference (a difference to the "left"):

$$\frac{u_i^{n+1} - u_i^n}{\Delta t} + c\frac{u_i^n - u_{i-1}^n}{\Delta x} = 0 \tag{6.61}$$

which is only first-order accurate in time and space, i.e., $O(\Delta t, \Delta x)$. However, when a von Neumann stability analysis is performed, the method is found to be stable for $0 \leq \upsilon \leq 1$, where $\upsilon = c\Delta t / \Delta x$, which is the *Courant number* (or *CFL number*) introduced in Section 6.6.2. This is the CFL condition that was mentioned earlier, and provides our first "working" numerical method (or *algorithm*), but also shows that the relationship between the time step and spatial step must be controlled for stability. This type of method is known as an *upstream* (or *windward*) difference method.

Computational Aerodynamics Concept Box

The Meaning of the CFL Number and Why the Upstream Method Worked

The CFL number for the wave equation using the upstream method was presented in Equations 6.51 and 6.52 as:

$$0 \le c \frac{\Delta t}{\Delta x} \le 1.$$

If you look at the CFL number for the wave, you will notice that it has two velocity terms: c, which is the physical wave speed, and $\Delta x / \Delta t$, which is the computational wave speed (how fast information is propagating within the grid). The CFL number is just the ratio of these two velocities: $c / (\Delta x / \Delta t)$. The CFL restriction says that this ratio cannot be greater than one or less than zero, or, in other words: the computational propagation speed cannot be less than the physical wave speed. That makes perfect sense; you do not want to allow your algorithm to propagate information slower than the physical problem you are trying to solve!

Now take the upstream method (from Equation 6.61) and solve for u_i^{n+1}:

$$u_i^{n+1} = u_i^n + c \underbrace{\frac{\Delta t}{\Delta x}}_{v} \left(u_{i-1}^n - u_i^n \right).$$

What happens to this algorithm when $v = 1$?

$$u_i^{n+1} = u_{i-1}^n.$$

What is this rather simple relation saying? When the computational speed is set exactly to the physical wave speed, the new value at a point is determined by the previous value at the point to the left (which is where the wave is coming from). If you compare this to the analytic solution for the wave equation given in Equation 6, you will find that (for a single wave traveling to the right with initial acceleration, $p(x) = u_t(x,0) = 0$, the upstream method gives the same result as the exact solution: the wave is merely translating to the right. This is a very fortuitous result that does not happen for other equations or algorithms, but is interesting and educational for the linearized wave equation.

6.7.4 Lax Method[26]

Since the central difference approach used in Section 6.7.2 was unconditionally unstable, it would be nice to find a way to make it stable since that method was second-order accurate in space. This is accomplished by replacing u_i^n in Equation 6.60 with $(u_{i+1}^n + u_{i-1}^n)/2$ to obtain:

$$\frac{u_i^{n+1} - (u_{i+1}^n + u_{i-1}^n)/2}{\Delta t} + c \frac{u_{i+1}^n - u_{i-1}^n}{2\Delta x} = 0 \qquad (6.62)$$

which has $O(\Delta t, \Delta x^2 / \Delta t)$ and is stable when a von Neumann stability analysis is performed if $|\upsilon| \leq 1$. This method also requires holding υ constant as $\Delta t, \Delta x \to 0$. This algorithm is quasi first-order-accurate, and has large *dissipation* errors (which will be described in Section 6.8), which further leads us on a search for higher-order accurate methods.

6.7.5 Lax-Wendroff Method[27]

Starting from a Taylor's series expansion in time we find that:

$$u_i^{n+1} = u_i^n + \Delta t u_t + \frac{1}{2}\Delta t^2 u_{tt} + O(\Delta t^3). \tag{6.63}$$

Our two wave equations (6.53 and 6.55) are:

$$u_{tt} = c^2 u_{xx} \quad u_t + c u_x = 0.$$

Substituting the two wave equations into Equation 6.63 yields:

$$u_i^{n+1} = u_i^n - c\Delta t u_x + \frac{1}{2}c^2 \Delta t^2 u_{xx} + O(\Delta t^3). \tag{6.64}$$

Using second-order central differencing for the x derivative terms gives:

$$u_i^{n+1} = (1 - \upsilon^2)u_i^n - \frac{1}{2}\upsilon(1 - \upsilon)u_{i+1}^n + \frac{1}{2}\upsilon(1 + \upsilon)u_{i-1}^n. \tag{6.65}$$

This representation has $O(\Delta t^2, \Delta x^2)$ and is stable for $|\upsilon| \leq 1$. The Lax-Wendroff method does not have the large dissipation error that the Lax method has and was one of the first useful algorithms for solving the wave equation.

6.7.6 MacCormack Method[28]

This is a two-step algorithm that has been widely used and that begins with our first approach from Section 6.7.1 and Equation 6.59 (which did not work):

$$\frac{u_i^{n+1} - u_i^n}{\Delta t} + c\frac{u_{i+1}^n - u_i^n}{\Delta x} = 0.$$

Now, solve for the unknown value, u_i^{n+1}, which will be called the "predictor" step (represented with a bar over the unknown value):

$$\overline{u_i^{n+1}} = u_i^n - \frac{c\Delta t}{\Delta x}(u_{i+1}^n - u_i^n). \tag{6.66}$$

The "corrector" step is given by:

$$u_i^{n+1} = \frac{1}{2}\left\{ u_i^n + \overline{u_i^{n+1}} - \frac{c\Delta t}{\Delta x}(\overline{u_i^{n+1}} - \overline{u_{i-1}^{n+1}}) \right\}. \tag{6.67}$$

We used a forward difference for u_x in the predictor step and a backward difference in the corrector step. This method also has an accuracy of $O(\Delta t^2, \Delta x^2)$ and is stable for $|v| \leq 1$. In fact, when applied to the wave equation, the Lax-Wendroff method and MacCormack's method can be shown to be identical! This method requires storing the predictor step results at all points prior to calculating the corrector step, which makes its memory requirements higher.

We now have four algorithms that can be used to solve the linearized wave equation. Not all of them are equally good, however, since each algorithm has a different level of accuracy and a different level of complexity (although these algorithms are all relatively straightforward to program).

Profiles in Computational Aerodynamics: Robert W. MacCormack

Bob MacCormack began his career by obtaining a BS in physics and mathematics from Brooklyn College, and then went to work as a research scientist in the Hypersonic Free Flight Branch at NASA Ames Research Center. During his time at NASA, he also attended Stanford University, where he obtained his MS degree in mathematics. While working on hypersonic research, he developed a second-order time-accurate explicit algorithm now known as MacCormack's explicit method, which was used to predict high-speed impact cratering (the method is presented in Section 6.7.6). He also developed implicit methods for predicting high-speed flows. His work in numerical methods led him to become the assistant chief of the Computational Fluid Dynamics Branch at NASA Ames Research Center, where he worked with Harvard Lomax and many other well-known CFD practitioners. His career at Ames culminated with his position as senior staff scientist in the Thermo– and Gas Dynamics Division. A few years later academia came calling, and Bob left NASA to become a faculty member at the University of Washington. In 1985 he was appointed as a professor in the Aeronautics and Astronautics Department at Stanford University, where he has worked since then. In recent years his research has focused on developing implicit procedures for three-dimensional flow and for hypersonic flows containing regions of chemical and thermal non-equilibrium, which requires developing numerical procedures for the solution of Maxwell's equations.

Bob is a member of the National Academy of Engineering, was awarded the NASA Medal for Exceptional Scientific Achievement and the AIAA Fluid Dynamics Award, and is a Fellow of the AIAA. He was a member of the U.S. Air Force Scientific Advisory Board from 2001 to 2005. He has authored or co-authored approximately 200 publications during his career.

6.8 Truncation Error Analysis of the Wave Equation: The Modified Equation

While the working methods of Section 6.7 are defined with an order of accuracy and stability restrictions, they do not all work in the same way. It will be useful to have a way to analyze their error in a more detailed fashion. We will do that starting with the linearized wave equation (Equation 6.55):

$$u_t + cu_x = 0$$

for $c > 0$.

Now apply the upstream differencing method from Section 6.7.3 to the equation and then analyze the truncation error:

$$\frac{u_i^{n+1} - u_i^n}{\Delta t} + c\frac{u_i^n - u_{i-1}^n}{\Delta x} = 0$$

which was found to be stable for $0 \le \upsilon \le 1$ where $\upsilon = c\Delta t / \Delta x$. If we substitute various Taylor's series for the terms in the equation and perform a great deal of algebra, we will obtain what is known as the *modified equation* for our algorithm:[5]

$$u_t + cu_x =$$
$$\frac{c\Delta x}{2}(1 - \upsilon)u_{xx} - \frac{c\Delta x^2}{6}(2\upsilon^2 - 3\upsilon + 1)u_{xxx} + O(\Delta x^3, \Delta x^2\Delta t, \Delta x\Delta t^2, \Delta t^3).$$

(6.68)

Since we truncated various terms when we formed the algorithm, we really are not solving the wave equation exactly. The modified equation shows us the PDE that we are solving exactly, and it is important to realize that the terms on the right-hand side of Equation 6.68 each represent a different type of error, which will be exhibited in different ways. The right-hand side of the equation is a representation of the truncation error of the algorithm, and the lowest order term tells us the order of the method (i.e., $O(\Delta x)$ in this case). If we set $\upsilon = 1$ the equation reduces to:

$$u_t + cu_x = 0 \qquad (6.69)$$

which means for this case the wave equation is solved exactly and the FDE is the same as the PDE. In the case where $\upsilon = 1$, the FDE becomes:

$$u_i^{n+1} - u_i^n = -\frac{c\Delta t}{\Delta x}(u_i^n - u_{i-1}^n) = -\upsilon(u_i^n - u_{i-1}^n) = -(u_i^n - u_{i-1}^n). \qquad (6.70)$$

Rewriting this expression give us:

$$u_i^{n+1} - u_i^n = -u_i^n + u_{i-1}^n \quad \text{or} \quad u_i^{n+1} = u_{i-1}^n \qquad (6.71)$$

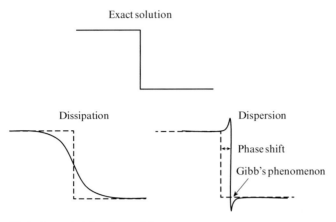

Exact solution

Dissipation   Dispersion

Phase shift

Gibb's phenomenon

Figure 6.10   Effects of numerical viscosity on solutions.

which is the characteristic equation. This algorithm fits the equation type very well and gives accurate results because of this – not all algorithms will be as good! A great deal more information can be obtained from the modified equation – interested readers are referred to Ref. 5 for more details.

Now we will look at the various truncation terms in Equation 6.68. The lowest-order term contains the partial derivative u_{xx}, which means the term behaves like a viscous term found in the Navier-Stokes equations, namely:

$$\frac{\partial \tau_{xx}}{\partial x} = \frac{4}{3}\mu u_{xx}. \tag{6.72}$$

Thus, for the equation when $\upsilon \neq 1$, the u_{xx} term (an even derivative term) is introduced, which adds what is called *artificial viscosity* or *numerical viscosity* to the results. The effect of the artificial viscosity is called *dissipation*, since it tends to smear out gradients (just like physical viscosity does), as seen in Fig. 6.10. Thus, when the leading-order term has an even derivative, the primary error will be dissipation.

If the first term in the modified equation is an odd derivative, then the numerical scheme is called *dispersive*. The primary effect of dispersion is to change the phase relationship between various waves, as well as creating the "blip" (or Gibb's phenomenon) at the point of high gradients (see Fig. 6.10). The combined effect of dissipation and dispersion is called *numerical diffusion*, which tends to spread out gradients that occur in the flowfield. Again, the predominant error is related to the leading term in the modified equation (Equation 6.68).

These terms should not be confused with explicitly added artificial viscosity, which is sometimes used in algorithms to maintain stability in the vicinity of shocks or other large flow gradients. This practice is often necessary, but care should always be taken to not add too much artificial viscosity, since it will smooth out all gradients, i.e., shocks will be smeared by the artificial viscosity.

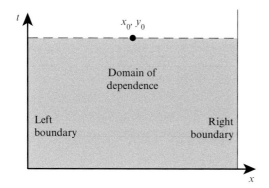

Figure 6.11 Domain of dependence for the heat equation.

6.9 The Heat Equation

Now that we have applied the finite difference approach to the wave equation, it would be nice to see if the same ideas will work for the heat equation, which we introduced previously (see Equation 6.31). The one-dimensional heat equation (also known as the diffusion equation) is given by Equation 3.93:

$$u_t = \alpha u_{xx} \tag{6.73}$$

which is a second-order PDE with coefficients $a = \alpha$, $b = 0$, $c = 0$, so $b^2 - 4ac = 0 - 4(\alpha)(0) = 0$. Therefore, the heat equation is a parabolic PDE. If we specify boundary and initial conditions

$$u(0,t) = u(1,t) = 0 \quad u(x,0) = f(x) \tag{6.74}$$

we can solve Equation 6.73 using separation of variables to obtain:

$$u(x,t) = \sum_{n=1}^{\infty} A_n e^{-\alpha k^2 t} \sin(kx) \tag{6.75}$$

$$A_n = 2\int_0^1 f(x)\sin(kx)dx \quad \text{for} \quad k = n\pi.$$

Whereas hyperbolic equations had a domain of dependence restricted in space, parabolic equations are much less restrictive. Specifically, where the domain of dependence for the wave equation was limited by the sloped characteristics lines (see Fig. 6.9), the dependence domain for a parabolic equation extends to all values of x, as shown in Fig. 6.11.

Because of the different nature of parabolic equations when compared with hyperbolic equations, we will want to choose numerical algorithms that take into account both the initial and boundary condition requirements, as

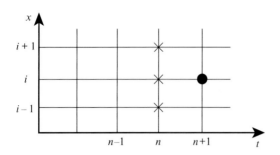

Figure 6.12 Grid points used in typical explicit calculation.

well as the domain of dependence issue which leads to the need to central difference in x. We will see that "implicit" algorithms may be a better choice for this type of PDE, but first we will evaluate an explicit approach.

6.9.1 Explicit Scheme

Consider the finite difference representation of the heat equation given in Equation 6.37. Using the notation shown in Fig. 6.12, we write the finite difference representation as:

$$\frac{u_i^{n+1} - u_i^n}{\Delta t} = \frac{\alpha}{\left(\Delta x\right)^2}\left(u_{i+1}^n - 2u_i^n + u_{i-1}^n\right) \tag{6.76}$$

where the solution at time level (n) is known. At time level $(n+1)$, there is only one unknown, which can be solved for explicitly.

Solving for the value of u at the $(n+1)$ time step gives:

$$u_i^{n+1} = u_i^n + \alpha\left(\frac{\Delta t}{\left(\Delta x\right)^2}\right)\left(u_{i+1}^n - 2u_i^n + u_{i-1}^n\right) \tag{6.77}$$

and thus at each i on $(n+1)$ we can solve for u_i^{n+1} algebraically, without solving a system of equations. This means that we can solve for each new value explicitly in terms of known values from the previous time step. These types of algorithm are known as explicit schemes, and they are a very straightforward procedures with some important features:

- Some good news: the algebra is simple.
- Some bad news: stability requirements require very small step sizes, which results in many iterations being needed to find a solution.
- More good news: this scheme is easily vectorized (see Chapter 2 for a discussion of vector processing) and is a natural for massively parallel computation.

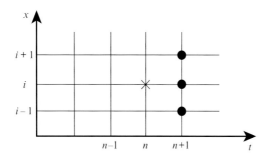

Figure 6.13 Grid points used in typical implicit calculation.

6.9.2 Implicit Scheme

Now consider an alternate finite difference representation of the heat equation given in Equation 6.37. Use the notation shown in Fig. 6.13 to define the location of grid points used to define the finite difference representation (notice how different this is from the "stencil" in Fig. 6.12).

Now we write the finite difference representation as:

$$\frac{u_i^{n+1} - u_i^n}{\Delta t} = \frac{\alpha}{(\Delta x)^2}\left(u_{i+1}^{n+1} - 2u_i^{n+1} + u_{i-1}^{n+1}\right) \tag{6.78}$$

where we use the spatial derivative at time $(n+1)$. By doing this, we obtain a system where, at each grid point on $(n+1)$, u_i^{n+1} depends on all other values at $(n+1)$, which is why the method is called implicit. Thus we need to find the values along $(n+1)$ simultaneously, which leads to a system of algebraic equations that must be solved.

For our model problem, this system of equations is linear, which we can see more clearly by rearranging Equation 6.78. First we will define:

$$\lambda = \alpha\frac{\Delta t}{(\Delta x)^2} \tag{6.79}$$

and rewrite Equation 6.78 (after some minor algebra) as:

$$-\lambda u_{i-1}^{n+1} + (1+2\lambda)u_i^{n+1} - \lambda u_{i+1}^{n+1} = u_i^n \qquad \text{for } i = 1,...,N. \tag{6.80}$$

This can be put into a matrix form to show that it has a particularly simple form:

$$\begin{bmatrix} (1+2\lambda) & -\lambda & 0 & \cdots & 0 & 0 & 0 \\ -\lambda & (1+2\lambda) & -\lambda & 0 & \cdots & 0 & 0 \\ 0 & \ddots & \ddots & \ddots & 0 & \cdots & 0 \\ 0 & 0 & \ddots & \ddots & \ddots & 0 & 0 \\ 0 & \cdots & 0 & \ddots & \ddots & \ddots & 0 \\ 0 & 0 & \cdots & 0 & -\lambda & (1+2\lambda) & -\lambda \\ 0 & 0 & 0 & \cdots & 0 & -\lambda & (1+2\lambda) \end{bmatrix} \begin{bmatrix} u_1^{n+1} \\ u_2^{n+1} \\ \vdots \\ u_i^{n+1} \\ \vdots \\ u_{N-1}^{n+1} \\ u_N^{n+1} \end{bmatrix} = \begin{bmatrix} u_1^n \\ u_2^n \\ \vdots \\ u_i^n \\ \vdots \\ u_{N-1}^n \\ u_N^n \end{bmatrix} \tag{6.81}$$

Equation 6.81 is a special type of matrix form known as a *tridiagonal form*, since the only non-zero elements lie along or immediately next to the diagonal of the matrix. A particularly easy solution technique is available to solve this form, which is known as the Thomas algorithm. The details of the Thomas algorithm are described in Section 6.14 and a routine called *TRIDAG* that implements the Thomas algorithm is described in Appendix D. Many numerical methods are tailored to be able to produce this form (note that the tridiagonal form results when only points at i, $i + 1$, and $i - 1$ are used). If higher-order methods are used that require values at points far away from i (like $i + 2$, $i - 2$, etc.), then a different matrix would result and a new matrix solution method would be needed.

The approach that leads to the formulation of a problem requiring the simultaneous solution of a system of equations is known as an *implicit* scheme. To summarize:

- The solution of a system of equations is required at each step.
- The good news: the lack of a stability restriction allows for a large step size.
- The not-so-good news: this scheme is harder to vectorize/parallelize.

A common feature for both explicit and implicit methods for parabolic and hyperbolic equations:

- A large number of mesh points can be treated; you only need the values at a small number of marching stations at any particular stage in the solution. This means you can obtain the solution with a large number of grid points using a relatively small amount of memory. Curiously, some recent codes don't take advantage of this last fact.

Computational Aerodynamics Concept Box

The Difference between Explicit and Implicit Methods

In describing the two numerical method approaches available for CFD solutions (explicit and implicit), we have left out a basic understanding about why these two methods were developed and how they are used today.

As we have said throughout the book, computational aerodynamics has challenged computer capabilities ever since there were computers. As soon as a new computer came into production, CFD researchers increased the complexity and size of their simulations, which required more grid points and larger computers. But getting a new computer is not always the most feasible way to make your code run faster (it can become quite expensive!). When explicit methods (such as MacCormack's method) were being used on complicated flow simulations in the 1970s, the size of the computers limited how large the grids could be. If no new computer was going to be available for a year or two, then researchers were left with the challenge of making their codes run faster. Implicit methods were a way to do that, but why? When we looked at explicit methods, you may have noticed that they always had a stability restriction. That is, you could not take any step size you wanted in order to obtain a solution, since your code would blow up if you did that. So, you would have to choose a very small step size for your relatively simple algorithm, which meant that the solution took a long time to complete.

We can better understand this using the analogy of walking across a room from one wall to the opposite wall; we can assume the room is 10m across for this example. If you take regular steps (1m per step) at a normal rate (1 step per second) you will usually cross the room quite quickly (for this example you would cross the room in 10 seconds). What if someone told you to take the same step rate (1 step per second) but limit your steps to a very small size (say 1mm per step). It would then take you 1,000 times longer to cross the room (in this case 10,000 seconds or about 2.8 hours). That would be an outrageous amount of time to walk across a room! This is exactly what explicit methods do – they allow you to take steps fairly quickly (the algorithms are fast for each iteration) but limit how big a step you can take due to the stability restriction. This generally makes explicit methods quite slow overall for a reasonable number of grid points.

Here is where implicit methods come in. Implicit methods are complicated and spend a great deal of time per iteration due to the need to perform a variety of matrix calculations. That is the equivalent of taking a step across the room at a very slow rate (say 1 step per minute). However, implicit methods do not have stability restrictions, so you can take much larger steps (2m for example). Now you have crossed the room in 5 minutes, compared with the 2.8 hours for the explicit method. Of course, implicit methods still have temporal accuracy levels that must be taken into account, but if you are trying to attain a steady-state solution, implicit methods will be much faster than explicit methods, which is why they are typically used today.

6.10 Laplace's Equation

We use Laplace's equation as the model problem for elliptic PDE's:

$$\phi_{xx} + \phi_{yy} = 0 \tag{6.82}$$

where $a = 1$, $b = 0$, $c = 1$, $b^2 - 4ac = 0 - 4 < 0$, therefore the equation is elliptic. Elliptic PDEs require information to be passed in all directions and there must be boundary conditions enclosing the solution region – we must model that feature with our algorithm. So, given the grid shown in Fig. 6.14, we will use information from all sides of each point to allow information to propagate in all directions equally.

Use the second-order accurate central difference formulas at (i, j):

$$\phi_{xx} = \frac{\phi_{i+1,j} - 2\phi_{i,j} + \phi_{i-1,j}}{(\Delta x)^2} + O(\Delta x)^2 \tag{6.83}$$

$$\phi_{yy} = \frac{\phi_{i,j+1} - 2\phi_{i,j} + \phi_{i,j-1}}{(\Delta y)^2} + O(\Delta y)^2 \tag{6.84}$$

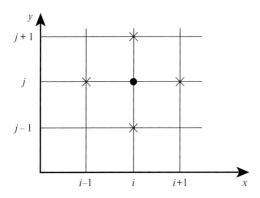

Figure 6.14   Grid points used in a typical representation of an elliptic equation.

and substitute these expressions into the governing equation:

$$\frac{\phi_{i+1,j} - 2\phi_{i,j} + \phi_{i-1,j}}{(\Delta x)^2} + \frac{\phi_{i,j+1} - 2\phi_{i,j} + \phi_{i,j-1}}{(\Delta y)^2} = 0. \tag{6.85}$$

Solve this equation for $\phi_{i,j}$:

$$\phi_{i,j} = \frac{(\Delta y)^2}{2\left[(\Delta x)^2 + (\Delta y)^2\right]}\left(\phi_{i+1,j} + \phi_{i-1,j}\right) + \frac{(\Delta x)^2}{2\left[(\Delta x)^2 + (\Delta y)^2\right]}\left(\phi_{i,j+1} + \phi_{i,j-1}\right) \tag{6.86}$$

And for the case where $\Delta x = \Delta y$, Equation 6.86 simplifies to:

$$\phi_{i,j} = \frac{1}{4}\left(\phi_{i+1,j} + \phi_{i-1,j} + \phi_{i,j+1} + \phi_{i,j-1}\right) \tag{6.87}$$

which is a simple average of the four points around the point where calculations are being performed. This expression illustrates the essential physics of flows governed by elliptic PDE's:

- $\phi_{i,j}$ depends on all the values around it
- all values of ϕ must be found simultaneously
- computer storage requirements are much greater than those required for parabolic/hyperbolic PDE's

Because of the large number of mesh points required to resolve the flowfield details, it is generally not practical to solve the system of equations arising from applying the preceding equation at each mesh point directly. Instead, an iterative procedure is usually employed. In this procedure an initial "guess" for the solution is made and then each mesh point in the flowfield is updated repeatedly until the values satisfy the governing equation. This iterative procedure can be thought of as having a time-like quality, which, as illustrated, has been exploited in many solution schemes in order to find the steady solution.

6.11 The Finite Volume Method

Finite difference methods are historically the most well-known methods in CFD, however, other methods have proven to be successful. One method in particular, the finite volume technique, actually forms the basis for most current successful codes. Instead of discretizing the PDE, the finite volume method uses the integral form of the governing equations (recall from Chapter 3 that each conservation law has both differential and integral statements). The integral form of the governing equations is more fundamental since it does not depend on the existence of continuous partial derivatives.

Consider the inviscid form of the general conservation equation (following the approach of Fletcher[2] and using Equation 3.63 in two dimensions for this example analysis):

$$\frac{\partial Q}{\partial t} + \frac{\partial F}{\partial x} + \frac{\partial G}{\partial y} = 0. \tag{6.88}$$

We can choose the particular equation used in the vector form to be the conservation of mass, where:

$$\begin{aligned} Q &= \rho \\ F &= \rho u \\ G &= \rho v \end{aligned} \tag{6.89}$$

and recall that this conservation law could also come from the integral form, Equation 3.7:

$$\frac{\partial}{\partial t} \iiint_{V} \rho dV = -\iint_{S} \rho \vec{V} \cdot \hat{n} dS. \tag{6.90}$$

Introducing the notation defined earlier and assuming two-dimensional flow, the conservation law can be rewritten as:

$$\frac{\partial}{\partial t} \iint Q dV + \int \vec{H} \cdot \hat{n} dS = 0 \tag{6.91}$$

where

$$\vec{H} = (F, G) = \rho \vec{V} \tag{6.92}$$

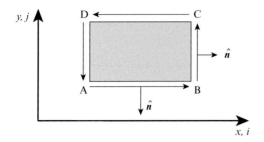

Figure 6.15   Basic nomenclature for finite volume analysis in Cartesian coordinates.

and

$$H_x = F = \rho u$$
$$H_y = G = \rho v. \tag{6.93}$$

Using the definition of $\hat{n}$ in Cartesian coordinates, and considering for illustration the Cartesian system given in Fig. 6.15, we can write:

$$\vec{H} \cdot \hat{n} dS = \left(H_x \hat{i} + H_y \hat{j} \right) \cdot \hat{n} dS$$
$$= \left(F\hat{i} + G\hat{j} \right) \cdot \hat{n} dS \tag{6.94}$$

along AB, $n = -j$, $dS = dx$, and:

$$\vec{H} \cdot \hat{n} dS = -G dx \tag{6.95}$$

along BC, $n = i$, $dS = dy$, and

$$\vec{H} \cdot \hat{n} dS = F dy \tag{6.96}$$

or, in general:

$$\vec{H} \cdot \hat{n} dS = F dy - G dx. \tag{6.97}$$

Using the general grid shown in Fig. 6.16, our integral statement, Equation 6.91, can be written as:

$$\frac{\partial}{\partial t} \left(A Q_{j,k} \right) + \sum_{AB}^{DA} \left(F \Delta y - G \Delta x \right) = 0. \tag{6.98}$$

Here, A is the area of the quadrilateral $ABCD$, and $Q_{i,j}$ is the average value of Q over $ABCD$, which is a cell-centered approach.

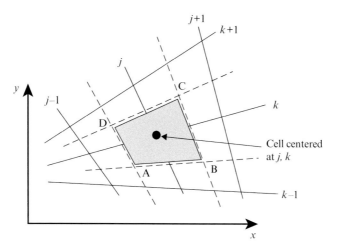

Figure 6.16 Circuit in a general grid system.

Now, define the flux quantities over each face. For illustration purposes consider AB:

$$
\begin{aligned}
\Delta y_{AB} &= y_B - y_A \\
\Delta x_{AB} &= x_B - x_A \\
F_{AB} &= \frac{1}{2}\left(F_{j,k-1} + F_{j,k}\right) \\
G_{AB} &= \frac{1}{2}\left(G_{j,k+1} + G_{j,k}\right)
\end{aligned}
\tag{6.99}
$$

and so on, over the other cell faces. Assuming A is not a function of time, and combining, we obtain:

$$
\begin{aligned}
A\frac{\partial Q_{j,k}}{\partial t} &+ \frac{1}{2}\left(F_{j,k-1} + F_{j,k}\right)\Delta y_{AB} - \frac{1}{2}\left(G_{j,k-1} + G_{j,k}\right)\Delta x_{AB} \\
&+ \frac{1}{2}\left(F_{j,k} + F_{j+1,k}\right)\Delta y_{BC} - \frac{1}{2}\left(G_{j,k} + G_{j+1,k}\right)\Delta x_{BC} \\
&+ \frac{1}{2}\left(F_{j,k} + F_{j,k+1}\right)\Delta y_{CD} - \frac{1}{2}\left(G_{j,k} + G_{j,k+1}\right)\Delta x_{CD}. \\
&+ \frac{1}{2}\left(F_{j-1,k} + F_{j,k}\right)\Delta y_{DA} - \frac{1}{2}\left(G_{j-1,k} + G_{j,k}\right)\Delta x_{DA} \\
&= 0
\end{aligned}
\tag{6.100}
$$

Suppose the grid is regular Cartesian as shown in Fig. 6.17. Then $A = \Delta x \Delta y$, and along:

$$
\begin{array}{lll}
AB: & \Delta y = 0, & \Delta x_{AB} = \Delta x \\
BC: & \Delta x = 0, & \Delta y_{BC} = \Delta y \\
CD: & \Delta y = 0, & \Delta x_{CD} = -\Delta x \\
DA: & \Delta x = 0, & \Delta y_{DA} = -\Delta y
\end{array}
\tag{6.101}
$$

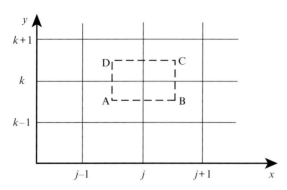

General finite volume grid applied in Cartesian coordinates.

Thus, from Equation 6.100, we are left with:

$$\Delta x \Delta y \frac{\partial Q_{j,k}}{\partial t} - \frac{1}{2}\left(G_{j,k-1} + G_{j,k}\right)\Delta x + \frac{1}{2}\left(F_{j,k} + F_{j+1,k}\right)\Delta y$$
$$+ \frac{1}{2}\left(G_{j,k} + G_{j,k+1}\right)\Delta x - \frac{1}{2}\left(F_{j-1,k} + F_{j,k}\right)\Delta y = 0.$$

(6.102)

Collecting terms:

$$\frac{\partial Q_{j,k}}{\partial t} + \underbrace{\frac{F_{j+1,k} - F_{j-1,k}}{2\Delta x} + \frac{G_{j,k+1} - G_{j,k-1}}{2\Delta y}}_{\substack{\text{for this reversion to Cartesian} \\ \text{coordinates, the equation reduces} \\ \text{to simple central differences of the original} \\ \text{partial differential equation}}} = 0$$

(6.103)

or

$$\frac{\partial Q}{\partial t} + \frac{\partial F}{\partial x} + \frac{\partial G}{\partial y} = 0.$$

(6.104)

Thus, and at first glance remarkably, the results of the finite volume approach can lead to the exact same equations to solve as the finite difference method on a simple Cartesian mesh. However, the interpretation is different:[vi]

- Finite difference: approximates the governing equation at a point
- Finite volume: approximates the governing equation over a volume
- Finite volume is the most physical representation for fluid mechanics codes and is actually used in most codes today.
- Finite difference methods were developed earlier; the analysis of methods is easier and better developed.

Both the finite difference and finite volume methods are very similar; however, there are subtle but important differences. We cite three points in

---

[vi] Summarized from Professor B. Grossman's unpublished CFD notes

favor of the finite volume method when compared with the finite difference method:

- Good conservation of mass, momentum, and energy using integrals when mesh is finite size
- Easier to treat complicated domains (integral discretization [averaging] easier to figure out, implement, and interpret)
- Average integral concept is a better approach when the solution has shock waves (i.e., the partial differential equations assume continuous partial derivatives)

Finally, special considerations are needed to implement some of the boundary conditions in this method. The references, in particular Fletcher,[2] should be consulted for more details.

Computational Aerodynamics Concept Box

Cell-Centered versus Cell Vertex (or node-centered) Methods

Unstructured mesh generation (which will be discussed in Chapter 7) has come a long way in the past twenty years, with rapid advances in generation schemes. However, one issue that has not been resolved is how to use the unstructured cells in finite volume calculations. There are two basic approaches to doing this: cell-centered and cell vertex, as shown in the figure.

The two approaches handle the cells in very different ways. The cell-centered approach defines unknown variables at the centroid of each cell (as shown in figure (a), with the "+" sign), and the cells created by the mesh generation system are used as the control volume (the triangle around the "+" sign). The number of control volumes for the cell-centered approach is equal to the number of cells in the input mesh. Typically the number of control volumes/unknowns for the cell-centered approach is much higher when compared to the cell

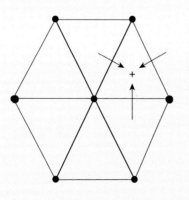

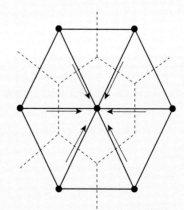

(a) Cell-centered method   (b) Cell vertex method with dual mesh vertex approach (typically about a factor of six), which partly expains why the cell-centered approach is more accurate in many cases (assuming the same input mesh is used).

The cell vertex approach starts with the same unstructured mesh but creates a "dual mesh" within the field (the dashed lines in figure (b)). The control volumes are arbitrary polyhedrons (the dashed hexagon in the figure) surrounding the vertices of the original mesh. The number of control volumes for the cell vertex approach is equal to the number of vertices in the input mesh, and the cell vertex approach is more efficient due to the smaller number of unknowns.

Both approaches have to find fluxes across the boundaries of the control volume; therefore, flow variables have to be reconstructed. This can be accomplished in different ways, such as averaging the values between two adjacent control volumes. For the cell vertex approach, a simple average between the vertices of the original mesh is sufficient. A simple average might not be sufficient for the cell-centered approach because, in general, the distances from the centers of two adjacent cells to the common boundary face are different.

Nevertheless, it is quite difficult to find a one-to-one comparison since the cell vertex approach is really using a different mesh for performing calculations. This leaves a great deal of room for debate about which method is better. To quote Dimitri Mavriplis of the University of Wyoming, "after twenty years of debate, the determination of which discretization approach, cell-based or vertex-based, is most effective for unstructured mesh methods, has still not been resolved definitely."

6.12 Time Integration and Differences

So far we have discussed various discrete forms of derivatives and applied them to spatial derivatives. This same procedure can be applied to time derivatives with one caveat: time derivatives are usually one-sided since we don't know future information! In fact, all of our previous temporal differencing used a first-order-accurate backward difference approximation for the time derivative. This has presented a great deal of difficulty for CFD researchers over the years, since the most obvious way to increase time integration accuracy involves storing multiple solutions at multiple time steps in order to take advantage of higher-order differencing. Unfortunately, computer memory limitations rarely make this a viable option. Therefore, a variety of more ingenious methods for obtaining higher orders of time accuracy have been developed.

There are two basic approaches to time integration, just as there were for spatial differencing: explicit methods, and implicit methods. The choice of method is highly dependent on the flow being solved for, primarily if the flow is steady or unsteady, but also if the solution is modeling a grid with motion (such as for a store separation or dynamic maneuver). Tom Pulliam of NASA Ames Research Center observes, "Implicit numerical schemes are usually chosen because we wish to obtain solutions which require fine grid spacing for numerical resolution, and we do not want to limit the time steps by employing a conditionally stable explicit scheme ... The extra work required for an implicit scheme is usually offset by the advantages obtained by the increased stability limits."[29]

In addition, there are various methods available to increase the accuracy of the results as the time integration is taking place, which usually involves some type of sub-iteration approach. Although surveying the full breadth and width of time integration methods is beyond the scope of this book, we will cover some of the basic approaches and provide references for more advanced methods.

6.12.1 Explicit Time Integration

The most basic time integration methods are the explicit approaches, which solve for the next time level only using information from the current (known) time level. These approaches are fairly straightforward to understand and are also easy to code. However, they often lack the accuracy required for more advanced unsteady/moving mesh calculations, and explicit methods have the stability restrictions that require a much longer time to compute their results due to the time step size limitations.

6.12.1.1 FIRST-ORDER TIME ACCURACY

Consider the conservation of mass equation in two dimensions (see Equation 3.4):

$$\frac{\partial \rho}{\partial t} + \frac{\partial \rho u}{\partial x} + \frac{\partial \rho v}{\partial y} = 0. \tag{3.4}$$

We can replace the time term with a first-order forward difference and the spatial terms with second-order central difference spatial derivatives to obtain:

$$\frac{\rho_{i,j}^{n+1} - \rho_{i,j}^{n}}{\Delta t} + \frac{(\rho u)_{i+1,j}^{n} - (\rho u)_{i-1,j}^{n}}{2\Delta x} + \frac{(\rho v)_{i,j+1}^{n} - (\rho v)_{i,j-1}^{n}}{2\Delta y} + O(\Delta t, \Delta x^2, \Delta y^2) = 0 \tag{6.105}$$

We need to "choose" what time level our spatial derivatives would be evaluated for [in this case we chose the current time level, (n)]. We can solve for the new time level $(n+1)$ by solving for the unknown value:

$$\rho_{i,j}^{n+1} = \rho_{i,j}^{n} - \Delta t \left[\frac{(\rho u)_{i+1,j}^{n} - (\rho u)_{i-1,j}^{n}}{2\Delta x} + \frac{(\rho v)_{i,j+1}^{n} - (\rho v)_{i,j-1}^{n}}{2\Delta y} \right] + O(\Delta t, \Delta x^2, \Delta y^2) \tag{6.106}$$

This is referred to as *Backward Euler* time differencing and is first-order accurate in time. We can apply this method to the full set of governing equations (mass, momentum, energy) to determine the set of unknowns at each time level. This type of differencing would be all that is required if we were trying to simulate a steady flow problem, since we could just continue iterating in time until the solution converged to the final steady-state solution. However, if we want to model unsteady flow, we will need higher orders of time accuracy or unrealistically small time step sizes.

6.12.1.2 SECOND-ORDER TIME ACCURACY

The obvious extension to the previous method is to use a second-order accurate time difference

$$
\frac{3\rho_{i,j}^{n+1} - 4\rho_{i,j}^{n} + \rho_{i,j}^{n-1}}{\Delta t} + \frac{(\rho u)_{i+1,j}^{n} - (\rho u)_{i-1,j}^{n}}{2\Delta x} + \frac{(\rho v)_{i,j+1}^{n} - (\rho v)_{i,j-1}^{n}}{2\Delta y}
$$
$$
+ O(\Delta t^2, \Delta x^2, \Delta y^2) = 0 \tag{6.107}
$$

or, solving for the density at the next time level, we have

$$
\rho_{i,j}^{n+1} = \rho_{i,j}^{n} + \frac{1}{3}(\rho_{i,j}^{n} - \rho_{i,j}^{n-1})
$$
$$
- \frac{2}{3}\Delta t \left[\frac{(\rho u)_{i+1,j}^{n} - (\rho u)_{i-1,j}^{n}}{2\Delta x} + \frac{(\rho v)_{i,j+1}^{n} - (\rho v)_{i,j-1}^{n}}{2\Delta y} \right] + O(\Delta t^2, \Delta x^2, \Delta y^2).
$$
$$
\tag{6.108}
$$

Notice that everything on the right-hand side of the equation is known (excluding the truncation term). However, the "penalty" for a higher order of accuracy is additional storage of the (*n-1*) time level of data, which can limit the size of the grid being used depending on the computer system capabilities. In addition, both the Lax-Wendroff algorithm and the MacCormack two-step algorithm mentioned in Section 6.7 are second-order accurate in time without the need to store information at three time levels.

6.12.1.3 GENERAL FORM OF BACKWARD TIME DIFFERENCE

Two of our explicit methods for time integration can actually be combined into a single equation with a "switch" to determine the time accuracy. Look at two of the previous ways we have to represent the time derivative:

$$
\frac{\partial \rho}{\partial t} = \frac{\rho^{n+1} - \rho^{n}}{\Delta t} + O(\Delta t) \tag{6.109}
$$

$$
\frac{\partial \rho}{\partial t} = \frac{3\rho^{n+1} - 4\rho^{n} + \rho^{n-1}}{2\Delta t} + O(\Delta t^2) \tag{6.110}
$$

Equations 6.109 and 6.110 can be combined with a switching parameter, ϕ, as:

$$
\frac{\partial \rho}{\partial t} = \frac{(1+\phi)\rho^{n+1} - (1+2\phi)\rho^{n} + \phi\rho^{n-1}}{\Delta t} + (1/2 - \phi)O(\Delta t) + O(\Delta t^2). \tag{6.111}
$$

where the switch can be set based on the time accuracy required:

$$\phi = 0 \rightarrow O(\Delta t)$$

$$\phi = 1/2 \rightarrow O(\Delta t^2)$$

This is done primarily to simplify the programming of the solution algorithm, since only one time integration equation is coded, but two time integration methods are available with an easily programmable switch.

6.12.1.4 RUNGE-KUTTA TIME INTEGRATION

Another method of time integration commonly used for higher orders of accuracy is the Runge-Kutta technique.[30] There are second-order approaches that use the Runge-Kutta approach, but we will demonstrate the fourth-order Runge-Kutta technique using the two dimensional conservation of mass equation, Equation 3.4:

$$\frac{\partial \rho}{\partial t} + \frac{\partial \rho u}{\partial x} + \frac{\partial \rho v}{\partial y} = 0. \tag{3.4}$$

We can rewrite this equation using partial derivative shorthand, and rearrange the equation to define a right-hand side term

$$\rho_t = -\left(\rho u_x + \rho v_y\right) \equiv R(\rho u, \rho v). \tag{6.112}$$

Using this notation we can express the fourth-order Runge-Kutta process in four steps:

Step 1: $\rho^{(1)} = \rho^n + \dfrac{\Delta t}{2} R^n$, where $R^n = R\left(\rho u^n, \rho v^n\right)$

Step 2: $\rho^{(2)} = \rho^n + \dfrac{\Delta t}{2} R^{(1)}$

Step 3: $\rho^{(3)} = \rho^n + \dfrac{\Delta t}{2} R^{(2)}$

Step 4: $\rho^{n+1} = \rho^n + \dfrac{\Delta t}{6}\left(R^n + R^{(1)} + R^{(2)} + R^{(3)}\right) + O(\Delta t^4, \Delta x^2, \Delta y^2)$

which assumes that second-order spatial derivatives are used. This process needs to be applied to the full set of governing equations (mass, momentum, energy) to determine the variables at each intermediate stage. Again, the increased time accuracy is obtained but, in this case, at added computational time and memory storage.

Table 6.2 Parameters for Implicit Time Integration Methods

Method	ϕ	θ	Error
Euler implicit	0	1	$O(\Delta t)$
Trapezoidal	0	½	$O(\Delta t)^2$
Three-point backward	½	1	$O(\Delta t)^2$

6.12.2 Implicit Time Integration

Implicit time integration is undertaken for similar reasons that implicit methods were used for CFD solutions in general: larger, more complex problems can be handled more efficiently and therefore faster. A general implicit time integration approach was developed by Beam and Warming[31] and can be written for the mass conservation equation, Equation 3.4, in a similar combined fashion as was used for the explicit approach in Equation 6.111:

$$\rho_{i,j}^{n+1} = \rho_{i,j}^n + \frac{\phi}{1+\phi}\left(\rho_{i,j}^n - \rho_{i,j}^{n-1}\right) + \frac{\theta\Delta t}{1+\phi}\left(\frac{\partial\rho_{i,j}^{n+1}}{\partial t} - \frac{\partial\rho_{i,j}^n}{\partial t}\right)$$
$$+ \frac{\Delta t}{1+\phi}\frac{\partial\rho_{i,j}^n}{\partial t} + O\left(\left(\theta - \phi - \frac{1}{2}\right)\Delta t^2 + \Delta t^3\right)$$

(6.113)

where

$$\frac{\partial\rho_{i,j}^n}{\partial t} = -\frac{(\rho u)_{i+1,j}^n - (\rho u)_{i-1,j}^n}{2\Delta x} + \frac{(\rho u)_{i,j+1}^n - (\rho u)_{i,j-1}^n}{2\Delta y}$$

(6.114)

$$\frac{\partial\rho_{i,j}^{n+1}}{\partial t} = -\frac{(\rho u)_{i+1,j}^{n+1} - (\rho u)_{i-1,j}^{n+1}}{2\Delta x} + \frac{(\rho u)_{i,j+1}^{n+1} - (\rho u)_{i,j-1}^{n+1}}{2\Delta y}.$$

(6.115)

The parameters θ and ϕ are chosen to create either first- or second-order accurate time integration approaches, as shown in Table 6.2.

The Euler implicit scheme is given by:

$$\rho_{i,j}^{n+1} = \rho_{i,j}^n - \Delta t\left(\frac{(\rho u)_{i+1,j}^{n+1} - (\rho u)_{i-1,j}^{n+1}}{2\Delta x} + \frac{(\rho u)_{i,j+1}^{n+1} - (\rho u)_{i,j-1}^{n+1}}{2\Delta y}\right)$$

(6.116)

which looks very similar to the explicit method from Equation 6.106, except that the variables on the right-hand side are evaluated at the next time level. That is what makes this an implicit method, since all of these "future" values will have to be solved for using a system of simultaneous equations.

6.12.3 Subiterations

Sometimes it is convenient to put an outer loop around the time integration scheme to eliminate errors caused by convergence acceleration schemes used in the flow solver. This allows multidisciplinary applications (such as aeroelasticity) to be used with the flow solver without completely rewriting the code. Look at a two-dimensional representation of the Euler equations (Equation 3.63), which neglects the viscous terms:

$$\frac{\partial Q}{\partial t} + \frac{\partial F}{\partial x} + \frac{\partial G}{\partial y} = 0 \tag{3.63}$$

where

$$Q = \begin{bmatrix} \rho \\ \rho u \\ \rho v \\ \rho e \end{bmatrix} \tag{3.34}$$

and F and G are the fluxes of Q. We can use discrete operators to obtain

$$\frac{(1+\phi)Q^{n+1} - (1+2\phi)Q^n + \phi Q^{n-1}}{\Delta t} + F_x^{n+1} + G_y^{n+1} = 0 \tag{6.117}$$

where F_x and G_x denote spatial differences in the x-direction. Now replace the unknown time level $n+1$ with a subiterative value $p+1$ such that, as $p \to \infty$, $Q^{p+1} \to Q^{n+1}$. We can also perform a Taylor's series expansion about the pth subiterate of the fluxes to obtain

$$F^{p+1} = F^p + \Delta Q \frac{\partial F^p}{\partial U} + O(\Delta Q^2) \tag{6.118}$$

$$G^{p+1} = G^p + \Delta Q \frac{\partial G^p}{\partial U} + O(\Delta Q^2) \tag{6.119}$$

where $\Delta Q^p = Q^{p+1} - Q^p$. These relations (Equations 6.118 and 6.119) can be combined with Equation 6.117 to obtain

$$\begin{aligned} &\frac{(1+\phi)Q^{p+1} - (1+2\phi)Q^n + \phi Q^{n-1}}{\Delta t} \\ &+ \frac{\partial F^p}{\partial x} + \frac{\partial}{\partial x}\left[\frac{\partial F^p}{\partial U}\right]\Delta Q^p + \frac{\partial G^p}{\partial y} + \frac{\partial}{\partial y}\left[\frac{\partial G^p}{\partial U}\right]\Delta Q^p = 0. \end{aligned} \tag{6.120}$$

Now add and subtract Q^p to obtain

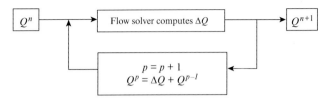

Block diagram for Newton subiteration concept.

$$(1+\phi)Q^{p+1} - (1+\phi)Q^{p} + (1+\phi)Q^{p} - (1+2\phi)Q^{n} + \phi Q^{n-1}$$
$$\overline{\Delta t}$$

$$+ \frac{\partial F^{p}}{\partial x} + \frac{\partial}{\partial x}\left[\frac{\partial F^{p}}{\partial Q}\right]\Delta Q^{p} + \frac{\partial G^{p}}{\partial y} + \frac{\partial}{\partial y}\left[\frac{\partial G^{p}}{\partial Q}\right]\Delta Q^{p} = 0 \qquad (6.121)$$

and rearrange as

$$\left[\frac{(1+\phi)}{\Delta t}I + \frac{\partial}{\partial x}\frac{\partial F^{p}}{\partial Q} + \frac{\partial}{\partial y}\frac{\partial G^{p}}{\partial Q}\right]\Delta Q^{p} =$$

$$-\left[\frac{(1+\phi)Q^{p+1} - (1+2\phi)Q^{n} + \phi Q^{n-1}}{\Delta t} + \frac{\partial F^{p}}{\partial x} + \frac{\partial G^{p}}{\partial y}\right] \qquad (6.122)$$

where the left-hand side of Eqn. 6.122 is the "solver" and the right-hand side is the governing equation.

As $p \to \infty$ then $\Delta Q^{p} \to 0$ and the right-hand side (RHS) goes to zero, which means the new time level has been obtained. This is an "implicit formulation," meaning the spatial fluxes are evaluated at the new time level (not lagged in time). Fig. 6.18 shows this subiteration method expressed in block diagram form.

6.12.4 Solution Method for Time Integration

Since there are basically only steady and unsteady flows, there are two basic ways to proceed with the time integration for a CFD solution. These two approaches may seem quite basic, but knowing how and when to integrate in time forms an important capability for obtaining good computational simulations. The two approaches to time integration are for:

Steady State – advance the initial conditions to the steady state using a time integration scheme. Time order is not important and sometimes we even use local time steps to accelerate to the steady-state solution. We must use a time step small enough to keep the algorithm numerically stable, typically by defining the *CFL number*, where

$$CFL = c\frac{\Delta t}{\Delta x}$$

Time Accurate – advance the initial conditions forward in time using a suitably accurate time integration scheme. Typically, second-order (or higher) accuracy is required. We must use a Δt that is small enough to "capture" the unsteady behavior, as well as maintain stability.

6.13 Boundary Conditions

So far we have obtained expressions for interior points on the mesh. However, the actual geometry of the flowfield we wish to analyze is introduced through the boundary conditions. Although there are many possible types of boundary conditions in modern codes (see the next concept box on the topic), we will focus only on farfield and solid wall (or nearfield) boundary conditions here.

6.13.1 Farfield Boundary Conditions

Consider the flow over a symmetric airfoil at zero angle of attack, as shown in Fig. 6.19. Because there is no lift, symmetry allows us to solve only the top half of the region. The boundary conditions representing the wall are discussed in more detail in Section 6.13.2. The specification of the conditions at the outer edges of the solution domain (known as farfield boundary conditions) are equally important. While the tasks of the boundary conditions at the extents of the domain are simple to describe, a valid implementation of the conditions in the code is much more difficult than you probably thought.

The farfield boundary condition has two main responsibilities. First, the freestream flow conditions must be established at those areas of the boundary where flow is entering the solution domain (similar to the inflow conditions at the front of the test section of a wind tunnel). This includes not only establishing the correct flow velocities but also the direction of the freestream flow as dictated by the desired angle of attack and angle of sideslip. Second, the farfield boundary must properly represent an infinite domain even though the boundaries are clearly a finite distance from the body. This means that any pressure waves or other flow disturbances must not be reflected back into the solution domain, most assuredly negatively impacting the quality of the simulation in the process. Moreover, it is desirable to have the farfield boundaries as close to the solid body as possible in order to reduce the computational expense of the simulation. An example of how to do this is shown in Fig. 6.20.

In Chapter 3 we discussed the importance of equation type on the method of solution. The equation type also has a big impact on applying the proper boundary conditions at the farfield. For example, in Chapter 3 we determined that in supersonic flow information travels downstream. Therefore at

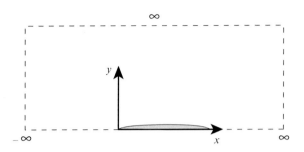

Figure 6.19   Example of boundary condition surfaces requiring consideration.

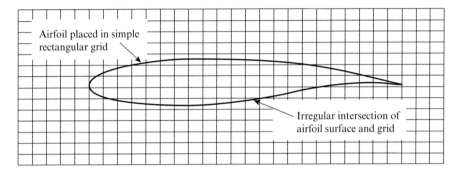

Figure 6.20   Surface passing through a general grid.

the inflow boundary of a supersonic simulation we need to specify all flow variables. In contrast to that, at the outflow of a supersonic flow we have to use information from inside the domain to set the flow variables on the boundary through extrapolation or other more sophisticated relationships.

Mathematically we can determine where to obtain boundary information (at the boundary or from within the flowfield domain) by looking at the characteristics of the governing equations. These characteristics can be determined by looking at the Eigenvalues of the matrix formed by the partial derivatives of the convective fluxes with respect to the flowfield variables. The sign of the eigenvalues determine the direction the information propagates and guides us in developing a consistent set of boundary conditions for all flow variables. Ignoring the concept of proper information propagation at the farfield boundary can have large consequences on simulation accuracy, stability, and how far away from the body we need to place the farfield boundaries. It should not be surprising that for inflow and outflow at supersonic speeds we find that all characteristics have the same sign and point into the domain at the inflow boundary and out of the domain at the outflow boundary consistent with our previous argument.

Characteristic analysis of subsonic flow produce one characteristic pointing upstream and the rest pointing downstream resulting in a mixed set of boundary conditions. At the inflow we can specify all but one of the flow variables and determine the last one from inside the domain. At the outflow, the opposite is true and we specify one condition (usually pressure) and extrapolate the rest from the interior of the domain. This requirement for a mixed set of boundary conditions is consistent with an elliptic set of equations having information propagation in all directions.

In actual calculations with complex domains and the possibility of arbitrary mesh motion we do not always know whether the flow is inflow or outflow or even supersonic or subsonic. For this reason, characteristic boundary conditions were developed that determine the direction of information propagation and modify the boundary conditions appropriately.

6.13.2 Solid Wall Boundary Conditions

The purpose of the solid wall (or nearfield) boundary conditions is to impose the presence of the geometry under study onto the solution domain. In reality, there is no such thing as a curved surface in a discretized geometry – curved surfaces are really just a collection of multiple straight-line segments (or planar elements in three dimensions). The degree to which the discretized geometry represents the actual geometry has a direct impact on how well the resulting flow solution will represent reality and thus how it will compare to any experimental data. Clearly, no mass is allowed to pass through the solid wall, and the fluid velocity tangential to the wall is a function of the wall geometry, the particular type of wall boundary condition, and any non-inertial motion of the wall. For viscous flows, heating effects must also be taken into account using known gradient information or by making simplifying assumptions.

There are several ways to approach the satisfaction of boundary conditions on the surface. These wall boundary conditions can be handled in at least two ways:

1. Use a standard grid and allow the surface to intersect grid lines in an irregular manner. Then, solve the equations with boundary conditions enforced between node points (Fig. 6.20 illustrates this approach). In the early days of CFD methodology development, this approach was not found to work well, and the approach discussed next was developed to handle these cases. However, using the finite volume method, an approach to treat boundary conditions imposed in this manner was successfully developed.

2. The most popular approach to enforcing surface boundary conditions is to use a coordinate system constructed such that the surface of the body is a coordinate surface in a curvilinear coordinate system; an example of this approach is shown in Fig. 6.21. This is currently the method of choice and by far the most popular approach employed in CFD, and it does work quite well. However, it complicates the problem formulation: to use

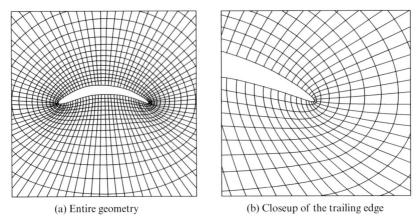

(a) Entire geometry   (b) Closeup of the trailing edge

Figure 6.21 Body conforming grid for easy application of BCs on curved surfaces (Ref. 34).

this approach, grid generation becomes even more crucial to the CFD process and leads to the need for more powerful grid generation codes, which will be discussed in Chapter 7.

Solid wall boundary conditions may be grouped into two main categories: slip and no-slip. (As discussed in the included concept box, symmetry plane boundaries are numerically identical to slip wall boundaries.) As mentioned earlier, the specification of velocity and temperature are the two main issues with solid wall boundary conditions. As such, the slip and no-slip walls typically only differ in how these two issues are handled.

Slip wall boundary conditions, as the name implies, are used to represent solid walls where flow is allowed to move tangentially along the surface but not normal to the surface since the wall is solid. The wall is effectively a streamline of the flowfield since the flow is allowed to slip along the wall, and no viscous effects are assumed to exist (e.g., no heating). The velocity at the boundary is specified by enforcing Equation 6.123:

$$\vec{V}_{wall} \cdot \vec{n}_{wall} = 0$$

(6.123)

In other words, the fluid is allowed to move parallel to the wall without restriction but may not move normal to the wall at the boundary since it is solid.

The more physically accurate wall boundary condition is the no-slip wall. As the name implies, the fluid at the wall is not allowed to move relative to the wall. That is, the fluid velocities at the wall are prescribed to be zero for non-moving walls. For moving walls (e.g., to model a car moving over a roadway), the no-slip condition is realized by specifying the fluid velocity be equal to the boundary velocity. Since the flow is not allowed to slip along the wall, the effects of viscosity must be taken into account. It is common to assume the wall is adiabatic (no temperature gradient or heat transfer) in which case the wall temperature is directly computed assuming constant internal energy relative to the boundary cell. For a constant-temperature, or isothermal, wall type, the wall temperature is specified directly. Finally, more general scenarios such as prescribed rates of heat transfer may be handled by incorporating the thermodynamic laws into the temperature calculation.

For both slip and no-slip wall types, the method by which pressure is determined can have drastic accuracy and stability effects on the solution. Often times, a zero static pressure gradient is assumed, and the wall pressure is simply taken as the computed pressure in the boundary element. However, the local element topology as well as geometry sometime invalidates this assumption, so it is common to utilize the so-called normal momentum relation introduced by Rizzi[32] in order to solve for the wall pressure:

$$\rho \vec{V}_{wall} \cdot (\vec{V}_{wall} \cdot \vec{\nabla}) \vec{n}_{wall} = \vec{n}_{wall} \vec{\nabla} p \tag{6.124}$$

This relation effectively balances the component of momentum normal to the wall with the pressure gradient at the wall, leading to a more accurate calculation for the pressure at the wall.

Computational Aerodynamics Concept Box

Common Boundary Conditions for CFD

One aspect of being a good practitioner of CA in general and CFD in particular is proper choice of boundary conditions (BCs). CFD codes in use today have a large variety of boundary conditions to choose from and this concept box seeks to define the most common BCs the user will encounter. However, this is not an exhaustive list by any means, and the mathematical descriptions of these BCs are left to the relevant references (Refs. 1–4). There are four main categories of BCs: Farfield, Internal, Solid Wall, and Source/Sink.

Farfield – CFD simulations of airfoils, wings, cars, submarines, and aircraft carriers all require specifying the conditions very far away from the body of interest where the flowfield is either essentially undisturbed or the conditions are completely known. These BCs are of the Farfield type and are dependent on whether the flow is supersonic or subsonic at the boundaries, since the equations change type with Mach number (see Chapter 3). The characteristics of the governing equation determine how many conditions can be specified at the boundary or must be determined from the domain interior without causing reflections of error back into the domain.

- Supersonic Inflow: all flow variables are specified at the boundary
- Supersonic Outflow: all flow variables are extrapolated from the interior of the domain
- Subsonic Inflow: specify all but one variable and extrapolate one from the interior
- Subsonic Outflow: specify one variable and extrapolate all of the others

- Characteristic: compute the characteristics "on the fly" to determine if flow is in or out
- Profile of flow conditions as a function of space and/or time: apply known conditions at the inflow such as a boundary layer profile or incoming vortex (important for matching wind tunnel data or flight in another aircraft's wake)

Internal – Sometimes boundary conditions are specified on the interior of the domain as a matter of efficiency, to aid in implementing the code, or to model a special type of flow.

- Linear, circumferential periodic: aids in efficiency by copying variables from one side of the domain to the other so the flowfield "feels" an infinite domain or axisymmetry
- Block interfaces, sliding interfaces: aids in decomposing the simulation with fictitious boundaries onto many CPUs for parallel processing or allows two bodies to move relative to one another, sliding along the fictitious boundary (e.g., engine rotor/stator)
- Actuator disks, momentum addition: allows the average effect of rotors (e.g. airplane propellers or helicopter blades) to be included in the interior of the domain

Solid Walls – CFD simulations are typically about computing flow around a body of interest. The solid wall boundaries contain all of the geometric shape information and therefore are critical to the accuracy of the simulation.

- No-slip wall: a solid wall with zero velocity relative to the wall (contravariant velocity for moving walls); temperature may be directly specified (isothermal), a zero temperature gradient may be used to determine the wall temperature (adiabatic), or a known heat flux may be used to determine the wall temperature; pressure is generally determined by either assuming a zero gradient and using the boundary cell pressure or by balancing the normal component of momentum with the pressure gradient in the normal direction and using the result to specify the wall pressure
- Slip wall: a solid wall that allows the flow to "slip" along it consistent with the inviscid assumption (e.g., Euler equations) by specifying the contravariant velocity normal to the wall as zero; pressure may be computed in the same way as with no-slip walls
- Symmetry plane: the symmetry plane condition allows only half of the domain to be computed if the geometry actually has symmetry about the plane of the boundary, mathematically this is the same as the slip wall condition

Source/Sink – Many CFD simulations require BCs that cause addition or subtraction of mass at the boundary referred to as source or sink BCs. The most prevalent use of these BCs is to simulate the flow leaving the domain into an engine compressor face and coming back into the domain after the turbine or converging-diverging nozzle. Another important use of these conditions is for flow control devices like leading edge suction/blowing (see Chapter 10 for an example).

- Maintain specified static/total properties: specifying static or total properties at a boundary results in flow either leaving the domain or entering the domain (i.e. mass flow subtraction or addition)
- Specified mass flow: this BC specifies directly the mass flow addition (source) or subtraction (sink)

6.13.3 Numerical Representation of Boundary Conditions

The manner in which boundary conditions are actually represented in the algorithm is highly dependent on the domain topology (structured, unstructured, Cartesian), the location of the solution points (cell-centered, vertex-centered), and the algorithm itself (finite difference, finite volume, etc.). Recall that there are normally two types of boundary conditions associated with a domain boundary: (1) the Dirichlet problem, where the independent variable values are specified on the boundary; and (2) the Neumann problem, where the derivative of the independent variables is specified. In the case of the Neumann problem, the particular boundary condition formulation may make use of the derivative value directly, or the derivative may be used to determine (interpolate) variable values (i.e. convert to an equivalent Dirichlet problem).

One of the most common ways to numerically represent boundary conditions is through the use of a so-called dummy row or ghost cells. In this case, a virtual solution point is defined inside the physical boundary (i.e., just outside of the discretized solution domain), and values are specified at the point such that any numerical scheme may solve up to the last interior domain point and simply access the neighboring virtual data to realize the effect of the boundary data. As an example (following the approach of Moran[33]), consider a case where the normal velocity, v, is set to zero at the outer boundary. The boundary is at grid line $j = NY$. Assume that another row is added at $j = NY + 1$, as indicated in Fig. 6.22.

The required boundary condition at $j = NY$ is:

$$\frac{\partial \phi}{\partial n} = 0 = \frac{\phi_{i,NY+1} - \phi_{i,NY-1}}{Y_{NY+1} - Y_{NY-1}} + O(\Delta Y)^2 \qquad (6.125)$$

and to ensure that the boundary condition is satisfied, simply define:

$$\phi_{i,NY+1} \equiv \phi_{i,NY-1}. \qquad (6.126)$$

The governing equations are then solved up to Y_{NY}, and whenever the value of ϕ at $NY+1$ is needed, simply use the value at $NY-1$. This approach can also be used at a solid surface.

Imposition of boundary conditions is sometimes more difficult than the analysis given here suggests. Specifically, both the surface and farfield boundary conditions for the pressure in the Navier-Stokes and Euler equations can be quite tricky. Care should be taken to implement boundary conditions correctly, and code users should understand what boundary conditions are implemented in codes that they use.

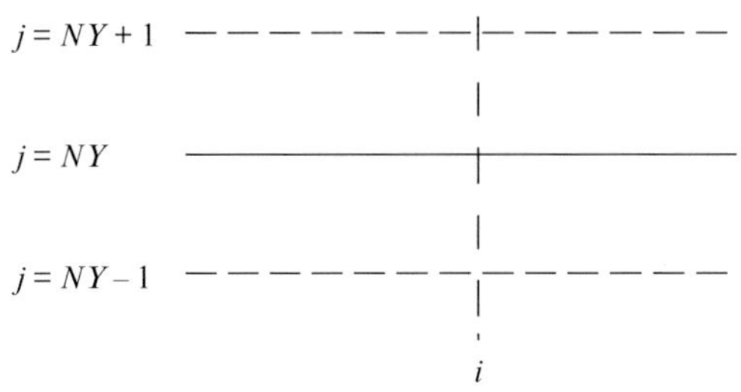

$j = NY + 1$ — — — — — — — |— — — — — —

$j = NY$

$j = NY - 1$ — — — — — — — |— — — — —

i

Figure 6.22   Boundary condition at farfield.

6.14 Solution of Algebraic Equations

We now know how to write down a representation of our partial differential equations at each grid point or for each cell. The next step is to solve the resulting system of equations. Recall that for finite differences, we have one algebraic equation for each grid point, and that the system of algebraic equations may, or may not, be linear. If they are nonlinear, the usual approach is to form an approximate linear system and then solve the system iteratively to obtain the solution of the original nonlinear system. The order of accuracy of the method being used dictates the number of grid points required to obtain the solution. Previously, we assumed that linear equation solution algorithms were available (as discussed in Chapter 3); however, the development of CFD methods requires knowledge of the forms of algebraic systems of equations being solved.

Recall that linear algebraic equations can be written in the standard form:

$$[A][x] = [b] \tag{6.127}$$

where $[A]$ is the coefficient matrix, $[x]$ is the vector of unknowns, and $[b]$ is the vector of constants (see Equation 6.81, for example). For an inviscid two-dimensional solution, a grid of 100×30 is typical. This is $N = 3000$ grid points and results in a matrix 3000×3000. In three dimensional inviscid flow, $250,000 \sim 300,000$ grid points are common, 500,000 points are not uncommon for some geometries, and millions of grid points are often required for complicated geometries; viscous flows require even more grid points. Clearly, you cannot expect to use classical direct linear equation solvers for systems of this size (for example, Cramer's rule, which requires $O(N^3)$ calculations, would take on the order of $3000^3 = 27 \times 10^6$ calculations for each solution step!).

Standard classification of algebraic equations depends on the characteristics of the elements in the matrix $[A]$. If $[A]$:

1. contains few or no zero coefficients, it is called dense,
2. contains many zero coefficients, it is called sparse,

3. contains many zero coefficients *and* the non-zero coefficients are close to the main diagonal; the $[A]$ matrix is called sparse and banded.

We will evaluate various approaches to solve each of these types of matrices.

6.14.1 Dense Matrix

For a dense matrix, direct methods are appropriate. Gaussian elimination is an example of the standard approach to solve these systems; LU decomposition[35] is another example of a standard method for solution of a dense matrix. These methods are not good for large matrices (> 200–400 equations), since the run time becomes long, and the results may be susceptible to roundoff error.

6.14.2 Sparse and Banded Matrix

Special forms of Gaussian elimination are available in many cases for sparse and banded matrices. The most famous banded matrix solution applies to the tridiagonal systems:

$$
\begin{bmatrix}
b_1 & c_1 & 0 & \cdots & 0 & 0 & 0 \\
a_2 & b_2 & c_2 & 0 & \cdots & 0 & 0 \\
0 & \ddots & \ddots & \ddots & 0 & \cdots & 0 \\
0 & 0 & a_i & b_i & c_i & 0 & 0 \\
0 & \cdots & 0 & \ddots & \ddots & \ddots & 0 \\
0 & 0 & \cdots & 0 & a_{N-1} & b_{N-1} & c_{N-1} \\
0 & 0 & 0 & \cdots & 0 & a_N & b_N
\end{bmatrix}
\begin{bmatrix}
x_1 \\
x_2 \\
\vdots \\
x_i \\
\vdots \\
x_{N-1} \\
x_N
\end{bmatrix}
=
\begin{bmatrix}
d_1 \\
d_2 \\
\vdots \\
d_i \\
\vdots \\
d_{N-1} \\
d_N
\end{bmatrix}
\qquad (6.128)
$$

The most common algorithm used to solve Equation 6.128 is known as the Thomas algorithm.[36] This algorithm is very easy to implement and is widely used (the Thomas algorithm is given in detail in the Tridiagonal Systems Concept Box, and a sample routine, *TRIDAG*, is described in Appendix D).

Computational Aerodynamics Concept Box

Solution of Tridiagonal Systems of Equations

The Thomas algorithm is a special form of Gaussian elimination that can be used to solve tridiagonal systems of equations. When the matrix is tridiagonal, the solution can be obtained in $O(N)$ operations, instead of $O(N^3/3)$. The form of the equation is:

$$a_i x_{i-1} + b_i x_i + c_i x_{i+1} = d_i \qquad i = 1, \cdots, N$$

where a_1 and c_n are zero. The solution algorithm starts with $k = 2,....,N$:

$$m = \frac{a_k}{b_{k-1}}$$
$$b_k = b_k - mc_{k-1}$$
$$d_k = d_k - md_{k-1}.$$

Then:

$$x_n = \frac{d_n}{b_n}$$

and finally, for $k = N - 1,...1$:

$$x_k = \frac{d_k - c_k x_{k+1}}{b_k}.$$

In CFD methods, this algorithm is usually coded directly into the solution procedure, unless machine optimized subroutines are employed on a specific computer.

6.14.3 General Sparse Matrix

These matrices are best treated with iterative methods. In this approach an initial estimate of the solution is specified (often simply 0), and the solution is then obtained by repeatedly updating the values of the solution vector until the equations are solved. This is also a natural method for solving nonlinear algebraic equations, where the equations are written in the linear equation form, and the coefficients of the $[A]$ matrix are changed as the solution develops during the iteration. Many methods are available, including Jacobi's method and the Gauss-Seidel method (discussed in the next section).

There is one basic requirement for iterative solutions to converge. The elements on the diagonal of the matrix should be large relative to the values off of the diagonal. The condition can be given mathematically as:

$$|a_{ii}| \geq \sum_{\substack{j=1 \\ j \neq i}}^{N} |a_{ij}| \tag{6.129}$$

and for at least one row:

$$|a_{ii}| > \sum_{\substack{j=1 \\ j \neq i}}^{N} |a_{ij}|. \tag{6.130}$$

A matrix that satisfies this condition is *diagonally dominant*, and, for an iterative method to converge, the matrix must be diagonally dominant. One example from aerodynamics of a matrix that arises that is not diagonally dominant is the matrix obtained in the monoplane equation formulation for the solution of the lifting-line theory problem.[37]

6.14.4 Point Jacobi and Point Gauss-Seidel

One class of iterative solution methods widely used in CFD is "relaxation." As an example, consider the solution of Laplace's equation using the discretized form from Equation 6.87. The iteration proceeds by solving the equation at each grid point i,j at an iteration $(n+1)$ using values found at iteration (n). Thus the solution at iteration $(n+1)$ is found from:

$$\phi_{i,j}^{n+1} = \frac{1}{4}\left[\phi_{i+1,j}^{n} + \phi_{i-1,j}^{n} + \phi_{i,j+1}^{n} + \phi_{i,j-1}^{n}\right]. \qquad (6.131)$$

The values of ϕ are computed repeatedly until they are no longer changing. The "relaxation" of the values of ϕ to final converged values is roughly analogous to determining the solution for an unsteady flow approaching a final steady-state value, where the iteration cycle is identified as a time-like step: this is an important analogy. Unfortunately, the idea of "iterating until the values stop changing" as an indication of convergence is not necessarily good enough. Instead, we must check to see if the finite difference representation of the partial differential equation using the current values of ϕ actually satisfies the partial differential equation. In this case, the value of the equation should be zero, and the actual value of the finite difference representation is known as the *residual*. When the residual is zero, the solution has converged. This is the value that should be monitored during the iterative process. Generally, as done in the program *THINFOIL* (described later) the maximum residual and its location in the grid and the average residual are computed and saved during the iterative process to examine the convergence history. Note that this method uses all old values of ϕ to get the new value of ϕ. This approach is known as *point Jacobi* iteration. This scheme requires that you save all old values of the array as well as the new values, and the procedure converges only very slowly to the final solution.

6.14.5 Gauss-Seidel and Successive Over-relaxation

A more natural approach to obtaining the solution is to use new estimates of the solution as soon as they become available. Figure 6.23 shows how this is

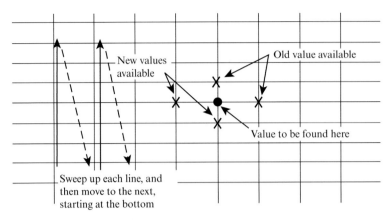

New values available

Old value available

Value to be found here

Sweep up each line, and then move to the next, starting at the bottom

Figure 6.23   Grid sweep approach to implement the Gauss-Seidel solution iteration scheme.

done using a simply programmed systematic sweep of the grid. With a conventional sweep of the grid this becomes:

$$\phi_{i,j}^{n+1} = \frac{1}{4} \left[\phi_{i+1,j}^{n} + \phi_{i-1,j}^{n+1} + \phi_{i,j+1}^{n} + \phi_{i,j-1}^{n+1} \right]. \tag{6.132}$$

This scheme is called the point Gauss-Seidel iteration. It also eliminates the need to store all the old iteration values as well as all the new iteration results, which was required with the point Jacobi method.

The point Gauss-Seidel iteration procedure also converges somewhat slowly. One method of speeding up the convergence is to make the change to the value larger than the change indicated by the normal Gauss-Seidel iteration. Since the methods that have been described are known as relaxation methods, the idea of increasing the change is known as successive over-relaxation (SOR). This is implemented by defining an intermediate value:

$$\hat{\phi}_{i,j}^{n+1} = \frac{1}{4} \left[\phi_{i+1,j}^{n} + \phi_{i-1,j}^{n+1} + \phi_{i,j+1}^{n} + \phi_{i,j-1}^{n+1} \right] \tag{6.133}$$

and then obtaining the new value as:

$$\phi_{i,j}^{n+1} = \phi_{i,j}^{n} + \omega \left(\hat{\phi}_{i,j}^{n+1} - \phi_{i,j}^{n} \right). \tag{6.134}$$

The parameter ω is a relaxation parameter. If it is unity, the basic Gauss-Seidel method is recovered. How large can we make the relaxation parameter and still obtain good results? For most model problems, a stability analysis indicates that $\omega < 2$ is required to obtain a converging iteration. The best value of ω depends on the grid and the actual equation being solved and is usually determined through numerical experimentation. Figure 6.24 presents

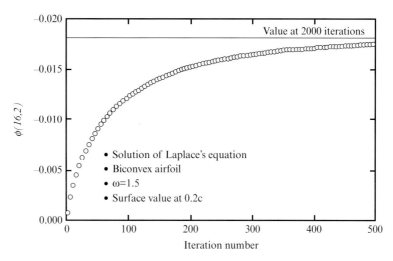

Figure 6.24   Typical variation of ϕ during solution iteration.

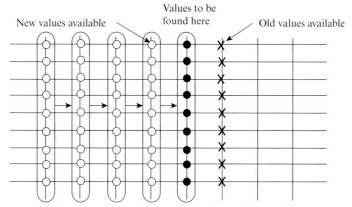

Starting upstream, move downstream, solving a line at a time.

Figure 6.25   Solution approach for SLOR.

an example of the manner in which the solution evolves with iterations; the value of ϕ after 2,000 iterations is approached very gradually. The figure also illustrates the time-like nature of the iteration.

6.14.6 Successive Line Over-Relaxation

Another way to speed up the iteration is to sweep the flowfield a "line" at a time rather than a point at a time. Applying over-relaxation to this process, the so-called successive line over-relaxation (SLOR) process is obtained. In this method, a system of equations must be solved at each line (Fig. 6.25 illustrates this approach). The method is formulated so that the system of equations is tridiagonal, and the solution is obtained very efficiently. This

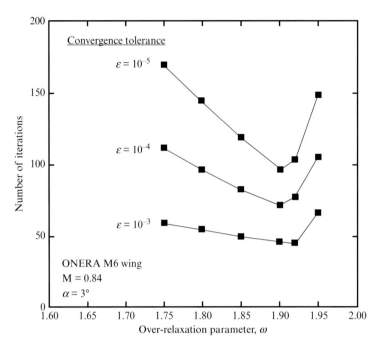

Figure 6.26 Effect of the value of ω on the number of iterations required to achieve various levels of convergence (Ref. 38).

approach provides a means of spreading the information from new values more quickly than the point-by-point sweep of the flowfield. However, all of these approaches result in a very slow approach to the final value during the iterations.

The effect of the value of the over-relaxation parameter on the convergence rate is shown in Figure 6.26. Here, the convergence level is compared for various values of ω. Notice that as convergence requirements are increased, the choice of ω becomes much more important. Unfortunately, the choice of ω may not only be dependent on the particular numerical method but also on the particular problem being solved.

Mathematically, the convergence rate of an iterative process depends on the value of the so-called *spectral radius* of the matrix relating the value of the unknowns at one iteration to the values of the unknowns at the previous iteration. The spectral radius is the absolute value of the largest Eigenvalue of the matrix.[5] The spectral radius must be less than one for the iterative process to converge, and, the smaller the value of the spectral radius, the faster the convergence.

6.14.7 Approximate Factorization

Another way to spread the information rapidly is to alternately sweep in both the x-and y-directions. This provides a means of obtaining the final

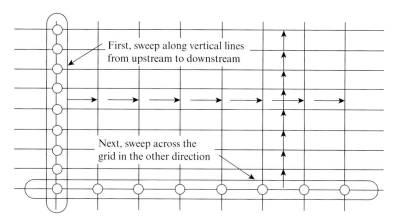

First, sweep along vertical lines from upstream to downstream

Next, sweep across the grid in the other direction

Figure 6.27   ADI scheme solution approach.

answers even more quickly and is known as an alternating direction implicit (ADI) method. Figure 6.27 illustrates the modification to the SLOR method that is used to implement an ADI scheme. Several different methods of carrying out the details of this iteration are available. The traditional approach for linear equations is known as the Peaceman-Rachford method and is described in standard textbooks (e.g., Ames[8] or Isaacson and Keller[39]). This approach is also known as an approximate factorization scheme (AF; it is typically referred to as AF1 because of the particular approach to the factorization of the operator). A discussion of ADI including a computer program is given in the *Numerical Recipes* book.[40]

Another approach has been found to be more robust for nonlinear partial differential equations, including the case of mixed sub- and supersonic flow. In this case the time-like nature of the approach to a final value is used explicitly to develop a robust and rapidly converging iteration scheme; this scheme is known as AF2. This method was first proposed for steady flows by Ballhaus et al.,[41] and Catherall[42] provided a theoretical foundation and results from numerical experiments. A key aspect of ADI or any AF scheme is the use of a sequence of relaxation parameters rather than a single value, as employed in the SOR and SLOR methods. Typically, the sequence repeats each eight to eleven iterations.

Holst[34] has given an excellent review and comparison of these methods. Figure 6.28 shows how the different methods use progressively "better" information at a point to find the solution with the fewest possible iterations. The advantage is shown graphically in Fig. 6.29 and is tabulated in Table 6.3. Program *THINFOIL*, described in Section 6.15, uses these methods, and Appendix D contains a description of the theoretical implementation of these methods.

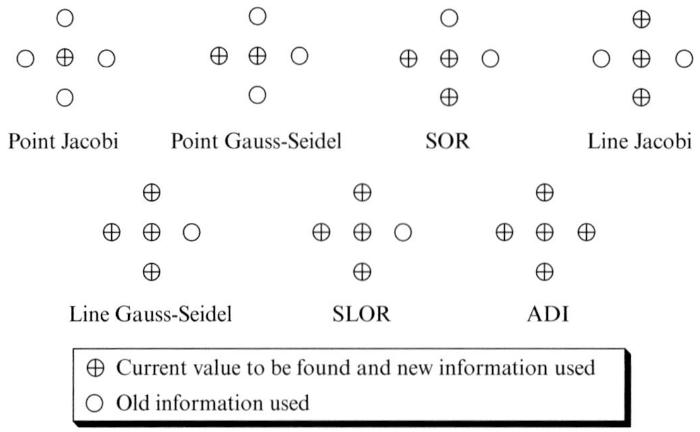

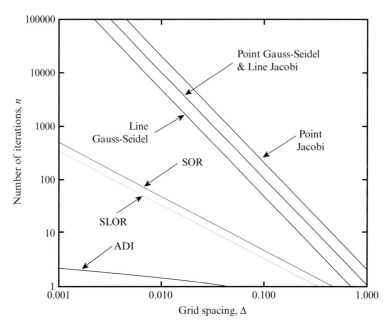

Figure 6.28 Stencil of information (Ref. 34).

Figure 6.29 Comparison of convergence rates of various relaxation schemes (Ref. 34). This is the number of iterations estimated to be required to reduce the residual by one order of magnitude.

Table 6.3 Convergence Rate Estimates for Various Relaxation Schemes (Ref. 34)

Algorithm	Number of iterations required for a one-order-of-magnitude reduction in error
Point-Jacobi	$2/\Delta^2$
Point-Gauss-Seidel	$1/\Delta^2$
SOR	$1/(2\Delta)$
Line-Jacobi	$1/\Delta^2$
Line-Gauss-Seidel	$1/(2\Delta^2)$
SLOR	$1/(2\sqrt{2}\Delta)$
ADI	$-\log(\Delta/2)/1.55$

Profiles in Computational Aerodynamics: Antony Jameson

Antony Jameson has authored or co-authored 400 scientific papers in a wide range of subject areas, including both control theory and aerodynamics, and is the principal developer of the well-known series of *FLO* and *SYN* codes, which have been used throughout the aerospace industry. The son of Brigadier Oscar Jameson and Olive Turney, he was born in Gillingham, Kent in 1934. Much of his early childhood was spent in India, where his father was stationed as a British Army officer. He first attended school at St. Edwards, Simla. Subsequently he was educated at Mowden School and Winchester College. He served as a lieutenant in the British Army in 1953–1955 and was sent to Malaya. After the army, he worked in the compressor design section of Bristol Aero-Engines in the summer of 1955, before studying engineering at Trinity Hall, Cambridge University, graduating with first class honors in 1958. Subsequently he stayed on at Cambridge to obtain a PhD in magnetohydrodynamics, and he was a Research Fellow of Trinity Hall from 1960 to 1963.

On leaving Cambridge he worked as an economist for the Trades Union Congress in 1964–1965. He then became chief mathematician at Hawker Siddeley Dynamics in Coventry. In 1966, he joined the aerodynamics section of the Grumman Aerospace Corporation in Bethpage, New York. During this period, his work was largely directed toward the application of automatic control theory to stability augmentation systems. Starting in 1970, he began to concentrate on the problem of predicting transonic flow. Existing numerical methods were not equal to the task, and it was clear that new methods would have to be developed. At that time limitations in computer capabilities also precluded any attempt to calculate the flow past a complete aircraft, but useful efforts could be made for simpler configurations such as airfoils and wings.

In 1972, he moved to the Courant Institute of Mathematical Sciences at New York University, where he continued his work on transonic flow. He was appointed Professor of Computer Science at New York University in 1974. He joined Princeton University in 1980, and in 1982 he was appointed James S. McDonnell Distinguished University Professor of Aerospace Engineering. He was director of the University's Program in Applied and Computational Mathematics from 1986 to 1988. During the last decade Professor Jameson devised a variety of new schemes for solving the Euler and Navier-Stokes equations for inviscid and viscous compressible flows and wrote a series of computer programs that have been widely used in the aircraft industry. He and his co-workers were finally able to realize their goal of calculating the flow past a complete aircraft in 1985, using his new finite element method. Subsequently, he re-focused his research on the problem of shape optimization for aerodynamic design. He is currently the Thomas V. Jones Professor of Engineering in the Department of Aeronautics and Astronautics at Stanford University, where he is actively involved in research into numerical methods, deformable meshes, and meshless schemes.

Professor Jameson has received numerous awards and honors over the course of his career. He received the NASA Medal for Exceptional Scientific Achievement in recognition of his earlier work on transonic potential flow. In 1988, he received the Gold Medal of the Royal Aeronautical Society for his outstanding contribution to the development of methods for the calculation of transonic flow over real aircraft configurations. In 1991, he was elected a Fellow of the American Institute of Aeronautics and Astronautics, and he was also elected an Honorary Fellow of Trinity Hall, Cambridge; in 1992 he was a W.R. Sears Distinguished Lecturer at Cornell University. In 1993, he was selected to receive the American Institute of Aeronautics and Astronautics Fluid Dynamics Award "in recognition of numerous contributions to computational fluid dynamics and the development of many widely used computer programs which have immeasurably improved the capability to analyze and understand complex flows." In 1996 he was selected to receive the Theodorsen Lectureship Award from ICASE/NASA Langley. In 1997 he was elected as a foreign associate to the National Academy of Engineering. He was awarded the degree Docteur Honoris Causa from the University of Paris in 2001, and in 2002 he received the degree Docteur Honoris Causa from Uppsala University. Both these degrees were in Applied Mathematics. In 2004, he became a fellow of the Royal Aeronautical Society, and in 2005 he was elected a Fellow of the Royal Academy of Engineering. He was also selected jointly by six engineering societies to receive the Elmer A. Sperry Award for Advancing the Art of Transportation, in recognition of his seminal and continuing contributions to the modern design of aircraft through his numerous algorithmic innovations and through the development of his *FLO*, *SYN*, and *AIRPLANE* series of computational fluid dynamics codes.

6.14.8 Multigrid Method

In addition to these methods, solutions can be obtained more rapidly by using so-called *multigrid methods*. These methods accelerate the convergence of iterative procedures by using a sequence of grids of different densities, and they have become one of the most important techniques used to solve field problems of all types. The overall levels of the solution are established by the solution on a crude grid, while the details of the solution are established on a series of finer grids. Typically, one iteration is made on each successively finer grid, until the finest grid is reached. Then, one iteration is made on each successively coarser grid. This process is repeated until the solution converges. This procedure can reduce the number of fine grid iterations from possibly thousands to less than about fifty iterations. This approach to the solution of partial differential equations was developed by Jameson[43] for the solution of computational aerodynamics problems. He used the multigrid approach together with an alternating direction method in an extremely efficient algorithm for the two-dimensional transonic flow over an airfoil.

The details of the multigrid method are beyond the scope of this chapter, and the reader should consult the standard literature for more details. This includes the original treatise on the subject by Brandt[44] (which includes an example program), another tutorial that includes a code,[45] and more recent presentations by Briggs[46] and Wesseling[47] (*Numerical Recipes*[35] also includes a brief description and sample program).

6.14.9 The Delta Form

To carry out the solution to large systems of equations, the standard numerical procedures require that the approach be generalized slightly from the one given here. Specifically, we define an operator, such that the partial differential equation is written as (continuing to use Laplace's equation as an example):

$$L\phi = 0 \tag{6.135}$$

where

$$L = \frac{\partial^2}{\partial x^2} + \frac{\partial^2}{\partial y^2}. \tag{6.136}$$

To solve this equation, we rewrite the iteration scheme expressions given in Equation 6.134 as:

$$N \underbrace{C_{i,j}^n}_{\substack{\text{n}^{\text{th}} \text{ iteration} \\ \text{correction}}} + \omega \underbrace{L\phi_{i,j}^n}_{\substack{\text{n}^{\text{th}} \text{ iteration} \\ \text{residual, $=0$ when} \\ \text{converged solution} \\ \text{is achieved}}} . \tag{6.137}$$

This form is known as the standard or *delta form*. The term C in Equation 6.137 is given by:

$$C_{i,j}^n = \phi_{i,j}^{n+1} - \phi_{i,j}^n, \tag{6.138}$$

and the actual form of the N operator depends on the specific scheme chosen to solve the problem.

6.15 Program *THINFOIL*

An example of the solution of Laplace's equation by finite differences is demonstrated in the program *THINFOIL*. This program offers the users options of SOR, SLOR, AF1, and AF2 to solve the system of algebraic equations for the flow over a biconvex airfoil at zero angle of attack. An unevenly spaced grid is used to concentrate grid points near the airfoil. The program and the theory are described in Appendix D. It can be used to study the effects of grid boundary location, number of grid points, and relaxation factor, ω.

Figure 6.30 provides the convergence history for the case for which the comparison with the exact solution is given next. Using SOR, this shows that hundreds of iterations are required to reduce the maximum change between iteration approximately three orders of magnitude. This is about the minimum level of convergence required for useful results. A check against results

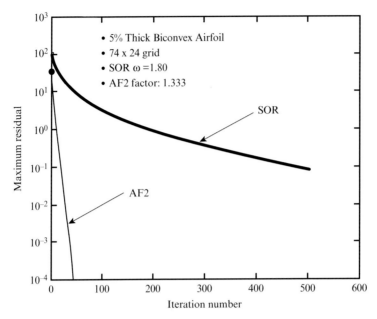

Figure 6.30   Convergence history during relaxation solution.

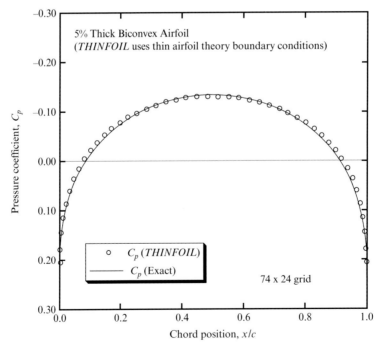

Figure 6.31   Comparison of numerical solution with analytic solution for a biconvex airfoil.

converged further should be made. The reader should compare this with the other iteration options.

The convergence history presented is actually the maximum residual of ϕ at each iteration. The solution is assumed to have converged when the residual goes to zero. Typical engineering practice is to consider the solution converged when the residual is reduced by three or four orders of magnitude. However, a check of the solution obtained at a conventional convergence level with a solution obtained at a much smaller residual (and higher cost) level should be made before conducting an extensive analysis for a particular study.

The solution for a 5% thick biconvex airfoil obtained with *THINFOIL* is presented in Fig. 6.31, together with the exact solution. For this case, the agreement with the exact solution is excellent. The exact solution for a biconvex airfoil is given by Van Dyke,[48] who cites Milne-Thompson[49] for the derivation.

Profiles in Computational Aerodynamics: Tom Pulliam

Thomas H. Pulliam started his career at NASA in the 1960s in the backyard of a grammar school friend (Larry Farrell) in an Apollo capsule made of an old refrigerator box and a TV antenna. He grew up on Long Island in New York and developed a keen interest in mathematics and a lifelong commitment to learn new things. This passion for learning and interest in mathematics and science led him to escape the suburbs of New York for the wilds of the Upper Peninsula of Michigan in 1969. There he attended Michigan Technological University, graduating with a degree in mathematics in 1973. It was in the embrace of the open wilderness of Upper Michigan that he met and married, in 1971, his lifelong companion, Carol Foster. During his tenure at Michigan Tech, he developed an interest in the mathematics of engineering, which led to graduate school in applied mechanics at Stanford University. During his first year at Stanford, he completed a master's in applied mechanics, worked with Professor Thomas Kane on bicycle dynamics, and began his thesis work with Professor John Sprieter on stratospheric circulation modeling. It was at this time that he began working at NASA Ames Research Center with two of the most influential people, both for his career and personality, Harvard Lomax and Joseph L. Steger. It was the early 1970s, and the new research area of Computational Fluid Dynamics (CFD) was emerging. Pulliam was integrated into a unique group of early CFDers, e.g., Robert MacCormack, William Ballhaus, Ronald Bailey, Barrett Baldwin, Robert Warming, and Richard Beam, to name a few. This group was working feverishly to develop the building blocks of theoretical and applied research in CFD that could take advantage of the new capabilities in computing power.

Pulliam's thesis was the development of a three-dimensional stratospheric circulation (the region of the atmosphere from 12 to 80 kilometers) CFD modeling code using semi-implicit methods. The approach and numerical techniques were at the leading edge of numerical method capabilities of that era, which by today's standards were rather limited. For example, the modeling covered the whole earth up to 80 km and used a 30 by 30 by 21 grid point mesh, the largest possible at that time. The results were obtained on a single core CDC 7600 computer, and to put that in perspective, recently Pulliam presented results for a rotor-craft configuration using over 3 billion grid points and 16,000 computer cores. Pulliam completed his thesis work in December 1976 and joined the then CFD Branch under Harvard Lomax as an NRC postdoctoral fellow until his hiring as a civil servant in 1978. Pulliam's post thesis work was the development of a three-dimensional Reynolds-Averaged Navier-Stokes code (*ARC3D*) working closely with Joseph L. Steger and Harv Lomax. *ARC3D* was one of the few three-dimensional codes used extensively in CFD through the early 1990s, both for numerical simulation in aerodynamics and as

a benchmark code for supercomputer development and assessment. In the early 1990s, many of the components of *ARC3D* were transitioned into a new multi-block structured code, *OVERFLOW*, which is the current state of the art for that class of code in CFD. The basis for a large segment of his research is the Beam-Warming Approximate Factorization Implicit scheme, which is computationally efficient for the Navier-Stokes equations in generalized geometries. The implicit nature of the scheme allows for stable numerical computations (especially on fine grids where the CFL restriction of explicit methods would be violated), while producing an algorithm which is computationally efficient. In multi-dimensions, an implicit method would produce a large linearized system of equations coupled across the multi-dimensions. The Beam-Warming algorithm factors into simpler one-coordinate dimension operators, which not only reduces the computational work, but also reduces memory requirements to manageable levels. A major contribution of Dr. Pulliam was the development of a diagonalized version of the Beam-Warming algorithm to further improve the computational efficiency of implicit methods. Dr. Pulliam's research activities include: numerical analysis, genetic algorithms and optimization, conventional aerodynamics, nonlinear dynamics, rotor-craft simulation, biologically inspired flight, and mathematics in general. He has published more than ninety publications as well as numerous presentations and lectures. Since 1990, he has been a consulting professor at Stanford University, teaching Introduction to CFD.

Dr. Pulliam has more than thirty years of service with NASA and, although he never got the cardboard Apollo capsule into orbit, he likes to think that his many contributions to NASA have helped others get there. In 1988 he received the Arthur S. Flemming Award for outstanding individual performance in the U.S. federal government, and in 1989 he was awarded the NASA Exceptional Scientific Achievement Medal. He has four children, Teela (PhD in particle physics), Troy (a restaurant manager), Merle Lynne (a car salesperson), and Richie (a golf professional). His many passions include hiking the high Sierras, skiing, biking, and zymurgy.

6.16 Modern Methods

Since the advent of digital computers, a large body of knowledge has been developed for solving systems of equations – these methods form the backbone of CFD. If you look at the methods described previously in this chapter, you will see that most of them were developed prior to 1970. These methods represent a major accomplishment, but as described in the biggest concept box, even these methods limited the size and complexity of problems that could be solved on computers at any given time. Efficient, but more complicated, methods were developed starting in the 1970s, which would allow researchers to maximize their potential for solving problems for a given computer capability. Although illustrative and simple to understand, most modern CFD codes do not use the methods described in this chapter up to this point. Both finite difference and finite volume codes use more advanced methods than these, and while the details of these methods are usually reserved for graduate courses, we will at least point out that the methods exist. If you have a greater interest in using any of these methods, you should consult the references that will be cited.

6.16.1 Finite Difference Methods

The most important class of modern numerical methods for finite difference CFD applications is based on the concepts developed by Richard Beam and Robert Warming at NASA Ames Research Center. These implicit methods have been described by Pulliam[50] (who was featured in the previous profile) and will be summarized briefly here. Of course, these algorithms start with the realization that the explicit algorithms presented in this chapter were just too slow for large CFD applications, and even the implicit methods were not fast enough. So, Beam and Warming improved the efficiency of the existing implicit methods in order to vastly improve the speed of performing these calculations. The trail that led them to their final algorithm took many twists and turns and involved a number of matrix procedures, including diagonalization, symmetrization, and upwind differencing. The last approach was important for aerodynamic flows because the Navier-Stokes equations are a mixed hyperbolic-elliptic set of equations which require changing from central difference for subsonic flow to upwind differencing for supersonic flows, as had been determined by Murman and Cole[20] in the early 1970s (see the discussion about this in Chapter 3).

The various requirements for solving the Navier-Stokes equations efficiently finally led Beam and Warming to approximate factorization methods. Just like the approximate (or iterative) methods mentioned in Section 6.14, approximate factorization gives solutions faster, but at the cost of introducing an error in the result, an error that can be reduced by iterating. The amount of iteration required depends on the level of accuracy desired, so while an improvement in computational speed can be achieved, it comes with potential problems. The resulting Beam-Warming implicit algorithm is best shown using the two-dimensional Euler equations for simplicity, Equation 3.63 without the viscous terms or z dimension included:

$$\frac{\partial Q}{\partial t} + \frac{\partial F(Q)}{\partial x} + \frac{\partial G(Q)}{\partial y} = 0 \tag{6.139}$$

where the dependence of the flux vectors on the conserved variables, Q, is shown to bring out the nonlinear nature of the flux terms in the equation. Beam and Warming started with an implicit second-order algorithm given by:

$$Q^{n+1} = Q^n + \frac{\Delta t}{2}\left[\frac{\partial Q^n}{\partial t} + \frac{\partial Q^{n+1}}{\partial t}\right] + O(\Delta t^3) \tag{6.140}$$

where the Q derivatives in Equation 6.140 are found from Equation 6.139 to give:

$$Q^{n+1} = Q^n - \frac{\Delta t}{2}\left[\left(\frac{\partial F}{\partial x} + \frac{\partial F}{\partial y}\right)^n + \left(\frac{\partial G}{\partial x} + \frac{\partial G}{\partial y}\right)^{n+1}\right] + O(\Delta t^3). \tag{6.141}$$

At this point, the method seems very similar to other implicit approaches spoken about earlier in the chapter, but the nonlinear nature of the equations presents some difficulties, since we need to find Q^{n+1} in terms of $F^{n+1} = F(Q^{n+1})$ and $G^{n+1} = G(Q^{n+1})$. Their next step was to perform a local Taylor series about Q^n to obtain:

$$\begin{aligned} F^{n+1} &= F^n + A^n\left(Q^{n+1} - Q^n\right) + O(\Delta t^2) \\ G^{n+1} &= G^n + B^n\left(Q^{n+1} - Q^n\right) + O(\Delta t^2) \end{aligned} \tag{6.142}$$

where $A = \partial F/\partial Q$ and $B = \partial G/\partial Q$ are called the Jacobians. If you substitute Equation 6.141 into Equation 6.142 you obtain the linear system (in terms of Q^{n+1}):

$$\begin{aligned} &\left[I + \frac{\Delta t}{2}\left(\delta_x A^n + \delta_y B^n\right)\right]Q^{n+1} = \left[I + \frac{\Delta t}{2}\left(\delta_x A^n + \delta_y B^n\right)\right]Q^n \\ &\quad - \Delta t\left(\delta_x F + \delta_y G\right)^n + O(\Delta t^3) \end{aligned} \tag{6.143}$$

where I is the identity matrix (1s on the diagonal and 0s elsewhere), and δ_x and δ_y are various possible finite difference formulas (for example a second-order central difference that would require values at $i-1$, i, and $i+1$). The problem with this approach is, according to Pulliam, that it requires "the inversion of a large block banded system of equations. It is well known that this direct inversion is, even today, prohibitively expensive in terms of computer operations and storage, and for the computers of the 1970s and 1980s it was an impossible task."[50] This led Beam and Warming to try an approximate factorization (AF) approach, which is similar to the ADI approach mentioned previously:

$$
\left[I + \frac{\Delta t}{2}\delta_x A^n\right]\left[I + \frac{\Delta t}{2}\delta_y B^n\right]Q^{n+1} - \frac{\Delta t^2}{4}\delta_x A^n \delta_y B^n Q^{n+1}
$$
$$
= \left[I + \frac{\Delta t}{2}\delta_x A^n\right]\left[I + \frac{\Delta t}{2}\delta_y B^n\right]Q^n - \frac{\Delta t^2}{4}\delta_x A^n \delta_y B^n Q^n \qquad (6.144)
$$
$$
-\Delta t\left(\delta_x F + \delta_y G\right)^n + O(\Delta t^3)
$$

Finally, they wrote the algorithm in Delta form as (after combining the extra terms with the error term):

$$
\left[I + \frac{\Delta t}{2}\delta_x A^n\right]\left[I + \frac{\Delta t}{2}\delta_y B^n\right]\Delta Q^n = -\Delta t\left(\delta_x F + \delta_y G\right)^n \qquad (6.145)
$$

where $\Delta Q^n = Q^{n+1} - Q^n$. To make the algorithm more general for a variety of time integration methods, Equation 6.144 was written as (see Section 6.12.2):

$$
\left[I + \frac{\Delta t}{2}\delta_x A^n\right]\left[I + \frac{\Delta t}{2}\delta_y B^n\right]\Delta Q^n
$$
$$
= -\frac{\Delta t}{1+\phi}\left(\delta_x F + \delta_y G\right)^n + \frac{\phi}{1+\phi}\Delta Q^{n-1} + \left(\theta - \phi - \frac{1}{2}\right)O(\Delta t^2) + O(\Delta t^3) \qquad (6.146)
$$

where θ and ϕ take on various values for different implicit schemes, as shown in Table 6.2 and discussed in Section 6.12.2.3.

The Beam-Warming algorithm then formed the basis for a number of improvements and alterations that would take place during the 1980s. These included the diagonalization of the method (to minimize the work involved in solving the block tridiagonal matrices)[51] and the introduction of flux split forms (flux difference and flux vector splitting to improve the solutions at supersonic speeds where shock waves form).[52]

These methods formed the basis of the development of *ARC2D* and *ARC3D* at NASA Ames, two important CFD codes used frequently in the 1980s. These codes formed the basis of *OVERFLOW*, which incorporates the overset grid scheme Chimera (which will be discussed in Chapter 7). According to Pulliam, *OVERFLOW* is "currently one of the most widely used general configuration codes."[53]

6.16.2 Higher-Order Methods

Remember the finite difference formulas we derived in Table 6.1? We used Taylor series expansions to approximate first and second derivatives to find formulas like the central-difference first derivative, which is second-order accurate:

$$\frac{\partial f}{\partial x} = \frac{f_{i+1} - f_{i-1}}{2\Delta x} + O(\Delta x^2). \tag{6.18}$$

The table also included a few fourth-order formulas, such as:

$$\frac{\partial f}{\partial x} = \frac{-f_{i+2} + 8f_{i+1} - 8f_{i-1} + f_{i-2}}{12\Delta x} + O(\Delta x^4).$$

You may have wondered why we do not use the higher-order formulas rather than the second-order formula shown in Equation 6.18. There are a number of reasons, especially relative to the numerical methods required (remember, the second-order central difference formulas led to a tridiagonal matrix, and the fourth-order formula above leads to a pentadiagonal matrix). Boundary conditions are also harder to implement since boundaries typically require one-sided differences that use large stencils (the footprint of the points required to perform the calculations), such as:

$$\frac{\partial f}{\partial x} = \frac{3f_{i-4} - 16f_{i-3} + 36f_{i-2} - 48f_{i-1} + 25f_i}{12\Delta x} + O(\Delta x^4). \tag{6.147}$$

As the order of the derivative goes up, the number of terms required also increases, as does the number of points required to evaluate the function. This led to matrix solutions that required more and more diagonals, making the solutions more expensive. Traditionally, this approach was prohibitive and was not pursued, so we used second-order schemes for decades.

Eventually, however, improved ways to use higher-order schemes with compact stencils were developed (higher-order refers to schemes that are greater than second-order accurate). A family of higher-order schemes were developed in "compact" form by Lele that were applied to finite difference grids with equal spacing.[54] He wrote general first-derivative and second-derivative 7-point stencils as:

$$\beta f'_{i-2} + \alpha f'_{i-1} + f'_i + \alpha f'_{i+2} + \beta f'_{i+2} = c\frac{f_{i+3} - f_{i-3}}{6\Delta x}$$
$$+ b\frac{f_{i+2} - f_{i-2}}{4\Delta x} + a\frac{f_{i+1} - f_{i-1}}{2\Delta x} \tag{6.148}$$

$$\beta f''_{i-2} + \alpha f''_{i-1} + f''_i + \alpha f''_{i+1} + \beta f''_{i+2} = c\frac{f_{i+3} - -2f_i + f_{i-3}}{6\Delta x}$$
$$+ b\frac{f_{i+2} - 2f_i + f_{i-2}}{4\Delta x} + a\frac{f_{i+1} - 2f_i + f_{i-1}}{2\Delta x} \tag{6.149}$$

Table 6.4 Coefficient Constraints for Compact Higher-order Formulas (Ref. 54)

Coefficient Relation	Order of Accuracy
$a+b+c=1+2\alpha+2\beta$	$O(\Delta x^2)$
$a+2^2 b+3^2 c = 2\dfrac{3!}{2!}(\alpha+2^2\beta)$	$O(\Delta x^4)$
$a+2^4 b+3^4 c = 2\dfrac{5!}{4!}(\alpha+2^4\beta)$	$O(\Delta x^6)$
$a+2^6 b+3^6 c = 2\dfrac{7!}{6!}(\alpha+2^6\beta)$	$O(\Delta x^8)$
$a+2^8 b+3^8 c = 2\dfrac{9!}{8!}(\alpha+2^8\beta)$	$O(\Delta x^{10})$

Table 6.5 Five-point Stencil Compact Higher-order Coefficients (Ref. 55)

Scheme	α	β	a	b	c
Fourth-order explicit	0	0	$\dfrac{4}{3}$	$-\dfrac{1}{3}$	0
Compact fourth-order	$\dfrac{1}{4}$	0	$\dfrac{3}{2}$	0	0
Compact sixth-order	$\dfrac{1}{3}$	0	$\dfrac{14}{9}$	$\dfrac{1}{9}$	0
Compact eighth-order (pentadiagonal matrix)	$\dfrac{4}{9}$	$\dfrac{1}{36}$	$\dfrac{40}{27}$	$\dfrac{25}{54}$	0
Compact eighth-order (tridiagonal matrix)	$\dfrac{3}{8}$	0	$\dfrac{75}{48}$	$\dfrac{1}{5}$	$-\dfrac{1}{80}$
Compact tenth-order	$\dfrac{1}{2}$	$\dfrac{1}{20}$	$\dfrac{17}{12}$	$\dfrac{101}{150}$	$\dfrac{1}{100}$

The values for α, β, a, b, c are chosen to create various difference schemes with different orders of accuracy, with both explicit ($a = b = 0$) and implicit approaches possible. Because these schemes retain the central-difference formulation, there is no dissipation error associated with any of them. Lele then used Taylor's series to find the relation between the coefficients for various orders of accuracy, as shown in Table 6.4.

Some of the common choices for central differencing coefficients with five-point stencils are shown in Table 6.5.[55] However, as mentioned earlier, "The primary difficulty in using higher order schemes is identification of stable boundary schemes that preserve their formal accuracy."[55] Once the

(a)

6th-Order

(b)

2nd-Order

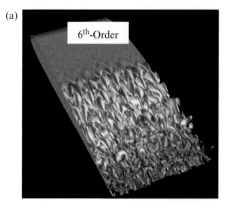
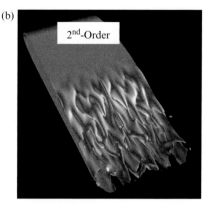

Figure 6.32 Transitional flow over an airfoil comparing a sixth-order scheme with a second-order scheme (Ref. 57; courtesy of Miguel Visbal of the Air Force Research Laboratory; a full color version of this image is available on the website: www.cambridge.org/aerodynamics).

boundary condition issue is addressed, however, accurate solutions can be obtained without significant increases in grid resolution.

"Based upon our experience, the tridiagonal subset of [Equation 6.148] increases the computational time by about a factor of two over that of a standard second-order explicit scheme for solution of [Equation 6.18]. But because of superior resolving capability, fewer computational resources need be expended with the high-order method, than are required with the standard approach, in order to attain the same level of resolution. Solution of the tridiagonal system of [Equation 6.148] is about 50% computationally less expensive than the pentadiagonal counterpart. Thus the fourth-order and sixth-order compact difference schemes provide a somewhat optimal balance between efficiency and accuracy."[56]

Another problem of higher-order schemes is the issues that arise when shocks are present in the flowfield, where central-difference schemes are unstable. Numerous upwind schemes have been employed, but upwind schemes have larger numerical dissipation, so finding the right balance of schemes and accuracy is challenging. A comparison of a sixth-order scheme with a second-order scheme on the same grid is shown in Fig. 6.32 for transitional flow over an airfoil.[57] You should notice that the sixth-order result shows a great deal more flow details and a different transition location than second-order result.

6.16.3 Finite Volume Methods

There is another class of modern methods based on the finite volume approach. These modern algorithms apply Equation 6.4 to known geometric shapes with planar sides (or polyhedron), called cells. By applying the

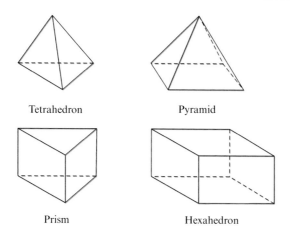

Tetrahedron   Pyramid

Prism   Hexahedron

Figure 6.33 Geometric shapes typically used as cells in finite volume algorithms.

governing equations to a known finite volume shape with planar sides, the triple integral over the cell volume can be evaluated directly and the double integral over all surfaces of the cell of interest can be evaluated with a summation. The most common three-dimensional geometric shapes are shown in Fig. 6.33. The structured mesh finite volume algorithms are a subset of the unstructured methods and only use hexahedra for all finite volume cells. The majority of unstructured methods use either all tetrahedral cells or a combination of prisms in the boundary-layer regions near solid surfaces that transition to tetrahedron away from the solid surfaces with pyramids used as transition elements between prisms and tetrahedra.

Computational Aerodynamics Concept Box

Advantages of Unstructured Mesh Finite Volume Methods

A few of the advantages of applying the finite volume approach to known cells (such as tetrahedra, pyramids, prisms, and hexahedra) to fill the volume surrounding a solid surface is the speed with which the volume mesh can be generated and the resulting quality of the mesh surrounding very complex geometries. This allows aircraft to be simulated with the majority of their details preserved, as in the C-17 cargo aircraft shown here.

Unstructured meshes typically have triangular or quadrilateral surface faces, allowing excellent control of the mesh to capture geometric features on the surface as in the C-17 engine nacelles, flap extensions, fuselage side doors, and side door aerodynamic diverters as seen in the figure (left). The volume mesh can usually be created from the surface mesh with very little input from the user, reducing the mesh generation task from months to weeks or even days for complex aircraft, as seen in the figure (right).

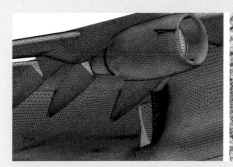

In addition to a reduction in the time to create the initial mesh and quality of the generated mesh, the mesh can be refined in regions of flow phenomena, such as shocks, massively separated flows, and vortices. This process is demonstrated in the image of a C-17 with refinement around the engine nacelles of the right inboard engine (image left).

These developments in unstructured mesh generation and the finite volume methods to solve on the created meshes has significantly increased the realism of the computed simulations in recent years. However, the expression "there is no free lunch" is applicable here as well. Unstructured finite volume methods have significantly higher memory and time per iteration requirements over structured mesh methods for a mesh with a similar number of cells.

We will now develop an unstructured finite volume solution algorithm starting with Equation 6.4 with the following assumptions: all finite volume cells are polyhedra, the polyhedra shape and locations are fixed in time, the primitive variables Q are solved for at the cell centroids, and the Q variables vary linearly in the cell. We will use the governing equations and these assumptions with some additional constraints noted later to develop a second-order temporally and spatially accurate algorithm for the Navier-Stokes equations. First, we write Equation 6.4 in terms of convective and viscous fluxes as

$$\frac{\partial}{\partial t}\left[\iiint_{V} Q\, dV\right] + \iint_{S}(\vec{F}_C - \vec{F}_V) \cdot \hat{n}\, dS = 0 \qquad (6.150)$$

where $\vec{F}_C$ and $\vec{F}_V$ are the convective and viscous fluxes, $\hat{n}$ is the vector normal to the surface of the volume, and S is the surface area. Next, we evaluate the triple integral over the volume with the assumption that the volume is one of our polyhedron with a fixed known volume

$$\frac{\partial}{\partial t}\left[\iiint_{V} Q\, dV\right] = V\frac{\partial Q}{\partial t}. \qquad (6.151)$$

Now, evaluating the surface integrals in the second term of Equation 6.150, we get

$$\iint_{S}(\vec{F}_C - \vec{F}_V) \cdot \hat{n}\, dS = \sum_{m=1}^{M}(\vec{F}_C{}^m - \vec{F}_V{}^m) \cdot \hat{n}^m S^m \qquad (6.152)$$

where m is the particular face of interest and M is the total number of faces. Substituting Equations 6.151 and 6.152 into Equation 6.150, we get

$$V\frac{\partial Q}{\partial t} + \sum_{m=1}^{M}\vec{F}_C{}^m \cdot \hat{n}^m S^m - \sum_{m=1}^{M}\vec{F}_V{}^m \cdot \hat{n}^m S^m = 0. \qquad (6.153)$$

Equation 6.153 is an expression of the Navier-Stokes equations applied to a particular cell of a fixed mesh.

We will next find a second-order temporal representation of the first term of Equation 6.153. Applying the same first- or second-order backward difference formula used in Equation 6.111 to our primitive variables Q at the $n+1$ time step, we get

$$\frac{\partial Q}{\partial t} = \frac{(1+\phi)Q^{n+1} - (1+2\phi)Q^n + \phi Q^{n-1}}{\Delta t} + (1/2 - \phi)O(\Delta t) + O(\Delta t^2) \qquad (6.154)$$

where ϕ is a parameter equal to 0 for a first-order formulation and ½ for a second-order formulation. As discussed in Section 6.12.3, many modern methods also include subiterations to improve temporal accuracy and to facilitate multidisciplinary coupling. We can include subiteration terms by adding and subtracting Q^p to Q^{n+1} in Equation 6.154, resulting in

$$
\frac{\partial Q}{\partial t} = \frac{(1+\phi)}{\Delta t}(Q^{n+1} - Q^p) + \frac{(1+\phi)Q^p - (1+2\phi)Q^n + \varphi Q^{n-1}}{\Delta t} + (1/2 - \phi)O(\Delta t) + O(\Delta t^2)
\tag{6.155}
$$

We can define the correction to the primitive variables for the current subiteration as

$$
\Delta Q^p = (Q^{n+1} - Q^p).
\tag{6.156}
$$

Substituting Equation 6.156 into Equation 6.155, we get an expression for the temporal derivative of Q in the Delta form:

$$
\frac{\partial Q}{\partial t} = \frac{(1+\phi)}{\Delta t}\Delta Q^p + \frac{(1+\phi)Q^p - (1+2\phi)Q^n + \phi Q^{n-1}}{\Delta t} + (1/2 - \phi)O(\Delta t) + O(\Delta t^2)
\tag{6.157}
$$

Next we use a Taylor series expansion to determine a representation of the convective flux at the $n+1$ time step on the mth face (where parentheses are used to delineate the temporal index from the index representing the cell faces)

$$
\vec{F}_C^{m,(n+1)} = \vec{F}_C^{m,(p)} + (Q^{n+1} - Q^p)\frac{\partial \vec{F}_C^{m,(p)}}{\partial Q} + O(\Delta Q^2)
\tag{6.158}
$$

where the partial differential term is the partial derivative of the fluxes with respect to the vector of primitive variables, commonly referred to as the Jacobian matrix in the literature. Substituting in Equation 6.156 we get the Delta form

$$
\vec{F}_C^{m,(n+1)} = \vec{F}_C^{m,(p)} + \Delta Q^p \frac{\partial \vec{F}_C^{m,(p)}}{\partial Q} + O(\Delta Q^2),
\tag{6.159}
$$

and similarly for the viscous flux in Delta form we get

$$
\vec{F}_V^{m,(n+1)} = \vec{F}_V^{m,(p)} + \Delta Q^p \frac{\partial \vec{F}_V^{m,(p)}}{\partial Q} + O(\Delta Q^2).
\tag{6.160}
$$

Substituting Equations 6.157, 6.158, and 6.159 into Equation 6.152 and applying it to the ith cell, we get

$$\mathcal{V}_i \frac{(1+\phi)}{\Delta t} \Delta Q_i^p + \mathcal{V}_i \frac{(1+\phi)Q_i^p - (1+2\phi)Q_i^n + \phi Q_i^{n-1}}{\Delta t}$$
$$+ \sum_{m=1}^{M} \left(\vec{F}_C^{m,(p)} \cdot \hat{n}^m S^m + \Delta Q_i^p \frac{\partial \vec{F}_C^{m,(p)}}{\partial Q} \cdot \hat{n}^m S^m \right) \qquad (6.161)$$
$$- \sum_{m=1}^{M} \left(\vec{F}_C^{m,(p)} \cdot \hat{n}^m S^m + \Delta Q_i^p \frac{\partial \vec{F}_C^{m,(p)}}{\partial Q} \cdot \hat{n}^m S^m \right) = 0$$

Grouping all terms of ΔQ_i^p on the left-hand side and all other terms on the right-hand side. we get the following:

$$\mathcal{V}_i \frac{(1+\phi)}{\Delta t} \left[I + \frac{\Delta t}{\mathcal{V}_i(1+\phi)} \sum_{m=1}^{M} \left(\frac{\partial \vec{F}_C^{m,(p)}}{\partial Q} - \frac{\partial \vec{F}_V^{m,(p)}}{\partial Q} \right) \cdot \hat{n}^m S^m \right] \Delta Q_i^p =$$
$$- \left[\mathcal{V}_i \frac{(1+\phi)Q_i^p - (1+2\phi)Q_i^n + \phi Q_i^{n-1}}{\Delta t} + \sum_{m=1}^{M} \left(\vec{F}_C^{m,(p)} - \vec{F}_V^{m,(p)} \right) \cdot \hat{n}^m S^m \right] \qquad (6.162)$$

or, by dividing through by the term in front of the square bracket on the left-hand side, we get the following system of equations for a particular cell:

$$\left[I + \frac{\Delta t}{\mathcal{V}_i(1+\phi)} \sum_{m=1}^{M} \left(\frac{\partial \vec{F}_C^{m,(p)}}{\partial Q} - \frac{\partial \vec{F}_V^{m,(p)}}{\partial Q} \right) \cdot \hat{n}^m S^m \right] \Delta Q_i^p =$$
$$- \frac{\Delta t}{\mathcal{V}_i(1+\phi)} \left[\mathcal{V}_i \frac{(1+\phi)Q_i^p - (1+2\phi)Q_i^n + \phi Q_i^{n-1}}{\Delta t} + \sum_{m=1}^{M} \left(\vec{F}_C^{m,(p)} - \vec{F}_V^{m,(p)} \right) \cdot \hat{n}^m S^m \right]$$

$$(6.163)$$

By applying Equation 6.163 to all cells in the domain, we have arrived at a coupled set of algebraic equations. Combining all cells into a single matrix results in the classic $Ax = b$ problem when considering the set of bracketed terms on the left-hand side as A, ΔQ_i^p as x, and the right-hand side as b. It is important to note that the bracketed term on the right-hand side of Equation 6.163, if set to zero, is an expression of Equation 6.152 when $Q_i^p \to Q_i^{n+1}$. In other words, if the system is convergent, successive solutions of Equation 6.163 will result in $\Delta Q_i^p \to 0$, which implies $Q_i^p \to Q_i^{n+1}$ and therefore, the governing equations will be satisfied. Convergence will depend on the time step Δt chosen, the number of subiterations p, and the method of solving the $Ax = b$ problem for ΔQ_i^p.

The A matrix resulting from applying Equation 6.163 to all cells in the domain is a sparse blocked matrix with block size equal to a 4×4 matrix for two-dimensional Euler or laminar Navier-Stokes calculations, a 5 × 5 matrix for three-dimensional Euler or laminar Navier-Stokes calculations, or larger blocks when coupling additional turbulence transport equations. There will be one block on the diagonal and additional blocks for each side of the polyhedral on the off-diagonals at a minimum. Because the mesh is unstructured, there is no fixed pattern of blocks on the off-diagonals (i.e., not block tridiagonal or penta-diagonal as in structured mesh solvers).

There are many ways to solve this $Ax = b$ problem that differentiate methods in the literature. Most methods approximate the A matrix to make the $Ax = b$ solution more efficient, resulting in inner iterations. The more complete the left-hand side, the fewer inner iterations are necessary but the higher the cost per iteration. One method in particular is called the "point implicit" method that uses a block Gauss-Seidel iteration process to solve the $Ax = b$ problem by sweeping through the matrix until there is no more change in ΔQ^p. During the sweeping process, the off-diagonal blocks are incorporated in a lagged fashion. This method requires inversion of only the diagonal blocks once per subiteration and many additional matrix vector multiplications.

We have now answered the question of "how" to solve the governing equations, but there are details buried in Equation 6.163 that still need to be defined to determine "what" we are solving. The biggest issue to deal with is that we only know information at the cell centroid but need information at the faces to compute the summation terms of Equation 6.163. We will first tackle the concept of "reconstruction." Reconstruction is the method of using data at the cell centroids to compute data at the cell face for use in the flux calculations. For example, a constant reconstruction results if we compute the data at the cell faces by using the cell centroid data "as is," giving us an overall first-order spatially accurate algorithm. If we follow our original assumption that the Q variables vary linearly in the cell, we can use a linear reconstruction technique and recover a second-order spatially accurate algorithm. Our linear reconstruction method will follow the approach that we can compute the Q variable anywhere near the centroid up to the boundaries of the cell by the following relationship:

$$Q_j = Q_i + \vec{\nabla} Q_i \cdot \vec{r}_{ij} \tag{6.164}$$

where Q_i is the primitive variable vector at the cell centroid, $\vec{r}_{ij}$ is the pointing vector from the cell centroid to the location of interest, $\vec{\nabla} Q_i$ is the gradient of the Q variables at the cell centroid, and Q_j is the resulting primitive variables at the location $\vec{r}_{ij}$ is pointing to in the cell. The gradient $\vec{\nabla} Q_i$ becomes a very important term that will impact both accuracy and robustness of the method.

There are many ways to calculate the gradient using the cell centroid Q data from the surrounding cells, but the most prevalent are either with a distance-weighted least squares approximation, or with a Gauss-gradient formulation.[4] There are advantages and disadvantages for each method detailed in Ref. 4. It is important to note that the choice of the gradient method can severely restrict the complexity of the solid surface geometry possible to compute due to the resulting volume mesh quality.

The next important concept in forming the Q variable data at the face is the impact of "extrema" on the robustness and accuracy of the solution. If face Q data are allowed to be either larger or smaller than both of the cell centroid data for cells sharing the face, we have created an extrema that can result in nonphysical oscillations in the solution. Extrema can be the result of inaccurate gradients, poor mesh quality, or using information inappropriately (e.g., downwind in a supersonic calculation) in the gradient calculation. A method of eliminating extrema is to use a limiter in the calculation of Equation 6.164. There are many limiters discussed in the literature, but two of the most prevalent in unstructured finite volume codes are the Barth-Jespersen[58] and Venkatakrishnan limiters.[59,60] These limiters both have advantages and disadvantages.[4] Most limiters use the current cell of interest centroid data and all neighbor cells' centroid data to determine a limit on the gradient at the cell that will result in no extrema for any of the faces on the cell. Mathematically we can express Equation 6.164 with a limiter as:

$$Q_j = Q_i + \Psi_i \vec{\nabla} Q_i \cdot \vec{r}_{ij} \qquad (6.165)$$

where Ψ_i in its simplest form is a scalar number between 0 and 1 computed to eliminate extrema at the faces and go to zero when strong discontinuities are encountered, such as shocks. More complicated formulations can have variations in Ψ_i for each face or for each term in Q. It is interesting to note that if the limiter is 0, we get a constant reconstruction and a first-order spatially accurate method in the convective terms, and if it is equal to 1, it is unlimited and a linear reconstruction results in a second-order spatially accurate method for the convective terms. The choice of method to compute Ψ_i can have the positive impact of increased robustness, but it can also have the negative impact of increased dissipation in the solution along with decreased accuracy and therefore must be considered carefully when simulating complex geometries.

Our desire is to compute the fluxes at the face of the cell using Equation 6.165. However, a face has two associated cells attached to it. This means that we can get different fluxes at the face depending on which cell we use for the reconstruction. We also know that the physical quantities that leave one cell must enter the other cell for conservation properties to hold. Therefore, we must compute the face quantities once using both the left cell reconstructed quantities Q_L and the right cell reconstructed quantities Q_R. Although it may be tempting to simply average the data from the left and right cells of the face, this has proven to be less accurate and robust than another approach that seeks to satisfy a physical relationship between the left and right cell states. Early developers of computational physics methods realized that the jump between the left and right states is similar to the problems solved by the Riemann method (e.g., the shock tube) and applied it to the cell faces. Most modern algorithms compute the face fluxes that both cells will use by solving the Riemann problem either exactly, as in the method of Gottlieb and Groth,[61] or approximately, as in the Roe scheme.[62] The Gottlieb and Groth method, while exact, requires more computational effort than the approximate methods. It also results in a face state Q_{face} that must be used to compute the fluxes at the face. The approximate Riemann solver of Roe results in the fluxes at the face and is widely used in current unstructured finite volume codes.

We have concentrated on the convective fluxes so far, but the viscous fluxes present their own issues to be solved. Construction of the viscous fluxes at the face requires us to know the component of the gradient of the primitive variables normal to the face at the face. As we mentioned earlier, only the Q variables and their gradients are known at the cell centroids. Most modern methods construct the gradient of Q at the face by a directed difference of the centroid Q values for the two cells on either side of the face of interest. One method of computing the directed difference gradient at the face was proposed by Crumpton et al.[63] and Weiss[64] and can be expressed as

$$\vec{\nabla} Q_{ij} = \frac{Q_j - Q_i}{\left| \vec{r}_{ij} \right|} \hat{r}_{ij} + \left[\overline{\nabla Q_{ij}} - \overline{\nabla Q_{ij}} \cdot \hat{r}_{ij} \right] \qquad (6.166)$$

where Q_j is the primitive variable at the centroid of the left cell, Q_i is the primitive variable at the centroid of the right cell, $\vec{r}_{ij}$ is the vector from the ith cell centroid to the jth cell centroid, $\left| \vec{r}_{ij} \right|$ is the magnitude of the vector between the ith cell centroid to the jth cell centroid, $\hat{r}_{ij}$ is the normalized $\vec{r}_{ij}$ vector, and $\overline{\nabla Q_{ij}}$ is the average of the i and j cell centroid gradients:

$$\overline{\nabla Q_{ij}} = \frac{1}{2} \left[\vec{\nabla} Q_i + \vec{\nabla} Q_j \right]. \qquad (6.167)$$

Equation 6.166 is a compact directed difference that uses the left and right Q variables to contribute the majority of the gradient at the face, and the averaged cell centroid gradients (Equation 6.167) are used to define the gradients normal to the $\vec{r}_{ij}$ vector. There are many modifications to the approach described by Equation 6.166 in the literature that seek to improve the accuracy and robustness of the algorithm for degraded cells in the volume mesh of complex geometries.

The algorithm presented in Equation 6.163 along with the supporting methods described in Equations 6.165 through 6.167 result in a point implicit, upwind-biased method of solving the Navier-Stokes equations on unstructured finite volume meshes. In addition to the algorithm, issues commonly encountered in the simulation of realistic air vehicles have been discussed to aid you as a future practitioner of computational aerodynamics when using codes with similar algorithms.

Summary of Best Practices

1. There are a number of approaches to discretizing the fluid dynamics equations of motion, all of which have strengths and weaknesses – you should understand the basic concepts of the approaches and know their strengths and weaknesses:
   a. Finite difference methods
   b. Finite volume methods
   c. Finite element methods/Pseudo-spectral methods
2. All numerical approaches have issues relative to numerical error that should be understood when using the methods, including:
   a. Order of accuracy
   b. Consistency
   c. Stability
   d. Convergence
   e. Numerical dissipation
3. Boundary conditions are essential concepts in computational aerodynamics, and there are many ways to discretize them. Having a full understanding of the boundary conditions used in a code can be crucial to obtaining good results.
4. There are many matrix solution algorithms used in computational aerodynamics, both for potential flow methods (discussed in Chapter 5) and CFD methods (discussed in Chapter 6) – you should understand the various methods and their strengths and weaknesses:
   a. Dense matrix methods
   b. Sparse and banded matrix methods
   c. General sparse matrix methods
   d. Exact vs. iterative methods

6.17 Projects

1. Finite Difference Project: Evaluate the accuracy of various derivative formulas using the Matlab program *Finite Differences*. Use the periodic function (shown here) to both analytically and numerically evaluate various derivatives.

$$u = -\sin(x) + \sin^2(2x)\cos(x)$$

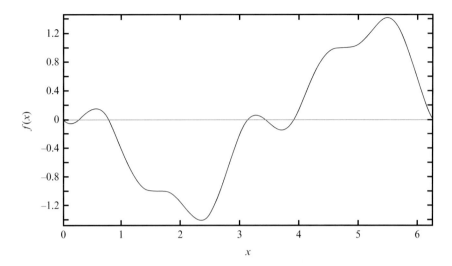

Evaluate both first and second derivatives using the following formulas:

- First Derivative
  - (a) 1st order backward
  - (b) 2nd order central
  - (c) 2nd order backward
  - (d) 4th order central
- Second Derivative
  - (a) 1st order backward
  - (b) 2nd order central
  - (c) 4th order central

Start with a Δx that yields five subdivisions for one period of the function, and then halve Δx until you reach "grid independence" (no change in the results). Analytically determine the derivatives from the given function. Compare the analytic results with the numerical results obtained using up to twelve levels of halving. The total error is defined as the maximum norm of the difference between the exact derivative and the numerical finite difference. Plot total error versus Δx on a log-log scale. The slope of the curves is the error order measure of the derivative. Write an explanation of what you observed. Perform a hand calculation of each of the derivatives for the first Δx at $x = 3.0$ and compare your answers with the results from the spreadsheet.

This project was created by Dr. Thomas Pulliam of NASA Ames Research Center and is used by permission.

2. Wave Equation Project: Solve the (linearized) wave equation

$$u_t + u_x = 0$$

on a computer for the initial conditions

$$u(x,0) = \sin\left[2n\pi\,\frac{x}{L}\right]$$

where $0 \le x \le L$ and for periodic boundary conditions, i.e.,

$$u_{m+1}^n = u_1^n.$$

Choose a grid of 41 mesh points with $\Delta x = 1/40$ and compute to $t = 18$. Solve for $n = 1$ and 3 and CFL = 1.0, 0.6, and 0.3 (CFL $= c\Delta t / \Delta x$, where c is the coefficient of u_x in the governing equation). Solve the equation using:

(a) the Upstream method
(b) the Lax method
(c) the Lax-Wendroff method
(d) the MacCormack method

Compare your solutions with the exact solution

$$u(x,0) = F(x)$$

$$u(x,t) = F(x - ct).$$

Comment on the results relative to stability, accuracy, type of error, number of grid points, method used, etc. Why did a CFL number of 1.0 give such good results?

This project is from Ref. 1 and is used by permission.

3. Heat Equation Project: Solve the heat equation

$$u_t = 0.2u_{xx}$$

on a computer using the simple explicit method for the initial conditions

$$u(x,0) = 100\sin\left[\frac{\pi x}{L}\right]$$

(where $L = 1$) and the boundary conditions

$$u(0,t) = u(L,t) = 0.$$

Be sure to define "pi" as $\pi = 4\tan^{-1}(1)$ for accuracy (it makes a difference on this project!).

Compute to $t = 1.5$ for the following cases (if possible):

Case	Number of Grid Points	CFL
1	19	0.25
2	9	0.50
3	19	0.50
4	19	? (find the CFL for instability)
5	19	2.00

where the CFL number is $\alpha \Delta t / \Delta x^2$ and for this problem $\alpha = 0.2$ (the coefficient of u_{xx} in the governing equation). Repeat the problem using Laasonen's simple implicit method. Compare all of your results with the exact solution:

$$u(x,t) = 100e^{-\alpha \pi^2 t} \sin(\pi x).$$

Comment on the results relative to stability, accuracy, type of error, number of grid points, method used, etc.

   This project is from Ref. 1 and is used by permission.

4. Get some experience with the solution of Laplace's equation using finite differences:
    *   Download a copy of *THINFOIL* from the book web page
    *   Study the program to understand the procedure:
    Pick as a baseline case: Xmin=-2.2, Xmax=3.2, Ymax=2.4, and NUP=14, NDOWN=14, NON=30, NABOVE=18
    *   Run SOR with $\omega = 1.6$ and see how many iterations to "convergence"
    *   Run with $\omega = 1.0, 1.50, 1.75, 1.90, 1.99$ (400 iterations max)
    *   Plot the convergence history as a function of iteration for each ω. Note that it is standard procedure to plot the log of the residual. (see examples in the text).
    *   For one ω, increase the number of grid points and compare (be careful of your dimensions):
        - the convergence rate with the same ω case above
        - the surface pressure distribution results for the two grids
    *   Draw conclusions about SOR as a numerical method for solving PDEs.
    *   Repeat the study using SLOR, AF1, and AF2. What do you conclude about the relative convergence times and solution accuracy?
5. Examine the effect of the number of grid points on the solution obtained using program *THINFOIL*. How many grid points are required for a grid converged solution?
6. Examine the effect of the location of the farfield boundary condition on the solution obtained using program *THINFOIL*. What do you conclude?

7. Change the farfield boundary condition in program THINFOIL to set $\phi = 0$, instead of $\partial\phi/\partial n = 0$. How does this affect the solution? the convergence rate?

8. Modify program *THINFOIL* to obtain the solution to the flow over an NACA 4-digit airfoil thickness shape. Address the following issues:
   (a) store the boundary condition values before the calculation begins instead of recomputing each time the BC needs the value
   (b) recognizing that the slope at the leading edge is infinite, assess two methods of avoiding numerical problems
   (c) place the leading edge between grid points
   (d) use Riegels' factor to modify the slope boundary condition, replacing df/dx by

$$\frac{df/dx}{\sqrt{1+(df/dx)^2}}.$$

6.18 **References**

1 Pletcher, R.H., Tannehill, J.C., and Anderson, D.A., *Computational Fluid Mechanics and Heat Transfer*, 3rd Ed., Boca Raton: CRC Press, 2013.

2 Fletcher, C.A.J., *Computational Techniques for Fluid Dynamics*, Vol. 1: "Fundamental and General Techniques," Vol. II, "Specific Techniques for Different Flow Categories," 2nd Ed. Berlin: Springer-Verlag, 1991.

3 Hirsch, C., *Numerical Computation of Internal and External Flows: The Fundamentals of Computational Fluid Dynamics*, 2nd Ed., Oxford: Butterworth-Heinemann, 2007.

4 Blazek, J., *Computational Fluid Dynamics: Principles and Applications*, 2nd Ed., Oxford: Elsevier, 2005.

5 Lomax, H., Pulliam, T.H., and Zingg, D.W., *Fundamentals of Computational Fluid Dynamics*, Berlin: Springer, 2001.

6 Morton, K.W., and Mayers, D.F., *Numerical Solution of Partial Differential Equations*, 2nd Ed., Cambridge: Cambridge University Press, 2005.

7 Smith, G.D., *Numerical Solution of Partial Differential Equations: Finite Difference Methods*, 3rd Ed., Oxford: Clarendon Press, 1985.

8 Ames, W.F., *Numerical Methods for Partial Differential Equations*, 3rd Ed., San Diego: Academic Press, 1992.

9 Versteeg, H.K., and Malalasekera, W., *An Introduction to Computational Fluid Dynamics: The Finite Volume Method*, 2nd Ed., Upper Saddle River: Prentice Hall, 2007.

10 LeVeque, R.J., *Finite Volume Methods for Hyperbolic Problems*, Cambridge: Cambridge University Press, 2002.

11 Ferziger, J.H., and Peric, M., *Computational Methods for Fluid Dynamics*, 3rd Ed., Berlin: Springer, 2002.

12 Peyret, R., and Taylor, T.D., *Computational Methods for Fluid Flow*, New York: Spring-Verlag, 1983.

13 Mavriplis, D.J., "Unstructured-Mesh Discretizations and Solvers for Computational Aerodynamics," *AIAA Journal*, Vol. 46, No. 6, 2008, pp. 1281–1298.

14 Baker, A.J., *Finite Element Computational Fluid Mechanics*, New York: Hemisphere Publishing, 1983.

15 Donea, J., and Huerta, A., *Finite Element Methods for Flow Problems*, West Sussex: John Wiley & Sons, 2003.

16 Löhner, R., *Applied CFD Techniques: An Introduction Based on Finite Element Methods*, West Sussex: John Wiley & Sons, 2001.

17 Gottlieb, D., and Orzsag, S.A., *Numerical Analysis of Spectral Methods*, Philadelphia: SIAM, 1977.

18 Peyret, R., *Spectral Methods for Incompressible Viscous Flow*, New York: Springer, 2002.

19 Canuto, C., Hussaini, M.Y., Quarteroni, A., and Zang, T.A., *Spectral Methods*, Berlin: Springer, 2007.

20 Murman, E.M., and Cole, J.D., "Calculation of Plane Steady Transonic Flows," *AIAA Journal*, Vol. 9, No. 1, 1971, pp. 114–121.

21 Lax, P.D., and Richtmyer, R.D., "Survey of the Stability of Linear Finite Difference Equations," *Communications on Pure and Applied Mathematics*, Vol. 9, 1956, pp. 267–293.

22 Morton, S.A., Cummings, R.M., and Kholodar, D.B., "High Resolution Turbulence Treatment of F/A-18 Tail Buffet," *Journal of Aircraft*, Vol. 44, No. 6, 2007, pp. 1769–1775.

23 Boelens, O.J., "CFD Analysis of the Flow Around the X-31 Aircraft at High Angle of Attack," AIAA Paper 2009–3628, June 2009.

24 Richtmyer, R.D., and Morton, K.W., *Difference Methods for Initial-Value Problems*, 2nd Ed., New York: Interscience, 1967.

25 Courant, R., Friedrichs, K.O., and Lewy, H., "Über die Partiellen Differenzengleichungen der Mathematischen Physik," *Mathematische Annalen*, Vol. 100, 1928, pp. 32–74.

26 Lax, P., "Weak Solutions of Nonlinear Hyperbolic Equations and Their Numerical Computation," *Communications on Pure and Applied Mathematics*, Vol. 7, Issue 1, 1954, pp. 159–193.

27 Lax, P., and Wendroff, B., "Systems of Conservation Laws," *Communications on Pure and Applied Mathematics*, Vol. 13, 1960, pp. 217–237.

28 MacCormack, R.W., "The Effect of Viscosity in Hypervelocity Impact Cratering," AIAA Paper 69– 0354, May 1969.

29 Pulliam, T.H., "Euler and Thin Layer Navier-Stokes Codes: ARC2D, ARC3D," Computational Fluid Dynamics User's Workshop, University of Tennessee Space Institute, March 1984.

30 Jameson, A., Schmidt, W., and Turkel, E., "Numerical Solution of the Euler Equations by Finite Volume Methods using Runge Kutta Time Stepping Schemes," AIAA Paper 81–1259, June 1981.

31 Beam, R., and Warming, R.F., "An Implicit Finite-Difference Algorithms for Hyperbolic Systems in Conservation Law Form," *Journal of Computational Physics*, Vol. 22, 1976, pp. 86–110.

32 Rizzi, A., "Numerical Implementation of Solid-Body Boundary Conditions for the Euler Equations," *Zeitschrift für Angewandte Mathematik und Mechanik*, Vol. 58, 1978, pp. 301–304.

33 Moran, J., *An Introduction to Theoretical and Computational Aerodynamics*, New York: John Wiley & Sons, 1984.

34 Holst, T.L., "Numerical Computation of Transonic Flow Governed by the Full-Potential Equation," VKI Lecture Series on Computational Fluid Dynamics, Rhode-St.-Genese, March 1983.

35 Press, W.H., Flannery, B.P., Teukolsky, S.A., and Vettering, W.T., *Numerical Recipes in FORTRAN: The Art of Scientific Computing*, 2nd Ed., Cambridge: Cambridge University Press, 1992.

36 Conte, S.D., and deBoor, C., *Elementary Numerical Analysis*, New York: McGraw-Hill, 1972.

37 Bertin, J.J., and Cummings, R.M., *Aerodynamics for Engineers*, 6th Ed., Upper Saddle River: Pearson, 2014.

38 Mason, W.H., MacKenzie, D., Stern, M., Ballhaus, W.F., and Frick, J., "An Automated Procedure for Computing the Three-Dimensional Transonic Flow Over Wing-Body Combinations, Including Viscous Effects," Vol. II Program User's Manual and Code Description, AFFDL-TR-77–122, February 1978.

39 Isaacson, E., and Keller, H.B., *Analysis of Numerical Methods*, New York: John Wiley & Sons, 1966.

40 Press, W.H., Flannery, B.P., Teukolsky, S.A., and Vettering, W.T., *Numerical Recipes: The Art of Scientific Computing (FORTRAN Version)*, Cambridge: Cambridge University Press, 1989.

41 Ballhaus, W.F., Jameson, A., and Albert, J., "Implicit Approximate-Factorization Schemes for the Efficient Solution of Steady Transonic Flow Problems," *AIAA Journal*, Vol. 16, No. 6, 1978, pp. 573–579.

42 Catherall, D., "Optimum Approximate-Factorisation Schemes for 2D Steady Potential Flows," AIAA Paper 81–1018, June 1981.

43 Jameson, A., "Acceleration of Transonic Potential Flow Calculations on Arbitrary Meshes by the Multiple Grid Method," AIAA Paper 79–1458, July 1979.

44 Brandt, A., "Multi-Level Adaptive Solutions to Boundary-Value Problems," *Mathematics of Computation*, Vol. 31, No. 138, 1977, pp. 333–390.

45 Stuben, K., and Trottenberg, U., "Multigrid Methods: Fundamental Algorithms, Model Problem Analysis and Applications," in *Multigrid*

Methods, Ed. W. Hackbusch and U. Trottenberg, Lecture Notes in Mathematics Vol. 960, Berlin: Springer-Verlag, Berlin, 1982. pp. 1–176.

46 Briggs, W.L., *A Multigrid Tutorial*, Philadelphia: SIAM, 1987.

47 Wesseling, P., *An Introduction to Multigrid Methods*, Chichester: John Wiley, 1992.

48 Van Dyke, M., *Perturbation Methods in Fluid Mechanics*, Annotated Edition, Stanford: Parabolic Press, 1975.

49 Milne-Thompson, L.M., *Theoretical Hydrodynamics*, 5th Ed., New York: Dover Publications, 1996.

50 Pulliam, T.H., "Early Development of Implicit Methods for Computational Fluid Dynamics at NASA Ames," *Computers and Fluids*, Vol. 38, Issue 3, 2009, pp. 491–495.

51 Pulliam, T.H., and Chaussee, D.S., "A Diagonal Form of an Implicit Approximate-Factorization Algorithm," *Journal of Computational Physics*, Vol. 29, Issue 2, 1981, pp. 347–363.

52 Steger, J.L., and Warming, R.F., "Flux Vector Splitting of the Inviscid Gas Dynamics Equations with Applications to Finite-Difference Methods," *Journal of Computational Physics*, Vol. 40, Issue 2, pp. 263–293.

53 Pulliam, T.H., "Development of Implicit Methods in CFD NASA Ames Research Center 1970s–1980s," *Computers and Fluids*, Vol. 41, Issue 1, 2011, pp. 65–71.

54 Lele, S.K., "Compact Finite Difference Schemes with Spectral-like Resolution," *Journal of Computational Physics*, Vol. 103, Issue 1, 1992, pp. 16–42.

55 Ekaterinaris, J.A., "High-Order Accurate, Low Numerical Diffusion Methods for Aerodynamics," *Progress in Aerospace Sciences*, Vol. 41, 2005, pp. 192–300.

56 Rizzetta, D.P., Visbal, M.R., and Morgan, P.E., "A High-Order Compact Finite-Difference Scheme for Large-Eddy Simulation of Active Flow Control," *Progress in Aerospace Sciences*, Vol. 44, 2008, pp. 397–426.

57 Garmann, D.J. and Visbal, M., "High-Order Solutions of Transitional Flow Over the SD7003 Airfoil Using Compact Finite-Differencing and Filtering," Case 3.3 Summary, 1st International Workshop on High-Order CFD Methods, January 7–8, 2012, Nashville, TN.

58 Barth, T.J., and Jespersen, D.C., "The Design and Application of Upwind Schemes on Unstructured Meshes," AIAA Paper 89–0366, January 1989.

59 Venkatakrishnan, V., "On the Accuracy of Limiters and Convergence to Steady State Solutions," AIAA Paper 93–0880, January 1993.

60 Venkatakrishnan, V., "Convergence to Steady State Solutions of the Euler Equations on Unstructured Grids with Limiters," *Journal of Computational Physics*, Vol. 118, Issue 1, 1995, pp. 120–130.

61 Gottlieb, J.J., and Groth, C.P.T., "Assessment of Riemann Solvers for Unsteady One-Dimensional Inviscid Flows of Perfect Gases," *Journal of Computational Physics*, Vol. 78, Issue 2, 1988, pp. 437–458.

62 Roe, P.L., "Approximate Riemann Solvers, Parameter Vectors, and Difference Schemes," *Journal of Computational Physics*, Vol. 43, Issue 2, 1981, pp. 357–372.

63 Crumpton, P.I., Moiner, P., and Giles, M.B., "An Unstructured Algorithm for High Reynolds Number Flows on Highly-Stretched Grids," 10th International Conference on Numerical Methods for Laminar Flows, Swansea, UK, July 21–25, 1997.

64 Weiss, J.M., Maruszewski, J.P., and Smith, W.A., "Implicit Solution of Preconditioned Navier-Stokes Equations Using Algebraic Multigrid," *AIAA Journal*, Vol. 37, No. 1, 1999, pp. 29–36.

7 Geometry and Grids: Key Considerations in Computational Aerodynamics

There I was, my grid was highly skewed and my solution was diverging.

From Don Kinsey, The Meshinger[1]

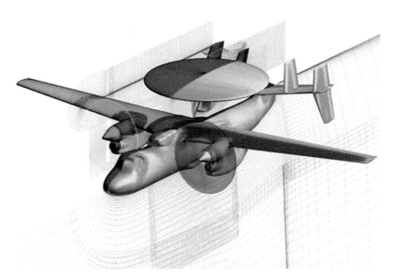

Block Structured Grid for an E-2C aircraft created by Warren H. Davis (Courtesy of Pointwise, Inc; a full color version of this image is available on the website: www.cambridge.org/aerodynamics).

LEARNING OBJECTIVE QUESTIONS

After reading this chapter, you should know the answers to the following questions:

- How does aircraft geometry become a surface representation that is usable in CA?
- How is a volume grid made from the aircraft surface representation?
- What are the differences between structured and unstructured grids?
- What are hybrid grids and how are they developed?
- What are some ways that grids are created around complex aircraft configurations?
- What are the ways that structured grids can be created? How about unstructured grids?
- What makes a grid "good" or "bad"? What are the specific ways that grids exhibit "goodness"?

- What is adaptive mesh refinement and why is it used?
- What level of grid spacing is required in a laminar boundary layer? How about for a turbulent boundary layer? Why are they different?
- What are the steps for conducting a grid sensitivity study?

7.1 Introduction

As difficult as it may be to believe, most aerodynamic analysis effort is 50 percent flowfield and 50 percent geometry. Although flowfield issues dominate the typical aerodynamics courses given in universities, the geometry issues often dominate real-world aerodynamic practice. The geometry issues can be even more daunting when it comes to performing computational analysis and design, where there are two distinct approaches to "the geometry problem" that require consideration. One approach is the creation of a grid system for the computation of a solution that can be used for analysis and/or design; the aerodynamicist achieves design objectives through the creation of geometry. Another approach is that practical aerodynamic results are created through the specification of a geometry that was defined specifically to produce a desired flowfield; the aerodynamicist specifies a flow and finds the geometry that creates it. Since most aerodynamicists use highly developed algorithms to perform their flow calculations, they typically spend most of their time defining geometry and creating grids to be used by their flowfield solvers. Thus, in practice, aerodynamic analysis and design are usually dominated by geometry considerations. This has led to the creation of software tools that can define the geometry and create grids about complex configurations – these tools are commonly referred to as *geometry modeling and grid generation* (GMGG) tools. We will attempt to provide an overview of the various approaches used to define geometry and create grids in this chapter.

7.2 Surface Shape Development: Lofting Techniques

The process of defining the external vehicle surface shape, or contour lines, is known as lofting; the basic ideas for lofting originated in the shipbuilding industry. To define the patterns for use in fabrication, a ship's hull shape was laid out in the lofts above the shop floor. Drafting curves appropriate for lofting are frequently known as "ship's curves," although many people call them aeronautical engineering curves since they were also used by aircraft designers (Raymer[2] provides examples and additional background on this subject). The purely visual art of lofting used on ships was replaced by the use of precise mathematical definitions for a surface in the aircraft industry in the 1930s. The particular mathematical form that was most popular at that time was the "conic section lofting," i.e., using parabolic, elliptic, and hyperbolic shapes to create the surface. This approach appears to have been formalized by Liming[3] at North American Aviation and was applied to the P-51 Mustang during World War II.

Often, aerodynamicists developed their own "desk drawer" geometry generators. Appendix A contains geometric details for numerous airfoils, fuselage type bodies, and wing geometries. These equations allow aerodynamicists to easily develop classic aerodynamic shapes, and computer programs are also available based on many of these definitions.[4] An aerodynamicist who developed simple methods to define geometric shapes is Ray Barger,[5,6] and, more recently, Brenda Kulfan at Boeing developed excellent geometry schemes suitable for designing aerodynamic shapes.[7,8] Many of these methods are often much more appropriate for aerodynamic design than the use of large general-purpose CAD systems. Various methods for smooth surface definition have also been developed for general applications to aircraft design; the book by Rogers and Adams[9] provides a description of general methodology used in CAD systems to develop surface contours, and de Berg et al. have recently released a new edition of their book on this subject.[10] Many engineers working in the aerospace industry doing geometry definition also use the book by Faux and Pratt (many computer lofting codes even specify the equation numbers being used from this book).[11] On a simpler level, an old data format developed at NASA and known as the "Craidon data" format[12] provides a common geometry that was used by many aerodynamic design codes.[13] However, that geometry definition format does not provide a convenient way to define smooth blending between components. The NASA-developed Vehicle Sketch Pad and AVID PAGE are examples of surface geometry programs that provide a smooth and flexible geometry definition for use in conceptual design as well as computational fluid dynamics. Boeing has also developed the Aero Grid and Paneling System (AGPS), which is a complete tool that "allows the engineer to create, manipulate, integrate, or visualize geometry of any type."[14] Thus, the aerodynamicist should either be able to "loft" his own shape or be able to work with the contour development group to obtain the desired shape. In this context, familiarity with conic lofting can be very useful. An example of the lofting process can be seen in Fig. 7.1, where the surface representation of the America's Cup yacht Stars & Stripes is shown.[9]

Today, modern *computer-aided design* (CAD) systems are used to define exterior geometry in what is known as *computer-aided geometry design* (CAGD). The group in an aircraft company that does the lofting is not called the lofting group anymore but is typically known as the contour development or master dimensions group. Surfaces are now defined using splines and other improved surface definition functions. Many of these advanced geometry features are available on various CAD systems including *Pro/Engineer*, *Inventor*, *SolidWorks*, *UGS NX*, *DesignModeler*, *Parasolid*, *ACIS*, *CAPRI*,

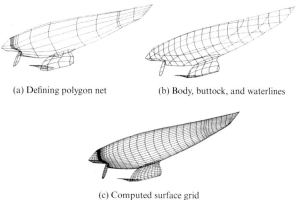

(a) Defining polygon net   (b) Body, buttock, and waterlines

(c) Computed surface grid

Figure 7.1 Basic lofting process for the America's Cup yacht Stars & Stripes (Ref. 10).

DXF/DWG, I-DEAS Viewer XML, Rhino3D, Workbench, and *CATIA.* The latter CAD system was developed in France by Dassault and is used at many airframe companies, including Boeing and Airbus, to design new airplanes. However, *CATIA* is a large, powerful program that requires considerable training to use and may not always be suitable for CFD since the detailed CAD descriptions usually take a great deal of effort to convert into useful surface geometries for computational aerodynamics. CAD descriptions of aircraft can be quite complex using the available programs, as is evident in the CAD representation of the F-16XL[15] shown in Fig. 7.2, which includes (among other things) a wing fence, an air dam (both located near the outboard wing crank), and a wing-tip missile (including the missile rail). This type of detail for CFD calculations was not possible even 10 years ago, but is becoming increasingly common with modern CAD systems and advanced computers.

Oftentimes, manufacturing considerations may restrict the shapes that aerodynamicists can use. Unless the aerodynamicist can demonstrate a severe performance penalty, the manufacturing group will urge that a "straight-line wrap" or "developable" surface be used. Using this approach, the wing will consist of straight-line wrap sections between a small number of production breaks. For highly tapered wings, this approach can lead to a poor aerodynamic twist distribution.[i] In addition, the axis about which wing twist is applied must be defined. Usually, fabrication is simplified if this axis is defined to coincide with the axis of the major control surface hinge line, since straight hinge lines simplify the manufacturing process. As an extreme example, the aerodynamicists designing the X-29 wanted a cambered and twisted canard. However, they were overruled, and a symmetric section untwisted canard was used so that only one mold was needed to fabricate the canard.

---

[i] See the exercise at the end of the chapter. Hint: If you think that the twist distribution between a root and tip incidence specification is linear, you would be wrong for a tapered wing. The same thing is true for the thickness distribution.

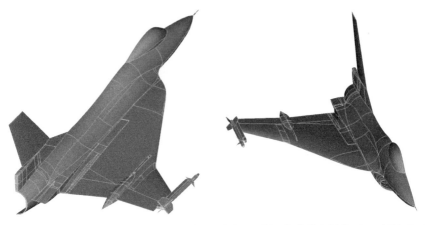

Figure 7.2 CAD representation of the F-16XL, including wing fence, air dam, and tip missile (Ref. 15; Courtesy of Okko Boelens of NLR).

Although the situation is much better today with the use of *computer-aided manufacturing* (CAM) systems, the aerodynamicist must always remain vigilant in case the contour development group makes shaping simplifications that may impact aerodynamic performance.

Profiles in Computational Aerodynamics: Joe Thompson

"A seventh-generation Mississippian, I grew up in Grenada, Mississippi, in a neighborhood of seven boys in four adjacent houses – all of whom went on to advanced degrees because we never knew college wasn't just automatically next down the road after high school. In early high school, I thought I might be a Presbyterian minister, but reading books on astronomy pulled me away from that. At graduation I briefly thought about majoring in English or music – both of which I now enjoy in retirement – but instead took the road at Mississippi State into chemical engineering because engineers had solid careers and I had had a good chemistry teacher in high school (as I had had in English and music). But halfway through college, I switched to physics, harking back to those astronomy readings. Graduation from college again posed a choice of roads and I initially continued in physics, only to be lured away into aerospace engineering by the emerging race to the moon. After getting my MS degree and joining NASA in the Apollo program, the road again forked, and I left the program before getting hooked into going all the way to the landing – to get my PhD and go for an academic career. That forty-five-year career was all at Mississippi State because of my inherent dedication to Mississippi and because MSU was a university large enough to have something happen and small enough that I could get the president on the phone.

"I got into CFD at NASA in the early '60s when I taught myself Fortran by comparing a paper with the equations with a computer printout modeling the pressurization of the liquid oxygen tank of the Saturn rockets. But my own CFD programming started with my dissertation at Georgia Tech, still with punched cards, the completion of which I think was determined by Tech wanting to get me off the computer. Then one day in the early 1970s, I was doing my monthly reading of the CFD literature when I picked up the idea of elliptic grid generation from a paper by W-H Chu of the Southwest Research Institute in San Antonio on sloshing in tanks. I still have a vivid memory of that day in my office – just back at MSU, freshly equipped with a PhD and looking for a key idea. As soon as I read Chu's paper, the idea leaped into reality. At that time, Frank Thames was at MSU looking for a dissertation topic. I gave an Open Forum presentation at the first AIAA CFD Conference in Palm Springs in 1973 on our preliminary results for potential flow, and met Bud Bobbitt and Jerry South of NASA Langley in the tram going up the mountain there. They funded the research that became Frank's dissertation in 1975 – with the grid around that cambered rock – and I was headed down my research road in grid generation. Our first paper came out in the *Journal of Computational Physics* in 1974, and it was at the 2nd AIAA CFD Conference in Hartford in 1975 that I first met most of the aerospace CFD crowd – the original generation.

"I continued in the grid generation part of CFD, and in the '80s, Dave Whitfield and I at MSU teamed for grids and CFD with Larry Lijewski's CFD group at Eglin Air Force Base. When Don Trotter brought microelectronics to MSU, we had the research nucleus to win an NSF Engineering Research Center award in 1990, dedicated essentially to CFD on high-performance computers. But that brought yet another fork in the road, taking me out of active research and teaching, and into research management for the second half of my career. Looking back, I'm glad for all the forks in the Yellow Wood, and I'm glad I stayed with my sense of place in Mississippi. My office wall was adorned with quotes from William Faulkner, my literary hero, and as he said, you always wear out life before you exhaust the possibilities of living. That being the case I walked away from engineering upon retirement, inspired by the muse to follow my interests in writing and music."

Dr. Thompson retired in 2009 as William L. Giles Distinguished Professor of Aerospace Engineering at Mississippi State University and director of the DoD Programming Environment & Training (PET) Center in the High Performance Computing Collaboratory (HPC²) at MSU. He received his BS in physics and MS in aerospace engineering from Mississippi State – and his PhD in aerospace engineering from Georgia Tech, where he has been named to the Academy of Distinguished Alumni. Dr. Thompson was the founding director of the NSF Engineering Research Center (ERC) for Computational Field Simulation at MSU and has been recognized by a number of university, state, and national awards for teaching, research, and service. He has authored and edited several books with colleagues, including the *Handbook of Grid Generation*, and has published numerous journal articles in the area of numerical grid generation applied in computational fluid dynamics. He received the AIAA Aerodynamics Award for this body of work and is a Fellow of the AIAA. Dr. Thompson was appointed by President Clinton to the President's Information Technology Advisory Committee (PITAC) and was reappointed by President Bush.

7.3 Computational Grid Overview

Once the surface shape is defined, the surface becomes the basis for the aerodynamic grid that will be used to make the calculations. In the early days of CFD, the algorithm developers created both the flowfield solver and the computational grid (often by themselves) using *research codes*. As time progressed, and the algorithms and applications became more and more complex, it quickly became apparent that both aspects of the CFD problem were very demanding. This led to highly specialized codes that concentrated on one aspect of the CFD problem – either grid generation or flow solution (as shown in Fig. 1.18). Therefore, the grid generation code output is typically an input to the flowfield code, and the solution from the flowfield code is an input to the flow visualization software (which will be discussed in Chapter 9). Once these more advanced tools were developed, it became possible for non-code developers to start using these methods to create grids. This has led to a more generalized (less specialized) approach to grid generation, with common characteristics among many software tools.

The basic grid generation approach consists of three fundamental aspects: the definition of the surface, as described in Section 7.2, the definition of a grid or mesh on the surface of the body being analyzed (the so-called *surface grid*), and the definition of the grid or mesh distribution off the surface (normally called the *volume grid*, as shown in Fig. 7.3). Each of these steps can be difficult and time-consuming for full, complex aircraft geometries. In fact, as we mentioned previously, the grid generation process has developed over the years and turned into a discipline in its own right. The initial book on the subject was written by Joe Thompson and his co-workers at Mississippi State University (this book is now freely available online at www.hpc.msstate.edu/publications/gridbook),[16] and they have also written another book on the subject.[17] A number of other good books have been written on grid generation, including those by Knupp and Steinberg,[18] Frey and George,[19] and Farrashkhalvat and Miles.[20] More generally, grid generation appears in many books emphasizing CFD, especially those where the flow solution algorithm and the grid structure and topology are closely connected. Grid generation has also been the subject of several NASA conferences,[21–23] as well as the symposium series on Computational Mechanics[24] and the Meshing Roundtable.[25]

While we will supply an overview of grid generation approaches in this chapter, readers interested in the details of the various methods should consult the references listed for more details.

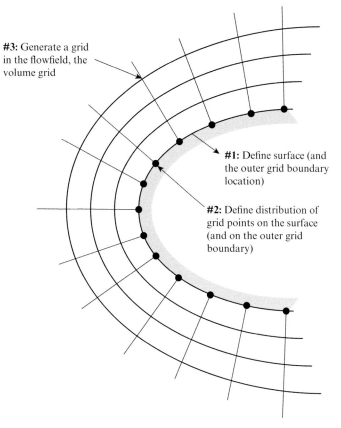

#3: Generate a grid in the flowfield, the volume grid

#1: Define surface (and the outer grid boundary location)

#2: Define distribution of grid points on the surface (and on the outer grid boundary)

Figure 7.3   Three parts required to generate a grid around a body.

7.4   Grid or Mesh Types

Just as there are various ways to numerically solve the equations of fluid motion (finite difference, finite volume, finite element, etc.), there are also many ways to create a grid around a surface. Perhaps the best way to discuss the various grid generation types is to see them all in relation to one another, as shown in Fig. 7.4. Some grids in use today, known as *structured grids*, are defined by a regular pattern of points. Because of the difficulty in generating grids for arbitrary geometries, another approach to grid generation has been developed that makes it easier to generate grids over arbitrary, complicated bodies. These grids use irregular patterns of tetrahedrons, prisms, and other grid cell types that allow the grid to be developed; the resulting mesh is known as an *unstructured grid*. Another potential advantage of unstructured grids is that the grid can be more easily modified to add resolution to the flowfield calculation as the solution evolves during an iteration, although this must be done with care in order to maintain grid quality.

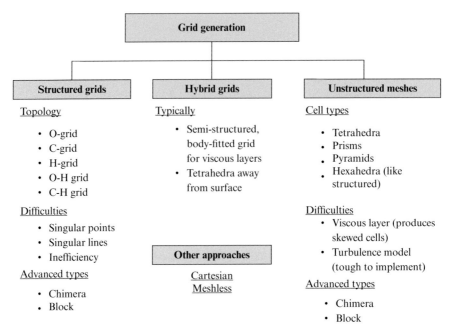

Various grid generation approaches.

A grid that is allowed to change during the solution is known as an *adaptive grid*, although there are also adaptive grid techniques that are used after a solution has been obtained. Grid adaptation is not yet in widespread use for general three-dimensional configurations, but various approaches to grid adaptation have been developed and will certainly play a larger role in future CFD approaches. Both approaches, structured and unstructured, have more advanced applications that will be discussed later in this chapter. As a side note, in general, structured formats are referred to as a *grid*, while unstructured formats are referred to as a *mesh*. Within the unstructured mesh, a single three-dimensional volume is often called a *cell*.

In practice we often use a combination of these two basic grid types. It has been found that the boundary layer flow over the surface can be computed more accurately using a semi-structured grid, while the rest of the flowfield can be easily defined using an unstructured mesh. Grids with both structured and unstructured regions are known as *hybrid grids*. Figure 7.5 illustrates the idea of a structured grid for a multi-element airfoil, while Fig. 7.6 shows an equivalent unstructured grid over a similar configuration (included in the review by Steinbrenner and Anderson[26]). Figure 7.7 shows an example of a hybrid grid over a wing, where a semi-structured grid (which is structured normal to the surface but unstructured on the surface) is used near the surface (usually to improve the quality of viscous flow predictions that use turbulence models), and an unstructured grid is used away from the surface (since they are typically quicker and easier to generate around complex geometries).[29]

When the aircraft geometry gets too complex for any single grid approach, several other approaches can be used. Figure 7.8 shows one of the methods

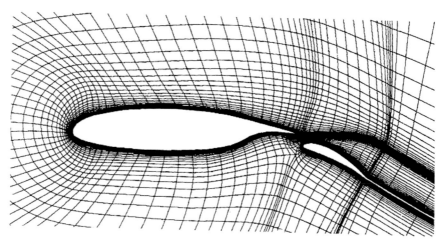

Figure 7.5 Structured grid example about a multi-element airfoil (Ref. 27; reprinted by permission of the American Institute of Aeronautics and Astronautics, Inc.).

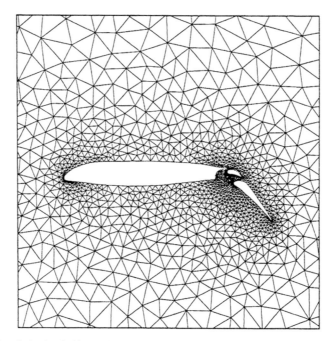

Figure 7.6 Unstructured grid example about a multi-element airfoil (Ref. 28; reprinted by permission of the American Institute of Aeronautics and Astronautics, Inc.).

commonly used for complex geometries which is known as a *Chimera grid* (or *overset grid*). A Chimera grid uses a background grid for the majority of the flowfield, and additional detailed grids around components where more grid resolution is required.[30] In the case of the high-lift airfoil shown in Fig. 7.8a, the fine overlapping grids are being used to provide adequate grid resolution in the vicinity of the leading-edge slat and the flap. The coarse and fine grids overlap each other, and the flow solver communicates information between them using various interpolation schemes. The Chimera approach

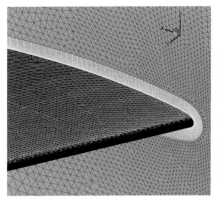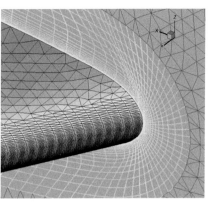

Figure 7.7   Hybrid grid example around an airfoil with a semi-structured grid near the surface and an unstructured grid away from the surface (Ref. 29).

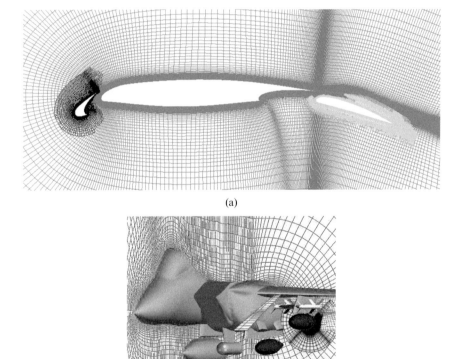

(a)

(b)

Figure 7.8   Examples of Chimera or overset grids: (a) high-lift airfoil (Ref. 31), and (b) F-15 with movable stores (Courtesy of USAF Seek Eagle Office; a full color version of this image is available on the website: www.cambridge.org/aerodynamics).

has proven to be extremely valuable for complex configurations in three dimensions, such as the wing-pylon-store configurations on the F-15 fighter shown in Fig. 7.8b, especially when simulations are required with moving objects, such as store separation.

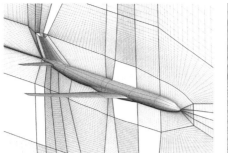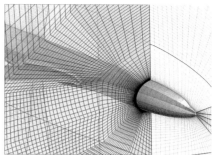

Figure 7.9   An example of a block structured grid around a passenger jet aircraft configuration (Ref. 32; Courtesy of Simao Marques and Ken Badcock of the University of Liverpool; a full color version of this image is available on the website: www.cambridge.org/aerodynamics).

Block structured grids are also very valuable in dealing with complex configurations, such as the passenger jet aircraft shown in Fig. 7.9.[32] Block structured grids create various zones (known as blocks) around an aircraft configuration and fill each block with a fairly simple structured grid. These structured grids are then connected to each other in a variety of ways, the simplest way being one-to-one point matching at the block interfaces. There are issues of numerical error and conservation (see Section 3.5.2 for a discussion on conservation form) that apply to the interfaces used in Chimera or block structured grids, but these issues are constantly being improved.[33]

Because of their relative simplicity, strictly *Cartesian grids* can also be very useful for complex configurations.[34] Cartesian grids fill the volume around a configuration with straight lines aligned with a Cartesian axis system. This has the advantage of being relatively easy to implement around complex configurations and readily allows for grid adaptation in regions of high flow gradients (rectangular cells are subdivided to form smaller cells where needed). The difficulty in using these grids is that the geometry of complex configurations is not aligned with the axis system (in general), and therefore the Cartesian grid must be refined heavily in the region of complex surface curvature using cut cells. Because of this, Cartesian grids show good results when solving the Euler equations for inviscid flow but are not as easy to apply to viscous flows. Figure 7.10 illustrates this type of gridding approach applied to the Space Shuttle during ascent and is compared with a Chimera grid over the same configuration. Notice how quickly the Cartesian grid becomes coarse away from the surface of the vehicle, and how regions of high flow gradients have finer grid density.

The Cartesian approach also lends itself to grid adaptation, since cells with high flow gradients are easily subdivided repeatedly until the flow gradients are adequately resolved. This is illustrated in Fig. 7.11, where the Cartesian grid of Fig. 7.10 has been refined. The original grid in Fig. 7.10a contains 96 million points, the Cartesian grid in Fig. 7.10b has 10 million cells, and the adapted grid in Fig. 7.11 contains 14 million cells.

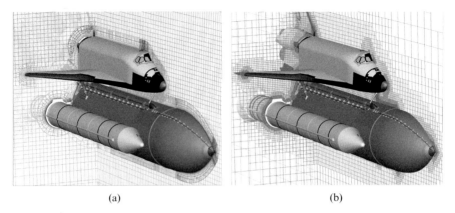

(a)   (b)

Figure 7.10 A Chimera grid around the Space Shuttle compared with a Cartesian grid (Courtesy of Reynaldo Gomez of NASA Johnson Space Center and Scott Murman of NASA Ames Research Center). (a) Chimera grid (picture shows every eighth point; 96 million points) (b) Cartesian grid (10 million cells)

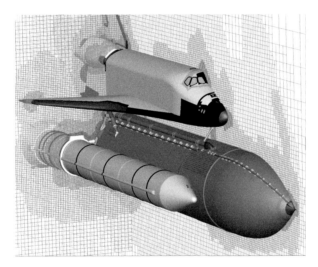

Figure 7.11 A Cartesian grid around the Space Shuttle with grid adaptation applied; 14 million cells (Courtesy of Reynaldo Gomez of NASA Johnson Space Center and Scott Murman of NASA Ames Research Center).

Cartesian grids can also adopt the basic concepts of the Chimera approach and allow relative motion between multiple sub-components within the grid, such as happens during store separation on tactical aircraft. Figure 7.12 shows a store separating from an F-18 using Cartesian grids, where a hole is cut from the main grid to allow the sub-munition grid to "fly through," which allows for the prediction of aerodynamics for the relative motion of the two components.[35]

Perhaps the most recent grid generation concept is one that really does not have a grid or mesh in the traditional sense: these are the "meshless" or "gridless" approaches. In this approach, a "cloud" of points is distributed

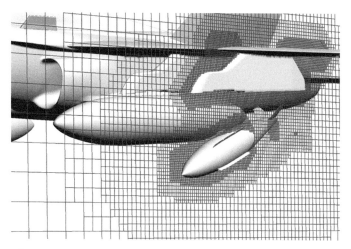

Figure 7.12 A Cartesian grid around the F/A-18 with a sub-munition hole cut (Ref. 35; Courtesy of Scott Murman of NASA Ames Research Center).

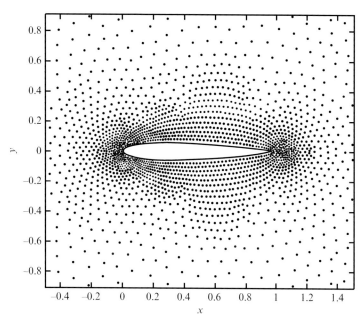

Figure 7.13 Global point cloud around an NACA 0012 airfoil (Ref. 36; reprinted by permission of the American Institute of Aeronautics and Astronautics, Inc.).

around a body, such as an airfoil or wing (see Fig. 7.13). The distribution of points could be random, but typically is based on a more complicated scheme for distributing points, so that the density of points is high in regions closest to the airfoil or wing and the points are conformal to the surface.[36] This distribution is known as the global point cloud, and notice that while the points that make up the cloud are defined, there are

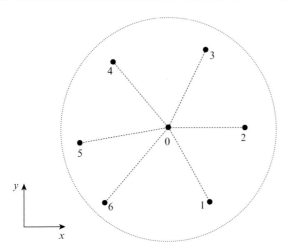

Figure 7.14 Local point cloud (Ref. 36; reprinted by permission of the American Institute of Aeronautics and Astronautics, Inc.).

no connecting lines between the points. This approach does not define relationships between points prior to performing any calculations (such as structured or unstructured grids), which makes creating these "grids" quite fast and relatively easy.

Once the global point cloud is created, calculations of flow quantities around the airfoil are performed by choosing a point in the field (such as Point 0 in Fig. 7.14) and finding the nearest neighbors to that point (in this case Points 1 through 6). This collection of points is known as the local point cloud. Once these points and their positions are defined, numerical representations of flow quantities and derivatives can be made. Some of the methods for performing these calculations include the Taylor series least squares method, the polynomial basis least squares method, and the radial basis function collocation method.[37] This allows for flow solutions to be obtained without spending a great deal of time creating grids, especially for grids near solid surfaces, something for which unstructured and Cartesian grids are not well suited.

An example of the power of the meshless approach can be seen in Fig. 7.15. Here, the meshless concept is applied in a hybrid fashion, in this case in conjunction with a Cartesian grid.[38] A body-conforming gridless method is used for accurate implementation of the boundary conditions on the body surfaces. It is relatively easy for the meshless region to move through the Cartesian grid with the store, allowing for relative motion calculations at relatively low computational cost.

These approaches to grid generation, as well as the other approaches shown in Fig. 7.4, can be quite complicated and could easily be described in their own chapters. Although we will provide an overview of how some of these

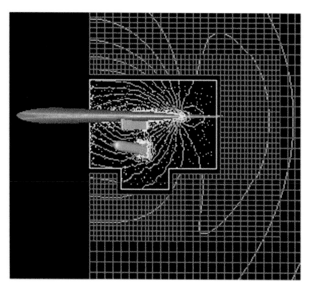

Figure 7.15   Meshless region containing a store separating from a wing, all contained inside a Cartesian grid (Ref. 38; reprinted by permission of the American Institute of Aeronautics and Astronautics, Inc.).

methods work, the details of these meshless methods are beyond the scope of this book.

7.5 Structured Grids

In a regular structured grid topology, each point (or node) in the network is connected with two neighbors along one or more dimensions, as shown in Fig. 7.16. This "structure" allows for relatively easy use of finite difference schemes, especially since they are derived using neighboring points (see Section 6.2.1 for more details and the definitions used for the index notation). The structured grid approach has been used to varying degrees over the years. In the early days of CFD, most grids used various structured topologies, but unstructured meshes have become quite popular in recent years. Improvements in block structured grid approaches have revitalized structured grids, and higher-order methods are easier and more efficient when applied to structured grids. So while unstructured grids have grown in popularity, structured grids are still widely used in various CFD applications.

7.5.1 Topologies

There are three basic topologies (or shapes) used for structured grids: O-grids, C-grids, and H-grids. Each of these grids is named after the shape the grid

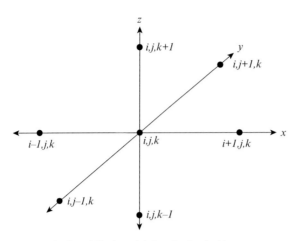

Figure 7.16   Index notation for neighboring points in a structured grid.

takes in a two-dimensional plane, as seen in Fig. 7.17. These grids are created using some type of grid generation scheme, and then, in order to perform computations on the grid, they are transformed into a computational Cartesian grid.[39]

Three-dimensional grids generally use combinations of these two-dimensional topologies. For example, Fig. 7.18 shows an H-H grid topology used on a commercial transport aircraft. One of the disadvantages of an H topology is that regions of grid clustering (locations where grid points are closer together) must extend out into the domain away from the vehicle, which leads to some inefficiency (for example, see the region above and below the wing in the centerline view). The advantage of an H topology, however, is the relative ease of creation. For example, an H-H grid around a wing can be made simply by stacking the airfoil H grid in various spanwise locations to create the three-dimensional shape.

Another example of how to make a three-dimensional wing is with a C-O grid, as shown in Fig. 7.19. Notice that the airfoil C grid has a similar difficulty as the H grid previously discussed, namely that the grid density required around the airfoil surface is continued into the wake (which may actually be helpful), but the high grid density over the surface of the wing is also continued above the wing, creating a large number of points where the flow gradients are small.

In general, the O grid uses points most efficiently, and the H grid uses points least efficiently, as mentioned previously. Eriksson[40] published a quantitative estimate of the relative number of grid points required to achieve the same accuracy on an isolated wing with different grid topologies, as shown in Fig. 7.20. Nearly a factor of four more grid points are required when using

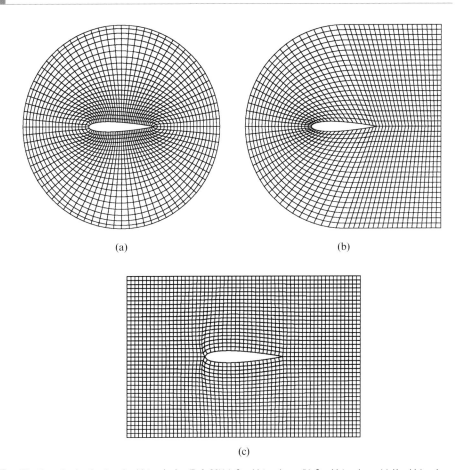

(a)   (b)

(c)

Figure 7.17   The three basic structured grid topologies.(Ref. 39)(a) O-grid topology; (b) C-grid topology; (c) H-grid topology (Courtesy of James Forsythe).

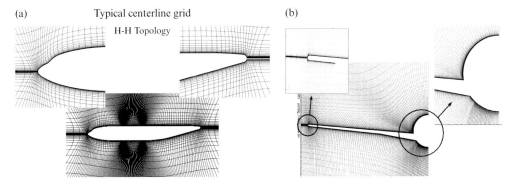

Figure 7.18   Example of H-H grid topology (Courtesy of Edward N. Tinoco). (a) aircraft centerline view; (b) spanwise view.

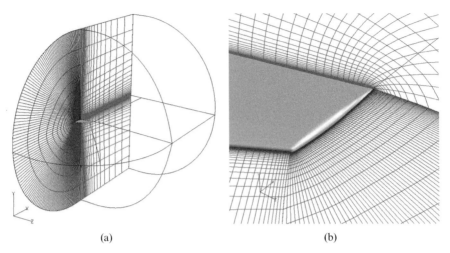

(a)   (b)

Figure 7.19   Example of C-O grid topology (Courtesy of NASA Glenn Research Center) (a) wing centerline view; (b) wingtip view

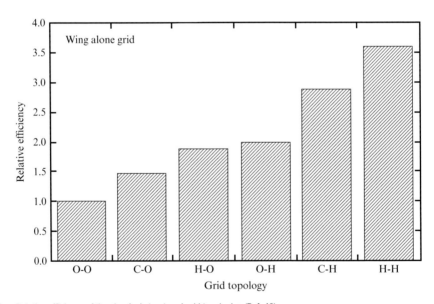

Figure 7.20   Relative efficiency of the classical structured grid topologies (Ref. 40).

an H-H grid topology as compared to an O-O grid topology. Clearly, it can be difficult to generate a single grid that conforms to the body and has all the characteristics of good grids (which we will discuss later). The bottom line is that all grid topologies have advantages and disadvantages, both in ease of creation and relative efficiency for calculations. These difficulties are what led to the Chimera and block-structured grid approaches described in Section 7.4. Unstructured grids, described in Section 7.7, also provide a solution for some of these difficulties.

Flow Visualization Box

How to Look at Grids

Start the FieldView Demo version and open the F-18 solution that you first saw in Chapter 3. Turn off the streamlines and iso-surfaces so you just see the surface geometry as shown in the figure.

Now add the surface mesh by clicking *Show Mesh* in the *Boundary Surface* window to obtain the following view.

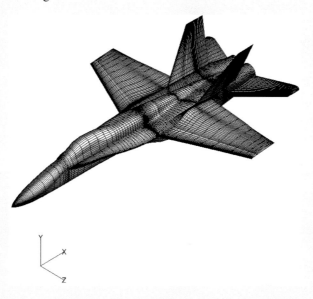

Can you tell what kind of grid this is? Structured, unstructured, viscous, inviscid? What topology does it have? If you are not sure, maybe we should add a *Coordinate Plane* to show what the grid looks like on the centerline.

Now what do you think? Yes, this is a structured grid with C topology on the centerline. You can also see that there is very little clustering near the surface (no resolved boundary layer) so this is an Euler grid used for inviscid calculations. Feel free to continue exploring this grid to determine the topology around the wing tips (an x-plane view), how far the outer boundaries are located from the aircraft, and other topics we are learning about.

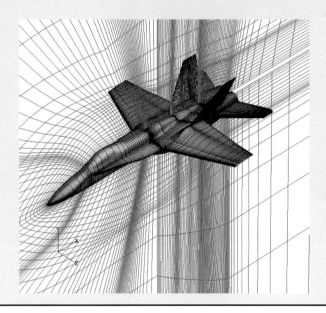

7.6 Methods for Creating Structured Grids

Several methods are used to systematically generate structured grids, depending on the complexity of the geometry and the requirements of the flow solver. These methods have developed over many years and a great deal of effort has been undertaken to improve the various structured grid generation approaches.

The basic methods used to generate structured grids are usually categorized into three basic approaches:

- algebraic grid generation
- conformal transformation through analytic mapping
- differential methods using partial differential equation concepts

Each of these methods has an important place in the history of CA, and have led to the modern grid generation tools that are available today. Understanding how each of these approaches works can be a valuable lesson,

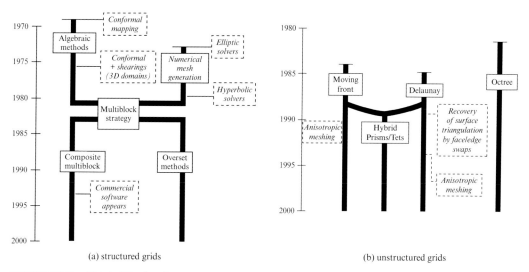

(a) structured grids (b) unstructured grids

Figure 7.21 Historical timelines for the development of structured and unstructured grid concepts (Ref. 41; Courtesy of Luigi Martinelli of Princeton University).

so we will discuss each of these at some length. Figure 7.21 shows how these grids have developed over the past forty years, including the development of methods that combine grids using multiblock or overset approaches; unstructured grid development is also shown and will be discussed later. Also notice that commercial software for grid generation did not become available until long after most of these methods had been in place and widely used.

There are a variety of purposes for using structured grid generation, many of which are a direct result of the requirements for utilizing finite difference representations of the governing equations, including:

1. distributing points over the flowfield so that neighboring points can be identified
2. allowing communication among points
3. allowing for the accurate representation of continuous functions by discrete values
4. supplying information about the quality of the grid depending on the gradients in the flow solution

Classical structured grid methods usually "map" the physical domain into a computational domain such that a constant coordinate surface in the mapped plane becomes the surface of the body (look at Fig. 7.22 to see how this works). The result is that the computational domain appears to allow the computational grid to be constructed quite simply. In the example in Fig. 7.22, $\eta = 0$ is taken to be the body surface. We would like the mapping or "transformation" to produce a nearly orthogonal grid in the physical domain, and we would also like the grid to be orthogonal to the surface. We do not want the grid to be excessively skewed or to have extremely high or

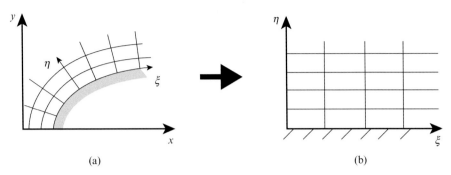

Figure 7.22   Example of mapping the physical domain to the computational domain. (a) physical domain; (b) computational domain.

low aspect ratios ($\Delta\eta/\Delta\xi$ too large or too small). In practice, achieving these objectives often proves to be quite difficult, especially in three dimensions. The transformed Cartesian (computational) space domain is where finite difference methods can be easily applied. After the governing equations are solved in the computational domain, the results can be mapped back to the physical domain.

One of the complications of the mapping approach is that it means the governing equations also have to be transformed, and we will see that this makes them more complicated. However, this is done so that the boundary conditions are easily and accurately imposed and so that the finite difference approach can be easily applied (see Appendix E for details).

We will discuss both the transformation process itself and issues that you need to watch out for when using mappings in the following section. In general, the various types of grid generation also allow for the following characteristics once the grid has been transformed into a computational domain:

1. allowing for grid clustering in regions of high flow gradients
2. creation of various grid topologies (C grid, O grid, etc.)
3. allowing for finite differencing in a rectangular mesh

A simple example of a mapping is the application of algebraic methods to a nozzle, as shown in Fig. 7.23. If the upper surface of the nozzle is defined as $h_{us}(x)$, and the lower surface is defined by $h_{ls}(x)$, then the mapping from the physical plane to the computational plane is simply:

$$\xi = x, \quad \eta = \frac{y - h_{ls}(x)}{h_{us}(x) - h_{ls}(x)} \tag{7.1}$$

where the upper nozzle wall corresponds to $\eta = 1$ and the lower nozzle wall is $\eta = 0$. The problem becomes much simpler computationally after this transformation is accomplished, since the resulting computational plane is Cartesian with equal grid spacing.

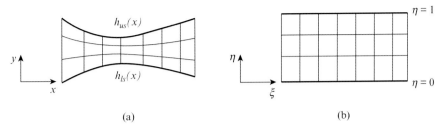

Figure 7.23   Example of mapping for a nozzle shape. (a) physical plane; (b) computational plane.

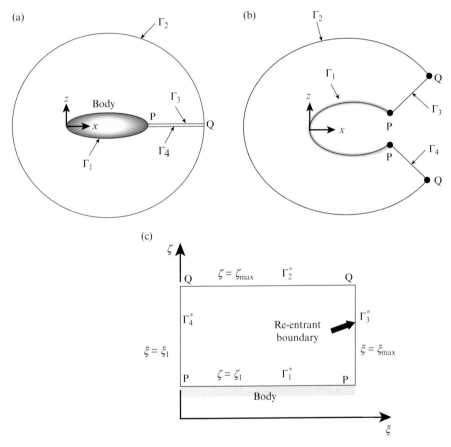

Figure 7.24   The "unwrapping" of a body grid: (a) physical domain (doubly connected region and branch cut); (b) unwrapping of the doubly connected region; and (c) computational domain (Ref. 42 Courtesy of Eddie Ly of RMIT.).

A more complicated example of a transformation relates to the airfoils or bodies we mentioned in Section 7.5. For example, an O-grid around a body is "unwrapped" so that the solid surface forms the bottom portion of the computational grid, while the outer boundary forms the top, and the wake cut forms the sides (see Fig. 7.24). Similar strategies for creating computational grids are possible for the other grid topologies shown in Fig. 7.17 as well.

Interested readers can learn more about grid generation in greater detail by reading the book by Thompson et al.[16] (available online at www.hpc.msstate.edu/publications/gridbook) and the introductory chapter of Ref. 18. Surveys by Steinbrenner and Anderson[26] and Thompson and Weatherill[43] also provide good details about the subject. The books cited previously[16,18] describe the methods, and briefer treatments are given in chapters in many CFD books. The book by Fletcher[44] is a good example and contains a good section devoted to grid generation, including a sample program.

7.6.1 Algebraic Grid Generation and Stretching/Clustering

The most basic way to create a structured grid is with *algebraic grid stretching* or *clustering*. Since in the mapped domain (the computational plane) the grid is Cartesian, we are mainly interested in how to divide the straight lines into reasonable increments, $\Delta \eta$ and $\Delta \xi$. This concept is straightforward to implement, in that the position of the grids is determined by various algebraic functions that create stretching according to their definition. Examples of functions that are commonly used for grid stretching include:

1. polynomials
2. sine and cosine
3. natural logarithm
4. hyperbolic sine and hyperbolic tangent
5. error function

To bunch grid points close to the areas of rapid flowfield change, a smoothly varying distribution of the grid spacing is developed using the mathematical functions listed previously (Fig. 7.25 illustrates the procedure using a hyperbolic sine function). Studies of clustering have been conducted to determine the best clustering formulas,[16] but, in general, good clustering minimizes the truncation error due to uneven spacing. A simple example of stretching commonly used in boundary-layer calculations is:[45]

$$\eta_j = \eta_{j-1} + h_j \tag{7.2}$$

where

$$h_j = K h_{j-1}. \tag{7.3}$$

Here, the value of K is taken to be constant, and the grid is defined by specifying K and the initial value of h. $K = 1$ results in a uniform spacing, and $K = 1.20$ opens up the grid spacing as you move away from the surface. This

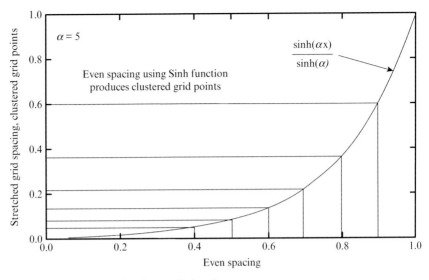

Figure 7.25 Algebraic grid example with clustering near the boundary.

defines a simple and systematic way to increase the grid spacing; however, it is not necessarily optimal. Another popular grid clustering function is the hyperbolic sine function:

$$y = \frac{\sinh(\alpha x)}{\sinh(\alpha)} \tag{7.4}$$

where an even spacing in x leads to a clustering of points in y, and the coefficient α is used to control the degree of clustering. A value of $\alpha = 5$ is a reasonable starting point for numerical investigations; the results from Equation (7.4) are illustrated in Fig. 7.25.

Another example of a common stretching function is:[45]

$$f(y) = B + \frac{1}{\tau}\sinh^{-1}\left[\left(\frac{y}{y_c} - 1\right)\sinh(\tau B)\right] \tag{7.5}$$

where

$$B = \frac{1}{2\tau}\ln\left[\frac{1 + (e^{\tau} - 1)(y_c / h)}{1 + (e^{-\tau} - 1)(y_c / h)}\right]. \tag{7.6}$$

This function creates grid clustering at a point in the middle of a grid as shown in Fig. 7.26.

When grid clustering is required in multiple dimensions, one-dimensional grid clustering can be accomplished in each coordinate direction independently (you would create the grids one dimension at a time, as shown in Fig. 7.27, where the error function, $erf(x)$, has been used to cluster the grid

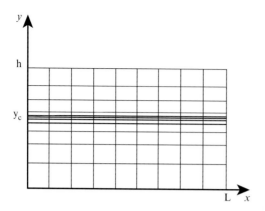

Figure 7.26 Hyperbolic sine grid clustering example; clustering in the middle of the region (Ref. 45).

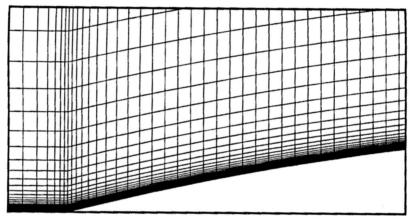

Figure 7.27 Example of algebraic stretching in two dimensions using the error function (Ref. 46).

in two directions). In general, algebraic grid generation is not as useful now as it once was due to major improvements in grid generation using the solution of partial differential equations, but algebraic clustering is still an easy, straightforward way to create a grid with high levels of control.

Two other techniques are widely used in conjunction with algebraic grid generation. The most common approach in applied aerodynamics is "transfinite interpolation," a technique developed by Gordon and Hall for application to automobiles[47] and made popular in aerodynamics by Eriksson.[40] In this approach, a simple two-dimensional interpolation scheme is used to create a smooth grid with varying levels of fineness. Most simple grid generation codes available in aerodynamics use this approach; an example for a delta wing is presented in Fig. 7.28.

To control orthogonality at the boundaries of the grid and at selected interior locations, a method developed by Eiseman[48] is also often used; this method is known as the multisurface method. The importance of

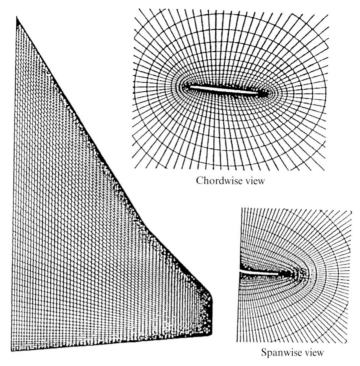

Chordwise view

Spanwise view

O-O type mesh with $160 \times 48 \times 80$ cells around a cranked delta wing

Figure 7.28   Algebraic grid generation with transfinite interpolation (Ref. 40).

boundary surface orthogonality will be discussed further in Section 7.11.2.3. Here, another surface is defined near the boundary, and the interior surface is used to ensure that the grid is locally orthogonal at the surface. An example of the multisurface method applied to an airfoil is presented in Fig. 7.29.

7.6.2 Conformal Transformation

Conformal transformation has its roots in the theoretical formulations for airfoil theory and is one of the original methods used to generate computational grids (see Fig. 7.21). The idea is to relate an easily constructed mesh over a simple geometry to a mesh around a general shape. Recalling that angles are preserved in conformal transformations, an orthogonal grid around the simple geometry will remain orthogonal around the general shape. In airfoil analysis, the method is closely related to the methods developed to analyze incompressible potential flow around arbitrary airfoils, and the Joukowski transformation was the original example, as we discussed in Section 4.4.1. The classic simple body shape used in potential flow aerodynamics is the circular cylinder. The grid around the cylinder or, in effect, the computational plane can then be transformed to create a grid in the physical plane (as shown in Fig. 7.30). This is the Circle Plane approach used by

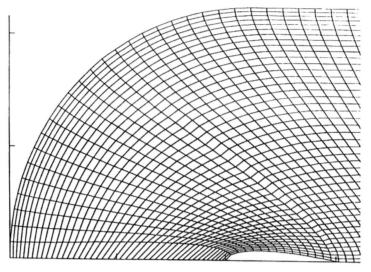

Figure 7.29   Algebraic grid generation concepts with the multisurface method (Ref. 44).

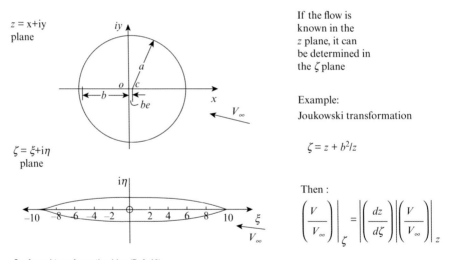

$z = x+iy$ plane

$\zeta = \xi+i\eta$ plane

If the flow is known in the z plane, it can be determined in the ζ plane

Example: Joukowski transformation

$$\zeta = z + b^2/z$$

Then :

$$\left.\left(\frac{V}{V_\infty}\right)\right|_\zeta = \left|\left(\frac{dz}{d\zeta}\right)\right|\left.\left|\left(\frac{V}{V_\infty}\right)\right|\right|_z$$

Figure 7.30   Conformal transformation idea (Ref. 49).

Jameson in his two-dimensional potential flow codes.[50] It is highly reliable and is one example where the grid is generated automatically in the flowfield solution code.

In the Circle Plane mapping, the computational domain is transformed from the region between the cylinder surface (normalized to a unit radius) and infinity to the region inside the cylinder. Here, "infinity" becomes the origin of the coordinate system, according to the original approach of Sells.[51] Figure 7.31 illustrates the idea, and Fig. 7.32 shows the flowfield inside the unit circle; the resulting grid in physical space is shown in Fig. 7.33. Note that the grid is naturally clustered near the leading and trailing edges using conformal transformations, and the grid is orthogonal to the surface of the

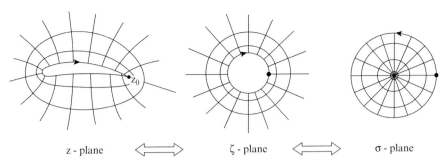

z - plane ⟺ ζ - plane ⟺ σ - plane

Figure 7.31   Circle plane mapping (Ref. 52).

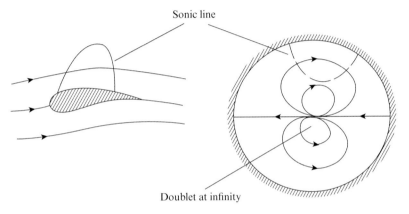

Sonic line

Doublet at infinity

Figure 7.32   Flowfield in circle plane (based on notes from Anthony Jameson).

airfoil. Additional details about the approach, and the extension of the concept to two-element airfoils, are described in the work by Ives.[53]

One problem with the method as presented here is that it is usually restricted to two-dimensional cases. A series of two-dimensional grids can be "stacked" together to create a three-dimensional grid (say, for a rectangular wing), or general three-dimensional shapes can be obtained by going through a series of transformations; the aft-fuselage mounted nacelle/pylon combination shown in Fig. 7.34 was created in this way.[54] The use of conformal transformation methods is a relatively sophisticated mathematical procedure, however, and for this reason this method is not often used in most current grid generation codes. Instead, different methods with greater flexibility and generality are usually used, such as methods based on PDE concepts.

7.6.3 Elliptic Grid Generation

Elliptic grid generation is accomplished by solving an elliptic PDE to create the grid. While it may sound strange to solve a PDE prior to finding the solution to a set of PDEs, this type of grid generation is quite popular,

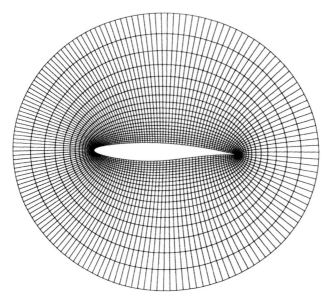

Figure 7.33 Grid in physical space generated with conformal mapping (Ref. 50; reprinted by permission of the American Institute of Aeronautics and Astronautics, Inc.).

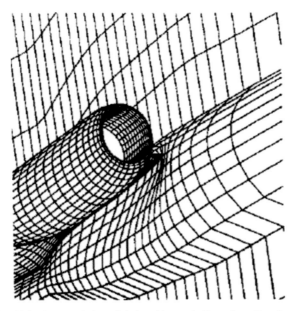

Figure 7.34 Aft-fuselage mounted nacelle/pylon grid generated by conformal transformations (Ref. 54; reprinted by permission of the American Institute of Aeronautics and Astronautics, Inc.).

especially now that computer speeds make these methods much faster and easier to use. Since the grid is a solution of a PDE, it often has the basic shape of the flowfield on which it will be solved, giving the grid good solution characteristics.

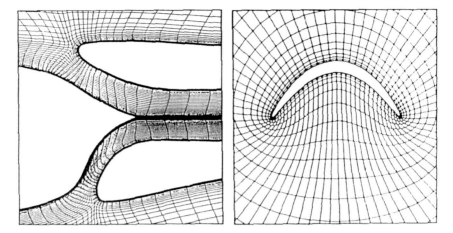

Figure 7.35 The elliptic PDE grid generation concept (Ref. 55).

The classic example of elliptic grid generation is to solve Poisson's equation numerically:

$$\Phi_{xx} + \Phi_{yy} = f(x, y). \tag{7.7}$$

In this case, we have written the equation in two dimensions to create a two-dimensional grid, but we could also create a three-dimensional grid using the three-dimensional version of the equation. Poisson's equation looks a great deal like Laplace's equation (Equation 3.130), and, in fact, the left-hand side is exactly the same (both Laplace's equation and Poisson's equation are elliptic PDEs). The function on the right-hand side of Equation 7.7, $f(x, y)$, is what makes this Poisson's equation. That function is a "control function," which is used to control the grid quality (including spacing, growth rates, etc.). Since this is an elliptic PDE, we must solve it in a region surrounded by boundaries with appropriate boundary conditions (which are also used to control the resulting grid). The solution process requires an initial grid to use as a starting solution, and this grid is normally provided by an algebraic grid generator. Then Laplace's equation is solved on the initial grid, with the appropriate boundary conditions. Since Laplace's equation was the basis for potential flow theory, solving this equation within a region tends to create grid lines that roughly follow flow directions, illustrated in Fig. 7.35. Notice that, in regions with confined boundaries, the grid lines compact (which is good for a region where flow gradients might be high), and the grid lines are roughly parallel to each surface. The resulting grids are often very good for computing flow over the boundary shapes, since the grid was created with an equation that models inviscid, incompressible flow.

Figure 7.36 provides a second example, in this case of a three-dimensional grid. Figure 7.36a shows the baseline grid created about the Space Shuttle

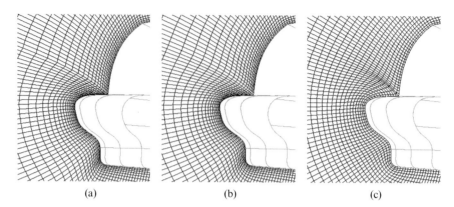

(a)   (b)   (c)

Figure 7.36   Importance of control functions with elliptic PDE grid generation (Ref. 56). (a) baseline grid; (b) grid created with fixed control functions; (c) grid created with control functions.

Orbiter using transfinite interpolation (see Section 7.6.1). The baseline grid then provides the starting solution for an elliptic PDE solver, specifically a Laplace equation solver with fixed control functions, $f(x, y, z) = 0$. The resulting grid, shown in Fig. 7.36b, has smoothed some of the irregular shapes in the grid, but the overall grid is very similar to the baseline grid, especially in the non-orthogonal nature of the cells near the body surface. Finally, a grid using Poisson's equation with control functions is shown in Fig. 7.36c. In this case the control functions were defined by Sorenson and Steger,[57] and the resulting grid is very smooth with nearly orthogonal interfaces with the solid surface.

Compared with algebraic and conformal grid generation concepts, elliptic PDE methods are very powerful and able to generate structured grids about complex configurations. The limitation is that they require multiple steps to create the final grid, and experience with the control functions becomes an important skill for the user.

7.6.4 Hyperbolic Grid Generation

Hyperbolic grid generation has found a great deal of favor among practitioners because hyperbolic grids can be "grown" away from an initial data surface with no need for multiple solutions to be obtained (which is usually the case with elliptic methods).[58,59] Since hyperbolic PDEs do not require boundaries enclosing the area where the grid is being created (see Chapter 3 for more details), the grid can often be created faster and with a higher level of control. A possible disadvantage is that the outer boundary shape cannot be controlled as easily, since it is the result of solving a hyperbolic PDE and marching outward in space.

Hyperbolic grid generation is accomplished with a set of steps in quite a different fashion from elliptic grid generation. First, the grid generator

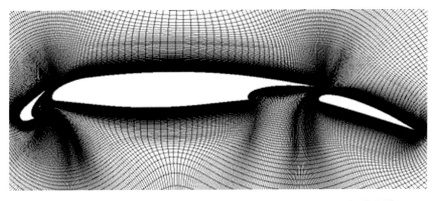

Figure 7.37   A single structured grid around a three-element airfoil created with hyperbolic grid generation (Ref. 59).

specifies that the body normal grid lines (the marching direction) must be orthogonal to the surface grid lines. Next, a surface normal direction is found, and, finally, a cell size is defined. This does not actually require the solution of a PDE, but, rather, creates a structured grid in a hyperbolic fashion similar to solving a PDE. An example of a hyperbolic grid is shown in Fig. 7.37, where a hyperbolic grid generator has created a single structured grid around a three-element airfoil. Note the control of grid spacing near the surface, in the cove region, and in the spaces between the elements. On the other hand, since the grid is "grown" from the surface outward, some irregular shapes occur in the grid, as can be seen in the wavy grid lines above and below the main element of the airfoil. Because of the high level of control, hyperbolic grid generators are currently quite popular in commercial grid generation software.

7.7 Unstructured Meshes

Unstructured meshes (or grids), as shown in Fig. 7.38, have become increasingly popular in recent years due to the relative ease of making meshes around complex configurations. While structured grids can take weeks (or even months) to create around a full aircraft, unstructured meshes can often be created in days or weeks. Unstructured meshes are not without their problems, however, as viscous computations requiring turbulence models are not easily applied to unstructured meshes. This is a major reason why hybrid grids are commonly used today, where quasi-structured grids are used near the surface of the body and unstructured meshes are used elsewhere (as shown in Fig. 7.7). There are a variety of methods for creating unstructured meshes, using different cell types, depending on the complexity of the configuration and whether the problem to be solved is two- or three-dimensional. An overview of the development of these unstructured grid types is shown in Fig. 7.21.

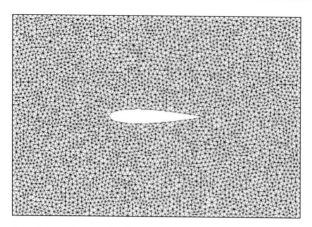

Figure 7.38 A simple isotropic unstructured mesh (Courtesy of James Forsythe).

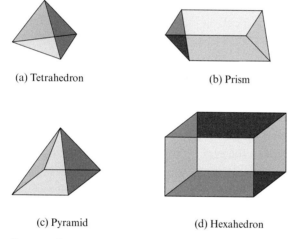

(a) Tetrahedron   (b) Prism

(c) Pyramid   (d) Hexahedron

Figure 7.39 Common cell types used in unstructured meshes.

7.7.1 Cell Types

In general, unstructured cell types can be almost any shape. Since unstructured meshes are most commonly used with finite volume formulations, which use the cell as a control volume for evaluating the equations of motion (as described in Section 6.11), any arbitrary volume could work. In general, however, there are a number of common cell types, as shown in Fig. 7.39, including:

1. tetrahedron – a four-sided cell (a three-dimensional triangle)
2. prism – a five-sided cell with triangular end faces connected by straight lines
3. pyramid – a polyhedron with one face usually having four sides and the other faces made of triangles
4. hexahedron – a six-sided cell with quadrilateral end faces connected by straight lines (essentially the same as used in structured grids)

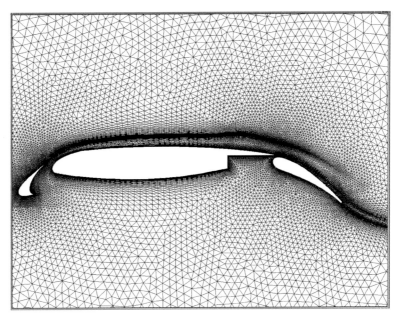

Figure 7.40 An unstructured mesh over a multi-element airfoil (Ref. 60); Courtesy of Dimitri Mavriplis of the University of Wyoming).

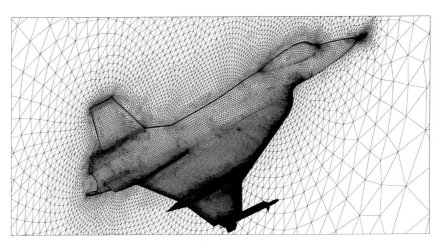

Figure 7.41 An unstructured mesh over an F-16XL aircraft (Ref. 61).

An example of an unstructured mesh for a multi-element airfoil is shown in Fig 7.40.[60] This shows the striking difference between structured and unstructured grids (compare Fig. 7.40 with Fig. 7.37, for example). Figure 7.41 illustrates the use of unstructured grids on an F-16XL aircraft with a wing-tip missile, which is a complex geometry with quite a bit of detail. The surface mesh has been generated with high fineness due to the complex flow topologies that exist over the wing, and was created from the CAD representation shown in Fig. 7.2.[61] For further details on unstructured meshes, see the surveys by Mavriplis[62,63] and Spragle, McGrory, and Fang.[64]

7.8 Methods for Creating Unstructured Meshes

Unstructured meshes have come of age over the past twenty years as CFD has been able to handle more and more complex aircraft configurations. The original unstructured meshes had several limitations, however, which had to be overcome before unstructured meshes became more useful. Some of these limitations were more perceived than actual, but, regardless of the reason for the skepticism about their use, these issues had to be resolved:[65]

1. unstructured (especially tetrahedral) meshes were computationally inefficient to generate/solve, and the corresponding flow solutions lack desired accuracy
2. the methodology was incapable of producing high-aspect-ratio cells suitable for resolving the boundary layer
3. even if tetrahedral "viscous" grids could be generated, they would be of such a large size that their routine application to realistic configurations would become impractical

These objections have been largely eliminated over the years, leading to a great improvement in unstructured meshing. This has led to a variety of approaches for generating unstructured meshes, with the three basic approaches being: Delaunay triangulation, advancing front/advancing layer methods, and Octree methods. Descriptions of these methods, including some examples of each, will be presented, including discussions of the strengths and weaknesses of each approach.

7.8.1 Delaunay Triangulation

Since unstructured meshes most often consist of triangles (or triangular faces), the geometry of mesh generation is linked to the geometry of triangles. Creating a triangle requires defining three points in space, and would seem quite straightforward, but the problems of unstructured mesh generation arise when trying to triangulate more than three points at the same time. Figure 7.42 shows five points (ABCDE) and the three ways to triangulate this set of two-dimensional points. The question for a CFD user relates to whether or not these three triangulations are equally as good for computing flow solutions, and the answer is definitely "No." The second and third triangulations create highly skewed triangular faces, which greatly increases the errors in the finite volume approach.

One way to obtain the "best" set of triangles for a given set of points is Delaunay triangulation, which was created by the Russian mathematician Boris Delaunay.[66] Delaunay triangulation is one of the best ways to draw triangles between a given sets of points in order to assure accuracy of the simulation. The basis of the approach is that each point is connected to its closest

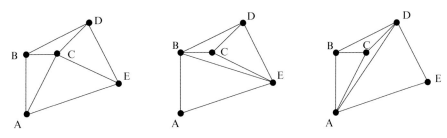

Figure 7.42 The three possible triangulations of points ABCDE in two dimensions.

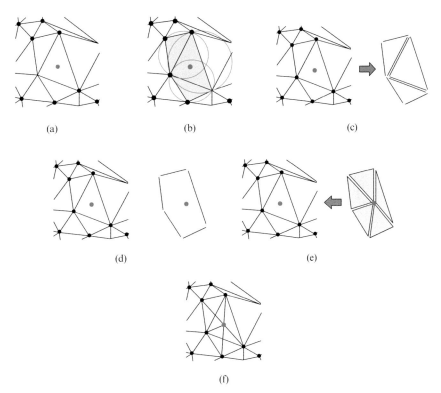

Figure 7.43 The creation of Delaunay triangulations (Ref. 67; Courtesy of Sjaak Priester). (a) vertex is added; (b) circumcircles drawn; (c) remove triangles that enclose vertex; (d) remove double edges; (e) form new triangles; (f) reform triangles.

neighbors, so that triangles that do not intersect one another are formed. All triangles in the Delaunay triangulation will have empty circumscribed circles, which is to say that no points lie in the interior of the circumcircle of any triangle. It can be shown that the Delaunay approach maximizes the minimum angle in a two-dimensional triangulation, but unfortunately the same property is not necessarily true in three dimensions.

The Delaunay triangulation process is best seen graphically in Fig. 7.43.[67] Say that we have a triangle where we want to increase the mesh density, so we insert a new vertex in the center of the triangle, as shown in Fig. 7.43a. Next, as shown in Fig. 7.43b, we find the triangles that have a circumscribed circle that encloses

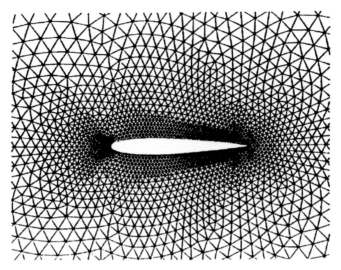

Figure 7.44 A Delauney triangulation mesh about an airfoil (Ref. 68; reprinted by permission of the American Institute of Aeronautics and Astronautics, Inc.).

the new vertex. Now we can remove the triangles that enclose the vertex, as shown in Fig. 7.43c, but, for future use, we must remember where the edges are located. Then we can remove any double edges, as shown in Fig. 7.43d, where now only unique edges are maintained in the construction. We can then form new triangles between the remaining edges and the vertex (Fig. 7.43e) and finally replace the newly formed triangles into the original mesh (Fig. 7.43f).

A Delaunay triangulation mesh for an airfoil is shown in Fig. 7.44.[68] Notice that the mesh is very uniform around the airfoil due to the positioning of the vertices in a uniform manner. In fact, one of the weaknesses of the Delaunay approach is that the user has to choose the points that will be used to create the triangles, and the initial choice of those points can have a large impact on the quality of the results.

7.8.2 Advancing Front/Advancing Layer

The advancing front[69] and advancing layer[70] methods (sometimes called moving front/layer methods) have become quite popular in recent years due to their relative efficiency and the quality of the resulting meshes. The idea behind this approach is actually quite straightforward; the process begins with a prescribed boundary and the surface meshing that was previously done on that boundary. The boundary would be a set of edges with points in two dimensions, or a set of triangular faces in three dimensions. "The boundary triangulation is regarded as a front on which a new layer of elements is built. The original front triangles become interior faces of the mesh and a new set of front faces is created, a process that continues until the entire domain has been filled."[70] The primary difficulty with the approach is

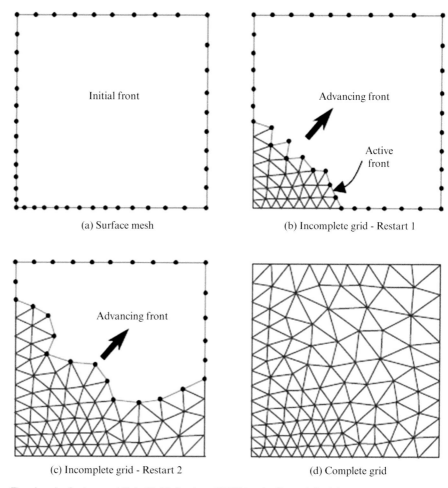

(a) Surface mesh

(b) Incomplete grid - Restart 1

(c) Incomplete grid - Restart 2

(d) Complete grid

Figure 7.45 The advancing front concept (Refs. 71, 72; Courtesy of NASA Langley Research Center).

the "end game," when the process reaches a pocket of cells that start to collapse on themselves before the final empty space is filled.

One example of the advancing front method (AFM) is shown in Fig. 7.45.[71,72] The starting edges and the surface mesh points are shown in Fig. 7.45a, and the process begins by creating triangles that connect the existing surface points one at a time. These triangles form an active front that "advances" across the domain, as shown in Fig. 7.45b. If the advancing front gets bogged down during the process, the front can be backed up and restarted using different triangles, as shown in Figures 7.45b and 7.45c. Finally, the entire region is filled with triangles to complete the mesh in Fig. 7.45d, where the size of the triangles is determined by the original spacing of points on the surface mesh.

While the advancing front approach was very useful in creating unstructured meshes for inviscid flow calculations, the necessity of using high aspect ratio cells close to the body surface (for turbulent flow computations) was another weakness of the original approach. This was overcome by the

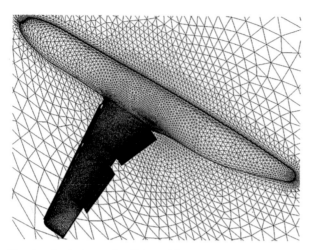

Figure 7.46   The advancing front concept applied to a civil transport with deflected flaps and slats (Courtesy of Dimitri Mavriplis of the University of Wyoming).

advancing layer method (ALM): "In contrast to the conventional method in which cells are added in no systematic sequence, the ALM advances one layer of cells at a time to reduce the complexity of generating high aspect ratio cells. The layers continue to advance in the field, while growing in thickness, until either (1) opposite fronts approach to within a cell width or (2) certain grid quality criteria dictated by a global background grid are locally satisfied."[73] In other words, the ALM method builds a layer of cells much like you might build up a brick wall – layer upon layer upon layer, until the wall reaches the height you are interested in. Then the AFM can continue to fill in the rest of the computational volume.

An example of an advancing front grid applied to a full transport configuration with flaps and slats extended is shown in Fig. 7.46. Notice the high level of control in changing mesh sizes that is evident on the surface of the wing and in the region of the flaps. This is one of the very nice features of this approach, and additional control of mesh density through the use of sources placed throughout the flowfield makes this method very popular today.

7.8.3 Octree

The Octree method for creating unstructured three-dimensional meshes is quite different from the previous approaches (the two-dimensional equivalent is known as the Quadtree method). In this method the region surrounding the surface of interest is divided into "a collection of rectangles followed by a division of rectangles into triangles. A rectangle can be further subdivided into four new rectangles. For a rectangle that intersects the boundary, this subdivision can be repeated until a sufficiently fine resolution has been achieved. Rectangles that intersect the boundary and are sufficiently small are then replaced by a polygon consisting of the part of the rectangle lying

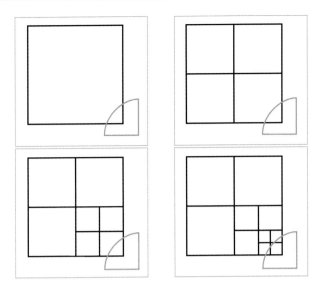

Figure 7.47   The Octree concept in the vicinity of a non-Cartesian boundary (Ref. 74; Courtesy of W.N. Dawes of Cambridge University).

inside the domain together with the part of the boundary that lies inside the rectangle. A further division of rectangles and boundary polygons into triangles creates a valid triangulation of the domain."[41]

The basic approach to the two-dimensional Quadtree concept can be easily understood by viewing Fig. 7.47.[74] The gray line is the surface of the boundary, so we continue subdividing cells until the resolution is fine enough for the particular region in the flow. Once a boundary surface is reached, as shown in Fig. 7.47, the regions are divided into smaller and smaller cells until the boundary is filled with polygons to complete the meshing.

Once the Octree cells are deemed fine enough, they can be triangulated to create the unstructured mesh, as shown in Fig. 7.48, although this is not necessary to the overall Octree method. An example of an Octree mesh that has not been triangulated is shown in Fig. 7.49, where the wing and nacelle of an aircraft are shown with the subdivided cells around it.[75] One of the unusual aspects of an Octree mesh is that it may look unusual, but good results can be obtained with the method, and the overall generation of the mesh is easily parallelized for faster results.

7.9 Cartesian Grids

Cartesian grids are both similar to and different from any of the grid concepts mentioned so far. These grids start from a very different premise, namely, that the boundary surfaces are not conformal to the grid. If you look at the examples of both structured and unstructured grids shown previously, they

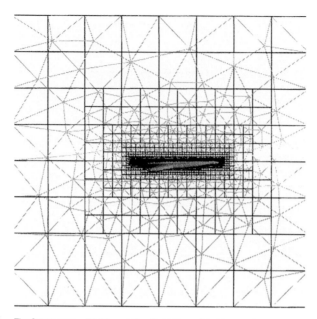

Figure 7.48 The Octree mesh with triangulation (Ref. 75; reprinted by permission of the American Institute of Aeronautics and Astronautics, Inc).

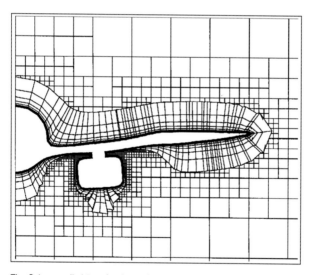

Figure 7.49 The Octree applied to a fuselage-wing-nacelle configuration (Ref. 75; reprinted by permission of the American Institute of Aeronautics and Astronautics, Inc.).

start from the surface of a vehicle in some fashion and then "grow" outward. Cartesian grids refer to the method of automatically intersecting or dividing the body-fluid interface into regular hexahedra (or quadrilaterals in two dimensions). There are many data structures that can be used to implement this, including unstructured meshes, Octrees, structured blocks, etc., so the Cartesian approach can use many of the methods already discussed. People often confuse the data structures of the implementation of Cartesian meshes with the methodology itself, but the implementation really comes from how you intend to handle different length scales (resolution), not the

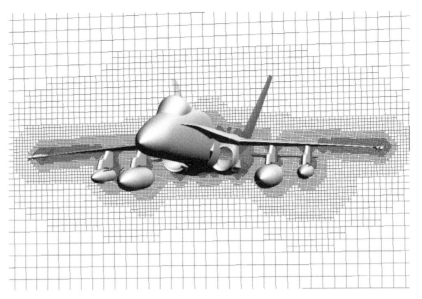

A Cartesian grid around the F-18 fighter (Ref. 78; Courtesy of Scott Murman of NASA Ames Research Center).

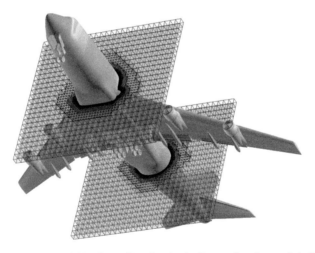

A Boeing 747 in full landing configuration showing the overall mesh generated with viscous layers (Ref. 74; Courtesy of Cambridge Flow Solutions Ltd; a full color version of this image is available on the website: www.cambridge.org/aerodynamics).

underlying numerical algorithm, which is independent of scale.[76,77] While the resulting grids look very simple, the mathematics behind the concept can be quite complicated. An example of a Cartesian grid around the F-18 fighter is shown in Fig. 7.50, where the Cartesian cells are quite small in the vicinity of the aircraft and gradually grow in size until reaching a relatively large size within one wing chord of the aircraft.[78]

An example of an octree-based Cartesian grid about a full 747 in landing configuration (including the deployment of the landing gear) is shown in Fig. 7.51.[74] In this case, viscous layers have been generated near the aircraft surface for solution with a Navier-Stokes solver.

Profiles in Computational Aerodynamics: Marsha Berger

"I am a computational scientist, although that field wasn't yet named when I got my PhD at Stanford in 1982. My PhD was actually from the computer science department, which is where Scientific Computing and Numerical Analysis research was located at that time. My BS degree, which I got in 1974 from SUNY Binghamton, was in math, since a computer science major wasn't a possibility then either. Luckily, it has turned out that all the different kinds of things I like to do can be put together and used in the field of computational science. I get to invent new algorithms, implement and test them on serial and parallel computers, and use them (mostly in collaboration with others) to solve problems in a variety of interesting application areas. And as a professor at the Courant Institute of New York University, you have a lot of freedom to choose what to work on.

"Most of the applications I work on are in computational fluid dynamics. With my collaborator Michael Aftosmis at NASA Ames (where I work during the summers) we are devising algorithms to simulate compressible flow in complicated geometries using Cartesian meshes. The goal is to simulate the flow accurately and automatically enough so it can be useful as a design tool, and reduces the need for wind tunnel testing. We use Cartesian cells that are cut by the embedded boundaries used to represent the geometry (for example the space shuttle), thus easing the problem of body-fitted grid generation around complicated objects. This involves mathematical work, for example, making up stable and accurate methods for the cut cells; and computer science, for example, using approaches found in computational geometry and computer graphics to generate the meshes. So again, I get to work on problems using many different approaches. I really like it when the results of my research are actually used by other people. That makes working at NASA Ames especially fun.

"Another project I am working on with Randy LeVeque (University of Washington) and David George (USGS) is simulating tsunamis. One of the difficulties here is that the fluid flow simulation (in this case water) has to cover hundreds of kilometers in the ocean, down to meters of resolution when the tsunami hits the shoreline. This is tough to simulate accurately, but if you want to know how high the wave will reach to help plan an evacuation route, this kind of resolution in the simulation is essential.

"My earliest work was in adaptive mesh refinement – which is one of the key ingredients in the types of simulations I just mentioned. When you solve a time-dependent fluid flow problem, the waves move, and the mesh resolution needed to accurately simulate the waves should vary in time too. It is hard to predict where this extra resolution will be needed, so mathematical algorithms are needed to estimate where the mesh needs to be refined. In the case of structured adaptive mesh refinement, the cells needing refinement are grouped into structured rectangular grid patches that try to efficiently cover the regions with high error, without overly refining the mesh, which would be too expensive. For this part I used pattern recognition algorithms to generate the grid patches.

"Finally, special formulas are needed to integrate the solution on a mesh with hanging nodes. My thesis proved a theorem about the stability and conservation properties of these formulas. So even in my thesis research this interdisciplinary approach turned out to be useful. In a sense I shouldn't be surprised that I like interdisciplinary work. When I applied to graduate school, after working for two years following completion of my undergraduate degree, I applied to six different departments in six schools. Now I'm working in the intersection of three of those areas."

Marsha Berger is the Silver Professor of Computer Science and Mathematics at the Courant Institute of New York University. The focus of her research is computational fluid dynamics. She received her BS in math from SUNY Binghamton in 1974. She worked for two years at Argonne National Laboratory as a scientific programmer in the Environmental Systems Division, mostly on projects modeling the circulation of Lake Michigan. She returned to school at Stanford University for her MS (1978) and PhD (1982). In 2000 she was elected to the National Academy of Sciences and in 2005, to the National Academy of Engineering. She is a SIAM (Society of Industrial and Applied Mathematics) Fellow, and currently serves on the SIAM Board of Trustees. She was a member of the team that won the NASA Software of the Year award in 2002 for *CART3D*.

7.10 Grid Adaptation

We have talked about many grid/mesh types in the previous sections, but all of the approaches have one issue in common: how do you know if the grid/mesh is fine enough in regions where high flow gradients exist (as discussed by Prof. Berger in the preceding profile)? It would be very useful if you could obtain a solution on a given grid and have the solution "tell" you (in one way or another) that it needed more resolution/accuracy, and therefore create a more refined mesh in certain regions – this is known as grid adaptation or adaptive mesh refinement (AMR).

Adapting a grid to the solution is a powerful tool in computational aerodynamics that will become increasingly important as users perform more and more complex simulations for detailed aerodynamic configurations. As the geometric complexity increases, the computational cost and time required also increases, which makes it unreasonable to perform global mesh refinement studies to satisfy flow gradient requirements. Hence, there is a potential for large savings in time (both CPU time and wall clock time) through the use of optimized mesh resolution methods.[79] Mavriplis has noted that grid adaptation is well suited for problems with a large range of scales, and allows for the possibility of automatic error estimation and control, but requires a very tight coupling to CAD (surface points) in order to insure that the adaptation is based in geometric reality and not on the previously discretized surface.[80]

There are various well-known methods available for adapting *isotropic meshes* (meshes where the cells have no directional bias), including:

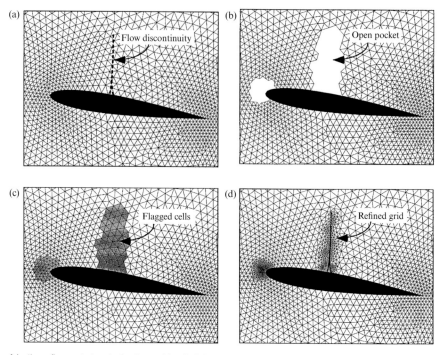

Figure 7.52 Adaptive refinement steps by local remeshing (Ref. 80; courtesy of NASA Langley Research Center).

- local remeshing
- Delaunay point insertion/retriangulation
- edge-face swapping
- element subdivision (both mixed elements [non-simplicial] and anisotropic subdivision required in transition regions)

Perhaps the most straightforward of these adaptation methods is local remeshing, where a region is determined to have high gradients of some flow quantity (perhaps vorticity or a pressure discontinuity at a shock) and a hole is cut around that region (as shown in Fig. 7.52a). The hole region (labeled "open pocket" in Fig. 7.52b) then flags the cells within that region (Fig. 7.52c) for remeshing, where the cells are subdivided using one of a variety of strategies (Fig. 7.52d).[80] The resulting mesh now has higher density in the region of the flow discontinuity and should result in an improved solution. Remeshing can also be accomplished using the Delaunay point insertion approach discussed in Section 7.8.1.

An example of adaptive mesh refinement applied to a delta wing at high angle of attack is shown in Fig. 7.53. The AMR method used was developed by Pirzadeh based on a tetrahedral unstructured grid technology originated at NASA Langley Research Center.[81] A large improvement of the adapted solutions in capturing vortex flow structures over the conventional unadapted results was demonstrated by comparisons with wind tunnel data. Pirzadeh's

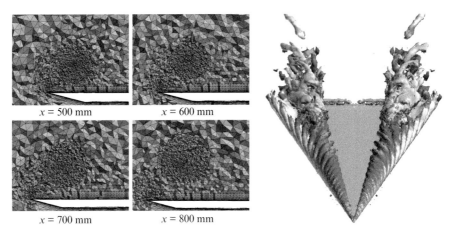

Figure 7.53 Adaptive refinement of a delta wing mesh for flow at 27 degrees angle of attack (Refs. 82, 83; a full color version of this image is available on the website: www.cambridge.org/aerodynamics).

method is applied to the delta-wing configuration after a steady-state flow solution was computed for a grid with relatively coarse surface resolution, and then the resulting flowfield was used to create a refined grid by eliminating all cells within an isosurface of vorticity at a particular level. The grid was then regrown inside of the isosurface with a scale factor of 0.5; this procedure was performed twice to create a vortex core and shear layer with one-quarter the cell sizes (in all coordinate directions) of the original grid. The new grid was then used to compute unsteady Detached-Eddy Simulations of the flowfield.[82,83] The resulting flowfield within the vortex core showed excellent agreement with experimental data for turbulent kinetic energy, and the refined mesh was able to capture the physics of the vortex with a substantially smaller grid than could be obtained by global refinement.

Another interesting example of adaptive mesh refinement is shown in Fig. 7.54, where the F-5E shock wave patterns are simulated numerically and compared with the Shaped Sonic Boom Demonstrator (SSBD) aircraft.[84] The SSBD was an aircraft designed by Northrop Grumman to minimize the shock wave strength and patterns of the F-5E by modifying the geometry of the vehicle, especially the nose section.[85] The simulations used Cartesian grids, as shown in Figures 7.54a and 7.54b, and the grids were refined with an adjoint-based approach using a "pressure sensor" placed 80 feet below and behind the aircraft to approximately match available flight test data. The aircraft was flying at $M = 1.4$, and the adapted meshes each contained approximately 10 million cells. For improved accuracy, the Cartesian mesh in this case has been rotated in order to be aligned with the shock waves. The resulting pressure fields for the two aircraft are shown in Figures 7.54c and 7.54d, where the reduced shock strengths for the SSBD are evident at the location of the pressure sensor. The predicted pressures at these locations matched the flight test data quite well and showed that the SSBD reduced the sonic boom signature when compared with the F-5E.

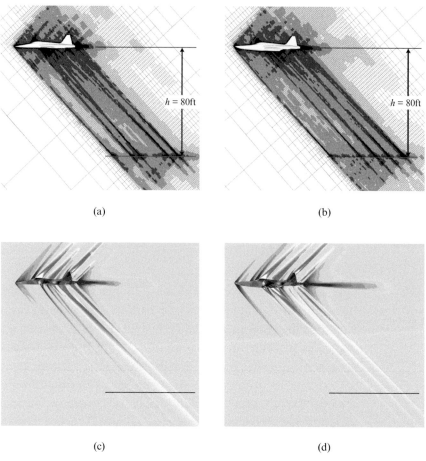

(a) (b)

(c) (d)

Figure 7.54 Grid adaptation applied to the Shaped Sonic Boom Demonstrator (Ref. 85; a full color version of this image is available on the website: www.cambridge.org/aerodynamics). (a) F-5E adapted mesh with 11 million cells; (b) SSBD adapted mesh with 9.2 million cells; (c) F-5E Mach and surface C_p contours; (d) SSBD Mach and surface C_p contours.

While mesh adaptation is a powerful tool, there are several weaknesses to the approach that are currently being investigated in an effort to try to improve the approach. Perhaps the most obvious weakness is that the refinement techniques are fairly good at finding regions of localized high-pressure gradients, but have a much harder time in refining for nonlocal (i.e., convective) flow features like wakes. Another area where AMR would be helpful is in multi-disciplinary optimization applications, where numerical approaches are used to perform design optimization studies. In this case, however, the AMR techniques may refine the grid of one configuration in a better way than it does for another configuration, which would lead to a bias in the results of design parameter sweeps and thereby "contaminate" the optimization study. As research is continued into the AMR approach, improved predictions of complex flowfields will be possible with smaller grids than global refinement approaches would yield, and the prediction of aircraft aerodynamics will be improved.

7.11 Grid Properties that Affect Solution Accuracy

A number of details about the grid and how it is created can have a significant impact on the results obtained using CFD. Many of the "problems" that people have in accepting CFD results are because of a lack of care by users in creating and checking grids. We cannot emphasize enough the importance of thinking about the size, shape, and type of grid to be used prior to obtaining solutions – the grid influences the results, so spend time making very good grids!

7.11.1 Outer Boundary Size

The location of the outer boundary of the grid can have a large impact on the results. Making the choice of outer boundaries even more difficult is the difference between subsonic and supersonic flows. Remember that subsonic flow propagates information in all directions, making the placement of the outer boundary (ahead, behind, above, and below the surface of interest, as shown in Fig. 7.55) crucial to adequately simulating a true subsonic flowfield. Supersonic flow can only propagate information downstream within the Mach cone coming from any point in the flow, so upstream boundaries can be significantly closer to the surface. We will call the left boundary the *inflow boundary*, the upper and lower boundaries *inflow/outflow boundaries* (because we do not know what the flow will do along these boundaries), and the right boundary the *outflow boundary*. Now we can look at a sample grid layout that does not take into account some of these important boundary concepts. Let us say that we are trying to simulate the subsonic flow over an airfoil and we use the grid shown in Fig. 7.56.

If we were to place the outer boundaries this close to the airfoil surface and then apply freestream boundary conditions ahead, above, and below

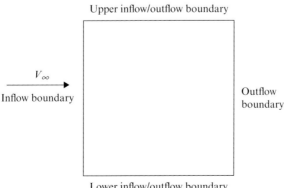

Upper inflow/outflow boundary

V_∞

Inflow boundary

Outflow boundary

Lower inflow/outflow boundary

Figure 7.55 Nomenclature for grid outer boundaries.

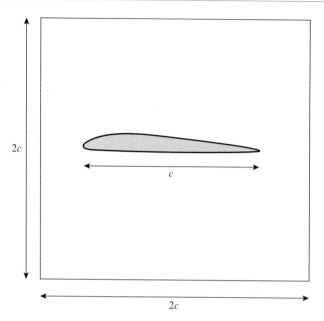

Figure 7.56   Poor airfoil grid layout for subsonic flow.

the airfoil, we would be requiring that the solution be obtained for the case where the flow is undisturbed from freestream conditions very close to the airfoil, which is definitely not true in subsonic flow! In fact, subsonic flow disturbances (theoretically) would impact conditions at a very large distance from the surface, so we need to place our boundaries very far away from the surface. But how far is far enough? Well, we could create a grid with all boundaries two chords away from the airfoil and obtain a solution, and then increase the outer boundary to three chords, and then four chords, until we found an outer boundary distance that did not impact the results on the surface of the airfoil. This is one form of a *grid sensitivity study*, and in this case we would be determining the sensitivity of the solution to the position of the outer boundary. However, over the years, CFD researchers have found a "rule of thumb" for minimum outer boundary distances, as shown in Fig. 7.57.

This grid allows for the boundaries where freestream conditions are prescribed to be at least ten chords away from the surface and for the wake of the airfoil to travel twenty chords downstream prior to interacting with the outflow boundary. Depending on the solution being attempted, however, these boundaries may not be far enough away, so *users should always determine their own boundary positions for the problem at hand*.

The actual boundary conditions used also depends on the type of flow and the type of boundary being modeled (as discussed for Fig. 7.55). For example, freestream conditions are usually specified at the inflow boundary, but there are a variety of ways to do that, some of which actually

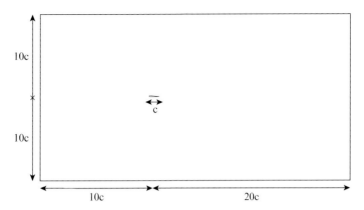

Figure 7.57 Minimum airfoil grid outer boundary distances for subsonic flow.

overspecify the freestream conditions. Two possible common inflow boundary conditions are:

1. fixed condition – you specify a Mach number, total pressure, and total temperature at the inflow and then calculate other required properties.
2. Riemann invariants – used a great deal for supersonic flows, this condition finds the characteristics of the flow in the vicinity of the boundary and uses them to find correct boundary conditions.

Once you have established these outer boundary positions for a certain class of flow, you usually do not have to re-invent the wheel each time you compute a similar flowfield, but rather use the experience you have gained in the previous calculations to place your boundaries for a new case.

7.11.2 Structured Cell Geometry

Another factor affecting the accuracy of our solutions is the local size and shape of the grid. These factors include whether or not the grid is ordered correctly and how the individual cells are shaped and stretched. Many of these "goodness" measures apply to either structured or unstructured grids, but some only apply to structured grids, such as the grid Jacobian.

7.11.2.1 JACOBIAN

Structured grids are almost always transformed from the physical space to a computational space, where the finite differencing and equation solution is obtained. The results are then transformed back to the physical space to complete the solution process. That means that each cell must have a geometric transformation from an arbitrary shape in the physical domain to a rectangular shape in the computational domain, as shown in Fig. 7.58. As long as the cells in the physical domain are "nice," the mathematical transformation can take place. However, once in awhile a cell exists in the physical

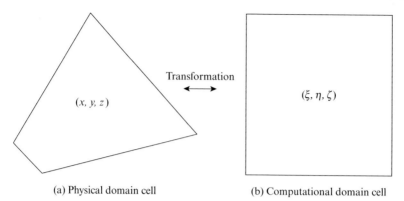

(a) Physical domain cell   (b) Computational domain cell

Figure 7.58   Transformation of a physical cell into a computational cell.

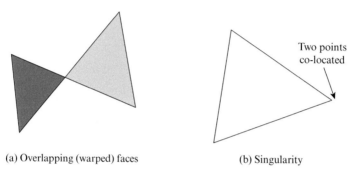

(a) Overlapping (warped) faces   (b) Singularity

Figure 7.59   Examples of cells with Jacobian problems.

domain that has a mathematical singularity that cannot be transformed, causing problems for generating the grid or obtaining a solution.

The function that determines the property of the transformation is called the *Jacobian* of the transformation. The Jacobian is defined as:

$$J = \left| \frac{\partial x_i}{\partial \xi_j} \right| \tag{7.8}$$

which simply relates the area (or volume) of the cell in the physical plane to the area (or volume) of the cell in the computational plane (see the three dimensional form in Appendix E). Problems arise when cells have overlapping faces or singularities (multiple vertices at the same location), as shown in Fig. 7.59. These problems can be avoided if the cell Jacobians are positive for all cells in the solution space.

7.11.2.2 CELL SHAPE

Since the cell forms the basis for all numerical approximations being performed (whether the code uses a finite difference or finite volume approach), the size and shape of the cell will have some impact on the quality of the

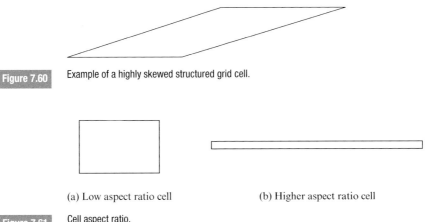

Figure 7.60 Example of a highly skewed structured grid cell.

(a) Low aspect ratio cell   (b) Higher aspect ratio cell

Figure 7.61 Cell aspect ratio.

approximation. Some of the geometric parameters that impact the solution accuracy are the cell skew angle, rotation angle, and aspect ratio. Highly skewed cells, as shown in Fig. 7.60, can cause the introduction of errors in numerical approximations due to inaccuracies in the transformation process from a highly skewed cell to a rectangular cell. The closer the cell is to being rectangular, the smaller the transformation error. In general, all angles within the cell should be approximately equal in order to reduce the error due to skewness. These errors are due to the truncation error issues we discussed in Chapter 6, since our derivative approximations assume that grid points are equally distributed and form rectangular cells. Even when grid transformations are used (as defined in Appendix E), the transformation derivatives (the metrics of the transformation, also in Appendix E) must be approximated numerically, and non-rectangular cells will create additional errors in the calculation of each derivative.

The aspect ratio of the cell is also important when looking at the quality of the grid. The aspect ratio is essentially the ratio of the length and width of a cell (which is similar to how aspect ratio is defined for an airplane wing), as shown in Fig. 7.61. In order to insure good grid quality (and low truncation error), the aspect ratio should be on the order of one.

Unfortunately, cells within boundary layers near solid surfaces often have exceedingly high aspect ratios (on the order of ten thousand or more), which is the result of needing very small grid resolution normal to the wall but not along the wall. These type of high aspect ratio cells are unavoidable in the boundary layer, but as long as the cells are nearly orthogonal to the surface there are few problems associated with them.

7.11.2.3 CELL ORTHOGONALITY AT A SURFACE BOUNDARY

While the preceding examples of cell shapes are very important, many researchers have found that one of the most important features for a grid is that it be nearly orthogonal to solid surfaces. "Structured grid flow solvers

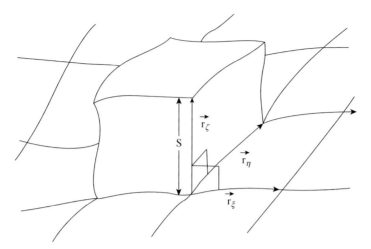

Figure 7.62 Cell near-orthgonality and spacing near a surface (Ref. 17; Courtesy of Joe Thompson of Mississippi State University).

give most accurate results when the grid cells are orthogonal since this mini-mizes the truncation error associated with the discretisation"[86] (see Chapter 6 for details on truncation error). In fact, "there are three geometric constraints which might be imposed at each boundary face of /a/ cube – two orthogo-nality relations and a specification of the spacing normal to the boundary"[17] (see Fig. 7.62). Not enforcing the orthogonality requirement near a surface (or any other topology junction) can degrade the quality of the flow predic-tions in this region.[87]

7.11.2.4 CELL STRETCHING

Another important grid quality measure is cell stretching. Most of the previ-ous grid quality measures have related to a single cell, but grid stretching has to do with the relationship of adjacent cells. Figure 7.63 shows an example of adjacent cells with poor cell stretching as well as an example of reason-able cell stretching. This error also relates directly to the finite difference formulations we discussed in Chapter 6 (see the Taylor's series expansion result in Equation 6.11). Our derivative approximations assume an evenly space grid system, which is true if there is no stretching, but stretched grids have higher error (as we found in Equation 6.27). However, most stretched structured grid systems are transformed into a Cartesian grid, as we showed in Section 7.6. The more we stretch the grid from one cell to the next, the larger the truncation error within our transformed equations (as discussed in Appendix E). Keeping a smooth and regular stretching is essential for reduc-ing errors in the solution.

The stretching ratio has to do with how a transition is made from one cell spacing distance to another. In Fig. 7.63a, the transition is made suddenly, where one cell spacing is used at the bottom of the grid and a different cell spacing is used at the top of the grid. This type of spacing change introduces

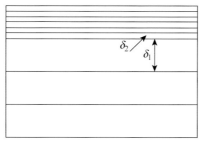
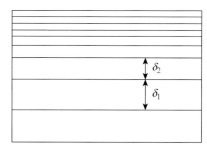

(a) Poor stretching ratio   (b) More reasonable stretching ratio

Figure 7.63   Cell stretching ratio examples.

errors in the numerical formulation of derivatives using adjacent points or cells. In order to reduce these numerical errors, a smooth transition must be made from one cell size to another, as shown in Fig. 7.63b. A rule of thumb for cell stretching is that any two adjacent cells should never have lengths that differ by more than 20 percent, or in equation form

$$0.8 \le \delta_1 / \delta_2 \le 1.2. \tag{7.9}$$

If this ratio is maintained in all directions and for all regions of the grid, then the grid quality should be quite high in terms of cell stretching. We should note that some people use slightly different values than 20 percent for cell stretching, such as 25 percent or 30 percent. Regardless of the exact number, the concept is the same: do not stretch your cells too much or you will incur numerical errors in your solution.

Flow Visualization Box

Grid Quality Investigation

Let's look at the NACA 0012 airfoil grid and see if the grid quality issues we are discussing are visible in that grid. First create a coordinate surface of the grid to view it from a distance.

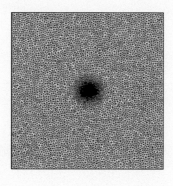

Wow, where is the airfoil? Well, that dark spot in the middle of the picture is the region of grid clustering around the airfoil, but we're so far away we can't see any details. How far away are the outer boundaries from the airfoil in this grid? What kind of grid is this? Let's see if zooming in a little helps.

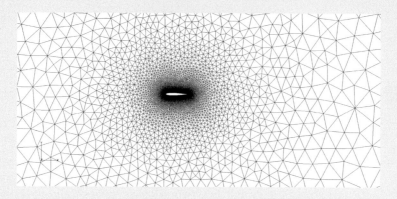

We can definitely see that this is an unstructured grid, and the airfoil is now visible, but is this a viscous or inviscid grid? Let's zoom in further to find out.

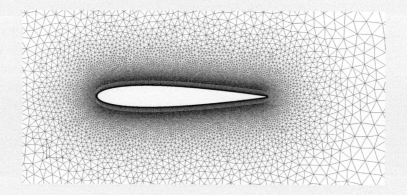

Now we can see that this is a hybrid grid (unstructured away from the surface and structured near the surface. Perhaps zooming in again will help us see everything clearly.

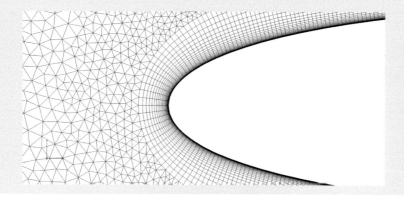

Now we can see the viscous clustering near the surface, which means that this is a viscous grid. Are the cells near the surface aligned with the surface? Are the unstructured cells nearly isotropic? Do the structured cells have high aspect ratios? Does the stretching follow the rules we have discussed? Investigate this grid in more detail and see if you can find the answers to these questions.

7.11.3 Unstructured Cell Geometry

While there are two basic numerical approaches for dealing with unstructured cell computations, the cell-vertex and cell-centered approaches (as discussed in Chapter 6), we will only discuss the cell requirements for the cell-centered approach. Some of the requirements for cell quality of unstructured meshes are similar to those for structured grids, but there are also several measures that are quite different. In general, keeping cells with relatively low aspect ratios and small levels of skew is still important for unstructured cells, but there are several other measures of quality that are important as well.

First, however, it is important to note there are two "pass/fail" criteria that must be satisfied in order for a cell-centered flow solution to continue without serious issues. First, as found for structured grids, every cell in the domain must have a positive volume that is above some minimum threshold (usually a few orders of magnitude above *machine zero*). Second, an element may not be "tangled" or inverted. In other words, none of the faces of the cell may be "folded" – a term used to describe the situation when the centroids of both adjacent cells are on the same side of the face.[88] It is impossible to maintain the conservation laws in this situation, since flow leaving one element would not be defined as entering the adjacent cell. An interesting complexity arises in this case, since it is possible for a cell to be tangled/inverted but still have a positive volume (in the case of non-simplex elements like prisms and hexahedrons). Conversely, a cell may be free of any folded faces but still have a positive volume that is below a minimum threshold.

7.11.3.1 FLOW ALIGNMENT AND BOUNDARY LAYER GRADIENTS

The presence of the solid walls in the flow solution of a hybrid mesh dictates a couple of obvious but important flow characteristics in the boundary-layer region (see Fig. 7.64); this requirement also applies to structured grids. We will begin by assuming the flow is nominally aligned with the solid wall. For viscous flows, the gradients in the streamwise direction are orders of magnitude lower than those in the transverse direction due to the no-slip viscous boundary condition at the wall. Since the solution needed to resolve the flow states on either side of a cell face is locally one-dimensional, the geometry most favorable for an accurate solution in the midst of these large gradients

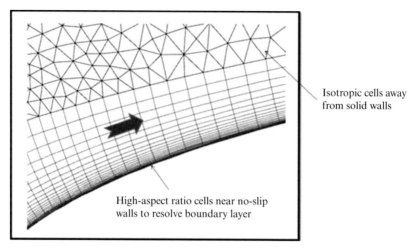

Isotropic cells away from solid walls

High-aspect ratio cells near no-slip walls to resolve boundary layer

Figure 7.64   Flow alignment in the boundary-layer region of a viscous mesh.

is one where the element faces are aligned parallel to the wall boundary. This is one reason why structured grids are often used near the surface of an aircraft, as shown in Fig. 7.64.

A way to measure how much cell faces are aligned with the tangential flow direction near solid wall boundaries is to find the angle between the cell face outward normal vector and a vector normal to the wall (as shown in Figure 7.65). Doing this for all four faces and comparing the result to 90 degrees gives a measure of quality of the cell alignment. Cell alignment with the flow near a solid surface is an important issue for viscous meshes, since alignment in the boundary-layer region keeps the larger face areas associated with the more accurate gradient projections due to the tight element spacing in the wall-normal direction. Notice that there is no preference in the formulation as to whether a particular face is aligned parallel or perpendicular to the flow direction. Also, it is only possible for quadrilateral, prismatic, and hexahedral elements to produce a perfect value for this measure of quality. Figure 7.65 shows the value of the flow alignment metric for various orientations of a quadrilateral element relative to a nearby wall boundary. It is clear the metric is insensitive to element shape and produces the worst value when all element faces are most out of alignment with the parallel or perpendicular wall directions (45 degrees off-axis).

7.11.3.2 CELL PLANARNESS

One important assumption of most cell-centered flow solvers is that the quadrilateral faces on non-simplex elements are planar. In the discrete implementation, the areas of the control volume faces (computed by subdividing into triangles if needed) are multiplied by a "wisely-computed" face normal to compute the area vector needed for the flux calculations. When non-planar faces are present, slight errors are introduced, resulting in a loss

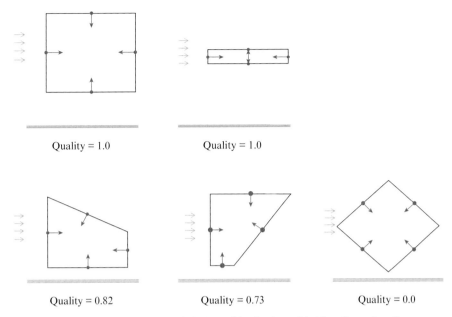

Quality = 1.0   Quality = 1.0

Quality = 0.82   Quality = 0.73   Quality = 0.0

Figure 7.65   Various quadrilateral element orientations relative to a solid wall and associated flow alignment quality measure.

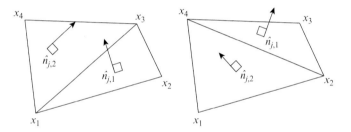

Figure 7.66   Definition of normal vectors of a quadrilateral face in planarness measure.

of flow conservation and possible stability issues. Therefore, all quadrilateral mesh faces should be kept as planar as possible. As discussed by Mavriplis,[89] this is not an issue with most cell-vertex flow solvers because of the way the dual control volumes are formed around the nodes.

The goal is to have any quadrilateral faces be as planar as possible, as shown in Fig. 7.66. Cell planarness is obviously not applicable in two dimensions or for faces with less than four nodes in three dimensions (since, by definition, three points form a plane).

The measure of planarness is found by dividing each quadrilateral cell into two triangles, as shown in Fig. 7.66. The normal vectors for each triangle are then found and a measure of their alignment is made. These alignment measures are then used to compute an area-weighted average over *all* faces in the cell. Since different values of alignment are possible, depending on which diagonal is used, the area-weighted measurement can be computed for each diagonal (diagonal $x_1 - x_3$ or $x_2 - x_4$), and the minimum value is

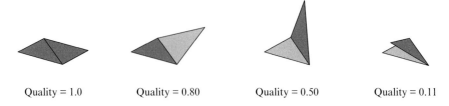

Quality = 1.0   Quality = 0.80   Quality = 0.50   Quality = 0.11

Figure 7.67   Planarness quality measure for a quadrilateral face.

used. Figure 7.67 gives the planarness quality measure for a single quadrilateral face in various "folded" configurations. The planarness quality measure can then be given an acceptable range, and each quadrilateral face can be required to fall within this range (planarness quality must be between 0.9 and 1.1, for example).

7.11.3.3 CELL SKEW AND SMOOTHNESS

One of the important issues for accurate unstructured mesh calculation is for the location of the flux calculation (face centroid) to be centrally located with respect to the solution points at the neighboring cell centroids (the place where flow properties are known). If this requirement is not met, the averaging process used to find the flow properties at the center of the face will introduce errors in the result. This characteristic is addressed as a measure of the skew and smoothness between adjacent elements in the mesh. Area (2D) or volume (3D) ratios between adjacent cells are commonly used to quantify the smoothness in a mesh. However, it is easy to create a manifold geometry that exhibits similar areas/volumes between adjacent elements but with drastically different projection distances between the centroids of adjacent cells and the included face centroid. Smoothness in these gradient projections, as well as the location of the face centroid relative to the line connecting the adjacent cell centroids, is what this metric evaluates. The skew-smoothness metric for the the jth face in the ith cell is defined using the vectors from the adjoining left and right cell centroids, respectively, to the centroid of the jth face of the cell; these definitions are shown in Fig. 7.68 for a simple two-dimensional case.

The degree of smoothness across a particular face in the mesh is quantified by the ratio of distances between the cell centroids and the included face centroid, $\tilde{v}_{left,j} / \tilde{v}_{right,j}$. If these distances are the same, the adjacent triangles are considered smooth. Also important is where the vector connecting two cell centroids crosses the included face relative to the face centroid location.[90] In Fig. 7.68, this vector would be quite far from the face centroid, leading to a fairly poor measure of skew. The skew-smoothness quality will only be "perfect" when the neighboring cell centroids are equidistant from the included face centroid *and* the included face centroid lies on a line connecting the two neighboring cell centroids.

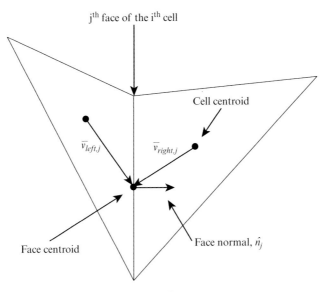

Figure 7.68   Terminology for the skew-smoothness quality measure.

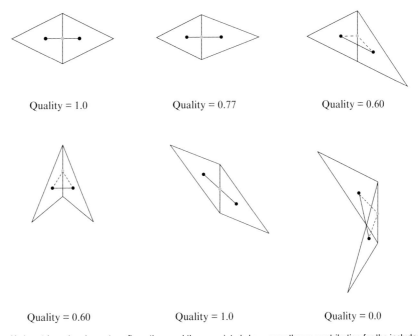

Quality = 1.0   Quality = 0.77   Quality = 0.60

Quality = 0.60   Quality = 1.0   Quality = 0.0

Figure 7.69   Various triangular element configurations and the associated skew-smoothness contribution for the included face.

Figure 7.69 shows the skew-smoothness quality value for a face adjoined by two triangular elements in various orientations. Notice that a parallelogram in two dimensions results in a perfect measure of quality, which minimizes finite volume truncation error for a midpoint-type evaluation of the solution values at the included face centroid.[91] Additionally, the adjacent volumes and areas are equal in this configuration, which gives merit to

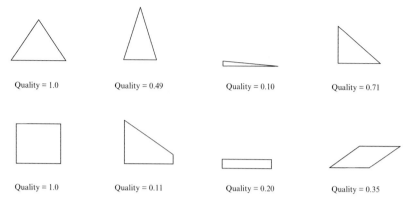

Quality = 1.0   Quality = 0.49   Quality = 0.10   Quality = 0.71

Quality = 1.0   Quality = 0.11   Quality = 0.20   Quality = 0.35

Figure 7.70 Isotropic quality metric for various 2D elements.

the common goal of having smooth element areas/volumes throughout the mesh. However, these comparisons break down when non-simplex elements are involved.

Just as you saw for structured grid cells, you should avoid highly skewed unstructured cells. For hexahedral cells or prisms, the included angles between the cell lines in 2D (faces in 3D) should be optimized in such a way that the angles are approximately 90 degrees. Angles measuring less than 40 degrees, or more than 140 degrees, often show a deterioration in the results or lead to numerical instabilities, especially in the case of transient simulations.[92] Away from boundaries, ensure that the aspect ratio (the ratio of the sides of the elements) is not too large, typically not larger than 20; near walls this restriction may be relaxed and indeed can be beneficial.

With regard to the shape of the computational cells, hexahedra are usually preferable to tetrahedra, as the former are known to introduce smaller truncation errors, and the latter require specialized solvers since they cannot use methods like ADI (see Chapter 6).[93] When you couple these requirements with the orthogonality requirement presented in Section 7.11.2.3, tetrahedral grids should use prismatic cells at the wall with tetrahedral cells away from the wall. For example, Fothergill et al. found improved results for a prismatic/tetrahedral grid compared to a purely tetrahedral grid,[94] and you should use this approach for unstructured grids.

7.11.3.4 CELL ISOTROPY AND SPACING

Uniform element arrangements are desired in the flow region beyond the boundary layer for various turbulence model applications (which will be discussed in Chapter 8). In general, more isotropic elements inherently imply a smoother mesh. A measure of deviation from an "equilateral" element shape is computed for general elements as a min/max ratio of edge lengths. This metric is independent of scaling, rotation, or translation since it is based solely on local distance measurements. Figure 7.70 shows the quality values

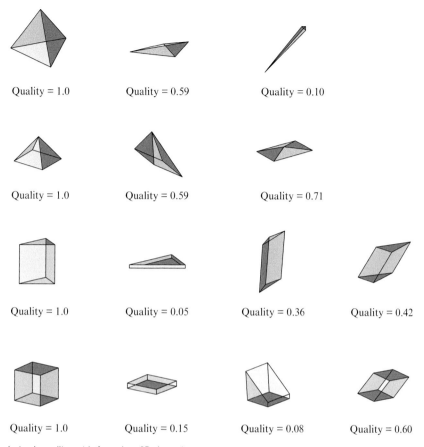

Quality = 1.0   Quality = 0.59   Quality = 0.10

Quality = 1.0   Quality = 0.59   Quality = 0.71

Quality = 1.0   Quality = 0.05   Quality = 0.36   Quality = 0.42

Quality = 1.0   Quality = 0.15   Quality = 0.08   Quality = 0.60

Figure 7.71   Isotropic quality metric for various 3D elements.

for different triangular and quadrilateral element shapes, and Fig. 7.71 gives results for various 3D element shapes. Note that this is not a measure of "degeneracy" but of "uniformity." Therefore, some element shapes may appear "better" than others, but give a lower isotropy metric value due to the large variation in length scales.

Applied globally, the spacing measure dictates a push toward an isotropic mesh. Since isotropic cells are directly related to the intended solver characteristic, we will use isotropy as a global measure formulation here instead of the isotropy metric defined previously. However, the isotropy metric is useful for static mesh analysis to identify regions of a mesh with very "poorly shaped elements." Also, note that it is locally possible for a skewed element to have perfect spacing, as shown in Fig. 7.72.

7.11.4 Viscous Grid Requirements

Viscous calculations create special challenges for grid generation, since any grid must provide enough grid points or cells to accurately represent the fluid processes taking place in the region of the points or cells. Look at the velocity

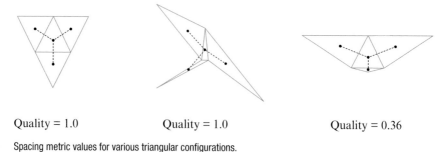

Quality = 1.0   Quality = 1.0   Quality = 0.36

Figure 7.72   Spacing metric values for various triangular configurations.

profile of a turbulent boundary layer shown in Fig. 4.12. The definitions for the wall variables used in the figure are:

$$y^+ = \frac{yu_\tau}{\upsilon} \qquad u_\tau = \sqrt{\frac{\tau_w}{\rho}} \qquad u^+ = \frac{u}{u_\tau}. \tag{7.10}$$

The turbulent boundary layer is made up of two regions, the inner region and the outer region, and the inner region has three sublayers: the viscous (or laminar) sublayer, the transitional (or log) layer, and the fully turbulent layer. A good CFD representation of this boundary layer must have points/cells within each of these regions and sublayers in order to allow a turbulence model to work properly (wall layer models are exceptions to these approximations; see Chapter 8 for more information on turbulence models). A good rule of thumb is that there must be more than two or three points/cells in each layer. The most difficult layer to grid is the viscous sublayer, since it is so small and so close to the wall. The viscous sublayer starts at the wall and continues to $y^+ \approx 10$.

So, how can a viscous grid properly model this boundary layer? This is even more difficult than it first appears, since the variables in Equation 7.10 depend on the solution to the boundary layer, which in most cases has not been obtained prior to creating the grid. If you couple the information about boundary layers from Fig. 4.12 with the cell stretching requirements from Equation 7.9, a good method for properly modeling a boundary layer arises. First, place the initial grid point at $y^+ \approx 1$ and then follow the cell stretching relationship (no points/cell should be more than 20 percent further away than the previous points/cells). This will insure that the various layers of the boundary layer have enough points and will also insure that *at least* 20 points/cells are used to model the boundary layer thickness, which is another good rule of thumb to follow when creating a viscous grid.

The key to achieving this cell quality measure is to find the placement of the first grid point/cell away from the wall. The relations in Equations 7.11 and 7.12 should provide good estimates for the spacing away from the surface to achieve a desired y^+ average for baseline calculations. However, after your runs are complete, you should verify that the actual y^+ values are acceptable and change your grid spacing if needed. The laminar and turbulent y^+ equations are:

$$\text{Laminar}: \Delta_{\text{lam}} = L\frac{1.3016\, y^+_{ave}}{Re_L^{0.75}} \tag{7.11}$$

$$\text{Turbulent}: \Delta_{\text{turb}} = L\frac{(13.1463 y^+_{ave})^{0.875}}{Re_L^{0.90}} \tag{7.12}$$

where L is the length of the body being computed (the chord of an airfoil or the length of a fuselage, etc.), Re_L is the Reynolds number at that location, and $y^+_{ave} \approx 1$. The resulting grid spacing, Δ, is the distance from the wall where the first grid point should be located to obtain an average value along the body length of $y^+_{ave} \approx 1$. These relationships were derived from flow conservation principles, and others have derived similar formulas. In fact, NASA Langley has an online calculator for estimating y^+ which uses a very similar formulation.[95]

7.12 Grid Sensitivity Studies

One of the most important aspects of CFD calculations is the grid sensitivity study. No matter how well you follow all of the rules of thumb, you can still create a grid that will not provide accurate results. How can this problem be fixed? That is where the grid sensitivity study comes into play. The idea is very simple: keep increasing the number of points/cells within the grid until the results no longer are affected by the addition of more points/cells (which is known as global refinement). Some type of sensitivity study is usually required for publication of work in papers or journal articles as well as for making design decisions, so you should understand the requirements of a grid sensitivity study and be sure to perform one for each class of flows being attempted.

In some cases, however, simply adding points where you "think" they should go and monitoring aerodynamic forces may not be enough. Since forces and moments acting on a flight vehicle are the integration of surface pressures and shear stresses, it is possible for a force to mask details within the flowfield, especially if the high gradients are isolated to one particular region of the solution space. In this case, grid sensitivity studies might need to look at flow quantities on the surface of the aircraft like pressure or off-surface quantities like vorticity or turbulent kinetic energy. In addition, varying global grid parameters, such as the total number of grid points, may not be as important as you might think, so some local grid parameters may need to be varied within the study (such as the number of points in a specific region or the fineness of the grid in the boundary layer).

Table 7.1 Grid Sensitivity Study, $M = 0.85$, $\alpha = 3$ deg. (Ref. 96)

J_{max}	y^+_{max}	C_L	$C_{D,total}$	$C_{D,pressure}$	$C_{D,friction}$
60	40	0.366	0.0247	0.0244	0.00037
60	13	0.409	0.0285	0.0249	0.00365
60	5	0.414	0.0327	0.0250	0.00764
60	1	0.416	0.0330	0.0254	0.00763
120	1	0.421	0.0318	0.0241	0.00767

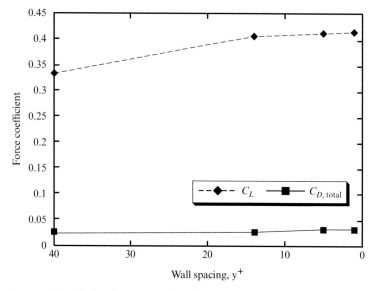

Figure 7.73   Force coefficients for four fineness levels (Ref. 96).

For example, a grid sensitivity study for a blended wing body was carried out by varying the average value of the first point off the surface in the boundary layer, as shown in Table 7.1.[96] The lift and drag coefficients were monitored for the various grids, with most of the grids having the same number of grid points normal to the surface ($J_{max} = 60$), while the values of y^+ were decreased from 40 to 1. Notice how the lift and drag coefficients both continue to change substantially until a value of $y^+ = 1$ is reached, as shown in Fig. 7.73. In addition, the number of points normal to the surface was then doubled (from $J_{max} = 60$ to $J_{max} = 120$). These results showed some differences, which would tend to imply that total grid convergence had not yet been obtained. At this point, it is really up to you to decide what aerodynamic features you are really interested in obtaining. For example, the lift coefficient (to two significant digits) could be obtained by the $J_{max} = 60$, $y^+ = 1$ grid at a lower cost than the $J_{max} = 120$, $y^+ = 1$ grid. But if drag coefficient within three significant digits is the parameter of interest, you may want to continue refining the grid before using the results.

Table 7.2 Grid Sensitivity Parameters (Ref. 97)

Grid	N1	N2	N3	N4	Total Points
Coarse	C	C	C	C	13 million
Medium 1	C	F	F	F	16 million
Medium 2	F	C	F	F	17 million
Medium 3	F	F	C	F	20 million
Medium 4	F	F	F	C	23 million
Fine	F	F	F	F	24 million

C = Coarse, F = Fine

Another example, in this case for a sonic boom prediction study, is shown next.[97] In this case the goal of the project was to predict the pressure waves emanating from an aircraft, so using forces and moments in a grid sensitivity study did not make sense. This was also a case where the boundaries of the computational region were very important. First of all, the computational region study was conducted before the grid sensitivity study. The computational region is inclined roughly at the Mach angle for the flow to increase the computational accuracy. The primary region of interest in this case was the region below and behind the aircraft in order to estimate the ground-level sonic boom.

Once the boundaries were determined, a grid sensitivity study for predicting near-field pressure signatures was investigated. A grid (the coarse grid) with 13 million grid points was generated as a baseline grid (shown in Fig. 7.74a, where every third point is shown for clarity). Several medium grids were generated by increasing the grid points in various regions, as defined in Fig. 7.74a (N1, N2, N3, and N4). Finally, a fine grid was generated by increasing the grid points in all four regions (Fig. 7.74b). Table 7.2 shows the list of the computational grids used for this grid sensitivity study, and Fig. 7.75 shows the CFD results for each grid. The fine grid captures the detailed change in the pressure signature at $h/L = 2.0$ below the airplane, where h represents a distance from the fuselage axis and L represents the length of the airplane. Another interesting result is that the signature with the Medium 2 grid (17 million points) is better than that with the Medium 3 grid (20 million points), which shows that the grid density in the N3 region is more important than that in the N2 region.

The proper approach to a grid sensitivity study can take many forms, but the essential point is that you know that your grid is resolving the flow features of interest in the best possible way. If you have not assured yourself of this important fact, then you should not proceed with detailed flow predictions for your configuration.

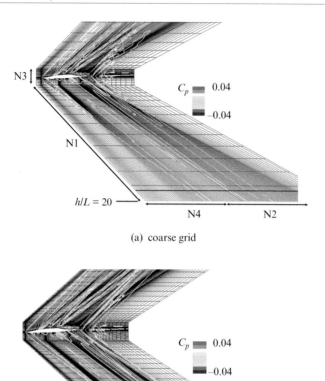

(a) coarse grid

(b) fine grid

Figure 7.74   CFD grid for near-field sonic boom prediction (Ref. 97; a full color version of this image is available on the website: www.cambridge.org/aerodynamics). (a) coarse grid; (b) fine grid.

7.13 Examples of Grids for Complex Geometries

While we cannot present every grid generation approach here, there are a number of examples of grids for complex aircraft geometries that should be informative. Each of the following grids has some unique feature that has been discussed earlier and that should prove interesting. Readers in search of additional details should consult the references listed with each example.

7.13.1 Ranger Jet Aircraft Inviscid Block-structured Grid

The Ranger model was simulated using the University of Liverpool block-structured solver PMB (Parallel MultiBlock), using a structured multiblock Euler grid generated using $ICEM\ ANSYS$.[98] Both full and half configurations

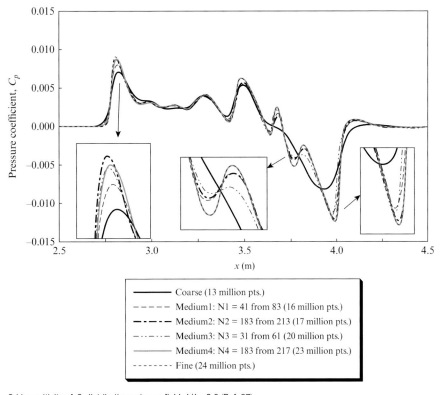

Coarse (13 million pts.)
Medium1: N1 = 41 from 83 (16 million pts.)
Medium2: N2 = 183 from 213 (17 million pts.)
Medium3: N3 = 31 from 61 (20 million pts.)
Medium4: N4 = 183 from 217 (23 million pts.)
Fine (24 million pts.)

Figure 7.75   Grid sensitivity of C_p distributions at near field, h/L= 2.0 (Ref. 97).

were generated, the latter to save on computing costs in the simulations of the longitudinal flight dynamics. The lateral flight dynamics tables, on the other hand, required the full model flow prediction. The model consists of a 14.5 million-point mesh for the half configuration arranged in 2028 blocks, which allow the calculations to be run evenly-balanced on up to 256 processors. The full configuration allows for twice the amount of processors since these characteristics are doubled.

The main characteristics of the blocking topology are shown in Fig. 7.76. Figure 7.76a shows the overall view of the meshed geometry with the shaded blocks on the surface of the model and a gray meshed symmetry plane. Figure 7.76b shows the diamond-shaped tip topology used to accommodate the cells around the main wing and tail-plane. The wing has an H-type topology around the leading edge to improve the cell quality in the wing-engine-fuselage junction, while the horizontal tail-plane blocking consists of a C-type around the leading edge. Most of the control surfaces present in the aircraft were also implemented in this grid model in order to simulate the required maneuvers for the flight dynamics analysis (these were the elevator, aileron, and rudder which can be deflected for steady-state calculations and during time-accurate simulations).

(a)   (b)

Figure 7.76   Block structured grid around the Ranger jet aircraft (Courtesy of David Vallespin and Ken Badcock of the University of Liverpool; a full color version of this image is available on the website: www.cambridge.org/aerodynamics).

Figure 7.77   X-31 aircraft in flight (Courtesy of NASA Dryden Flight Research Center).

7.13.2 X-31 Viscous Block Structured Grid

The X-31 enhanced fighter maneuverability (EFM) experimental aircraft obtained data to aid in the design of highly maneuverable next-generation fighters. "The goals of the flight program were to demonstrate EFM technologies, investigate close-in-combat exchange ratios, develop design requirements, build a database for application to future fighter aircraft, and develop and validate low-cost prototype concepts."[99] The X-31 had a fairly complex configuration, including a delta wing with leading and trailing edge devices, a canard, and several strakes (see Fig. 7.77 to see some of these details).

The approach to creating a grid around a complex configuration starts with an abstraction of the geometry description, as shown in Fig. 7.78a. This low-level abstraction represents the basic geometric features of the aircraft with a set of Cartesian blocks. The low-level abstraction is then projected

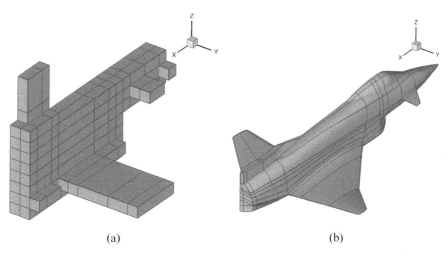

(a)   (b)

Figure 7.78 X-31 geometry abstraction (Courtesy of Okko Boelens of NLR; a full color version of this image is available on the website: www.cambridge.org/aerodynamics). (a) abstraction of the surface geometry; (b) projected abstraction.

onto the real geometry description, as shown in Fig. 7.78b, in order to create the various blocks that will make up the grid. Next, the Navier-Stokes blocks (which are the first layer of blocks around the surface) and the field blocks in physical space are generated automatically. "In the resulting topology, the grid dimensions are set manually and the edges are connected automatically. Finally, the grid quality is improved by applying an elliptic smoothing algorithm and the resolution in the Navier-Stokes blocks is improved such that the desired boundary layer resolution is obtained."[100]

The original grid generated for the X-31 configuration did not have leading-edge flap gaps (designated grid G3), which are the narrow gaps between the various segments of the leading-edge devices. These gaps have an impact on the flowfield, especially on the formation of vortices along the leading edge of the wing. So, in this case, the leading-edge gaps (both longitudinal and spanwise) were also simulated in the grid. Based on grid G3, the grids around the X-31 configuration with only the longitudinal leading-edge flap gaps (G2) and the X-31 configuration with all leading-edge flap gaps (G1) were constructed by inserting the respective blocks. Generating the grids in this manner has the advantage that the grids are virtually identical, except for the flap gaps.

Using this approach, grids around complex configurations can be generated within relatively short times. For the X-31 configuration with no leading-edge flap gaps (G3), the grid was completed in slightly more than one week, whereas including the longitudinal flap gaps and the spanwise flap gaps only took one more day each. The resulting surface grid and grid on the symmetry plane for the X-31 configuration with all leading-edge flap gaps (G1) are shown in Fig. 7.79. Details for all three grids can be found in Table 7.3.

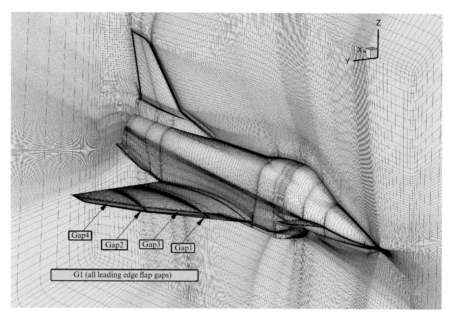

Figure 7.79 X-31 block structured grid topology (Courtesy of Okko Boelens of NLR; a full color version of this image is available on the website: www.cambridge.org/aerodynamics).

Table 7.3 Details of X-31 Block Structured Grids (Ref. 100)

Grid	G1 (all leading-edge flap gaps)	G2 (longitudinal leading-edge flap gaps)	G3 (no leading-edge flap gaps)
Number of blocks	1,317	1,307	1,299
Number of cells	24,899,072	24,737,792	24,651,776
Number of grid points	27,836,365	27,641,739	27,542,211
Number of blocks after merging	207	197	189

7.13.3 Block Structured Grid for Helicopter with Sliding Interface for Rotor

Figure 7.80 shows an example of multi-block structured fuselage and rotor meshes that have non-matching topologies along the sliding-mesh interface separating the two domains.[101] Non-matching cell faces have been used in CFD by many researchers; the fundamentals of interface conditions for non-matching cell faces have been explored.[102,103] These non-matching methods are commonly used in turbo-machinery simulations,[104] where non-matching and rotating cell faces are used for the simulation of the flow between adjacent blade-rows of aero engines.

7.13.4 Unstructured High Lift Commercial Transport

Now, we have an example of an unstructured mesh applied to a commercial transport with high-lift flap and slat deflections.[105] Figure 7.81 shows the

(a)   (b)

(c)   (d)

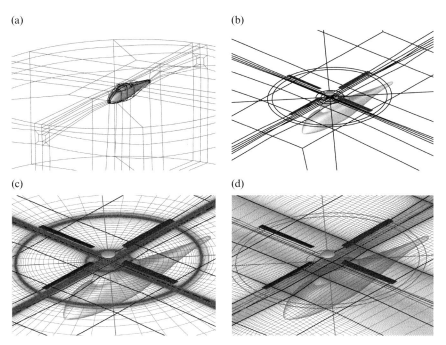

Figure 7.80 Block structured helicopter grid with sliding interfaces for rotor (Ref. 104; Courtesy of George Barakos of the University of Liverpool; a full color version of this image is available on the website: www.cambridge.org/aerodynamics).

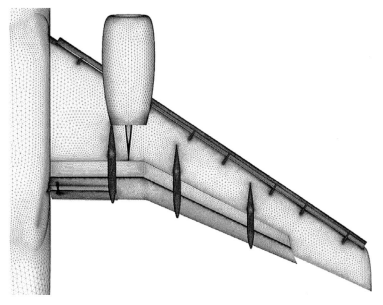

Figure 7.81 Commercial transport with extended leading-edge and trailing-edge devices (Ref. 105; reprinted by permission of the American Institute of Aeronautics and Astronautics, Inc.).

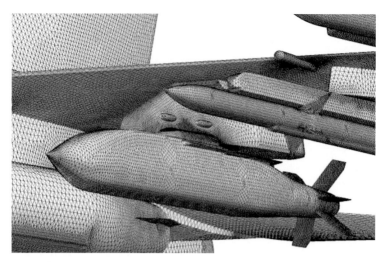

Figure 7.82 F-16C aircraft with stores (Ref. 106).

computational unstructured mesh with flap track fairings (FTF) and brackets to support the slats and aft-flap. The unstructured meshes for the case without FTF and with FTF have approximately 5.8 and 7.4 million mesh points, respectively. The minimum grid spacing in the normal direction to the wing surface is $0.02 / \sqrt{Re}$. Only one or two cells are placed on the blunt trailing edges.

7.13.5 Unstructured Mesh for Aircraft with Stores

Predicting the aerodynamics of aircraft with under-wing stores is extremely important for military applications, especially since the stores can drastically change the flowfield in the vicinity of the aircraft.[106] Being able to properly model the presence of stores (both statically and dynamically) is a crucial capability for store certification programs. An F-16C with a variety of under-wing stores is shown in Fig. 7.82. All grids are unstructured and were created with *SolidMesh*,[107] a solid modeling and unstructured grid generation system and the AFLR grid generation tool (developed at Mississippi State University).[108,109] Full-span grid sizes ranged from 13 million cells for a clean F-16C to 30+ million cells for a fully loaded aircraft with stores. All grids are unstructured mixed element grids containing tetrahedral and five- and six-sided elements. An initial boundary-layer spacing of $y^+ = 1$ was specified for all grids as well.

7.13.6 C-130 Unstructured Chimera Mesh with Ring-Slot Parachute

The C-130H performs airdrops at speeds between 130 and 140 knots (indicated airspeed) and at altitudes ranging from 500 ft to 20,000 feet. The Reynolds number based on the chord used in these calculations will therefore

Figure 7.83   C-130 with a ring-slot parachute; note the propeller disk simulation (Ref. 110).

range from 8.6 million to 19.6 million. The grids were designed to provide an average surface normal spacing in wall units (y^+) less than 1 for the first cell off the wall for the whole aircraft.[110] Grids were developed using the software programs *Gridtool*[111] (to develop the surface point distributions and other background information) and *VGRIDns*[112] (to grow the volume grid). Grids were typically created with concentration of points near the surface in the viscous region, as shown in Fig. 7.83. After growing the tetrahedral grid with *VGRID*, a *Cobalt* utility code called Blacksmith was used to "weld" the tetrahedra in the viscous region into prisms. Using prisms in the viscous region improves the aspect ratio of the cells, and a side benefit is a reduction in the overall number of cells.

All of the grids in this study consist of an inner region of approximately twenty layers of prisms for the boundary layer, with a wall normal spacing in viscous wall units less than 1 and an outer region of tetrahedra. The prism dimensions on the surface were a factor of approximately 200 times larger than the wall normal dimension for all grids. The C-130H has four turbo-shaft engines turning four blade propellers. These propellers are all turning in the same direction at 1020 RPM creating 3000 lb of thrust for each engine. The propellers were modeled by actuator disk boundary conditions with peak loading set to 90 percent of the propeller radius, with a linear load distribution assumed. The C-130H with the extraction chute configuration was developed with a 20-ft ringslot parachute on a 60-ft extraction line resulting in the chute at 50 ft aft of the cargo ramp. Overset cases include positioning the ringslot parachute 50 ft, 25 ft, and 12.5 ft aft of the

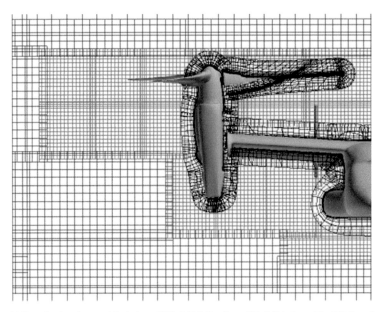

V-22 unstructured, overset Cartesian grid (Ref. 113; Courtesy of Mark Potsdam of the U.S. Army Aeroflightdynamics Directorate; a full color version of this image is available on the website: www.cambridge.org/aerodynamics).

cargo ramp to determine the effect of aircraft proximity to the loading on the ringslot parachute.

7.13.7 V-22 Rotorcraft with Cartesian Overset Grids

Finally, we show an example of a Cartesian grid, in this case, one using the Chimera overset capability in order to model the rotors of the V-22, as shown in Fig. 7.84.[113] In the Chimera methodology, overset, structured near-body grids are generated about the geometry. They extend approximately one tip chord away from the body and include sufficient resolution to capture boundary-layer viscous effects. They use C-mesh topology for the blades and tip caps. The first four points away from the blade surfaces have a constant spacing, verified to produce a $y+ \leq 1$. These spacing parameters closely match those determined for accurate drag prediction on transonic fixed wings using *OVERFLOW*.[114] Off-body Cartesian grid generation is automatically performed by *OVERFLOW-D*. The finest off-body spacing for the baseline grid is 0.10 of the tip chord. This level-1 grid surrounds the blades and is manually specified in order to contain the entire wake. A total of five progressively coarser levels are generated out to the farfield boundary, which is placed at five times the rotor radius in all directions from the center of the domain. The grid spacings differ by a factor of two between each mesh level. The baseline grid contains 15.9 million points: 6.2 million near-body and 9.7 million off-body. Where grid points fall inside the geometry, hole cutting is employed to blank out these points. A cut through the grid system in Fig. 7.84 shows the near- and off-body grids, hole cuts, and overlap.

7.14 Current Grid Generation Software and Data Structures

As we said earlier, grid generation has certainly been one of the bottlenecks in computational aerodynamics. Creating accurate geometric shapes that represent realistic airplanes, and then defining surface grids and volume grids, is a time-consuming job. In fact, Joe Steger, a pioneer in CFD grid generation, said, "grid generation remains the key pacing item in making CFD useful to most engineering applications in aerodynamics."[115] Unfortunately, he said that in 1991, and the statement is still true today! We hope that, as time goes on, the software tools for grid generation will evolve to help make this task easier, but, for now, it is challenging and requires a certain level of expertise.

Several summaries of grid generation methods are available to help define the issues and challenges in grid generation. Included in these overviews are the results of a workshop held at NASA Langley in 1992,[22] a review of grid generation capabilities,[116] an overview of the enabling technology requirements for CFD,[117] a review of grid generation requirements for design and optimization,[118] and a discussion of the grid generation requirements for hypersonic aerothermodynamics.[119] Although many, many systems are widely used, we will identify a few codes as being representative of various capabilities that exist. Some widely used grid generation systems are:

- *GRAPE* (www.pdas.com/grape.html): This set of elliptic PDE grid generation codes originates at NASA Ames, beginning with the original code by Reese Sorenson.[120]
- *Gridgen/Pointwise* (www.pointwise.com/): This program was originally developed at General Dynamics in Fort Worth for the Air Force Wright Laboratories[121] and is now a commercially available code.
- *VGRID* (tetruss.larc.nasa.gov/vgrid/): unstructured tetrahedral grids with the capability of generating thin-layered "viscous" grids and multidirectional anisotropic grid stretching.
- *SolidMesh* and *GUMBO* from Mississippi State University (www.simcenter.msstate.edu/docs/solidmesh/): unstructured grid generation system that quickly generates high-quality unstructured grids for complex 2D and 3D geometries.
- *ICEM CFD* (www.ansys.com/products/icemcfd.asp): hexa mesh (structured or unstructured) with grid generation control, extended mesh diagnostics, and advanced, interactive mesh editing.

- *CentaurSoft* (www.centaursoft.com/): Uses hybrid (prismatic/hexahedral/pyramidal/tetrahedral) meshing strategy with prismatic and hexahedral elements used in regions of high solution gradients, and tetrahedra are used elsewhere with pyramids used in some locations to allow for a transition between the prisms/hexahedra and the tetrahedra.
- *Cubes* from *CART3D* (people.nas.nasa.gov/~aftosmis/cart3d/): Produces topologically unstructured, adaptively refined, Cartesian meshes around any geometry or configuration that may be described by a collection of simplicial polyhedral.
- *blockMesh* from *OpenFOAM* (Open Field Operation And Manipulation) (www.openfoam.com/): Handles unstructured meshes of mixed polyhedra with any number of faces: hexahedra, tetrahedra, and degenerate cells.
- *OVERFLOW* (aaac.larc.nasa.gov/~buning/codes.html): A comprehensive program from NASA for creating grids, including overset grids, for complex geometries.

Many of these codes are currently under continual development, with the primary emphasis on improving the user interface and automation.

In addition to the various software programs available for creating grids, there is a great deal of effort underway to define grid data structures … but why? Over the years a variety of programs have been written to create grids or post-process data. Each of these programs used its own definitions (called formatting) for how data should be input and output from various programs. This has led to a wide variety of data formats, and while many codes try to read (and write) as many data formats as possible, not all codes can utilize all data formats. One of the oldest formats commonly used for structured grids is *PLOT3D*,[122] a post-processing tool developed at NASA Ames Research Center. Not only does *PLOT3D* define how output data would be written, but it also defines how input information is formatted, including the grids being used by various NASA Ames Research Center flow solvers (*ARC2D*, *ARC3D*, and *OVERFLOW*). Many programs, therefore, are able to read and write the *PLOT3D* format for grids and post-processing data. But there are numerous data formats, and not every program can read every format, which can cause a great deal of confusion for people trying to read various grids into their flow solver.

This situation led a number of researchers to suggest the creation of a standard data format for CFD, a format that would also be proposed as an international standard (meeting International Organization for Standardization, ISO, requirements if possible). The format that was created, and is still evolving, is called the CFD General Notation System (CGNS).[123,124] The development was by "a public forum made up of international representatives from government and private industry, responsible for the development, evolution, support, and promotion of [CGNS]."[125] CGNS allows for a great deal of generality and has as its goal being the de facto standard for grid formats. While this is a laudable goal, complete generality also leads to complexity and the need for a large library of translators and "checkers." This generality, therefore, can also be seen as being cumbersome, possibly taking a great deal of time to input and output large data files.

Summary of Best Practices

1. The grid/mesh must be capable of capturing all of the flow physics of interest, which is a challenging task that should not be underestimated by the user.
2. To obtain good (and dependable) results from CFD codes, you must have a thorough understanding of the creation, purposes, and relative strengths and weaknesses of the following grids/meshes:
   a. structured grids (including algebraic stretching, conformal transformations, elliptic and hyperbolic PDE grid generation, etc.)
   b. unstructured meshes
   c. hybrid meshes
   d. Cartesian grids
3. The selected grid/mesh must fit the geometry and requirements of your application, including the requirements of your flow solver.
4. Grid farfield boundaries should be far enough away from the surfaces of interest so they do not impact the solution. Remember:
   a. subsonic flows require large outer boundary distances: at least ten chords in front and above/below the vehicle and usually at least twenty chords behind the vehicle
   b. supersonic and hypersonic flows do not require the inflow boundary to be as far away from the vehicle, due to the way information is propagated
5. In general, use finer grids in regions of high gradients in order to capture the details of the flow physics of interest: use grid refinement studies and/or adaptive mesh refinement to accomplish this goal.
6. Viscous grids should have grid "support" in the boundary layer for turbulence models to work well. Elements of adequate grid support include:
   a. average $y^+ \leq 1$
   b. stretching growth rates no greater than 1.2
   c. at least 20 points/cells through the boundary layer
   d. at least two or three points in each sublayer, although more points are better
7. Grid/mesh cells should:
   a. avoid being highly skewed or warped
   b. avoid high aspect ratios (less than 20 away from a wall, although higher values close to a wall are usually unavoidable)
   c. have stretching ratios in all directions no greater than 1.2
   d. use nearly orthogonal cells near boundaries
   e. have appropriate planarness, skew-smoothness, and isotropic quality
8. A grid sensitivity study should be carried out for each class of problem with at least three grids of significantly different size.

7.15 Projects

Overview: perform a viscous CFD study of the 4-digit NACA airfoil section you chose in Chapter 1. You will complete all four steps in the CFD solution process several times to accomplish this work: generate airfoil geometry, create several grids and perform a grid refinement study, generate computational lift-curve slopes and drag polars, compare those with experiment, and run a post-stall case.

Part 1. Generate a 2D contour of your assigned NACA airfoil section.

Use the airfoil/wing geometry program to generate contour coordinates for your airfoil section and export a data file that can be imported into grid-generation software. Generate approximately 200 points on both the upper and lower surfaces for this database geometry. Be sure to distribute the points appropriately along the surface of the airfoil in order to accurately define the surface regions with high curvature (cosine spacing), to prepare for the grid-generation process. *NOTE: These points are not grid points, they are just geometry.*

Part 2. Generate computational grids for your airfoil shape.

Once you read the database file into a grid generation software program, for more accurate results you may choose to modify the database to give a sharp trailing edge for the airfoil; note the chord length and be sure it matches the value for your reference area. Generate a grid fine enough to accurately capture the flowfield phenomena near the airfoil surface but not so fine that it includes unnecessary grid points in the farfield regions (that is, a hybrid grid: structured close to airfoil to properly resolve the boundary layer, unstructured away from the airfoil for better computing efficiency). Be sure the outer domain boundaries are far enough away from the airfoil boundary to accurately capture the entire flowfield; e.g., you may obtain a more accurate answer for a subsonic airfoil by using more than 10 chord lengths above, below, and upstream, and more than 20 chord lengths downstream. You should use a high decay rate for the unstructured portion of the domain, something like 0.99. Also, provide cell clustering in the wake, using one or several wake connectors. Exactly how you do this is up to you. For viscous calculations, the grid spacing at the no-slip wall (airfoil surface) should be tight enough to give an average $y^+ \leq 1$. After your runs are complete, you should verify the actual y^+ values are acceptable and change grid spacing if needed. *Include in your observations the y^+ values and cell counts for each grid.* The initial spacing should be found from

$$\Delta_{turbulent} = L \frac{(13.1463 y_{ave}^+)^{0.875}}{Re_L^{0.90}}$$

where L is the length at the point of interest (the chord for this project) and Re_L is the Reynolds number based on that length. Your first grid should be the "medium grid," and 201 grid points on both the lower and upper surface is recommended. Once you have created a suitable grid, create two more grids: one grid with half the number of points and another grid with double the number of points. You should now have three grids with a consistent number of points that can be evaluated for grid sensitivity.

Part 3. Using a flow solver, solve the 2D flow around the airfoil *at an α in the linear regime* for all three grids.

For this turbulent Reynolds number and non-stalled α, run the SA turbulence model. Flowfield conditions (p, T, and M) need to give $Re_c = 3,000,000$ (or whatever the Reynolds number is for your data) and also be in the incompressible regime ($0.2 < M < 0.3$). Use the following relationships to determine p, T, and M. (*Hint: assume M and T, then calculate the rest. Be consistent with units!*)

$$Re_c = \frac{\rho_\infty U_\infty c}{\mu_\infty}$$

$$p_\infty = \rho_\infty R T_\infty$$

$$M_\infty = \frac{U_\infty}{a_\infty},$$

$$\mu(T) = \frac{C_1 T^{3/2}}{T + C_2}$$

where the viscosity is determined using Southerland's law from Eqn. 3.32 and $C_1 = 1.458 \times 10^{-6}$ N-s/(m^2-K$^{1/2}$) and $C_2 = 110.4$ K (see Table 3.1). Run the simulation for a sufficient number of iterations to ensure convergence of forces and moments. *Be sure to include at least one convergence plot in your reporting.* Perform a grid sensitivity study and use that information to choose a single grid for the rest of the project. You may have to make another grid, but make no more than four, even if you don't completely reach grid independence. Be sure to verify that all solutions are converged. Now solve the 2D flow around the airfoil at two additional equally spaced α's in the linear region of the C_l vs. α curve using a steady-state assumption. Also compute one point *well* into the post-stall region (but within the bounds of your available data for later comparison), *still using the steady-state assumption.* Reference the experimental data for your airfoil section to make sure your selected angles of attack are in the linear region and one point in the post-stall region of the lift curve.

Perform an *unsteady analysis* of your airfoil at the α WELL into the post-stall region (same α as selected earlier). Solve the flowfield for this unsteady case with second-order temporal accuracy at an appropriate Reynolds number and for a nondimensional time step of:

$$\Delta t^* = \frac{\Delta t\, U_\infty}{c} = 0.01.$$

Note: codes usually require Δt, NOT Δt^*, so you must solve for Δt using the preceding relationship. Capture enough data for each of these runs to ensure that no transients exist. You may use the steady case as the starting point for the unsteady computations: for example, run steady for 2,000 iterations at $\alpha = 20°$ for Part 3 of this project, then change to unsteady and run for at least 4,000 iterations. Compute the mean C_l, C_d, and $C_{mc/4}$ and add these data points to the steady-state plots.

Part 4. Compare CFD results to experimental data.

Collate the computed lift, drag, and pitching moment coefficient data from your output files and create comparison plots for C_l, C_d, and $C_{mc/4}$. To provide a more precise comparison with experimental data, you should "digitize" the data using the digitizing software for your particular airfoil and plot your CFD data along with the digitized data (only lift and drag comparisons are shown here). Comment on how well the CFD predictions match the experimental data and how the unsteady calculations changed the post-stall results.

7.16 References

1   Kinsey, D., "The Meshenger: For the Grid Generation Enthusiast," USAF Wright Laboratory, Aeronautical Systems Division, Wright-Patterson Air Force Base, Ohio, January 1991.

2   Raymer, Daniel P., *Aircraft Design: A Conceptual Approach*, 4th Ed., Reston: AIAA, 2006.

3   Liming, Roy A., *Practical Analytic Geometry with Applications to Aircraft*, New York: MacMillan, 1944.

4   www.cambridge.org/aerodynamics.

5   Barger, R.L., "An Analytical Procedure for Computing Smooth Transitions Between Two Specified Cross Sections with Applications to Blended Wing-Body Configurations," NASA TP 2012, May 1982.

6   Barger, R.L., and Adams, M.S., "Semianalytic Modeling of Aerodynamic Shapes," NASA NP 2413, April 1985.

7   Kulfan, B.M., "Universal Parametric Geometry Representation Method – 'CST'," *Journal of Aircraft*, Vol. 45, No. 1, 2008, pp. 142–158.

8   Kulfan, B.M., "Recent Extensions and Applications of the 'CST' Universal Parametric Geometry Representation Method," AIAA Paper 2007–7709, September 2007.

9   Rogers, D.F., and Adams, J.A., *Mathematical Elements of Computer Graphics*, 2nd Ed., New York: McGraw-Hill, 1990.

10   de Berg, M., Cheong, O., van Kreveld, M., and Overmars, M., *Computational Geometry: Algorithms and Applications*, 3rd Ed., Berlin: Springer, 2008.

11   Faux, I.D., and Pratt, M.J., *Computational Geometry for Design and Manufacture*, Chichester: Ellis Horwood Limited, 1979.

12   Craidon, C.B., "Description of a Digital Computer Program for Airplane Configuration Plots," NASA TM X-2074, 1970.

13   Mason, W.H., and Rosen, B.S., "The COREL and W12SC3 Computer Programs for Supersonic Wing Design and Analysis," NASA CR-3676, December 1983.

14   Capron, W.K., and Smit, K.L., "Advanced Aerodynamic Applications of an Interactive Geometry and Visualization System," AIAA Paper 91–0800, January 1991.

15   Boelens, O.J., Görtz, S., Morton, S., Fritz, W., and Lamar, J.E., "Description of the F-16XL Geometry and Computational Grids Used in CAWAPI," AIAA Paper 2007-0488, January 2007.

16   Thompson, J.F., Warsi, Z.U.A., and Mastin, C.W., *Numerical Grid Generation*, New York: North-Holland, 1985.

17   Thompson, J.F., Soni, B.K., and Weatherill, N.P., eds., *Handbook of Grid Generation*, Boca Raton: CRC Press, 1999.

18   Knupp, P., and Steinberg, S., *Fundamentals of Grid Generation*, Boca Raton: CRC Press, 1993.

19   Frey, P.J., and George, P.L., *Mesh Generation*, London: Wiley, 2008.

20   Farrashkhalvat, M., and Miles, J.P., *Basic Structured Grid Generation: With an Introduction to Unstructured Grid Generation*, Oxford: Butterworth-Heinemann, 2003.

21   Smith, R.E., "Numerical Grid Generation Techniques," NASA CP 2166, 1980.

22   Smith, R.E., "Software Systems for Surface Modeling and Grid Generation," NASA CP 3143, 1992.

23   Posenau, M.-A., "Unstructured Grid Generation Techniques and Software," NASA CP 10119, September 1993.

24   www.usnccm.org.

25   www.imr.sandia.gov.

26   Steinbrenner, J.P., and Anderson, D.A., "Grid-Generation Methodology in Applied Aerodynamics," in *Applied Computational Aerodynamics*, P.A. Henne, Ed., AIAA Progress in Astronautics and Aeronautics, Vol. 125, 1990, pp. 91–130.

27   Schuster, D.M., and Brickelbaw, L.D., "Numerical Computation of Viscous Flowfields about Multiple Component Airfoils, AIAA Paper 85–0167, January 1985.

28   Fang, J., and Kennon, S.R., "Unstructured Grid Generation of Non-Convex Domains," AIAA Paper 89–1983, June 1989.

29   van der Burg, J.W., de Cock, K.M.J., and van der Pijl, S.P., "Demonstration of Viscous Flow Computations on Hybrid (Prismatic/Tetrahedral) Grids," National Aerospace Laboratory (The Netherlands), NLR-TP-2000–378, 2000.

30   Steger, J.L., Dougherty, C.F., and Benek, J.A., "A Chimera Grid Scheme," in *Advances in Grid Generation*, ASME, 1983, pp. 59–69.

31   Nichols, R.H., and Buning, P.G., "User's Manual for OVERFLOW 2.1," August 2008.

32   Badcock, K.J., Richards, B.E., and Woodgate, M.A., "Elements of Computational Fluid Dynamics on Block Structured Grids Using Implicit Solvers," *Progress in Aerospace Sciences*, Vol. 36, 2000, pp. 351–392.

33   Wang, Z.J., Hariharan, N., and Chen, R., "Recent Development on the Conservation Property of Chimera," *International Journal of Computational Fluid Dynamics*, Vol. 15, No. 4, 2001, pp. 265–278.

34   Aftosmis, M.J., Berger, M.J., and Melton, J.E., "Robust and Efficient Cartesian Mesh Generation for Component-Based Geometry," *AIAA Journal*, Vol. 36, No. 6, 1998, pp. 952–960.

35   Murman, S.M., Aftosmis, M.J., and Berger, M.J., "Simulations of Store Separation from an F/A-18 with a Cartesian Method," *Journal of Aircraft*, Vol. 41, No. 4, 2004, pp. 870–878.

36   Katz, A., and Jameson, A., "Meshless Scheme Based on Alignment Constraints," *AIAA Journal*, Vol. 48, No. 11, 2010, pp. 2501–2511.

37   Löhner, R., Sacco, C., Oñate, E. and Idelsohn, S., "A Finite Point Method for Compressible Flow," *International Journal for Numerical Methods Engineering*, Vol. 53, 2002, pp. 1765–1779.

38   Tang, L., Yang, J., and Lee, J., "Hybrid Cartesian Grid/Gridless Algorithm for Store Separation Prediction," AIAA Paper 2010-0508, January 2010.

39   Filipiak, M., *Mesh Generation*, Edinburgh Parallel Computing Centre, University of Edinburgh, November 1996.

40   Eriksson, L.E., "Generation of Boundary-Conforming Grids Around Wing-Body Configurations Using Transfinite Interpolation," *AIAA Journal*, Vol. 20, No. 10, 1982, pp. 1313–1320.

41   Baker, T.J., "Three Decades of Meshing: A Retrospective View," AIAA Paper 2003–3563, June 2003.

42   Ly, E., and Norrison, D., "Automatic Elliptic Grid Generation by an Approximate Factorisation Algorithm," *Australian and New Zealand Industrial and Applied Mathematics Journal*, Vol. 48, 2007, pp. C188–C202.

43   Thompson, J.F., and Weatherill, N.P., "Aspects of Numerical Grid Generation: Current Science and Art," AIAA Paper 93–3539, August 1993.

44   Fletcher, C.A.J., *Computational Techniques for Fluid Dynamics*, Vol. II. "Specific Techniques for Fluid Dynamics," Berlin: Springer-Verlag, 1988.

45   Pletcher, R.H., Tannehill, J.C., and Anderson, D.A., *Computational Fluid Mechanics and Heat Transfer*, 3rd Ed., Boca Raton: CRC Press, 2013.

46   Cummings, R.M., Yang, H.T., and Oh, Y.H., "Supersonic, Turbulent Flow Computation and Drag Optimization for Axisymmetric Afterbodies," *Computers and Fluids*, Vol. 24, No. 4, 1995, pp. 487–507.

47   Gordon, W.J., and Hall, C.A., "Construction of Curvilinear Co-ordinate Systems and Applications to Mesh Generation," *International Journal for Numerical Methods in Engineering*, Vol. 7, 1973, pp. 461–477.

48   Eiseman, P.R., "Grid Generation for Fluid Mechanics Computations," *Annual Review of Fluid Mechanics*, Vol. 17, 1985, pp. 487–522.

49   Houghton, E.L., and Brock, A.E., *Aerodynamics for Engineering Students*, London: Edward Arnold, 1960.

50   Jameson, A., "Full-Potential, Euler, and Navier-Stokes Schemes," in *Applied Computational Aerodynamics*, ed. P.A. Henne, AIAA Progress in Astronautics and Aeronautics, Vol. 125, 1990, pp. 39–88.

51   Sells, C.C.L., "Plane Subcritical Flow Past a Lifting Aerofoil," *Proceedings of the Royal Society London*, Vol. 308A, 1968, pp. 377–401.

52   Curle, N., and Davies, H.J., *Modern Fluid Dynamics, Vol. 2, Compressible Flow*, London: Van Nostrand Reinhold, 1971.

53   Ives, D.C., "A Modern Look at Conformal Mapping Including Multiple Connected Regions," *AIAA Journal*, Vol. 14, No. 8, 1976, pp. 1006–1011.

54   Halsey, N.D., "Use of Conformal Mapping in Grid Generation for Complex Three-Dimensional Configurations," *AIAA Journal*, Vol. 25, No. 10, 1987, pp. 1286–1291.

55   Sorenson, R.L., "A Computer Program to Generate Two-Dimensional Grids About Airfoils and Other Shapes by the Use of Poisson's Equation," NASA TM 81198, 1980.

56   Steinbrenner, J.P., and Chawner, J.R., "Gridgen's Implementation of Partial Differential Equation Based Structured Grid Generation Methods," 8th International Meshing Roundtable, Lake Tahoe, CA, October 1999.

57   Sorenson, R.L., and Steger, J.L., "Grid Generation in Three Dimensions by Poisson Equations with Control of Cell Size and Skewness at Boundary Surfaces," *Advances in Grid Generation*, Vol. 5, Ed. K.N. Ghia, and U. Ghia, ASME Applied Mechanics, Bioengineering, and Fluids Engineering Conference, 1983.

58   Steger, J.L., and Chaussee, D.S., "Generation of Body-Fitted Coordinates Using Hyperbolic Partial Differential Equations," *SIAM Journal of Scientific and Statistical Computing*, Vol. 1, No. 4, 1980, pp. 431–437.

59   Chan, W.H., "Hyperbolic Methods for Surface and Field Grid Generation," in *Handbook of Grid Generation*, Ed. Thompson, J.F., et al., Boca Raton: CRC Press, 1999, pp. 5-1–5-26.

60   Mavriplis, D.J., and Pirzadeh, S., "Large-Scale Parallel Unstructured Mesh Computations for Three-Dimensional High-Lift Analysis," *Journal of Aircraft*, Vol. 36, No. 6, 1999, pp. 987–998.

61   Görtz, S., Jirasek, A., Morton, S.A., McDaniel, D.R., Cummings, R.M., Lamar, J.E., and Abdol-Hamid, K.S., "Standard Unstructured Grid Solutions for Cranked Arrow Wing Aerodynamics Project International F-16XL," *Journal of Aircraft*, Vol. 46, No. 2, 2009, pp. 385–408.

62   Mavriplis, D.J., "Unstructured Mesh Algorithms for Aerodynamic Calculations," ICASE Report No. 92-35 (also NASA CR 189685), July 1982.

63   Mavriplis, D.J., "Unstructured Mesh Generation and Adaptivity," Lecture notes, 28th VKI CFD Lecture Series, April 1995.

64   Spragle, G.S., McGrory, W.R., and Fang, J., "Comparison of 2D Unstructured Grid Generation Techniques," AIAA Paper 91-0726, January 1991.

65   tetruss.larc.nasa.gov/vgrid/grid_gen.html

66   Delaunay, B., "Sur la Sphère Vide," *Otdelenie Matematicheskikh i Estestvennykh Nauk*, Vol. 7, 1934, pp. 793–800.

67   www.codeguru.com/cpp/cpp/algorithms/general/article.php/c8901/

68   Venkatakrishnan, V., and Barth, T.J., "Application of Direct Solvers to Unstructured Meshes for the Euler and Navier-Stokes Equations Using Upwind Schemes," AIAA Paper 89-0364, January 1989.

69   Löhner, R., and Parikh, P., "Generation of Three-Dimensional Unstructured Grids by the Advancing-Front Method," *International Journal for Numerical Methods in Fluids*, Vol. 8, 1988, pp. 1135–1149.

70   Pirzadeh, S., "Unstructured Viscous Grid Generation by Advancing-Layer Methods," AIAA Paper 1993-3453, July 1993.

71   Pirzadeh, S., "Recent Progress in Unstructured Grid Generation," AIAA Paper 92-0445, January 1992.

72   Pirzadeh, S.Z., and Zagaris, G., "Domain Decomposition By the Advancing-Partition Method for Parallel Unstructured Grid Generation," AIAA Paper 2009-0979, January 2009.

73   Pirzadeh, S., "Viscous Unstructured Three-Dimensional Grids by the Advancing-Layers Method," AIAA Paper 94-0417, January 1994.

74   Dawes, W.N., Harvey, S.A., Fellows, S., Eccles, N., Jaeggi, D., and Kellar, W.P., "A Practical Demonstration of Scalable, Parallel Mesh Generation," AIAA Paper 2009-0981, January 2009.

75   McMorris, H., and Kallinderis, Y., "Octree-Advancing Front Method for Generation of Unstructured Surface and Volume Meshes," *AIAA Journal*, Vol. 35, No. 6, 1997, pp. 976–984.

76   Aftosmis, M.J., Berger, M.J., and Melton, J.E., "Robust and Efficient Cartesian Mesh Generation for Component-Based Geometry," AIAA Paper 97-0196, January 1997.

77   Aftosmis, M.J., Berger, M.J., and Murman, S.M., "Applications of Space-Filling Curves to Cartesian Methods for CFD," AIAA Paper 2004-1232, January 2004.

78   Murman, S.M., Aftosmis, M.J., and Berger, M.J., "Simulations of Store Separation from an F/A-18 with a Cartesian Method," *Journal of Aircraft*, Vol. 41, No. 4, 2004, pp. 870–878.

79 Mavriplis, D.J., "Unstructured Mesh Related Issues in Computational Fluid Dynamics (CFD)-Based Analysis and Design," *11th International Meshing Roundtable*, September 2002.

80 Pirzadeh, S.Z., "A Solution-Adaptive Unstructured Grid Method by Grid Subdivision and Local Remeshing," *Journal of Aircraft*, Vol. 37, No. 5, 2000, pp. 818–824.

81 Pirzadeh, S., "Vortical Flow Prediction Using an Adaptive Unstructured Grid Method," Symposium on Advanced Flow Management. Part A: Vortex Flow and High Angle of Attack, NATO RTO-MP-069-13, May 2001.

82 Mitchell, A.M., Morton, S.A., Forsythe, J.R., and Cummings, R.M., "Analysis of Delta-Wing Vortical Substructures Using Detached-Eddy Simulation," *AIAA Journal*, Vol. 44, No. 5, 2006, pp. 964–972.

83 Morton, S.A., "Detached-Eddy Simulations of Vortex Breakdown over a 70-Degree Delta Wing," *Journal of Aircraft*, Vol. 46, No. 3, 2009, pp. 746–755.

84 Wintzer, M., Nemec, M., and Aftosmis, M.J., "Adjoint-Based Adaptive Mesh Refinement for Sonic Boom Prediction," AIAA Paper 2008–6593, August 2008.

85 Pawlowski, J., Graham, D., Boccadoro, C., Coen, P., and Maglieri, D., "Origins and Overview of the Shaped Sonic Boom Demonstration Program," AIAA Paper 2005-0005, January 2005.

86 Gribben, B.J., Badcock, K.J., and Richards, B.E., "Towards Automatic Multiblock Topology Generation," AIAA Paper 99–3299, June 1999.

87 Casey, M., and Wintergerste, T., Eds., "Quality and Trust in Industrial CFD: Best Practice Guidelines," ERCOFTAC, 2000.

88 Strang, W.Z., Tomaro, R.F., and Grismer, M.J., "The Defining Methods of Cobalt$_{60}$: A Parallel, Implicit, Unstructured Euler/Navier-Stokes Flow Solver," AIAA Paper 1999-0786, January 1999.

89 Mavriplis, D.J., "Unstructured Mesh Discretizations and Solvers for Computational Aerodynamics," AIAA Paper 2007–3955, June 2007.

90 Ferziger, J.H., and Perić, M., *Computational Methods for Fluid Dynamics*, 3rd Ed., Berlin: Springer Verlag, 2002.

91 Mastin, C.W., "Truncation Error for Finite Volume Approximations on Unstructured Grids," 8th Mississippi State University of Alabama at Birmingham Conference on Differential Equations and Computational Simulations, May 2009.

92 Anonymous, "Best Practice Guidelines for Marine Applications of Computational Fluid Dynamics," MARNet CFD, Final Report, 2003.

93 Hirsch C., Bouffioux, V., and Wilquem F., "CFD simulation of the impact of new buildings on wind comfort in an urban area," Impact of Wind and Storm on City Life and Built Environment, Proceedings of the Workshop, pp. 164–171, 2002.

94   Fothergill, C.E., Roberts, P.T., and Packwood, A.R., "Flow and Dispersion Around Storage Tanks. A Comparison Between Numerical and Wind Tunnel Simulations," *Wind and Structures*, Vol. 5, No. 2–4, 2002, pp. 89–100.

95   geolab.larc.nasa.gov/APPS/YPlus/

96   Qin, N., Vavalle, A., and Le Moigne, A., "Spanwise Lift Distribution for Blended Wing Body Aircraft," *Journal of Aircraft*, Vol. 42, No. 2, 2005, pp. 356–365.

97   Ishikawa, H., Makino, Y., Ito, T., and Kuroda, F., "Sonic Boom Prediction Using Multi-Block Structured Grids CFD Code Considering Jet-On Effects," AIAA Paper 2009–3508, June 2009.

98   Ghoreyshi, M., Vallespin, D., Badcock, K.J., Da Ronch, A., Vos, J., and Hitzel, S., "Simulation of Aircraft Manoeuvres Based on Computational Fluid Dynamics," AIAA Paper 2010-8239, August 2010.

99   Fischer, D.F., Cobleigh, B.R., Banks, D.W., Hall, R.M., and Wahls, R.A., "Reynolds Number Effects at High Angles of Attack," AIAA Paper 98–2879, June 1998.

100   Boelens, O.J., "CFD Analysis of the Flow Around the X-31 Aircraft at High Angle of Attack," AIAA Paper 2009–3628, June 2009.

101   Steijl, R., and Barakos, G., "Sliding Mesh Algorithm for CFD Analysis of Helicopter Rotor–Fuselage Aerodynamics," *International Journal for Numerical Methods in Fluids*, Vol. 58, No. 5, 2008, pp. 527–549.

102   Rai, M., "A Conservative Treatment of Zonal Boundaries for Euler Equation Calculations," *Journal of Computational Physics*, Vol. 62, 1986, pp. 472–503.

103   Rai, M., "A Relaxation Approach to Patched-Grid Calculations with the Euler Equations," *Journal of Computational Physics*, Vol. 66, 1986, pp. 99–131.

104   Barakos, G., Vahdati, M., Sayma, A.I., Breard, C., Imregun, M., "A Fully Distributed Unstructured Navier-Stokes Solver for Large-Scale Aeroelasticity Computations," *Aeronautical Journal*, Vol. 105, 2001, pp. 419–426.

105   Murayama, M., Yokokawa, Y., and Yamamoto, K., "CFD Validation Study for a High-Lift Configuration of a Civil Aircraft Model," AIAA Paper 2007–3924, June 2007.

106   Dean, P.D., Clifton, J.D., Bodkin, D.J., Morton, S.A., and McDaniel, D.R., "Determining the Applicability and Effectiveness of Current CFD Methods in Store Certification Activities," AIAA Paper 2010–1231, January 2010.

107   Gaither, J.A., Marcum, D.L., and Mitchell, B., "SolidMesh: A Solid Modeling Approach to Unstructured Grid Generation," 7th International Conference on Numerical Grid Generation in Computational Field Simulations, September.

108   Marcum, D.L., and Weatherill, N.P., "Unstructured Grid Generation Using Iterative Point Insertion and Local Reconnection," *AIAA Journal*, Vol. 33, No. 9, 1995, pp. 1619–1625.

109 Marcum, D.L., "Unstructured Grid Generation Using Automatic Point Insertion and Local Reconnection," *The Handbook of Grid Generation*, Ed. J.F. Thompson, B. Soni, and N.P. Weatherill, Boca Raton: CRC Press, p. 18–1, 1998.

110 Morton, S.A., Tomaro, R.F., and Noack, R.W., "An Overset Unstructured Grid Methodology Applied to a C-130 With a Cargo Pallet and Extraction Parachute," AIAA Paper 2006-0461, January 2006.

111 Samareh, J., "Gridtool: A Surface Modeling and Grid Generation Tool," Proceedings of the Workshop on Surface Modeling, Grid Generation, and Related Issues in CFD Solution, NASA CP-3291, May 1995.

112 Pirzadeh, S., "Progress toward a User-Oriented Unstructured Viscous Grid Generator," AIAA Paper 96-0031, January 1996.

113 Potsdam, M.A., and Strawn, R.C., "CFD Simulations of Tiltrotor Configurations in Hover," *Journal of the American Helicopter Society*, Vol. 50, No. 1, 2005, pp. 82–94.

114 Vassberg, J. C., Buning, P. G., and Rumsey, C. L., "Drag Prediction for the DLR-F4 Wing/Body Using OVERFLOW and CFL3D on an Overset Mesh," AIAA Paper 2002-0840, January 2002.

115 Steger, J.L., "Technical Evaluation Report: AGARD Fluid Dynamics Panel Specialty Meeting on Applications of Mesh Generation to Complex 3-D Configurations," AGARD-AR-268, 1991.

116 Thompson, J.F., and Weatherill, N.P., "Aspects of Numerical Grid Generation: Current Science and Art," AIAA Paper 93-3539, July 1993.

117 Soni, B., Cheng, G., Koomullil, R., Shih, A., Luke, E., and Thompson, D., "Enabling Technologies for Complex CFD Applications," AIAA Paper 2004-3987, July 2004.

118 Samareh, J.A., "Status and Future of Geometry Modeling and Grid Generation for Design and Optimization," *Journal of Aircraft*, Vol. 36, No. 1, 1999, pp. 97–104.

119 Gnoffo, P.A., Weilmuenster, K.J., Hamilton, H.H., Olynick, D.R., and Venkatapathy, E., "Computational Aerothermodynamic Design Issues for Hypersonic Vehicles," *Journal of Spacecraft and Rockets*, Vol. 36, No. 1, 1999, pp. 21–43.

120 Sorenson, R.L., "A Computer Program to Generate Two-Dimensional Grids About Airfoils and Other Shapes by the Use of Poisson's Equation," NASA TM 81198, 1980.

121 Steinbrenner, J.P., Chawner, J.R., and Fouts, C.L., "The GRIDGEN 3D Multiple Block Grid Generation System," WRDC-TR-90–3022, Vol. I: Final Report, July 1990, Vol. II: User's Manual, February 1991.

122 Walatka, P.P., Buning, P.G., Pierce, L., and Elson, P.A., "PLOT3D User's Manual," NASA TM 101067, March 1990.

123 Rumsey, C.L., Poirier, D.M.A., Bush, R.H., and Towne, C.E., "A User's Guide to CGNS," NASA TM 2001–211236, October 2001.

124 Legensky, S.M., Edwards, D.E., Bush, R.H., Poirier, D.M.A., Rumsey, C.L., and Towne, C.E., "CFD General Notation System (CGNS): Status and Future Directions," AIAA Paper 2002-0752, January 2002.

125 Bush, R.H., Cosner, R., Poirier, D.M.A., Hall, E., Darian, A., Dominik, D., and Rumsey, C.L., "CGNS Steering Committee Charter," Version 3.1.2, www.grc.nasa.gov/WWW/cgns/CGNS_docs_current/charter/index.html.

8 Viscosity and Turbulence Modeling

I consider turbulence as the single most intractable problem in physics.[1]

Steven Weinberg, Nobel Prize winner in Physics

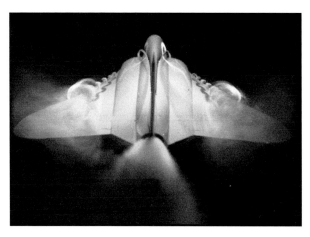

Vortex breakdown on an aircraft with a delta wing (Photograph by Henri Werlé © ONERA, the French Aerospace Lab; a full color version of this image is available on the website: www.cambridge.org/aerodynamics).

LEARNING OBJECTIVE QUESTIONS

After reading this chapter, you should know the answers to the following questions:

- What differences do laminar and turbulent boundary layers exhibit relative to skin friction and flow separation?
- How are laminar boundary layers maintained in flight?
- How is laminar viscosity modeled in computational simulations?
- What are some of the flow disturbances that make a boundary layer "receptive" to transition?
- What are some of the characteristics of turbulent flow?
- What are the regions and sublayers of a turbulent boundary layer? How many grid points are required in each sublayer? In the entire boundary layer? How close should the first grid point be to a solid surface?
- What are the differences (in equation complexity, level of modeling, and computation time) between DNS, LES, RANS, and hybrid RANS/LES?

8.1 Introduction

Another important piece in our computational puzzle is understanding how modern CFD simulations model viscosity. Both laminar and turbulent flows require some way to model viscosity, although turbulent flow is far more difficult to model than laminar flow. In fact, turbulence is one of the most complex, difficult concepts in engineering, and the approximate modeling of turbulence is a great challenge. By developing a basic physical understanding of turbulence you will be better prepared to understand the various models and how they work. We have already covered some of the building-block concepts required to create turbulence models, including the derivation of the governing equations of fluid flow and the effects of turbulence on aerodynamics. In Chapter 3 we derived the governing equations of fluid motion and saw that many of the terms involved fluid viscosity, μ. In Chapter 4, however, we showed how many of the basic concepts of the aerodynamics of streamlined vehicles could be developed assuming that the flow was essentially inviscid, with the viscous effects confined to a thin region adjacent to the surfaces – the boundary layer. In spite of this, the effects of viscosity are critically important in aerodynamics. For example, viscosity is the key for airfoils to generate lift! The Kutta condition allows us to incorporate the effects of viscosity in an otherwise inviscid flow model and gives us the ability to obtain lift from our theories. But viscosity and the no-slip boundary condition also lead to skin friction, and hence drag. The ability of the boundary layer to withstand the "load" imposed by the pressure distribution required for high lift without separating is the limiting factor for the maximum lift coefficient, which impacts takeoff and landing speeds and distances. So, viscous effects are critical in aerodynamics (they are even responsible for the aerodynamic heating on hypersonic vehicles).

The goal of this chapter is to illustrate the wide variety of types of viscous effects and flow separation that aerodynamicists deal with and to provide an introduction to viscous flow modeling, including laminar, turbulent, and transitional flows. We will need to spend most of our time understanding turbulent flows and the need to use a "turbulence model" when making aerodynamic calculations. In general, users of computational aerodynamics will need to choose a turbulence model, but making that choice can be quite challenging. Turbulence models have been developed for a wide range of applications, and finding the appropriate model is essential to making good predictions of aerodynamics.

8.2 Types of Viscous Effects

The boundary-layer concept is an excellent model for viscous effects in attached flow, but there are a variety of ways viscosity affects aerodynamics.

The key viscous effect facing aerodynamicists is flow separation. Flow separation leads to additional drag on aircraft, although controlled flow separation provides a way of achieving enhanced performance in some cases. For example, the strakes on the F-16 generate vortices that enhance lift, and the leading-edge vortex on the Concorde, with the additional "vortex lift" at takeoff and landing, was responsible for making the aircraft feasible. Another important viscous effect is skin friction drag, which is a function of the boundary-layer state (laminar or turbulent) and where transition takes place. All told, viscous effects play a major role in the analysis and design of aircraft.

The water tunnel flow visualization pictures in Fig. 8.1 illustrate a few of the different types of viscous effects that might arise in an aerodynamic analysis. They were made by Henri Werlé at ONERA, the French Aerospace Lab. Figure 8.1a illustrates the case of attached flow over an airfoil, where the inviscid flow/boundary-layer approach provides an excellent flow model. In Fig. 8.1b the angle of attack of the airfoil is increased and the flow separates over the aft portion of the upper surface. While in the first case a simple turbulence model is all that is required to make the calculation, if the flow is separated on the aft upper surface (as shown in Fig. 8.1b), a more complicated turbulence model must be used. The case of mixed attached-separated flow is one of the more difficult cases to compute, as shown in Fig. 8.1b for an airfoil and in Figures 8.1c–d, which illustrate bluff body separation. In the first case, the flow is turbulent when it separates, and the separated wake is small compared to the wake in Fig. 8.1d, where the flow is laminar at separation and the resulting wake is much larger. Looking at Fig. 8.1d, you get a sense of the flow unsteadiness in the separated flow region. These cases represent massive flow separation. Also notice the difference in separation points between the two cases, with the laminar boundary layer separating earlier (as was discussed in Chapter 4). Figures 8.1e and 8.1f illustrate examples of three-dimensional flow separation. Figure 8.1e shows the core of a leading-edge vortex on a delta wing. Figure 8.1f illustrates one flow separation pattern on an axisymmetric body of revolution at a relatively high angle of attack. When looking at these figures of steady vortical flow over wings and bodies (Figures 8.1e and 8.1f), we see that flow separation in three dimensions can be quite different from flow separation in two dimensions (compare Figures 8.1c and 8.1d, and recall D'Alembert's paradox from Section 4.3!). We will see that the steady vortical flows require one level of modeling, while the unsteady, chaotic flow requires a completely different degree of modeling.

Another example of a viscous flow case typical in aerodynamics is shown in Fig. 8.2, which is a photograph of a wing being tested in a transonic wind tunnel. The wing has been coated with "oil" so that the surface flowfield can be seen (this concept will be discussed in greater detail in Chapter 9). The purpose of the test is to see if any separation is occurring, since separation

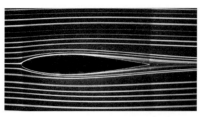

(a) Attached flow over an NACA 64A015
airfoil at $\alpha=0°$; laminar flow

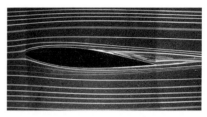

(b) Separated flow over an NACA 64A015
airfoil at $\alpha=5°$; laminar flow

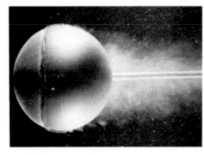

(c) Bluff body separation, flow over a sphere,
tripped to produce turbulent flow separation

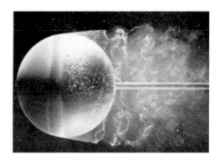

(d) Bluff body separation, flow over a
sphere, with laminar flow separation
(quickly becoming turbulent)

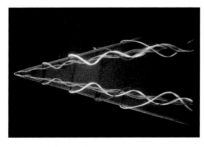

(e) Leading edge vortex flow type separation,
dye injected along the leading edge, and
showing the formation of a vortex core
above the wing; under some flow conditions
the vortex will "breakdown"

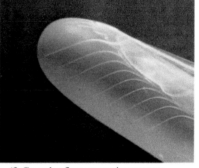

(f) Complex flow separation pattern
over a Rankine Ogive at 30°
(other flow patterns possible)

Figure 8.1   Some types of separation of interest to computational aerodynamicists (Ref. 2; Photographs by Henri Werlé © ONERA, the French Aerospace Lab).

will cause additional drag: the photo shows that there is no separation since all of the surface streamlines flow straight back over the wing (the flow in the wind tunnel is from the top of the picture to the bottom). Two other features of the test and flowfield are also of interest. The surface is smooth close to the leading edge because the flow is laminar and there is not enough shear force to move the oil. At about 15 percent of the chord, the surface streamlines suddenly start to be visible. This is because there is a boundary layer trip strip on the wing to force the flow to transition to turbulent flow. This is the classic problem of the wind tunnel not being able to simulate full-scale Reynolds number, hence the need to use artificial means to force the flow to

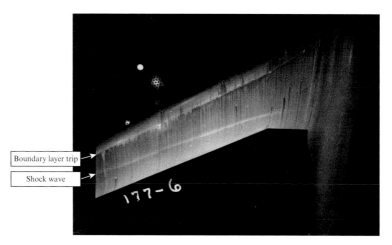

Figure 8.2 Oil flow photo of wing being tested in a wind tunnel at $M = 0.825$, $\alpha = 4°$, $C_L = 0.516$ (Courtesy of Pres Henne of Gulfstream Aerospace Corporation).

be turbulent. The other feature of interest occurs aft of the mid chord where the oil shows the position of the shock wave, since the shock thickens the oil locally. We can also see that the shock unsweeps slightly as it moves outboard, which is a way to determine how well the wing was designed.

8.3 Laminar Flow

In Chapter 4, we described how viscous effects are largely determined by the state of the boundary layer: laminar or turbulent. At low Reynolds numbers, the flow is typically laminar, the boundary layer is thin, and the skin friction is low. Controlling whether the flow is laminar or turbulent is part of modern aerodynamic design. Maintaining laminar flow over an entire aircraft to reduce skin friction has been called the "holy grail" of aerodynamics. The reason can be seen by comparing the skin friction formulae for laminar flow, Equation 4.47, and turbulent flow, Equation 4.51:

$$C_{F_{laminar}} = \frac{1.328}{\sqrt{Re_L}} \quad C_{F_{turbulent}} = \frac{0.074}{\left(Re_L\right)^{0.2}}$$

For a surface with a Reynolds number of 1 million, the laminar skin friction coefficient is $C_{F_{laminar}} = 0.001328$ and the turbulent skin friction coefficient is $C_{F_{turbulent}} = 0.004669$: the turbulent flow skin friction is 3.5 times higher than the laminar flow skin friction! As the price of aircraft fuel continues to increase, there will be new efforts to obtain significant portions of laminar flow on aircraft.

Aerodynamic design features that promote extensive runs of laminar flow include favorable pressure gradients (so that the flow is continually

 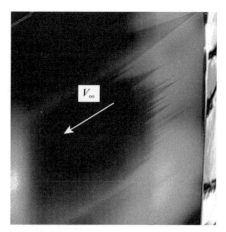

Figure 8.3 In-flight transition measurement using IR flow-visualization technique. (Ref. 4; Courtesy of William S. Saric of Texas A&M University; a full color version of this image is available on the website: www.cambridge.org/aerodynamics).

accelerating) and low wing sweep. Designing to promote natural laminar flow is known as passive laminar flow control. Today's sailplanes, many modern light airplanes, and the new Honda Jet[3] achieve significant regions of laminar flow naturally. Figure 8.3 shows the laminar and turbulent flow portions of a test wing mounted on a Cessna O-2A Skymaster using infra-red (IR) flow visualization.[4] Notice the jagged transition front behind the leading edge, which we will discuss shortly. The wing was designed so it could also be tested in the low-disturbance wind tunnel at Texas A&M University in order to compare flight test and wind tunnel results.

In contrast, active laminar flow control may also be used. Typically, this means that suction is applied to the wing to keep the boundary layer thin, which delays the onset of transition. The U.S. Air Force experimented with this idea by using suction through slots running outboard on the wing in the X-21 program. The first flight occurred in 1963, and the program was eventually successful. A more recent application of active laminar flow control is the F-16XL, which was fitted with a perforated glove through which air could be drawn in order to control transition at supersonic speeds (see Fig. 8.4)[5,6]

The computation of laminar flow is relatively straightforward when compared to turbulent flow. Since Sutherland's law relates laminar viscosity to temperature (see Equation 3.32 and Table 3.1), that relation can be used directly in the fluid equations of motion, and laminar flow predictions can be made directly. For example, the Unmanned Combat Air Vehicle (UCAV) shown in Fig. 8.5 was simulated using the Navier-Stokes equations solved with laminar viscosity only, in order to simulate results from a low-speed wind tunnel test. The details of the flowfield are apparent in the visualization, including the leading-edge vortex formation and the breakdown of the vortex at approximately 60 percent of the length of the vehicle.[7]

Figure 8.4   The F-16XL aircraft fitted with an active laminar flow control glove on the left wing (courtesy of NASA Dryden Flight Research Center).

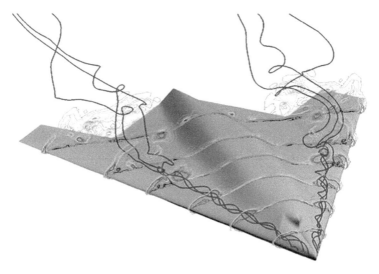

Figure 8.5   Laminar flow over a UCAV at 15-degree angle of attack showing vortical flow structures and vortex breakdown (Ref. 7; a full color version of this image is available on the website: www.cambridge.org/aerodynamics).

8.4 **Transition**

Another key aspect of computational simulation is the estimation of transition. For a low-speed flow with no pressure gradients and relatively high freestream turbulence levels, we said in Chapter 4 that an approximate location for transition could be estimated to be where the local Reynolds number

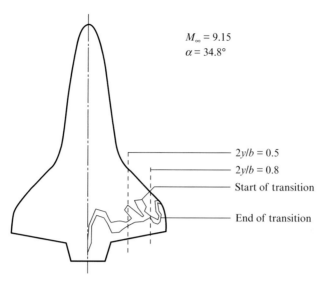

$M_\infty = 9.15$
$\alpha = 34.8°$

$2y/b = 0.5$
$2y/b = 0.8$
Start of transition
End of transition

Figure 8.6 Transition location on the Space Shuttle Orbiter measured during flight (Ref. 9; reprinted by permission of the American Institute of Aeronautics and Astronautics, Inc.).

is 500,000. However, a laminar boundary layer is sensitive to a number of factors that affect the transition location. In addition to the Reynolds number and maintaining favorable pressure gradients or suction as described, the boundary layer is also affected by surface roughness[i] (recall the trip shown in Fig. 8.2 used to simulate high Reynolds number flow in a low Reynolds number wind tunnel test). Other factors that affect transition include surface sweep, freestream turbulence, and noise. Mach number and surface temperature also affect the transition location, with supersonic and hypersonic transition taking place at higher Reynolds numbers than subsonic transition. The basic physics of transition and methods used to analytically estimate where transition will occur are described in Schlichting et al.[8] Since adverse pressure gradients hasten transition, it is usually safe to assume that transition occurs rapidly if an airfoil has a pressure distribution that reaches a pronounced peak followed by a strong adverse pressure gradient (such as in Fig. 4.24). If the pressure gradient is favorable, there is much more uncertainty in the estimation of the transition location.

Transition occurs over a fairly short distance at low speeds and is often assumed to occur instantly in CFD simulations. However, at hypersonic speeds transition occurs over a significant distance, and, in fact, the uncertainty associated with transition has been a major issue in the development of hypersonic vehicles. Figure 8.6 shows the irregular location of transition

---

[i] The North American P-51 Mustang was designed with a laminar flow airfoil. The slight gaps and steps that occurred in the manufacturing process, as well as maintenance required to keep the surface clean and smooth in operation, resulted in a loss of most of the laminar flow. Today, composite skin surfaces can be built so smoothly that significant laminar flow is possible.

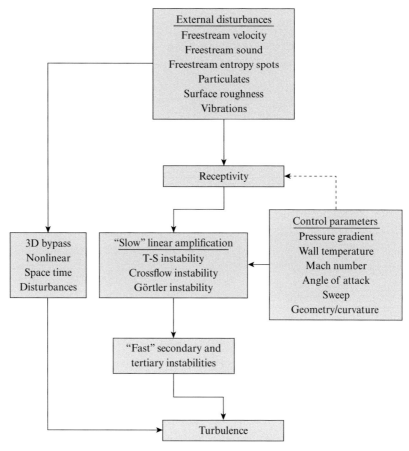

Figure 8.7   Transition flow chart (Ref. 10; reprinted by permission of the American Institute of Aeronautics and Astronautics, Inc.).

on one Space Shuttle flight as well as the beginning and end of the transition region.[9] There is a large difference in the aerodynamic heating at high speeds, depending on whether the flow is laminar or turbulent, so predicting transition is important for aircraft design considerations.

Boundary-layer transition is composed of several physical processes, as described in Fig. 8.7.[10] The transition process begins by introducing external disturbances into the boundary layer through a process known as "receptivity."[11] Some of these external disturbances include freestream vorticity, surface roughness, vibrations, and sound. Identifying and defining the initialization of these external disturbances for a given problem is the basis for the prediction of transition. The initial disturbance is a function of the aircraft and its environment, and therefore is not usually known in advance.[12] The disturbances in the boundary layer eventually amplify and can be modeled by linear stability theory (see Ref. 8 for example). The normal modes responsible for the amplification of these disturbances in boundary-layer flow are the viscous Tollmein-Schlichting (T-S) waves, inflectional Rayleigh waves (i.e., instabilities due to crossflow), and Görtler vortices for curved

Figure 8.8 Flow visualization of the transition process on a flat plate (Courtesy of A.S.W. Thomas, Lockheed Aeronautical Systems Company).

streamlines. Once the amplifications are large enough, nonlinearity sets in through secondary and tertiary instabilities and the flow becomes "transitional." It should be noted that the nonlinear portion of the flow is small compared to the linear region and can still often be approximated by linear stability theory for preliminary designs.

Stability theory predicts, and experiments verify, that the initial instability is usually in the form of two-dimensional T-S waves that travels in the mean flow direction. Flows that are not two-dimensional can also have crossflow instabilities as well. Even though the mean flow is two-dimensional, unstable three-dimensional waves and hairpin eddies soon develop as the T-S waves begin to show spanwise variations. The experimental verification of the transition process is illustrated in the photograph of Fig. 8.8. A vibrating ribbon perturbs the low-speed flow upstream of the left side of the photograph. Smoke accumulation in the small recirculation regions associated with the T-S waves can be seen at the left edge of the photograph. The sudden appearance of three-dimensionality is associated with the nonlinear growth region of the laminar instability. In the advanced stages of the transition process, intense local fluctuations occur at various times and locations in the viscous layer. From these local intensities, true turbulence bursts forth and grows into a turbulent spot. Downstream of the region where the spots first form, the flow becomes fully turbulent. Transition-promoting phenomena, such as an adverse pressure gradient and finite surface roughness, may short-circuit the transition process, eliminating one or more of the five transitional regions described previously.

When one or more of the transitional regions are bypassed, we term the cause (e.g., roughness) a bypass mechanism.

Although efforts are currently under way (and have been ongoing for years) to develop methods to estimate transition that can be incorporated into CFD codes, the usual practice is to require the user to specify the transition location. Various approaches to defining the transition "line" are used in various codes, but many users just assume the flow is either fully laminar or fully turbulent.[13] More information on the estimation of transition using computational methods is available in the book by Cebeci,[14] the article by Krumbein et al.,[15] and the review paper by Cheng et al.[16] The problem of transition at hypersonic speeds continues to be an important research area, as indicated by a 2008 special section on transition in the *Journal of Spacecraft and Rockets*; interested readers should obtain the articles by Lin[17] and Schneider[18] for more details.

Computational Aerodynamics Concept Box

Crossflow Instability and Transition on Swept Wings

Most commercial and military jet aircraft used today have swept wings, as is true for both the KC-10 and F-22 shown here. Unfortunately, flight test experience "has shown that on swept wings the transition point is considerably farther forward than on unswept wings."[19] This was found to be caused by an unstable boundary-layer profile in the spanwise direction (toward the wing tip) and in a plane tangential to the wing surface. "The crossflow is strongest where there is a high curvature of the streamlines in the tangential planes, i.e., in the immediate neighborhood of the leading edge and in the region of the rear pressure rise."[19]

Since the streamlines have both a freestream and crossflow (outboard) component, the flow near the surface of the wing at the leading edge can be quite complex, as shown.

KC-10 refueling an F-22 (U.S. Air Force photo).

Perhaps the best explanation of how this flow pattern leads to early boundary-layer transition was given by William S. Saric, Helen Reed, and Edward White: "The crossflow instability occurs in regions of pressure gradient on swept surfaces or on rotating disks. In the inviscid region outside the boundary layer, the combined influences of sweep and pressure gradient produce curved streamlines at the boundary-layer edge. Inside the boundary layer, the streamwise velocity is reduced, but the pressure gradient is unchanged. Thus, the balance between centripetal acceleration and pressure gradient does not exist. This imbalance results in a secondary flow in the boundary layer, called crossflow, that is perpendicular to the direction of the inviscid streamline."[20] The three-dimensional profile and resolved streamwise and crossflow boundary-layer profiles are shown in the figure.

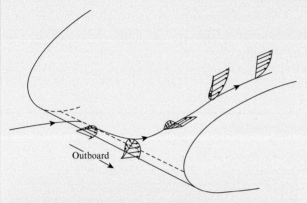

Boundary layer velocity profiles on a swept wing

"Because the crossflow velocity must vanish at the wall and at the edge of the boundary layer, an inflection point exists and provides a source of an inviscid instability. The instability appears as co-rotating vortices whose axes are aligned to within a few degrees of the local inviscid streamlines."[20] This instability can cause transition to take place at locations on the wing far ahead of where natural transition might occur, which means our most straightforward methods for estimating transition locations will be in error.

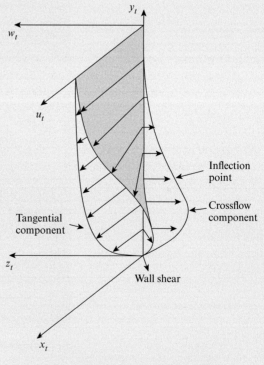

Inflection points in a 3D boundary layer velocity profile

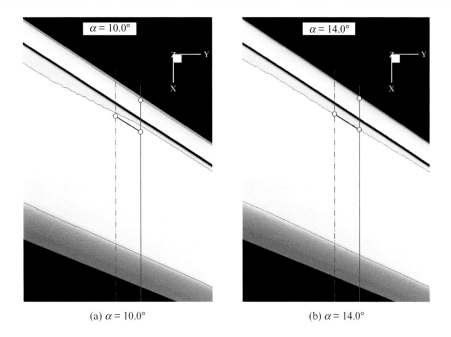

(a) $\alpha = 10.0°$ (b) $\alpha = 14.0°$

Figure 8.9 Transition comparisons for the DLR F-11 wing with slats and flaps extended in a landing configuration. Experimental transition locations using hot films are shown with the solid line, and numerically predicted transition regions are shown by the gray laminar flow region. (a) $\alpha = 10°$ (b) $\alpha = 14°$ (Ref. 21; Courtesy of Andreas Krumbein of DLR; a full color version of this image is available on the website: www.cambridge.org/aerodynamics).

Recent applications of coupled Navier-Stokes/linear stability theory codes have provided good results in predicting transition for transport aircraft, as shown in Fig. 8.9. Experimental hot films are shown from an experiment where the transition region is visible, and the predicted location for transition (at two angles of attack) is shown to agree very well with the hot film data.

When this approach is applied to a full transport aircraft at transonic speeds, the results show important differences between the fully turbulent prediction (the common approach used in most CFD simulations) and the flow with transition, as shown in Fig. 8.10. While the flow over the fuselage, nacelle, and pylon are largely the same, the wing shows significant differences. The fully turbulent case shows much higher levels of skin friction coefficient than the laminar case, which indicates that the turbulent flow wing would have higher skin friction drag. The prediction of transition will continue to be an important area of research in aerodynamics, and a large amount of work remains to be done in order to accurately predict transition for all flight conditions.

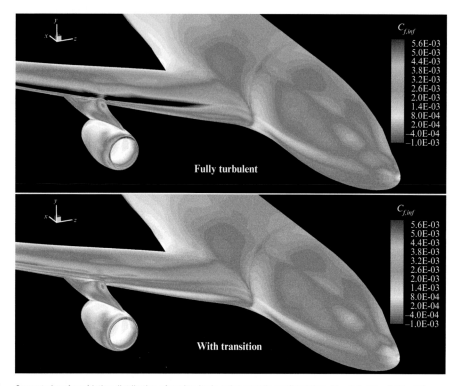

Computed surface friction distribution of a wing-body-pylon-nacelle configuration, $M = 0.8$, $Re = 9.97 \cdot 10^6$, $\alpha = 2.2°$, k-ω model (fully turbulent or with fixed transition) (Ref. 22; Courtesy of Andreas Krumbein of DLR; a full color version of this image is available in the color insert pages of this text as well as on the website: www.cambridge .org/aerodynamics).

Profiles in Computational Aerodynamics: Dimitri Mavriplis

"I am a researcher and developer of computational fluid dynamics algorithms and software for aerospace applications. I obtained my PhD in mechanical and aerospace engineering at Princeton University in 1987. In the 1980s, my PhD advisor Antony Jameson had perfected fast solvers for the Euler equations and had written a series of codes for the simulation of inviscid transonic flow. Although these codes were widely used in the aircraft industry, they were based on structured meshes and proved difficult to use for complex geometries other than simple wing or wing-fuselage configurations. At that time it was realized that unstructured meshes held the best potential for enabling the simulation of flows over more complex configurations and, as a PhD project, I was tasked with devising an efficient Euler solver using unstructured meshes. Using some of the most powerful supercomputers at the time, I devised an unstructured multigrid solver for the Euler equations that solved two-dimensional single airfoil and multi-element airfoil problems in less than one hour. Today, a derivative of this same solver runs in minutes on a cell phone.

"After graduating from Princeton, I joined ICASE (Institute for Computer Applications in Science and Engineering) located at NASA Langley in Hampton, Virginia, where I continued my work on unstructured mesh technology. The principal goals were to extend these methods from the 2D Euler equations to the 3D Reynolds-averaged Navier-Stokes equations, a feat that took no less than a decade. In those days, unstructured mesh methods were novel and everything had to be done in-house, including mesh generation, solver development, and even writing our own visualization software. One of the driving applications in the aerospace industry was the accurate simulation of high-lift problems, i.e., aircraft with deployed flap systems. Structured mesh methods had evolved to where they could be used for complete aircraft configurations including engines in cruise conditions, but the complex geometry of wings with flaps, slats, and support track fairings still proved to be a stumbling block for these methods.

"Today, unstructured mesh methods are commonly used to simulate very detailed and complete geometries in aerospace, automotive, and various other industries. Working on the development of these methods has been, and continues to be, a truly interdisciplinary endeavor. For example, work in unstructured mesh generation involves concepts such as Delaunay triangulation and draws heavily on computational geometry. Solver development is firmly rooted in applied mathematics, while the need to simulate very complex configurations requires porting these methods to the latest high-performance computing hardware, involving the field of computer science. Finally, since our ultimate goal is the simulation and improvement of aerodynamic characteristics, a good knowledge of aerospace engineering is obviously required as well.

"In 2003, I joined the University of Wyoming as a professor in mechanical engineering. Current research directions are focused on pushing the boundaries of unstructured mesh CFD technology to more complex and multidisciplinary simulations. Time-dependent simulations using overlapping, moving, and deforming meshes for rotorcraft and wind energy applications, has been an important thrust of ours in recent years. The inclusion of aeroelastic effects, through the coupled simulation of fluid and structural models, is also becoming more important for both fixed and rotary wing aircraft. At the same time, there is a need to provide higher accuracy simulation results, thus driving the development and use of adaptive meshing methods, as well as higher order discretization methods (such as discontinuous Galerkin methods), both of which are ideally suited for use on unstructured meshes.

"In summary, my career in CFD research and development continues to be truly exciting and rewarding, not only because of the interdisciplinary nature of this field, but also because CFD is continually advancing so quickly and has applications in so many fields of engineering."

Dimitri Mavriplis is the Max Castagne Professor of Mechanical Engineering at the University of Wyoming, where he is the director of the computational fluid dynamics laboratory. He received his Bachelor of Engineering and Master's of Engineering degrees in Mechanical Engineering from McGill University, and his PhD degree in Mechanical and Aerospace Engineering from Princeton University. He is an AIAA Associate Fellow, SIAM member, and AHS member. He has been a participant and committee member for the AIAA-sponsored Drag Prediction, High-Lift Prediction and Aeroelastic Prediction Workshop series.

8.5 Turbulent Flow

Turbulent flow is fundamentally different from any other aspect of fluid dynamics, which is why a Nobel laureate would make the statement found at the beginning of the chapter. The boundary layers over fighter airplanes and transonic jet transports are almost entirely turbulent, and, in some cases, turbulent flow may be considered to be "bad" since it causes higher skin friction drag. However, turbulence also delays the onset of flow separation, which could be considered to be "good" (see Section 4.6.8 for details). So, turbulence is a mixed blessing with applications in which it causes problems and applications in which it can be beneficial. For example, effective turbulent mixing to spread the fuel in a scramjet engine is an enabling technology for future advances in hypersonic flight. Thus, in many cases, turbulent flows can be advantageous and can even be necessary.

The special methods used to estimate turbulent flows in CFD are generally termed *turbulence models*, and we will discover why this terminology is used. Turbulent flows are important in all aspects of fluid mechanics and gas dynamics; in aerodynamics, we are especially interested in high Reynolds number flows (in the millions at least, although there are exceptions) and we are mainly interested in external flows. As shown from the flow examples described in Fig. 8.1, part of the challenge is the need to compute attached flows, mixed attached/separated flows, and massively separated flows, sometimes all on one airplane at the same time.

Modeling turbulence is challenging and complicated. When you start to model any physical behavior, the first requirement is to gain some understanding of the behavior. Unfortunately, turbulence is one of the most difficult unsolved problems in physics.[23] What variables define the phenomenon? Does the phenomenon change character at certain critical values of those variables? Is the phenomenon linear or nonlinear? The list of questions is endless, and the list of answers is all too short! Many people have tried to model the behavior of turbulent boundary layers, starting with Prandtl's mixing length concept, which was developed in the 1920s.[24] The problem with finding a useful model of turbulence is that turbulent flows have many characteristics that are difficult to quantify (in addition to the prediction of transition), including turbulence variation with Reynolds number, variation with compressibility (Mach number), and the multilayer structure of a turbulent boundary layer (see Fig. 4.12), and the list can go on from there.[25] Before delving into the specific details of turbulence models, therefore, a brief review of turbulence will be presented to help you gain a better physical understanding of the nature of turbulence. Then, we will better appreciate why turbulence models do not always work well, and we will become better able to intelligently use these models in simulations.

8.6 **Characteristics of Turbulence**

Understanding turbulence is difficult, and predicting turbulent flow behavior is even more difficult. Perhaps a fairly simple flow example will help to illustrate this point. Figure 8.11 shows a subsonic jet transitioning to turbulence and provides a good image to describe turbulence. A turbulent flow has several characteristics:[26]

- Chaotic and irregular – fundamentally unsteady and three-dimensional
- Enhanced mixing of momentum, heat, and mass
- Large Reynolds numbers
- Three-dimensional vorticity fluctuations

Consider these characteristics as you look at the jet flow shown in Fig. 8.11. As the jet exits the nozzle, there are instabilities that cause the shear layer to roll up. Initially, these instabilities are highly regular and, therefore, not turbulent. Only when these irregularities break down and become chaotic can the flow be considered turbulent; prior to the onset of chaotic behavior the flow is considered laminar. In between the laminar and turbulent regimes, the flow is considered transitional (changing state from laminar to turbulent). Something else to note from Fig. 8.11 is that the jet is spreading more rapidly (becoming wider) after it becomes turbulent, which is due to the enhanced mixing that the turbulence provides.

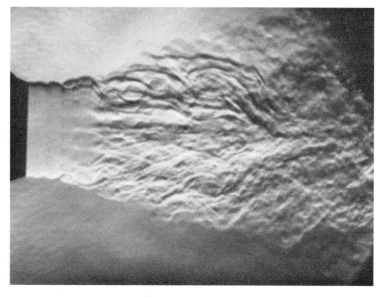

Figure 8.11   A jet of air from a nozzle into ambient air at subsonic speed (Ref. 2; Reprinted with Permission. Courtesy of Cambridge University Press).

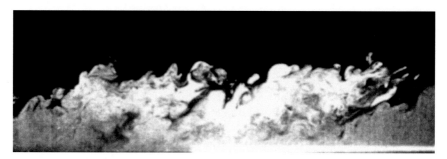

Figure 8.12 Turbulent boundary layer (Ref. 27; Reprinted with permission from "Coherent motions in the outer region of turbulent boundary layers," *Physics of Fluids*, Vol. 20, © 1977, AIP Publishing LLC).

Other important features of turbulence are the length and time scales that are found within turbulent flows. Figure 8.12 shows a turbulent boundary layer as it flows over a flat plate. The boundary layer is highly unsteady, with flow quantities (such as velocity, density, pressure, and temperature) fluctuating at high frequencies. In addition to the fluctuations of the flow properties, there are both large- and small-scale structures in turbulent flow. The *large-scale structures* are visible in Fig. 8.12, appearing like waves convecting along with the flow, having a length scale on the order of the boundary-layer thickness. These are normally called eddies, and are in effect clumps of fluid that appear to be rolling along together in the flow. The large-scale structures are essentially "grabbing" kinetic energy from the freestream flow and dragging it down into the inner layers of the boundary layer (remember, Fig. 8.12 is an instant in time of the unsteady turbulent flow process occurring in an otherwise steady flow). Perhaps this can best be appreciated by watching the smoke coming from a smokestack. The edges of the plume are in constant, irregular motion and, at any instant, will be very similar to the photo in Fig. 8.12. The large-scale structures also eject low-momentum fluid from the inner layer to the outer layer. This mixing effect significantly alters the average velocity profile compared to a laminar boundary layer. Underlying the large-scale turbulence is an isotropic *small-scale turbulence*, meaning it is essentially the same everywhere within the boundary layer and acts in all directions equally. The small-scale turbulence converts the kinetic energy of the large turbulent structures into other forms of energy, including friction and heat.[ii]

Because the turbulence length scales can be extremely small, they pose a problem for numerical simulation of turbulence. To properly simulate these flows, the numerical grid must be small enough to resolve the small scales in the turbulence, and this requires a grid within the boundary layer that is so

[ii] The smallest length scale found in turbulence is known as the *Kolmogorov length scale*. It turns out to be nearly microscopic at Reynolds numbers of interest in aerodynamics but is still much larger than the size of molecules (see Ref. 26 for details).

small that we probably would not have enough computer memory to contain our results unless we were simulating a very simple flow (like flow over a flat plate, or in a channel, at low Reynolds numbers).

An additional difficulty for those interested in modeling a turbulent boundary layer is the various sublayers that make up the full boundary layer. Refer back to Fig. 4.12 to see the velocity variation within a turbulent boundary layer in both linear and logarithmic scaling. The boundary layer is actually made up of two primary regions, the *inner region* and the *outer region*, with the inner region being made up of three sublayers: the laminar (or viscous) sublayer, the transitional (or buffer) region, and the fully turbulent region (also called the log-law region). The inner region is much thinner than the outer region (usually about 10–20 percent of the boundary-layer thickness), even though it looks larger in Fig. 4.12 due to the use of the logarithmic scale. Because the regions are so thin, we use variables that arise naturally in boundary-layer theory, the *wall units*, y^+ and u^+, when describing the boundary layer near the wall. These were defined in Chapter 7 as:

$$y^+ = \frac{yu_\tau}{v} \tag{8.1}$$

$$u_\tau = \sqrt{\frac{\tau_w}{\rho}} \tag{8.2}$$

$$u^+ = u / u_\tau \tag{8.3}$$

where y^+ is the nondimensional distance normal to the wall (and can be thought of as a local Reynolds number) and u^+ is the streamwise velocity nondimensionalized by the turbulent velocity and length scales, u_τ, known as the *friction velocity*. The reason the laminar sublayer exists is that the presence of the wall damps velocity fluctuations due to the no-slip condition. So, just at the wall the flow is laminar, and in wall units the velocity profile is $u^+ = y^+$. In the fully turbulent region the velocity profile can be described by:

$$u^+ = \frac{1}{\kappa}\log y^+ + 5.0 \tag{8.4}$$

which is called the *Law of the Wall* for a turbulent boundary layer, where κ, the Kármán constant, is given as $\kappa = 0.41$. The Law of the Wall can be used as a near wall turbulence model in order to reduce the grid clustering required to compute turbulent flows, saving large numbers of grid points in the boundary layer. Other velocity variations occur for each sublayer within the boundary layer, as shown in Fig. 4.12. Any boundary-layer textbook, e.g., Schlichting et al.[8] or Schetz and Bowersox,[28] will provide complete details

    Karman vortex street behind two cylinders (photo courtesy of NASA).

of the theoretical derivation and nomenclature used for turbulent boundary layers.

So why do turbulent flows occur at high Reynolds numbers? It should be kept in mind that the Reynolds number (see Section 3.5.3) is a ratio of inertial effects to viscous effects:

$$Re \equiv \frac{\rho_\infty V_\infty L}{\mu}. \tag{8.5}$$

Rearranging terms and arguing that the velocity gradient would depend on V_∞/L (i.e., $\partial V/\partial y \approx V_\infty/L$) we obtain:

$$\frac{\rho_\infty V_\infty L}{\mu} = \frac{\rho_\infty V_\infty^2}{\mu \dfrac{V_\infty}{L}} \approx \frac{\rho_\infty V_\infty^2}{\mu \dfrac{\partial V_\infty}{\partial L}} \tag{8.6}$$

where $\mu \partial V_\infty/\partial L$ represents viscous shear stress. So a large Reynolds number means high inertial effects (the numerator) compared to viscous effects (the denominator). If the Reynolds number is low, viscous effects are relatively large and would tend to stabilize the flow by smoothing out any velocity gradients that could cause instabilities. This can be seen in the flow over a cylinder, where, at Reynolds numbers less than 40, the flow is steady. As the Reynolds number increases to about 70, vortices begin to shed in an alternating unsteady pattern known as the Kármán vortex street (see Fig. 8.13). Although the flow is unsteady, it is still laminar, since the shedding is orderly. As the Reynolds number increases further ($Re_D > 3 \times 10^5$), the wake begins to break down into a turbulent motion, since the disturbances are carried downstream in the flow and grow until turbulence is established.

The final property of turbulence listed previously is that it is chaotic and irregular. The fact that the flow is chaotic makes prediction quite challenging, since chaos has sensitive dependence on initial conditions. A slight change in initial conditions can have a drastic impact on the flowfield at a later time, which is sometimes referred to as the "butterfly effect."[29] So, predicting the exact details of the instantaneous flow shown in Figures 8.11 or 8.12 is virtually impossible, since we would require an almost exact initial flowfield to start the calculation. Luckily, we are generally concerned with statistical quantities that are not sensitive to initial conditions since they involve long time averages. For example, we may want the time-averaged lift on an aircraft, which can be obtained if we can calculate the time-averaged pressure and velocity field over the aircraft (the velocity field is required to obtain skin-friction). Even if we want unsteady information, it is usually from a statistical standpoint (such as the standard deviation of pressure fluctuations about a mean). If more detailed information is required, then more elaborate (and more computationally expensive) turbulence methods will have to be used. The accompanying Computational Aerodynamics Concept Box discusses the difference between instantaneous and average flowfields.

Computational Aerodynamics Concept Box

The Difference between "Instantaneous" and "Average"

It is quite common for people to get confused when talking about unsteady physical properties, like fluid flows. A simple example can help to illustrate this: you are driving in your car and measuring instantaneous speed (recording readings from your speedometer), but you also can estimate your average speed (how many miles you traveled and how long it took you to complete the trip). If you are driving through a city while doing this, you will see that your instantaneous speed goes up and down a great deal (as you start and stop the car), but your average speed (total trip distance divided by total trip length) does not contain any of this information.

This is also true for unsteady flow predictions with CFD. For example, two flow visualizations of the velocities around the University of New South Wales Sunswift solar car are shown in the figures: one is instantaneous (like a picture taken at one moment in time) and one is averaged over a long period of time. Can you tell which one is which? For most of the flow around the car, there is no difference between the two, because most of the flow is steady (as if you were driving your car at a constant speed – there would be no difference between your instantaneous velocity and your average velocity). However, there is one region where the flow is quite different between the two pictures: immediately behind the car. The first picture is an instantaneous flowfield, where the details of the wake behind the car are clear. The second picture is the averaged picture, where the wake only shows gross features, but no instantaneous details. Understanding the difference between these two pictures, and the reason behind the difference, is important for CFD.

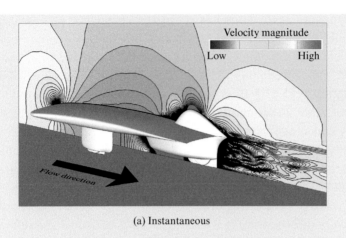

(a) Instantaneous

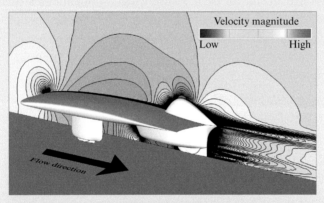

(b) Time averaged

Simulated flow over a solar car (a) instantaneous (b) time averaged (Courtesy of Tracie Barber, University of New South Wales; a full color version of this image is available on the website: www.cambridge.org/aerodynamics).

8.7 Turbulence Modeling Approaches

Because of its importance in aerodynamics, methods to simulate turbulent flow have been proposed for more than one hundred years. Most of the approaches focus on approximating the inherently unsteady turbulent fluctuations for use in numerical simulations – the approximation is known as a *turbulence model*. Initially, the focus was on developing turbulence models for solving the boundary-layer equations. Two of the most important aerodynamic researchers, Prandtl and von Kármán, made key contributions to the development of turbulence models for those types of flows. A modern overview of the various approaches has been given by Spalart for the numerical prediction of turbulent flows.[30] Tennekes and Lumley also include a good philosophical debate on the many choices for numerically modeling turbulence.[26] Current approaches to turbulence modeling include the *Reynolds-Averaged Navier-Stokes* (RANS), *Large Eddy Simulation*

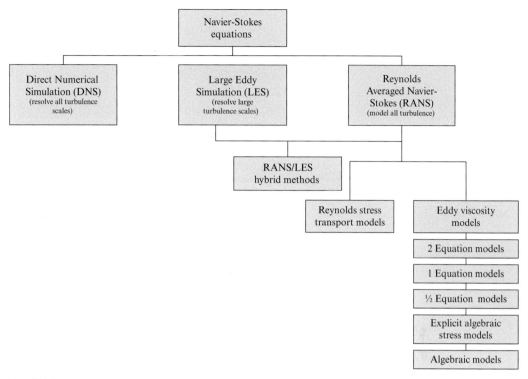

Figure 8.14 Turbulence modeling approaches.

(LES), and Direct Numerical Simulation (DNS) approaches, as categorized in Fig. 8.14.

Osborne Reynolds, of Reynolds number fame, also made an important contribution to turbulence modeling by developing a method to convert the fundamentally unsteady turbulent flow equations into a time-averaged version of the Navier-Stokes equations that provides an opportunity to capture much of the physics of turbulent flow with a steady calculation. Thus, the *Reynolds-Averaged Navier-Stokes* approach (RANS) is used to solve for time-averaged flow. This means that all scales of turbulence must be modeled and requires that empirical information based on experimental observation be used in the turbulence models. The RANS approach can provide accurate results for attached boundary-layer flows (and even steady separated flows) *with minimal grid requirements*. However, since the large scales for unsteady separated flows are highly dependent on the geometry, RANS-based turbulence models often fail to provide accurate results for flows with large regions of unsteady separated flows. One approach to dealing with these unsteady flows is to use *Unsteady Reynolds-Averaged Navier-Stokes* (URANS) or, as some have called it, Very Large Eddy Simulation (VLES). Of course, the first question that is often asked is how can a time-averaged set of equations, RANS, be used to make unsteady calculations? The apparent answer lies in just how the time averaging is done, and we will discuss that in greater detail later in the next

Table 8.1 Turbulence Model Methods Classified According to Needs, Performance, and Readiness (Ref. 30)

Method	Aim*	Grid Re Dependence	Empiricism	Grid Size	Time Steps	Readiness
2D URANS	Numerical	Weak	Strong	10^5	$10^{3.5}$	1980
3D URANS	Numerical	Weak	Strong	10^7	$10^{3.5}$	1995
DES	Hybrid	Weak	Strong	10^8	10^4	2000
LES	Hybrid	Weak	Weak	$10^{11.5}$	$10^{6.7}$	2045
DNS	Numerical	Strong	None	10^{16}	$10^{7.7}$	2080

* "Numerical" means that adding grid points does not add any new physics to the solution but improves numerical accuracy. "Hybrid" means that both physical and numerical improvements occur with grid refinement.

section. Also notice that the RANS models are broken into two groups: the Reynolds Stress models (RSM) and the Eddy Viscosity models (EVM); EVM RANS models can also come in linear or nonlinear variations.

The Direct Numerical Simulation (DNS) approach resolves *all* scales of turbulence directly, without relying on any empirical data. Because DNS must model all scales from the largest to the smallest, the grid resolution requirements are very high and increase drastically with Reynolds number. The Large Eddy Simulation (LES) approach attempts to model the smaller, more homogeneous scales of turbulence, while resolving the larger, energy-containing scales numerically. This reduces the grid requirements for LES compared to DNS. To accurately resolve the boundary layer, however, LES must accurately resolve the energy-containing eddies in the boundary layer, which requires very small streamwise grid spacing. To circumvent this problem, Spalart et al. proposed a new method, *Detached-Eddy Simulation* (DES), that combines the RANS and LES approaches.[30,31] The DES hybrid model was proposed in 1997 and has been applied a great deal since then. An extension of DES was proposed in 2006, known as *Delayed Detached-Eddy Simulation* (DDES), that is also being used a great deal in CFD simulations.[32] In fact, many other hybrid models have been developed in the past decade because of the success of DES and DDES.

Table 8.1 compares these various approaches based on the application, requirements, and readiness of each method to compute full aircraft aerodynamics at flight Reynolds numbers. This comparison was performed by Spalart and is very valuable in determining how to approach various applications. We will describe the RANS approach first, then LES, followed by descriptions of DES and DNS later in the chapter. The important thing to notice about the table, however, is that LES and DNS solutions for full aircraft at flight Reynolds numbers will not be available for decades! Even with the fastest supercomputers of today (remember Chapter 2), we still do not have adequate computer memory and speed to handle these problems.

8.8 **Reynolds-Averaged Navier-Stokes (RANS)**

To obtain the RANS equations from the Navier-Stokes equations, we will divide the flow variables into mean (time-averaged values denoted by $\bar{f}$) and fluctuating (time-varying values denoted by f') quantities, as shown in Fig. 8.15. For example:

$$f(\vec{x},t) = \bar{f}(\vec{x}) + f'(\vec{x},t) \tag{8.7}$$

where $\overline{f'} = 0$. Mathematically, the time averaging process is expressed as:

$$\bar{f}(\vec{x}) = \lim_{T \to \infty} \frac{1}{T} \int_{t}^{t+T} f(\vec{x}, \tau) d\tau \tag{8.8}$$

where the time constant, T, in Equation 8.8 should be large compared to the period of the random fluctuations associated with turbulence, represented by T_1, but small compared to the characteristic time scale of the vehicle being simulated, represented by T_2 (the time required for a fluid particle to travel the length of the body, which is known as the *convective time scale*), as shown in Fig. 8.15.

When this averaging process is applied to the Navier-Stokes equations, the result is an equation for the mean quantities with extra terms involving the fluctuating quantities – the *Reynolds stress* terms. For example, if you start with the dimensional form of the boundary-layer momentum equation, given previously as Equation 3.137, and apply the Reynolds averaging process, we see explicitly where the new Reynolds stress term appears (see the Computational Aerodynamics Concept Box for the details of how this happens for two-dimensional, incompressible flow):

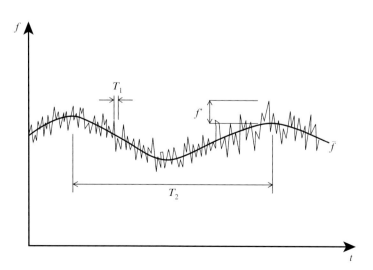

Figure 8.15   Reynolds averaging process for turbulent flow quantities (Ref. 33.).

$$\rho\bar{u}\frac{\partial\bar{u}}{\partial x}+\rho\bar{v}\frac{\partial\bar{u}}{\partial y}=-\frac{\partial\bar{p}}{\partial x}+\frac{\partial}{\partial y}\overbrace{\left(\underbrace{\mu\frac{\partial\bar{u}}{\partial y}-\underbrace{\rho\overline{u'v'}}_{\substack{\text{Reynolds}\\\text{stress}}}}\right)}^{\tau}. \tag{8.9}$$

As is customary, the bars over the time-averaged terms are dropped for simplicity to yield the Reynolds-averaged equation:

$$\rho u\frac{\partial u}{\partial x}+\rho v\frac{\partial u}{\partial y}=-\frac{\partial p}{\partial x}+\frac{\partial}{\partial y}\left[\mu\frac{\partial u}{\partial y}-\rho\overline{u'v'}\right]. \tag{8.10}$$

If Equation 8.9 had been written out in three dimensions, but still using the incompressible flow assumption, the Reynolds-averaged x-momentum equation would be:

$$\rho\frac{\partial u}{\partial t}+\rho u\frac{\partial u}{\partial x}+\rho v\frac{\partial u}{\partial y}+\rho w\frac{\partial u}{\partial z}$$
$$=-\frac{\partial p}{\partial x}+\frac{\partial}{\partial x}\left(\tau_{xx}-\rho\overline{u'u'}\right)+\frac{\partial}{\partial y}\left(\tau_{yx}-\rho\overline{u'v'}\right)+\frac{\partial}{\partial z}\left(\tau_{yz}-\rho\overline{u'w'}\right). \tag{8.11}$$

The Reynolds stresses take into account the transfer of momentum by the turbulent fluctuations, and are typically written as:

$$\tau_{ij}^{r}=-\rho\overline{u_i'u_j'} \tag{8.12}$$

where the subscripts i and j in general take the values of 1, 2, and 3 to account for all three directions, x, y, and z (which is known as index notation). For example, $\tau_{11}=\tau_{xx}=-\rho\overline{u_1'u_1'}=-\rho\overline{u'u'}$ and $\tau_{12}=\tau_{xy}=-\rho\overline{u_1'u_2'}=-\rho\overline{u'v'}$. Because we time-averaged the equations, the averaging process introduced these additional unknowns in the governing equations, which is where the requirement for *turbulence models* arises. In fact, for the Navier-Stokes equations we started with four equations in four unknowns and now we have four equations in ten unknowns (the various Reynolds stress terms). Having to provide equations to model these additional terms is commonly referred to as the *closure problem* (we need to close the set of equations by defining the Reynolds stress terms). As described earlier, the modeling of the terms represented in Equation 8.12 requires empirical information (which usually shows up as constants in the model), and is why the models are never a completely accurate representation of turbulence for all cases of interest (recall the wide range of flows with viscous effects illustrated in Fig. 8.1). This process can also lead to an equation for turbulent kinetic energy and viscous dissipation which would also need to be modeled with PDEs.

Finally, we can now begin to understand the meaning of URANS discussed earlier. Since Equation 8.7 has both upper and lower limits on the time scale of the motion, it could be argued that the time averaging is only averaging out small-scale time motions (such as occur in turbulence), but that large time scales are left un-averaged (such as vortex shedding or aircraft motion). Therefore, it is possible to obtain unsteady results with a RANS model, even though the turbulence model does not have any temporal component. The limitation here is that the temporal scales of the fluid motion being predicted should not be too large, or the URANS approach might give very unusual results. Or, as Phillipe Spalart said, "It is simple to describe how to conduct a URANS, but the physical implications give pause."[30]

Computational Aerodynamics Concept Box

How the Reynolds Averaging Process Creates the Reynolds Stress Terms

For an example of how the Reynolds averaging process creates the Reynolds stress term, we will start with the continuity equation and the x-momentum boundary-layer equation (Equation 3.137) for steady, two-dimensional, incompressible flow:

$$\frac{\partial u}{\partial x} + \frac{\partial v}{\partial y} = 0 \quad \rho u \frac{\partial u}{\partial x} + \rho v \frac{\partial u}{\partial y} = -\frac{\partial p}{\partial x} + \mu \frac{\partial^2 u}{\partial y^2}.$$

Now decompose the velocity and pressure terms, remembering that the density is constant:

$$u = \bar{u} + u' \quad v = \bar{v} + v' \quad p = \bar{p} + p'$$

substitute into the momentum equation to obtain:

$$\rho(\bar{u} + u') \frac{\partial}{\partial x}(\bar{u} + u') + \rho(\bar{v} + v') \frac{\partial}{\partial y}(\bar{u} + u') = -\frac{\partial}{\partial x}(\bar{p} + p') + \mu \frac{\partial^2}{\partial y^2}(\bar{u} + u')$$

and multiply out the terms to obtain:

$$\rho \bar{u} \frac{\partial \bar{u}}{\partial x} + \rho \left[\bar{u} \frac{\partial \bar{u}}{\partial x} + u' \frac{\partial \bar{u}}{\partial x} + \bar{u} \frac{\partial u'}{\partial x} + u' \frac{\partial u'}{\partial x} \right] + \rho \left[\bar{v} \frac{\partial \bar{u}}{\partial y} + v' \frac{\partial \bar{u}}{\partial y} + \bar{v} \frac{\partial u'}{\partial y} + v' \frac{\partial u'}{\partial y} \right]$$

$$= -\left[\frac{\partial \bar{p}}{\partial x} + \frac{\partial p'}{\partial x} \right] + \mu \left[\frac{\partial^2 \bar{u}}{\partial y^2} + \frac{\partial^2 u'}{\partial y^2} \right].$$

Now, time-average the entire equation, remembering that, if you average an already averaged term, you get the same term, and, if you average a fluctuating term, you get zero ($\bar{\bar{u}} = \bar{u}$ and $\overline{u'} = 0$):

$$\rho \bar{u} \frac{\partial \bar{u}}{\partial x} + \rho \bar{v} \frac{\partial \bar{u}}{\partial y} + \rho \overline{u' \frac{\partial u'}{\partial x}} + \rho \overline{v' \frac{\partial u'}{\partial y}} = -\frac{\partial \bar{p}}{\partial x} + \mu \frac{\partial^2 \bar{u}}{\partial y^2}.$$

We cannot get rid of the third and fourth terms in the equation because they are the product of fluctuating values, and the time average of such a term is not necessarily zero. So we will analyze those terms using the continuity equation. Applying the same process to the continuity equation yields two relationships:

$$\frac{\partial \bar{u}}{\partial x} + \frac{\partial \bar{v}}{\partial y} = 0 \qquad \frac{\partial u'}{\partial x} + \frac{\partial v'}{\partial y} = 0.$$

If we multiply the second equation by $\rho(\bar{u} + u')$ and time average we obtain:

$$\overline{\rho u' \frac{\partial u'}{\partial x}} + \overline{\rho u' \frac{\partial v'}{\partial y}} = 0.$$

Now add this equation to the momentum equation and neglect $\partial/\partial x(u'^2)$ as being small in a boundary layer to obtain:

$$\rho \bar{u} \frac{\partial \bar{u}}{\partial x} + \rho \bar{v} \frac{\partial \bar{u}}{\partial y} = -\frac{\partial \bar{p}}{\partial x} + \frac{\partial}{\partial y}\left[\mu \frac{\partial \bar{u}}{\partial y} - \overline{\rho u' v'} \right]$$

which is Equation 8.9.

8.8.1 Mass-Weighted Averaging

For compressible flows, the density becomes a variable, and it also must be decomposed into mean and fluctuating components, just like the other quantities:

$$\rho(\vec{x},t) = \bar{\rho}(\vec{x}) + \rho'(\vec{x},t). \tag{8.13}$$

However, if the velocity is time averaged in the same way, the resulting RANS equations become quite complex. Instead, a more concise set of equations can be obtained, if the density and pressure are time averaged and the other variables are *mass averaged*;[34] (which is also known as Favre averaging) according to:

$$\tilde{f}(\vec{x}) = \frac{1}{\bar{\rho}} \lim_{T \to \infty} \frac{1}{T} \int_{t}^{t+T} \rho(\vec{x},t) f(\vec{x},\tau) d\tau = \frac{\overline{\rho f}}{\bar{\rho}} \tag{8.14}$$

where $\overline{\rho f''} = 0$. The variables are then decomposed as

$$f(\vec{x},t) = \tilde{f}(\vec{x}) + f''(\vec{x},t). \tag{8.15}$$

This approach to averaging yields a mass conservation equation that takes the same form as the non-averaged equation (compare with Equation 3.5):

$$\frac{\partial \overline{\rho}}{\partial t} + \frac{\partial \overline{\rho}\tilde{u}}{\partial x} + \frac{\partial \overline{\rho}\tilde{v}}{\partial y} + \frac{\partial \overline{\rho}\tilde{w}}{\partial z} = 0 \tag{8.16}$$

which is why this form of averaging is typically chosen, since it insures that the average flux of mass across a streamline vanishes if mass is conserved. The mass-averaged x-momentum equation becomes:

$$\begin{aligned}
\frac{\partial \overline{\rho}\tilde{u}}{\partial t} &+ \frac{\partial \overline{\rho}\tilde{u}\tilde{u}}{\partial x} + \frac{\partial \overline{\rho}\tilde{u}\tilde{v}}{\partial y} + \frac{\partial \overline{\rho}\tilde{u}\tilde{w}}{\partial z} \\
&= -\frac{\partial \overline{p}}{\partial x} + \frac{\partial}{\partial x}\left(\overline{\tau}_{xx} - \overline{\rho u''u''}\right) + \frac{\partial}{\partial y}\left(\overline{\tau}_{yx} - \overline{\rho u''v''}\right) + \frac{\partial}{\partial z}\left(\overline{\tau}_{zx} - \overline{\rho u''w''}\right)
\end{aligned} \tag{8.17}$$

which is very similar to the Reynolds-averaged x-momentum equation in Equation 8.11. Note that averaging the energy equation also leads to an additional term. The turbulence model used to approximate the Reynolds stress could be an algebraic relation, an ordinary differential equation, or additional partial differential equations, as categorized in Fig. 8.14.

8.8.2 Taxonomy of Turbulence Models

Turbulence models that solve for the Reynolds stresses directly are known as *Reynolds stress models* (RSM). Since there are six unique components of the Reynolds stress tensor (Equation 8.12), as well as the turbulence dissipation, these models often contain seven equations – one for each stress component and one for the dissipation term. Models such as these are usually quite elaborate and complex and will not be covered in this chapter.

A simpler approach is to assume that the Reynolds stress is proportional to the mean strain rate, or

$$\tau'_{ij} = \mu_t \left(\frac{\partial \overline{u}_i}{\partial x_j} + \frac{\partial \overline{u}_j}{\partial x_i}\right) - \frac{2}{3}\overline{\rho} k \delta_{ij} \tag{8.18}$$

that can be used to model the turbulence. This is known as the *Boussinesq eddy viscosity approximation* and can be generalized to compressible flows to provide the Reynolds stress

$$\overline{\rho u_i''u_j''} = \mu_t \left(\frac{\partial \tilde{u}_i}{\partial x_j} + \frac{\partial \tilde{u}_j}{\partial x_i} - \frac{2}{3}\frac{\partial \tilde{u}_k}{\partial x_k}\delta_{ij}\right) - \frac{2}{3}\overline{\rho} k \delta_{ij} \tag{8.19}$$

where μ_t is the unknown turbulent viscosity and must be provided by the turbulence model.[iii] Boussinesq originally assumed that the turbulent viscosity

---

[iii] The last term in Eqn. 8.19 guarantees that the trace of the Reynolds stress tensor returns $2\overline{\rho} k$ where k is: $k = \frac{1}{2}\left(\overline{u'^2} + \overline{v'^2} + \overline{w'^2}\right)$, the kinetic energy of the turbulent fluctuations. This ensures that Eqn. 8.17 is satisfied.

Table 8.2 Taxonomy of Typical Eddy-Viscosity Models

Number of PDEs	Common Names	Examples
0	Algebraic model Zero-equation model	Cebeci-Smith[35] Baldwin-Lomax[36]
1/2	Half-equation model	Johnson-King[37]
1	One-equation model	Spalart-Allmaras[38] Baldwin-Barth[39]
2	Two-equation model	$k - \varepsilon$[40] $k - \omega$[33] Menter's $k - \omega$ models[41] Menter's SST model, [42] Explicit Algebraic Stress Model (EASM)[43]

would be some large constant multiple of the basic fluid viscosity. This turned out to be wrong, but the use of a "turbulent viscosity" concept allowed computations to mimic laminar flow calculations with the modification that μ was replaced by $\mu + \mu_t$, and μ_t would require modeling. The eddy viscosity is analogous to the kinematic viscosity which has dimensions of *length* x *velocity*, and is defined as $v_t = \mu_t / \rho$.[iv] It is easier to visualize the idea in terms of the boundary-layer equation, where the eddy viscosity is cast in the same form as the fluid viscosity (although it is not a fluid property and varies throughout the flow). Eddy-viscosity models can be used to calculate μ_t, and the models vary in complexity and accuracy. They can be roughly broken down into categories based on the number of additional partial differential equations that must be solved in order to complete the estimation of the eddy viscosity. Table 8.2 shows examples of various eddy-viscosity models and the number of PDEs they require (the *half-equation model* is a model that solves a single ordinary differential equation, and a *zero-equation model* is an algebraic model).

Flow Visualization Box

What is your turbulence model doing?

One of the important things we just talked about was making sure our grid resolution in the boundary layer was fine enough for our flow. Specifically, the $y+$ value should be less than 1 for most regions in the boundary layer. Shown here is our airfoil (see in three dimensions from above) where $y+$ has been shaded onto the airfoil surface. Notice that there are values of 2 near the leading edge, but this is where the boundary layer is very thin and it is nearly impossible to make the grid fine enough in this region. As the boundary layer thickens further downstream, however, the value is close to (or less than) 1 over the majority of the airfoil.

---

iv Various authors use different nomenclatures, so take care to make sure you understand the definitions.

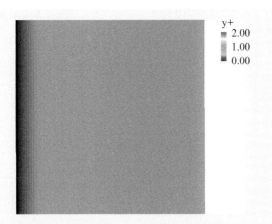

The next thing we can look for is where our turbulence model is adding "eddy viscosity." The figure here shows eddy viscosity around the airfoil, where dark shading represents no eddy viscosity (no turbulent viscosity being predicted) and light shading represents higher values of eddy viscosity. Notice that the boundary layer has eddy viscosity being added, and there are large values being added near the trailing edge and in the wake (there is a small amount of separated flow in this region). Does this make sense? Gaining a physical feeling for turbulence (and where it happens) will help you to ensure that your CFD code is working properly (a full color version of these images are available on the website: www.cambridge.org/aerodynamics).

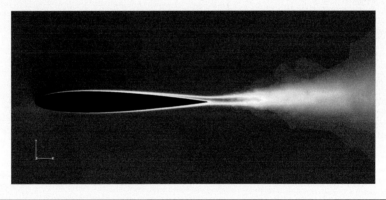

8.8.3 Prandtl's Mixing-Length Theory – An Example of a Zero-Equation Model

It will be useful to see how one of the original algebraic turbulence models was developed – the algebraic model developed by Prandtl. We start by considering how the fluid passes along the wall in turbulent motion, as shown in Fig. 8.12: fluid particles coalesce into lumps (eddies), which move and cling together over a given traversed length, both in the longitudinal and wall-normal directions, retaining their momentum parallel to the x direction. Now, assume that the lump of fluid is displaced a distance in the transverse

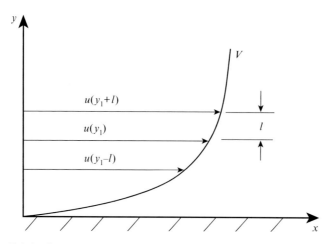

Figure 8.16   Turbulent flow exchange within a flat plate boundary layer.

direction, l, which is known as *Prandtl's mixing length*. We can visualize the process and define the terms for the analysis using Fig. 8.16, where l is the transverse direction traveled by the lump (either up or down), and the velocities within the turbulent boundary layer are shown for each position: $y_1 - l$, y_1, and $y_1 + l$.

If we expand the velocities shown in Fig. 8.16 in a Taylor's series expansion for $v' > 0$ (which represents a fluid lump moving up),

$$\bar{u}(y_1 - l) = \bar{u}(y_1) + \left[(y_1 - l) - y_1 \right] \left(\frac{d\bar{u}}{dy} \right)_1 + \cdots \tag{8.20}$$

or

$$\bar{u}(y_1) - \bar{u}(y_1 - l) = l \left(\frac{d\bar{u}}{dy} \right)_1 + \cdots. \tag{8.21}$$

So, the difference in velocities is

$$\Delta u_1 = \bar{u}(y_1) - \bar{u}(y_1 - l) \approx l \left(\frac{d\bar{u}}{dy} \right)_1. \tag{8.22}$$

In the same way, a lump of fluid from $y_1 + l$ has a greater velocity ($v' < 0$)

$$\Delta u_2 = \bar{u}(y_1 + l) - \bar{u}(y_1) \approx l \left(\frac{d\bar{u}}{dy} \right)_1. \tag{8.23}$$

The velocity differences caused by the motion represent the turbulent velocity at y_1, and the difference is obtained by taking the average of the two velocity magnitudes:

$$\left|\overline{u'}\right| = \frac{1}{2}\left(\left|\Delta u_1\right| + \left|\Delta u_2\right|\right) = l\left|\left(\frac{d\overline{u}}{dy}\right)_1\right|. \tag{8.24}$$

So, the level of turbulence is related to the distance the lumps must travel (this idea came from the mean free path concept in the kinetic theory of gases).

Further, Prandtl assumed that the wall-normal component of turbulent velocity could be represented by a constant percentage of the longitudinal component of turbulent velocity:

$$\left|\overline{v'}\right| = \text{constant}\left|\overline{u'}\right| = \text{constant}\, l\left|\frac{d\overline{u}}{dy}\right|. \tag{8.25}$$

In order to finish the model, we need $\overline{u'v'}$ to define the Reynolds stress term in Equation 8.9, so Prandtl assumed

$$\overline{u'v'} = -c\left|\overline{u'}\right| \cdot \left|\overline{u'}\right| \tag{8.26}$$

or

$$\overline{u'v'} = -\text{constant}\, l^2\left(\frac{d\overline{u}}{dy}\right)^2 \tag{8.27}$$

or we can remove the constant by redefining the mixing length, and Equation 8.27 becomes:

$$\overline{u'v'} = -l^2\left(\frac{d\overline{u}}{dy}\right)^2 \tag{8.28}$$

and the shear stress (Reynolds stress) is then:

$$\tau_t = -\rho\overline{u'v'} = \rho l^2\left(\frac{d\overline{u}}{dy}\right)^2. \tag{8.29}$$

To take into account the correct sense of the direction of the flow (either forward or backward) we can write:

$$\tau_t = \rho l^2 \left|\frac{d\overline{u}}{dy}\right|\left(\frac{d\overline{u}}{dy}\right). \tag{8.30}$$

So, our algebraic turbulent viscosity model is

$$\mu_t = \rho l^2 \left|\frac{d\overline{u}}{dy}\right| \tag{8.31}$$

where μ_t can be used directly in the averaged governing equations, as we will see shortly. The turbulent eddy viscosity can then be defined as:

$$v_t = \frac{\mu_t}{\rho} = l^2 \left|\frac{d\bar{u}}{dy}\right|. \tag{8.32}$$

In order for the model to be "closed," some relation for the mixing length, l, would have to be defined. Here is where the semi-empirical nature of turbulence modeling arises. The mixing length would normally be defined using some empirical knowledge or data, which is why turbulence models do not work for all types of flow.

8.8.4 Examples of the Use of Various RANS Models

It might seem impossible to decide which RANS turbulence model to use at this point. There have been so many different models created over the past decades, and finding out how well they work for various flows would seem to be a daunting task. Luckily for us, however, a large number of people have done some of this work for us. While you should not just depend on the work of others when deciding which turbulence model to use, you certainly can make a great deal of progress by looking at the work of others.

The effect of turbulence in the RANS equations is incorporated using the turbulent viscosity, μ_t, that accounts for the increased mixing in turbulent flow. To repeat, μ_t is not a constant, but varies throughout the flowfield, according to the specific model being used. Additionally, a turbulent Prandtl number can also be specified as: $\text{Pr}_t = C_p \mu_t / k_t$, where the turbulent thermal conductivity coefficient, k_t, is then calculated from the turbulent viscosity and turbulent Prandtl number. For example, in the case of a two-dimensional formulation:

$$\tau_{xx} = (\mu + \mu_t)\left(\frac{4}{3}\frac{\partial u}{\partial x} - \frac{2}{3}\frac{\partial v}{\partial y}\right) \tag{8.33}$$

$$\tau_{yy} = (\mu + \mu_t)\left(\frac{4}{3}\frac{\partial v}{\partial y} - \frac{2}{3}\frac{\partial u}{\partial x}\right) \tag{8.34}$$

$$\tau_{xy} = \tau_{yx} = (\mu + \mu_t)\left(\frac{\partial u}{\partial y} + \frac{\partial v}{\partial x}\right) \tag{8.35}$$

$$q_x = -(k + k_t)\frac{\partial T}{\partial x} \tag{8.36}$$

$$q_x = -(k + k_t)\frac{\partial T}{\partial y} \tag{8.37}$$

Table 8.3 Commonly Used Eddy-Viscosity Turbulence Models

Model Name	PDEs	Year	Abbreviation
Cebeci-Smith	0	1974	CS
Baldwin-Lomax	0	1978	BL
Johnson-King	½	1985	JK
Baldwin-Barth	1	1991	BB
Spalart-Allmaras	1	1994	SA
Jones-Launder k-ε	2	1972	k-ε
Launder-Spalding k-ω	2	1972	k-ω
Menter's k-ε/k-ω	2	1991	k-ε/k-ω
Menter's Shear Stress Transport	2	1994	SST
Wilcox k-ω	2	1998	Wilcox k-ω
Explicit Algebraic Stress Model	2	2003	EASM

and an experimentally determined value for the Prandtl number, $Pr_t = 0.9$, is often used to find k_t.

Several commonly used eddy-viscosity RANS turbulence models are presented in Appendix F, including some of the models listed in Table 8.3. Many of the models have become known by the names of their developers (and by the abbreviations of their names), or by the PDE variables that are defined by the model (like k – turbulent kinetic energy, ε – turbulent dissipation, or ω – turbulent dissipation rate). Deciding which model to use for which type of flow has been the subject of a great deal of research over the past thirty years. We will look at some of these results to develop a sense of how well the various models predict typical flows (an excellent collection of turbulence model information is also available at the NASA Langley Research Center Turbulence Modeling Resource, http://turbmodels.larc.nasa.gov/index.html).

An example of a comparison between algebraic and ½-equation turbulence models is shown in a set of predictions made for flow over a prolate spheroid (an ellipse of revolution shown in Fig. 8.17).[44] The goal of the study was to determine the ability of various turbulence models to predict the steady vortical flows that take place on these bodies at high angles of attack (in this case, $\alpha = 30°$). The turbulence models tested were labeled as follows: A = Baldwin-Lomax, B = Baldwin-Lomax with Degani-Schiff modification, C = Johnson-King, and D = Johnson-King with Degani-Schiff modification. The structured grid consisted of 121 axial, 53 circumferential, and 65 normal points for a total of more than 400,000 points, which is fairly coarse by today's standards. The surface geometry and predicted surface streamlines are also shown compared with experimental data in Fig. 8.17.

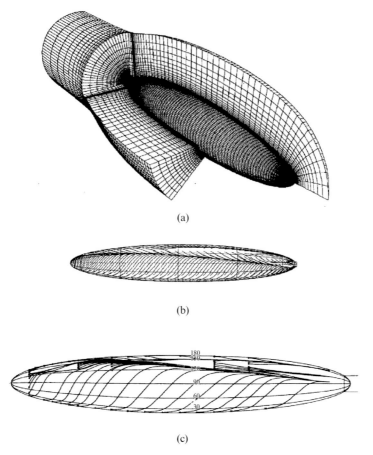

(a)

(b)

(c)

Figure 8.17 Surface geometry and streamlines: $M = 0.25$; $\alpha = 30$ deg; $Re = 7.2 \times 10^6$ (Ref. 44). (a) surface geometry and grid. (b) surface streamlines using Model B (viewed from the side). (c) experimental surface streamlines.

The surface streamline patterns shown in Fig. 8.17b (the flow is going from left to right) show primary vortex separation on the side of the body and a secondary vortex separation line near the top of the geometry. These predictions compare very well with the experimental surface streamlines shown in Fig. 8.17c. These types of separated flow regions are what caused the basic turbulence models to have great difficulty in accurately predicting the flowfield. More details on interpreting surface and off-surface flow topology, as seen in Figures 8.17b and 8.17c, can be found in the review article by Tobak and Peake,[45] as well as in the article by Dallmann (which will be discussed in Chapter 9).[46]

One of the issues being investigated in this study was how well the turbulence models, which were created for attached flow, would function when the flow separated. Degani and Schiff believed that these turbulence models would overpredict the turbulent viscosity, μ_t, and devised a way to modify the model in the presence of vortices.[47] The resulting values for the four turbulence models are presented in Fig. 8.18 at two longitudinal ($x/L = 0.22$ and

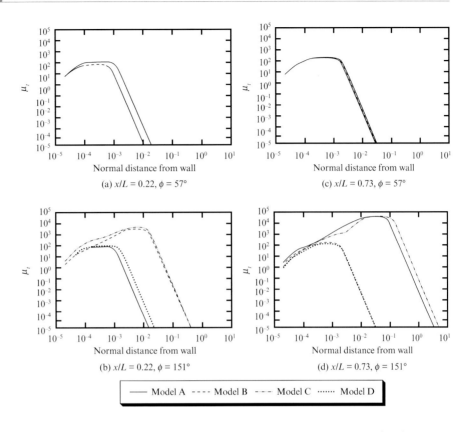

Figure 8.18 Profiles of μ_t at two leeward and two windward stations: $M = 0.25$; $\alpha = 30$ deg; $Re = 7.2 \times 10^6$. (Ref. 44)

0.73) and two windward positions ($\phi=57$ and 151 deg, where $\phi=0$ deg is at the bottom of the prolate spheroid where the flow is attached and $\phi=180$ deg is at the top where the flow is separated). Notice that, at the attached flow positions ($\phi=57$ deg), both at the front and rear of the body, the four models are yielding very similar results. That means that, in this case, the models work well for attached flow. However, the original (unmodified) models are giving much higher values of turbulent viscosity at $\phi=151$ deg where the flow is separated than are the models using the Degani-Schiff modification (in fact, several orders of magnitude higher values). When these higher values are put into the shear stress terms (Equations 8.33 through 8.35), the shear stress is extremely high, resulting in "smeared out" (or dissipated) flow features. Adding the modification greatly improved the vortical flow predictions and surface pressure variations, resulting in an overall assessment that the unmodified algebraic and ½-equation models could not adequately predict steady separated flow.

Another excellent example showing the ability of RANS turbulence models was an extensive set of comparisons that were made for the NACA 0012 airfoil at the Viscous Transonic Airfoil Workshop, which was reported by Holst.[48] This study evaluated the abilities of three eddy-viscosity turbulence

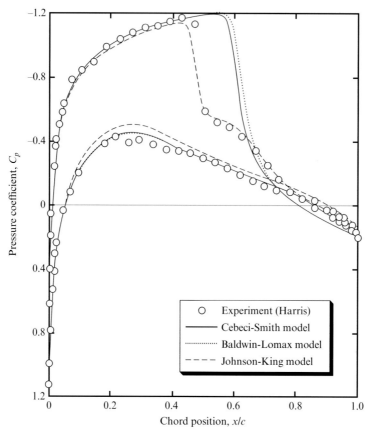

Experiment (Harris)
Cebeci-Smith model
Baldwin-Lomax model
Johnson-King model

Figure 8.19 Comparisons for pressure distributions using different turbulence models for an NACA 0012 airfoil at $M = 0.80$, $\alpha = 2.26°$, and $Re_c = 9 \times 10^6$ (Ref. 48.).

models: Cebeci-Smith, Baldwin-Lomax, and Johnson-King. Experimental data for transonic airfoils have their own uncertainties, but, of all the transonic tests of NACA 0012 airfoils, a study by McCroskey[49] suggested that the data by Harris used in this study was the highest quality data available.[50] Figure 8.19 provides a comparison of pressure distributions for the NACA 0012 airfoil, where the primary difference in the predictions is the location of the shock. The Johnson-King model produced the best agreement between the experimental and predicted positions of the shock wave at $M = 0.80$. However, even though the shock position is predicted well, the lower surface pressure distribution is not predicted well. Figure 8.20 provides an example of the difficulty in choosing a turbulence model by showing the drag predictions using the Baldwin-Lomax and Johnson-King models for $M = 0.7$ (the equivalent pressure and force results were not presented for this comparison at the same Mach number). Here, the Baldwin-Lomax model produces the best results, which shows why the CFD user must investigate the validity of predictions carefully before using a specific turbulence model and code in real work.

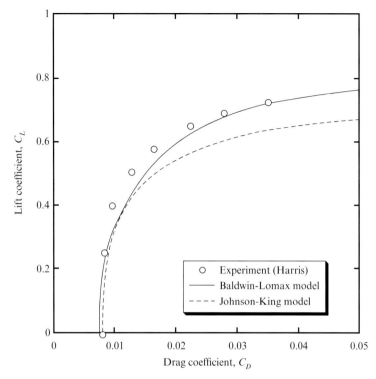

Figure 8.20   Comparisons of drag predictions using different turbulence models for an NACA 0012 airfoil at $M = 0.70$, and $Re_c = 9 \times 10^6$ (Ref. 48.).

Another turbulence model comparison study was conducted to see the difference between an algebraic turbulence model and a two-equation model (the Wilcox $k\text{-}\omega$ model). The three-dimensional configuration used in the study was the DLR F4 wing-body configuration.[51] Experimental data for this configuration were available from three wind tunnels, and the experimental results were reported by Redeker et al.[52] The computations were carried out on a grid with 128 equal sized blocks with a total of 6.3 million cells, which was generated by Airbus.

Figure 8.21a shows pressure distributions in two chordwise sections near the engine for $M = 0.75$, $\alpha = 0.98$ deg, and $Re = 3 \times 10^6$, which shows no significant differences in the numerical prediction between the Wilcox $k\text{-}\omega$ and Baldwin-Lomax models. However, the $k\text{-}\omega$ model did a better job predicting the lift, as shown in Fig. 8.21b, as the Baldwin-Lomax model results are consistently too high. The Wilcox $k\text{-}\omega$ computations for the drag polar also agree well with the experiment; in particular, the prediction of drag was within 5 percent of the measurements. "The differences are in part a result of the modeling of the nearly separated flow at the trailing edge of the wing, which is more accurately represented by the Wilcox $k\text{-}\omega$ model. This, in turn, has an expectedly strong effect on the shock position and lift coefficient"[51]

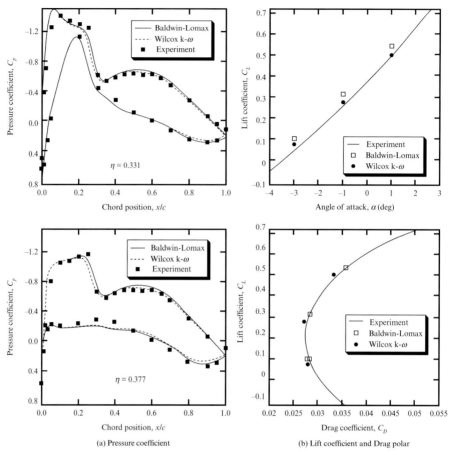

(a) Pressure coefficient
(b) Lift coefficient and Drag polar

Figure 8.21   Comparisons of surface pressures, lift, and drag polars using different turbulence models for the DLR F4 wing-body at $M = 0.75$, $\alpha = 0.93°$, and $Re_c = 3 \times 10^6$ (Ref. 51). (a) pressure coefficient; (b) lift coefficient and drag polar.

Now we can look at how various turbulence models perform for the flow over the Aérospatiale A airfoil at maximum lift, with an angle of attack of 13.3 deg at $M = 0.15$ and a Reynolds number of $Re = 2 \times 10^6$.[53] Transition in the experiment was induced by tripping the flow near the leading edge at $x/c = 0.12$ on the upper surface of the airfoil and at $x/c = 0.30$ on the lower surface of the airfoil. "The computational treatment of these transition locations was accounted for by multiplying the entire source term (production, redistribution and dissipation) in the equations governing the Reynolds stresses and dissipation rate by an appropriate function providing its zero value in the laminar flow part and a unit value in the fully turbulent flow region."[53] Two-dimensional calculations were performed using a mesh with 512×128 cells. The turbulence models evaluated include the Baldwin-Lomax (BL), Spalart-Allmaras (SA), k-ω, and the SSG/LRR-ω Reynolds stress models. Figure 8.22 shows the comparison of the computed pressure coefficient distribution with available experimental data. The highest positive pressure coefficient is seen at the stagnation point ($C_p = 1.0$), and the highest

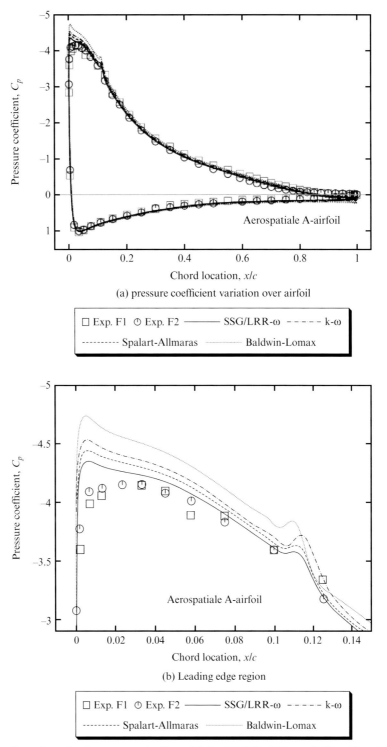

(a) pressure coefficient variation over airfoil

□ Exp. F1   ☉ Exp. F2 ——— SSG/LRR-ω − − − − − k-ω
--------- Spalart-Allmaras ·············· Baldwin-Lomax

(b) Leading edge region

□ Exp. F1   ☉ Exp. F2 ——— SSG/LRR-ω − − − − − k-ω
--------- Spalart-Allmaras ·············· Baldwin-Lomax

Figure 8.22 Comparisons of surface pressures for Aérospatiale A airfoil at $M = 0.15$, $\alpha = 13.3°$, and $Re_c = 2 \times 10^6$ (Ref. 53; reprinted by permission of the American Institute of Aeronautics and Astronautics, Inc.). (a) pressure coefficient variation over airfoil; (b) leading-edge region; (c) trailing-edge region.

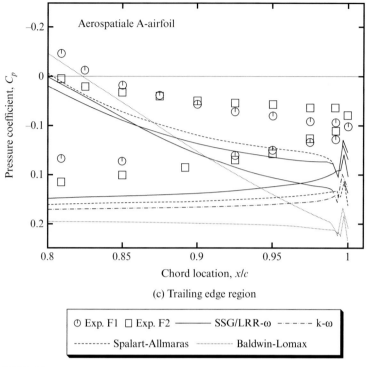

(c) Trailing edge region

○ Exp. F1 □ Exp. F2 ———— SSG/LRR-ω - - - - - k-ω

---------- Spalart-Allmaras .... Baldwin-Lomax

Figure 8.22 (*continued*)

negative pressure coefficient takes place on the upper surface at the suction peak ($4.0 \leq -C_p \leq 5.0$). While all models compute the pressure distribution well over the majority of the airfoil, the largest differences are seen near the leading-edge and the trailing-edge separation region (see Fig. 8.22b for the leading-edge region and Fig. 8.22c for the trailing-edge region). Notice that the BL model is overpredicting the leading-edge suction more than any of the other models and is underpredicting the trailing-edge separation less. While none of the models exactly matches the experimental data in these regions, the Reynolds stress model is doing better than the rest of the models on the same grid.

Therefore, we can conclude that the algebraic models do well for situations similar to the conditions for which they were created: attached boundary-layer flow over a relatively flat surface. However, more complex models are required in regions with large pressure gradients or where flow separation or recirculation exists.

Next, we will evaluate the ability of the Wilson k-ω RANS model to be useful in predicting separated flows with shocks over a delta wing, as shown in Fig. 8.23.[54] The results upstream of the vortex breakdown shows good agreement with the experimental data (Fig. 8.23a). However, the results downstream of breakdown shows significant discrepancies arising from the premature prediction of vortex breakdown, which is a highly unsteady

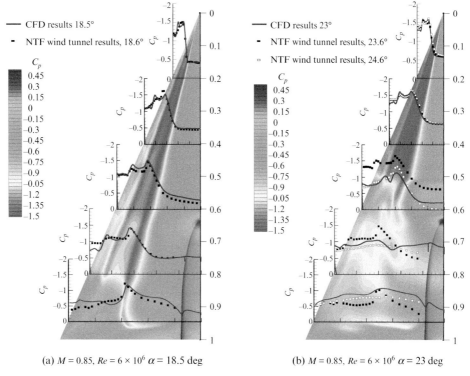

(a) $M = 0.85$, $Re = 6 \times 10^6$ $\alpha = 18.5$ deg   (b) $M = 0.85$, $Re = 6 \times 10^6$ $\alpha = 23$ deg

Figure 8.23   Comparisons of predictions of surface pressure for a sharp leading-edge delta wing at $M = 0.85$ and $Re_c = 6 \times 10^6$ using the Wilcox k-ω model (Ref. 54; a full color version of this image is available in the color insert pages of this text as well as on the website: www.cambridge.org/aerodynamics).

phenomenon. In fact, the sudden movement of breakdown is predicted about 3 deg earlier for the CFD when compared with the measurements. Notice that, in Fig. 8.23b, experimental results are included at 23.6 deg and 24.6 deg, and the large difference in pressure aft of $x/c = 0.5$ indicates that breakdown has moved rapidly up the wing between these two angles of attack.

Finally, we will look at one case where unsteady RANS (URANS) predictions are made for supersonic flow in the base region of a missile body.[55] The base flow region is shown in Fig. 8.24 with vorticity contours highlighting the recirculation region and showing the unsteadiness in the flow. The base pressure variation across the axisymmetric base is shown in Fig. 8.25a. The one-equation Spalart-Allmaras (SA) model predicts a base pressure that is too low and with a slight variation across the base. The experimental data are fairly constant along the base, which was determined to be because of the highly unsteady effects in the base region. The SA model is also used with a compressibility correction (SA-CC), which not only shows a strong effect as the results are much closer to the experiments, but also introduces an even larger radial variation. The SST model without the compressibility corrections does about as well as SA with the correction and yields a flatter radial profile. The compressibility correction applied to the SST (SST-CC) model further improves the pressure level but, again, introduces more radial variation.

Missile base flow recirculation region at $M = 2.46$ and $Re_d = 2.858 \times 10^6$ (Ref. 55; a full color version of this image is available on the website: www.cambridge.org/aerodynamics).

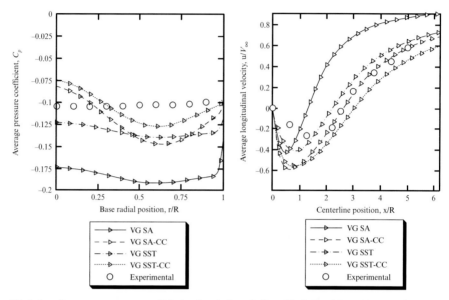

Missile base flow average pressure coefficient and centerline velocity at $M = 2.46$ and $Re_d = 2.858 \times 10^6$ (Spalart-Allmaras model with and without compressibility corrections on *VGRIDns* grid are VG SA and VG SA-CC, respectively; shear stress transport model with and without compressibility corrections on VGRIDns grid are VG SST and VG SST-CC, respectively) (Ref. 55).

The centerline velocity behind the base is shown in Fig. 8.25b. The SA model greatly underpredicts the shear layer reattachment location when compared with the experimental data. In general, the peak reverse velocity is overpredicted by the models with compressibility corrections, which helps explain the increased variation in pressure along the base. Streamlines flowing along the centerline toward the base stagnate on the center of the base, leading to the high pressure seen there. The large reduction in turbulent eddy viscosity has the effect of increasing the recirculation region size, which makes the turning angle at the base more realistic, but allows a larger reverse velocity, which leads to a larger variation in pressure. The SST model starts with much lower turbulent viscosity than SA, which allows for the larger recirculation region, as seen in Fig. 8.25b. The compressibility correction further reduces the levels of eddy viscosity, increasing the size of the recirculation region further and increasing the peak reverse velocity. These results show that, while URANS predictions are possible, and often even yield quite reasonable results, they do not always accurately model the flow well and can lead to results with physically dubious meaning.

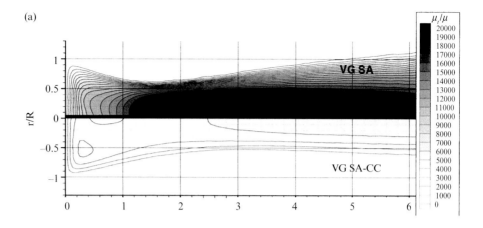

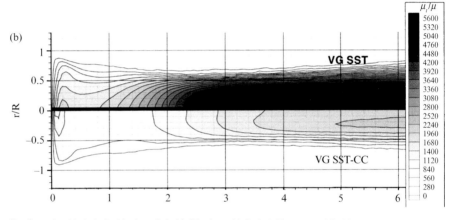

Figure 8.26 Nondimensional turbulent eddy viscosity behind the base. (a) Spalart-Allmaras model with and without compressibility corrections on *VGRIDns* grid (VG SA and VG SA-CC, respectively); (b) shear stress transport model with and without compressibility corrections on VGRIDns grid (VG SST and VG SST-CC, respectively) (Ref. 55; a full color version of this image is available on the website: www.cambridge.org/aerodynamics).

The nondimensional eddy viscosity for the SA, SA-CC, SST, and SST-CC turbulence models is shown in Fig. 8.26 (all shown using the fine grid). The important lesson here is how different the results are for each of the models, and also how much difference the compressibility correction makes in the results. For both models, the compressibility correction significantly reduces the eddy viscosity, which helps to explain some of the differences seen in Fig. 8.25. Also, notice that none of the models accurately captures the pressure distribution on the base of the missile, leading us to believe that these models do not work well in this highly unsteady, separated flow region.

8.8.5 FLOMANIA Project Results

A European Union-sponsored program that compared the abilities of various turbulence models for URANS predictions was conducted from 2002 through 2004. This project, called FLOMANIA (for FLOw Physics

Modelling – An Integrated Approach), had seventeen different partners who used a total of forty-seven turbulence models (or model variations) on a total of twenty test cases of varying complexity.[56] The test cases included six main application categories, including airfoils, wings/delta wings, wing-bodies, bluff bodies, transonic compressor rotors, and internal flows. Some of the fluid dynamic complexities that were simulated included shock/boundary-layer interactions, shock/shock interactions, pressure and shock-induced separation, high angle of attack flow, vortical/vortex flows, vortex-boundary-layer interactions, jets, mixing layers, control devices (such as vortex genera-tors), and Reynolds number scaling. As an example of the capabilities being assessed, Fig. 8.27 shows a comparison of the lift coefficient on a pitching airfoil using four turbulence models (including SA and BB). Notice that the older algebraic model (Baldwin Barth) did not do as well as the newer one- or two-equation models in predicting the flow.

To summarize their results with the various turbulence models used, the participants rated each model for the various flow types using the following performance descriptors: recommended, applicable, not recommended. The ratings were then tallied for each flow type to determine an overall rating of how the turbulence model performed for the flow type. The participants warned against blindly following their recommendations, however, by stating that "the potential user must compare their own applications closely with the FLOMANIA application to determine the extent to which this information can help in choosing an appropriate model." In other words, you should not just blindly follow their recommendations but, rather, intelligently and dili-gently determine how well any model may work for a type of flow prediction. For basic airfoil applications, most of the models were rated as "applicable," while the SST model was rated as "recommended." Wing and delta wing problems saw "applicable" ratings for the SA, k-ω, and SST, with only full Reynolds stress models receiving a "recommended" rating.[57]

Flomania participants also began using hybrid turbulence models (such as DES) in some of their predictions, which resulted in some very good results when compared with URANS predictions. In fact, they stated that DES "was found to deliver clear improvements in predictive accuracy for flows dominated by massive flow separation (such as bluff bodies) com-pared to unsteady RANS." We will discuss DES in more detail later in the chapter.

8.8.6 AIAA Drag Prediction Workshop Results

Five Drag Prediction Workshops (DPW) have now been conducted (or cur-rently are being conducted) by AIAA. The purpose of these workshops is to assess the ability of CFD to predict drag on realistic aircraft configurations. This requires the use of good turbulence models and good grids (as we dis-cussed in Chapter 7), both of which are crucial to being able to accurately

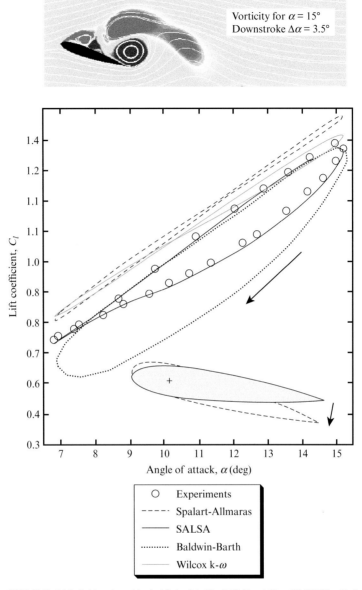

Figure 8.27 NACA 0015 airfoil pitching at $\alpha = 11 +/-4.2$, $k = 0.1$, $M = 0.29$, $Re = 1.95 \times 10^6$ (SALSA = Strain Adaptive Formulation of Spalart-Allmaras One-Equation Model) (Reprinted from Ref. 57, © 2002, with permission from Springer)

predict drag. Remember from Chapter 4 that the drag of an airplane is made up of four components: skin friction drag (which requires predicting the boundary-layer profile correctly), pressure drag (which requires predicting flow separation correctly), vortex drag (which requires predicting the shed vorticity of the aircraft correctly), and wave drag (which requires predicting the shock wave strength and position correctly). This presents quite a challenge to most flow solvers, turbulence models, and grid generation approaches.

Some of the five workshops "assigned" a configuration to the participants that had high-quality experimental data available. The workshop participants

Table 8.4 AIAA Drag Prediction Workshop III Submissions (Ref. 58)

Tag	Code	Grid Type	Turbulence Model	Submitter
A	PAB3D	Multiblock	Girimaji EASM	ASM
B	PAB3D	Multiblock	k-ε	ASM
C	PAB3D	Multiblock	SZL EASM	ASM
D	STAR-CCM+	Multiblock	Wilcox k-ω	QinetiQ
E	UPACS	Multiblock	Modified SA	JAXA
F	UPACS	Multiblock	Modified SA	JAXA
G	CFL3D-Thin	Multiblock	SA	Boeing
H	CFL3D-Thin	Multiblock	SST	Boeing
I	CFL3D-Full	Multiblock	SA	Boeing
J	CFL3D-Full	Multiblock	SST	Boeing
K	CFL3D-Full	Overset	SST	LaRC
L	CFL3D-Full	Overset	SST	LaRC
M	OVERFLOW	Overset	SA	Boeing
N	TAU	Hybrid	Edwards SA	DLR
O	EDGE	Hybrid	Hellsten EARSM	FOI
P	FUN3D	Unstructured	SA	LaRC
Q	NSU3D	Hybrid	SA	U. Wy.
R	CFD++	Hybrid	SA	Boeing
S	BCFD	Hybrid	SA	Boeing
T	UPACS	Multiblock	SST	JAXA
U	OVERFLOW	Overset	SST	LaRC
V	FLUENT	Unstructured	k-ε	Fluent
W	STAR-CCM+	Unstructured	SST	CD-Adapco
X	USM3D	Unstructured	SA/WF	Raytheon
Y	BCFD	Hybrid	SST	Boeing
Z	TAS	Unstructured	Modified SA	JAXA

were then asked to simulate the flow around the configuration, using their best practices, and report their results to the workshop without knowing what the experimental data looked like (which is called a "blind" test). The Third DPW used as one of its test cases the DLR F6 wing-body geometry, which is a representative commercial transport configuration.[58] The configuration was tested in the NASA National Transonic Facility (NTF) at $M = 0.75$ and $Re = 5 \times 10^6$. The participants were asked to perform a grid sensitivity study at a lift coefficient of $C_L = 0.5$, and then to perform an angle of attack sweep from $-3°$ to $+1.5°$. A list of participating organizations, shown in Table 8.4, includes a wide variety of experienced CFD users from industry, government, and university organizations.

When the wind tunnel test was conducted, a separation region was detected at the side of the body near the trailing edge of the wing, as shown with surface

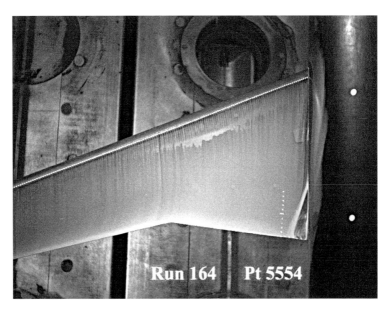

Figure 8.28   Side-of-body separation bubble on the baseline DLR F6 in the National Transonic Facility (Ref. 58; reprinted by permission of the American Institute of Aeronautics and Astronautics, Inc.).

oil flow visualization in Fig. 8.28 (compare this with Fig. 8.2, where there is no flow separation). The presence of this separated flow region added difficulty to the predictions and challenged the participants to use all of their experience to obtain a good prediction. A great deal of variation in the size and location of the separated flow region was seen by the participants,[59] leading to very different drag predictions (even when grid convergence issues were taken into account). A fairing was also added to the configuration in the experiment (designated FX2B); there was no separation of flow for the configuration with the fairing. The participants predicted the drag of both configurations (with and without the fairing) and compared their results with the wind tunnel data.

The results for the CFD drag predictions for the various participants are shown in Fig. 8.29. Notice that the vertical axis of the bar chart is the drag coefficient on a greatly expanded scale, which was necessary to determine the accuracy of each prediction. The drag coefficients were extrapolated from the grid sensitivity studies using Richardson extrapolation[60] to approximate grid-independent results. The wind tunnel experimental data values from the NTF are also included in the bar chart. Notice that most of the predictions were within 10 drag counts (which is less than 5%; recall from Chapter 4 that a drag count is $C_D = 0.0001$) of the experimental data, and some of the predictions were extremely close to the data. The workshop participants concluded that "there is a set of CFD codes whose members all seem to agree relatively well with each other, and do so over all of the test cases spanning the DPW Series. Most noteworthy about this core set of codes is that it is comprised of flow solvers that are based on all types of grids."[59]; In other words, good CFD practitioners using best practices can obtain good results

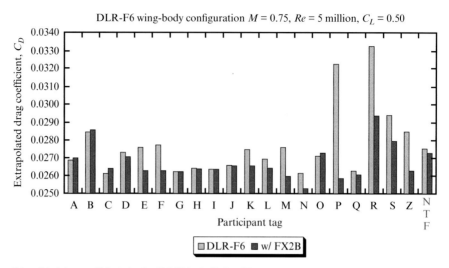

Figure 8.29 Extrapolated drag coefficients for the DLR F6 in the National Transonic Facility with and without the FX2B fairing (Ref. 59; reprinted by permission of the American Institute of Aeronautics and Astronautics, Inc.).

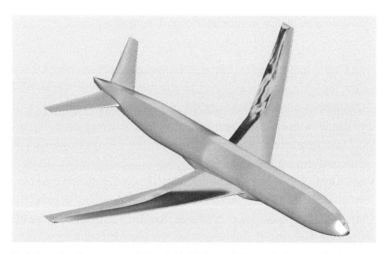

Figure 8.30 Predictions of surface pressures (left) and skin friction coefficient (right) for the NASA CRM at $M = 0.87$, $\alpha = 4.0°$, and $Re_c = 5 \times 10^6$ (Courtesy of Dimitri Mavriplis, University of Wyoming; a full color version of this image is available on the website: www.cambridge.org/aerodynamics).

using any type of grid, whether it is structured or unstructured, multiblock or overset.

A Fourth and Fifth DPW are using a body-wing-horizontal tail geometry called the NASA Common Research Model (CRM). These workshops are examining the ability of the various codes to predict downwash on the horizontal tail, Reynolds number effects on the configuration, buffet on the wing, and the impact of turbulence models on these predictions. Some predictions for a high subsonic Mach number case ($M = 0.87$) using the SA turbulence model are shown in Fig. 8.30. Details of the predictions and comparisons with wind tunnel data have been presented at recent AIAA meetings.

Profiles in Computational Aerodynamics: Christopher Rumsey

"My interest in fluid dynamics began only after starting my undergraduate degree in aeronautical engineering at Rensselaer Polytechnic Institute (RPI). Although I was fascinated with all aspects of flight, I was particularly enchanted by the fact that the science of fluid dynamics was so broad, and that air itself was a "fluid." I worked as a co-op very briefly at Sikorsky in the early 1980s. There, I was introduced to the challenge of using computers to predict the behavior of rotor blades. My master's thesis at RPI was on non-steady aerodynamic characteristics of helicopter rotors in forward flight. My chief memory about this research was the fact that I spent nearly my entire computer budget on a series of computer runs that had a bug in the code. This bug made most of the results obsolete, and I had to beg for more computer hours. I have had a love-hate relationship with computers ever since.

"In 1983 I joined NASA Langley Research Center as an engineer in a newly formed computational fluid dynamics group. At the time, CFD was a very dynamic, expanding field, so much of my effort was spent with on-the-job training. At first I was primarily involved with helping to develop and test the code CFL3D, which, despite its age, is still around and useful today. Many of my early studies involved unsteady flows, such as pitching airfoils.

"In the late 1980s to early 1990s I took advantage of an opportunity to complete my PhD at the University of Michigan with Bram van Leer as my advisor. The subject was research into a grid-independent approximate Riemann solver. Although the particular method I employed never became a practical tool, it was a fascinating foray into theoretical and creative computational methodologies.

"Since that time, I have been involved in many different aspects of CFD, most notably turbulence modeling. The challenge of turbulence (impossible to solve) is simply too big to resist. I have had the privilege to work with several very talented turbulence modelers, and I have helped to implement their ideas into CFD codes. For example, we spent a lot of time working to develop advanced models beyond simple one- and two-equation models that rely on the Boussinesq assumption. It is a difficult task, because more complex models are often less robust, so their benefits in terms of accuracy in some instances are often outweighed by their disadvantages.

"Of course, turbulence modeling plays a part in most aerodynamic flows of interest to NASA and the aircraft industry. My career has been dedicated to computing and validating against many different types of flows, such as high-lift airfoils and wings, buffet onset, and flow control applications. I have been involved in many workshops and other efforts to use CFD collectively to (1) learn what methods and models work best, (2) achieve consistent results, and (3) quantify the error and uncertainty. I have also been a dedicated member of a long-standing team to bring a universal CFD storage and interchange capability called CFD General Notation System (CGNS) to fruition. One of the goals of CGNS is to make it easier for different CFD groups to work together by providing a common file format that is adaptable and will stand the test of time as well.

"I am continuing to work with turbulence models today, with a primary focus in areas such as separated flows, where they often perform poorly. Although computer hardware advances are making simulations that directly resolve many of the larger turbulence scales more affordable, the need for turbulence modeling improvements – both for Reynolds averaging and for sub-grid scales – will no doubt exist for a long time."

Chris Rumsey received his BS and MS degrees from Rennselaer Polytechnic Institute and his PhD from the University of Michigan. He works as a CFD Research Scientist at NASA Langley Research Center, where he is a specialist in numerical algorithms and code development for fluid dynamics, turbulence modeling development and implementation, and complex fluid and aerodynamic applications. He is an AIAA Fellow and the recipient of the NASA Superior Accomplishment Award for his outstanding systematic assessment of state-of-the-art CFD uncertainties for prediction of buffet onset for a transport aircraft. He was also deeply involved in the development of CGNS, the CFD General Notation System.

8.9 Large Eddy Simulation

Large Eddy Simulation (LES) can be seen as a compromise between DNS (described more fully in Section 8.11) and RANS. The basis of LES is that the flow is decomposed into small- and large length scales (or eddy sizes); the small scales are modeled, while the large scales (large eddies) are computed (or resolved) numerically. See Fig. 8.31 for a comparison of the energy spectrum modeling comparisons between DNS and LES approaches, which

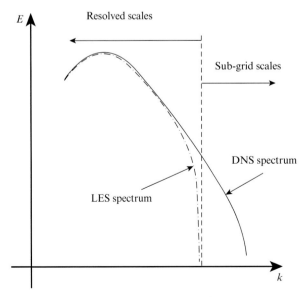

Figure 8.31 Energy spectrum – LES vs. DNS (Courtesy of James Forsythe).

results in the cost of computation for LES being greater than for RANS but less than for DNS. Since the small scales are modeled, the Reynolds number restriction of DNS is eased, so LES can be used to solve engineering flows that DNS cannot. LES can also be used to provide detailed flow information for the creation of RANS or other models in situations where DNS is too expensive (for example high Reynolds number flows, complex geometries, etc.).

The theory behind large eddy simulation is that the large scales of motion should be responsible for most of the energy and momentum transport, but they are highly dependent on the flow being considered. Small scales, on the other hand, are more homogeneous and universal and should be easier to model, but they are mostly responsible for energy dissipation and need to be taken into account. Breuer provides a brief outline of the history of LES,[61] including the first three-dimensional LES simulations performed by Smagorinsky in 1963 for meteorology (weather forecast), and the first engineering application of LES by Deardorff in 1970 for a turbulent channel flow. LES was almost forced out of favor in the 1980s by DNS, but has regained favor recently. In the last decade, significant progress has been made on LES in the engineering field. Some examples are the application of LES to compressible flows, the development of the dynamic model, and the use of LES in modeling chemically reacting flows. There is still much work to be accomplished, however, before LES will be fully mature.

8.9.1 Spatial Filtering

The backbone of LES is spatial filtering, which is how the small-scale fluctuations are removed from the flow. If f is a quantity of interest (velocity, density, etc.), the spatial filter of this quantity (denoted by a bar) is given by:

$$\bar{f}(\vec{x}) = \int G(\vec{x} - \vec{x}')f(x')dx'. \tag{8.38}$$

The function G is the filter function. It has a characteristic length scale of Δ over which its value drops to zero. It should be constructed so that mean values are preserved. An example is the one-dimensional Top Hat filter:

$$G(x) = \begin{cases} \frac{1}{\Delta} & ; \quad |x| < \frac{\Delta}{2} \\ 0 & ; \quad \text{otherwise} \end{cases} \tag{8.39}$$

There are several filter functions that can be used. Two others are the Gaussian filter and the sharp Fourier cutoff filter. The type of filter used is generally implied by the numerical scheme applied. For example, in finite differences, discretizing the quantities onto a grid implies that you have averaged over $\Delta/2$ (where Δ is the grid spacing and the length scale of the filter), implying the use of the Top Hat filter. For spectral methods, the Fourier cutoff filter is implied. This filter simply removes all Fourier frequencies with wavelengths shorter than Δ. The length scale of the filter is generally taken to be the same as the grid spacing. This does not need to be the case, although the length scale cannot be smaller than the grid spacing.

The LES equations are obtained by applying the spatial filter to the Navier-Stokes equations. This process returns the original equations in terms of the filtered variables, with an additional term:

$$\tau_{ij}^S = -\rho\left(\overline{u_i u_j} - \overline{u}_i \overline{u}_j\right). \tag{8.40}$$

This term is known as the subgrid scale (SGS) Reynolds stress, and it accounts for the length scales smaller than the cutoff length scale. Since it involves correlations between unfiltered velocities, it is unknown and must be modeled.

8.9.2 Subgrid Scale Models

The ability of LES to make accurate predictions of the flow is highly dependent on the model for the small scales of turbulence. There are many models available, including the Smagorinsky model, two-point closure models, Renormalization Group theory models, dynamic models, scale-similar models, and one-equation models. The Smagorinsky model is of great historical significance (it was applied as early as 1963) and is the most commonly used model. The dynamic model is more contemporary (1990) and shows promise for the future. Piomelli gives a good overview of many of these models, which will be of interest to the reader looking for more details.[62] The Smagorinsky model is based on the eddy viscosity concept. This concept relates the subgrid-scale stresses to the large-scale (resolved) stresses by a relation that is very similar to RANS eddy viscosity models. The difference is that the small-scale turbulent stresses (unresolved) are assumed to be proportional to the large-scale stresses (resolved). With RANS, the turbulent stresses (stresses from the fluctuating component of the velocity) are assumed to be proportional to the stresses from the time-averaged velocities.

For a description of some of these approaches, and to determine the constants of the model, refer to Piomelli.[62] The Smagorinsky model unadjusted does not accurately predict flows in the near wall region. The presence of the viscous sublayer, where the effects of turbulence decrease, is not accurately modeled. One means of adjusting the model is to reduce the model constant in the near wall region by using van Driest damping[63] (as in RANS).

The Smagorinsky model has provided good results for a number of flows but has several drawbacks. One is that the model constant may be different for different flows (e.g., shear flows), and may not be able to be determined a priori. The second disadvantage is the need to reduce the model constant in the near wall region. In complex geometries, the van Driest damping may not provide good results. Also, in the near wall region, stretched grids are often used. In this case, it may be difficult to define Δ. Third, since the model constant is fixed, the subgrid stresses do not vanish in laminar flow, as they should. The final problem is that the model is purely dissipative. The subgrid stresses always decrease the energy of the flow. Although this is generally true on the average in a given flow, there are often regions of backscatter where energy is transferred from the small scales to the large scales. This backscatter can be important to the dynamics of the flow.

8.9.3 Example LES Applications

The application problem is the challenge of computing the flowfield for a V/STOL airplane at touchdown. The jet exhaust impinging on the ground leads to a complicated flowfield, where the interaction of the flowfield on the airplane can have a profound, and quite possibly adverse, effect on the safety of the landing. This occurs because, if any of the hot exhaust gases are re-ingested, the engine will lose thrust and possibly also surge. Figure 8.32, which illustrates the complexity of the flowfield, is from an investigation of a model problem[64] before using the method to investigate a realistic case.[65] These types of flowfields are dominated by viscous effects and are well beyond the situations that the RANS equations are capable of accurately computing.

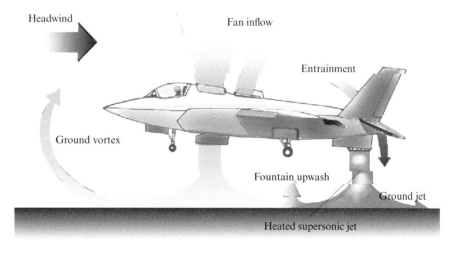

Figure 8.32   Vertical landing aircraft flowfield (Ref. 64; Courtesy of Gary J. Page of Loughborough University).

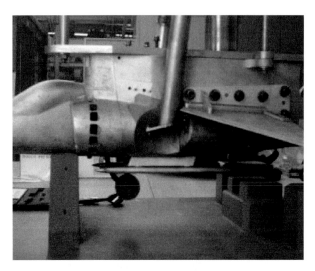

Figure 8.33   Harrier geometry used in LES simulation (Ref. 65; Courtesy of Gary J. Page of Loughborough University).

The geometry for the calculation of interest is shown in Fig. 8.33. This is a 1/15 scale AV-8B Harrier, and the lower surface geometry is accurately modeled, while the geometry is modified at the top of the plane to provide the air supply and to be able to extract the air from the inlets. The rear jet is hot (700 K), and the Reynolds number of the jets is about 8 million, based on the jet exit conditions. Finally, there is a crossflow of 6m/sec. This is the amount of detail required to simulate the flowfield accurately.

The turbulence structures found in the calculation are shown in Fig. 8.34. These are the cores of vortical structures, shaded by temperature. The results are found after 100,000 time steps, where the time-step value was 5×10^{-8} seconds. The calculation was done on an HPCx IBM Power5 system with 128 processors and represents the equivalent of 43,000 CPU hours. However, although the key flow structures were identified, the authors state that insufficient samples were generated to obtain a statistically valid mean.

Another example of LES applied to aerodynamics flowfields is shown in Fig. 8.35 for a wing-tip vortex simulation.[66] The main computational grid consists of $1536 \times 128 \times 128$ points, with the near-field wake simulated with 25 million grid points. The simulation modeled wind tunnel test conditions of $Re = 4.6 \times 10^6$, $M = 0.15$, and $\alpha = 10$ deg. The nondimensional time step based on the freestream velocity is approximately $9 \times 10^{-5} \, c/V_{\infty}$, where c is the chord length and V_{∞} is the freestream velocity. The time-dependent vorticity contours behind the wing tip where the wing-tip vortex convects is clear and easily identified, and the LES predictions agree qualitatively with the experimental data.

Figure 8.34   Turbulent structures found from an LES calculation (Ref. 65; Courtesy of Gary J. Page of Loughborough University; a full color version of this image is available on the website: www.cambridge.org/aerodynamics).

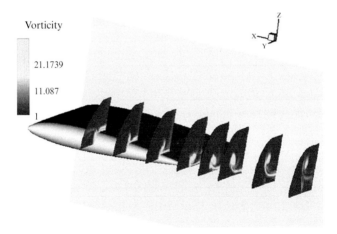

Figure 8.35   Time-averaged vorticity contours showing the unsteady roll-up of the wing-tip vortex (Ref. 66; a full color version of this image is available on the website: www.cambridge.org/aerodynamics).

8.10 Hybrid Approach (RANS/LES)

While there are a number of hybrid turbulence models in use today, the vast majority of them are variations of the Detached-Eddy Simulation approach developed by Philippe Spalart and his associates. The next section will concentrate on the DES models, but interested readers can easily find results for a variety of hybrid approaches in the literature.

8.10.1 Detached-Eddy Simulation

Spalart et al. estimated the cost in terms of grid size for DNS and LES computations of a full-scale aircraft at flight Reynolds numbers.[31] Their conclusion, assuming that Moore's law continues to hold (see Table 8.1), was that DNS computations would not be feasible for full aircraft at flight Reynolds numbers until approximately the year 2080, and that LES computations would not be possible until 2045. This estimate motivated the formulation of *Detached-Eddy Simulation*, which combines the advantages of LES and RANS into one model that is usable with today's computers.[31] The model uses RANS in the boundary layer, where RANS performs well with much lower grid requirements than LES. LES is then used in separated flow regions, where it can resolve large-scale turbulence motion well. Shur et al. calibrated the model for isotropic turbulence and applied it to an NACA 0012 airfoil section.[67] The model agreed well with drag predictions all the way to 90 degrees angle of attack. Constantinescu and Squires applied DES to the turbulent flow over a sphere at several Reynolds numbers.[68] Issues of grid resolution, numerical accuracy, and values of the model constant were examined, and the model was compared to LES and RANS predictions. These successes with the hybrid turbulence model approach led Salas to state that "DES and other RANS-LES hybrids appear to have the greatest promise at this time for unsteady separated flows."[69]

Specifically, the DES model was originally based on the Spalart-Allmaras (SA) one equation RANS turbulence model[38] detailed in Appendix F. The wall destruction term is proportional to $\left(\tilde{v}/d\right)^{2}$, where d is the distance to the closest wall. When this term is balanced with the production term, the eddy viscosity becomes $\tilde{v} \propto S d^{2}$, where S is the local strain rate. The Smagorinsky LES model varies its subgrid scale (SGS) eddy viscosity with the local strain rate and the grid spacing: Δ, i.e., $v_{t} \propto S\Delta^{2}$. If, therefore, d is replaced by Δ in the wall destruction term, the SA model will act as a Smagorinsky LES model.

To exhibit both RANS and LES behavior, d in the SA model is replaced by:

$$\tilde{d} = \min\left(d, C_{DES}\Delta\right). \tag{8.41}$$

When $d \ll \Delta$, the model acts as a RANS model. When $d \gg \Delta$, the model acts as a Smagorinsky LES model. The thickness of the RANS region is therefore determined by the grid spacing and the value of C_{DES}. The model can be "switched" to the LES mode by locally refining the grid. In an attached boundary layer, a RANS simulation will have highly stretched grids in the streamwise direction. To retain RANS behavior in this case, Δ is taken as

(a) SST   (b) SA   (c) SA-DES

Figure 8.36 RANS and DES simulations of flow over an F/A-18C at α = 30 deg, Rec = 13 × 10^6, leading-edge flaps set to –33 deg, trailing-edge flaps set to 0 deg, with no diverter slot present (Ref. 70; a full color version of this image is available in the color insert pages of this text as well as on the website: www.cambridge.org/aerodynamics).

the largest spacing of any direction ($\Delta = \max\left(\Delta x, \Delta y, \Delta z\right)$). The role of the RANS region is to accurately capture boundary-layer characteristics, such as skin friction and boundary-layer thickness, and to predict the separation location. At the separation point, a wealth of scales is allowed to grow rapidly by the model's LES characteristics outside the boundary layer.

Figure 8.36 shows comparisons between two RANS models (SST and SA) and DES for the F/A-18C at high angles of attack.[70] The flowfield is unsteady and would best be viewed in a movie, but Fig. 8.36 depicts snapshots of solutions for each method with the surface shaded by pressure and an iso-surface of vorticity shown. The chosen vorticity level for the isosurface and the pressure color map are held fixed. Although the snapshots are not necessarily synchronized in time, the overall differences are striking. The SA-DES solution (Fig. 8.36c) produces a much more detailed view of the simulation because it is able to capture much finer flowfield scales. The SST (Fig. 8.36a) and SA (Fig. 8.36b) models are unable to capture the proper post–breakdown behavior or the leading-edge separation regions of the wing, horizontal and vertical tails. It is also apparent that the SST leading-edge extension vortex pressure footprint on the surface is significantly different from either the SA or SA-DES solutions. The low-pressure region represented by a dark shade is greatly reduced in size on the SST solution. The SA-DES solution is also capturing the vortical substructures around the primary vortex.

Since DES utilizes LES in separated flow regions, the model can produce very good unsteady (and three-dimensional) flow results. When the F-18E/F started experiencing abrupt wing stall, a large computational and experimental investigation was initiated to understand this dynamic phenomenon.[71] The vortex produced by the snag (leading-edge discontinuity) interacts with the normal shock above the wing to create stall on one wing, which rolls the aircraft at rapid rates.[72] Figure 8.37 shows the prediction of abrupt wing stall on the F-18E/F, which occurs at transonic Mach numbers. The figure shows isosurfaces of vorticity, shaded by pressure, and is an indicator of the separation regions. Flow visualizations are provided for this calculation

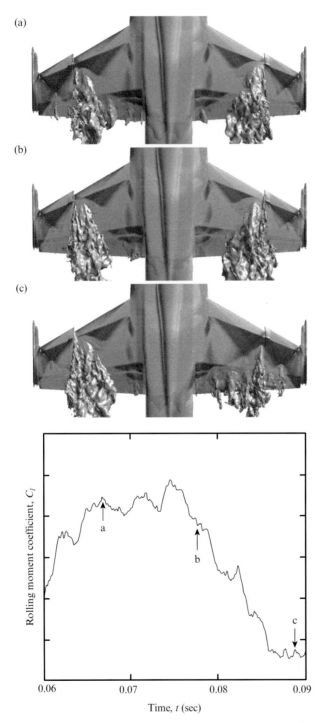

Figure 8.37   Roll moment vs. time and flow visualizations at specific times, whole aircraft without tails; flow visualizations are isosurfaces of vorticity colored by pressure (Ref. 72; a full color version of this image is available on the website: www.cambridge.org/aerodynamics).

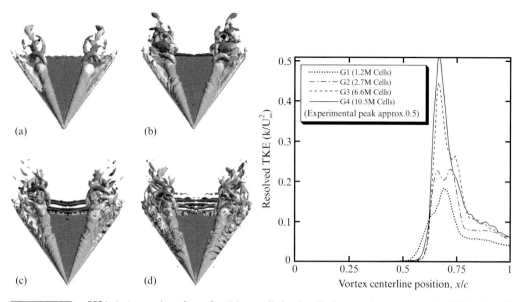

Figure 8.38 DES instantaneous isosurfaces of vorticity magnitude colored by the spanwise component of vorticity for four grids: (a) G1 (1.2×10^6 cells), (b) G2 (2.7×10^6 cells), (c) G3 (6.7×10^6 cells), and (d) G4 (10.5×10^6 cells). The flowfield conditions are $\alpha = 27$ deg, $M = 0.069$, and $Re_c = 1.56 \times 10^6$ (Ref. 73; a full color version of this image is available on the website: www.cambridge.org/aerodynamics).

with a corresponding region of the roll-moment plot. These isosurfaces of zero streamwise velocity are an indicator of the separated region. The shock on the left side starts farther back in Fig. 8.37a, giving a large, positive, roll moment. As this shock moves forward, the roll moment moves toward zero in Fig. 8.37b. Then, the right shock moves aft in Fig. 8.37c, giving a large, negative roll moment.

Another example that demonstrates the power of DES to resolve separated flow regions can be seen with the flow over a delta wing, shown in Fig. 8.38. Morton[73] used increasingly finer grids (labeled G1 through G4, which ranged from 1.2 million cells to 10.5 million cells) and showed that DES could resolve the turbulent kinetic energy in the vortex core to levels that match experimental data.

8.10.2 Delayed Detached-Eddy Simulation

One of the problems with the DES model occurs with highly refined grids, which activates the LES model in the boundary layer, since the switch between the RANS turbulence model and LES is controlled by the grid spacing. While correct grid spacing could solve this problem for many flows,[74] in cases where the separation is quite shallow (such as a separation bubble over an airfoil), the model switches in a nonphysical manner. An example of this is shown in Fig. 8.39, where a RANS model and DES are used to predict flow separation on an airfoil. Notice that the DES model predicts separation

(a)   (b)

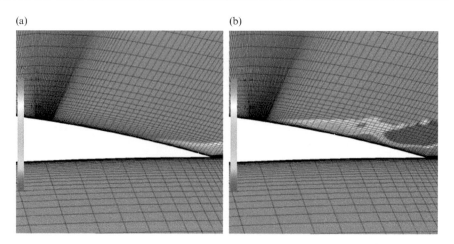

Figure 8.39 Vorticity contours over an airfoil: (a) Reynolds-Averaged Navier-Stokes; and (b) Detached-Eddy Simulation with grid-induced flow separation. (Reprinted from Ref. 75, © 2002, with permission from Springer).

occurring in the region of high grid density, which is called grid-induced separation (GIS).[75]

Delayed Detached-Eddy Simulation (DDES)[32] was developed to correct this problem exhibited by DES. DDES has a switch based on the location of the outer edge of the boundary layer, so that RANS is always used within the boundary layer and LES is used outside of the boundary layer. Early results from DDES have shown that the model works very well for flows that are massively or shallowly separated.

For example, DDES was applied to the high angle of attack aerodynamics of the Modular Transonic Vortex Interaction (MTVI) configuration, which exhibits a sharp nonlinearity in roll moment at small sideslip angles.[76] Figure 8.40 captures the time-accurate vortex breakdown when the sideslip angle is instantaneously increased from $\beta = 1$ to 2 deg. Plotted in Fig. 8.40a is the time history of the roll-moment response. Plotted in Figures 40b–40f are isosurfaces of Q (see Section 9.5.2.4) contoured by axial velocity and surface pressure coefficient at $t = 0.03, 0.17, 0.18, 0.19$, and 0.30 s. From $t = 0.03$ to 0.17 s, there is a significant negative increase in rolling moment coefficient from –0.012 to –0.028. Inspection of Figures 40b and 40c indicates that this increase is due to an increased suction on the upper surface of the windward wing and a decrease in suction on the upper surface of the leeward wing. This change in surface pressure distribution is due to an increased interaction of the windward vortex system and a small upward and outboard movement of the leeward vortex system. From $t = 0.17$ to 0.30 s, there is a rapid increase in the roll-moment coefficient to its steady value of 0.018. DDES is able to capture the dynamic vortex breakdown of the windward vortex system during this time period in a way that matches the available wind tunnel data very well. Comparisons to RANS models for the same flow conditions

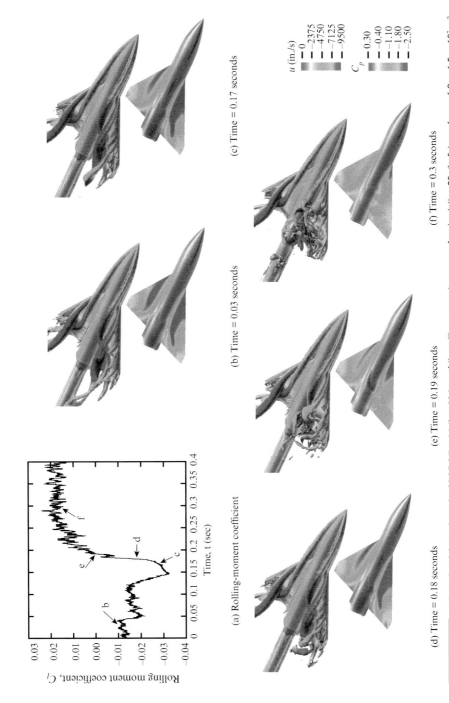

(a) Rolling-moment coefficient

(b) Time = 0.03 seconds

(c) Time = 0.17 seconds

(d) Time = 0.18 seconds

(e) Time = 0.19 seconds

(f) Time = 0.3 seconds

Figure 8.40 Vortex breakdown dynamics highlighting (a) time history of the rolling-moment response for simulation 23; (b–f) isosurfaces of $Q = 1.5 \times 10^7 \ s^{-2}$ contoured by axial velocity and surface pressure coefficients at various times; $\alpha = 30$ deg, $\beta = 2$ deg, $Re_c = 2.68 \times 10^6$, and $M = 0.40$. (Ref. 76; a full color version of this image is available on the website: www.cambridge.org/aerodynamics)..

showed that the RANS model could predict the non-linear trend reasonably well but it was not able to predict the sideslip angles where the nonlinearity occurred, and could not predict the magnitudes of the rolling moment as well as the DDES model.

8.10.3 Improved Delayed Detached-Eddy Simulation

Improved Delayed Detached-Eddy Simulation (IDDES) was proposed by Shur et al.[77] to overcome some additional observed weaknesses of DES. DES does not work as well when it is used as a wall-layer model for LES (i.e., LES is used inside the boundary layer where the boundary layer-sized eddies are resolved, while the near wall smaller scale turbulence is modeled). This was not the original intent of DES, but having it be used as a wall-layer model extends its applicability. Because, in this case, two models are being used within the boundary layer, they overlap in the logarithmic layer (or fully turbulent region) of the boundary layer (see Fig. 4.12). Unfortunately, the two models give different values in the logarithmic-layer and therefore do not match each other.

IDDES resolves these problems by using both local grid characteristics and a wall-length scale in the model that reduces or eliminates the mismatch, making the model suitable for use as a wall-modeled LES (WMLES) model. Again, if the LES mode is activated in a simulation with insufficient upstream resolved turbulence content, then there needs to be a way to switch off the WMLES, and retain the RANS behavior in the boundary layer, which is done in IDDES through a blending function that makes it behave like DDES.[78] IDDES uses empirical functions, "which address log-layer mismatch and the bridge between wall-resolved and wall-modeled DES."[79] All these modifications are only necessary when grid resolution becomes fine enough in the boundary layer to start resolving the large scales, which was not the original intent of DES but is important to extend the range of applicability as computer speeds increase.

IDDES is currently being evaluated for a variety of flows that were a challenge for DES, such as flow over a backward-facing step, where there are a number of wall distances and regions of attached and separated flow. A comparison with experimental data is shown in Fig. 8.41 for a backward-facing step with a Reynolds number of 28,000 (based on step height) and a channel expansion ratio of 5/4. The figure shows a comparison of the mean skin friction coefficients on the lower wall of the channel downstream of the step. "In the vicinity of this wall IDDES operates in WMLES mode, and provides a better resolution of the fine turbulent structures, which is especially important for the region of recovery of the reattached boundary layer known to be the most challenging for both RANS and DDES. This leads to a more accurate prediction of the mean flow characteristics by both SA- and

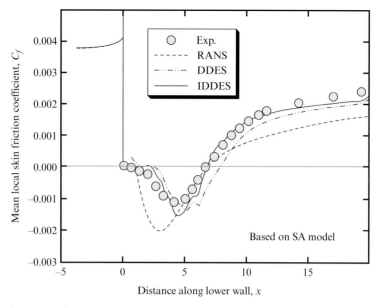

Figure 8.41   Comparison of mean friction coefficient distributions in backward-facing step flow predicted by RANS, DDES, and IDDES based on SA model and compared with experiment (Ref. 79).

MSST-based [Menter's SST model] versions of IDDES not only versus the corresponding background RANS models but, more importantly, versus the corresponding DDES versions as well."[79]

8.10.4 DESider Results

DESider was another European Union project that compared the ability of various hybrid turbulence models to predict unsteady, separated flow.[80] This project, which took place between 2004 and 2007, had eighteen primary participants and five observer organizations, including representatives from the aerospace industry, research labs, and universities. The project had as its stated objectives: strengthening competitiveness, improving environmental impact by reducing noise and emissions, improving aircraft safety and security, and increasing operational capacity and safety of air transport systems. This was accomplished via a comprehensive work plan (see Figure 8.42) that included experimental measurements to create an improved data base, improvements to modeling for URANS, DES, and LES approaches, and advancing the ability of CFD to predict aerodynamics and multidisciplinary applications, including aeroelasticity and aeroacoustics.

The experiments used to collect detailed turbulence data for basic flows included flow over a backward-facing 35 degree ramp and flow over a circular cylinder. Other experiments used to evaluate URANS, DES, and LES predictions included a delta wing, a NACA 0021 airfoil at 60 degrees angle of attack, an oscillating NACA 0012 airfoil, the Ahmed car body with 25- and

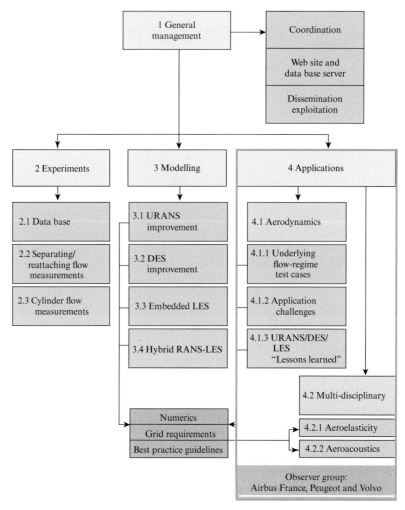

Figure 8.42   The DESider work plan overview (Reprinted from Ref. 80, © 2009, with permission from Springer).

35-degree slant angles, a multi-element airfoil including a slat and a flap, the Simpson 3D hill, a circular cylinder on a ground plane, flow in a fully developed channel, a bump in a square channel, a supersonic base flow, a transonic cavity flow, separated flow behind an airfoil trailing edge, a fuel cell, a simplified landing gear, the FA-5 aircraft, and the EC145 helicopter fuselage.

As an example of the results obtained during the DESider program, we will look in detail at the predictions for the FA-5 fighter type aircraft, which has a clipped delta wing and canards. A 1:15 scale model, shown in Fig. 8.43, was tested at the Technical University of Munich low-speed wind tunnel at $M_\infty = 0.15$, $\alpha = 15°$, and $Re = 2.8 \times 10^6$. Detailed flowfield data (including mean velocity, RMS velocity, Reynolds stresses, vorticity, and turbulent kinetic energy) was collected at twelve experimental crossflow planes ranging from 20 percent to 120 percent of the wing chord. Computations were performed using the following codes by the organizations shown: *CFX* (ANSYS), *TAU* (DLR), *FLOWer* and *TAU* (EADS), and *ELAN* (TU Berlin).

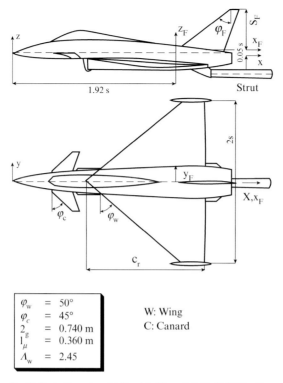

$$\varphi_w = 50°$$
$$\varphi_c = 45°$$
$$2_g = 0.740 \text{ m}$$
$$1_\mu = 0.360 \text{ m}$$
$$\Lambda_w = 2.45$$

W: Wing
C: Canard

Figure 8.43   The FA-5 wind tunnel model (Reprinted from Ref. 80, © 2009, with permission from Springer).

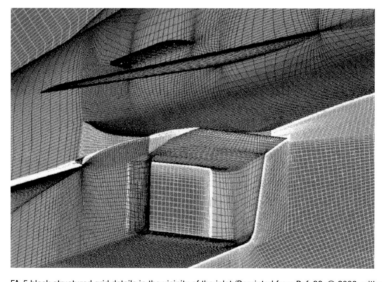

Figure 8.44   FA-5 block structured grid details in the vicinity of the inlet (Reprinted from Ref. 80, © 2009, with permission from Springer.

Both block-structured and -unstructured grids were used, with grid refinement and time-step studies being carried out for each. An example of the block-structured grid in the vicinity of the engine inlet is shown in Fig. 8.44, including details of the canard, a strake (which was not included on the wind

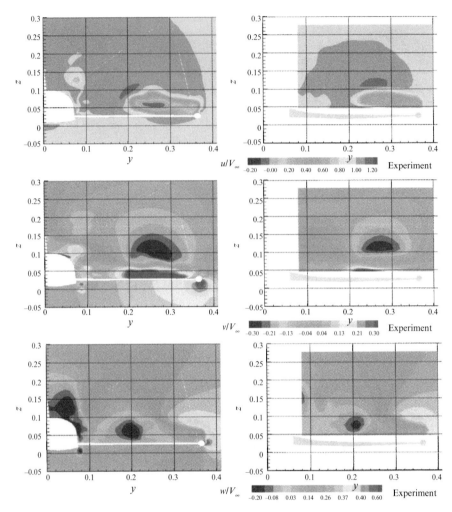

Figure 8.45   Comparison of EASM-DES mean velocities at $x/c = 0.90$ with experimental data for the FA-5 aircraft (CFD predictions are on the left and experimental data is on the right) (Reprinted from Ref. 80, © 2009, with permission from Springer; a full color version of this image is available on the website: www.cambridge.org/aerodynamics).

tunnel model), the boundary-layer diverter, and the inlet lip. The grid had 196 blocks with a total size of 11.3 million points.

The prediction of mean u, v, and w velocity components at $x/c = 0.90$ are shown in Fig. 8.45 for the EASM-DES model. These predictions compared very well with the experimental data, especially for the size of the region of reversed flow within the vortex. These results showed "that URANS is not able to reproduce the unsteadiness of this kind of flow correctly, regardless of the size of the time step."[80] The researchers for this case (from TU Berlin using the *ELAN* code) also found that the results were highly dependent on the RANS model used within the DES formulation. Similar results were also obtained for the RMS values of the velocity components.

The summary for the DESider program included a detailed list of drawbacks that hybrid models suffer from. These drawbacks include the following:[80]

1. RANS and LES Reynolds stresses are derived in very different ways, yet are often dealt with as though they were equivalent.
2. RANS models were developed and validated for steady flow but are now applied to many situations with highly unsteady flowfields and were certainly never envisioned to be used in conjunction with LES for only a portion of the flowfield, so great care must be taken when using RANS models for separated flowfields.
3. Selection of an adequate grid and an appropriate time step is very important for accurately predicting separated flow when using hybrid models, and the use of isotropic grid refinement in all directions within a focus region is essential to obtaining good results, even though the computational cost will be quite high.
4. The accurate prediction of mean flow quantities for vortical flows requires high grid resolution in the vortex core.
5. The blending approach from RANS to LES in hybrid models is crucial to their general usefulness, but the ability of the user to provide adequate grid resolution without knowing in advance where the blending region is located leads to a difficult challenge, usually leading to the need for automatic grid adaption methods.
6. Obtaining good results for airfoil cases requires an adequate spanwise domain and the need for very long time samples.

In other words, using advanced turbulence models requires a well-rounded knowledge of the model as well as other best practices (such as grid generation) for CFD.

8.11 Direct Numerical Simulation

Direct Numerical Simulation overcomes the shortcomings of RANS and LES by fully resolving all turbulent length scales. DNS solves the full Navier-Stokes equations without any averaging or models. To do this, DNS must use grids that resolve the smallest scales of turbulence (the Kolmogorov length and time scales) as well as the largest, as shown in Fig. 8.46.

Figure 8.46 shows a typical energy spectrum of a turbulent flow, where k represents the wave number and E is the energy contained at that wave number. The lower wave numbers (or larger length scales) contain most of the energy. In the high wave number range (small length scales) – the viscous range – molecular viscosity dominates (since length scales are small, the gradients are large and the viscous shear stress is large), dissipating the energy of the flow. Between these two ranges is the inertial sub-range, where energy is transferred from the large scales to the small. The largest length scales are highly dependent on the flow geometry and are represented by L. The

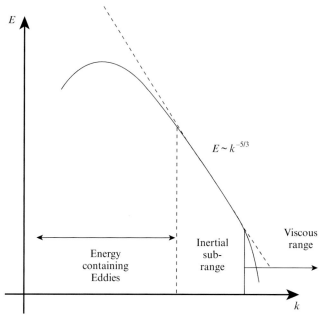

Figure 8.46 Typical energy spectrum for a turbulent flow on a log-log scale (Courtesy of James Forsythe).

smallest scales are known as the Kolmogorov microscales for length (η) and time (τ) and are given by $\eta = (\upsilon^3/\varepsilon)^{1/4}$ and $\tau = (\upsilon/\varepsilon)^{1/2}$ where υ is the kinematic viscosity and ε is the turbulence dissipation rate per unit mass.[26] Since these relationships define the spatial and temporal scales of turbulence, they set the requirements for grid spacing and time steps for DNS. For example, the grid spacing must be less than the turbulence length scale, and the time step must be less than the turbulence time scale. The ratio of these two scales, L/η, is proportional to $Re^{3/4}$, and determines the number of grid points in each direction. Thus, in three dimensions the number of grid points is this ratio cubed ($Re^{9/4}$). As the number of grid points increases, the time step taken must decrease. All combined, this leads to the cost (in terms of computation time) of DNS being roughly proportional to Re^3. In other words, a full-scale aircraft with a DNS grid of 10^7 grid points would require approximately 10^{23} time steps, a completely unreasonable calculation with current computer technology. A more detailed analysis can be found in references by Ferziger and Peric,[81] as well as Piomelli.[62] Although DNS makes up for all the disadvantages of RANS, it lacks the great advantage of RANS – speed. The high computation cost currently limits DNS to fairly simple low Reynolds number flows. This limit is not likely to disappear anytime soon, even though computer speeds are constantly increasing (see Fig. 2.22). An analysis of the future capabilities of DNS has been written by Nieuwstadt et al.[82]

The advantage of DNS is that it provides very detailed information about turbulence that is difficult or impossible to obtain in experiments. In fact, there are many instances where DNS and experiments disagree. After

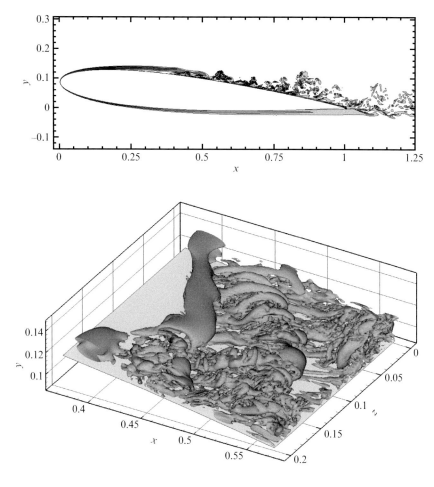

Figure 8.47   DNS simulation of flow over an NACA 0012 airfoil; $M = 0.4$, $\alpha = 5°$, $Re = 5 \times 10^4$ (Ref. 84; Courtesy of Neil Sandham of the University of Southampton; a full color version of this image is available on the website: www.cambridge.org/aerodynamics).

rechecking both methods, it has often been found that the experiments were in error, and, after correction, they agreed with DNS. This extensive amount of detail can be used to obtain insight into turbulence to develop and test models for RANS or other techniques.[83]

An example of a calculation using DNS is shown for the low Reynolds number flow over an NACA 0012 airfoil in Fig. 8.47.[84] Numerical simulations were performed on a NACA 0012 airfoil at an attack angle of 5 degrees, $Re = 5 \times 10^4$, and $M = 0.4$. The computational domain upstream boundary is approximately seven chord lengths away from the leading edge of the airfoil, the upper and lower boundaries are also 7 chord lengths from the solid surface, and the outflow boundary is 5 chord lengths downstream of the trailing edge. There are approximately 1,000 grid points on the surface of the airfoil and nearly 700 grid points in the normal direction from the surface to the outer boundary. The airfoil is regarded as infinite in the spanwise direction by using a periodic boundary condition at the spanwise boundaries with 96 grid planes used in the spanwise direction. Notice that, over the majority of the airfoil, there is no evidence of significant vorticity, other than within the largely steady boundary layer. However, at the 40 percent chord location on the upper surface, the boundary layer separates and an unsteady wake region is evident. It is this level of turbulent detail that DNS can provide; we look forward to the day when we can apply DNS to full aircraft.

Summary of Best Practices

1. You should know the various types of turbulence models and have a basic understanding of how they work.
2. You should understand the differences between DNS, LES, RANS, and hybrid RANS/LES approaches to computing turbulence effects and understand when and where these computations are currently possible.
3. You should understand that there is no universally valid general model of turbulence that is accurate for all classes of flows.
4. Check the literature to understand known weaknesses of the model(s) you are using.
5. Validation of a turbulence model you intend to use is essential for all applications and should be done without altering the model or constants.
6. If possible, you should examine the effect and sensitivity of results to the turbulence model by trying different models on problems of interest.
7. The relevance of turbulence modeling becomes significant in CFD simulations only when other sources of error, in particular the numerical and convergence errors, have been removed or properly controlled.
8. Grid sensitivity studies become crucial for all turbulence model computations.
9. Check that the grid is consistent with the near wall modeling used and that it follows the grid best practices outlined in Chapter 7.

8.12 References

1   Browne, M.W., "Science Squints at a Future Clogged by Chaotic Uncertainty," *New York Times*, September 22, 1998, p. F4.

2   Van Dyke, M., *An Album of Fluid Motion*, Stanford: Parabolic Press, 1982.

3   Fujino, M., Yoshizaki, Y., and Kawamura, Y., "Natural-Laminar-Airfoil Development for a Lightweight Business Jet," *Journal of Aircraft*, Vol. 40, No. 4, 2003, pp. 609–615.

4   Crawford, B.K., Duncan, G.T., West, D.E., and Saric, W.S., "Laminar-Turbulent Boundary Layer Transition Imaging Using IR Thermography," *Optics and Photonics Journal*, Vol. 3, No. 3, 2013, pp. 233–239.

5   Joslin, R.D., "Overview of Laminar Flow Control," NASA TP-1998–208705, October 1998.

6   Braslow, A.L., *A History of Suction-Type Laminar-Flow Control with Emphasis on Flight Research*, NASA Monographs in Aerospace History, Number 13, 1999.

7   Cummings, R.M., Morton, S.A., and Siegel, S.G., "Numerical Prediction and Wind Tunnel Experiment for a Pitching Unmanned Combat Air Vehicle," *Aerospace Science and Technology*, Vol. 12, Issue 5, 2008, pp. 355–364.

8   Schlichting, H., Gersten, K., Krause, E., and Oertel, H., *Boundary-Layer Theory*, 8th Ed., New York: Springer, 2004.

9   DeJarnette, F.R., Hamilton, H.H., Weilmuenster, K.J., and Cheatwood, F.M., "A Review of Some Approximate Methods Used in Aerodynamic Heating Analyses," *Journal of Thermodynamics and Heat Transfer*, Vol. 1, No. 1, 1987, pp. 5–12.

10 Malik, M.R., "Stability Theory for Laminar Flow Control Design," in *Viscous Drag Reduction in Boundary Layers*, AIAA Progress in Astronautics and Aeronautics, Vol. 123, 1990, pp. 3–46.

11 Morkovin, M.V., "Critical Evaluation of Transition from Laminar to Turbulent Shear Layer with Emphasis on Hypersonically Traveling Bodies," Air Force Flight Dynamics Laboratory, AFFDL-TR-68–149, 1969.

12 Garcia, J.A., Tu, E.L., and Cummings, R.M., "A Parametric Study of Supersonic Laminar Flow for Swept Wings using Linear Stability Analysis," AIAA Paper 95-2277, June 1995.

13 Cheng, G., Nichols, R., Kshitij, Neroorkar, D., and Radhamony, P.G., "Validation and Assessment of Turbulence Transition Models," AIAA Paper 2009-1141, January 2009.

14 Cebeci, T., *Stability and Transition: Theory and Application*, Long Beach: Horizons Publishing, January 2004.

15 Krumbein, A., Krimmelbein, N., and Schrauf, G., "Automatic Transition Prediction in Hybrid Flow Solver, Part 1: Methodology and Sensitivities," *Journal of Aircraft*, Vol. 46, No. 4, 2009, pp. 1176–1190.

16 Cheng, G., Nichols, R., Neroorkar, K.D., and Radhomony, P.G., "Validation and Assessment of Turbulence Models," AIAA Paper 2009-1141, January 2009.

17 Lin, T.C., "Influence of Laminar Boundary-Layer Transition on Entry Vehicle Design," *Journal of Spacecraft and Rockets*, Vol. 45, No. 2, 2008, pp. 165–175.

18 Schneider, S., "Effects of Roughness on Hypersonic Boundary-Layer Transition," *Journal of Spacecraft and Rockets*, Vol. 45, No. 2, 2008, pp. 193–209.

19 Kosin, R.E., "Laminar Flow Control by Suction as Applied to the X-21A Airplane," *Journal of Aircraft*, Vol. 2, No. 1, 1965, pp. 384–390.

20   Saric, W.S., Reed, H.L., and White, E.B., "Stability and Transition of Three-Dimensional Boundary Layers," *Annual Review of Fluid Mechanics*, Vol. 35, 2003, pp. 413–440.

21   Krumbein, A., Krimmelbein, N., and Schrauf, G., "Automatic Transition Prediction in Hybrid Flow Solver, Part 2: Practical Application," *Journal of Aircraft*, Vol. 46, No. 4, 2009, pp. 1191–1199.

22   Krumbein, A., "Transition Modeling in FLOWer – Transition Prescription and Prediction," in *MEGAFLOW – Numerical Flow Simulation for Aircraft Design*, Berlin: Springer, 2005, pp. 45–62.

23   Strahlman, B.I., and Siggia, E.D., "Scalar Turbulence," *Nature*, Vol. 405, 2000, pp. 639–646.

24   Prandtl, L., "Report on Investigation of Developed Turbulence," NACA TM 1231, September 1949 (translation of "Bericht über Untersuchungen zur ausgebildeten Turbulenz," *Zeitschrift für Angewandte Mathematik und Mechanik*, Vol. 5, No. 2, 1925).

25   Liepmann, H.W., "The Rise and Fall of Ideas in Turbulence," *American Scientist*, Vol. 67, 1979, pp. 221–228.

26   Tennekes, H., and Lumley, J.L., *A First Course in Turbulence*, Cambridge: MIT Press, 1972.

27   Falco, R.E., "Coherent Motions in the Outer Region of Turbulent Boundary Layers," *Physics of Fluids*, Vol. 20, 1977, pp. S124–S132.

28   Schetz, J.A. and Bowersox, R.D.W., *Boundary Layer Analysis*, 2nd Ed., Reston: AIAA, 2011.

29   Lorenz, E., "Predictability: Does the Flap of a Butterfly's Wings in Brazil Set Off a Tornado in Texas," Annual Meeting of the American Association for the Advancement of Science, Washington, DC, December 1979.

30   Spalart, P.R., "Strategies for Turbulence Modelling and Simulations," *International Journal of Heat and Fluid Flow*, Vol. 21, Issue 3, 2000, pp. 252–263.

31   Spalart, P.R., Jou, W-H., Strelets, M., and Allmaras, S.R., "Comments on the Feasibility of LES for Wings, and on a Hybrid RANS/LES Approach," *Advances in DNS/LES*, 1st AFOSR International Conference on DNS/LES, Greyden Press, Columbus OH, August 4–8, 1997.

32   Spalart, P.R., Deck, S., Shur, M.L., Squires, K.D., Strelets, M.K., and Travin, A., "A New Version of Detached-Eddy Simulation, Resistant to Ambiguous Grid Densities," *Theoretical and Computational Fluid Dynamics*, Vol. 20, 2006, pp. 181–195.

33   Wilcox, D.C., *Turbulence Modeling for CFD*, 3rd Ed., La Cañada: DCW Industries, Inc., 2006.

34   Favre, A., "Equations des Gaz Turbulents Compressibles," *Journal de Mécanique*, Vol. 4, 1965, pp. 361–390.

35   Cebeci, T., and Smith, A.M.O., *Analysis of Turbulent Boundary Layers*, New York: Academic Press, 1974.

36 Baldwin, B., and Lomax, H., "Thin-Layer Approximation and Algebraic Model for Separated Turbulent Flows," AIAA Paper 1978-257, January 1978.

37 Johnson, D.A., and King, L.S., "A Mathematically Simple Turbulence Closure Model for Attached and Separated Turbulent Boundary Layers," *AIAA Journal*, Vol. 23, No. 11, 1985, pp. 1684–1692.

38 Spalart, P.R., and Allmaras, S.R., "A One Equation Turbulence Model for Aerodynamic Flows," *La Recherche Aerospatiale*, Vol. 1, 1994, pp. 5–21.

39 Baldwin, B.S., and Barth, T.J., "A One-Equation Turbulence Transport Model for High Reynolds Number Wall-Bounded Flows," AIAA Paper 1991-0610, January 1991.

40 Jones, W.P., and Launder, B.E., "The Calculation of Low-Reynolds-Number Phenomena with a Two-Equation Model of Turbulence," *International Journal of Heat and Mass Transfer*, Vol. 16, Issue 6, 1973, pp. 1119–1130.

41 Menter, F.R., "Influence of Freestream Values on $k - \omega$ Turbulence Model Predictions," *AIAA Journal*, Vol. 30, No. 6, 1991, pp. 1657–1659.

42 Menter, F.R., "Two-Equation Eddy-Viscosity Turbulence Models for Engineering Applications," *AIAA Journal*, Vol. 32, No. 8, 1994, pp. 1598–1605.

43 Rumsey, C.L., and Gatski, T.B., "Recent Turbulence Model Advances Applied to Multielement Airfoil Computations," *Journal of Aircraft*, Vol. 38, No. 5, 2001, pp. 904–910.

44 Gee, K., Cummings, R.M., and Schiff, L.B., "Turbulence Model Effects on Separated Flow About a Prolate Spheroid," *AIAA Journal*, Vol. 30, No. 3, 1992, pp. 655–664.

45 Tobak, M., and Peake, D.J., "Topology of Three-Dimensional Separated Flows," *Annual Review of Fluid Mechanics*, Vol. 14, 1982, pp. 61–85.

46 Dallmann, U., "Three-Dimensional Vortex Structures and Vorticity Topology," *Fluid Dynamics Research*, Vol. 3, 1988, pp. 183–189.

47 Degani, D., and Schiff, L.B., "Computation of Turbulent Supersonic Flows Around Pointed Bodies Having Crossflow Separation," *Journal of Computational Physics*, Vol. 66, No. 1, 1986, pp. 173–196.

48 Holst, T., "Viscous Transonic Airfoil Workshop: Compendium of Results," *Journal of Aircraft*, Vol. 25, No. 12, 1988, pp. 1073–1097.

49 McCroskey, W.J., "A Critical Assessment of Wind Tunnel Results for the NACA 0012 Airfoil," NASA TM 100019, October 1987.

50 Harris, C.D., "Two-Dimensional Aerodynamic Characteristics of the NACA 0012 Airfoil in the Langley 8-Foot Transonic Pressure Tunnel," NASA TM 81927, April 1981.

51 Monsen, E., Franke, M., Rung, T., Aumann, P., and Ronzheimer, A., "Assessment of Advanced Transport-Equation Turbulence Models for Aircraft Aerodynamics Performance Prediction," AIAA Paper 99-3701, July 1999.

52 Redeker, G., Müller, R., Ashill, P.R., Elsenaar, A., and Schmitt, V., "Experiments on the DFVLR-F4 Wing-Body Configuration in Several European Windtunnels," AGARD-CP 429 Symposium, September 1987.

53 Jakirlic, S., Eisfeld, B., and Basara, B., "Performance Assessment of Some Popular RANS Models by Relevance to High-lift Aerodynamics," AIAA Paper 2009-0049, January 2009.

54 Schiavetta, L.S., Boelens, O.J., Crippa, S., Cummings, R.M., Fritz, W., and Badcock, K.J., "Shock Effects on Delta Wing Vortex Breakdown," *Journal of Aircraft*, Vol. 46, No. 3, 2009, pp. 903–914.

55 Forsythe, J.R., Hoffman, K.A., Cummings, R.M., and Squires, K.D., "Detached-Eddy Simulation With Compressibility Corrections Applied to a Supersonic Axisymmetric Base Flow," *Journal of Fluids Engineering*, Vol. 124, No. 4, 2002, pp. 911–923.

56 Bunge, U., Mockett, C., Aupoix, B., Haase, W., Menter, F., Schwamborn, D., and Weinman, K., "FLOMANIA – A European Initiative on Flow Physics Modelling," Chapter 5, Summary of Experience, *Notes on Numerical Fluid Mechanics and Multidisciplinary Design*, Vol. 94, Berlin: Springer, 2004.

57 Bunge, U., Rung, T., and Thiele, F., "Methods Used and Highlight Results Achieved in UNSI," in Computational Fluid-Structure Interaction, *Notes on Numerical Fluid Mechanics and Multidisciplinary Design*, Vol. 81, 2002, pp. 139–154.

58 Vassberg, J.C. Tinoco, E.N., Mani, M., Levy, D., Zickurh, T., Mavriplis, D.J., Wahls, R.A., Morrison, J.H., Brodersen, O.P., Eisfeld, B., and Murayama, M., "Comparison of NTF Experimental Data with CFD Predictions from the Third AIAA CFD Drag Prediction Workshop," AIAA Paper 2008-6918, August 2008.

59 Vassberg, J.C., Tinoco, E.N., Mani, M., Broderson, O.P., Eisfeld, B., Wahls, R.A., Morrison, J.H., Zickuhr, T., Laflin, K.R., and Mavriplis, D.J., "Abridged Summary of the Third AIAA Computational Fluid Dynamics Drag Prediction Workshop," *Journal of Aircraft*, Vol. 45, No. 3, 2008, pp. 781–798.

60 Roy, C.J., "Review of Code and Solution Verification Procedures for Computational Simulation," *Journal of Computational Physics*, Vol. 205, Issue 1, 2005, pp. 131–156.

61 Breuer, M., "Numerical and Modeling Influences on Large Eddy Simulations for the Flow Past a Circular Cylinder," *International Journal of Heat and Fluid Flow*, Vol. 19, No. 5, 1998, pp. 512–521.

62 Piomelli, U., "Large Eddy Simulation of Turbulent Flows," TAM Report No. 767, September 1994.

63 Van Driest, E.R., "On Turbulent Flow Near a Wall," *Journal of the Aerospace Sciences*, Vol. 23, No. 1, 1956, pp. 1007–1011.

64 Page, G.J., Li, Q., and McGuirk, J. J., "LES of Impinging Jet Flows Relevant to Vertical Landing Aircraft," AIAA Paper 2005-5226, June 2005.

65   Page, G.J., Li, Q., McGuirk, J.J., and Richardson, G.A., "Large Eddy Simulation of a Harrier Aircraft at Touch Down," AIAA Paper 2007-4294, June 2007.

66   Liu, C., "High Performance Computation for DNS/LES," *Applied Mathematical Modelling*, Vol. 30, 2006, pp. 1143–1165.

67   Shur, M., Spalart, P.R., Strelets, M., and Travin, A., "Detached-Eddy Simulation of an Airfoil at High Angle of Attack," Proceedings of the Fourth International Symposium on Engineering Turbulence Modelling and Measurements, 24–26 May 1999, Corsica, Elsevier, Amsterdam.

68   Constantinescu, G.S., and Squires, K.D., "LES and DES Investigations of Turbulent Flow over a Sphere," AIAA Paper 2000-0540, January 2000.

69   Salas, M.D., "Digital Flight: The Last CFD Aeronautical Grand Challenge," *Journal of Scientific Computing*, Vol. 28, No. 2/3, 2006, pp. 479–505.

70   Morton, S.A., Cummings, R.M., and Kholodar, D.B., "High Resolution Turbulence Treatment of F/A-18 Tail Buffet," *Journal of Aircraft*, Vol. 44, No. 6, 2007, pp. 1769–1775.

71   Hall, R.M., and Woodson, S.H., "Introduction to the Abrupt Wing Stall Program," *Journal of Aircraft*, Vol. 41, No. 3, 2004, pp. 425–435.

72   Forsythe, J.R., and Woodson, S.H., "Unsteady Computations of Abrupt Wing Stall using Detached-Eddy Simulation," *Journal of Aircraft*, Vol. 42, No. 3, 2005, pp. 606–616.

73   Morton, S.A., "Detached-Eddy Simulations of Vortex Breakdown over a 70-Degree Delta Wing," *Journal of Aircraft*, Vol. 46, No. 3, 2009, pp. 746–755.

74   Spalart, P.R., "Young Person's Guide to Detached-Eddy Simulation Grids," NASA CR 2001–211032, July 2001.

75   Menter, F.R., and Kuntz, M., "Adaptation of Eddy-Viscosity Turbulence Models to Unsteady Separated Flow Behind Vehicles," *Proceedings of the Conference on The Aerodynamics of Heavy Vehicles: Trucks, Busses and Trains*, Asilomar, CA, 2002.

76   Jeans, T.L., McDaniel, D.R., Cummings, R.M., and Mason, W.H., "Aerodynamic Analysis of a Generic Fighter Using Delayed Detached-Eddy Simulation," *Journal of Aircraft*, Vol. 46, No. 4, 2009, pp. 1326–1339.

77   Shur, M.L., Spalart, P.R., Strelets, M.K., Travin, A.K., "A Hybrid RANS-LES Approach with Delayed-DES and Wall-modeled LES Capabilities," *International Journal of Heat and Fluid Flow*, Vol. 29, Issue 6, 2008, pp. 1638–1649.

78   Forsythe, J.R., personal communication.

79   Spalart, P.R., "Detached-Eddy Simulation," *Annual Review of Fluid Mechanics*, Vol. 41, 2009, pp. 181–202.

80   Haase, W., Braza, M., and Revell, A., "DESider – A European Effort on Hybrid RANS-LES Modelling," *Notes on Numerical Fluid Mechanics and Multidisciplinary Design*, Vol. 103, Berlin: Springer, 2007.

81 Ferziger, J.H., and Peric, M., *Computational Methods for Fluid Dynamics*, New York: Springer-Verlag, 1997.

82 Nieuwstadt, F.T.M., Eggels, J.G.M., Janssen, R.J.A., and Pourquie, M.B.J.M., "Direct and Large-Eddy Simulations of Turbulence in Fluids," *Future Generation Computer Systems*, Vol. 10, Issue 2–3, 1994, pp. 189–205.

83 Moin, P., and Mahesh, K., "Direct Numerical Simulation: A Tool in Turbulence Research," *Annual Review of Fluid Mechanics*, Vol. 30, 1998, pp. 539–578.

84 Jones, L.E., Sandberg, R.D., and Sandham, N.D., "Direct Numerical Simulations of Forced and Unforced Separation Bubbles on an Airfoil at Incidence," *Journal of Fluid Mechanics*, Vol. 602, 2008, pp. 175–207.

9 Flow Visualization: The Art of Computational Aerodynamics

The greatest value of a picture is when it forces us to notice what we never expected to see.[1]

John W. Tukey, mathematician

CFD Simulation of F/A-18 at high angle of attack (a full color version of this image is available in the color insert pages of this text as well as on the website: www.cambridge.org/aerodynamics).

LEARNING OBJECTIVE QUESTIONS

After reading this chapter you should know the answers to the following questions:

- What is flow visualization and why is it important?
- What are five of the ways to "say nothing" with flow visualization?
- What are some of the ways that early CFD results were visualized?
- Why was the advent of color graphics workstations so important for the development of flow visualization?
- What are a node and a focus and how are they different from each other?
- What are some of the errors that can be found in flow visualization?
- What are the pros and cons of using scalar or vector properties for visualizing various flow features?
- Why are feature extraction methods important today?

Figure 9.1   Time-dependent particle traces for the V-22 in hover, colored by particle release time: lighter – earliest, darker – latest (Ref. 2; Courtesy of Mark Potsdam of the U.S. Army Aeroflightdynamics Center; a full color version of this image is available in the color insert pages of this text as well as on the website: www.cambridge.org/aerodynamics).

9.1 Introduction

Flow visualization (often referred to as "flow viz") is perhaps the most interesting and enjoyable, yet often frustrating, aspect of computational aerodynamics. Modern computational tools make visualization very easy to do, yet being able to correctly "see" and "show" the details of a flow solution can be quite challenging. Three-dimensional flows around aircraft configurations can lead to many complex flow features, which are not always easy to understand by looking at a two-dimensional computer terminal. For example, the simulation of the V-22 tilt-rotor in hover results in a complex flowfield, as shown in Fig. 9.1.[2] The flowfield complexity includes the interaction of each rotor blade with the wake of the preceding blade, the interaction of the rotors with the fuselage, and the interaction of the rotor wakes with one another and the freestream flow. But this visualization only shows one geometric plane of the solution volume, and it would be informative to see more of the results. However, it would be very difficult to comprehend too much more information on one figure. And of course, there may be many other ways to "see" this flow, some of them more instructive and informative than others. It might also be informative to put such a flowfield into a movie so that time-dependent flow features could be seen and understood. Doing these things in an intelligent, accurate, and useful way is the challenge of flow visualization.

There might also be ways to make a simulation look better than it really is, or to "hide" imperfections in the flowfield simulation, or to just plain overlook important details of a solution. In fact, some people over the years have referred to CFD as "colorized fluid dynamics." It is certainly true that computational aerodynamics uses computer visualization techniques that require a great deal of color to view the results, but there may be some truth in these accusations, since it is altogether too easy to concentrate on how good your results *look* rather than how good your results *are*. Realizing this, two researchers came up with a list of things that many CFD researchers do (whether intentionally or unintentionally) that make their results essentially *useless*.[3,4] The Scientific Visualization Concept Box discusses these items in a humorous way – we hope you realize that these are things that you DO NOT want to do! Rather, you should enjoy the humor of these observations, and then do just the opposite in order to take full advantage of the flow visualization capabilities that exist today.

Computational Aerodynamics Concept Box

How to Say Nothing with Scientific Visualization (Or how NOT to present visualization results) (based on Refs. 3 and 4)

1. Never Include a Color Legend
As you shall see, it is common in CFD to assign colors to various scalar values (for example blue for cold and red for hot). If you don't want others to know what your results are, then just leave out the color legend where the values are defined. *However, if you want people to truly understand your results (qualitatively and quantitatively), then you should always include a color legend with your figures.*

2. Avoid Annotation
Many fields of science and engineering present results with images that are annotated (they use arrows and explanations to highlight regions of interest within the results). If you don't want anyone to understand your results, you can just leave out these annotations. *You probably want people to have a full understanding of your images (a picture is worth a thousand words), so annotations may prove worthwhile.*

3. Never Mention Error Characteristics
If scientists using visualization software were aware that visualization techniques might introduce error, they might not be properly impressed by our masterworks. Therefore, never imply by word or deed that your algorithm introduces any error whatsoever. After all, if the picture looks good, it must be correct. *On second thought, it might be good to be up front about errors and issues with your visualization so that others have the fullest possible understanding of your work.*

4. Never Compare Your Results with Other Visualization Techniques
There are many flow visualization software systems available today, and all of them have strengths and weaknesses. Comparing results with other visualization techniques could show problems with your results, so you might want to avoid doing that. *Conversely, if you want to have confidence that your visualization software is giving you the best possible results, it may be a good idea to compare results with other approaches.*

5. Avoid Visualization Systems
Come to think of it, writing your own flow visualization software (or using one from a co-worker) is probably much better than using modern flow visualization software that allows for defining new techniques within a proven program structure. You can then provide results without knowing if the visualizations are correct! *Well, maybe that isn't such a good idea after all – modern flow visualization systems are so advanced and flexible that you probably will never be able to match their capability with your own code.*

6. Quietly Use Stop-Frame Video Techniques
Each frame of a scientific video usually takes seconds, minutes, or even hours to produce. To achieve smooth animation it is usually necessary to generate video frames one at a time

to a movie file which is played back at 30 or even 60 frames a second. The magic is lost, however, if you are so foolish as to tell anyone what you're doing. *Actually, fully describing the process used to make a movie will add understanding of your results, including showing time values or other information for each frame.*

7. Never Cite References for the Data

If you cite a reference describing the data used to generate images, someone may read the paper and discover that your visualization bears no relationship to the key elements the original experiment was meant to elucidate. This will detract from your picture's appeal and should be avoided. *Actually, that is rather bad advice! Always cite sources of data so that others will understand your results in context.*

8. Use Viewing Angle to Hide Blemishes

Many otherwise excellent algorithms produce 3D objects containing unsightly blemishes. Avoid carelessly choosing viewing angles that expose such flaws. If a suitable angle cannot be found, try another dataset. *Actually, being honest and open about troubling issues within your results is what true science is all about – sometimes the problems help to lead you to the right answer!*

9. If Viewing Angle Fails, Try Specularity or Shadows

Sometimes every possible viewing angle is marred by some small ugliness. In these cases, try adding shadows or brilliant highlights in appropriate places. However, never resort to using a paint program to touch up your image; that wouldn't be scientific. *While you certainly can use shadows and other visualization techniques to aid in viewing complex results, you should never use these devices to hide results.*

10. Never Learn Anything About the Data or Scientific Discipline

Scientific visualization software is much more difficult if you are worried about producing correct results. Irritating details like accurate interpolation techniques get in the way; in many cases, ad-hoc interpolation techniques can produce much prettier pictures with significantly less work. As we all know, beauty is the higher truth. *Correction: showing your results accurately and correctly is key to developing understanding about the flow, so always do your best to show the key features in your results.*

Throughout this book we have tried to use good examples of flow visualizations that illustrate how to present results graphically – you should strive to make your results usable and professional.

9.2 Flow Visualization Background

When computational aerodynamics was in its early stages, researchers could look at tabulated computer printouts to see the results of their simulations. By tabulating the relatively small amount of data (maybe surface pressures at a handful of points on an airfoil) and graphing it by hand (remember, there were no spreadsheets back then), a picture of the flowfield could be developed

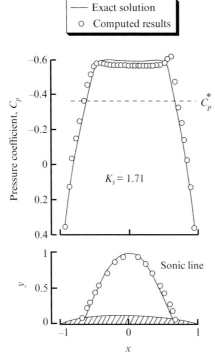

Figure 9.2 Transonic results for a 10.76% Nieuwland airfoil at $M = 0.8257$, c. 1970 (Ref. 5; reprinted by permission of the American Institute of Aeronautics and Astronautics, Inc.).

by the user, so that analysis could take place. If the researcher was going to make a presentation or write a paper, they would have a graphic artist create professional looking plots, as shown in Fig. 9.2.[5] As researchers started producing results for three-dimensional shapes like wings, rudimentary plotting packages were developed that used pen plotters, as shown in Fig. 9.3.[6] In both of these cases, however, only surface scalar quantities (pressure) were being evaluated and graphed; little or no off-surface information was visualized.

As computers got larger and the number of grid points increased, it was no longer possible to just tabulate a few numbers and make basic graphs, so more complicated software had to be developed to visualize the flow solutions. Programs like *PLOT3D* initially made it possible to make black

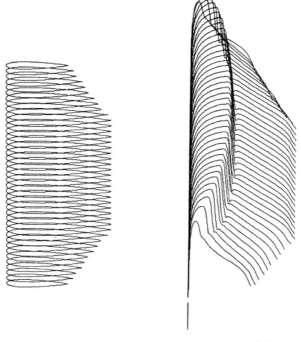

Wing shape   Pressure coefficient

Figure 9.3 Transonic wing prediction with NACA 0012 airfoil section; $AR = 8.9$, $M = 0.75$, $\alpha = 3.0$ deg, c. 1973 (Ref. 6; reprinted by permission of the American Institute of Aeronautics and Astronautics, Inc.) (a) wing shape (b) pressure coefficient.

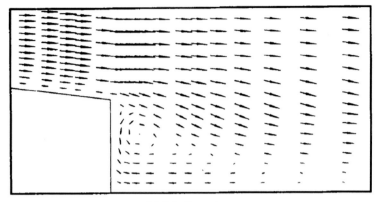

(a) Velocity vectors

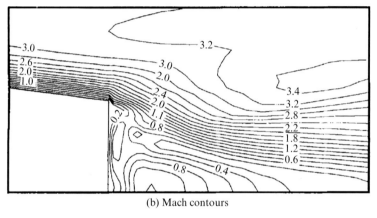

(b) Mach contours

Figure 9.4 Velocity vectors and Mach contours in the base region of an axisymmetric missile body at $M = 3.0$ using *PLOT3D*, c. 1986 (Ref. 7).

and white contour or vector plots (stored as graphics files) in order to visualize off-surface flow variables, as shown in Fig. 9.4.[7] However, these rudimentary approaches to flow visualization would become relatively useless as the size of the datasets continued to increase in size.

As the CFD solutions produced more and more information, it was no longer possible to fully sift through all of the results using simple approaches. For example, a structured three-dimensional grid for viscous flow computations with 1,000,000 points will produce at least 6,000,000 pieces of data for every time step (density, velocities, pressure, and temperature at every point!). If the flow is unsteady and you are computing in time-accurate mode with a fairly small time step, you can easily have hundreds or thousands of sets of data to comprehend at the same time. That could mean you would need to understand (in one way or another) billions (or more) pieces of information in some logical, straightforward way, which is why flow visualization is so important.

We need fairly quick and easy post-processing tools to help us "see" the results, and how you look at the data depends on what you're looking for!

Figure 9.5   Silicon Graphics (SGI) Personal Iris Workstation, c. 1988 (Ref. 8; Courtesy of Silicon Graphics International Corp.).

Knowing what to look for often requires some basic aerodynamic knowledge and understanding, which is why you should refer to Chapter 4 (and other aerodynamics sources) while you are performing your simulations. Remember, too, that what you are looking for often depends on the speed regime, angle of attack, type of vehicle being simulated, etc.

Luckily, these increased needs for flow visualization (which were due to the increased capabilities of mainframe computers) were met with the improved technology of color graphics computer workstations. These workstations (an example of one from the late 1980s is shown in Fig. 9.5) were designed to perform high-quality graphics data manipulation, but no commercial software existed to take full advantage of them. By the way, these workstations were also used to make the initial computer-generated imagery (CGI) for movies and were considered very futuristic – to put this into context, personal computers had only recently been introduced and they did not have color monitors!

These color graphics workstations allowed researchers to develop software that could make color images of the flowfield and allowed the use of various

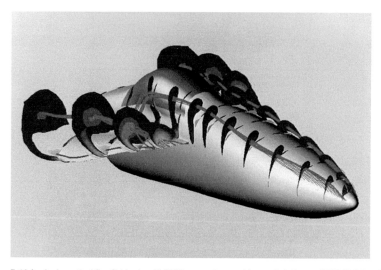

Figure 9.6   F-18 forebody vortical flowfield using *PLOT3D* on a color graphics workstation, c. 1989 (Ref. 9; a full color version of this image is available on the website: www.cambridge.org/aerodynamics).

colors to represent parameter values rather than using contour lines (as seen in Fig. 9.4). Once the computer hardware had been created, programs developed at NASA Ames Research Center, such as *PLOT3D* and *FAST* (Flow Analysis Software Toolkit), expanded in capability to take advantage of the new workstations (as shown in Fig. 9.6[9]). Eventually, commercial flow visualization packages (like *Fieldview*®, *Tecplot*®, and *Ensight*®) became available, initially requiring dedicated hardware systems such as upgraded versions of the workstation shown in Fig. 9.5. The advent of these workstations and graphics software enabled the CFD user to begin to visualize flows as never before by using color.

The increased computer power of PC's, coupled with massive improvements in affordable graphics power, enabled high-level flow visualization software to be available to the average user at affordable prices. This, in turn, led to an explosion of research and concepts for flow visualization of scientific information. These color graphics–based concepts have now become common in the CFD world and require users to understand the various ways that numerical predictions can be visualized with modern computer systems.

As flow visualization systems have developed over the past 10–15 years, many of them possess capabilities that mimic the basic fluid dynamic concepts that experimentalists and theoreticians understand. This requires having a basic understanding of these flow features before delving too deeply into the world of flow visualization, so it may be helpful to take some time and understand basic flow topology (see the accompanying Computational Aerodynamics Concept Box).[10,11,12] Flow topology concepts and classification can be very helpful when trying to understand what you are seeing with your flow visualization software.

Computational Aerodynamics Concept Box

Flow Topology Basics (Refs. 10–12)

Topology is the mathematical study of geometric surfaces and features in multi-dimensional spaces; flow topology is the application of these concepts to the various flow features that occur in aerodynamics. For example, the F-16 shown here is flying at a fairly high angle of attack, which creates natural visualization of the vortices on the strake of the aircraft via condensation of the water vapor in the air. Notice that the wing-tip vortices are also visible in the picture.

Vortices visible on a maneuvering F-16 (U.S. Air Force photo)

If a numerical simulation of this flowfield were being performed, there might be a variety of ways to visualize the vortices present around the aircraft, including representation of the flow near or on the surface as well as visualization of the off-surface flow. Both the surface and off-surface flow features can be described by various topological properties which they possess, such as nodes of attachment or separation (in 2D on a surface), lines of attachment or separation (in 3D in the fluid), or general descriptions of flow features like vortices. While the general study of topology is beyond the scope of this book, a few basic concepts can come in handy when performing flow visualizations.

For example knowledge of various critical points in surface flow topology can help you understand what is happening off the surface. The pictures included here show some surface nodes and flow lines in the vicinity of a flow separation line, which is a line that surface streamlines approach and into which they "feed." The next picture shows some surface flow lines in the vicinity of a flow attachment line, which is a line from which surface streamlines come, since the flow is coming from off the surface and attaching to a line on the surface.

Some examples of flow topology are shown for high angle-of-attack flow over a sharp leading-edge delta wing. Since the leading edges are sharp, the flow separates and rolls up into two counter-rotating primary vortices, as shown in the schematic. These vortices induce outboard flow near the surface of the delta wing (shown as skin friction lines in the schematic), which eventually separate into secondary vortices. The various surface flow patterns can be used to deduce the presence of the vortices.

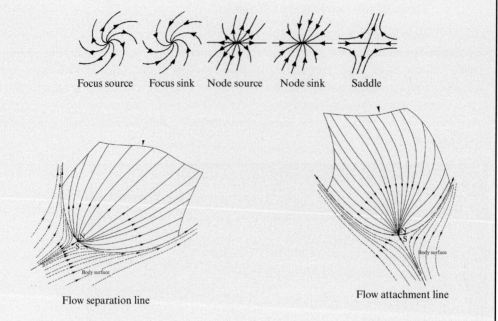

Focus source   Focus sink   Node source   Node sink   Saddle

Flow separation line

Flow attachment line

Flow topology on the upper surface of a thin delta wing at angle of attack.

An experimental flow visualization of the primary and secondary vortices is shown in the photograph. The flow topology is mapped in the sketch below the photograph. All of these flow features can be visualized numerically, if you know what to look for.

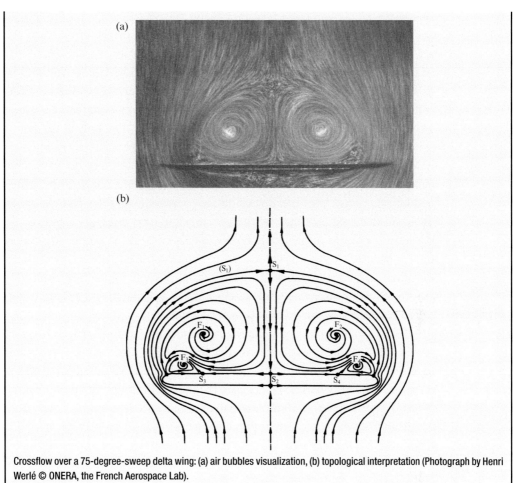

Crossflow over a 75-degree-sweep delta wing: (a) air bubbles visualization, (b) topological interpretation (Photograph by Henri Werlé © ONERA, the French Aerospace Lab).

9.3 How Flow Visualization Works

Flow visualization software, like the computational aerodynamics codes that calculate the results, requires the use of numerical methods. The first thing you should realize, therefore, is that flow visualization is subject to errors, since there are no perfectly accurate numerical methods (see Rule 3 in the scientific visualization box at the beginning of the chapter). The important thing to know is how well your software performs and which flow quantities require numerical integration or differentiation, both of which introduce errors in your results. These are key concepts to understand while using modern flow visualization software.

Most computational aerodynamics codes use primitive variables (the variables used by the numerical method) to perform calculations. Panel methods and vortex lattice codes usually obtain results in terms of the pressure coefficient, C_p. If you use a post-processing flow visualization tool to look at any quantity other than those derived from the flow solver output, then you are "processing" the results and introducing some error into the answers. CFD codes typically perform calculations on the conserved variables of fluid flux, as given in Equation 3.64. Specifically, the output variables for most CFD codes are typically: ρ, ρu, ρv, ρw, and E_t. These are the only variables that are known exactly when using flow visualization (and then only at the points where they were calculated). Unfortunately, you probably will only rarely want to look at any of these variables directly (with the exception of density). That means all other values you may want to see, like pressure, temperature, Mach number, etc., must be calculated using the conserved variables. In most cases, the equations used are obtained theoretically and have no error ... but you do not always want to know about flow variables only at the points where they were calculated. These variables need to be interpolated between grid points (or cell centers), and the type of interpolation used will impact the results (for example, a flow solver might use a cell-centered approach, and a flow visualization package may use a vertex-centered approach).

Let's say you want to know the Mach number in a region between two grid points. While it may be perfectly reasonable to assume that the conserved quantities can be interpolated accurately between points, it does not mean that anything else can easily be known between the points. Buning states that "making the assumption that the solution variables are linear between grid points does not mean that Mach number, or pressure, or even u, v, and w vary linearly!"[13] It is possible that the particular flow solver you are using automatically computes some derived quantities at the solution points or interpolates the solution from cell-centers to vertices (or vice versa). This still does not mean the derived and/ or interpolated quantities are of the same accuracy as the computed solution. You must be keenly aware of any post-processing the CFD code is automatically accomplishing separate from the actual flow solution.

As graphics post processing programs were originally developed, they needed to address the numerical issues that were going to arise when users started creating "pictures" of their computer flowfields. The types of issues include: producing symmetric plots of symmetric functions (that does not sound hard, but it was!), drawing smooth contour lines, shading, smoothing, making vector plots, creating streamlines via numerical integration, and calculating various flow functions from the conserved variables. Some of the issues will be addressed here because they still impact graphics in modern flow visualization programs, and users should be aware of these issues when they look at their results.

9.3.1 Smooth Contour Lines

Early flow visualization software was black and white, so "seeing" the flow was quite challenging. This led to the use of contour plots for visualization purposes (for example, see Fig. 9.4b). As the color workstations became

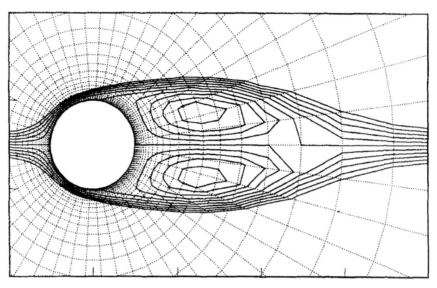

Figure 9.7   Asymmetric plot of stream function contours for a symmetric flow about a circular cylinder (Ref. 13; Courtesy of Pieter G. Buning of NASA Langley Research Center).

available, the contours could be color coded (rather than marked with numbers), but problems still can arise with this approach.

According to Buning, "two types of problems arise when computing contour lines. The first is a classic ambiguity, while the second is caused by a mismatch between the order of accuracy of the contouring scheme versus the numerical method of the flow solver."[13] The first problem is caused by the type of interpolation used between solution points to determine the location for a constant value for a flow variable. The interpolation scheme can easily determine where the constant value is located between solution points, but the scheme has no way of knowing where that constant value should extend elsewhere. Therefore, the program needs some algorithm to determine how to draw the contour line through the cell. Depending on how this is done, the results can lead to problems, as shown in Fig. 9.7, where a symmetric flow solution behind the circular cylinder results in an asymmetric set of contour plots. This example was performed on a coarse grid to show the problem more clearly, but, no matter how fine a grid is, this problem could occur if care is not taken.

The other common problem with contour plots is "scalloping." Scalloping occurs when the contours are being drawn on a highly stretched grid and the algorithm uses interpolation to find the contour lines, but they appear crinkled due to the grid cell orientation (see Fig. 9.8).

Modern flow visualization software allows for contours to be shaded with colors, creating a continuous picture, but distortions can still take place. In Fig. 9.9, an inviscid solution for the supersonic flow over a diamond wedge airfoil is shown. The unstructured mesh (Fig. 9.9a) seems reasonably fine for a Euler solution, but, because oblique shocks and expansion waves are being computed through this mesh (both of which will be straight oblique lines), the

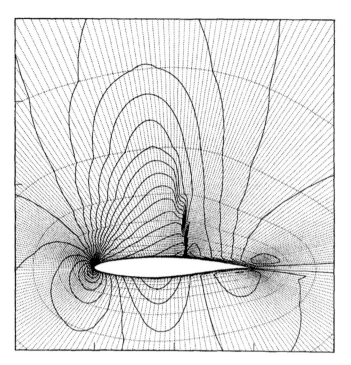

Figure 9.8   Density contours around an airfoil showing scalloping on a highly stretched grid (Ref. 13; Courtesy of Pieter G. Buning of NASA Langley Research Center).

resulting flow does not look reasonable on the mesh (Fig. 9.9b). Looking at contour lines of pressure (Fig. 9.9c) there is a great deal of oscillation, which is still evident when looking at the smooth surface of pressure (Fig. 9.9d). This emphasizes the importance of having a good grid for the flow you are trying to compute (in this case shocks and expansion waves), because no amount of reasonable post-processing can make the results look better. In other words, you should not expect more from the solver and mesh than what is possible, which is why good mesh refinement is so important (as discussed in Chapter 7).

9.3.2 Three-Dimensional Vector Plots

Vector plots are perhaps the most basic flow visualization tool available (for example, see Fig. 9.4a). The output variables from your flow code are easily converted into vectors (like velocity vectors), so graphing the velocity vectors at each point of the grid or mesh is relatively straightforward. The problem with vector plots is that you can only look at them in two dimensions on your computer screen, yet vectors are typically three-dimensional. When you look at a three-dimensional vector in two dimensions, you lose information (you only see a projection of the vector on your two-dimensional computer screen), which means you also lose the ability to completely "see" the flow. Most flow visualization software allows you to rotate your view of the flow, and doing this can begin to fill in your mental picture of the flow more fully, although you may still

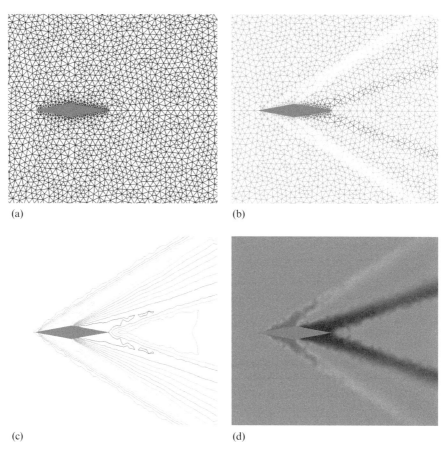

(a)   (b)

(c)   (d)

Figure 9.9 Pressure field around a diamond wedge airfoil in supersonic flow showing the impact of the mesh on the smoothness of the results. (a) airfoil surface and coarse mesh. (b) pressure field superimposed on mesh. (c) pressure contours around airfoil. (d) smooth shaded pressure contours (a full color version of this image is available on the website: www.cambridge.org/aerodynamics).

have trouble seeing everything in which you are interested. Another way to help see the three-dimensional nature of the vector is to color code each vector with some scalar property, such as the magnitude of velocity or pressure.

9.3.3 Streamlines

Calculating streamlines (also called particle tracing) is an important capability of computational flow visualization. Seeing a two-dimensional map of a flow quantity such as pressure or density often leaves you unsure of the exact flow behavior: streamlines often help to fill in the flow picture in your mind.

Streamlines are obtained by integrating a very simple ordinary differential equation:

$$\vec{V} = \frac{d\vec{x}}{dt}$$

$$(9.1)$$

that assumes the particle being tracked follows the flow as if it is a part of the fluid. The problem that arises is that numerical time integration produces errors, as does the process of finding the velocity vector at any location within the flow solution. Particle paths are also integrated throughout a flow solution space, so the total error can accumulate as the integration is being done (as you progress along the streamline). Not only can this calculation take a long time, it will become more inaccurate the longer the path of the particle. As recommended by Buning: "one can check the magnitude of the error by integrating a particle trace forward in time to the outflow boundary, then backwards from there to the inflow, and comparing the two paths."[13] If you do this and get two very different streamlines, then there is a relatively large error in your integration and interpolation methods. Remember: you often do not know exactly what your flow visualization software is doing (does it integrate with a first-order Euler explicit method, a second-order Runge-Kutta method, etc.?), so checking some of these things can be a valuable exercise to complete prior to wondering why your results do not look correct.

9.3.4 Flow Function Computation

Perhaps the best example of the difficulty in computing flow functions is how we compute vorticity. As described in Section 4.2.1, vorticity is the curl of the velocity vector field, and, for a finite difference formulation in two dimensions may be computed from:

$$\omega_z = \frac{\partial v}{\partial x} - \frac{\partial u}{\partial y} \tag{9.2}$$

Likewise, for a finite volume calculation in two dimensions the vorticity could be found from:

$$\oint_C \vec{V} \cdot d\vec{l} = \iint_A \vec{\omega} \cdot d\vec{A} \approx \omega_z A. \tag{9.3}$$

Notice that the finite difference formulation requires differentiation, while the finite volume formulation requires integration. Mathematically, these two approaches can be shown to be identical in the limit. However, when you take these approaches and approximate them with a numerical method, you may get different results, depending on the fineness of the grid and the method used. The derivatives in Equation 9.2, for example, could be approximated the same way as derivatives were numerically evaluated in Chapter 6. That means that whatever method we choose will have a truncation error dependent on the grid size. Our flow visualization software is now being required to mimic some of the capabilities of our flow solver, which also means the calculations could be quite time consuming, in addition to introducing errors. Some flow solvers give the user options for which variables are output to visualization files (for example, you may ask for primitive variables only, but then have to calculate variables such as vorticity in the post-processing software). It often will be better to have the flow solver output various flow variables using the correct schemes and methods, rather than trying to re-create those variables after the fact.

Profiles in Computational Aerodynamics: Scott Murman

"Growing up in suburban Detroit, I frustrated my parents by disassembling my new toys to find out how they worked. I can remember sitting in our basement taking apart my toy pistols and race cars, examining the springs and linkages that animated them. Occasionally, I was even able to get them back together again.

"As I got older I found new ways to frustrate my parents, but the passion for discovery and learning how things work never left me, and I knew I wanted a career in science and engineering. I enrolled in the aerospace engineering department at the University of Michigan. While I certainly was inspired by the space program and jet airplanes, one of the main reasons I chose aerospace engineering is the broad spectrum of disciplines aerospace includes. One studies fluid dynamics, structural mechanics, electrical and computer engineering, chemical propulsion, systems engineering, etc. This diversity still appeals to me today as I continue to work in the aerospace field.

"During my final year at Michigan, I discovered computational fluid dynamics and have pursued it ever since. This discipline combines many elements that I enjoy: physics, mathematics, and computer science. While I don't get to work with my hands as much as I'd like, this is counterbalanced by the limitless construction possibilities of computer software.

"After Michigan, I pursued a Master's degree at Cal Poly, San Luis Obispo. I was seeking to complement the strong academic training I'd received as an undergraduate with a more applied focus. I wanted to take things apart again. Cal Poly is a smaller school in a great location, known for hands-on, student-led projects. As part of this hands-on study, I began a graduate research project in CFD at NASA Ames Research Center in Silicon Valley. I haven't left since.

"At NASA Ames I transitioned from student to professional, and began working at the NASA Advanced Supercomputing (NAS) facility. NAS provides high-end computing capability to support NASA missions and science partnerships, and has consistently maintained one of the top ten supercomputers in the world. I also completed a doctorate at nearby Stanford University through a NASA grants program, while continuing to work part-time. Working directly with the professors at Stanford and the senior researchers at NASA provided invaluable experience performing and directing independent research. This is experience I now find myself passing on as I work with younger researchers.

"When I began working at NASA, my goal was to develop the capability to perform physics-based simulations of maneuvering aircraft, e.g., a CFD flight simulator. I've worked on many projects that build toward this ultimate goal: unsteady physics and turbulence modeling, moving-body algorithms, automation and robustness, etc. Each project moves us closer to a comprehensive CFD capability for maneuvering vehicles.

"The dynamic CFD capability we've been developing led to our support of the Space Shuttle Columbia accident investigation. This was a difficult time for us, as one of the astronauts, Dr. Kalpana Chawla, worked with us at the NAS facility for many years. The CFD support provided insight into the velocity of the foam debris, guiding the ground testing which confirmed the cause of the accident. We continued to support shuttle operations with high-fidelity CFD simulations for debris risks until it was retired.

"I still have the same curiosity and passion for discovery that guided me toward engineering as a child, and have no doubt it will continue. I supplement my scientific research with hands-on woodworking, metalworking, and electronics as hobbies. I'm also now teaching my two young girls to take apart their toys and discover for themselves what is inside."

9.4 How to View Scalar Properties

The most straightforward way to "see" a numerical flowfield is to look at scalar properties. These include the traditional fluid dynamic properties such as density, pressure, and temperature, but also include the magnitudes of vector properties like velocity and vorticity. The viewing can be done on the solid surfaces that make up the geometry of interest (like an airplane wing, for example) or on surfaces created within the flowfield (such as coordinate planes or isosurfaces of other variables). These views of scalar properties often help you initially understand what the flowfield looks like and begin to develop a deeper understanding of what is happening in your solution.

A comparison of the surface pressure coefficients for a generic fighter configuration at $\alpha = 30°$ is shown in Fig. 9.10.[14] The pressure coefficient, C_p (defined in Equation 4.19 and used in Fig. 9.10), is a very useful way to compare pressures between different solutions since it is nondimensionalized by

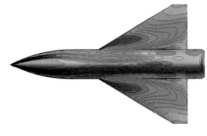

C_p

— 0.300
— −0.633
— −1.567
— −2.500

Figure 9.10   Surface pressures for a generic fighter at $\alpha = 30°$, $\beta = 2°$ (Ref. 14; a full color version of this image is available on the website: www.cambridge.org/aerodynamics).

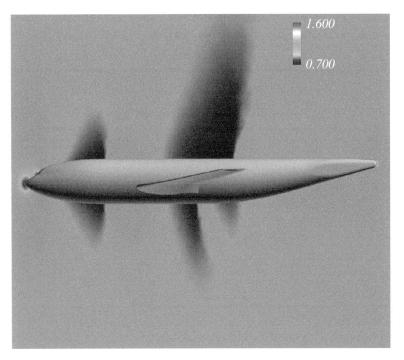

Figure 9.11 Centerline plane showing pressure coefficients around a generic commercial transport; note regions of high-speed flow (dark), quickly followed by lower speed flow (light), which helps to identify shock locations (Ref. 15; reprinted by permission of the American Institute of Aeronautics and Astronautics, Inc; a full color version of this image is available on the website: www.cambridge.org/aerodynamics).

the freestream dynamic pressure. Now that you can see the surface pressures, you have to determine what you are seeing. For example, the scale states that positive pressure coefficients are represented by red shades, while very negative pressure coefficients are represented by blue. Therefore, dark blue regions have very low pressures compared with the freestream dynamic pressure. What has caused these pressures to be so low? In this case, the low pressures are caused by a vortex over the wing, which we will visualize in greater detail later.

Another way to use scalar visualization is to create a fictitious planar surface within the flowfield and show a scalar quantity on that surface. Figure 9.11 shows this being done around a transport aircraft, where the vertical geometric plane along the centerline shows pressure coefficients. In this case, the pressure coefficient is showing the presence of transonic shocks, since shocks produce a nearly discontinuous increase in pressure (see Chapter 4 for more details on this).[15] There are various ways to visualize a shock, but it is not easy to "see" a shock in three dimensions; it is usually easier to view a shock in two dimensions, as has been done here. Traditionally, shocks are shown on surface pressure coefficient graphs, where the discontinuity is usually obvious. Pictorially, shocks are also easy to see as abrupt changes in color on pressure or velocity contour pictures. Notice that the shocks occur at several locations above and below the fuselage, which are seen where the colors rapidly change from blue (lower pressure) to green (higher pressure).

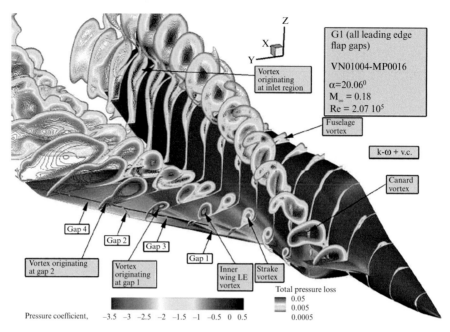

Pressure coefficient,   −3.5 −3 −2.5 −2 −1.5 −1 −0.5 0 0.5

Total pressure loss
0.05
0.005
0.0005

Figure 9.12   Multiple crossflow planes showing total pressure loss with pressure coefficients on the surface of the X-31 (Ref. 16; Courtesy of Okko Boelens of NLR; a full color version of this image is available in the color insert pages of this text as well as on the website: www.cambridge.org/aerodynamics).

Also, notice the stagnation regions at the front of the airplane, which are bright red (very dark) to represent high pressure.

A very helpful way to use plotting planes is shown in Fig. 9.12. A detailed view of the complexity of the flowfield for the X-31 is shown using multiple crossflow planes which are colored with total pressure loss (the difference between the freestream total pressure and the total pressure in the plotting plane). The surface of the X-31 is also colored by the static pressure coefficient.[16] Notice how easy it is to see the various vortices that have formed around the aircraft and how they interact with each other. This is especially true in the vicinity of the wing leading edge, where multiple vortices are formed by the gaps between the leading-edge slats, all of which interact with each other and the vortices coming from the strake and the wing leading edge. While this type of visualization is time consuming, it is especially valuable in improving your understanding of the flow. Fortunately, most modern visualization packages have some sort of "restart" capability so the time consuming actions do not have to be repeated for every unique solution.

Figure 9.13 (repeated from Figure 8.28) gives a good example of combining surface scalar flow visualization with quantitative comparisons of pressure from numerical simulations and wind tunnel experiments.[17] This type of detail, combining the computed flow visualization and experimental data, can be very informative to help you quickly decide how well a CFD simulation models the details of the experimental flowfield. Again, while this type of

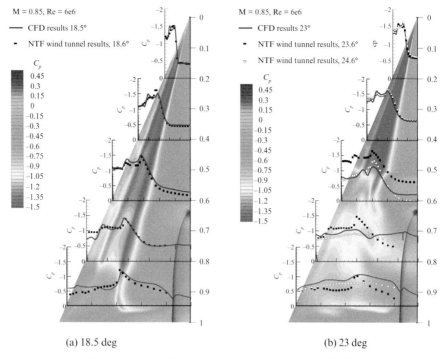

(a) 18.5 deg   (b) 23 deg

Figure 9.13   Flowfield above a delta wing visualized with surface pressure coefficient and overlaid with graphs comparing the prediction with experimental data (Ref. 17; Courtesy of Kenneth Badcock of the University of Liverpool; a full color version of this image is available in the color insert pages of this text (see Fig. 8.23) as well as on the website: www.cambridge.org/aerodynamics).

visualization can be quite time consuming to create, it is extremely valuable in understanding a great deal about the flow in a very compact format.

9.5 How to View Vector Properties

As we mentioned in Section 9.3, vector arrows are relatively easy to visualize since they are directly available from the solution; you need only create the vector from the three velocity components and determine how to draw the vector. The problem arises when the vectors are not primarily within one plane of the flow, since a two-dimensional view of three-dimensional vectors can hide information. This section will discuss different ways to view vector properties in light of these difficulties.

9.5.1 Commonly Used Vectors in Fluid Dynamics and Flow Visualization

Before we look at examples of flow visualization using vectors and their magnitudes, we will review some of the commonly used vectors in fluid dynamics. You will know the most common vector, velocity, where $\vec{V} = u\hat{i} + v\hat{j} + w\hat{k}$. However, other vectors are also commonly used in flow visualization, some of which may be more familiar to you than others.

9.5.1.1 VORTICITY

As discussed in Chapter 4 and Appendix C, vorticity is a measure of the rotation and shear of a fluid element. Mathematically, vorticity is the curl of the velocity vector, or, from Equation 4.3,

$$
\begin{aligned}
\vec{\omega} &= \vec{\nabla} \times \vec{V} \\
&= \begin{vmatrix} \hat{i} & \hat{j} & \hat{k} \\ \dfrac{\partial}{\partial x} & \dfrac{\partial}{\partial y} & \dfrac{\partial}{\partial z} \\ u & v & w \end{vmatrix} \\
&= \left(\dfrac{\partial w}{\partial y} - \dfrac{\partial v}{\partial z} \right)\hat{i} - \left(\dfrac{\partial w}{\partial x} - \dfrac{\partial u}{\partial z} \right)\hat{j} + \left(\dfrac{\partial v}{\partial x} - \dfrac{\partial u}{\partial y} \right)\hat{k} \\
&= \omega_x \hat{i} + \omega_y \hat{j} + \omega_z \hat{k}
\end{aligned}
\tag{9.4}
$$

Vorticity can sometimes be used directly in flow visualization, but more commonly the various magnitudes of the vorticity vector, ω_x, ω_y, ω_z, and $|\vec{\omega}| = \sqrt{\omega_x^2 + \omega_y^2 + \omega_z^2}$, are used. Magnitudes are often easier to visualize since the scalar methods described in Section 9.4 can be used directly, rather than trying to deal with vector quantities. When using vorticity magnitudes, it will help you to remember that each quantity is the amount of rotation around the individual axis (ω_x is the rotation around the x-axis, etc.).

9.5.1.2 HELICITY DENSITY AND RELATIVE HELICITY

Helicity density is the scalar product of the velocity and vorticity vectors, as shown in Equation 9.5.[18]

$$
H = \vec{V} \cdot \vec{\omega} = \vec{V} \cdot \left(\nabla \times \vec{V} \right)
\tag{9.5}
$$

The main value in using relative helicity instead of helicity density is that you will be able to roughly nondimensionalize the helicity density to manageable values, since helicity density can have very large numerical values in many cases. Relative helicity is defined as:

Helicity density is a scalar quantity which can show a sense of rotation (clockwise or counterclockwise) and magnitude, which is often useful for visualizing multiple vortices in close proximity to each other, especially when some of them are rotating in opposite directions. An example of this is the primary and secondary vortices over a delta wing, as shown in the Flow Topology Concept Box:

$$
\frac{H}{|\vec{V}||\vec{\omega}|}.
\tag{9.6}
$$

One of the drawbacks to helicity is how the function behaves for vortices that are not aligned with the local velocity vector. As the angle between the velocity and vorticity vectors varies away from zero, the helicity decreases to zero. For many situations (like vortices over a delta wing), this is not a problem, since the vortices are aligned with the flow direction. For a situation like the tip vortices from a helicopter rotor, however, helicity does not show the vortex very well at all positions of the rotor, since, in some cases, the scalar product is large (when the rotor is at each side of the vehicle), while in other cases the value is very small (when the rotor is ahead of or behind the vehicle).

9.5.1.3 *Q*-CRITERION

The *Q*-criterion is a relatively new variable for visualizing vortices.[19] Specifically, the *Q*-criterion is defined as:

$$Q = \left(\left| \vec{\xi} \right|^2 - \left| \vec{S} \right|^2 \right) / 2 \qquad (9.7)$$

where $\vec{\xi} = \left(\nabla \vec{V} - \nabla \vec{V}^T \right) / 2$ is the rate of rotation, and $\vec{S} = \left(\nabla \vec{V} + \nabla \vec{V}^T \right) / 2$ is the rate of shearing. In two dimensions this would be:

$$\left| \vec{\xi} \right| = \left(\frac{\partial v}{\partial x} - \frac{\partial u}{\partial y} \right) / 2 \ \text{ and } \ \left| \vec{S} \right| = \left(\frac{\partial v}{\partial x} + \frac{\partial u}{\partial y} \right) / 2. \qquad (9.8)$$

The *Q*-criterion is an attempt to determine when a vortex exists by stating $Q \geq 0$, since this would require that the rate of rotation be greater than the rate of shearing at some point in the flow. Since this criterion does not depend on local alignment of the vorticity and velocity vectors, nearly all vortices are relatively easy to visualize with Q.

Other vector magnitude criteria are now commonly used for visualizing vortices, some of which can be fairly complicated to calculate and understand. Two other relatively recent ways to visualize vortices include the Δ-criterion[20] and the λ_2-criterion.[21] In addition, there are methods based on Eigenvalues and Eigenvectors of the velocity gradient tensor in the vicinity of the critical points of the flow (see Flow Topology Concept Box).[22,23]

9.5.1.4 SHEAR STRESS VECTOR

Shear stress is the force per unit area acting parallel to a solid surface due to the friction of the fluid. Shear stress for a two-dimensional boundary-layer flow over a flat plate is defined as:

$$\tau_w = \mu \frac{du}{dy}\bigg|_w \qquad (9.9)$$

where μ is the coefficient of viscosity of the fluid and du/dy is the velocity shear at the surface (or wall). The shear stress is often represented as a non-dimensional skin friction coefficient (see Section 4.5.1 for more details):

$$C_f = \frac{\tau_w}{q_\infty}. \qquad (9.10)$$

In general three-dimensional flows, however, the surface shear stress is a vector with both magnitude and direction, depending on the direction of the local flow above the surface (see Section 3.3.2.2 for details on the stress definition). This vector also serves as a way to determine the direction of the flow directly above the surface and can be used in flow visualization techniques such as simulating surface oil flows.

9.5.2 Examples of Vector Flow Visualization

Now that we have examined many of the most common vector quantities used in fluid dynamics, we will look at examples of flow visualizations using some of these quantities. Of course the most common use of vectors is for vortex visualization. If you remember the two-dimensional vortex from Chapter 4, some of the important characteristics of a vortex are the high velocities near the center (which correspond to low pressures) and the high vorticity in the vicinity of the core. These physical measures form the basis for visualization of vortices, but each approach has its own particular limitations, which is why a large number of methods have been developed for vortex visualization.[24]

9.5.2.1 VECTOR ARROWS

An example of using velocity vector arrows on a cutting plane in a flow is shown in Fig. 9.14. In this case we are looking at the flow around a delta wing at a high angle of attack (similar to the flow discussed in the Flow Topology Concept Box) in the vicinity of the leading edge.[25] The velocity vectors are projected onto a plane and then colored with local static pressure to help indicate regions of high pressure and low pressure. The flow is separating at the leading edge, rolling up into a large primary vortex (visible at the upper left corner of the figure), and creating a counter-rotating secondary vortex (the region with the low pressure vectors).

9.5.2.2 **STREAMLINES/STREAM RIBBONS**

Flow streamlines are obtained by integrating the trajectory of a fluid particle through a velocity field. If the flow is unsteady, various types of lines, including streaklines, pathlines, and streamlines, can occur. For steady flow all of these lines are the same. While streamlines are valuable flow visualization tools, sometimes it is difficult to see if a streamline is rotating (such as within a vortex), so stream ribbons can be used for that purpose. Figure 9.15 shows the use of streamlines and stream ribbons to visualize the vortical flow over a UCAV.[26] In this case, the streamlines are used in conjunction with other

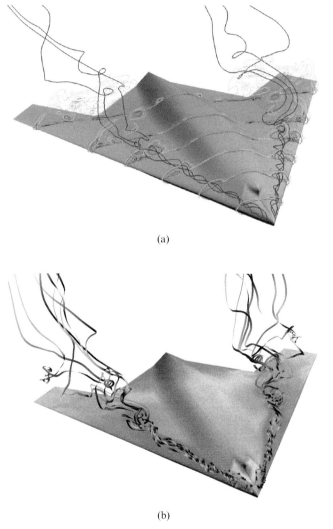

(a)

(b)

Figure 9.15   Flow around a UCAV at angle of attack: (a) leading-edge vortex streamlines and cross-sections of vorticity, (b) leading-edge vortex stream ribbons and surface colored by pressure (Ref. 26; a full color version of this image is available on the website: www.cambridge.org/aerodynamics).

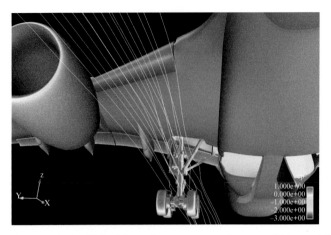

Figure 9.16 Streamlines in the vicinity of the landing gear on a generic commercial transport (Courtesy of Airbus; a full color version of this image is available in the color insert pages of this text as well as on the website: www.cambridge. org/aerodynamics).

flow visualization techniques, such as cross sections of vorticity (Fig. 9.15a) or coloring the surface with pressure (Fig. 9.15b). Notice in Fig. 9.15b that the vortices lie directly above a region with low pressure (visible from surface shadings), since vortices have high velocities and induce low pressure footprints on the surface of the vehicle.

Figure 9.16 shows a case where streamlines are used to determine the flow patterns in the vicinity of a landing gear design on a generic commercial transport. The aircraft surface is colored by the pressure coefficient to show any unusual impact of the flowfield on the aerodynamics of the wing. The airplane has its leading- and trailing-edge devices extended, and the impact of the landing gear on the flow under the wing is seen with streamlines. Notice that all of the streamlines shown flow past the landing gear without significant disruption or with no visible flow into the wheel well. This type of flow visualization is useful for determining the validity of design choices on the aerodynamics of a complex configuration.

A dynamic flowfield visualized using vorticity is shown in Fig. 9.17. Here, flow around an automobile is visualized with a number of techniques combined together, which gives an expanded view of what is happening around the car.[27] Streamlines and velocity vectors are shown with streaklines, ribbons, and glyphs (vector symbols showing the flow direction at points in the flowfield). A vertical plane along the length of the car is also shown to aid in the visualization process.

Another interesting visualization using vorticity is shown in Fig. 9.18. The visualization shows a liquid fuel spray that is widely used to improve combustion performance. However, the physical mechanism of liquid atomization has not been understood well because the phenomenon is highly multi-scale and turbulent. The figure shows a direct numerical simulation of a liquid fuel

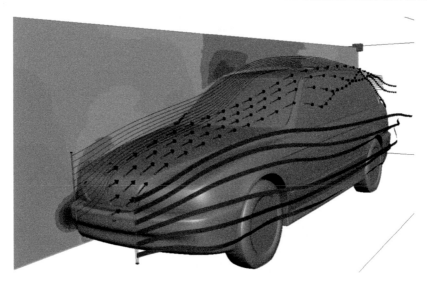

Figure 9.17   Streaklines, stream ribbons, and glyphs, used in conjunction with a cutting plane, to show the flow around an auto-mobile (Ref. 27; Courtesy of Wolf Bartelheimer of BMW; a full color version of this image is available in the color insert pages of this text as well as on the website: www.cambridge.org/aerodynamics).

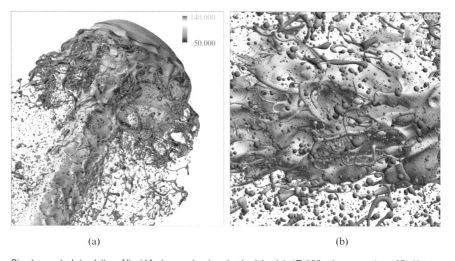

(a)   (b)

Figure 9.18   Direct numerical simulation of liquid fuel spray showing atomized droplets (*FieldView* image courtesy of Dr. Matsuo, Japan Aerospace Exploration Agency, JAXA; a full color version of this image is available on the website: www.cambridge.org/aerodynamics). (a) top view (b) side view.

jet with 6 billion grid points to better understand how the spray improves performance.

9.5.2.3 VORTEX VISUALIZATION USING VECTOR MAGNITUDES

As we mentioned earlier, magnitudes of velocity vectors, rather than the vectors themselves, are commonly used in flow visualization. An example of this is shown in Fig. 9.19, where a delta wing at a 27 degree angle of attack

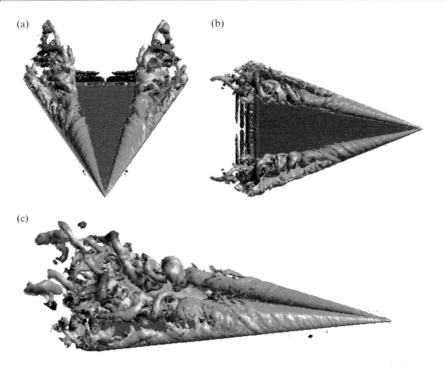

Figure 9.19 Instantaneous isosurfaces of vorticity magnitude colored by the spanwise component of vorticity for three views of a 70 deg delta wing; $\alpha = 27$ deg, $M = 0.069$, and $Re = 1.56 \times 10^6$. (Ref. 28; a full color version of this image is available in the color insert pages of this text as well as on the website: www.cambridge.org/aerodynamics).

has large vortices over the upper surface. The vortices are being visualized with a combination of methods previously described, in order to visualize the core of a vortex and compare the results with equivalent experimental data.[28] First, isosurfaces of vorticity magnitude ($|\vec{\omega}| = \sqrt{\omega_x^2 + \omega_y^2 + \omega_z^2}$) are created to see the size of the vortex. These isosurfaces are then colored by the spanwise component of vorticity (ω_y) to show details of the flow that otherwise might not be visible. Notice the complexity of the visualization that results, and the details of the vortex flow that are visible due to these choices of visualization.

Figure 9.20 shows the vortical structures and wake produced by a generic commercial transport when the slats and flaps are deflected, which happens during takeoff and landing. The wake structures are visualized with an isosurface of vorticity, but in this case the isosurface is not colored by some other quantity. This helps to show the size and extent of the wake structures to aid in design evaluation and to find the impact of the slat/flap system on the lift and drag of the airplane. Notice that there are vortical structures coming from the edges of the slats and flaps, the jet engine, and the wing tips. Additionally, notice the "crinkle" effect on the isosurfaces which results from the visualization algorithm identifying complete mesh elements within the specified isosurface threshold instead of (inappropriately) trying to interpolate a surface within the elements.

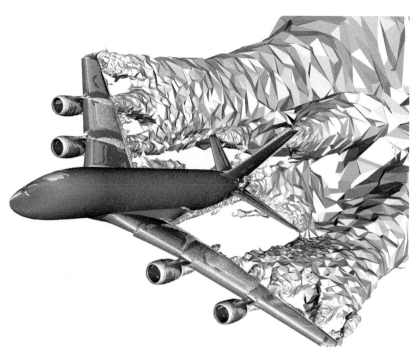

Figure 9.20 Isosurfaces of vorticity showing the wakes of the flaps and slats on a generic commercial transport (Courtesy of Airbus; a full color version of this image is available on the website: www.cambridge.org/aerodynamics).

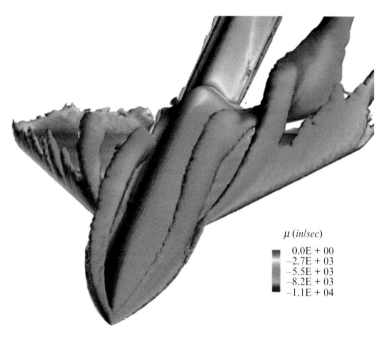

μ (*in/sec*)

0.0E + 00
−2.7E + 03
−5.5E + 03
−8.2E + 03
−1.1E + 04

Figure 9.21 MTVI vortex breakdown visualized with isosurfaces of $Q = 1.0 \times 10^7$ s^2 colored by axial velocity; $\alpha = 30$ deg, $\beta = 2$ deg, $Re = 2.68 \times 10^6$, and $M = 0.40$ (Ref. 14; Courtesy of Tiger Jeans of the USAFA High Performance Research Center; a full color version of this image is available on the website: www.cambridge.org/aerodynamics).

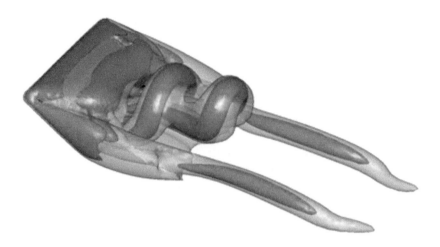

Figure 9.22 Separated flows behind low-aspect-ratio wings have been of interest for the development of micro air vehicles as well as for understanding the flight mechanism of insect wings. RMS of $|\omega|$ and Q (Ref. 29; Courtesy of Kunihiko Taira of Florida A&M/Florida State University and Tim Colonius of Caltech; a full color version of this image is available on the website: www.cambridge.org/aerodynamics).

9.5.2.4 VORTEX VISUALIZATION USING THE Q-CRITERION

Figure 9.21 shows the complex vortical flowfield above a generic fighter aircraft. This geometry was part of a large research program called MTVI (Modular Transonic Vortex Interaction) coordinated by NASA Langley Research Center.[14] The aircraft is at 27 degree angle of attack and a sideslip angle of 2 degrees. In this case the Q-criterion has been used to visualize the vortices, and the surfaces created are also colored by the axial velocity, u. The surface of the aircraft is colored by pressure to aid in a more complete visualization. Notice that the Q-criterion results in a very clear vortex, and the contour coloring helps you to see regions of high and low velocity, especially near a point where vortex breakdown is taking place.

Another example of the use of the Q-criterion is shown in Fig. 9.22. In this case, which is the flow over a low aspect ratio wing at low speeds, the complex vortex field includes vortices that are not aligned with the nominal freestream direction.[29] Therefore, there are two visualization tools being used: the magnitude of the vorticity vector (since this flow is unsteady, this quantity is actually a root-mean-square value) and the Q-criterion. In order to see the isosurfaces of both of these visualization tools at the same time, the Q-criterion has been made transparent so that the vorticity isosurfaces are also visible. Making some surfaces transparent is another useful tool in flow visualization.

A very complex flowfield is created by rotors on a helicopter, especially during hover or near-hover conditions. Figure 9.23 shows the V-22 Osprey rotor simulation using a highly refined mesh. This unsteady simulation was

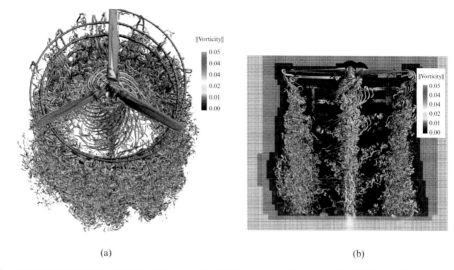

(a)   (b)

Figure 9.23 Isosurfaces of vorticity magnitude in the wake of a V-22 Osprey rotor. The complex flowfield includes the rotor tip vortex as well as the shear layers formed by the rotor (Ref. 30; Courtesy of Neal Chaderjian of NASA Ames Research Center; a full color version of this image is available in the color insert pages of this text as well as on the website: www.cambridge.org/aerodynamics). (a) top view (b) side view showing grid refinement.

post-processed on a large-scale workstation to show results from a grid with 700 million cells and more than 14,000 zones in an overset AMR configuration. "Accurately predicting aeromechanic performance and noise production for rotorcraft is very challenging and requires a multi-disciplinary approach to account for rotor blade aerodynamics, blade flexibility, and blade motion for trimmed flight. Moreover, rotor blades encounter the tip vortices of other rotor blades resulting in very complex blade-vortex interactions and vortex wake structures."[30]

9.5.2.5 VORTEX VISUALIZATION USING VORTEX TRACKING

Vortex tracking is another relatively new approach in flow visualization. In this case, where a set of vortices exists over the wing of the F-16XL aircraft, it might be advantageous to have an automatic way to find the vortices and show them – this is known as vortex tracking.[31] Vortex tracking uses one of several vortex identification methods to find and track the center of a vortex throughout the flowfield. The results, as shown in Fig. 9.24, identify multiple vortices coming from the leading edge, the crank of the leading edge, and the tip missile of the aircraft. This approach can make a complex flowfield, with many features, relatively easy to visualize and understand.

9.5.3 Skin Friction Lines

A flow visualization technique common in wind tunnel testing is to determine the off-surface flowfield using oil on the surface of the wind tunnel

Figure 9.24 Vortex tracking on the F-16XL CAWAPI aircraft simulation (Ref. 31; reprinted by permission of the American Institute of Aeronautics and Astronautics, Inc. a full color version of this image is available on the website: www.cambridge. org/aerodynamics).

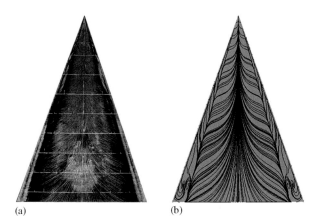

(a)   (b)

Figure 9.25 Surface oil flow on a delta wing: (a) wind tunnel results (Ref. 32), (b) CFD simulation using SA-DES turbulence model; $M = 0.069$, $\alpha = 27$ degrees, $Re = 1.56 \times 10^6$ (Ref. 28; a full color version of this image is available on the website: www.cambridge.org/aerodynamics).

model, as shown in Fig. 9.25a. Prior to turning on the wind tunnel, a thin film of oil is placed on the surface of the model. Then, the tunnel is turned on with the model at some pre-determined angle of attack, 27 degrees in this case. Once the flow has been established for a short period of time, the tunnel is turned off and a picture of the model is taken. Engineers use flow topology mathematics to determine separation lines, attachment lines, saddles, etc., as we discussed earlier. All of these surface flow features point to off-surface flow structures such as vortices or other separated flow phenomena. The equivalent of a surface oil flow in a wind tunnel is a surface oil flow simulation using CFD. Once the flow solution has been obtained, a post-processing software tool can be used to "release" particles just above the surface of the geometry that constrain the particle to move only within the region immediately above the surface. This produces results such as those shown in Fig. 9.25b, which closely resembles the wind tunnel results of Fig. 9.25a.[28]

Profiles in Computational Aerodynamics: Wei Shyy

"I was born and grew up in Taiwan. The school system there was, and still is, largely built on a competitive exam system. After taking multiple exams in my youth, I found myself in National Tsing-Hua University in Hsinchu. My major was power mechanical engineering, which basically was a hybrid between mechanical and electrical engineering, a very "multidisciplinary" program.

"With no attendance taken by professors, and away from home with many newly found classmates staying in the same dormitory for four years, the Tsing-Hua experience was unique to its students and has had a lifelong impact on me. After some doubt about my major in freshman and sophomore years, I picked up fluid dynamics enthusiastically. I was fortunate to have excellent and caring teachers, who interacted with us in and out of the classroom. In my senior year, I took a two-semester sequence of numerical methods in the mathematics department. These and fluid mechanics topics were my favorites in Tsing-Hua.

"After college, like others, I served in the military for about two years to fulfill a citizen's duty. I was a lieutenant in the armor division. It was a demanding experience with little opportunity for leisure time. Toward the end of the military service, what I learned in college was mostly a fuzzy memory. After I finished military service, I was fortunate to be offered a scholarship to do graduate study in aerospace engineering at the University of Michigan. I arrived in Ann Arbor in August 1979, shortly after I left my M48 tank. I was extremely lucky to have come to Ann Arbor as my first stop in the United States. I entered an extremely scholarly and friendly department, and had a good fortune to study under Professor Tom Adamson, Jr.

"My PhD thesis was on the analysis of hydrocarbon emissions from an internal combustion engine. The research was largely analytical, based on singular perturbation and matched asymptotic expansion techniques. At the end of analytical development, I needed to solve nonlinear, time-dependent partial differential equations with undefined boundary location. That was my first taste of doing original computation for research. After I received my PhD degree in summer 1982, I spent some time as a consultant helping incorporate my HC emissions model into Ford's engine simulation code.

"I joined the GE Research and Development Center in Schenectady, New York, in 1983. There, I had substantial freedom and support to develop computational capabilities with the goal of computing gas turbine combustor flows. With little existing tools and many open and basic issues waiting to be addressed, it was a great time to be in GE. Besides turbulent reacting flows, I was involved in numerous activities at GE, including turbomachinery for hydraulic and steam turbines, high-pressure discharge lamp, materials processing, etc. I was also encouraged to do basic research and publish. My association with GE in many ways strongly influenced my approach to do research. In other words, I am interested in addressing issues encompassing both fundamental research and engineering application, including:

- development of original numerical and modeling techniques for problems related to thermo-fluid dynamics and their interaction with other disciplines;
- advancement of computational tools to a point that they form a comprehensive capability to tackle physical and engineering issues;
- consistent emphasis on close collaboration between theory/computation and experiment;
- extension of scientific research to address engineering issues arising from optimization, assessment, and design tool development.

"I joined the University of Florida in 1988, and stayed there until I returned to my alma mater, University of Michigan, in January 2005. During this period, I had opportunities to broaden my interest in areas related to multiphase flows, bio-inspired flight, fluid-structure interactions, battery technologies, design optimization, interfacial transport and moving boundaries, combustion, materials processing, and micro-scale biofluid dynamics. I enjoy seeing the common computational techniques applicable to seemingly very different physical problems, while in the meantime appreciate that in spite of the similar mathematical structures, the physical implications and mechanisms associated with these problems are really quite distinct."

Professor Shyy was named provost of the Hong Kong University of Science and Technology in Fall 2010.

9.6 Newer Flow Visualization Approaches

Over the past few years, a number of new approaches for performing flow visualization have been developed. Each of these approaches offers a way to see specific flow features in ways that mimic experimental methods (such as particle image velocimetry or Schlieren).

9.6.1 Line Integral Convolution

The line integral convolution (LIC) approach combines a vector field, which is sampled on a uniform rectilinear grid (or sometimes on a structured

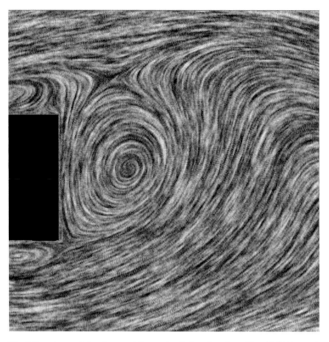

Figure 9.26 Line integral convolution image of the unsteady flow for a base flow (Ref. 34; courtesy of the DoD High Performance Computing Modernization Program.).

curvilinear grid) with a texture map image (often a random noise field). The texture map image is approximately the same dimension as the original uniform grid, which produces a result that looks blurred by the flowfield. The pixels in the output image are produced by the one-dimensional convolution of the texture pixels along a streamline with a filter kernel.[33] Streamlines are started at each grid cell, then the input texture map is smeared in the direction of the vector field. An example of this for a base flow is shown in Fig. 9.26.[34]

9.6.2 Numerical Schlieren

Schlieren photography of high-speed flowfields has been conducted for decades, but reproducing the look of a Schlieren picture has not always been easy. Schlieren photography (explained in the accompanying Concept Box) shows the gradient of the density field, $|\nabla\rho|$. When this quantity is reproduced in flow visualization software, the results can be quite striking, especially when visualizing shocks and other high-speed waves, as shown in Fig. 9.27.[35]

Notice that the shocks at the nose of the projectile are clearly visible, and so are the expansion waves at the junction of the nose and the cylindrical body. The shocks and expansion waves, including the wake flow structures, are also clearly visible in the base region of the projectile.

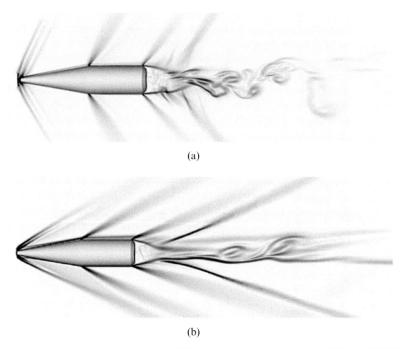

(a)

(b)

Figure 9.27 Numerical Schlieren visualizations of the flowfield for a projectile: instantaneous flowfield from RANS/LES simulation at $\alpha = 3°$ for (a) Mach 1.2, and (b) Mach 2.5 (Ref. 35; Courtesy of James DeSpirito, U.S. Army Research Laboratory).

Computational Aerodynamics Concept Box

Schlieren Photography

Schlieren photography has been around for centuries, but really came of age during the twentieth century with the growth of supersonic flow research. The Schlieren concept takes advantage of light refracting through density gradients in the flowfield by passing parallel

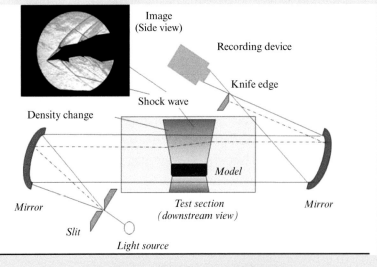

Layout of a Schlieren System (Courtesy of NASA Glenn Research Center)

light waves through the test section of a wind tunnel. Those waves are then focused and a knife edge is placed just to the side of the focal point. Refracted light waves that have been "bent" by density gradients in shocks and expansion waves are blocked by the knife edge, creating dark regions that make the flow structures visible.

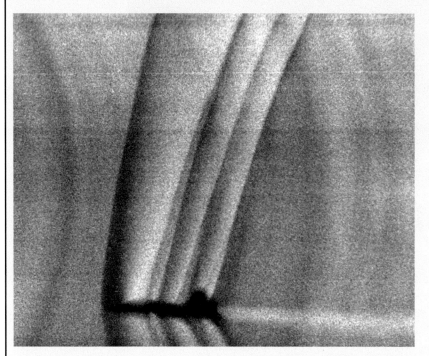

Schlieren photograph of a T-38 at Mach 1.1 and 13,700 feet altitude (Courtesy of NASA Dryden Flight Research Center)

A "natural" Schlieren photograph of a T-38 flying at Mach 1.1 and an altitude of 13,700 feet is shown here (photo courtesy of NASA Dryden Flight Research Center). Notice the multiple shock wave and expansion wave systems, as well as the wake of the aircraft.

9.6.3 Feature Extraction

Often, datasets are so large, and flow features are so complex, that it is difficult to find all important flow features. The old saying, "it's like looking for a needle in a hay stack," comes to mind for these situations. Some more modern flow visualization features have been developed specifically to take care of this problem, which "... lifts the visualization to a higher level of abstraction, by extracting physically meaningful patterns from the dataset."[36]

In general, feature extraction examines a flow solution dataset and finds what you are looking for (which, of course, means the user has to be involved!). If you are looking at the flow over a delta wing and are concerned primarily about separation and attachment lines (which determine vortex location, as shown in Fig. 9.28), you could create a feature

Figure 9.28 Separation and attachment lines on a delta wing using feature extraction (Ref. 37; Courtesy of NASA Ames Research Center; a full color version of this image is available on the website: www.cambridge.org/aerodynamics).

extraction algorithm to automatically find these flow features and visualize them.[37] One important benefit of feature extraction is that since it is automated by definition, it may typically be accomplished in-line with the flow simulation or at a minimum on the same (remote) machine as the flow simulation. This drastically decreases the time-to-visualization for very large datasets.

As feature extraction has evolved, it has become more and more apparent that some flow features are more difficult to visualize than others. For example, CFD is often used to predict details about turbulence, and seeing the details of turbulent flows can be quite difficult (see Fig. 8.12 for example). So, in addition to finding specific flow features, researchers have begun to come up with creative ways to show the features. For example, Fig. 9.29 shows a region of turbulent vortex structures, which are visualized with isosurfaces in the shape of ellipsoids, where the ellipsoids are outlined by red markers to make their shape and relative position more obvious to the user.

Another way to visualize turbulent flow structures is with solid surfaces in the shape of ellipsoids or "stick" or "skeleton" icons, as shown in Fig. 9.30. Notice that both pictures show the same flowfield, but the visualization takes on a slightly different appearance, depending on how the features are marked.

A simple way to see the power of feature extraction is to apply it to a flow that we already understand. Figure 9.31 shows the flow behind a circular cylinder as computed with CFD. This flow should have a Karman vortex street of alternating vortices, with each vortex rotating in the opposite direction (see Fig. 8.13 for an example). In this case, the vortices are extracted and then marked with colored ellipsoids, where the color represents the sense

Figure 9.29   Vortices in a dataset with turbulent vortex structures, visualized using isosurfaces and ellipsoids (Ref. 36; Courtesy of Frits Post and the TU Delft Visualization Group; a full color version of this image is available in the color insert pages of this text as well as on the website: www.cambridge.org/aerodynamics).

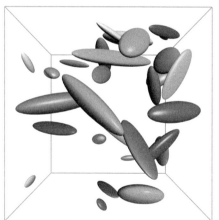 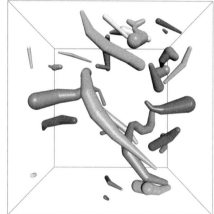

Figure 9.30   Turbulent vortex structures represented by ellipsoid icons (left) and skeleton icons (right) (Ref. 36; Courtesy of Frits Post and the TU Delft Visualization Group; a full color version of this image is available on the website: www.cambridge.org/aerodynamics).

of rotation. These types of visualization tools can be very important when looking at a dataset for the first time, since they can show you where interesting flow features are taking place or where additional grid refinement may be necessary due to unexpected flow features.

Finally, an example of Eigenvector feature extraction is shown in Fig. 9.32. Here, flow over the F/A-18 at a high angle of attack is visualized using feature extraction. The motivation for doing this is explained: "For large data

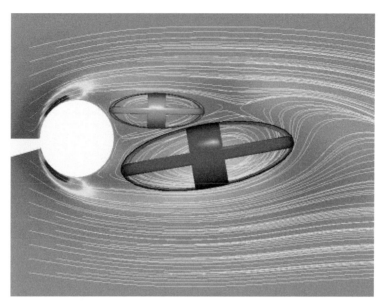

Figure 9.31 Vortices behind a tapered cylinder visualized with ellipsoids (the color of the ellipsoids represents the rotational direction) (Ref. 36; Courtesy of Frits Post and the TU Delft Visualization Group; a full color version of this image is available on the website: www.cambridge.org/aerodynamics).

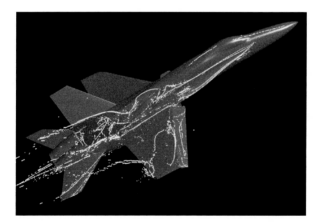

Figure 9.32 Vortices over the F-18 aircraft, showing the leading-edge extension vortex, as well as vortices over the wing (Ref. 37; Courtesy of NASA Ames Research Center; a full color version of this image is available in the color insert pages of this text as well as on the website: www.cambridge.org/aerodynamics).

sets like this, the computations for the particle traces can take several hours for each set of seed locations. Finding the right set of seed points to highlight a particular vortex can therefore take days."[38] In this case, the feature extraction shows the core of vortices, including vortices that have experienced breakdown. Also, "The vortex cores extracted from the F/A-18 data set revealed many other interesting features, such as the vortices separating off the leading edge flap and the vortices entering the engine intakes. Although irrelevant to this study, these features are important to aircraft designers in that they could have a negative impact on maneuverability and on the engine's performance."[38]

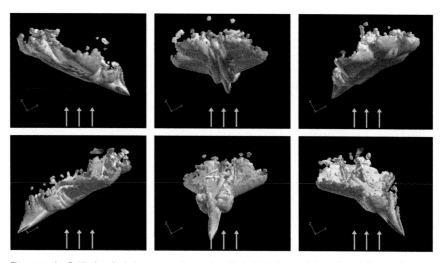

Figure 9.33   Flow over the F-15 aircraft during a prescribed spin, with individual snapshots of the solution serving to give the impression of motion (arrows from the bottom show the direction of the flow) (Ref. 39; Courtesy of James Forsythe of the U.S. Air Force Academy; a full color version of this image is available in the color insert pages of this text as well as on the website: www.cambridge.org/aerodynamics).

9.6.4 Unsteady Flow and Movies

Perhaps some of the more difficult flows to visualize are flows that are highly unsteady. The unsteadiness can come from massively separated flow (such as stall on a wing or flow over a blunt body) or from the vehicle of interest moving during the simulation (such as a maneuvering aircraft). "Seeing" the flow in two dimensions becomes a problem due to the four-dimensional nature of the flowfield (three spatial dimensions and one temporal dimension).

An example of an unsteady flowfield is shown in Fig. 9.33, where a simulation of an F-15 undergoing a spin has been conducted.[39] In this case, the flow is unsteady both for the vehicle dynamics (the spin) and the flow unsteadiness (due to separation and turbulence). The figure shows the spin during one complete cycle of revolution, but at only six positions. The surface of the F-15 is colored by pressure, and isosurfaces of vorticity are used to show the separated flow region.

While this type of visualization can show many of the flow features of interest (such as the flow separating from the leading and trailing edges of the wing and the resulting complex flow features that form above the wing), a great deal of information is also lost. In order to show the flow in more detail, a large number of these "snapshots" of the flow can be added together to form the frames of a movie. The resulting movie is not easily shown on the pages of the book but can be viewed at www.cambridge.com/aerodynamics.

Another interesting way to visualize unsteady flow, in this case, the flow around a maneuvering aircraft, is to superimpose individual images to show the aircraft as it moves (see Fig. 9.34).[40] In this case, the Ranger aircraft is performing an Immelmann turn, simulated with CFD. The surface of the

Figure 9.34   Ranger aircraft simulation undergoing an Immelman turn; the aircraft size to the flight path scale is 50:1 for clarity (Ref. 40; Courtesy of Mehdi Ghoreyshi of the University of Liverpool; a full color version of this image is available on the website: www.cambridge.org/aerodynamics).

aircraft is shaded by pressure coefficient to show the impact of the maneuver on the surface pressures throughout the turn.

An additional word of caution is appropriate as we conclude our study of flow visualization. Never forget that flow visualization is a tool. Being able to "see" the flow can add a great deal of understanding and greatly aid in design and modification of a configuration. Flow visualization cannot, however, answer all questions, nor should it be used to make precise comparisons for verification and validation purposes. Our good friend Ed Tinoco told us, "While flow visualization pictures are very useful for understanding the solutions, they do not take the place of the ordinary 2-D scalar plots. The FAA is not going to certify an airplane on the basis of a fancy color plot, but will rely on well annotated 2-D scalar plots." As with all concepts we have discussed in this book, flow visualization is another tool in the aerodynamicist's collection of capabilities that should be used with judgment and experience.

Summary of Best Practices

1. Choose the best way to show the details of the flow, even if the results are not very flattering (do not hide poor results with clever pictures).
2. Always include a color scale for each quantity being visualized.
3. Always cite references for any data you are comparing with.

4. Know your flow visualization software, and know what it can and cannot do.
5. Understand the basic flow visualization options, such as scalar or vector quantities for various flows.
6. Try various ways to show the same flow feature to aid in the viewer's understanding.
7. Unsteady flows may require multiple "snapshot" images or a video to convey the details of the flow.

9.7 Projects

Using the airfoil simulations provided at www.cambridge.org/aerodynamics, try as many of the flow visualization approaches described in this chapter as possible.

9.8 References

1 Tukey, J.W., *Exploratory Data Analysis*, Reading: Addison-Wesley, 1977.

2 Potsdam, M.A., and Strawn, R.C., "CFD Simulations of Tiltrotor Configuration in Hover," *Journal of the American Helicopter Society*, Vol. 50, No. 1, 2005, pp. 82–94.

3 Globus, A., and Raible, E., "13 Ways to Say Nothing with Scientific Visualization," NAS Report RNR-92-006, February 1992.

4 Globus, A., and Raible, E., "14 Ways to Say Nothing with Scientific Visualization," *Computer*, Vol. 27, No. 7, 1994, pp. 86–88.

5 Murman, E.M., and Cole, J.D., "Calculation of Plane Steady Transonic Flows," *AIAA Journal*, Vol. 9, No. 1, 1971, pp. 114–121.

6 Jameson, A., "Numerical Calculation of the Three Dimensional Transonic Flow over a Yawed Wing," AIAA Paper 73-3002, July 1973.

7 Cummings, R.M., Oh, Y.H., and Yang, H.T., "Supersonic, Turbulent Flow Computation and Drag Optimization for Axisymmetric Afterbodies," *Computers and Fluids*, Vol. 24, No. 4, 1995, pp. 487–507.

8 http://www.futuretech.blinkenlights.nl/pcw9-90pi4d25.html

9 Schiff, L.B., Cummings, R.M., Sorenson, R.L., and Rizk, Y.M., "Simulation of High-Incidence Flow About the F-18 Fuselage Forebody," *Journal of Aircraft*, Vol. 28, No. 10, 1991, pp. 609–617.

10 Tobak, M., and Peake, D.J., "Topology of Three Dimensional Separated Flows," *Annual Review of Fluid Mechanics*, Vol. 14, 1982, pp. 61–85.

11 Délery, J.M., "Robert Legendre and Henri Werlé: Toward the Elucidation of Three-Dimensional Separation," *Annual Review of Fluid Mechanics*, Vol. 33, 2001, pp. 129–154.

12 Pobitzer, A., Peikert, R., Fuchs, R., Schindler, B., Kuhn, A., Theisel, H., Matkovic, K., and Hauser, H., "On the Way towards Topology-Based Visualization of Unsteady Flow – the State of the Art," Eurographics STAR Proceedings, 2010, pp. 137–154.

13 Buning, P.G., "Sources of Error in the Graphical Analysis of CFD Results," *Journal of Scientific Computing*, Vol. 3, No. 2, 1988, pp. 149–164.

14 Jeans, T.L., McDaniel, D.R., Cummings, R.M., and Mason, W.M., "Aerodynamic Analysis of a Generic Fighter Using Delayed Detached-Eddy Simulation," *Journal of Aircraft*, Vol. 46, No. 4, 2009, pp. 1326–1339.

15 Watanabe, T., Yamazaki, W., Matsushima, K., and Nakahashi, K., "Wing Shape Optimization of a Near-Sonic Passenger Plane," AIAA Paper 2007-4168, June 2007.

16 Boelens, O.J., "CFD Analysis of the Flow around the X-31 Aircraft at High Angle of Attack," AIAA Paper 2009-3628, June 2009.

17 Schiavetta, L.A., Boelens, O.J., Crippa, S., Cummings, R.M., Fritz, W., and Badcock, K.J., "Shock Effects on Delta Wing Vortex Breakdown," *Journal of Aircraft*, Vol. 46, No. 3, 2009, pp. 903–914.

18 Levy, Y., Degani, D., and Seginer, A., "Graphical Visualization of Vortical Flows by Means of Helicity," *AIAA Journal*, Vol. 28, No. 8, 1990, pp. 1347–1352.

19 Dubief, Y., and Delcayre, F., "On Coherent-Vortex Identification in Turbulence," *Journal of Turbulence*, Vol. 1, No. 11, 2000, pp. 1–22.

20 Chong, M.S., Perry, A.E., and Cantwell, B.J., "A General Classification of Three-Dimensional Flowfields," *Physics of Fluids A*, 1990, Vol. 2, No. 5, pp. 765–777.

21 Jeong, J., and Hussain, F., "On the Identification of a Vortex," *Journal of Fluid Mechanics*, 1995, Vol. 285, pp. 69–94.

22 Berdahl, C.H., and Thompson, D.S., "Education of Swirling Structure Using the Velocity Gradient Tensor," *AIAA Journal*, Vol. 31, No. 1, 1993, pp. 97–103.

23 Kenwright, D.N., and Haimes, R., "Automatic Vortex Core Detection," *IEEE Computer Graphics and Applications*, Vol. 18, No, 4, 1998, pp. 70–74.

24 Jiang, M., Machiraju, R., and Thompson, D., "Detection and Visualization of Vortices," in *The Visualization Handbook*, Oxford: Elsevier, 2004, pp. 295–309.

25 Cummings, R.M., Morton, S.A., and Siegel, S.G., "Computational Simulation and Experimental Measurements for a Delta Wing with Periodic Suction and Blowing," *Journal of Aircraft*, Vol. 40, No. 5, 2003, pp. 923–931.

26 Cummings, R.M., Morton, S.A., and Siegel, S.G., "Numerical Prediction and Wind Tunnel Experiment for a Pitching Unmanned Combat Air Vehicle," *Aerospace Science and Technology*, Vol. 12, No. 5, 2008, pp. 355–364.

27 Schulz, M., Reck, F., Bartelheimer, W., and Ertl, T., "Interactive Visualization of Fluid Dynamics Simulations in Locally Refined Cartesian Grids," Proceedings of IEEE Visualization '99, October 1999, pp. 413–416.

28 Morton, S.A., "Detached-Eddy Simulations of Vortex Breakdown over a 70-Degree Delta Wing," *Journal of Aircraft*, Vol. 46, No. 3, 2009, pp. 746–755.

29 Taira, K., and Colonius, T., "Three-Dimensional Flows Around Low-Aspect-Ratio Flat-Plate Wings at Low Reynolds Numbers," *Journal of Fluid Mechanics*, Vol. 623, 2009, pp 187–207.

30 Chaderjian, N.M., and Buning, P.G., "High Resolution Navier-Stokes Simulation of Rotor Wakes," Proceedings of the American Helicopter Society 67th Annual Forum, Virginia Beach, VA, May 3–5, 2011.

31 Jankun-Kelly, M., Thompson, D., Jiang, M., Shannahan, B., and Machiraju, R., "Vortex Characterization for Engineering Applications," AIAA Paper 2008-0929, January 2008.

32 Mitchell, A.M., Barberis, D., Molton, P., and Délery, J., "Oscillation of Vortex Breakdown Location and Control of the Time-Averaged Location by Blowing," *AIAA Journal*, Vol. 38, No. 5, 2000, pp. 793–803.

33 Laramee, R.S., Hauser, H., Doleisch, H., Vrolijk, B., Post, F.H., and Weiskopf, D., "The State of the Art in Flow Visualization: Dense and Texture-Based Techniques," *Computer Graphics Forum*, Vol. 23, No. 2, 2004, pp. 203–221.

34 http://daac.hpc.mil/gettingStarted/Line_Integral_Convolution.html

35 DeSpirito, J., and Plostins, P., "CFD Prediction of M910 Projectile Aerodynamics: Unsteady Wake Effect on Magnus Moment," AIAA Paper 2007-6580, August 2007.

36 Post, F.H., Vrolijk, B., Hauser, H., Laramee, R.S., and Doleisch, H., "The State of the Art in Flow Visualization: Feature Extraction and Tracking," *Computer Graphics Forum*, Vol. 22, No. 4, 2003, pp. 775–592.

37 Kenwright, D.N., Henze, C., and Levit, C., "Feature Extraction of Separation and Attachment Lines," *IEEE Transactions on Visualization and Computer Graphics*, Vol. 5, No. 2, 1999, pp. 135–144.

38 Kenwright, D.N., and Haimes, R., "Automatic Vortex Core Detection," *IEEE Computer Graphics and Application*, Vol. 18, No. 4, 1998, pp. 70–74.

39 Forsythe, J.R., Strang W.Z., and Squires, K.D., "Six Degree of Freedom Computation of the F-15E Entering a Spin," AIAA Paper 2006-0858, January 2006.

40 Ghoreyshi, M., Vallespin, D., DaRonch, A., Badcock., K.J., Vos, J., and Hitzel, S., "Simulation of Aircraft Manoeuvres Based On Computational Fluid Dynamics," AIAA Paper 2010-8239, August 2010.

10 Applications of Computational Aerodynamics

"Would you tell me, please, which way I ought to go from here?"
"That depends a good deal on where you want to get to," said the Cat.[1]

Lewis Carroll, *Alice's Adventures in Wonderland*

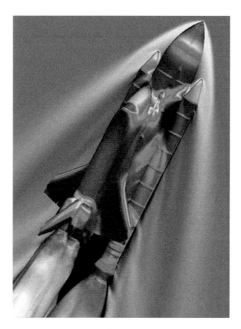

CFD simulation of Space Shuttle during ascent (Courtesy of NASA; a full color version of this image is available in the color insert pages of this text as well as on the website: www.cambridge.org/aerodynamics).

LEARNING OBJECTIVE QUESTIONS

After reading this chapter you should know the answers to the following questions:

- What are the challenges and approaches for computing transonic aerodynamics?
- What are the challenges and approaches for computing supersonic aerodynamics?
- What are the challenges and approaches for computing hypersonic aerodynamics?
- What is multidisciplinary design optimization and how is it useful in aircraft design?
- What are the potential benefits of performing aerodynamic studies in an integrated fashion, where CFD, EFD, and TFD are done together?

- Are potential flow methods still being developed, and if so, what are they being used for?
- What is the future of Computational Aerodynamics?

10.1 Introduction

Since the earliest uses of computational aerodynamics in the 1960s, the breadth and complexity of the aerodynamic problems being solved using computers have continued to grow and mature. From two-dimensional airfoil flows using panel methods to complex aircraft flying at high angles of attack, the march of computational aerodynamics progress has continuously expanded the notion of what was possible. This chapter will outline a few (and only a few) ways that computational aerodynamics continues to grow into more complex and challenging regimes. Many of these challenging areas are applications that we, the authors of the book, have been involved in, but there are many other applications to understand and evaluate. Truly, where you go from here depends on "where you want to get to."

10.2 Getting to Know Flowfields

Have you ever found yourself watching water flowing along in a stream, or staring at waves crashing at the seashore? There is something intriguing about the motion of fluids; in fact Nicholas Sparks said, "It is life, I think, to watch the water. A man can learn so many things." And while air is a fluid, it is definitely not easy to see, but we might have an equal sense of awe if we could easily see the motion of air around us. That is why we added this section to a book on computational aerodynamics, because the computer makes it possible to see the flow of air.

Remember the picture from Chapter 1, shown again in Fig. 10.1? We are showing it again because it demonstrates how little we see of what is happening in a flowfield. The CFD simulation of the F-16 is showing significant, important flow features (such as the strake vortices) that also exist on the real F-16, but we cannot see them on the actual airplane. The inability to see the flow around an airplane is one of the reasons that aerodynamics is so challenging to most of us – we typically do not see it, so we really do not understand what is happening.

In fact, there are many such situations where significant aerodynamic flow patterns exist around aircraft that can be adequately simulated with CFD; until we see the CFD prediction we may not even know that the flow patterns are there. Take a look at Fig. 10.2a, where the strake vortices on an F-16 are clearly visible due to air-water vapor condensation. The picture shows even

Figure 10.1 F-16 fighter in formation flight with a computational aerodynamic simulation showing strake vortices, surface pressures, and the exhaust of the engine (courtesy of Stefan Görtz and the USAFA High Performance Computing Research Center; a full color version of this image is available in the color insert pages of this text (see Fig. 1.1) as well as on the website: www.cambridge.org/aerodynamics).

(a) Thunderbird F-16 in a high-g turn   (b) CFD simulation of F-16

Figure 10.2 Comparison of real life and CFD: F-16 in a high-g turn with: (a) strake vortices visible from air-water vapor conden-sation, and (b) strake vortices visible from vorticity contours (U.S. Air Force photo by Staff Sergeant Larry E. Reid Jr.; USAFA High Performance Computing Research Center; a full color version of this image is available in the color insert pages of this text (see Fig. 1.4) as well as on the website: www.cambridge.org/aerodynamics).

more detail than you might have noticed. If you look closely at the vortices, you can see that they appear to have helical structures wrapping around the vortex, almost like a braided rope. Even relatively fine details such as these can be predicted by CFD, as shown in Fig. 10.2b, but how will you know what they are if you do not spend time getting to know about various flow features and how/why/when they appear?

We would love to answer that question by providing a wide variety of flow visualizations from real life, coupled with similar CFD predictions of the same flow, but that could fill an entire book by itself. So, instead, we recommend to you that in order to be a good practitioner of computational aero-dynamics, you should spend time learning about fluid flow (something else we mentioned in Chapter 1).

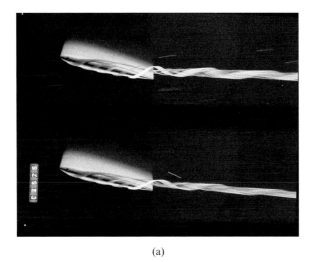

(a)

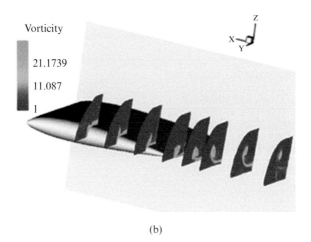

(b)

Figure 10.3   Comparison of physical and CFD simulation of wing-tip vortices. (a) Wing-tip vortices where the sense of rotation, diameter, and pitch of the helical streamlines of the vortex are revealed by liquid tracers of various colors (Photograph by Henri Werlé © ONERA, the French Aerospace Lab (a full color version of this image is available on the website: www.cambridge.org/aerodynamics). (b) Time-averaged vorticity contours showing the unsteady roll-up of the wing-tip vortex (a full color version of this image is available on the website: www.cambridge.org/aerodynamics as Figure 8.35).

One of the true geniuses of visualization flowfields was Henri Werlé of ONERA in France. He was able to use smoke (in air) or dye (in water) and "see" what we often cannot see, the flow features around aircraft. For example, look at the flow in the vicinity of a wing-tip shown in Fig. 10.3a. Multicolor liquid tracers were used to see how the vortex rolls up in the vicinity of the wing, which is a very important flow feature on aircraft. Fig. 10.3b shows the DNS simulation of a similar flow showing the formation of the wing-tip vortices.

Rather than continue to show various aerodynamic flow features, we suggest that you take some time to investigate various fascinating websites that will show you a number of beautiful flow images. Try to do more than enjoy the images, also try to understand what is happening and why. You may need to spend time reviewing the aerodynamic concepts presented in Chapter 4, but the end result will be a better understanding of how fluids flow and the impact of these flows on aerodynamics.

Now that we have you looking and thinking about fluid flows in general, and aerodynamics specifically, we will turn our attention to various areas of application for CA that have not already been mentioned. We will not concentrate on subsonic aerodynamics in this chapter, since most of the examples and applications shown up to this point have been for subsonic flow; rather, we will discuss transonic, supersonic, and hypersonic aerodynamics, followed by a variety of specialty applications where computational aerodynamics is making great strides at the present time. Finally, we will take a look into the future to see where CA might be going next.

Computational Aerodynamics Concept Box

Where to Find Fluid Flow Visualizations

A wide variety of fluid flow picture galleries are available on the World Wide Web; here is a small sample set.

efluids image gallery:
http://www.efluids.com/efluids/pages/gallery.htm

C-17 with smoke showing wing-tip vortices (U.S. Air Force photo)

Physics of Fluids (American Institute of Physics):
http://pof.aip.org/gallery_of_fluid_motion

Turbulent boundary layer (Courtesy of American Institute of Physics; a full color version of this image is available on the website: www.cambridge.org/aerodynamics).

Flow Visualization: A Course in the Physics & Art of Fluid Flow (University of Colorado, Boulder):
http://www.colorado.edu/MCEN/flowvis/
Gallery of Fluid Dynamics (Virginia Tech):
http://www.fluids.eng.vt.edu/msc/gallery/gall.htm

Flow over a triangular cylinder (Courtesy of Jean Hertzberg of the University of Colorado; a full color version of this image is available on the website: www.cambridge.org/aerodynamics).

B-2 at transonic conditions (Courtesy of Virginia Tech Gallery of Fluid Dynamics)

Gallery of Fluid Motion (American Physical Society):
http://www.aps.org/units/dfd/pressroom/gallery/

Vortices in the wake of a dual-step cylinder (Courtesy of American Physical Society; a full color version of this image is available on the website: www.cambridge.org/aerodynamics).

Books or papers with collections of fluid photos:

Milton van Dyke, *An Album of Fluid Motion*, 14th Ed., Parabolic Press, 2012.

M. Samimy, K.S. Breuer, L.G. Leal, and P.H. Steen, *A Gallery of Fluid Motion*, Cambridge University Press, 2004.

J.F. Campbell and J.R. Chambers, *Patterns in the Sky: Natural Visualization of Aircraft Flowfields*, NASA SP-514, 1994.

H. Werlé, "On the Flow of Fluids Made Visible," *Leonardo*, Vol. 8, No. 4, 1975, pp. 329–331.

10.3 Transonic Aerodynamics Prediction

Transonic aerodynamics was one of the most mysterious areas for aerodynamicists prior to the development of CFD. As we defined in Chapter 4, transonic flowfields contain regions of both subsonic and supersonic flow. This means that the governing equations are nonlinear even at the highest level of approximation. As a result, very few analytic theories are available to guide designers. In addition, until the concept of a vented wind tunnel was developed in the late 1940s, it was not possible to perform reliable wind tunnel tests at transonic speeds.[2] Also, numerous aerodynamic problems had been discovered in transonic flight for the piston-powered fighters of World War II, which led to the use of jet engines and the resulting swept wings that enabled higher-speed flight. Finally, although inviscid flow models were often sufficient for making aerodynamic estimates, the abrupt adverse pressure gradients that shocks impose on the boundary-layer and the asymmetric boundary-layer development on supercritical airfoils (due to the aft camber) meant that viscous effects also had to be included in the analysis (see Section 4.8 for more details). As a result, transonic aerodynamics depended on advances in CFD more than any other flow regime. In this section we illustrate the capabilities of CFD with a very few of the literally thousands of examples available in the literature (for example, see the survey paper by Vos et al.[3]).

10.3.1 Brief Review of Methodology Development for Transonic Flow Calculations

In the late 1960s computers had become powerful enough (by the standards of the day) to try to compute transonic flow, leading many researchers to try to develop methods to compute the transonic flow over an airfoil. In 1970, the key breakthrough was made by Murman and Cole,[4] who presented the solution for the transonic small disturbance equation. In their method, a

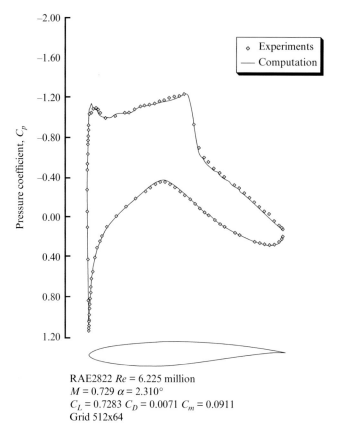

RAE2822 Re = 6.225 million
M = 0.729 α = 2.310°
C_L = 0.7283 C_D = 0.0071 C_m = 0.0911
Grid 512x64

Figure 10.4   The RAE 2822 airfoil at M = 0.729 and α = 2.31° (Ref. 7, attributed to Luigi Martinelli).

shock wave emerged during the iterative solution, and their inviscid solution agreed well with experimental data. Apparently, because the flow model was considered simplistic and results were presented for a nonlifting circular arc airfoil, the achievement was not widely appreciated. However, Antony Jameson realized the importance of their work and quickly repeated the calculations. He then extended the method to solve the full potential equation (Equation 3.102) for arbitrarily thick lifting airfoils. Jameson and his co-workers have written numerous survey articles that tell the story of the development of transonic computational methodology in detail.[5,6]

10.3.2 Airfoils

Today, we can reliably find the transonic flowfield over an airfoil using CFD. Figure 10.4 shows a typical solution for a classic transonic test case, the RAE 2822 airfoil.[7]

Several aspects of the figure for this "supercritical" airfoil are worth discussing. Perhaps the first feature of interest in the figure is that the airfoil is shown below the pressure distribution, which was the standard presentation format. This format allows you to develop a feel for the relation between the shape and the pressure distribution in a visual manner. Next, notice the good agreement between the experimental data and the computations. On the upper surface, we can see expansion around the leading edge, the slight acceleration of the flow up to the shock wave, and the essentially linear recompression to the trailing edge. As long as the shock is not too strong so that the flow does not separate, the presence of the shock is acceptable. On the lower surface, the "bulge" on the bottom of the airfoil leads to an acceleration and reduced pressure. This is followed by the shaping of the aft surface, known as "the cove" region, with the resulting camber producing lift on the aft portion of the airfoil. Finally, the flow accelerates approaching the trailing edge. The difference in pressure distributions between the upper and lower surface results in the asymmetric development of the boundary layer, which is why it became important to include viscous effects in the calculation of supercritical airfoils.

Another example of modern transonic airfoil development is the advanced airfoil developed for the new HondaJet. This airplane has an essentially unswept wing and achieves reduced drag by designing the airfoil to have a significant percentage of laminar flow, with the associated reduced skin friction drag. This is achieved without any boundary-layer suction and thus is a "natural" laminar flow airfoil. This is one of many ideas employed to achieve excellent aerodynamic characteristics,[8] shown in Fig. 10.5.

The airfoil designed for this airplane has a favorable pressure gradient (flow accelerating) as far along the chord as possible to encourage laminar flow. The airfoil was developed using a combination of CFD methods.[9] The

Figure 10.5   The HondaJet (Courtesy of Honda Aircraft Company).

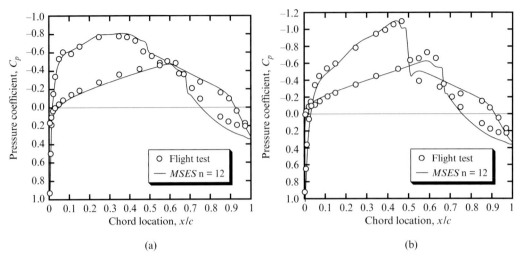

(a)   (b)

Figure 10.6   Comparison of HondaJet natural laminar-flow airfoils between CFD and flight test obtained on a modified T-33 (Ref. 9; Courtesy of Honda Aircraft Company). (a) $M = 0.62$, $Re = 13.6 \times 10^6$ (b) $M = 0.72$, $Re = 16.2 \times 10^6$

pressure distributions shown in Fig. 10.6 illustrate the excellent agreement between computational predictions and flight test pressure distributions. It is rare to find these types of comparisons, and they illustrate the success of CFD during the design effort. Because laminar flow is highly Reynolds number dependent, it was important to measure the pressure distributions at the full-scale Reynolds number of approximately 15×10^6.

10.3.3 Wings

Next, we will examine some examples of CFD applied to transonic wings. The first example is a comparison of a classic transonic test case, the ONERA M6 wing, which has been used by countless code developers. It has a symmetric airfoil, a wing with an aspect ratio of 3.86, a taper ratio of 0.5377, and a leading edge sweep of 30°, as shown in the ONERA S2MA wind tunnel in Fig. 10.7. This case is used primarily for pressure distribution comparisons. The case typically used for comparison is the $M = 0.84$, $\alpha = 3.06°$ case. Streamwise pressure distributions were measured at seven span stations, but several of these span station results are frequently omitted in the comparisons. The results shown in Fig. 10.8 contain detailed pressure distribution comparisons between CFD and wind tunnel data (as presented by Strang et al.[10]), and a planform showing the surface isobar pattern (presented by Holst[11]). This case has fairly strong shocks, including both a forward and aft shock on the chord merging just outboard of the 80 percent semi-span station. In general, the agreement between the RANS simulation and the data is very good. Because it has proven very difficult to predict the double shock at the 80 percent span station, almost all code developers omit this station when comparing their results with the wind tunnel data (although the results

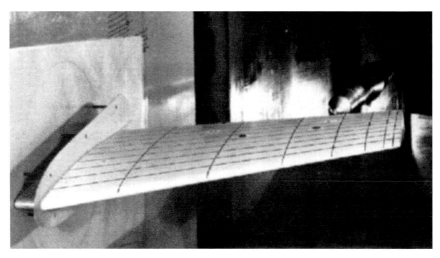

Figure 10.7 M6 Wing in the ONEAR S2MA Wind Tunnel (Courtesy of Patrick Champigny of ONERA).

shown here include this difficult aspect of the flowfield). The pressure isobars shown on the surface of the wing are valuable in illustrating the shock sweeps and merger. This wing was not intended to be an example of a modern, well-designed transport wing, but rather provides a good test case because of this complex shock pattern. We will present the results of a modern design exercise later in the chapter, where the isobars are much more well behaved (meaning that they are swept along the element lines of the wing).

10.3.4 Drag Prediction

We have already looked at drag prediction from the perspective of turbulence modeling in Chapter 8. Now we want to look at some of the results from the perspective of how well we can predict drag, since most users are ultimately interested in the predictions of forces and moments. In this regard, the aerodynamicist uses pressure distributions to infer aerodynamic performance. Examining the flowfield provides so-called flow diagnostics to help understand the force and moment results. In this section, we provide an example of the ability to predict drag for transonic transport type configurations. In these *Drag Prediction Workshops*, participants were given geometry and flowfield conditions but not the data from the wind tunnel tests. Each participant predicted the drag, and the organizers compiled the results. The outcome of this series of workshops showed that accurate drag prediction remains difficult and points out the importance of skill of the engineers performing the calculations. Rumsey et al.[12] and Vassberg et al.[13] provide summaries of the workshop results. Here we have chosen to present results from Langtry et al.[14] using one of the test cases.

Figure 10.9 shows the geometry used to examine the effects of the nacelle on the aerodynamics and the comparison with the lift and drag results

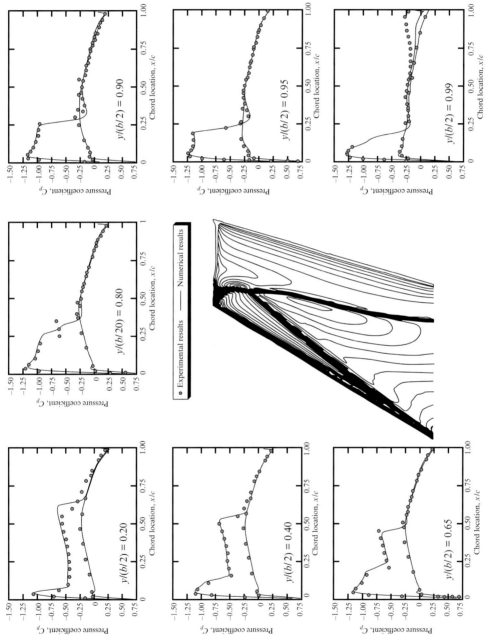

Figure 10.8

The famous ONERA M6 case (Ref. 10); $M = 0.84$, $\alpha = 3.06°$.

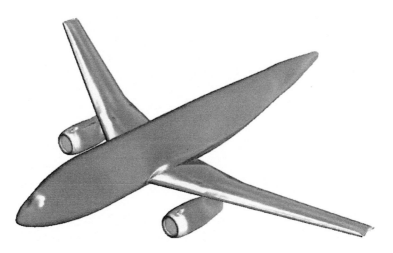

Figure 10.9   DLR-F6 configuration used to evaluate CFD drag prediction capability (Ref. 14 courtesy of Florian Menter of ANSYS, Inc.; a full color version of this image is available on the website: www.cambridge.org/aerodynamics); $M = 0.75$, $Re = 3 \times 10^6$.

obtained in a wind tunnel test. The configuration is known as the DLR-F6 configuration. The wind tunnel results were obtained in the French ONERA S2MA wind tunnel (the same tunnel used for the M6 wing); the comparison cases are for a Mach number of 0.75. The test Reynolds number was 3×10^6, and transition strips were used to trip the boundary layer, producing turbulent flow intended to simulate full-scale flight conditions.

The calculation results shown in Figs. 10.10 and 10.11 used the CFX-5 RANS code. The Menter SST model was used to model the turbulence, and the grid contained 5.8 million nodes. The results of calculations for cases with and without the nacelle and pylon are shown in Figs. 10.10 (lift coefficient) and 10.11 (drag polar). Figure 10.10 illustrates the good agreement between the calculations and the wind tunnel results for the lift as a function of angle of attack. In this case, the design C_L was 0.5. Although the angle of attack range seems small, it is typical of transonic transport testing. Notice that including the nacelle and pylon results in a loss of lift and that the computational results overpredict the drag effect compared to the wind tunnel results. Figure 10.11 shows the comparison of wind tunnel results and CFD calculations for this case. The agreement is good, and the addition of the extra surface area is evident, leading to more skin friction drag when the nacelle pylon is included.

The last example for a transonic transport wing illustrates the results from a wing design carried out by Jameson, Pierce, and Martinelli.[15] Previously, we showed the strong shock waves that typically arise on the wing at transonic speeds. Now, we show the result of contouring the wing to reduce or eliminate shocks and maintain a wing isobar pattern that does not unsweep; Fig. 10.12 illustrates the results. At the design point of $M = 0.86$, the wing is essentially shock free, and the isobars over the key portions of the wing are swept in line with the wing element line.

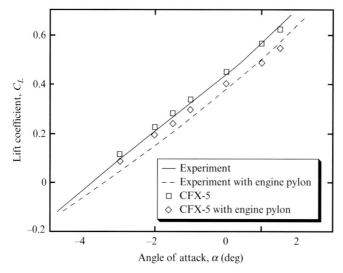

Figure 10.10 Lift results for the DLR-F6 geometry (Ref. 14; courtesy of Florian Menter of ANSYS, Inc.).

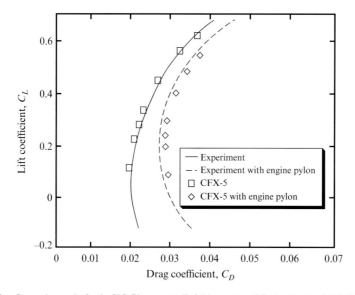

Figure 10.11 Drag polar results for the DLR-F6 geometry (Ref. 14; courtesy of Florian Menter of ANSYS, Inc.).

10.3.5 Fighter Aircraft Design

The situation for fighter wings is more complicated than that for transonic transport wings, especially since there are multiple design points for the aircraft, as shown in Fig. 10.13 from Bradley.[16] The various design points include both attached and separated flow conditions. Figure 10.14 shows Bradley's notion of where these flows will occur in the design space.

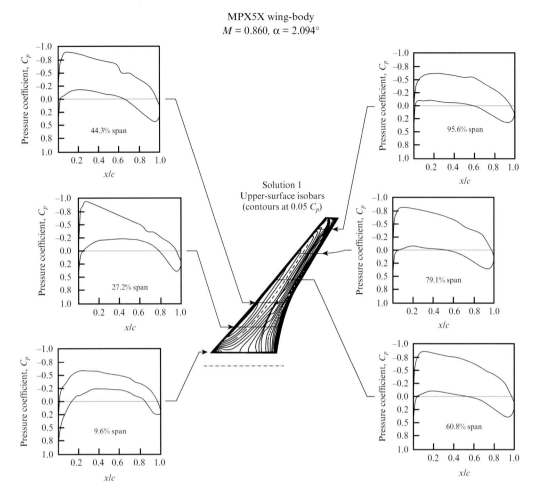

Figure 10.12 A wing designed for efficient transonic cruise (Ref. 15; reprinted by permission of the American Institute of Aeronautics and Astronautics, Inc.).

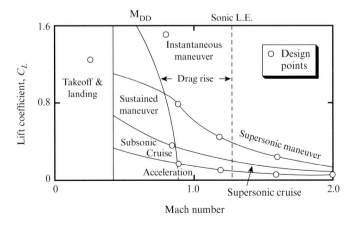

Figure 10.13 Typical fighter aerodynamic design points (Ref. 16; reprinted by permission of the American Institute of Aeronautics and Astronautics, Inc.).

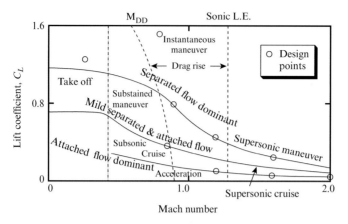

Figure 10.14 Typical boundaries for attached and separated flow regions for fighter design (Ref. 16; reprinted by permission of the American Institute of Aeronautics and Astronautics, Inc.).

From Fig. 10.14 it is clear that the aerodynamicist has to deal with cruise conditions where the flow should be attached, fully separated flows, and, perhaps most difficult, mixed attached and separated flows. In the United States, sustained maneuver design with nominally attached flow was used by the F-15 and later the X-29. Another way to attain acceptable maneuver design conditions was to introduce "hybrid" flow concepts, where a LEX (leading-edge extension) on the F-18 or a chine on the F-16 could be used to generate a stable vortex flow providing additional aerodynamic lift. More recently, the F-22 and F-35 have forebody chines to reduce radar cross section and also produce vortices to improve aerodynamic characteristics.

Figure 10.15 shows the most recent U.S. fighter, the F-35, which is being developed in three variations. The F-35A is the U.S. Air Force version, the F-35B is the version capable of short takeoff and vertical landing being developed for the U.S. Marines, and the F-35C is the carrier-capable version for the U.S. Navy. The story of the F-35 is presented by Bevilaqua in the 2009 AIAA Wright Brothers Lecture.[17]

The chine is clearly visible in Fig. 10.15, and notice that there is also a distinct leading-edge device. At transonic maneuver conditions, the flow over the airplane is much more complicated, as shown in Fig. 10.16.[18] The figure includes both the off-surface vorticity distribution and surface pressure distribution, indicated by the color pattern, as well as so-called computational surface "oil flows" showing the surface streamlines. Under these conditions, it is important to examine the flowfield both off and on the surface. Clearly, these conditions are among the most important and complicated flowfield calculations made using CFD.

A comparison between the computed pressure distribution and results from wind tunnel tests is shown in Fig. 10.17. These results are for a mid-

Figure 10.15 The F-35 Lightning II (U.S. Air Force photo)

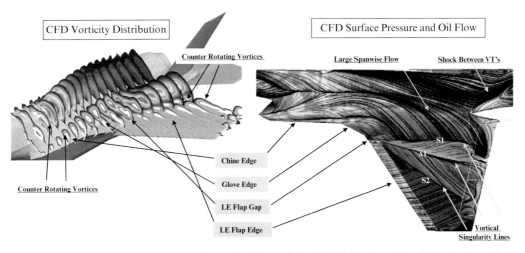

Figure 10.16 F-35 upper surface on- and off-body flowfield visualization (Ref. 18; Courtesy of Brian Smith of Lockheed Martin; a full color version of this image is available in the color insert pages of this text as well as on the website: www.cambridge.org/ aerodynamics).

span station at an angle of attack of 14°. Under these maneuver conditions, the pressures are very different from the transonic transport cases shown previously. Also, the effect of the deflection of the leading-edge device is quite apparent. The circle on the lower surface was drawn to point out the effect of a flap attachment device on the wind tunnel model. The agreement between the calculations and the test results should be judged to be excellent.

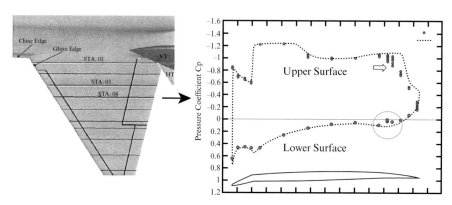

Figure 10.17   Mid-span pressure distribution comparisons between wind tunnel and CFD calculations for the F-35 at transonic maneuver conditions (Ref. 18; Courtesy of Brian Smith of Lockheed Martin).

10.4 Supersonic Aerodynamics Prediction

In recent years supersonic aerodynamic analysis and design have focused on the possibility of developing a new commercial supersonic airplane. The Concorde was an extraordinary achievement, but in the end it was not commercially viable; the last commercial flights of the Concorde were made in 2003. After the Air France crash in 2000, the plane never fully recovered to successful service, in part because the additional weight required to seal the fuel tanks resulted in a reduced passenger load. Another problem was that it was too loud around airports (the so-called community noise problem), and most countries did not allow supersonic flight over land because of the sonic boom created by the aircraft. The range of the Concorde was marginal, barely able to fly nonstop from London or Paris to New York (supersonic $L/D_{\max}$ was usually cited as 7.4).

In the 1990s, the United States studied the possibility of a new supersonic transport, the High Speed Civil Transport (HSCT). The goal was to design an aircraft capable of trans-Pacific range (the range from San Francisco to Hong Kong, for example, is approximately 7,000 miles). The problem turned out to be too challenging, and the program was terminated. An offshoot of the HSCT program was the Boeing Sonic Cruiser, a near Mach 1 airplane that was also abandoned in favor of the Boeing 787 Dreamliner. The new, highly efficient transonic transport used most of the advanced technologies that had been explored for use in the HSCT and Sonic Cruiser programs.

10.4.1 Initial Application of CFD at Supersonic Speeds

Before using a CFD code in a design application, the user should evaluate the code and his/her ability to produce accurate results. We will show that supersonic designs can be extremely sensitive to accurate drag predictions.

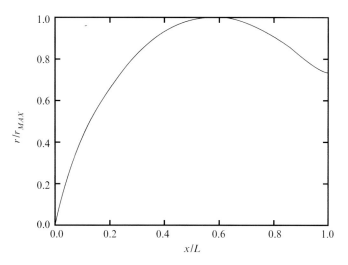

Figure 10.18 Haack-Adams Body of Revolution for *Sbase/Smax* = 0.523 (Ref. 19).

In this section, we provide several examples of simple test cases that should be used to gain an understanding of supersonic aerodynamics. We will start with the prediction of wave drag. The initial examples were completed some years ago in preparation for MDO studies of the HSCT.[19]

The first case we will examine is the wave drag of an axisymmetric body. The original wave drag estimation method was based on the slender body integral that is a double integral of the second derivative of the cross-sectional area distribution. The code that became the standard for this integration was developed by The Boeing Company for NASA,[20] which was developed under the monitoring of Roy Harris, and the code is generally known as "The Harris Code." To evaluate the code, a wind tunnel test at NASA Langley was conducted to provide experimental data.[21] Several bodies of revolution were tested, and the results we use here are for the so-called Haack-Adams body with fineness ratios of 7 and 10 and a closure ratio (which is the ratio of the base area and the maximum cross-sectional area, S_{base}/S_{max}) of 0.523. The surface pressures were measured experimentally and then integrated to obtain the experimental drag. Figure 10.18 shows the radius distribution for the body, which has a maximum radius at 60 percent of the length.

Figure 10.19 shows the wave drag results for the two fineness ratios mentioned earlier. Both plots compare the drag obtained in the wind tunnel with the results from the Harris code and a CFD code (in this case GASP was used to obtain the CFD results – GASP[22] is a standard CFD code). This case was computed using the Parabolized Navier-Stokes (PNS) approach described in Chapter 6. The results shown in Fig. 10.19 are generally good for this type of prediction. The CFD results are close to the data, while the classic Harris code results generally overpredict the drag. For the higher fineness ratio (more slender), the agreement with the Harris code is better, as

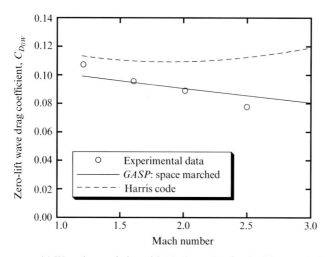

(a) Wave drag variation with Mach number for the l/d_{max} = 7 body

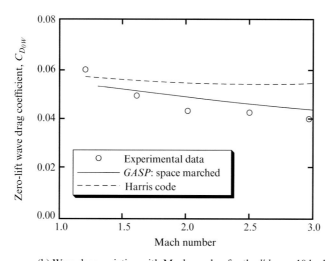

(b) Wave drag variation with Mach number for the l/d_{max} = 10 body

Figure 10.19   Wave drag results for the Haack-Adams Bodies (Ref. 19). (a) Wave drag variation with Mach number for the l/d_{max} = 7 body. (b) Wave drag variation with Mach number for the l/d_{max} = 10 body.

you might expect, since the Harris code assumes that the body is slender and works better for very thin fuselage shapes.

The next case we will look at is a symmetric wing at zero degrees angle of attack. A general geometry with wind tunnel data is available: the so-called Squire Wing,[23] as shown in Fig. 10.20. The geometry is defined analytically, making it easy to model with CAD or other computer-generated geometry approaches. In this case the wing planform is a delta wing, and the center section is a parabolic biconvex airfoil with a 9% thickness ratio. The cross sections are elliptic and fit the centerline section. The experimental wave drag values were obtained by integrating surface pressures from the wind tunnel tests.

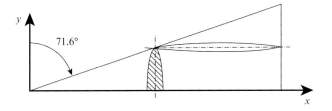

Figure 10.20 Squire wing geometry (Ref. 23).

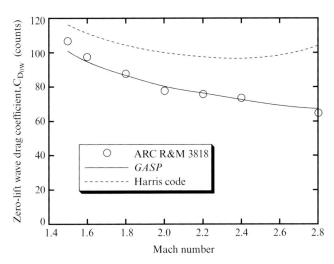

Figure 10.21 Squire wing wave drag predictions and wind tunnel results (Ref. 23).

The wave drag results are given in Fig. 10.21 and, again, the agreement between experiment and CFD is excellent, while the Harris code overpredicts the wave drag (as we also saw for the body of revolution).

10.4.2 Application of CFD to a Supersonic Configuration

Having established the capability of CFD to predict volumetric wave drag, we can now look at a more realistic configuration, a generic High Speed Civil Transport (HSCT) aircraft.[19] Figure 10.22 shows both the type of configuration and the computational grid. Notice that when computational design methods using numerical optimization are employed, the construction of a grid for each shape change must be made automatically (for example, using an approach from NASA Langley Research Center[24,25]). This is an important consideration in establishing your approach to computational design.

We will now look at the comparison between predicted lift, moment, and drag at the design Mach number of 2.4 for the HSCT-class configuration. We will compare classical linear theory with Euler and Parabolized Navier-Stokes predictions. The wing is cambered using the modified linear theory of Carlson and Walkley.[26] In these cases, the skin friction estimates were

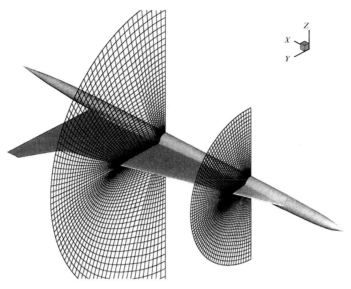

Figure 10.22 HSCT-class wing geometry and grid (Ref. 19).

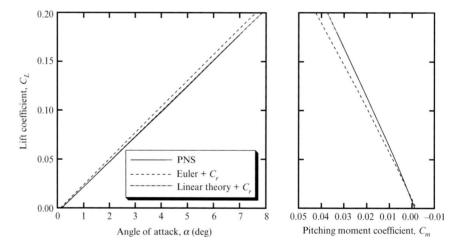

Figure 10.23 HSCT-class wing supersonic lift and pitching moments (Ref 19).

added to the linear theory and Euler calculations (following the recommendation of Hopkins and Inouye,[27] we used the Van Driest II method for friction drag). Figure 10.23 shows the comparisons for lift and pitching moment and Fig. 10.24 shows the related drag polar.

The linear theory lift curve slope is slightly higher than the PNS and Euler predictions. In addition, the slope of the pitching moment differs between the liner theory and the Euler and Navier-Stokes values.

The drag estimates are also interesting, as shown in Fig. 10.24, where the linear theory drag-due-to-lift is consistently low. The study shows that the Harris wave drag estimates are within 2 counts of the CFD value, and the

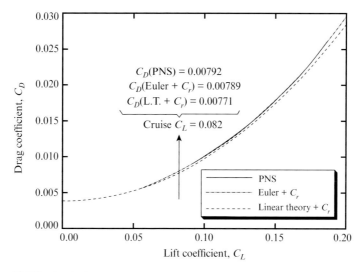

Figure 10.24 HSCT Cambered wing drag polar estimates (Ref 19).

skin friction estimate is slightly larger than the PNS value, but still within 1 count. While it appears that the agreement between the estimation methods is excellent, and it is, these types of configurations are unbelievably sensitive to the drag. As an example, Brenda Kulfan of Boeing quotes a sensitivity to drag of 10,400 pounds per count for one configuration concept she studied.[28] This type of sensitivity is not too unusual for supersonic and hypersonic concepts and illustrates why the full accuracy of CFD is needed for drag prediction, which we discussed in Chapter 8.

10.4.3 Application to Low Sonic Boom Aircraft Designs

Having examined the basic ability of CFD to predict supersonic aerodynamics, we will now highlight the primary consideration for current supersonic vehicle design, which is reduction of the strength of the sonic boom. Figure 10.25 from Aronstein and Schueler[29] illustrates the crux of the problem. The typical sonic boom is characterized by the strength of an N-wave. The term N-wave is used because the pressure distribution along the length of the body forms an "N" shape, since the pressure variation consists of an initial and final very sudden change in pressure due to shocks. The goal of advanced supersonic design shown in Fig. 10.25 is to reduce the strength and tailor the shape of the N-wave in order to decrease the sonic boom the shocks make as the aircraft flies overhead.

Designs carried out with the goal of reducing the sonic boom are apparently feasible: Fig. 10.26 shows both the baseline and a low-boom configuration. In order to obtain a lower boom, the aircraft had to be lengthened, and the cross-sectional area distribution variation was made extremely smooth. For these types of designs, the longitudinal lift distribution must also be

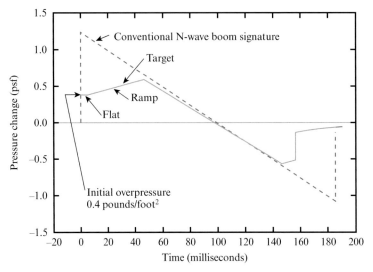

Figure 10.25   Target sonic boom signature (Ref 29; reprinted by permission of the American Institute of Aeronautics and Astronautics, Inc.).

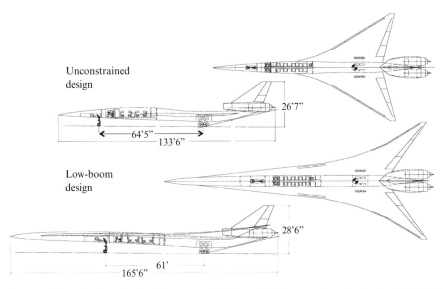

Figure 10.26   Unconstrained and low-boom business jet concepts (Ref. 29; reprinted by permission of the American Institute of Aeronautics and Astronautics, Inc.).

carefully tailored. In addition, the wetted area that arises naturally because of the increased length had to be reduced as much as possible. The result of the study on reducing the N-wave is shown in Fig. 10.27, where a significant reduction in the estimated sonic boom strength was achieved.

Progress on reducing the strength of the sonic boom while maintaining reasonably efficient drag performance is continuing in a number of NASA studies. An example of the use of CFD calculations for use in sonic boom calculations is shown in Figures 10.28 and 10.29.[30] In this NASA study

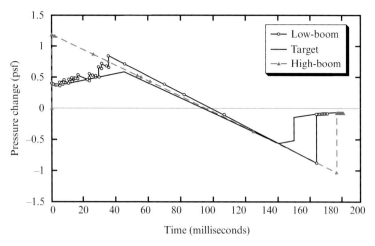

Figure 10.27 Sonic boom estimates of the concepts (Ref. 29; reprinted by permission of the American Institute of Aeronautics and Astronautics, Inc.).

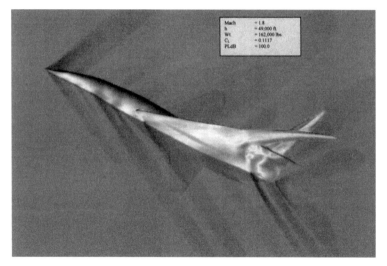

Figure 10.28 *CART3D* pressure prediction for low sonic boom design at *M* = 1.8 (Ref. 30; a full color version of this image is available on the website: www.cambridge.org/aerodynamics).

conducted by Boeing and Georgia Tech, a so-called N+2 (second-generation) technology goal was assumed. The specific concept shown here is designed for thirty passengers and a trans-Atlantic range of 4000nm, and emphasis was placed on low sonic boom strength. The figure shows both surface pressure variations and the initial shock wave field, which is then used to estimate the sonic boom strength. In this case the CFD code was CART3D,[31,32] an inviscid Euler code. The estimated sonic boom strength is shown in Fig. 10.29, and, once again, the strength of the boom has been significantly reduced.

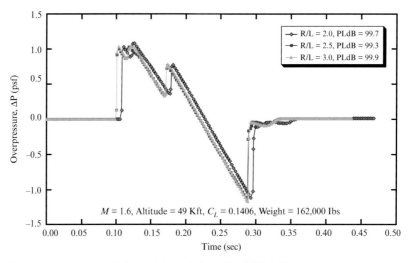

Figure 10.29   Resulting ground signature for low sonic boom design at $M = 1.8$ (Ref. 30).

Profiles in Computational Aerodynamics: Mark Lewis

"I was seven years old when Neil Armstrong and Buzz Aldrin landed on the moon. Like many in my generation caught in the excitement of Apollo, I knew then that I wanted to pursue a career in aerospace. Throughout my childhood, I read every book on space and airplanes that I could get my hands on, built more model airplanes and rockets than I could count, and dragged my parents to science museums, airports, planetariums, any place that touched on space and atmospheric flight. I was also an enthusiastic science fiction fan, gravitating especially towards the hard science writers. In the summer after ninth grade I spent a week visiting with my older brother, who was a physics major at MIT. Thumbing through his course catalog, I was amazed to find that there were university courses on such topics as rocket propulsion, aerodynamics, and space systems design. By the fall of 1980 I began my freshman year at MIT, planning to major in Aeronautics and Astronautics with a second degree in Earth and Planetary Science. At MIT I worked on research projects in the Space Systems Laboratory, including some early efforts on space robotics and aerobraking, and in the Gas Turbine Laboratory. Undergraduate research work in the Gas Turbine Lab ultimately led to a master's degree doing experimental flow visualization in a turbine stage.

"My master's work explored the use of a laser-induced fluorescence to directly measure both temperature and density of the gas moving through a turbine passage. Though experimental, we did this work in parallel with a computational effort, trying to match our flow visualization results to CFD solutions. Flow through a jet engine includes some of the most challenging problems in gas dynamics, and that was an early use of CFD to solve complex problems in moving blade systems and matching to real data.

"As I was finishing my master's thesis, I began to look for a doctoral research topic. The United States had just begun a national program called the X-30 National AeroSpace Plane (NASP). The plan was to build a so-called "hypersonic" aircraft that could take off from any runway, accelerate through the atmosphere beyond 25 times the speed of sound, and reach low earth orbit in a single stage. MIT was under contract with the Draper Labs to explore basic propulsion challenges, and I became one the first graduate students to join that effort, sponsored by a fellowship from the Office of Naval Research.

"I began my doctoral work with an analytical and computational study of the interaction between the forebody of a hypersonic vehicle and its engine. It was understood that a hypersonic aircraft, powered by a so-called "scramjet" – a special ramjet engine that burned its fuel in supersonic conditions- would be highly integrated. The aircraft would essentially be a flying engine, with its front part serving as the inlet, the aftbody functioning as a nozzle, and all the parts interacting. It was also known that there were no experimental facilities that could adequately test a hypersonic craft, so much of the design work would rely on fundamental understanding and computation. In fact, some engineers in the NASP program were proclaiming that their vehicle would be designed completely with modeling and simulation. In retrospect, this was perhaps naïve, but presaged the increasing importance of CFD in designing future aircraft.

"I left MIT after a total of eight years with four degrees and immediately began life as a faculty member in the Department of Aerospace Engineering at the University of Maryland. I was drawn to a faculty career because of the flexibility – being able to work on whatever research project I wished, as long as I could secure funding – and also the excitement of teaching and working with students. Much of my research at Maryland has built on my early interests in hypersonic engine/airframe integration, though my activities have run the gamut from some very basic fundamental investigations to very applied. On the basic side, I have worked on a range of subjects from basic physics of high-speed unsteady flows, sharp leading edges, and shock interactions, including shocks intersection boundary layers and other shocks. In many cases, these are problems that cannot be solved analytically, so my students and I have used CFD tools as a "numerical wind tunnel." My work has also extended to the rarefied flow regimes, the chemistry of heat shield ablation, and plasma effects at high speed.

"Even when I am working on a fundamental problem, I always like to have some relevant application in mind. I have been especially interested in optimal design of hypersonic craft – using analytical and computational methods to identify vehicles with good aerodynamics, good packaging, and good engine flowfields. A large portion of my work has included the use of so-called inverse design techniques, where a desired flowfield is first calculated, then the shape that creates that flow is determined. Among these configurations is a category of vehicle known as "waveriders" – designed so that their shockwaves are attached to their leading edge. This offers excellent aerodynamic performance as well as engine integration at on-design conditions. However, there has always been a question about off-design performance, and so we have used CFD to determine solutions to inverse configurations at varying Mach numbers, attitudes, etc.

"Though I had embarked on what I expected to be a very traditional academic career, I have found myself taking some detours along the way. In 2000 I joined the USAF Scientific Advisory Board, which is a collection of scientists and engineers who advise the Secretary of the Air Force and Chief of Staff. As my four year term was ending, I was asked to become the Chief Scientist of the USAF. Unlike the Advisory Board, the Chief Scientist position is a full-time (and then some) job, so I took four years of leave from the Maryland campus to work in the Pentagon. I had never taken any type of leave or sabbatical from campus before, but I found this time away to be a highly valuable, broadening experience. Being Chief Scientist impressed upon me the value of relevance in our research, even for very basic research topics. The Air Force sponsors billions of dollars worth of research each year in laboratories and universities around the world, including some of the most cutting edge efforts in physics, chemistry, and biology. Indeed, much of the work described in this book came about as a result of Air Force sponsorship. But for each project, there is always the question of how the results of that work can be used to benefit the mission of the Air Force. To that end, one of my greatest professional joys was participating in the first flight of the Air Force's X-51 test vehicle in May 2010. X-51 was a hydrocarbon-fueled scramjet-powered vehicle, using a waverider forebody. That was a practical culmination of much of the work I have done as a researcher and faculty mentor."

Dr. Mark J. Lewis was the Willis Young, Jr. Professor and Chair of the Department of Aerospace Engineering at the University of Maryland. He is currently the Director of the Science and Technology Policy Institute at IDA (Institute for Defense Analyses). A recipient of both the Meritorious and Exceptional Civilian Service Award, Dr. Lewis also received the IECEC/AIAA Lifetime Achievement Award, and was named an Aviation Week and Space Technology Laureate in 2007. He is a Fellow of the American Society of Mechanical Engineers, the American Institute of Aeronautics and Astronautics, and the Royal Aeronautical Society. He also served a term as the President of AIAA.

10.5 Hypersonic Aerodynamics Prediction[33]

Hypersonic CFD is one of the most challenging and difficult areas of aerodynamic simulation. You only need to look at the typical assumptions made when deriving the equations of fluid motion (as we did in Chapter 3) to understand why hypersonic flow prediction has been more difficult than lower-speed simulations. Typically, the equations solved in CFD programs are based on the assumptions that the fluid is a continuum, that the perfect-gas law applies, and that the only forces are due to pressure and viscous effects. The continuum assumption can invalidate the equations for rarefied gas flows at high altitudes when the mean-free path of the fluid molecules is on the order of the length of the vehicle; the other assumptions can lead to additional difficulties. These problem areas are due to the complex flow features that can occur in hypersonic flow: thin shock layers (high compression), entropy layers caused by highly swept and curved shock waves, viscous/inviscid interactions, and real gas effects, including dissociation, ionization (high temperatures), and rarefication at high altitudes.[34]

Figure 10.30 shows many of these fluid dynamic challenges in a computation for the Hyper-X vehicle at $M_\infty = 7$. The thin shock layer coming from

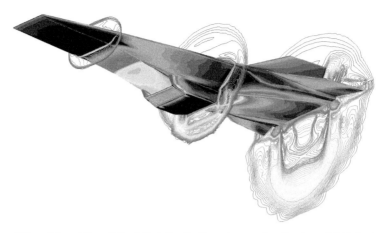

Figure 10.30   CFD prediction of Hyper-X flowfield at $M = 7$ with engine operating (Courtesy of NASA Dryden Flight Research Center; a full color version of this image is available in the color insert pages of this text as well as on the website: www.cambridge.org/aerodynamics).

the forebody remains very close to the vehicle surface for a large distance down the body length, various shock-shock and shock-boundary layer interactions occur in the vicinity of the ramp inlet, flow interactions with the engine exhaust and lower surface body contouring create a complex flowfield under the back half of the vehicle, and all of this could take place at flight conditions where chemical reactions could be important. This makes for one of CFD's most challenging problems, made even more difficult by our limited ability to adequately represent hypersonic flow experimentally.

H. K. Cheng, expanding on the difficulties in predicting viscous/inviscid interactions, stated that "the fluid dynamics of hypersonic flows is complicated by the interaction of the boundary layer and the shear layer with shock waves, leading to flow separation and instability not amenable to straightforward analysis."[35] In addition, the key issues for CFD related to hypersonic aerodynamic design, such as real-gas effects, heating, turbulence, and viscous interactions, are not currently modeled well by CFD.[36]

From a simulation perspective, at least five factors affect the accuracy of flow predictions for hypersonic vehicles. These factors include the grid density, the ability to model either equilibrium or nonequilibrium flows, the ability to model transition and turbulence, and the effort involved in accurately modeling the vehicle geometry (especially in regions of shock interactions).[37] While it is possible for CFD to accurately predict the aerodynamic environment of hypersonic vehicles (including surface pressures and forces/moments), CFD is severely limited in accurately predicting the aerothermodynamic environment of these vehicles (including heat transfer and skin friction). Additional challenges exist for predicting flows in deflected flaps and rudders or in gaps between tiles, high-enthalpy flows (including radiation cooling), wall catalycity, turbulence modeling, and jet/airflow interaction. Longo[38] quantified the effect of these simulation factors on the prediction

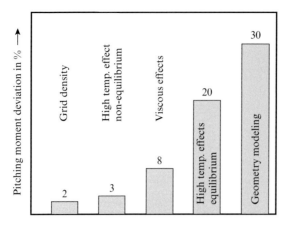

Figure 10.31   Numerical modeling aspect impact on predicting pitching moment at hypersonic speeds (Ref. 38; The original version of this material was published by the North Atlantic Treaty Organization Scientific and Technology Organization, Collaboration and Support Office [NATO/STO/CSO] in Technical Report RTO-EN-AVT-116 on Critical Technologies for Hypersonic Vehicle Development published in 2004).

of pitching moments, as shown in Fig. 10.31. Note that the largest source of error is geometry modeling, something that certainly needs to be addressed by the simulation community. In fact, the geometry itself changes during hypersonic flight due to the high temperatures interacting with the structure, and must be accounted for in the simulation to achieve reasonable accuracy for some flight conditions.

A variety of important numerical algorithms has been developed over the years specifically to deal with the shocks that are formed at supersonic and hypersonic speeds. Most methods fall into the "shock capturing" approach, where the equations of fluid motion are solved to determine shock locations and strengths. These approaches were described in Chapter 6, including the explicit methods of Lax and Wendroff and the predictor-corrector scheme of MacCormack. An alternative approach for predicting shocks is "shock fitting" as proposed by Moretti, which uses the Rankine-Hugoniot shock relations to determine shock jump conditions, once the position of the shock is known.[39] In spite of the appeal of the shock fitting method, the vast majority of algorithms developed over the years have been of the shock capturing variety.

Shocks cause instabilities in most numerical methods and must be treated with algorithms that can deal with the instability either through some sort of upwind differencing, or through the explicit addition of numerical smoothing. Unfortunately, the numerical dissipation needed to dampen the shock often degrades the prediction of viscous phenomena such as boundary layers or vortical flows. Significant improvements in predicting shocks have been made over the years using either the flux vector splitting or flux difference splitting approaches, including Van Leer's second-order extension to Godunov's method,[40,41] the Steger-Warming flux vector splitting approach,[42] the characteristics-based schemes of Roe,[43] and Harten's high-resolution total variation diminishing (TVD) schemes[44] (several of these methods are described in Chapter 6). These advances in computational algorithms, coupled with the amazing growth of computer processor speeds, have led to improvements in the computational prediction of high-speed flows, but further progress is still needed.

Closely coupled with the numerical algorithm that is being used is the ability to create quality grids during the design and analysis of a vehicle. Grid generation has made great strides over the past decades, with a number of commercially available programs that do a reasonably good job of creating grids around complex geometries, but the difficult requirements of hypersonic flow can tax even the best grid generation software. It can still take weeks or even months to create complex grids, especially when a variety of configurations are being considered during the conceptual design phase.

Surface geometry definition and grid generation still remain the biggest bottlenecks in the CFD process. One research group described it this way: "the factors that limit the use of CFD in the early design phase are how quickly the grids can be generated and how quickly 'sufficiently' accurate CFD simulations can be provided. For complex shapes, 10–100 CFD solutions with marginal fidelity but a very fast turn-around time would allow CFD to be part of this stage."[45] In addition, the grid density "support" required in regions of high flow gradients can be difficult to supply, especially without knowing where all of those regions are located a priori. The orientation of the grid relative to a shock can also be important, and numerical schemes need to address how to handle shocks with the grid in different orientations to the shock. Adaptive mesh refinement methods mentioned in Chapter 7 may also prove useful for improving the grid in regions of high flow gradients.

Near the surface of a vehicle, grid resolution can be crucial to the aerothermodynamic predictions being made. Papadopoulos et al.[45] showed that "a computational mesh and a level of convergence which result in accurate surface pressure and shear stress values do not guarantee accurate heat-transfer values." In addition, they found that both the level of convergence required for accurate predictions and the numerical algorithm used (including the flux-splitting method and the limiter used) could impact the results of a flow simulation at hypersonic speeds. Papadopoulos et al.[45] believe that accurate predictions require careful monitoring of both the cell Reynolds number and the temperature jump near the wall. The cell Reynolds number, given by:

$$Re_{cell} = \frac{\rho_{wall} a_{wall} \Delta y}{\mu_{wall}} \qquad (10.1)$$

helps to insure that the initial grid spacing near the wall is small enough to accurately deal with the viscous and heat-transfer effects at the surface. In Equation 10.1, Re_{cell} is the cell Reynolds number, ρ_{wall} is the density at the wall, a_{wall} is the speed of sound at the wall, Δy is the y-step size of the cell, and μ_{wall} is the viscosity at the wall. A grid sensitivity study is especially important for accurate flow prediction at hypersonic speeds, and the cell Reynolds number is especially important to the results: "even a small value of order 10 of Re_{cell} might result in 100% error in the vicinity of the stagnation point, and the heat flux converges when the value of Re_{cell} is less than or equal to 3."[46]

The accurate prediction of transition and turbulence is another crucial area in hypersonic flow simulation, but accurate transition models may not be available for a long time due to the difficulties in understanding the transition process in hypersonic flow.[47] Although numerical predictions of transition in high-speed boundary layers have been conducted (see Fasel et al. for example),[48] it remains a daunting challenge, so only turbulence modeling will be reviewed here. One thing is certain: Few, if any, of the models were developed specifically with high-speed flows in mind. Compressible dissipation and pressure dilatation corrections, such as those proposed by Suzen and Hoffmann,[49] will need to be included to properly model these high-speed flows. Forsythe et al.[50] showed the impact of compressibility corrections on the prediction of supersonic flows; detailed analysis of these turbulence models and corrections for hypersonic flows is certainly needed.

Showing special promise for predicting turbulent flows are the hybrid RANS/LES turbulence models, such as detached-eddy simulation (DES) by Spalart et al., which was discussed in Section 8.10.1. The DES model takes advantage of the RANS approach within the attached boundary layer, where RANS models work quite well. Once the flow separates, an LES approach is used, which is computationally affordable using current computers. Again, this approach has shown promise for massively separated flowfields only, including at supersonic speeds as shown by Forsythe et al. in Chapter 8,[50] but work is needed to verify that these models work well in hypersonic flows.

Chemically reacting flows exist when the flowfield temperature reaches a level that causes the constitutive parts of the atmosphere to start reacting and changing their basic molecular state. For example, as temperatures reach 2000 K (3600 R) O_2 begins to dissociate, at 4000 K (7200 R) O_2 dissociation is complete and N_2 dissociation begins, and at 9000 K (16,200 R) N_2 dissociation is complete and O and N begin to ionize.[51] Taking these chemical reactions into account is essential to fully analyzing the viscous and heating effects of the flow. According to Edwards,[52] "Several levels of approximation are available to model hypersonic flows. These include the widely-used perfect gas model, along with the more general equilibrium, nonequilibrium, and frozen chemistry assumptions." Cheng and Emanuel[53] summarize the difficulty in accurately modeling these chemical reactions: "Current research developments in high-temperature flow physics still do not possess a methodology base with unquestioned certainty. A rational way to derive an equation set for a nonequilibrium flow is to write the time rate of population change of atoms and molecules, with a specific energy state, as the difference between the sum of rates of all collisional and radiative transitions that populate a given state and the sum of rates that depopulate the state." They go on to explain that "this hypothesis, it should be noted, still requires rather bold assumptions regarding the interaction potentials and the collision mechanisms." In other words, even if we could describe the exact chemical reactions, defining the reaction rates and interactions is something that still requires a fairly high level of modeling.

Perhaps the most commonly used nonequilibrium model in CFD simulations is Park's two-temperature model.[54] This model uses translational and vibrational temperatures, along with various assumptions about the translational, rotational, vibrational, and electronic excitation modes of the gas species. The impact of using the nonequilibrium model in predicting the flow around an airfoil at hypersonic speeds showed that the lift and pitching moment changed by 10 percent and 20 percent, respectively.[55] Nompelis et al.[56] found that numerical predictions of heat-transfer rates for a double-cone geometry could differ by 20 percent from experimental data, with the nonequilibrium model, the surface no-slip condition, grid density, and non-uniform tunnel flow conditions accounting for the difference.

Other rate-limiting issues for chemically reacting flows are detailed by Sarma.[57] These include, but are not limited to, the availability of accurate flight-test data for heat fluxes, forces, and moments, the availability of high-fidelity, ground-based experimental data with defined and controlled experimental conditions, the applicability of the Navier-Stokes equations and traditional boundary conditions, and non-Boltzmann effects. Sarma concludes that, "it is obvious that we have still a long way to go before fully validated predictive hypersonic CFD codes can emerge out of current efforts."[57]

In spite of these difficulties, reasonable flow predictions can be made for quantities such as heat transfer when modeling issues are carefully considered, as shown in Fig. 10.32. While the overall heat fluxes are well predicted, there remain regions where differences between the numerical predictions and experimental data are fairly large, which could lead to design problems for a future vehicle.

The ability to simulate hypersonic flow has not kept stride with the advances made for lower-speed, less-complex flows – it should be apparent that a great deal of work still needs to be done before computational simulations of hypersonic flow are of a caliber that could be used for analysis and design of hypersonic vehicles for the entire flight regime.

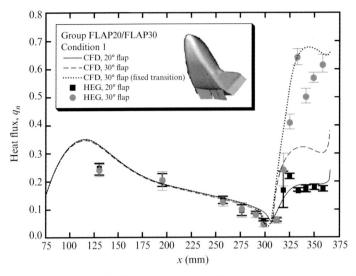

Figure 10.32 Numerical predictions for X-38 heat flux distributions (HEG = High-Enthalpy Shock Tunnel Göttingen) (Ref. 38; The original version of this material was published by the North Atlantic Treaty Organization Scientific and Technology Organization, Collaboration and Support Office [NATO/STO/CSO] in Technical Report RTO-EN-AVT-116 on Critical Technologies for Hypersonic Vehicle Development published in 2004).

10.6 Aerodynamic Design and MDO

Computational aerodynamics can be much more than a wind tunnel in a computer. Originally, computational aerodynamics was used primarily for the *analysis* of a configuration. However, it quickly became clear that it would be desirable to use computational methods to find the shape of configurations that achieve an "optimum" design directly, rather than have the user employ a "cut and dry" approach simulating a wind tunnel development program. Computational *design* methods began to emerge shortly after computational analysis became available.[58]

The design problem is a more sophisticated one than the analysis problem because the designer needs to decide what constitutes a good design and translate this desire into a mathematical statement suitable for computational methodology. Unfortunately, it is challenging to define the design goals and constraints to be used. Originally, aerodynamicists considered the design problem as a requirement to specify the pressure distribution rather than the geometry. Thinking in two dimensions, the questions became: what pressure distribution do we want on the airfoil and what are the constraints? This led to the development of various "inverse" methods that started with the flow and found the geometry that could create that flow. Since the pressure distribution is prescribed, the lift and pitching moment requirement can be incorporated directly. However, it is possible to prescribe a pressure distribution for which no geometry exists or, if it does exist, it may not satisfy some thickness constraint. Also, multiple pressure distributions could exist that results in the same integrated loads (lift, drag, and/or pitching moment). Furthermore, the airfoil is not useful if it does not have good characteristics over a range of Mach numbers and lift coefficients (a design that is good at one flow condition might not be good at a closely related flow condition). Nevertheless, inverse methods are useful when used by experienced aerodynamicists.[59]

Once computational methods started to be used for configuration design, the problem became much broader. That is because a configuration depends on many other disciplines besides aerodynamics (such as structures, propulsion, electro-mechanical systems, etc.). While the aerodynamicist might be thinking in terms of maximizing the lift-to-drag ratio of the design, curiously, this will not produce the best overall design. Especially for commercial transports the goal is the lowest cost, and "cost" can be viewed from many points of view. The manufacturer may want to sell a design that costs the least to build. The operator may be interested primarily in the design that costs the least to operate. Some insight into the cost issue can be gained by studying the paper by Jensen et al.,[60] who compared the designs that resulted from using a variety of criteria in an optimization study. In optimization terminology this criterion is known as a "figure of merit" or objective function (or cost function). A similar, more modern study has been carried out by Gur et al.[61] In these types of studies, it becomes evident that the most attractive configuration is the one that reflects aerodynamics and structures together, either by minimizing the vehicle takeoff gross weight or minimizing the fuel used and emissions. This means that both the structures and the aerodynamics have to be considered simultaneously in the computational design. Other disciplines such as propulsion and stability and control also need to be included. This approach to computational design has become known as Multidisciplinary Design Optimization (MDO). The initial proponent of MDO in the United States was Jarek Sobieski, who spearheaded the development of systematic computational methods for MDO.[62] Use of MDO is essential to cope with advanced concepts where the disciplinary interactions are tightly coupled.

As an active area of research, there are numerous approaches being investigated for MDO, but experience has shown that no single approach is best for all situations. For aerodynamic design, there are two key issues that must be made efficient for MDO to be practical. The first issue is the representation of the geometry. The geometry needs to be defined by a parametric model that can represent a broad range of shapes with a small number of "parameters." In the design process, these parameters will become design variables used to define the optimized design. In general, it has been found that full-strength CAD packages are not appropriate for this approach, and many geometry schemes developed specifically for MDO have been proposed.[63-65] Careful consideration should be given to the selection of a geometry model for a specific problem. The second issue is the choice of the appropriate level of fidelity to be used for the aerodynamics and how to represent the aerodynamics in the MDO process. In some cases linear theory is appropriate; however, in many cases Euler or Navier-Stokes (likely RANS) equations will be required. This will influence how the problem is approached and how expensive (in time and resources) the computation may be.

By the very definition of MDO, some sort of optimization method will be involved in the design process.[66] For the nonlinear problems in which we are interested, optimization methods require repeated executions of the aerodynamics code in order to estimate trends or gradients. The choice of the level of fidelity required from the aerodynamics is important because of the computational cost of the calculation. If linear aerodynamics is acceptable, an analysis method may be coupled directly to an optimizer. However, higher fidelity methods require a different approach. We have found that it is best to develop a database of solutions and interpolate the database to obtain the aerodynamics required during the optimization. There are several reasons why this is a good approach. First, many thousands of aerodynamic cases will be required during the optimization, so it is impractical to do these calculations in a serial fashion. We can take advantage of today's parallel computers by running the cases in parallel before the optimization. The cases to be run are various combinations of the design variables selected using design of experiments theory. Taking this approach can speed up the process time by orders of magnitude. Another reason for this approach is the opportunity to filter artificial numerical noise arising from the discretized calculations (shocks jumping between grid lines, etc.).

Once the database is assembled, it can be used repeatedly without incurring an additional cost. The process just described is known in statistics as developing a response surface. Once we have the response surface, we develop a model of the surface, and the model can be used quickly and inexpensively. The response surface model can be constructed in many ways, the simplest approach being to represent the surface with a low-order polynomial; many other approaches can be used. The result of this procedure produces a "surrogate" for the expensive problem that is almost free in terms of computation.[67] The ideas described here are simple, but success in a particular application requires an understanding of both the process described here and the problem being solved. See the paper by Giunta et al.[68] for more details and examples of this approach.

Key issues to consider when developing an MDO approach for design include:

- Computational grids: the grid required for evaluation of each set of design variables must be generated automatically. In many cases this is still difficult, and innovative approaches are required for each problem.
- Gradients of the objective function and constraints: Although there are optimization methods that do not require gradients with respect to the

objective function and constraints, gradient-based methods are among the most efficient approaches. Originally, these gradients were found using finite differences of the solution with respect to each design variable. Although relatively simple to implement, this approach is not particularly accurate and is very expensive computationally. Instead, the gradients, also known as sensitivities, can be found using methods borrowed from control theory. These are known as adjoint methods and provide a means of finding sensitivities at a computational cost only modestly higher than the basic analysis; Jameson pioneered this approach.[69]

- Local optima: When trying to find the best design using optimization methods, it is possible to obtain a local rather than a global optimum. The problem is, in general, nonlinear, and there is no way to guarantee that a global minimum has been found. Since optimizations typically start with a baseline design, the usual approach is to perform the process several times, each using a different baseline design or starting point. The resulting designs may be the same or slightly different, and the user will select the best outcome from this process.

This section concludes with two examples: the high-speed civil transport studied in the 1990s and a strut-braced transonic transport that remains under current study as a future fuel-efficient airliner.

10.6.1 High Speed Civil Transport Example[70]

The HSCT problem studied by NASA and Boeing/McDonnell Douglas in the 1990s pushed the available technology to the limits. It was very sensitive to aerodynamic drag, and eventually studies at Virginia Tech used the Euler equations to model the aerodynamics (with skin friction formulas used to estimate the viscous drag; see Chapter 4 for examples). As many as twenty-eight design variables and fifty constraints were used to define the problem. The automatic grid generation issue described previously was resolved because the HSCT was a slender, smoothly varying shape, and an automatic grid generation tool appropriate for these geometries was available.[24] Figure 10.22 shows a typical configuration with the corresponding grid.

In this case, the key tradeoffs were between the wing thickness and planform sweep resulting in low structural weight, sufficient fuel volume, and low aerodynamic drag. The figure of merit was chosen to be the takeoff gross weight (TOGW). The response surfaces were constructed using analysis of variance methods, where thousands of linear theory aerodynamic analyses were examined to find out which design variables were the most important, and Euler calculations using this reduced set of design variables were made using coarse-grained parallel processing. An advantage of the response surface approach is that it can also be used to examine the behavior of the

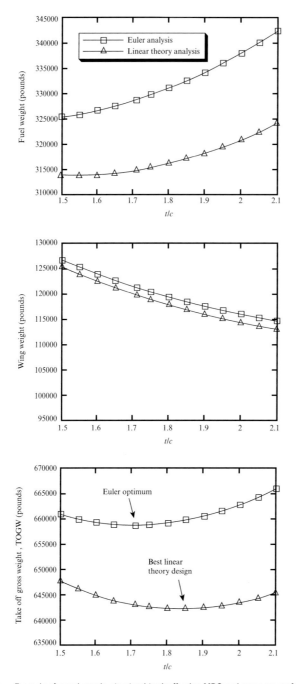

Figure 10.33 Example of aerodynamic-structural tradeoff using MDO and response surfaces (Ref. 70).

design around the optima. Figure 10.33 illustrates the insight that can be obtained from this approach. The TOGW variation with t/c is shown for both the Euler and linear theory aerodynamic models. The Euler level of aerodynamic fidelity results in a TOGW that is higher than that found using linear theory, and the Euler optimum t/c is slightly less than the linear theory

Figure 10.34   Typical truss-braced wing concept for a transonic commercial transport (Ref. 72).

result. The figure also includes both the wing weight variation with t/c and the fuel weight (aerodynamic drag) change with t/c. These types of sensitivity curves aid designers in choosing various geometric parameters for their configuration. As various parameters are established through this process, the shape and dimensions of the aircraft begin to develop.

10.6.2 Truss-Braced Wing Example[71]

Struts are traditionally used on high-wing general aviation airplanes to reduce the wing weight by reducing the wing root bending moment. In fact, general aviation wings are pinned at the wing root. In contrast, transonic commercial transports use cantilever wings without struts to reduce the drag that would be caused by the strut. If we assume that modern CFD methods can be used to reduce or eliminate the aerodynamic interference of the strut-wing junction, then struts would be advantageous at transonic speeds also. This allows the wing to have a greater span (reducing induced drag), be thinner (reducing wave drag), and even allows some reduction in wing sweep. However, the highly coupled nature of the problem requires the use of MDO to explore the potential of transonic strut and truss-braced wing concept. An MDO approach was used to find the benefits of various strut and truss configurations. In this case, most of the analysis methods were coupled directly to the optimizer, with the exception of a response surface used to represent the transonic interference drag at the location of the strut-wing junction. Figure 10.34 illustrates a typical conceptual representation of a truss-braced wing concept for use in an MDO study.[72]

The results of numerous studies have shown anywhere from a 10 percent to a 20 percent reduction in TOGW compared with an equivalent cantilever wing design, depending on the specific concept and problem details. shows the various configuration topologies investigated in a typical study.[73] The results of the MDO study are shown in Figs. 10.35 and 10.36. In this case, the fuel weight is minimized and the potential fuel savings resulting from the truss configuration are clear.

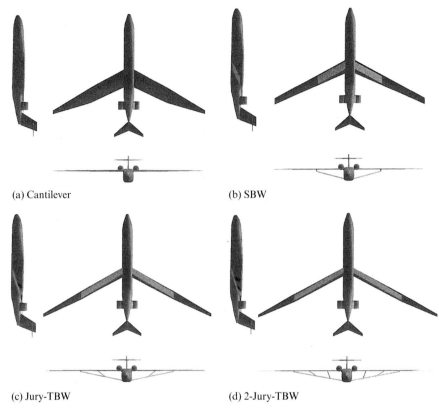

(a) Cantilever

(b) SBW

(c) Jury-TBW

(d) 2-Jury-TBW

Figure 10.35   Configuration topologies studied using MDO (Ref. 73).

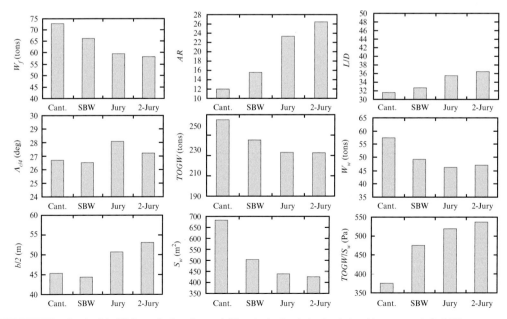

Figure 10.36   Result of the MDO examination of several different wing topologies for strut and truss concepts (Ref. 73).

Profiles in Computational Aerodynamics: William Mason

"I grew up in Winchester, Virginia, where I built and flew model airplanes, competing in control-line model airplane contests in the Virginia-Maryland area. The main lesson from those days was the importance of being able to start the engine and build (and re-build) a strong structure. Proper balance and control were also critical. Really good aerodynamics was only important afterward! I was able to go to National Airport in the days when you could still stand outside above the gates on the observation deck and watch (and smell) the prop airplanes. I learned about the large surface area of airplanes compared to cars when I washed airplanes at the local airport in exchange for rides.

"Attending Virginia Tech, I got a co-op job at McDonnell Aircraft in Saint Louis before completing my freshman year. Over the next few years I had a variety of assignments, including flight test department jobs (2 ½ F-4 Phantoms were being built each day!) and low-speed wind tunnel testing (including a swing-wing version of the F-4). Typically these jobs required lots of hand plotting of data. Although this was mainly flight test and wind tunnel data analysis, I also got an assignment to Spacecraft Advanced Design. There, I helped with computational analysis of ascent and entry trajectories of low L/D spacecraft vehicles and got to find allowable re-entry angles (skip or burnup limits) and available landing footprints. The aerodynamic characteristics were estimated by the aerodynamicists using Newtonian flow theory. I got to walk by the white room where Gemini spacecraft were being assembled on the way to my desk.

"The summer before my senior year I got a job as an engineering aide at Edwards Air Force Base – but was assigned to an Army unit that tested helicopters, including the HueyCobra. Flight test data reduction was a peculiar combination of hand and computer analysis. They assigned the summer aides to digitize oscillograph traces by hand for vibration analysis.

"Attending graduate school, I did a Master's degree studying aircraft trailing vortices, both experimentally and theoretically. For my PhD work, I studied transonic normal-shock boundary-layer interaction, developing an "analytic" theory. The work required extensive use of classical numerical methods. However, CFD was just starting to be developed for transonic flows, and for a graduate numerical methods class, I programmed the transonic flow solution over a circular arc airfoil using the method in the landmark paper written by Murman and Cole.

"I went to work at Grumman after graduation and my initial assignment was to the flutter group doing unsteady aerodynamics. However, I was quickly reassigned to the aerodynamics group, where I worked on a software development team that added both viscous and fuselage effects to an early 3D transonic code by Ballhaus and co-workers at NASA Ames for the Air Force. This gave me the opportunity to meet many of the early pioneers of CFD. I did other aerodynamics methodology projects, including work contributing to the design of the X-29, and the design of several wings for transonic and supersonic conditions. Then I applied a unique application of transonic methodology at supersonic speeds, leading to the design and test of a supersonic maneuver wing with supercritical conical camber, SC3. The final years at Grumman were spent on several fighter design projects, including supersonic cruise, supersonic maneuver, transonic maneuver, and low speed–high alpha configuration development. Most of the work at Grumman involved using supercomputers, including the NASA CDC 7600 and various Cray computers. However, I also got to work in several major wind tunnels, including a wind tunnel test program for high angle of attack aerodynamics in the 30 × 60 foot test section Full Scale Wind Tunnel at NASA Langley.

"Shortly after Grumman lost the competition for the ATA (A-12), I returned to Virginia Tech to teach a new course on applied computational aerodynamics and to advise senior aircraft design students. Coaching design teams was a lot of fun. My primary research work was associated with using computational methods for aerodynamic applications, either individually or as part of the development and application of MDO methods to airplane design. Key applications included HSCT design using MDO and then transonic strut-based wing transports. The HSCT work exploited the use of parallel computing. Since my retirement, I continue to try to learn aerodynamics and teach classes in configuration aerodynamics and applied computational aerodynamics."

10.7 Integration of Computational and Experimental Work

We believe it is imperative that people conducting joint aerodynamics analyses should be well-versed in the advantages and disadvantages of the computational, experimental, and theoretical approaches, including the errors and/or assumptions of each. Without this knowledge, collaboration is more difficult, and a great deal of time can be wasted trying to resolve discrepancies in results. We will briefly describe some of the advantages and disadvantages of both experiments and computations (theoretical predictions come with their own assumptions and limitations, as long as you remember them!).

10.7.1 Pros and Cons of Experiments

While we all could make our own lists of the advantages and disadvantages of experimental methods, a number of issues come to mind immediately. Certainly, any list has its own deficiencies, so we invite you to add your own ideas to these lists.

Some of the strong advantages of wind tunnel testing include:

- Well known and understood capabilities
- Usually easy to set and verify freestream conditions
- Forces and moments are relatively easy to obtain
- Flowfield properties are readily available (from probes, hot wire, LDV, PIV, PSP, etc.)

Some of the disadvantages of wind tunnel testing include:

- Many measurements are intrusive and modify the flow
- Wall corrections are often required and can be difficult to make
- Support system corrections are often required and are difficult to make
- Blockage issues must be addressed
- Model fidelity is often a challenge
- Matching flight conditions can be difficult (Reynolds number, transition, etc.)
- Transonic flow is especially troublesome due to nearly normal shocks

10.7.2 Pros and Cons of Computations

Perhaps the biggest disadvantage of CFD predictions is the over-optimism of the earliest users, as described in Section 2.3.2. In fact, the non-acceptance of CFD by many people led to a common lament among CFD practitioners: "No one believes CFD results except the person who ran the code, and everyone believes wind tunnel results except the person who conducted the test." There is certainly a great deal of truth to this, but in reality there are a number of advantages and some important disadvantages to CFD.

Some of the advantages of CFD include:

- Complete flowfield prediction (all properties are predicted throughout the flow)
- Matching flight conditions is fairly straightforward
- Nonintrusive flowfield "measurements" can be made
- Steady or time-accurate results are possible
- Flow visualization is straight forward

Some of the disadvantages of CFD include:

- Turbulence models
- Transition prediction
- Numerical dissipation
- Numerical error
- "Black box" syndrome (GIGO is still a common CFD problem)

Without understanding the strengths and weaknesses of each approach, engineers are left to "grope" in the dark to gain understanding about various aerodynamic phenomena; using both approaches is often enlightening and beneficial to understanding. Several examples of collaboration will be shown that detail how experiments and CFD can be used together and how an evolution is taking place that utilizes both approaches to their fullest capability in aerodynamic design and analysis.

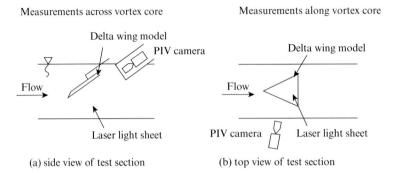

Measurements across vortex core   Measurements along vortex core

(a) side view of test section   (b) top view of test section

Figure 10.37 PIV measurements of delta wing in water tunnel (Refs. 74, 75).

10.7.3 Delta Wing with Periodic Suction and Blowing for Flow Control

The first example of a situation where close collaboration between experimentalists and computationalists paid dividends is a delta wing study conducted at the U.S. Air Force Academy. The purpose was to determine the feasibility of using periodic suction and blowing (PSB) along the leading edge of the wing.[74,75] Researchers tested the 70° delta wing configuration in the U.S. Air Force Academy water tunnel at $\alpha = 35°$ and $Re_c = 40,700$ (see Fig. 10.37). Two-dimensional PIV measurements were taken of the flow over the upper surface of the delta wing, but no force and moment data were taken.

To perturb the shear layer originating at the leading edge of the delta wing, a semi-spherical rubber cap was used as an oscillatory blowing and suction flow actuator. It was moved back and forth by a connecting rod, eccentrically mounted on a disk that was driven by a 560W DC motor. The water displacement produced by the moving cap was channeled through a tube 2 cm in diameter to the hollow wing and to the length of the slot in its leading edge. With this setup, as with any oscillatory flow control method, fluid is drawn into the actuator over half of the sinusoidal cycle and ejected over the other half ($V = V_o \sin \omega t$). The phase during the forcing cycle is determined by the position of the rotating disk flywheel, which features an adjustable optical pickup to synchronize the data acquisition with a particular phase of the forcing cycle. A forcing cycle starts at 0° with the blowing phase, which extends to 180°. The suction portion between 180° and 360° completes the cycle.

One of the problems encountered during the experimental phase of the investigation was that the suction phase did not appear to be as effective as the blowing phase of the periodic cycle. While this observation was important to the experiment, no direct reasons for the apparent anomaly were known, leaving the experimentalists to wonder if their apparatus was operating correctly, or if there was some fluid dynamic interaction at work. Another difficulty realized by the researchers was that, while they knew the impact of the

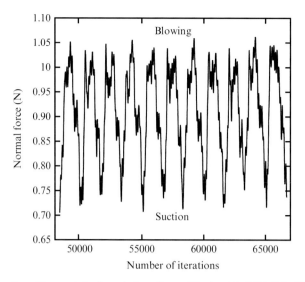

Figure 10.38 Normal force variation for periodic suction and blowing, $\Delta t = 0.006$ (Refs. 74, 75).

suction and blowing on the flowfield, they did not know the impact on the aerodynamic forces of the delta wing.

In order to help answer these questions a companion CFD study was conducted. The study allowed for the modeling of the PSB, but also allowed for keeping track of the forces and moments during the oscillations. The various frequencies for the time-converged PSB case are visible in Fig. 10.38, which shows the normal force variation for 17,000 iterations (over 10 cycles of the suction and blowing). The suction and blowing frequency is obvious, but, overlaid on that frequency is the shear-layer instability frequency, constantly oscillating around the lower frequency. Also notice that the blowing portion of the suction/blowing cycle is more effective, as evidenced by the amount of time the normal force remains at the highest levels. When the suction cycle takes place, decreasing the normal force, the force spikes to a minimum value but then quickly rises again as the suction phase ends. This verified what was seen during the experimental portion of the work, where it appeared that the suction was incomplete (or possibly not working correctly). Even the numerical simulation clearly shows that the suction phase is not as effective in altering the normal force acting on the delta wing. There is also a slight dwell as suction begins, which was also not explained by the PIV results.

Again, the CFD results were able to answer some of these questions due to the ability to interrogate the flowfield at all locations and at all times. Figure 10.39 shows the velocity vectors in the vicinity of the delta wing leading edge at the 60% chord location. During the blowing phase (90 deg.), Fig. 10.39a shows that the fluid is able to expel directly into the surrounding flow and have a direct impact on the shear layer region. However, during the suction phase (270 deg.), Fig. 10.39b shows that the flow in the vicinity of

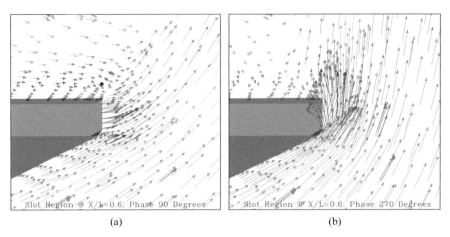

(a)   (b)

Figure 10.39   Velocity vectors near delta wing leading edge showing difference between suction and blowing (Refs. 74, 75; a full color version of this image is available on the website: www.cambridge.org/aerodynamics). (a) Blowing phase, 90 deg. (b) Suction phase, 270 deg.

the leading edge of the delta wing is not able to turn the sharp corner and be fully brought into the PSB channel. This explains the difference seen in between suction and blowing and may also explain the dwell during the blowing phase, since the flow is attached and fully formed during this phase.

10.7.4 Pitching UCAV Configuration

A full-scale model for the Boeing 1301 UCAV configuration is shown in Fig. 10.40; the configuration has many similar features to the X-45A. The 1301 configuration has a straight, 50° sweep leading edge, an aspect ratio of 3.1, rounded leading edges, a top-mounted engine inlet, and a blended wing/body planform. A 1:46.2 scale model of the configuration was tested in the USAF Academy 3 ft × 3 ft (0.914 m × 0.914 m) open return low-speed wind tunnel.[76,77] The scaled model has a mean aerodynamic chord of 5.24 in. (0.133 m) and a reference area (wing planform area) of 46.82 in.^{2} (302.1 cm^2). The tunnel has less than 0.05 percent freestream turbulence levels at all speeds. The test was conducted at a freestream velocity of 65.4 ft/s (20 m/s), which corresponds to a chord-based Reynolds number of 1.42 × 10^5. The model was sting-mounted from the rear, and forces and moments were measure with a six-component force balance. Both static and dynamic testing was done; forces during the dynamic runs were obtained by subtracting the force history with the tunnel off from the dynamic data. The dynamic pitching was done with a shifted cosine oscillation, starting at a certain angle of attack and pitching up to twice the peak amplitude of the cosine wave, then back to the original angle of attack.

$$\alpha(t) = \alpha_\circ + m^\circ - m^\circ \cos(\omega t) \qquad (10.2)$$

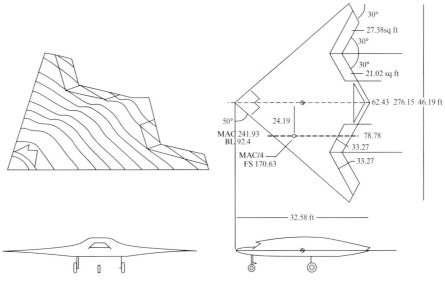

Figure 10.40   Boeing 1301 UCAV configuration (Refs. 76, 77).

where α_o and m were varied to obtain results for $0° \leq \alpha \leq 45°$ in three pitching cycles. This pitching function was used because it produces a motion without any discontinuities in acceleration or velocity at the beginning and end of the motion and thus is easier to implement in both an experiment and a CFD code.

One of the limitations encountered during the experiment was that the experiment only was able to measure forces. This is a common problem during wind tunnel tests, where tests are either of the force and moment variety or of the flowfield property variety, but experiments cannot always include both sets of measurements. Because of this, researchers are usually unsure of the fluid dynamic causes of various results, being left to make educated guesses about unusual or unexpected results. For example, in the case of the UCAV wind tunnel tests, the vehicle lift coefficient showed linear lift characteristics up to 10° to 12° angle of attack, as shown in Fig. 10.41. Wing stall was evident at about 20° angle of attack, with the lift being re-established up to 32°, after which an abrupt loss of lift takes place. But what is the cause of the poor lift characteristics? Are the results a direct effect of leading-edge vortices and vortex breakdown? The experimentalist is left to hypothesize and wonder, but the numerical researcher can add insight into the problem.

Also shown in Fig. 10.41 are the results of the CFD predictions. Perhaps the most important result of the CFD simulation was the realization of just how unsteady the flowfields in the post-stall region were. Time-accurate results matched the experiment fairly well, with fairly good modeling of the flowfield, including drag, up to $\alpha = 45°$. However, there was a difference in lift from $\alpha = 20°$ to $\alpha = 30°$, which could have been caused by the presence of the sting, surface roughness, transition, or a host of other phenomena.

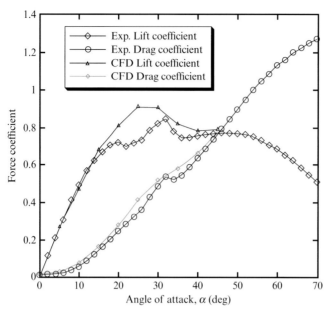

Figure 10.41 Numerical (time accurate) and experimental (static) force coefficient comparison (Refs. 76, 77).

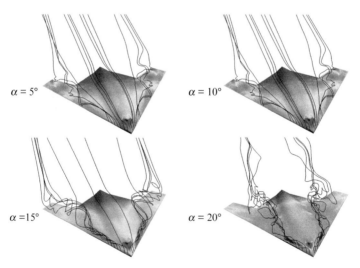

Figure 10.42 Numerical flowfield predictions for various angles of attack (surface colored by pressure) (Refs. 76, 77; a full color version of this image is available on the website: www.cambridge.org/aerodynamics).

Figure 10.42 shows representative numerical simulations of the configuration at $\alpha = 5°$, $10°$, $15°$, and $20°$, with the flowfield being visualized with streamlines and the surface coloured with pressure. The leading-edge vortices are clearly visible closely following the $50°$ sweep, until approximately $x/l = 0.40$ when vortex breakdown is evident. Low surface pressures are visible beneath the vortex prior to breakdown; these low pressures account for the lift on the configuration at $\alpha = 20°$. After breakdown, the vortex wake quickly moves up and behind the leading edge, leading to higher pressures on the upper surface of the wing. The vortices are very wide compared with their

Figure 10.43   C-130 wind tunnel model part showing decomposition of surface shape (Ref. 78).

height, most likely due to the rounded leading edges of the wing. Secondary vortices are also visible beneath the primary vortices. The primary vortex is seen splitting into two flow structures after the breakdown location.

These numerical simulations help to answer some of the questions raised by the wind tunnel tests. The rounded leading edges and mid-range leading-edge sweep yield weak leading-edge vortices that do not produce very much lift. The vortices are only just beginning to form (and are very weak) when breakdown takes place, and their contribution to lift is reduced. These are common characteristics of "lambda" type wings, but seeing the CFD simulation increases understanding of the wind tunnel tests data.

10.7.5 C-130 Airdrop Configuration

This is a case where CFD and experiments were being conducted in a collaborative fashion. A wind tunnel model of a C-130 was being tested at various wind tunnels, with different types of flowfield tests being conducted at each tunnel (force and moments, surface flow visualization, etc.).[78] As the wind tunnel tests proceeded, the results were in disagreement with the CFD simulations being performed, leading to a great deal of hand wringing and consternation. Finally, after a lot of hard work, it was discovered that the wind tunnel model had been degrading in shape as the various tests were being performed (see Fig. 10.43), leading to more and more configuration mismatch between the model and the original CAD description that had been used to create the CFD grids. This led to a study of the model material

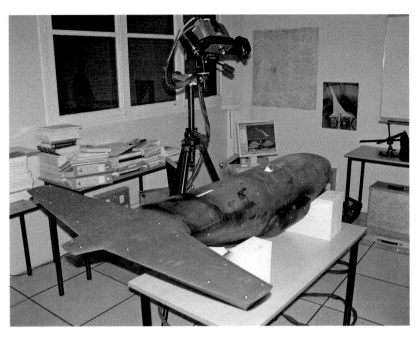

Figure 10.44   3D optical digitizing of C-130 wind tunnel model (Ref. 78).

and their chemical compatibility with products to be used in the wind tunnel tests, perhaps including:

- Fluorescent viscous wall coatings
- Acetone
- Black paint
- Fog generating liquid, etc.
- Filling of holes and gaps with putty, etc.

A three-dimensional optical digitizing of the wind tunnel model was performed at ENSICA (achieved by GOM Company, as shown in Fig. 10.44) in Toulouse, and comparisons were made with the original CAD geometry. Figure 10.45 shows the differences between the original CAD geometry and the actual model being tested in the ENSICA wind tunnels, with large variations evident at various locations around the fuselage and horizontal tails. In fact, the tips of the horizontal tails appear to be bent as much as 2mm away from the original CAD shape. This shows another example of how CFD and experiments can be used together to insure accuracy of results in aerodynamic evaluation.

10.7.6 Closed-Loop Flow Control

A novel combination of numerical and experimental evaluation is being conducted to show the effect of feedback flow control on the wake of a circular

Figure 10.45 Differences between C-130 wind tunnel model and original CAD definition: color scale: −2 to +2 mm (Ref. 78; a full color version of this image is available in the color insert pages of this text as well as on the website: www.cambridge.org/aerodynamics).

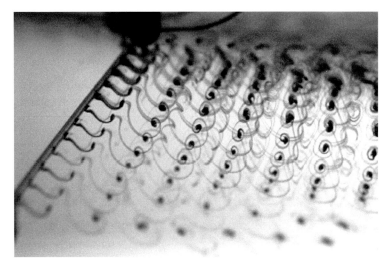

Figure 10.46 Flow visualization of the cylinder wake at Re = 120, forced at the natural shedding frequency with an amplitude of 30% of the cylinder diameter (Refs. 79–81; a full color version of this image is available on the website: www.cambridge.org/aerodynamics).

cylinder at a Reynolds number of 120, as shown in Fig. 10.46.[79–81] An initial 2D numerical simulation of the laminar flow was investigated in direct numerical simulation using proper orthogonal decomposition (POD, a nonlinear time-continuous reduced-order modeling approach) by placing sensors at various locations downstream of the cylinder as shown in Fig. 10.47. The flow was computed using the commercial Navier-Stokes solver Cobalt, and the POD analysis was done with MatLab. Also shown in is the feedback loop after information from MatLab is used to oscillate the cylinder normal to the freestream flow in order to excite or dissipate the vortex wake. The CFD was used to determine optimal number and location of the sensors in order to accurately (to required levels) describe the flow.

In the unforced flow, the vortices roll up between 1 and 2 diameters downstream of the cylinder, while, in the feedback-controlled situation the rollup

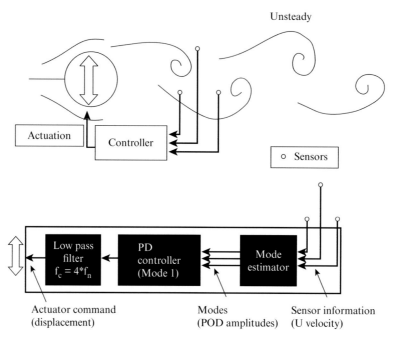

Unsteady

Actuation

Controller

Sensors

Low pass filter
$f_c = 4*f_n$

PD controller
(Mode 1)

Mode estimator

Actuator command
(displacement)

Modes
(POD amplitudes)

Sensor information
(U velocity)

Figure 10.47   Flow geometry around a circular cylinder including sensor placement and control concept (Refs. 79–81).

occurs between 3 and 4 diameters downstream, as shown in Fig. 10.48. The engineers observed a reduction in the vortex shedding frequency simultaneous with the lengthening of the recirculation zone. In the low drag state, the near wake is entirely steady, while the far wake exhibits vortex shedding at a reduced intensity. The forced case achieved a drag reduction of close to 90 percent of the vortex-induced drag and lowered the unsteady lift force by the same amount.

This success of the low dimensional feedback control of the circular cylinder wake in the two-dimensional CFD simulation led to the implementation of the control approach in a water tunnel experiment. An in-house developed real time PIV system was used to provide sensor information at the same downstream and flow normal locations used in the CFD simulation, using a grid of 35 off-body sensors. The main difference was that the CFD simulation was two-dimensional, while the water tunnel model features a three-dimensional model and flowfield with an aspect ratio of more than 40. With these experimental findings, three-dimensional numerical simulations were performed to gather quantitative data along the span of the model, which is not possible with current state-of-the-art experimental measurement techniques. The simulation setup shown in Fig. 10.49 resembles the water tunnel experiment in terms of aspect ratio, Reynolds number, and feedback control method employed. This approach showcases the highest level of integration of CFD and experiments that the authors have seen and shows what can be accomplished when the best attributes of each approach are used together.

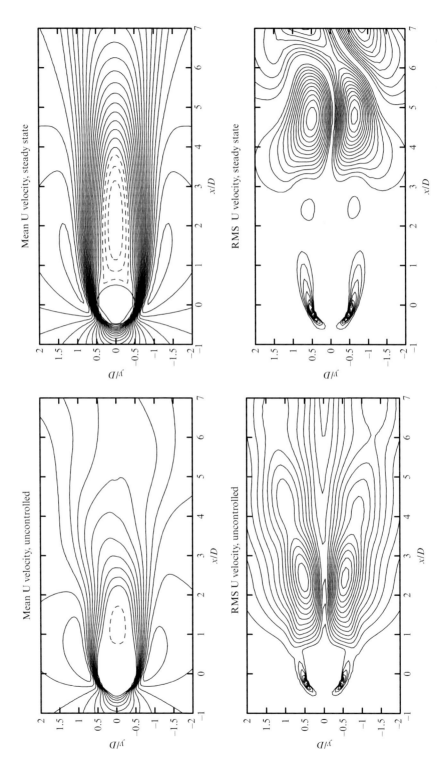

Figure 10.48 Mean flow (top) and RMS velocity distributions (bottom). Left, uncontrolled, right, controlled case. The cylinder is centered at (0,0) and of diameter 1, flow from left to right. Negative isocontours are dashed; positive isocontours are solid lines (Refs. 79–81).

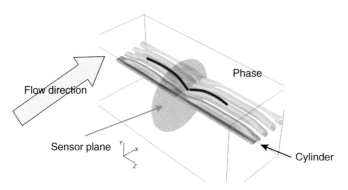

Figure 10.49   Numerical simulation setup for the three-dimensional case (Refs. 79–81).

10.7.7 Data Assimilation

There are many applications in geo-sciences and engineering where "sophisticated mathematical models have been developed and for which significant observational data are also available. Blending the predictive capability of the mathematical models with the information contained in the data—known as data assimilation—is a very important task and has enormous potential pay-back. Many challenging mathematical questions arise naturally in the context of data assimilation. These are related to the strongly nonlinear dynamical systems governing the evolution of key quantities, the non-Gaussian behaviour of statistical estimates of the system, and the multi-scale nature of the processes involved, and the high dimensionality of computational algorithms for the problem."[82]

Data assimilation has been widely used in weather prediction by NOAA and the National Weather Service.[83] MADIS, NOAAs Meteorological Assimilation Data Ingest System (http://madis.noaa.gov), uses massively parallel computer systems to collect, integrate, perform quality control, and distribute observations from NOAA and non-NOAA organizations to perform short-term weather prediction. Highly nonlinear mathematical models are required to blend the experimental data (which all have different levels of uncertainty and come from numerous sources including satellites, ground stations, and aircraft) with numerical predictions. The end result is a prediction of the weather that can be compared with the real weather that happens only a few hours later. The improvements in this approach over the past decade have allowed weather forecasters to increase their forecast time from hours/days to days/weeks.

What does all of this have to do with aerodynamic predictions? Some engineers have started to use data assimilation techniques to create an aerodynamic "truth" model that combines experimental data from numerous sources (and with varying degrees of uncertainty such as wind tunnel testing and flight testing) with numerical predictions. This is a very new and interesting idea, and can be quite controversial since the traditional views of "truth" in aerodynamics can be challenged, forcing us to rethink how we perform aerodynamic predictions.

Profiles in Computational Aerodynamics: Tracie Barber

"I grew up in New Zealand and from an early age I planned to be an ornithologist and part-time truck driver, as I very much liked birds and I also liked mechanical things. However, when I was at high school I realized that I didn't really enjoy biology and was more attracted to courses such as physics, so ornithology was not going to be a good option. My interest in flight turned from birds to aircraft, helped along by the *Top Gun* movie appearing just as I was at the age of choosing a possible career, so I decided to try (unsuccessfully) to join the New Zealand Air Force. Fortunately for me the NZ Air Force ceased to exist a few years later anyway. In the meantime, I decided to study mechanical engineering. During my first year I found some guides to Australian universities, and discovered I could study aerospace engineering if I moved, so I surprised my parents and moved to Sydney, where a handy job at an ice cream shop supported me during my studies.

"During my undergraduate degree my favorite courses were fluid mechanics and aerodynamics, and when the opportunity came to study a PhD in aerodynamics, I very happily stayed on at the university. My PhD was a computational fluid dynamics (CFD) study of wings in ground effect, with the main application being wing-in-ground effect vehicles, such as the large Russian Ekranoplans of the 1970s. I completed a small amount of experimental work using a Particle Imaging Velocimetry system and a very primitive moving ground wind tunnel. After completing the PhD in 1999, I worked for the then-distributor of the CFD software Fluent. This gave me an excellent overview of other applications of CFD as I was called on to support the local users in various industry and university groups.

"After a couple of years I missed research and my PhD supervisor, Professor Eddie Leonardi, suggested I become a lecturer at the University of New South Wales. I began to be more interested in experimental fluid dynamics (EFD) and the research I was doing in ground effect aerodynamics demanded a better version of my primitive moving ground wind tunnel. With the help of some talented PhD students, a new wind tunnel was constructed and a moving ground plane was designed. Because of the strong CFD background of my research group, we used CFD to entirely design the wind tunnel before anything was built, resulting in an excellent facility and the start of my very keen interest in integrating CFD and EFD in my research. As most of my PhD students were interested in cars, the wings turned upside down and we started to focus much more on race-car aerodynamics, considering such applications as wing/wheel interaction, flow separation on downforce wings, delaying separation using dimpled surfaces on downforce wings. and wing-tip vortex movement from downforce wings. We also acquired a three-dimensional Laser Doppler Anemometry system and built our own PIV system from an existing Nd-YAG laser. Other lasers were soon purchased for the use of flow visualization. The students in my group quickly became used to working in both CFD and EFD and are skilled in both. This made any experiments very easy to simulate in CFD as they were designed for just this purpose.

"My involvement in the experimental side led to my role as laboratory director, where I now manage the school's laboratories (the largest mechanical engineering complex in Australia). In the last few years, I have started to work more in the field of vascular fluid dynamics, where the knowledge I have gained in EFD and CFD is useful, but the challenges and unknowns are significant. It is an interesting change and I am enjoying learning the specifics of a new field while making use of everything I have learned in my ground effect aerodynamics work. The courses I teach include fluid mechanics and two CFD courses. In my spare time, when not reading interesting fluid mechanics books, I spend time with my engineer husband Ashton, our two children, and our two big hairy dogs. We live near the beach in Sydney."

Tracie Barber is associate professor of mechanical and manufacturing engineering at the University of New South Wales in Sydney, Australia. She has published nearly thirty journal articles on a variety of applied aerodynamics topics, including race car aerodynamics, high-speed aerodynamics, and biological fluid dynamics. She was the recipient of the Zonta International Amelia Earhart Fellowship in 1997 and the University of New South Wales Vice-Chancellor's Teaching Award in 2005.

10.8 Current Applications of Potential Flow Codes

In spite of the fact that in Chapter 5 we said very few new panel method or vortex lattice codes were still being developed, the fact remains that these methods are still very valuable for conceptual and preliminary design, and various improvements and extensions to these approaches are still being made. We will review some of these recent developments here, although there are many more approaches still being worked on by various people, especially now that these methods can often be used on desktop or laptop computers. We will also review some new approaches to performing numerical simulations of the potential flow equations.

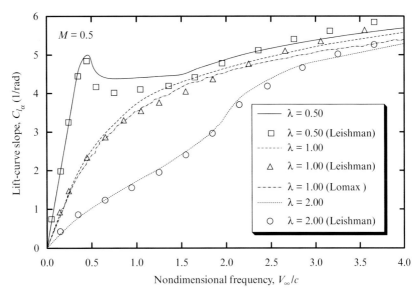

Sharp edge gust predictions compared with theories of Leishman and Lomax for $M = 0.5$ (Ref. 84; reprinted by permission of the American Institute of Aeronautics and Astronautics, Inc.)).

10.8.1 Compressible Vortex Lattice Method

Vortex Lattice Methods (VLM) were covered in Chapter 5, but the methods shown and the results presented were for incompressible flow conditions. Extensions within those codes for subsonic compressibility were often made with the Prandtl-Glauert compressibility correction, but no formal compressible versions of the codes were presented. The ability to perform unsteady and/or compressible simulations using the VLM approach requires recasting the equations governing the flowfield and coming up with improved approaches for finding solutions to the equations. A method capable of performing these calculations was used to perform unsteady gust simulations of the NACA 0012 airfoil at $M = 0.5$.[84] The lift-curve slope results in Fig. 10.50 are compared with the theoretical estimates due to Leishmann and due to Lomax. Since the compressible flow equations allow for subsonic or supersonic solutions, predictions were also made for $M = 2.0$. Notice that the predictions compare very well with the two theories for most nondimensional frequencies and gust velocity ratios.

10.8.2 Transonic Lifting-Line Method

Classic lifting-line theory was described in Chapter 4. However, this theory is based on incompressible flow concepts and is only valid for low subsonic Mach numbers. Lifting-line theory requires knowing the basic aerodynamics of the airfoils being used along the span of the wing, and a new approach has been developed at Georgia Tech that uses compressible airfoil panel methods, such as *XFOIL* or *MSES*, to supplement lifting-line theory, creating

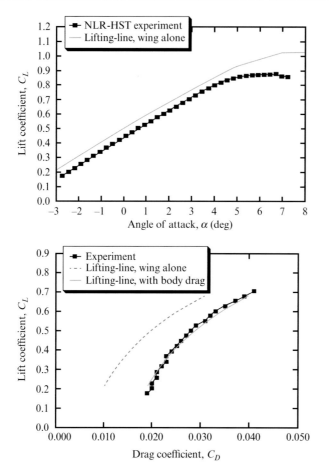

Figure 10.51   Lift and drag polar predictions for the DLR F-4 wing-body configuration at $M = 0.6$ (Ref. 85; reprinted by permission of the American Institute of Aeronautics and Astronautics, Inc.).

a semi-empirical approach for estimating the characteristics of wings and tandem lifting surfaces.[85] Programs like *XFOIL* and MSES are also able to estimate viscous effects, including skin friction and stall characteristics, so approximations of the wing-body skin friction and stall region are included in the prediction. A prediction for the DLR F-4 wing-body configuration that was used for the First Drag Prediction Workshop is shown in Fig. 10.51. The prediction shown is for $M = 0.6$, but predictions were also made at $M = 0.75$. The results were compared with wind tunnel data and show that the approach makes reasonable predictions for drag that allow for use in conceptual design and optimization approaches.

10.8.3 Unsteady Vortex-Lattice Method for Aeroelasticity

An unsteady Vortex-Lattice Method was described in Chapter 5 that was used for estimating the aerodynamics of flapping wing configurations. Another approach uses these methods to estimate aeroelastic effects on a

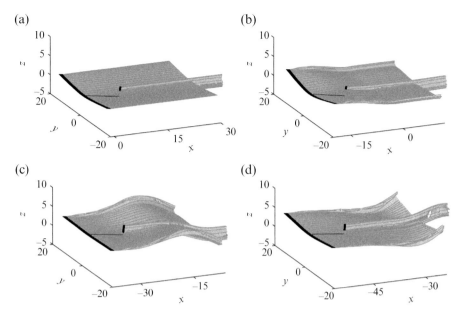

Figure 10.52 Snapshots (a–d) of the aircraft and the wake as they pass through a discrete "1-cos" gust (wing span is 32m) (Reprinted from Ref. 86, © 2012, with permission from Elsevier; a full color version of this image is available on the website: www.cambridge.org/aerodynamics).

glider configuration, including the fuselage and empennage sections of the aircraft.[86] The structural dynamics of the aircraft are simulated using nonlinear composite beam modeling, and the results provide detailed information about the flight dynamic characteristics of the aircraft (see Fig. 10.52). This approach provides valuable information with affordable methods that are capable of running on laptop computer technology.

10.8.4 Meshless Full Potential Solver

An approach for solving the Full Potential Equation (Equation 3.104) using finite difference approaches has been developed that takes advantage of the meshless grid generation approach described in Chapter 7.[87] A baseline node distribution is shown in Fig. 10.53, and the pressure coefficient distribution for the NACA 0012 airfoil at $\alpha = 2°$ and $M = 0.63$ is shown in Fig. 10.54. The results are favorably compared with the predictions of German[88] and the experimental data of Lock.[89] Efforts are under way to extend these results to three dimensions and fully transonic predictions.

10.9 The Future of Computational Aerodynamics

The future of computational aerodynamics, as we have described throughout this book, is especially bright. The continued progress of Moore's law,

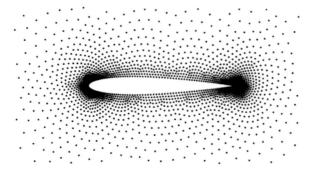

Figure 10.53   Close-up view of baseline airfoil node distribution (Ref. 87; Courtesy of Alejandro Ramos and Rob McDonald of Cal Poly).

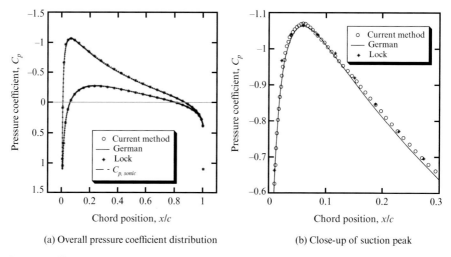

(a) Overall pressure coefficient distribution   (b) Close-up of suction peak

Figure 10.54   Pressure coefficient distribution for NACA 0012 airfoil, $\alpha = 2°$, $M = 0.63$ (Ref. 87; Courtesy of Alejandro Ramos and Rob McDonald of Cal Poly and Brian German of Georgia Tech). (a) overall pressure coefficient distribution; (b) close-up of suction peak.

coupled with advancements in flow solver technology and ever-improving turbulence models, means that CA is ready to follow wherever aerodynamic requirements lead. Remember back in Chapter 2 we discussed how aerodynamic design requires a variety of tools and capabilities? That will continue to be true into the future as new challenges and capabilities are required to design future aircraft.

For example, NASA is currently working on developing the technology that will be required for future aircraft (N+2 or N+3). They foresee that these aircraft will be required to have significantly less emissions, greatly reduced noise, and significant drag reduction for improved fuel burn. These goals will be achieved through a variety of technical developments, many of which will include CA capabilities we have described. These goals will require significant improvements in propulsion integration (how the propulsion system

Figure 10.55   Boeing's X-48B blended wing body (Courtesy of NASA Dryden Flight Research Center).

is built into the aircraft, aerodynamic drag reduction, and multidisciplinary optimization [MDO]).

Many of the improvements required for future aircraft can be significantly improved with CA as part of an integrated design approach that uses computations, wind tunnel testing, and flight testing in an intelligent way to greatly improve aircraft. For example, the goals of engine integration and drag reductions will require higher fidelity flow physics modeling, including transition prediction and improved turbulence modeling. Some of the aircraft will operate at the edges of current flight envelope capabilities (higher speeds and altitudes, for example), which will require improved validation of computational capabilities. Of course, integrating aerodynamics with structural and systems modeling through MDO tools will be an essential part of this progress. Finally, as we mentioned in this chapter, intelligent design will take advantage of fully integrating computational capabilities with wind tunnel and flight testing in a way that utilizes the strengths of all three while avoiding the weaknesses. An example of applying some of these concepts to aircraft design is the Blended Wing Body concept being tested by Boeing and NASA in the X-48 program, as shown in Fig. 10.55. This aircraft serves as a test bed for new and advancing CA technology as aircraft evolve and improve in the coming years. So, where should you be going? As the Cheshire Cat said in *Alice's Adventures in Wonderland*: "That depends a good deal on where you want to get to."[1] Enjoy the journey now that you have acquired some of the tools and knowledge necessary to proceed down the path.

Summary of Best Practices

1. An intelligent user of computational aerodynamics should have a basic understanding of how real flows look, so take time to absorb flow visualizations that are readily available.
2. Transonic, supersonic, and hypersonic flows all create unique challenges for CA; be sure to understand these challenges and know how to deal with them as you simulate these flows.
3. MDO is a valuable and ever-expanding area of CA, but obtaining a valid optimized solution is challenging; know how to test MDO software to insure you are obtaining a physically realistic optimized solution.
4. Up until the present time, integration of CA with other areas of fluid dynamics has not been done as much as it should have been. Learn how to enjoy all three areas of fluid dynamics (CFD, EFD, and TFD) so you will be a well-rounded aerodynamicist.
5. Potential flow codes are still valuable tools to the modern aerodynamicist, and many new codes and applications should be examined to make sure the best tool is being used to accomplish each task (as we discussed in Chapter 2).

10.10 Projects

Go and start exploring computational aerodynamics on your own. Find an area within CA that challenges and interests you and become an expert at it. We know that this book will serve as a resource as you start your explorations, and we wish you all the best in your journey!

10.11 References

1 Carroll, L., *Alice's Adventures in Wonderland*, London: Macmillan and Co., 1865.
2 Becker, J.V., "Transonic Wind Tunnel Development (1940–1950)," in *The High-Speed Frontier, Case Histories of Four NACA Programs 1920–1950*, NASA SP-445, 1980.
3 Vos, J.B., Rizzi, A., Darracq, D., and Hirschel, E.H., "Navier-Stokes Solvers in European Aircraft Design," *Progress in Aerospace Sciences*, Vol. 38, 2002, pp. 601–697.
4 Murman, E.M., and Cole, J.D., "Calculation of Plane Steady Transonic Flows," *AIAA Journal*, Vol. 9, No. 1, 1971, pp. 114–121.
5 Caughey D.A., and Jameson, A., "Development of Computational Techniques for Transonic Flows: An Historical Perspective," *IUTAM Symposium Transsonicum IV*, H. Sobieczky, Ed., Dordrecht: Kluwer, 2003, pp. 183–194.

6 Jameson, A., and Ou, K., "50 Years of Transonic Aircraft Design," *Progress in Aerospace Sciences*, Vol. 47, No. 5, 2011, pp. 308–318.

7 Jameson, A., "A Perspective on Computational Algorithms for Aerodynamic Analysis and Design," *Progress in Aerospace Sciences*, Vol. 37, 2001, pp. 197–243.

8 Fujino. M., "Design and Development of the HondaJet," *Journal of Aircraft*, Vol. 42, No. 3, 2005, pp. 755–764.

9 Fujino, M., Yoshizaki, Y., and Kawamura, Y., "Natural-Laminar-Flow Airfoil Development for a Lightweight Business Jet," *Journal of Aircraft*, Vol. 40, No. 4, 2003, pp. 609–615.

10 Strang, W.Z., Tomaro, R.F., and Grismer, M.J., "The Defining Methods of Cobalt$_{60}$: A Parallel, Implicit, Unstructured Euler/Navier-Stokes Flow Solver," AIAA Paper 99-0786, January 1999.

11 Holst, T.L., "On Approximate Factorization Schemes for Solving the Full Potential Equation," NASA TM 110435, February 1997.

12 Rumsey, C.L., Rivers, S.M., and Morrison, J.H., "Study of CFD Variation on Transport Configurations from the Second Drag Prediction Workshop," *Computers & Fluids*, Vol. 34, 2005, pp. 785–816.

13 Vassberg, J.C., Tinoco, E.N., Mani, M., Rider, B., Zickuhr, T., Levy, D.W., Brodersen, O.P., Eisfeld, B., Crippa, S., Wahls, R.A., Morrison, J.H., Mavriplis, D.J., and Murayama, M., "Summary of the Fourth AIAA CFD Drag Prediction Workshop," AIAA Paper 2010–4547, June 2010.

14 Langtry, R.B., Kuntz, M., and Menter, F.R., "Drag Prediction of Engine-Airframe Interference Effects with CFX 5," *Journal of Aircraft*, Vol. 42, No. 6, 2005, pp. 1523–1529.

15 Jameson, A., Pierce, N.A., and Martinelli, L., "Optimum Aerodynamic Design Using the Navier-Stokes Equation," AIAA Paper 97-0101, January 1997.

16 Bradley, R.G., "Practical Aerodynamic Problems – Military Aircraft," in *Transonic Aerodynamics*, D. Nixon, Ed., AIAA Progress in Astronautics and Aeronautics Series, Vol. 81, New York: AIAA, 1982.

17 Bevilaqua, P.M., "Genesis of the F-35 Joint Strike Fighter," *Journal of Aircraft*, Vol. 46, No. 6, 2009, pp. 1825–1836.

18 Wooden, P.A., Smith, B.R., and Azevedo, J.J., "CFD Predictions of Wing Pressure Distributions on the F-35 at Angles-of-Attack for Transonic Maneuvers," AIAA Paper 2007-4433, June 2007.

19 Knill, D.L., Balabanov, V., Golovidov, O., Grossman, B., Mason, W.H., Haftka, R.T., and Watson, L.T., "Accuracy of Aerodynamic Predictions and Its Effects on Supersonic Transport Design," Virginia Tech MAD Center Report 96-12-01, December 1996.

20 Harris, R.V., "An Analysis and Correlation of Aircraft Wave Drag," NASA TM X-947, March 1964.

21 Harris, R.V., and Landrum, E.J., "Drag Characteristics of a Series of Low-Drag Bodies of Revolution at Mach Numbers from 0.6 to 4.0," NASA TN D-3163, December 1965.

22 McGrory, W.D., Slack, D.C., Applebaum, M.P., and Walters, R.W., GASP Version 2.2 Users Manual, Aerosoft, Inc., 1993 (http://www.aerosoftinc. com).

23 Weber, J., and King, C., "Analysis of the Zero-Lift Wave Drag Measured on Delta Wings," Aeronautical Research Council Reports and Memoranda No. 3818, 1978.

24 Barger, R.L., Adams, M.S., and Krishnam, R.R., "Automatic Computation of Euler-Marching and Subsonic Grids for Wing-Fuselage Configuration," NASA TM 4573, July 1994.

25 Barger, R.L., and Adams, M.S., "Automatic Computation of Wing-Fuselage Intersection Lines and Fillet Inserts with Fixed-Area Constraints," NASA TM 4406, March 1993.

26 Carlson, H.W., and Walkley, K.B., "Numerical Methods and a Computer Program for Subsonic and Supersonic Aerodynamic Design and Analysis of Wings with Attainable Thrust Corrections," NASA CR-3808, 1984.

27 Hopkins, E.J., and Inouye, M., "An Evaluation of Theories Predicting Turbulent Skin Friction and Heat Transfer on Flat Plates at Supersonic and Hypersonic Mach Numbers," *AIAA Journal*, Vol. 9, No. 6, 1971, pp. 993–1003.

28 Kulfan, B.K., "Fundamentals of Supersonic Wave Drag," Fourth International Conference on Flow Dynamics, Sendai, Japan, September 2007.

29 Aronstein, D.C., and Schueler, K.L., "Conceptual Design of a Sonic Boom Constrained Supersonic Business Aircraft," AIAA Paper 2004-0697, January 2004.

30 Welge, H.R., Bonet, J., Magee, T., Chen, D., Hollowell, S., Kutzmann, A., Mortlock, A., Stengle, J., Nelson, C., Adamson, E., Baughcum, S., Britt, R.T., Miller, G., and Tai, J., "N+2 Supersonic Concept Development and Systems Integration," NASA CR-2010–216842, August 2010.

31 http://people.nas.nasa.gov/~aftosmis/cart3d/

32 Hahn, A., "Application of CART3D to Complex Propulsion-Airframe Integration with Vehicle Sketch Pad," AIAA Paper 2012-0547, January 2012.

33 Bertin, J.J., and Cummings, R.M., "Critical Hypersonic Aerothermodynamic Phenomena," *Annual Review of Fluid Mechanics*, Vol. 38, 2006, pp. 129–157.

34 Bertin, J.J., and Cummings, R.M., "Fifty Years of Hypersonics: Where We've Been, Where We're Going," *Progress in Aerospace Sciences*, Vol. 39, 2003, pp. 511–536.

35 Cheng, H.K., "Perspectives on Hypersonic Viscous Flow Research," *Annual Review of Fluid Mechanics*, Vol. 25, 1993, pp. 455–484.

36 Longo, J.M.A., "Aerothermodynamics – A Critical Review at DLR," *Aerospace Science and Technology*, Vol. 7, 2003, pp. 429–438.

37 Longo, J.M.A., Orlowski, M., and Brück S., "Consideration on CFD Modeling for the Design of Re-entry Vehicles," *Aerospace Science and Technology*, Vol. 4, 2000, pp. 337–345.

38 Longo, J.M.A., "Modelling of Hypersonic Flow Phenomena," NATO RTO-EN-AVT-161, Paper 6, 2004.

39 Moretti, G., "On the Matter of Shock Fitting," *Lecture Notes Physics*, Vol. 35, 1974, pp. 287–292.

40 Godunov, S.K., "Finite-Difference Method for Numerical Computation of Discontinuous Solutions of the Equations of Fluid Dynamics," *Matematicheskii Sbornik*, Vol. 47, 1959, pp. 271–306.

41 Van Leer, B., "Towards the Ultimate Conservative Difference Scheme," *Journal of Computational Physics*, Vol. 32, 1979, pp. 101–136.

42 Steger, J.L., and Warming, R.F., "Flux Vector Splitting of the Inviscid Gasdynamic Equations with Application to Finite Difference Methods," *Journal of Computational Physics*, Vol. 40, 1981, pp. 263–293.

43 Roe, P.L., "Characteristic-Based Schemes for the Euler Equations," *Annual Review of Fluid Mechanics*, Vol. 18, 1986, pp. 337–365.

44 Harten, A., "High Resolution Schemes for Hyperbolic Conservation Laws," *Journal of Computational Physics*, Vol. 49, 1983, pp. 357–393.

45 Papadopoulos, P., Venkatapathy, E., Prabhu, D., Loomis, M.P., and Olynick, D., "Current Grid-Generation Strategies and Future Requirements in Hypersonic Vehicle Design, Analysis, and Testing," *Applied Mathematical Modeling*, Vol. 23, 1999, pp. 705–735.

46 Men'shov, I.S. and Nakamura, Y., "Numerical Simulations and Experimental Comparison for High-Speed Nonequilibrium Air Flows," *Fluid Dynamics Research*, Vol. 27, 2000, pp. 305–334.

47 Schneider, S.P., "Hypersonic Laminar-Turbulent Transition on Circular Cones and Scramjet Forebodies," *Progress in Aerospace Sciences*, Vol. 40, 2004, pp. 1–50.

48 Fasel, H., Thumm, A., and Bestel, H. "Direct Numerical Simulation of Transition in Supersonic Boundary Layer: Oblique Breakdown," In *Transitional and Turbulent Compressible Flows*, Ed. L.D. Kral and T.A. Zang, pp. 77–92, New York: ASME, 1993.

49 Suzen, Y.B., and Hoffmann, K.A., "Investigation of Supersonic Jet Exhaust Flows by One- and Two-equation Turbulence Models," AIAA Paper 98-0322, January 1998.

50 Forsythe, J.R., Hoffmann, K.A., Cummings, R.M., and Squires, K.D., Detached-Eddy Simulation with Compressibility Corrections Applied to a Supersonic Axisymmetric Base Flow," *Journal of Fluids Engineering*, Vol. 124, 2002, pp. 911–923.

51 Anderson, J.D., *Hypersonic and High Temperature Gas Dynamics*, New York: McGraw-Hill, 1989.

52 Edwards, T.A., "Fluid/chemistry Modeling for Hypersonic Flight Analysis," *Computers and Mathematics with Applications*, Vol. 24, 1992, pp. 25–36.

53 Cheng, H.K., and Emanuel, G., "Perspective on Hypersonic Nonequilibrium Flow," *AIAA Journal*, Vol. 33, No. 3, 1995, pp. 385–400.

54 Park, C., "Assessment of Two-Temperature Kinetic Model for Ionizing Flow," *Journal of Thermophysics and Heat Transfer*, Vol. 3, 1989, pp. 233–244.

55 Park, C., and Yoon, S., "Fully Coupled Implicit Method for Thermochemical Nonequilibrium Air at Suborbital Flight Speeds," *Journal of Spacecraft Rockets*, Vol. 28, No. 1, 1991, pp. 31–39.

56 Nompelis, I., Candler, G.V., and Holden, M.S., "Effect of Vibrational Nonequilibrium on Hypersonic Double-Cone Experiments," *AIAA Journal*, Vol. 41, No. 11, 2003, pp. 2162–2169.

57 Sarma, G.S.R., "Physico-chemical Modeling in Hypersonic Flow Simulation," *Progress in Aerospace Sciences*, Vol. 36, 2000, pp. 281–349.

58 Keane, A.J., and Nair, P.B., *Computational Approaches for Aerospace Design: The Pursuit of Excellence*, Chichester: John Wiley & Sons, 2005.

59 Volpe, G., "Inverse Airfoil Design: A Classical Approach Updated for Transonic Applications," in *Applied Computational Aerodynamics*, P.A. Henne, Ed., Washington, DC: AIAA, 1990.

60 Jensen, S.C., Rettie, I.H., and Barber, E.A., "Role of Figures of Merit in Design Optimization and Technology Assessment," *Journal of Aircraft*, Vol. 18, No. 2, 1981, pp. 76–81.

61 Gur, O., Bhatia, M., Schetz, J.A., Mason, W.H., Kapania, R., and Mavris, D., "Design Optimization of a Truss-Braced-Wing Transonic Transport Aircraft," *Journal of Aircraft*, Vol. 47, No. 6, 2010, pp. 1907–1917.

62 Sobieszczanski-Sobieski, J., and Haftka, R.T., "Multidisciplinary Aerospace Design Optimization: Survey of Recent Developments," *Structural Optimization*, Vol. 14, 1997, pp. 1–23.

63 Rodriguez, D.L., and Sturdza, P., "A Rapid Geometry Engine for Preliminary Aircraft Design," AIAA Paper 2006-0929, 2006.

64 Kulfan, B., "Universal Parametric Geometry Representation Method," *Journal of Aircraft*, Vol. 45, No. 1, 2008, pp. 142–158.

65 Hahn, A., "Vehicle Sketch Pad: A Parametric Geometry Modeler for Conceptual Aircraft Design," AIAA Paper 2010-0657, 2010.

66 Arora, J., *Introduction to Optimum Design*, 3rd Ed., Waltham, MA: Academic Press, 2011.

67 Forrester, A.I.J., Sobester, A., and Keane, A.J., *Engineering Design via Surrogate Modelling: A Practical Guide*, Reston: AIAA, 2008.

68 Giunta, A.A., Golovidov, O., Knill, D.L., Grossman, B., Mason, W.H., Watson, L.T., and Haftka, R.T., "Multidisciplinary Design Optimization of Advanced Aircraft Configurations," *Fifteenth International Conference on Numerical Methods in Fluid Dynamics*, P. Kutler, J. Flores, and J.-J. Chattot, eds., in Lecture Notes in Physics, Vol. 490, Berlin: Springer-Verlag, 1997, pp. 14–34.

69 Jameson, A., and Vassberg, J.C., "Computational Fluid Dynamics for Aerodynamic Design: Its Current and Future Impact," AIAA Paper 2001-0538, January 2001.

70 Knill, D.L., Giunta, A.A., Baker, C.A., Grossman, B., Mason, W.H., Haftka, R.T., and Watson, L.T., "Response Surface Models Combining Linear and Euler Aerodynamics for Supersonic Transport Design," *Journal of Aircraft*, Vol. 36, No. 1, 1999, pp. 75–86.

71 Gern, F., Ko, A., Grossman, B., Haftka, R.T., Kapania, R.K., and Mason, W.H., "Transport Weight Reduction through MDO: The Strut-Braced Wing Transonic Transport," AIAA Paper 2005-4667, June 2005.

72 Gur, O., Bhatia, M., Mason, W.H., Schetz, J.A., Kapania, R.K., and Nam, T., "Development of a Framework for Truss-Braced Wing Conceptual MDO," *Structural and Multidisciplinary Optimization*, Vol. 44, No. 2, 2011, pp. 277–298.

73 Gur, O., Schetz, J.A., and Mason, W.H., "Aerodynamic Considerations in the Design of Truss-Braced Wing Aircraft," *Journal of Aircraft*, Vol. 48, No. 3, 2011, pp. 919–939.

74 Cummings, R.M., Morton, S.A., and Siegel, S.G., "Computational Simulation and PIV Measurements of the Laminar Vortical Flowfield for a Delta Wing at High Angle of Attack," AIAA Paper 2003-1102, January 2003.

75 Cummings, R.M., Morton, S.A., and Siegel, S.G., "Computational Simulation and Experimental Measurements for a Delta Wing eith Periodic Suction and Blowing," *Journal of Aircraft*, Vol. 40, No. 5, 2003, pp. 923–931.

76 Cummings, R.M., Morton, S.A., Siegel, S.A., and Bosscher, S., "Numerical Prediction and Wind Tunnel Experiment for a Pitching Unmanned Combat Air Vehicle," AIAA Paper 2003-417, January 2003.

77 Cummings, R.M., Morton, S.A., and Siegel, S.G., "Numerical Prediction and Wind Tunnel Experiment for a Pitching Unmanned Combat Air Vehicle," *Aerospace Science and Technology*, Vol. 12, 2008, pp. 355–364.

78 Cummings, R.M., and Morton, S.A., "Continuing Evolution of Aerodynamic Concept Development Using Collaborative Numerical and Experimental Evaluations," *Journal of Aerospace Engineering*, Vol. 220, No. 6, 2006, pp. 545–557.

79 Siegel, S., Cohen, K., and McLaughlin, T., "Feedback Control of a Circular Cylinder Wake in Experiment and Simulation," AIAA Paper 2003-2569, June 2003.

80 Siegel, S., Cohen, K., and McLaughlin, T., "Experimental Variable Gain Feedback Control of a Circular Cylinder Wake," AIAA Paper 2004-2611, June 2004.

81 Seidel, J., Siegel, S., Cohen, K., and McLaughlin, T., "Three-Dimensional Simulations of a Feedback Controlled Circular Cylinder Wake," AIAA Paper 2005-0295, January 2005.

82 http://www.icfd.rdg.ac.uk/Workshops/Mathematics%20and%20Applications%20of%20Data%20Assimilation/Mathematics%20and%20Applications%20of%20Data%20Assimilation.htm (accessed 16 August 2014)

83 Kalnay, E., *Atmospheric Modeling, Data Assimilation and Predictability*, Cambridge: Cambridge University Press, 2001.

84 Hernandes, F., and Soviero, P.A.O., "Unsteady Aerodynamic Coefficients Obtained by a Compressible Vortex Lattice Method," *Journal of Aircraft*, Vol. 46, No. 4, 2009, pp. 1291–1301.

85 Jacobs, R.B., Ran, H., Kirby, M.R., and Mavris, D.N., "Extension of a Modern Lifting-Line Method to Transonic Speeds and Application to Multiple-Lifting-Surface Configurations," AIAA Paper 2012-2889, June 2012.

86 Murua, J., Palacios, R., and Graham, J.M.R., "Applications of the Unsteady Vortex-Lattice Method in Aircraft Aeroelasticity and Flight Dynamics," *Progress in Aerospace Sciences*, Vol. 55, 2012, pp. 46–72.

87 Ramos, A., and McDonald, R.A., "A Meshless Finite Difference Scheme for Compressible Potential Flows," AIAA Paper 2011-0654, January 2011.

88 German, B.J., "A Riemannian Geometric Mapping Technique for Identifying Incompressible Equivalents to Subsonic Potential Flows," PhD thesis, Georgia Institute of Technology, 2007.

89 Lock, R.C., "Test Cases for Numerical Methods in Two-Dimensional Transonic Flows," AGARD Tech. Rep. 575, 1970.

Geometry for Aerodynamicists

Aerodynamicists control the flowfield through geometry definition and are always interested in possible geometric shapes that would be useful in design. This appendix provides the detailed definition of many of the classic shapes frequently specified in aerodynamics. It is not intended to be encyclopedic, but will provide a good starting point for where to obtain geometric definitions for aerodynamic shapes.

A.1 Airfoil Geometry

A.1.1 The NACA Airfoils

The NACA (National Advisory Committee for Aeronautics) airfoils were designed during the period from 1929 through 1947 under the direction of Eastman Jacobs at the NACA's Langley Field Laboratory (now NASA Langley Research Center). Most of the airfoils were based on simple geometrical descriptions of the section shape, although the 6 and 6A series were developed using theoretical analysis and don't have simple shape definitions. Although a new generation of airfoils has emerged as a result of improved understanding of airfoil performance and the ability to design new airfoils using computational methods, the NACA airfoils are still useful in many aerodynamic design applications. A number of references have been included to allow the reader to study both the older NACA literature and the new airfoil design ideas. Taken together, this literature provides a means of obtaining a rather complete understanding of the ways in which airfoils can be shaped to obtain desired performance characteristics.

The NACA airfoils are constructed by combining a thickness envelope with a camber or mean line. The equations that describe this procedure are:

$$\begin{aligned} x_u &= x - y_t(x)\sin\theta \\ y_u &= y_c(x) + y_t(x)\cos\theta \end{aligned}$$

(A.1)

and

$$x_l = x + y_t(x)\sin\theta$$
$$y_l = y_c(x) - y_t(x)\cos\theta \qquad (A.2)$$

where $y_t(x)$ is the thickness function, $y_c(x)$ is the camber line function, and

$$\theta = \tan^{-1}\left(\frac{dy_c}{dx}\right) \qquad (A.3)$$

is the camber line slope. It is not unusual to neglect the camber line slope terms in Equations A.1 and A.2. This simplifies the equations and makes the reverse problem of extracting the thickness envelope and mean line for a given airfoil straightforward. The primary reference volume for all the NACA subsonic airfoil studies remains *Theory of Wing Sections* by Abbot and von Doenhoff.[1]

Table A.1 provides a brief history of the development of the NACA Airfoils. References to the development of the NASA advanced airfoils, which were developed from 1966 to approximately 1977, are also provided here. The original NACA Reports are available from the NASA Technical Report Server (http://ntrs.nasa.gov/) and can also be obtained from the Cranfield University NACA mirror site (http://naca.central.cranfield.ac.uk/); these reports represent a wealth of data and experience and are well worth reading.

Historical accounts of the NACA airfoil program are contained in "Airfoils: Significance and Early Development by Abbott,[3] and "Recollections From an Earlier Period in American Aeronautics" by Jones.[4] NASA has published two reports describing computer programs that produce the NACA airfoil ordinates, both by Ladson and Brooks: "Development of a Computer Program to Obtain Ordinates for the NACA 4-Digit, 4-Digit Modified, 5-Digit, and 16-Series Airfoils"[5] and "Development of a Computer Program to Obtain Ordinates for the NACA 6- and 6A-Series Airfoils."[6] This program is included in the utility programs described in Appendix D, as *LADSON*, although it is not extremely accurate for sections less than 6 percent thick or greater than 15 percent thick.

Various websites can also create the NACA geometries. A spreadsheet is available at the book website which can also create airfoil and wing geometries (www.cambridge.org/aerodynamics) for 4- and 5-digit airfoils. An extensive and excellent survey of the older airfoils is contained in a German book available in English translation by Riegels.[7] Finally, NASA supercritical airfoil development is described in reports by Whitcomb,[8] Harris,[9] and Becker.[10] These airfoils, in virtually any combination of camber lines and thickness envelopes, can be constructed using program *FOILGEN*, described in Appendix D.

Table A.1 NACA Airfoil Families

Evolution of the NACA Airfoils	Primary NACA Report	Authors	Date
The 4-digit airfoils: According to Abbott, Pinkerton found that the thickness distribution of the Clark Y and Gottingen 398 airfoils were similar, and Jacobs selected a function to describe this thickness distribution. The mean lines were selected to be described by two parabolic arcs tangent at the position of maximum camber.	R-460	Jacobs, Ward, and Pinkerton	1933
The 4-digit modified airfoils: The camber lines were identical to the 4-digit series, and a more general thickness distribution was defined, which allowed variations in the leading-edge radius and position of maximum thickness to be investigated.	R-492	Stack and von Doenhoff	1934
The 5-digit airfoils: The thickness distribution was kept identical to the 4-digit series, and a new camber line was defined which allowed for camber to be concentrated near the leading edge. A reflexed camber line was designed to produce zero pitching moment, but has generally not been used. These foils were derived to get good high lift with minimum Cm_0.	R-537 R-610	Jacobs, Pinkerton, and Greenberg	1935 1937
The 6-series airfoils: The airfoils were designed to maintain laminar flow over a large portion of the chord by delaying the adverse pressure gradient. The thickness envelope was obtained using exact airfoil theory, and no simple formulas are available to describe the shapes. The camber lines were designed using thin airfoil theory, and simple formulas are available that describe their shape.	R-824*	Abbott, von Doenhoff, and Stivers	1945
The 6A-series airfoils: To improve the trailing edge structurally, the 6-series airfoils were redesigned to provide sections with simple (nearly straight) surface geometry near the trailing edge while maintaining the same general properties as the original airfoils. The camber line can be described by a simple alteration of the standard 6-series mean line.	R-903*	Loftin	1948

*Additional section data are contained in a report by Patterson and Braslow.[2]

A.1.2 Other Airfoil Definition Procedures

Interest in defining airfoils by a small number of parameters for use in numerical optimization has led to several parametric representations that might be useful. In particular, the work by Verhoff and his co-workers at McDonnell Douglas uses Chebyshev functions to obtain relations that can represent very general airfoil shapes with from 5 to 20 coefficients required. This work is described in two papers by Verhoff, Tookesberry, and Cain.[11,12]

Another approach using Bezier methods frequently used in CAD surface representation software has been described in two papers by Ventkataraman that uses 14 design variables to represent the airfoil.[13,14] Smith and co-workers at NASA Langley Research Center have used a similar approach based on non-uniform rational B-splines (NURBS).[15]

A.1.3 The NACA 4-Digit Airfoil

The numbering system for these airfoils is defined by:

$$\text{NACA MPXX}$$

where

        M is the maximum camber in percent of chord,
        P is the chordwise position of the maximum camber in tenths of chord,
        XX is the maximum thickness, t/c, in percent chord.

Note that, although the numbering system implies integer values, the equations can provide 4-digit foils for arbitrary values of M, P, and XX.

An example: NACA 2412 • a 12% thick airfoil,

                                • a max value of the camber line of 0.02, at $x/c = 0.4$.

The NACA 4-digit thickness distribution is given by:

$$\frac{y_t}{c} = \left(\frac{t}{c}\right)\left[a_0\sqrt{x/c} - a_1(x/c) - a_2(x/c)^2 + a_3(x/c)^3 - a_4(x/c)^4\right] \quad (A.4)$$

where:

$$a_0 = 1.4845 \qquad a_2 = 1.7580 \qquad a_4 = 0.5075$$
$$a_1 = 0.6300 \qquad a_3 = 1.4215$$

Note that this definition results in a small but finite trailing-edge thickness. Many computational methods require a zero-thickness trailing edge, and the coefficient definitions are frequently modified to produce a zero-thickness trailing edge. This can lead to the wrong value of drag from calculations. Van Dam cites a value of wave drag 15.3% too high for the 0012 modified to zero trailing-edge thickness at $M = 0.78$.[16]

The maximum thickness occurs at $x/c = 0.30$, and the leading-edge radius is

$$\left(\frac{r_{LE}}{c}\right) = 1.1019\left(\frac{t}{c}\right)^2. \quad (A.5)$$

The included angle of the trailing edge is:

$$\delta_{TE} = 2\tan^{-1}\left\{1.16925\left(\frac{t}{c}\right)\right\}. \quad (A.6)$$

As noted earlier, it is important to realize that the airfoil has a finite thickness at the trailing edge.

The camber line is given by:

$$\left.\begin{array}{l} \dfrac{y_c}{c} = \dfrac{M}{P^2}\left[2P\left(x/c\right)-\left(x/c\right)^2\right] \\[2mm] \dfrac{dy_c}{dx} = \dfrac{2M}{P^2}\left(P-\left(x/c\right)\right) \end{array}\right\} \quad \left(\dfrac{x}{c}\right) < P \qquad (A.7)$$

and

$$\left.\begin{array}{l} \dfrac{y_c}{c} = \dfrac{M}{\left(1-P\right)^2}\left[1-2P+2P\left(x/c\right)-\left(x/c\right)^2\right] \\[2mm] \dfrac{dy_c}{dx} = \dfrac{2M}{\left(1-P\right)^2}\left(P-\left(x/c\right)\right) \end{array}\right\} \quad \left(\dfrac{x}{c}\right) \ge P. \qquad (A.8)$$

The camber line slope is found from Equation A.3 using Equations A.7 and A.8, and the upper and lower surface ordinates resulting from the combination of thickness and camber are then computed using Equations A.1 and A.2.

A.1.4 The NACA 5-Digit Airfoil

This airfoil is an extension of the 4-digit series that provides additional camber lines. The numbering system for these airfoils is defined by:

NACA LPQXX

where:

> L is the amount of camber; the design lift coefficient is 3/2 L, in tenths
>
> P is the designator for the position of maximum camber, x_f, where $x_f = P/2$, and P is given in tenths of the chord
>
> Q = 0; standard 5-digit foil camber
>
>   = 1; "reflexed" camber
>
> XX is the maximum thickness, t/c, in percent chord

An example: the NACA 23012 airfoil is 12% thick, the design lift coefficient is 0.3, the position of maximum camber is located at $x/c = 0.15$, and the "standard" 5-digit foil camber line is used.

The thickness distribution is the same as the NACA 4-digit airfoil thickness distribution described in Equation A.4. The standard 5-digit series camber line is given by:

$$\left.\begin{array}{l} \dfrac{y_c}{c} = \dfrac{K_1}{6}\left[\left(x/c\right)^3 - 3m\left(x/c\right)^2 + m^2\left(3-m\right)\left(x/c\right)\right] \\[2mm] \dfrac{dy_c}{dx} = \dfrac{K_1}{6}\left[3\left(x/c\right)^2 - 6m\left(x/c\right)+m^2\left(3-m\right)\right] \end{array}\right\} \quad 0 \le \left(x/c\right) \le m \qquad (A.9)$$

and

$$\left.\begin{array}{l} \dfrac{y_c}{c} = \dfrac{K_1}{6} m^3 \left[1 - (x/c) \right] \\[3mm] \dfrac{dy_c}{dx} = -\dfrac{K_1}{6} m^3 \end{array}\right\} \quad m < (x/c) \le 1 \qquad (A.10)$$

where m is not the position of maximum camber, but is related to the maximum camber position by:

$$x_f = m\left(1 - \sqrt{\dfrac{m}{3}} \right) \qquad (A.11)$$

and m is found from a simple fixed-point iteration for a given x_f. K_1 is defined to avoid the leading-edge singularity for a prescribed ideal lift C_{l_i} and m:

$$K_1 = \dfrac{6C_{l_i}}{Q} \qquad (A.12)$$

where:

$$Q = \dfrac{3m - 7m^2 + 8m^3 - 4m^4}{\sqrt{m(1-m)}} - \dfrac{3}{2}(1-2m)\left[\dfrac{\pi}{2} - \sin^{-1}(1-2m) \right]. \quad (A.13)$$

Note that K_1 is a linear function of C_{l_i} and the K_1's were originally tabulated for $C_{l_i} = 0.3$. The tabulated K_1's are multiplied by $(C_{l_i}/0.3)$ to get values at other C_{l_i}. To compute the camber line, the values of Q and K_1 must be determined. In some cases the computed values of K_1 and Q differ slightly from the official tabulated values (remember these were computed in the 1930s). The tabulated values should be used to reproduce the official ordinates; Table A.2 illustrates the differences. Once the camber line parameters are chosen, the airfoil is constructed using the preceding equations.

A.1.4.1 CAMBER LINES DESIGNED TO PRODUCE ZERO PITCHING MOMENT

The reflexed mean line equations were derived to produce zero pitching moment about the quarter chord.

$$\dfrac{y_c}{c} = \dfrac{K_1}{6}\left[\left\{ (x/c) - m \right\}^3 - \dfrac{K_2}{K_1}(1-m)^3(x/c) - m^3(x/c) + m^3 \right] \quad 0 \le (x/c) \le m \quad (A.14)$$

$$= \dfrac{K_1}{6}\left[\dfrac{K_2}{K_1}\left\{ (x/c) - m \right\}^3 - \dfrac{K_2}{K_1}(1-m)^3(x/c) - m^3(x/c) + m^3 \right] \quad m < (x/c) \le 1 \quad (A.15)$$

Table A.2 Table of Tabulated and Computed Values for m and K_1.

Mean Line	x_f	m (tabulated)	m (computed)	K_1 (tabulated)	K_1 (using tabulated m)	K_1 (using computed m)
210	0.05	0.0580	0.0581	361.4	351.56	350.332
220	0.10	0.1260	0.1257	51.65	51.318	51.578
230	0.15	0.2025	0.2027	15.957	15.955	15.920
240	0.20	0.2900	0.2903	6.643	6.641	6.624
250	0.25	0.3910	0.3913	3.230	3.230	3.223

Table A.3 Tabulated Values for Camber Lines.

Mean Line	(P/2) x_f	m	K_1	K_1/K_2
211	0.05	–	–	–
221	0.10	0.1300	51.99	0.000764
231	0.15	0.2170	15.793	0.006770
241	0.20	0.3180	6.520	0.030300
251	0.25	0.4410	3.191	0.13550

where

$$\frac{K_2}{K_1} = \frac{3\left(m - x_f\right)^2 - m^3}{\left(1 - m\right)^3}. \tag{A.16}$$

The parameters are defined as follows: (i) given x_f, find m to give $C_{mc/4} = 0$ from thin airfoil theory; (ii) given x_f and m, calculate K_1 to give $C_{l_i} = 0.3$. The tabulated values for these camber lines are given in Table A.3

A.1.5 The NACA Modified 4-Digit Airfoil

This airfoil is an extension of the 4-digit series to allow for a variation of leading-edge radius and location of maximum thickness. The numbering system is defined by:

NACA MPXX-IT

where **MPXX** is the standard 4-digit designation (see Section A.1.3) and the **IT** appended at the end describes the modification to the thickness distribution. They are defined as:

I is the designation of the leading-edge radius

T is the chordwise position of maximum thickness in tenths of chord

$$\frac{r_{le}}{c} = 1.1019 \left(\frac{I}{6} \cdot \frac{t}{c} \right)^2 \qquad \text{for } I \le 8 \qquad (A.17)$$

and

$$\frac{r_{le}}{c} = 3 \times 1.1019 \left(\frac{t}{c} \right)^2 \qquad \text{for } I = 9 \qquad (A.18)$$

$I = 6$ produces the leading-edge radius of the standard 4-digit airfoils.

An example: NACA 0012-74 denotes an uncambered 12% thick airfoil, with a maximum thickness at $x/c = 0.40$ and a leading-edge radius of 0.0216, which is 36 percent larger than the standard 4-digit value.

The NACA 16 series is a special case of the modified 4-digit airfoil with a leading-edge radius index of $I = 4$, and the maximum thickness located at $x/c = 0.5$ ($T = 5$). As an example, the NACA 16-012 is equivalent to a NACA 0012-45.

The thickness distribution is given by:

$$\frac{y_t}{c} = 5 \left(\frac{t}{c} \right) \left[a_0 \sqrt{\frac{x}{c}} + a_1 \left(\frac{x}{c} \right) + a_2 \left(\frac{x}{c} \right)^2 + a_3 \left(\frac{x}{c} \right)^3 \right] \qquad 0 < \frac{x}{c} < T \qquad (A.19)$$

and

$$\frac{y_t}{c} = 5 \left(\frac{t}{c} \right) \left[.002 + d_1 \left(1 - \frac{x}{c} \right) + d_2 \left(1 - \frac{x}{c} \right)^2 + d_3 \left(1 - \frac{x}{c} \right)^3 \right] \qquad T < \frac{x}{c} \le 1 \quad (A.20)$$

The coefficients are determined by solving for the d's first, based on the trailing-edge slope and the condition of maximum thickness at $x/c = T$. Once these coefficients are found, the a's are found by relating a_0 to the specified leading-edge radius, the maximum thickness at $x/c = T$, and the condition of continuity of curvature at $x/c = T$. These constants are all determined for $t/c = 0.2$ and then scaled to other t/c values by multiplying by $5(t/c)$. The value of d_1 controls the trailing-edge slope and was originally selected to avoid reversals of curvature. In addition to the tabulated values, Riegels has provided an interpolation formula. The official (tabulated) and Riegels (approximate) values of d_1 are given in Table A.4, where the Riegels approximation is given by:

$$d_1 \cong \frac{(2.24 - 5.42T + 12.3T^2)}{10(1 - 0.878T)}. \qquad (A.21)$$

Once the value of d_1 is known, d_2 and d_3 are found from the relations given by Riegels:

Table A.4 Tabulated and Approximate Values for d_1

T	Tabulated d_1	Approximate d_1
0.2	0.200	0.200
0.3	0.234	0.234
0.4	0.315	0.314
0.5	0.465	0.464
0.6	0.700	0.722

$$d_2 = \frac{0.294 - 2(1-T)d_1}{(1-T)^2} \tag{A.22}$$

and

$$d_3 = \frac{-0.196 + (1-T)d_1}{(1-T)^3}. \tag{A.23}$$

With the d's determined, the a's can be found. a_0 is based on the leading-edge radius:

$$a_0 = 0.296904 \cdot \chi_{LE} \tag{A.24}$$

where

$$\chi_{LE} = \frac{I}{6} \qquad \text{for } I \le 8 \\ = 10.3933 \qquad \text{for } I = 9. \tag{A.25}$$

Defining:

$$\rho_1 = \left(\frac{1}{5}\right) \frac{(1-T)^2}{[0.588 - 2d_1(1-T)]} \tag{A.26}$$

the rest of the a's can be found from:

$$a_1 = \frac{0.3}{T} - \frac{15}{8} \cdot \frac{a_0}{\sqrt{T}} - \frac{T}{10\rho_1} \tag{A.27}$$

$$a_2 = -\frac{0.3}{T^2} + \frac{5}{4} \cdot \frac{a_0}{T^{3/2}} + \frac{1}{5\rho_1} \tag{A.28}$$

$$a_3 = \frac{0.1}{T^3} - \frac{0.375 a_0}{T^{5/2}} - \frac{1}{10\rho_1 T}. \tag{A.29}$$

The camber lines are identical to the standard 4-digit airfoils described in Section A.1.3. The upper and lower ordinates are then computed using the standard equations.

A.1.6 The NACA 6 and 6A-Series Mean Lines[i]

The 6-series mean lines were designed using thin airfoil theory to produce a constant loading from the leading edge back to $x/c = a$, after which the loading decreases linearly to zero at the trailing edge. Theoretically, the loading at the leading edge must be either zero or infinite within the context of thin airfoil theory analysis. The violation of the theory by the assumed finite leading-edge loading is reflected by the presence of a weak singularity in the mean line at the leading edge, where the camber line has an infinite slope. Therefore, according to Abbott and von Doenhoff,[1] the 6-series airfoils were constructed by holding the slope of the mean line constant in front of $x/c = 0.005$, with the value at that point. For round leading edges the camberline values are essentially not used at points ahead of the origin of the leading edge radius. The theory is discussed by Abbott and von Doenhoff on pages 73–75, 113, and 120. Tabulated values are contained on pages 394–405. The derivation of this mean line is a good exercise in thin airfoil theory. By simply adding various mean lines together, other load distributions can be constructed.

From Abbott and von Doenhoff:[1]

> The NACA 6-series wing sections are usually designated by a six-digit number together with a statement showing the type of mean line used. For example, in the designation NACA 65,3-218, a = 0.5, the 6 is the series designation. The 5 denotes the chordwise position of minimum pressure in tenths of the chord behind the leading edge for the basic symmetrical section at zero lift. The 3 following the comma (sometimes this is a subscript or in parenthesis) gives the range of lift coefficient in tenths above and below the design lift coefficient in which favorable pressure gradients exist on both surfaces. The 2 following the dash gives the design lift coefficient in tenths. The last two digits indicate the thickness of the wing section in percent chord. The designation a = 0.5 shows the type of mean line used. When the mean line is not given, it is understood that the uniform-load mean line (a = 1.0) has been used.

The 6A series airfoils employed an empirical modification of the $a = 0.8$ camber line to allow the airfoil to be constructed of nearly straight line segments near the trailing edge. This camber line is described by Loftin.[17]

---

[i] Only the mean camber lines have analytical definitions. The thickness distributions are the result of numerical methods which produced tabulated coordinates. In addition to the values tabulated in the NACA reports, the closest approximation for the thickness distributions is available in program LADSON (see Appendix D).

Basic Camberline Equations

When $a = 1$ (uniform loading along the entire chord):

$$\frac{y}{c} = -\frac{C_{l_i}}{4\pi}\left[\left(1-\frac{x}{c}\right)\ln\left(1-\frac{x}{c}\right)+\frac{x}{c}\ln\left(\frac{x}{c}\right)\right] \tag{A.30}$$

and

$$\frac{dy}{dx} = \frac{C_{l_i}}{4\pi}\left[\ln\left(1-\frac{x}{c}\right)-\ln\left(\frac{x}{c}\right)\right] \tag{A.31}$$

where C_{l_i} is the "ideal" or design lift coefficient, which occurs at zero angle-of-attack.

For $a < 1$,

$$\frac{y}{c} = \frac{C_{l_i}}{2\pi(1+a)}\left\{\frac{1}{1-a}\left[\begin{array}{c}\frac{1}{2}\left(a-\frac{x}{c}\right)^2\ln\left|a-\frac{x}{c}\right|-\frac{1}{2}\left(1-\frac{x}{c}\right)^2\ln\left(1-\frac{x}{c}\right)\\ +\frac{1}{4}\left(1-\frac{x}{c}\right)^2-\frac{1}{4}\left(a-\frac{x}{c}\right)^2\\ -\frac{x}{c}\ln\left(\frac{x}{c}\right)+g-h\frac{x}{c}\end{array}\right]\right\} \tag{A.32}$$

with

$$g = \frac{-1}{(1-a)}\left[a^2\left(\frac{1}{2}\ln a-\frac{1}{4}\right)+\frac{1}{4}\right] \tag{A.33}$$

$$h = (1-a)\left[\frac{1}{2}\ln(1-a)-\frac{1}{4}\right]+g \tag{A.34}$$

and

$$\frac{dy}{dx} = \frac{C_{l_i}}{2\pi(1+a)}\left\{\frac{1}{1-a}\left[\left(1-\frac{x}{c}\right)\ln\left(1-\frac{x}{c}\right)-\left(a-\frac{x}{c}\right)\ln\left(a-\frac{x}{c}\right)\right]-\ln\left(\frac{x}{c}\right)-1-h\right\} \tag{A.35}$$

The associated angle of attack is:

$$\alpha_i = \frac{C_{l_i}h}{2\pi(1+a)} \tag{A.36}$$

a = 0.8 (modified), the 6A-series mean line

For $0 < x/c < .87437$, use the basic $a = .8$ camber line, but with a modified value of the ideal lift coefficient, $C_{l_{i}\text{mod}} = C_{l_i}/1.0209$. For $0.87437 < x/c < 1$, use the linear equation:

$$\frac{y_c/c}{C_{l_i}} = 0.0302164 - 0.245209\left(\frac{x}{c} - 0.87437\right) \tag{A.37}$$

and

$$\frac{dy}{dx} = -0.245209 C_{l_i}. \tag{A.38}$$

Note that at $x/c = 1$, the foregoing approximate relation gives $y/c = -0.000589$, indicating an α shift of $0.034°$ for $C_{l_i} = 1$.

A.1.7 Tabulated Airfoil Definitions and the Airfoil Library

Most modern airfoils are not described by equations but are defined by a table of coordinates. Frequently, these coordinates are the results of a computational aerodynamic design program, and simple algebraic formulas cannot be used to define the shape (this was the case with the NACA 6-series airfoils described earlier). The following table provides a list of the tabulated airfoils currently available at www.cambridge.org/aerodynamics. The subsequent tables, Table A.5 and A.6, provide a guide to these airfoils. The values are given from leading edge to trailing edge, with the top followed by the bottom.

	File Name	Comments
NACA 4 digit airfoils		
NACA 0010	N0010.DAT	
NACA 0010-35	N001035.DAT	Abbott and von Doenhoff
NACA 0012	N0012.DAT	
NACA 4412	N4412.DAT	
NACA 6 and 6A airfoils		
NACA 63(2)-215	N632215.DAT	NASA TM 78503
NACA 63(2)-215 mod B	N632215m.DAT	
NACA 64A010	N64A010.DAT	
NACA 64A410[ii]	N64A410.DAT	
NACA 64(3)-418	N643418.DAT	
NACA 65(1)-012	N651012.DAT	
NACA 65(1)-213	N651213.DAT	

[ii] Be aware that the coordinates for this airfoil, widely used as a test case, on page 356 of the book by Abbott and von Doenhoff, contain some typos. In particular, the value of y at $x = 7.5\%$ is supposed to be 2.805. Also, it is obvious that the x-value between 30

	File Name	Comments
NACA 65(1)A012	N65A012.DAT	
	N658299M.DAT	
	N658299R.DAT	
NACA 65(2)-215	N652215.DAT	
NACA 66(3)-018	N663018.DAT	
NACA 747A015	N747A015.DAT	NACA Report 824

NASA General Aviation Series

LS(1)-0417	GAW1.DAT	originally known as: GA(W)-1
LS(1)-0417 mod	LS10417M.DAT	
LS(1)-0413	GAW2.DAT	originally known as: GA(W)-2
LS(1)-0013	LS10013.DAT	

NASA Medium Speed Series

MS(1)-0313	MS10313.DAT
MS(1)-0317	MS10317.DAT

NASA Laminar Flow Series

NLF(1)-1215F	NL11215F.DAT	
NLF(1)-0414F	NL10414F.DAT	
NLF(1)-0416	NL10416.DAT	
NLF(2)-0415	NL20415.DAT	
HSNLF(1)-0213	HSN0213.DAT	
HSNLF(1)-0213mod	HSN0213D.DAT	drooped le

NASA Supercritical Airfoils[iii]

SC(2)-0402	SC20402.DAT
SC(2)-0403	SC20403.DAT
SC(2)-0503	SC20503.DAT
SC(2)-0404	SC20404.DAT
SC(2)-0406	SC20406.DAT
SC(2)-0606	SC20606.DAT
SC(2)-0706	SC20706.DAT

and 40 is 35. Finally, the table is mislabeled. The y typo was responsible for the widely circulated story that the coordinates contained in Abbott and von Doenhoff were not accurate enough to be used in modern computational airfoil codes.

[iii] These coordinates were typed in by students when NASA TP 2969 was issued. They are exactly the values printed in the TP. It turns out that those values were developed by taking the original set of coordinates and enhancing them to 100 points upper and lower using straight-line interpolation. As such, they are not really usable. Our experience is that we can identify the original values and weed out the straight-line interpolated values, which produces usable values.

NASA Supercritical Airfoils (cont.)

SC(2)-1006	SC21006.DAT	
SC(2)-0010	SC20010.DAT	
SC(2)-0410	SC20410.DAT	
SC(2)-0610	SC20610.DAT	
SC(2)-0710	SC20710.DAT	also known as Foil 33
SC(2)-1010	SC21010.DAT	
SC(2)-0012	SC20012.DAT	
SC(2)-0412	SC20412.DAT	
SC(2)-0612	SC20612.DAT	
SC(2)-0712	SC20712.DAT	
SC(3)-0712(B)	SC20712B.DAT	
SC(2)-0414	SC20414.DAT	
SC(2)-0614	SC20614.DAT	
SC(2)-0714	SC20714.DAT	Raymer, Ref. NASA TP 2890
SC(2)-0518	SC20518.DAT	
FOIL31	FOIL31.DAT	
SUPER11	SUPER11.DAT	11% thick
SUPER14	SUPER14.DAT	14% thick, NASA TM X-72712

NYU Airfoils (see Table A.7)

82-06-09	K820609.DAT	
79-03-12	K790312.DAT	
72-06-16	K720616.DAT	
71-08-14	K710814.DAT	
70-10-13	K701013.DAT	
65-14-08	K651408.DAT	
65-15-10	K651510.DAT	
75-06-12	KORN.DAT	the "Korn" Airfoil
75-07-15	K750715.DAT	

Miscellaneous Transonic Airfoils

CAST 7	CAST7.DAT	
DSMA 523	DSMA523.DAT	AIAA Paper 75-0880
NLR HT 731081	NLRHT731.DAT	AGARD AR-138
ONERA M6	ONERAM6.DAT	
RAE 2822	RAE2822.DAT	
WILBY B	WILBYB.DAT	
WILBY C	WILBYC.DAT	
WILBY R	WILBYR.DAT	
SUPER10	NASA10SC.DAT	AGARD AR-138
	MBB-A3.DAT	AGARD AR-138

Eppler Airfoils

EPPLER 662	EPP662.DAT	Raymer[18] NASA CP 2085
EPPLER 748	EPP748.DAT	Raymer[18] NASA CP 2085

Wortman Airfoils

FX-63-137-ESM	FX63137.DAT
FX-72-MS-150A	FX72M15A.DAT
FX-72-MS-150B	FX72M15B.DAT

Miscellaneous Foils

ClarkY	CLARKY.DAT	
Early Liebeck High Lift	RHLHILFT.DAT	
NLR-1	NLR1.DAT	Rotorcraft foil, NASA CP 2046
RAE 100	RAE100.DAT	
RAE 101	RAE101.DAT	
RAE 102	RAE102.DAT	
RAE 103	RAE103.DAT	
RAE 104	RAE104.DAT	

VariEze Airfoils

VariEze wing bl23	VEZBL32.DAT
VariEze winglet root	VEZWLTR.DAT
VariEze winglet tip	VEZWLTT.DAT
VariEze canard	VEZCAN.DAT

Human powered aircraft airfoils

DAE 11	DAE11.DAT	Daedalus airfoils
DAE 21	DAE21.DAT	
DAE 31	DAE31.DAT	
DAE 51	DAE51.DAT	
Lissaman 7769	LISS7769.DAT	Gossamer Condor airfoil

Many other airfoils are available on the web. In particular, the Applied Aerodynamics group at the University of Illinois, under the direction of Professor Michael Selig, has established a massive online source for airfoil definitions and includes data from wind tunnel tests on the airfoils. Their focus is directed toward airfoils designed for low speeds and low Reynolds numbers; see http://m-selig.ae.illinois.edu/ads.html. In addition, Richard Eppler has published an entire book of his airfoil shapes and experimental results.[19]

Table A.5 NASA Low Speed, Medium Speed, and Natural Laminar Flow Airfoil Chart

Airfoil Designation	Design Lift	Design Thickness	Design Mach	Test?	Ordinates in Airfoil Library?	Ref.	Comment
GA(W)-1	0.4/1.0	0.17		yes	yes	TN D-7428	Low Speed
LS(1)-0417mod		0.17			yes		"
GA(W)-2		0.13		yes	yes	TM X-72697	"
mod		0.13		yes	yes	TM X-74018	"
?		0.21		yes		TM 78650	"
LS(1)-0013		0.13		yes	yes	TM-4003	"
MS(1)-0313		0.13		yes	yes	TP-1498	Medium Speed
MS(1)-0317	0.30	0.17	0.68	yes	yes	TP-1786	"
mod		0.17		yes		TP-1919	"
NLF(1)-0215F	0.20?	0.15			yes	Raymer[18]	Natural Laminar Flow
NLF(1)-0414F					yes		"
NLF(1)-0416					yes		"
NLF(1)-0414F drooped L.E.					yes		"
NLF(2)-0415	0.40?	0.15?	?	yes	yes	Raymer[18]	"
HSNLF(1)-0213	0.20?	0.13?	?		yes	TM-87602	"
HSNLF(1)-0213 mod					yes		

Several transonic airfoils were developed by a group led by Paul Garabedian at New York University. Table A.7 provides a list of the airfoils they published. Their airfoils are included in Refs. 20–22.

A.2 Classic Bodies of Revolution

Bodies of revolution form the basis for a number of shapes used in aerodynamic design and are also often used in comparing computational methods. The bodies defined in this section are generally associated with supersonic aerodynamics.

A.2.1 Summary of Relations

The body radius r is given as a function of x, $r/l = f(x/l)$. Once r is known, a number of other values characterizing the shape can be determined. The cross-sectional area and derivatives are:

Table A.6 NASA Supercritical Airfoils – Phase 2, from NASA TP 2969, March 1990, by Charles D. Harris

Airfoil Designation	Design Lift	Design Thickness	Design Mach	Test?	Ordinates in Airfoil Library?	Ref.	Comment
SC(2)-0402	0.40	0.02			yes		
SC(2)-0403	0.40	0.03			yes		
SC(2)-0503	0.50	0.03			yes		
SC(2)-0404	0.40	0.04			yes		
SC(2)-0406	0.40	0.06			yes	unpubl.	
SC(2)-0606	0.60	0.06			yes		
SC(2)-0706	0.70	0.06	0.795	yes	yes	unpubl.	
SC(2)-1006	1.00	0.06		yes	yes	unpubl.	
SC(2)-0010	0.00	0.10			yes		
SC(2)-0410	0.40	0.10	0.785		yes		
SC(2)-0610	0.60	0.10	0.765		yes		
SC(2)-0710	0.70	0.10	0.755	yes	yes	TM X-72711	Airfoil 33
SC(2)-1010	1.00	0.10	0.70		yes		
SC(2)-0012	0.0	0.12		?	yes	TM-89102	
SC(2)-0412	0.40	0.12			yes		
SC(2)-0612	0.60	0.12			yes		
SC(2)-0712	0.70	0.12	0.735	?	yes	TM-86370	TM-86371
SC(2)-0414	0.40	0.14			yes		
SC(2)-0614	0.60	0.14			yes		
SC(2)-0714	0.70	0.14	0.715	Yes	yes	TM X-72712	Low Speed TM-81912
SC(2)-0518	1.00	0.18			yes		

Table A.7 Garabedian and Korn Airfoil Chart

Airfoil Designation	Design Lift	Design Thickness	Design Mach	Test?	Ordinates in Airfoil Library?	Pages in Ref.[21]	Comment
79-03-12	0.293	0.123	0.790		yes	37,41-43	
72-06-16	0.609	0.160	0.720		yes	48,52-54	
71-08-14	0.799	0.144	0.710		yes	55,59-61	
70-10-13	0.998	0.127	0.700		yes	62,66-68	
65-14-08	1.409	0.083	0.650		yes	73,77-79	
65-15-10	1.472	0.104	0.650		yes	80,84-86	
82-06-09	0.590	0.092	0.820		yes	91,95	
75-06-12	0.629	0.117	0.750	yes	yes	96,99-101	the "Korn"
75-07-15	0.668	0.151	0.750		yes	102,106	

$$S(x) = \pi r^2 \tag{A.39}$$

$$\frac{dS}{dx} = 2\pi r \frac{dr}{dx} \tag{A.40}$$

$$\frac{d^2S}{dx^2} = 2\pi \left[\left(\frac{dr}{dx} \right)^2 + r \frac{d^2r}{dx^2} \right]. \tag{A.41}$$

Basic integrals that are useful in calculations are:

Volume:

$$V = \int_0^l S(x)\,dx. \tag{A.42}$$

Surface area:

$$S_{wet} = 2\pi \int_0^l r(x)\,dx. \tag{A.43}$$

Length along the contour:

$$p(\bar{x}) = \int_0^l \sqrt{1 + \left(\frac{dr}{dx} \right)^2}\,dx. \tag{A.44}$$

Note that the incremental values can be found by changing the lower limit of the integrals. The local longitudinal radius of curvature is given by:

$$R(x) = \frac{\left[1 + \left(\frac{dr}{dx} \right)^2 \right]^{3/2}}{\left| \frac{d^2r}{dx^2} \right|}. \tag{A.45}$$

Several simple shapes are also of interest in addition to those presented in detail. They are:

Parabolic spindle:

$$\frac{r}{l} = 4\frac{r_{mid}}{l}\frac{x}{l}\left(1 - \frac{x}{l} \right). \tag{A.46}$$

Ellipsoid of revolution:

$$\frac{r}{l} = 2\frac{r_{mid}}{l}\sqrt{\frac{x}{l}\left(1 - \frac{x}{l} \right)}, \tag{A.47}$$

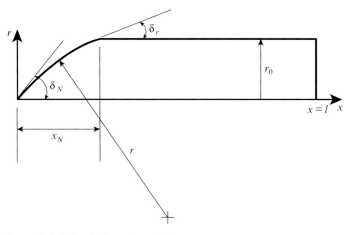

Figure A.1   Geometric definitions for tangent/secant ogive noses.

Power law body:

$$\frac{r}{l} = \frac{r_0}{l}\left(\frac{x}{x_N}\right)^n \tag{A.48}$$

where x_n is the nose length, and r_0 is the radius at $x = x_n$. The nose is blunt for $0 < n < 1$.

Another common shape is the spherical nose cap, discussed in detail in Krasnov.[22] References that discuss geometry of bodies of revolution in great detail are Refs. 23 and 24.

A.2.2 Tangent/Secant Ogives

The tangent or secant ogives are frequently used shapes in supersonic aerodynamics. The nomenclature is illustrated in Fig. A.1. Note that the ogive is actually the arc of a circle and when $\delta_r = 0$ the ogive ends tangent to the body, so that $\delta_r = 0$ represents the tangent ogive body. If $\delta_r = \delta_N$, the cone-cylinder is recovered. If $\delta_r = 0$ and $\delta_N = 90°$, the spherical cap case is obtained.

The expression for the radius r is determined using three basic constants for a particular case:

$$A = \frac{r_0}{l}\left(\frac{\cos\delta_N}{\cos\delta_r - \cos\delta_N}\right) \tag{A.49}$$

$$B = 2\frac{r_0}{l}\left(\frac{\sin\delta_N}{\cos\delta_r - \cos\delta_N}\right) \tag{A.50}$$

and

$$C = \frac{r_0}{l} \tag{A.51}$$

The radius is then given by:

$$\frac{r}{l} = \sqrt{A^2 + B\left(\frac{x}{l}\right) - \left(\frac{x}{l}\right)^2} - A \qquad 0 < \frac{x}{l} < \frac{x_N}{l} \tag{A.52}$$

$$= C \qquad \frac{x_N}{l} < \frac{x}{l} < 1$$

where x_N is found as follows.

For a tangent ogive ($\delta_r = 0$), the ogive can be defined by specifying either x_N/r_0 or δ_N. The other value can then be found following this procedure:

Given δ_N,

$$\frac{x_N}{r_0} = \frac{\sin \delta_N}{1 - \cos \delta_N} \tag{A.53}$$

Or given x_N/r_0,

$$\delta_N = \cos^{-1} \left[\frac{\left(\frac{x_N}{r_0}\right)^2 - 1}{\left(\frac{x_N}{r_0}\right)^2 + 1} \right]. \tag{A.54}$$

For the secant ogive, the simplest analytical procedure is to define the ogive in terms of δ_N and δ_r and then find x_N/l from:

$$\frac{x_N}{l} = \frac{r_0}{l} \left(\frac{\sin \delta_N - \sin \delta_r}{\cos \delta_r - \cos \delta_N} \right). \tag{A.55}$$

If x_N/l is not satisfactory, δ_N and δ_r can be adjusted by trial and error to obtain the desired nose length. A program can be set up to handle this process quite simply.

The first and second derivatives are then given by:

$$\frac{d(r/l)}{d(x/l)} = \frac{B - 2(x/l)}{2\left[(r/l) + A \right]} \tag{A.56}$$

and

$$\frac{d^2(r/l)}{d(x/l)^2} = -\frac{\left[B - 2(x/l)\right]^2}{4\left[(r/l) + A\right]^3} - \frac{1}{\left[(r/l) + A\right]}. \tag{A.57}$$

The relationships between radius and area derivatives given in Section A.2.1 are then used to complete the calculation.

A.2.3 The von Kármán Ogive

The von Kármán Ogive is the shape that produces minimum wave drag for a specified base area and length, according to slender body theory. This ogive has a very slightly blunted nose.[25] In this case, it is convenient to work with the cross-sectional area and a new independent variable:

$$\theta = \cos^{-1}\left[2\left(\frac{x}{x_N}\right) - 1\right] \tag{A.58}$$

or

$$\frac{x}{x_N} = \frac{1}{2}(1 + \cos\theta) \tag{A.59}$$

where the nose is at $\theta = \pi$, and the base is located at $\theta = 0$. Here we use x_N to denote the "nose length" or length of the ogive and allow this shape to be part of an ogive-cylinder geometry. The shape is then given as:

$$\frac{S(x)}{l^2} = \frac{S_B}{l^2}\left[1 - \frac{\theta}{\pi} + \frac{\sin 2\theta}{2\pi}\right] \tag{A.60}$$

and

$$\frac{r}{l} = \sqrt{\frac{S/l^2}{\pi}} \tag{A.61}$$

where S_B is the prescribed base area and l is the total length.

Defining

$$\bar{S} = \frac{S}{l^2}, \quad \bar{x} = \frac{x}{l}, \tag{A.62}$$

we have

$$\frac{d\bar{S}}{d\theta} = -\frac{\bar{S}_B}{\pi}\left[1 - \cos 2\theta\right] \tag{A.63}$$

$$\frac{d^2\overline{S}}{d\theta^2} = -\frac{2}{\pi}\overline{S}_B \sin 2\theta \tag{A.64}$$

and

$$\frac{d\overline{S}}{dx} = \overline{S}' = \frac{4}{\pi}\left(\frac{l}{x_N}\right)\overline{S}_B \sin\theta$$

$$\frac{d^2\overline{S}}{dx^2} = \overline{S}'' = -\frac{8}{\pi}\left(\frac{l}{x_N}\right)^2 \frac{\overline{S}_B}{\tan\theta}. \tag{A.65}$$

The radius derivatives are then computed by:

$$\frac{d\overline{r}}{dx} = \frac{\overline{S}'}{2\pi\overline{r}}, \qquad \frac{d^2\overline{r}}{dx^2} = \frac{\overline{S}''}{2\pi\overline{r}} - \frac{\overline{r}'^2}{\overline{r}}. \tag{A.66}$$

A.2.4 The Sears-Haack Body

This is the minimum wave drag shape for a given length and volume according to slender body theory. The body is closed at both ends and has a very slightly blunted nose and is symmetric about the midpoint.[25]

Although the notation used in Section A.2.3 for the von Kármán Ogive section could be used, it is more common to describe the Sears-Haack body in the manner presented here. This form uses the fineness ratio, $f = l/d_{max}$ to scale the shape. However, it is important to realize that the Sears-Haack shape is the minimum drag body for a specified volume and length, not for a specified fineness ratio. The minimum drag body for a specified fineness ratio is described in the next section.

Defining

$$\varsigma = 1 - 2\left(\frac{x}{l}\right), \tag{A.67}$$

the Sears-Haack body is defined as

$$\frac{r}{l} = \frac{1}{2f}\left(1 - \varsigma^2\right)^{3/4}. \tag{A.68}$$

The derivatives are given by:

$$\frac{d(r/l)}{d(x/l)} = \frac{3\varsigma}{1 - \varsigma^2}\left(\frac{r}{l}\right) \tag{A.69}$$

and

$$\frac{d^2(r/l)}{d(x/l)^2} = -\left(\frac{1}{1-\varsigma^2}\right)\left[\varsigma\frac{d(r/l)}{d(x/l)} + 6\left(\frac{r}{l}\right)\right]. \tag{A.70}$$

The fineness ratio is related to the length and volume, V, by:

$$f = \sqrt{\frac{3\pi^2}{64}\frac{l^3}{V}}. \tag{A.71}$$

In terms of f and either V or l, the other values can be found from the following:

Given f and l:

$$V = \frac{3\pi^2}{64}\frac{l^3}{f^2}. \tag{A.72}$$

Given f and V:

$$l = \left[V\frac{64}{3\pi^2}f^2\right]^{1/3}. \tag{A.73}$$

The relationships between radius and area derivatives given in Section A.2.1 are then used to complete the calculation.

A.2.5 The Haack-Adams Bodies

The Haack-Adams bodies define a number of minimum drag shapes.[26] These bodies correspond to the following cases:

  I. Given length, base area, and contour passing through a specifically located radius
 II. Given length, base area, and maximum area
III. Given length, base area, and volume

In case I, the specified radius will not necessarily be the maximum radius.

The notation used in Ref. 26 is employed in the equations, leading to the following definitions:

$$S = 4\frac{\overline{S}(x)}{l^2}, \qquad B = 4\frac{S_{BASE}}{l^2}, \qquad A = 4\frac{S_A}{l^2}, \qquad V = 8\left(\frac{\overline{V}}{l^3}\right) \tag{A.74}$$

where $S(x)$ is the area, S_A corresponds to either the specified area at a given location or the maximum area, and V is the volume. The independent variable is defined with its origin at the body midpoint:

$$\varsigma = 2\left(\frac{x}{l}\right) - 1 \tag{A.75}$$

and the location of the specified radius (Case I) and maximum radius (Case II) is designated c and given in ς coordinates.

The equation for each case can be written in a standard form:

Case I – Given S_{BASE}, S_A, c_x:

$$\frac{\pi S}{B} = \left[\frac{\pi A}{B} - \cos^{-1}(-c)\right]\frac{\sqrt{1-\varsigma^2}\,(1-c\varsigma)}{\left(1-c^2\right)^{3/2}} + \frac{\sqrt{1-\varsigma^2}\,(\varsigma - c)}{\left(1-c^2\right)}$$
$$+ \left[\frac{\pi A}{B} - \cos^{-1}(c) - c\sqrt{1-c^2}\right]\frac{(\varsigma - c)^2}{\left(1-c\right)^2}\ln N + \cos^{-1}(-\varsigma) \tag{A.76}$$

where

$$N = \frac{1 - c\varsigma - \sqrt{1-c^2}\,\sqrt{1-\varsigma^2}}{|\varsigma - c|}. \tag{A.77}$$

Case II – Given S_{BASE}, S_{MAX}:

First, find the location of the maximum thickness from the implicit relation

$$f(c) = 0 = \frac{\pi A}{B}c - \sqrt{1-c^2} - c\cos^{-1}(-c). \tag{A.78}$$

Use Newton's iteration method

$$c^{i+1} = c^i - \frac{f(c^i)}{f'(c^i)} \tag{A.79}$$

where

$$f'(c) = \frac{\pi A}{B} - \cos^{-1}(-c). \tag{A.80}$$

An initial guess of $c = 0$ is sufficient to start the iteration. Given c, the relation for the area is:

$$\frac{\pi S}{B} = \frac{\sqrt{1-\varsigma^2}}{c} + \frac{(\varsigma - c)^2}{c\sqrt{1-c^2}}\ln N + \cos^{-1}(-\varsigma) \tag{A.81}$$

where N is the same function as given in Case I.

Case III – Given S_{BASE} and V:

$$\frac{\pi S}{B} = \frac{8}{3}\left[\frac{V}{B} - 1\right]\left(1 - \varsigma^2\right)^{3/2} + \varsigma\sqrt{1 - \varsigma^2} + \cos^{-1}(-\varsigma). \qquad (A.82)$$

The maximum thickness for this case is located at:

$$e = \frac{1}{4(V \, / \, B - 1)} \qquad (A.83)$$

and in x coordinates

$$e_x = \frac{1}{2}(1 + e). \qquad (A.84)$$

Note that if $S_{BASE} = 0$, the Sears-Haack body is recovered.

A.3 Cross-Section Geometries for Bodies

The axisymmetric bodies described earlier can be used to define longitudinal lines for aerodynamic bodies. However, many aerodynamic bodies are not axisymmetric (the fuselage cross section is not round). In this section, we define a class of cross-section shapes that can be used to develop more realistic aerodynamic models. In particular, they have been used to study geometric shaping effects on forebody aerodynamic characteristics using an analytical forebody model with the ability to produce a wide variation of shapes. This generic model makes use of the equation of a super-ellipse to define cross-sectional geometry. The super-ellipse can recover a circular cross section, produce elliptical cross sections, and can also produce chine-shaped cross sections. Thus it can be used to define a variety of different cross-sectional shapes.

The super-ellipse equation for a cross section is:

$$\left(\frac{z}{b}\right)^{2+n} + \left(\frac{y}{a}\right)^{2+m} = 1 \qquad (A.85)$$

where n and m are adjustable coefficients that control the surface slopes at the top and bottom plane of symmetry and chine leading edge. The constants a and b correspond to the maximum half-breadth (the maximum width of the body) and the upper or lower center lines, respectively. Depending on the values of n and m, the equation can be made to produce all the shapes described earlier. The case $n = m = 0$ corresponds to the standard ellipse. The body is circular when $a = b$.

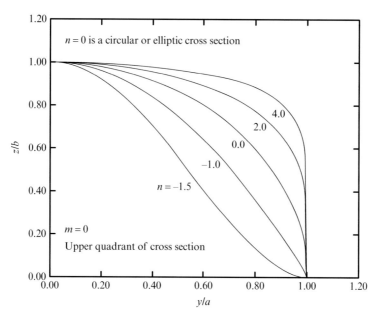

$n = 0$ is a circular or elliptic cross section

4.0

2.0

0.0

−1.0

$n = -1.5$

$m = 0$

Upper quadrant of cross section

y/a

z/b

Figure A.2 Super ellipse cross sections for various values of n.

When $n = -1$ the sidewall is linear at the maximum half breadth line, forming a distinct crease line. When $n < -1$ the body cross section takes on a cusped or chine-like shape. As n increases, the cross section starts to become rectangular.

The derivative of z/b with respect to y/a is:

$$\frac{d\bar{z}}{d\bar{y}} = -\frac{\left(\dfrac{2+m}{2+n}\right)}{\left[1-\bar{y}^{(2+m)}\right]^{\left(\frac{1+n}{2+n}\right)}} \qquad (A.86)$$

where $\bar{z} = z/b$ and $\bar{y} = y/a$. As $\bar{y} \to 1$, the slope becomes:

$$\frac{d\bar{z}}{d\bar{y}} = \begin{cases} \infty & n > -1 \\ 0 & n < -1 \\ -(2+m)\bar{y}^{1+m} & n = -1 \end{cases} \qquad (A.87)$$

Figure A.2 shows a quadrant of the cross section for various values of n ranging from a chine to a rectangle. Different cross sections can be used above and below the maximum half-breadth line. Even more generality can be provided by allowing n and m to be functions of the axial distance, x. The parameters a and b can also be functions of the planform shape and varied to study planform effects. Notice that when $n = -1$ the value of m can be used to control the slope of the sidewall at the crease line. Also, observe that

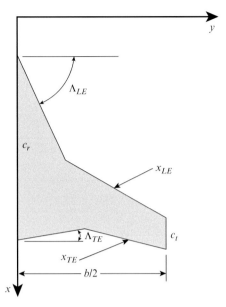

Figure A.3 Wing planform nomenclature.

large positive values of n drive the cross-section shape to approach a rectangular or square shape.

Connecting various cross-section shapes is part of the subject of lofting, described in Chapter 7. One of the few other textbook discussions is contained in Raymer in Chapter 7.[18] Dan Raymer worked at North American Aviation (Rockwell), where Liming[iv] literally wrote the book on the analytic definition of aircraft lines.[27]

A.4 Planform Analysis

Several local and integral planform properties are of interest in aerodynamic analysis; they are summarized in this section and Section 4.7.1. (Note: Biplanes use the total area of both wings as the reference area.) For a more complete presentation, see *DATCOM*.[28]

The local values are the leading- and trailing-edge locations, $x_{le}(y)$ and $x_{te}(y)$, the local chord, $c(y)$, and the leading- and trailing-edge sweep angles: $\Lambda_{le}(y)$ and $\Lambda_{te}(y)$. Figure A.3 illustrates the standard nomenclature.

Assuming the planform is symmetric, the integral properties are:

Planform Area, S:

$$S = 2\int_0^{b/2} c(y)\, dy. \tag{A.88}$$

[iv] The book emphasizes conic sections, and the examples in the book are for the P-51 Mustang.

Mean aerodynamic chord, *mac*:

$$\bar{c} = \frac{2}{S} \int_0^{b/2} c^2(y)\, dy. \tag{A.89}$$

X position of centroid of area, x_{cen}:

$$x_{cen} = \frac{2}{S} \int_0^{b/2} c(y) \left\{ x_{LE}(y) + \frac{c(y)}{2} \right\} dy. \tag{A.90}$$

Spanwise position of *mac*:

$$y_{mac} = \frac{2}{S} \int_0^{b/2} y\, c(y)\, dy. \tag{A.91}$$

Leading edge location of *mac*:

$$x_{LE_{mac}} = \frac{2}{S} \int_0^{b/2} x_{LE}(y) c(y)\, dy. \tag{A.92}$$

In addition, the following derived quantities are often of interest:

Aspect ratio:

$$AR = \frac{b^2}{S_{ref}}. \tag{A.93}$$

Average chord:

$$c_A = \frac{S_{ref}}{b}. \tag{A.94}$$

Taper ratio:

$$\lambda = \frac{c_t}{c_r}. \tag{A.95}$$

S_{ref} is usually chosen to be equal to the area of a basic reference trapezoidal planform, and thus, the actual planform area, *S*, may not equal S_{ref}.

When considering two areas, recall that the centroid of the combined surfaces is:

$$\begin{aligned}
S\bar{x} &= S_1\bar{x}_1 + S_2\bar{x}_2 \\
S\bar{y} &= S_1\bar{y}_1 + S_2\bar{y}_2.
\end{aligned} \tag{A.96}$$

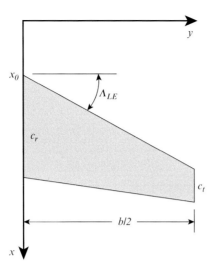

Trapezoidal wing planform nomenclature.

For a standard trapezoidal wing it is convenient to collect the following formulas, where Fig. A.4 shows the nomenclature:

$$
\begin{aligned}
x_{LE}(y) &= x_{LE_0} + y \tan \Lambda_{LE}(y) \\
x_{TE}(y) &= x_{TE_0} + y \tan \Lambda_{TE}(y)
\end{aligned}
\tag{A.97}
$$

and the local chord is:

$$
\frac{c(y)}{c_r} = 1 - (1 - \lambda)\eta
\tag{A.98}
$$

where:

$$
y = \frac{b}{2}\eta \quad \text{or} \quad \eta = \frac{y}{b/2} \quad \text{and} \quad \lambda = \frac{c_t}{c_r}.
\tag{A.99}
$$

The sweep at any element line can be found in terms of the sweep at any other by:

$$
\tan \Lambda_n = \tan \Lambda_m - \frac{4}{AR}\left[(n-m)\left(\frac{1-\lambda}{1+\lambda}\right)\right]
\tag{A.100}
$$

where n, m are fractions of the local chord. An alternate formula is available using the leading-edge and trailing-edge sweep angles:

$$
\tan \Lambda_n = (1 - n)\tan \Lambda_{LE} + n \tan \Lambda_{TE}.
\tag{A.101}
$$

The integral and other relations are given by:

$$S = \frac{b}{2} c_r (1 + \lambda)$$

$$c_{ave} = \frac{S}{b}$$

$$\frac{\bar{c}}{c_r} = \frac{2}{3}\left(\frac{1 + \lambda + \lambda^2}{1 + \lambda}\right)$$

$$AR = \frac{b^2}{S} = \frac{b/2}{c_r}\left(\frac{4}{1 + \lambda}\right)$$

$$y_{mac} = \frac{b}{6}\left(\frac{1 + 2\lambda}{1 + \lambda}\right)$$

$$\frac{x_{LE_{mac}}}{c_r} = \frac{x_{LE_0}}{c_r} + \left(\frac{1 + 2\lambda}{12}\right) AR \tan \Lambda_{LE} \tag{A.102}$$

$$x_{cen} = x_{LE_{mac}} + \frac{\bar{c}}{2}.$$

When computing the projected planform area of an entire configuration, the following formula is useful:

$$S = \sum_{k=1}^{k=N} (y_{k+1} + y_k)(x_{k+1} - x_k) \tag{A.103}$$

where Fig. A.5 defines the nomenclature.

At $k = N$, y_{k+1}, x_{k+1} refer to the initial points y_1, x_1. For normal planforms, $y_{n+1} = y_1 = 0$, so that the summation can be terminated at N-1. This formula assumes planform symmetry and provides the total planform area with only one side of the planform used in the computation.

A.4.1 A Note on Reference Chords and Aerodynamic Center Location

Looking at the aerodynamics literature, especially the older literature and textbooks, you find that the nomenclature for reference chords is not uniform. However, there is a standard in use today in the U.S. industry. The issue is what value to use when nondimensionalizing the pitching moment:

$$C_m = \frac{M}{q S_{ref} c_{ref}}. \tag{A.104}$$

The USAF *DATCOM*[28] defines the standard in the U.S. They define c_{ref} to be the *mean aerodynamic chord* or *mac*:

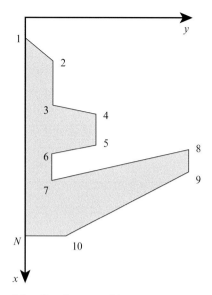

Figure A.5   Full configuration nomenclature.

$$\bar{c} = \frac{2}{S} \int_0^{b/2} c^2(y)dy \qquad (A.89)$$

which, for a straight tapered wing, is:

$$\frac{\bar{c}}{c_r} = \frac{2}{3}\left(\frac{1 + \lambda + \lambda^2}{1 + \lambda}\right). \qquad (A.102)$$

The use of the symbol $\bar{c}$ to represent the *mean aerodynamic chord* is also confusing because $\bar{c}$ is sometimes used to represent the average chord, ($c_a = S/b$ in other literature). For example, ESDU (Engineering Sciences Data Unit, discussed in Chapter 2) uses $\bar{c}$ to be the standard (or geometric) mean chord, which we have defined here as the average chord. ESDU uses $\bar{\bar{c}}$ as the *mac* defined by Equation A.89. The main problem occurs when the report or paper does not specifically define the terms used. For example, Lamar and Alford[29] use Equation A.89 to nondimensionalize the pitching moment, but they call it the mean geometric chord.

Another issue to be aware of is that the *mac* does not necessarily lie on the actual planform if the wing is not a simple straight tapered configuration.

A.5 Conical Camber

An important class of camber distributions is associated with the planform, and not the airfoil. Conical camber has been widely used in many forms; however, the NACA defined a specific type of conical camber that is known as NACA conical camber. The most recent example of NACA conical camber is the F-15 wing. It improves the drag characteristics of wings in the subsonic and transonic flow region even though it was developed to reduce the drag at supersonic speeds. A NACA report by Hall[30] provides the original mathematical definition of NACA conical camber and provides a large range of test conditions for which the camber was effective. A NASA report by Boyd[31] provides more details of the derivation of the formulas for NACA conical camber and corrects errors in the equations presented in the Hall report; additional experimental results were also presented. An advanced form of conical camber addressing supercritical crossflow cases is Supercritical Conical Camber (SC^3).[32,33]

A.6 Three-Dimensional Wing Geometry

Wing geometry is often defined by interpolating between airfoil sections specified at particular spanwise stations; however, some care should be taken to interpolate properly. A discussion about wing lofting is presented in Chapter 7, and Program *WNGLFT* is available on the book website. This program includes an example of a lofting scheme to provide wing ordinates at any desired location for a wide class of wings. The book website contains a spreadsheet that can loft wings for CA applications based on defining the root and tip airfoils: www.cambridge.org/aerodynamics.

A.7 Simple Computer Geometry Modeling

As we discussed in Chapter 7, there are complex computer-aided design (CAD) tools available for defining aerodynamics geometry. However, these programs can be quite challenging to learn to use and often are not valuable for conceptual design and development. However, alternate sources of geometric modeling have been developed recently that may be quite useful to basic CA geometry modeling: Vehicle Sketch Pad and CST.

A.7.1 Vehicle Sketch Pad

The Open Source version of Vehicle Sketch Pad, OpenVSP, is a parametric aircraft geometry tool that makes it possible to create a three-dimensional model of an airplane using defined common engineering parameters (basic analytic geometric shapes such as trapezoids, etc.).[34] This model can be processed into formats suitable for engineering analysis. Figure A.6 shows the GUI for OpenVSP, and the software can be obtained via the NASA Open Source Agreement at http://www.openvsp.org/.

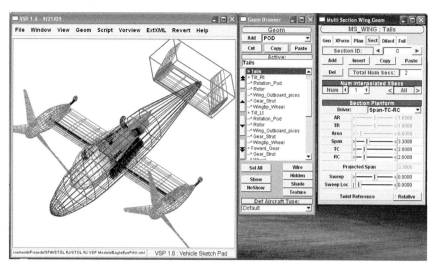

Figure A.6   Vehicle Sketch Pad GUI (Ref. 34; Courtesy of NASA).

A.7.2 Class function / Shape function Transformation (CST)

Brenda Kulfan of The Boeing Company created a geometric parameteriza-
tion approach, CST, that also uses various basic analytic functions to create more complex aircraft geometries. The resulting approach, called CST
(Class function/Shape function Transformation).[35] The class function is a
unit shape function that can define fundamental classes of airfoils, axisymmetric bodies, and axisymmetric nacelles geometries. The shape function
defines unique geometric shapes within each fundamental class. Special
polynomials are used to define the shapes for ease of use. Matlab files are
available to perform the geometry definition (for example, airfoil modeling is
available at: http://www.mathworks.com/matlabcentral/fileexchange/42239-
airfoil-generation-using-cst-parameterization-method).

A.8 References

1 Abbott, I.H., and von Doenhoff, A.E., *Theory of Wing Sections*, New York:
   Dover, 1959.

2 Patterson, E.W., and Braslow, A.L., "Ordinates and Theoretical Pressure-
   Distribution Data for NACA 6- and 6A-Series Airfoil Sections With
   Thicknesses from 2 to 21 and From 2 to 15 Percent Chord, Respectively,"
   NASA TR R-84, 1958.

3 Abbott, I.H., "Airfoils: Significance and Early Development," AIAA Paper
   80-3033, March 1980.

4 Jones, R.T., "Recollections from an Earlier Period in American Aeronautics,"
   Annual Review of Fluid Mechanics, Vol. 9, 1977, pp. 1–11.

5 Ladson, C.L., and Brooks, C.W., "Development of a Computer Program to Obtain Ordinates for the NACA 4-Digit, 4-Digit Modified, 5-Digit, and 16-Series Airfoils," NASA TM X-3284, November 1975.

6 Ladson, C.L., and Brooks, C.W., Jr., "Development of a Computer Program to Obtain Ordinates for the NACA 6- and 6A-Series Airfoils," NASA TM X-3069, September 1974.

7 Riegels, F.W., *Airfoil Sections*, London: Butterworths, 1961.

8 Whitcomb, R.T., "Review of NASA Supercritical Airfoils," ICAS Paper 74-10, August 1974.

9 Harris, C.D., "NASA Supercritical Airfoils," NASA TP 2969, March 1990.

10 Becker, J.V., "The High-Speed Airfoil Program," in *The High Speed Frontier*, NASA SP-445, 1980.

11 Verhoff, A., Stookesberry, D., and Cain, A., "An Efficient Approach to Optimal Aerodynamic Design. I – Analytic Geometry and Aerodynamic Sensitivities," AIAA Paper 93-0099, January 1993.

12 Stookesberry, D., Verhoff, A., and Cain, A., "An Efficient Approach to Optimal Aerodynamic Design. II – Implementation and Evaluation," AIAA Paper 93-0100, January 1993.

13 Ventkataraman, P., "A New Procedure for Airfoil Definition," AIAA Paper 95-1875, June 1985.

14 Ventkataraman, P., "Optimum Airfoil Design in Viscous Flows," AIAA Paper 95-1876, June 1985.

15 Sadrehaghighi, I., Smith, R.E., and Tiwari, S., "Grid and Design Variables Sensitivity Analysis for NACA Four-Digit Wing-Sections," AIAA Paper 93-0195, January 1993.

16 van Dam, C.P., "Recent Experiences with Different Methods of Drag Prediction," *Progress in Aerospace Science*, Vol. 35, 1999, pp. 751–798.

17 Loftin, L.K., "Theoretical and Experimental Data for a Number of NACA 6A-Series Airfoils," NACA TR-903, 1948.

18 Raymer, D.P., *Aircraft Design: A Conceptual Approach*, 5th Ed., Reston: AIAA, 2012.

19 Eppler, R., *Airfoil Design and Data*, Berlin: Springer-Verlag, 1990.

20 Bauer, F., Garabedian, P., and Korn, D., "A Theory of Supercritical Wing Sections with Computer Programs and Examples," *Lecture Notes in Economics and Mathematical Systems*, Vol. 66, Berlin: Springer-Verlag, 1972.

21 Bauer, F., Garabedian, P., Jameson, A., and Korn, D., "Supercritical Wing Sections II, A Handbook," *Lecture Notes in Economics and Mathematical Systems*, Vol. 108, Berlin: Springer-Verlag, 1975.

22 Bauer, F., Garabedian, P., and Korn, D., "Supercritical Wing Sections III," *Lecture Notes in Economics and Mathematical Systems*, Vol. 150, Berlin: Springer-Verlag, 1977.

23 Krasnov, N.F., *Aerodynamics of Bodies of Revolution*, edited and annotated by D.N. Morris, New York: American Elsevier, 1970.

24 *Handbook of Supersonic Aerodynamics*, Vol. 3, Section 8, "Bodies of Revolution," NAVWEPS Report 1488, October 1961.

25 Ashley, H., and Landahl, M., *Aerodynamics of Wings and Bodies*, Reading: Addison-Wesley, 1965, pp. 178–181.

26 Adams, M.C., "Determination of Shapes of Boattail Bodies of Revolution for Minimum Wave Drag," NACA TN 2550, November 1951.

27 Liming, R., *Practical Analytic Geometry with Applications to Aircraft*, New York: MacMillan, 1944.

28 Williams, J.E., and Vukelich, S.R., *The USAF Stability and Control Digital DATCOM*, AFFDL TR-79–3032, 1979.

29 Lamar, J.E., and Alford, W.J., "Aerodynamic-Center Considerations of Wings and Wing-Body Combinations," NASA TN D-3581, October 1966.

30 Hall, C.F., "Lift, Drag, and Pitching Moment of Low Aspect Ratio Wings at Subsonic and Supersonic Speeds," NACA RM A53A30, 1953.

31 Boyd, J.W., Migotsky, E., and Wetzel, B.E., "A Study of Conical Camber for Triangular and Swept Back Wings," NASA RM A55G19, November 1955.

32 Mason, W.H., "A Wing Concept for Supersonic Maneuvering," NASA CR 3763, December 1983.

33 Mason, W.H., "SC^3 – A Wing Concept for Supersonic Maneuvering," AIAA Paper 83-1858, July 1983.

34 Hahn, A.S., "Vehicle Sketch Pad: A Parametric Geometry Modeler for Conceptual Aircraft Design," AIAA Paper 2010-0657, January 2010.

35 Kulfan, B.M., "Universal Parametric Geometry Representation Method," *Journal of Aircraft*, Vol. 45, No. 1, 2008, pp. 142–158.

Sources of Experimental Data for Code Validation

Some sources of aerodynamic geometry and experimental data for use in code evaluation are listed here. They are invaluable for making sure that you are using a computational aerodynamics code correctly. Always check a code that is new to you against known results, which we already discussed in Chapter 2. Note that rigorous validation of codes requires very careful analysis and an understanding of possible experimental, as well as computational, error (which was also discussed in Chapter 10). Most of the NASA and NACA reports cited here are available from the NASA Technical Reports Server, http://ntrs.nasa.gov/; a mirror website for NACA reports is available from Cranfield University at http://naca.central.cranfield.ac.uk/. Some of the reports listed here will also be provided at the book website: http://www.cambridge.org/aerodynamics. Most of the results are presented graphically, so a utility such as *DataThief* or *Engauge* is needed to digitize the data for comparison with calculations.

B.1 Airfoil Data Sources

B.1.1 Books

Abbott, I.H. and von Doenhoff, A.E., *Theory of Airfoil Sections*, New York: Dover, 1959: This is a book every aerodynamicist should have. Look in the references for the original NACA airfoil reports. The aerodynamic descriptions contained in the reports are unsurpassed. However, beware of the actual data presented prior to 1939, which is when they discovered that they had to apply a different support interference correction to the measured results (see NACA Report 669 by Jacobs for details). Note that pressure distributions for airfoils are fairly rare. See also NACA Report 824, which is an earlier compendium similar to this book.

Riegels, F.W., *Airfoil Sections*, London: Butterworths, 1961.

Selig, M.S., Donovan, J.F., and Fraser, D.B. *Airfoils at Low Speeds*, Soartech Virginia Beach: H.A. Stokely, 1989 (most of this material is available at http://m-selig.ae.illinois.edu/ads.html).

Eppler, R. *Airfoil Design and Data*, Berlin: Springer-Verlag, 1990.

B.1.2 The NASA Low- and Medium-Speed Airfoils

These airfoil tests were essentially an outgrowth of the NASA Supercritical Airfoils designed for transonic flight; all follow a similar template and typically include both force and moment results and pressure distributions.

McGhee, R.J., and Beasley, W.D., "Low Speed Aerodynamic Characteristics of a 17-Percent-Thick Airfoil Section Designed for General Aviation Applications," NASA TN D-7428, 1973.

McGhee, R.J., Beasley, W.D., and Somers, D.M., "Low Speed Aerodynamic Characteristics of a 13-Percent-Thick Airfoil Section Designed for General Aviation Applications," NASA TM X-72697, 1975.

McGhee, R.J., and Beasley, W.D., "Effects of Thickness on the Aerodynamic Characteristics of an Initial Low-Speed Family of Airfoils for General Aviation Applications," NASA TM X-72843, 1976.

McGhee, R.J., and Beasley, W.D., "Low-Speed Wind-Tunnel Results for a Modified 13-Percent-Thick Airfoil," NASA TM X-74018, 1977.

Barnwell, R.W., Noonan, K.W., and McGhee, R.J., "Low Speed Aerodynamic Characteristics of a 16-Percent-Thick Variable Geometry Airfoil Designed for General Aviation Application," NASA TP-1324, 1978.

McGhee, R.J., and Beasley, W.D., "Wind-Tunnel Results for an Improved 21-Percent-Thick Low-Speed Airfoil Section," NASA TM-78650, 1978.

McGhee, R.J., Beasley, W.D., and Whitcomb, R.T., "NASA Low- and Medium-Speed Airfoil Development," NASA TM-78709, 1979.

McGhee, R.J., and Beasley, W.D., "Low-Speed Aerodynamic Characteristics of a 13-Percent-Thick Medium Speed Airfoil Designed for General Aviation Applications," NASA TP-1498, 1979.

McGhee, R.J., and Beasley, W.D., "Low Speed Aerodynamic Characteristics of a 17-Percent-Thick Medium Speed Airfoil Designed for General Aviation Applications," NASA TP-1786, 1980

McGhee, R.J., and Beasley, W.D., "Wind-Tunnel Results for a Modified 17-Percent Thick Low-Speed Airfoil Section, "NASA TP-1919, 1981.

Ferris, J.D., McGhee, R.J., and Barnwell, R.W., "Low Speed Wind-Tunnel Results for Symmetrical NASA LS(1)-0013 Airfoil," NASA TM-4003, 1987.

B.1.3 NASA Transonic Airfoils

Whitcomb, R.T., "Review of NASA Supercritical Airfoils," ICAS Paper 74-10, August 1974: This was the first public description of the supercritical airfoil concept.

Harris, C.D., "NASA Supercritical Airfoils," NASA TP 2969, March 1990: See the references contained in this report for the specific NASA report citations containing the experimental data for the various NASA Supercritical Airfoils; also a "must have" report for aerodynamicists.

B.1.4 Laminar Flow Airfoils

Somers, D.M., "Design and Experimental Results for a Flapped Natural-Laminar-Flow Airfoil for General Aviation Applications," NASA TP-1865, June 1981: NLF(1)-0215F, Lancair and Wheeler express airfoil.

McGhee, R.J., Viken, J.K., and Pfenninger, W.D., "Experimental Results for a Flapped Natural-Laminar Flow Airfoil with High Lift/Drag Ratio," NASA TM-85788, 1984: This report contains tab data as well as plots.

Sewell, W.G., McGhee, R.J., Viken, J.K., Waggoner, E.G., Walker, B.S., and Miller, B.F., "Wind Tunnel Results for a High-Speed, Natural Laminar Flow Airfoil Designed for General Aviation Aircraft," NASA TM 87602, November 1985: This report contains tab data as well as plots.

B.1.5 Other Low- and Medium-Speed Airfoils and Airfoil Data

Stack, J., Lindsey, W.F., and Littell, R.E., "The Compressibility Burble and the Effect of Compressibility on Pressures and Forces Acting on an Airfoil," NACA Report 646, 1939: The results are presented for an NACA 4412 airfoil.

Stivers, L.S., "Effects of Subsonic Mach Number on the Forces and Pressure Distributions on Four 64A-Series Airfoil Sections at Angles of Attack as High as 28°," NACA TN 3162, 1954: This report contains tab data as well as plots.

Bingham, G.J., and Chen, A.W.S., "Low Speed Aerodynamic Characteristics of an Airfoil Optimized for Maximum Lift Coefficient," NASA TN D-7071, December 1971.

Beasley, W.D., and McGhee, R.J., "Experimental and Theoretical Low-Speed Aerodynamic Characteristics of the NACA 65(1)-213, $a = 0.50$, Airfoil," NASA TMX-3160, February 1975.

Hicks, R.M., "A Recontoured Upper Surface Designed to Increase the Maximum Lift Coefficient of a Modified NACA 65(0.82) (9.9) Airfoil Section," NASA TM 85855, February 1984.

B.1.6 The NACA 0012 Airfoil

This is one of the most widely used airfoils for code validation; we list four important reports for this use.

Harris, C.D., "Two-Dimensional Aerodynamic Characteristics of the NACA 0012 Airfoil in the Langley 8-Foot Transonic Pressure Tunnel," NASA

TM 81927, April 1981. Recommended by McCroskey. Has both force and moment and pressure distribution data, but no tab data.

McCroskey, W.J., "A Critical Assessment of Wind Tunnel Results for the NACA 0012 Airfoil, NASA TM 100019, October 1987.

Ladson, C.L., Hill, A.S., and Johnson, W.G., "Pressure Distributions from High Reynolds Number Transonic Tests of an NACA 0012 Airfoil in the Langley 0.3-Meter Transonic Cryogenic Tunnel," NASA TM 100526, 1987: This report contains tab data of the pressures.

Ladson, C.L., "Effects of Independent Variation of Mach and Reynolds Numbers on the Low-Speed Aerodynamic Characteristics of the NACA 0012 Airfoil Section," NASA TM 4074, 1988: This report has tab data for the forces.

B.1.7 Low Reynolds Number Airfoils

The key source for low Reynolds number airfoils is available through the University of Illinois website: http://m-selig.ae.illinois.edu/ads.html.

McGhee, R.J., Walker, Betty S., and Millard, Betty F., "Experimental Results for the Eppler 387 Airfoil at Low Reynolds Numbers in the Langley Low-Turbulence Pressure Tunnel," NASA TM 4062, 1988: This report contains tab data.

B.1.8 Multi-element Airfoil Data

Weick, F.E., and Shortal, J.A., "The Effect of Multiple Fixed Slots and a Trailing-edge Flap on the Lift and Drag of a Clark Y Airfoil," NACA Report 427, 1932.

Wenzinger, C.J., and Delano, J., "Pressure Distribution Over an NACA 23012 Airfoil with a Slotted and Plain Flap," NACA R 633, 1938.

Harris, T.A., and Lowry, J.G., "Pressure Distribution over an NACA 23012 Airfoil with a Fixed Slot and a Slotted Flap," NACA R 732, 1942.

Kelly, J.A., and Hayter, N-L.F., "Lift and Pitching Moment at Low Speeds of the NACA 64A010 Airfoil Section Equipped with Various Combinations of Leading Edge Slat, Leading Edge Flap, Split Flap and Double-Slotted Flap," NACA TN 3007, September 1953: This report illustrates the use of experimental gap and overlap studies to locate the best flap position relative to the airfoil; no drag or pressure distributions are presented.

Axelson, J.A., and Stevens, G.L., "Investigation of a Slat in Several Different Positions on an NACA 64A010 Airfoil for a Wide Range of Subsonic Mach Numbers," NACA TN 3129, March 1954: This report contains tab data as well as plots.

Wentz, W.H., and Seetharam, H.C., "Development of a Fowler Flap System for a High Performance General Aviation Airfoil," NASA CR-2443, 1974.

Seetharam, H.C., and Wentz, W.H., "Experimental Studies of Flow Separation and Stalling on a Two-Dimensional Airfoil at Low Speeds," NASA CR-2560, 1975.

B.2 Three-Dimensional Wing Data Sources

B.2.1 Elementary Body Geometries

The NACA conducted many tests using geometries that are simple to model. Similar tests were also done in the early days of NASA. The NACA reports were classified at the time, but have since been declassified.

Loving, D.L., and Estabrooks, B.B., "Transonic Wing Investigation in the Langley Eight Foot High Speed Tunnel at High Subsonic Mach Numbers and at a Mach number of 1.2," NACA RM L51F07, 1951.

Williams, C.V., "An Investigation of the Effects of a Geometric Twist on the Aerodynamic Loading Characteristics of a 45° Sweptback Wing-Body Configuration at Transonic Speeds," NACA RM L54H18, 1954.

McDevitt, J.B., "An Experimental Investigation of Two Methods for Reducing Transonic Drag of Swept Wing and Body Combinations," NACA RM A 55B21, April 1955.

Keener, E.R., "Pressure Measurements Obtained in Flight at Transonic Speeds for a Conically Cambered Delta Wing," NASA TM X-48, October 1959: This is data from flight tests of the JF-102A airplane.

Runckel, J.F., and Lee, E.E., "Investigation at Transonic Speeds of the Loading Over a 45° Sweptback Wing Having an Aspect Ratio of 3, Taper Ratio of 0.2, and NACA 65A004 Airfoil Sections," NASA TN D-712, 1961: This report contains tab data for pressures.

Chu, J., and Luckring, J.M., "Experimental Surface Pressure Data Obtained on 65° Delta Wing Across Reynolds Number and Mach Number Range," NASA TM 4645, 1996, Vol. 1. Sharp Leading Edge, Vol. 2. Small-Radius Leading Edge, Vol. 3. Medium-Radius Leading Edge, Vol. 4. Large Leading Edge Radius. These reports contain the tab data as well as plots. Results were obtained for Re from 6 to 36 million at M – 0.85, and Mach numbers from 0.4 to 0.9 at Re of 6 million.

Storms, B.L., Ross, J.C., Horne, W.C., Hayes, J.A., Dougherty, R.P., Underbrink, J.R., Scharpf, D.F., and Moriarty, P.J., "An Aeroacoustic Study of an Unswept Wing with a Three-Dimensional High-Lift System," NASA TM 112222, February 1998.

The standard transonic test case: the ONERA M6 wing has been used in practically every transonic code validation calculation ever published. The data are contained in AGARD AR-138 cited below and one case (Test 2308) is available at the NPARC Alliance Validation Archive: http://www.grc.nasa.gov/WWW/wind/valid/m6wing/m6wing.html

B.2.2 Supercritical Wings

Harris, C.D. and Bartlett, D.W., "Tabulated Pressure Measurements on a NASA Supercritical-Wing Research Airplane Model With and Without Fuselage Area-Rule Additions at Mach 0.25 to 1.00," NASA TM X-2634, 1972: This is data for the supercritical wing that NASA put on an F-8 airplane.

Harris, C.D., "Wind-Tunnel Measurements of Aerodynamic Load Distribution on a NASA Supercritical-Wing Research Airplane Configuration," NASA TM X-2469, 1972: This is also data for the F-8 supercritical wing airplane.

Montoya, L.C., and Banner, R.D., "F-8 Supercritical Wing Flight Pressure, Boundary Layer and Wake Measurements and Comparisons with Wind Tunnel Data," NASA TM X-3544, March 1977: The report contains tab data.

Hinson, B.L., and Burdges, K.P., "Acquisition and Application of Transonic Wing and Far-Field Test Data for Three-Dimensional Computational Method Evaluation," AFOSR-TR-80–0421, March 1980, available from DTIC as AD A085 258: These are the Lockheed Wings A, B, and C.

Keener, E.R., "Pressure Distribution Measurements on a Transonic Low-Aspect Ratio Wing," NASA TM 86683, 1985: This is the so-called Lockheed Wing C.

Keener, E.R., "Boundary Layer Measurements on a Transonic Low-Aspect Ratio Wing," NASA TM 88214, 1986: This is the so-called Lockheed Wing C and represents a monumental effort.

B.2.3 Supersonic Wing Data

Miller, D.S., Landrum, E.J., Townsend, J.C., and Mason, W.H., "Pressure and Force Data for a Flat Wing and a Warped Conical Wing Having a Shockless Recompression at Mach 1.62," NASA TP 1759, April 1981: This report contains tab data.

Pittman, J.L., Miller, D.S., and Mason, W.H., "Supersonic, Nonlinear, Attached-Flow Wing Design for High Lift with Experimental Validation," NASA TP 2336, August 1984: This report contains tab data.

B.3 AGARD Test Cases

AGARD (Advisory Group for Aeronautics Research and Development; later changed to Research and Technology Organization, RTO, and now called Science and Technology Organization, STO) selected various test cases for CFD code validation. These cases are important because an attempt has been made to define the test conditions and any corrections required precisely enough for use in code validation work. This also means that the airfoil test coordinates and results are available in tabulated form in these reports. Some of these reports are available from the NATO Science and Technology Organization reports website: http://www.cso.nato.int/abstracts. aspx. The key test case reports are described next.

B.3.1 AGARD AR-138, "Experimental Data Base for Computer Program Assessment," May 1979

B.3.1.1 TWO-DIMENSIONAL TEST CASES

1. NACA 0012, over a range of subsonic Mach and angle of attack, both force and moment and pressure distributions;
2. NLR QE 0.11-0.75-1.375, a symmetrical airfoil designed to be shock-free at a transonic design point, Mach range from 0.30 to 0.85, all at zero angle of attack;
3. Supercritical airfoil CAST 7, pressure distributions over a range of Mach from 0.40 to 0.80, α from $-2°$ to $5°$, also boundary layer measurements. No force and moment data;
4. NLR7301, thick supercritical airfoil (16.5%), Mach from 0.30 to 0.85, α from $-4°$ to $+4°$, pressure, and force and moment;
5. SKF 1.1/with maneuver flap (French), Mach number from 0.50 to 1.2, force and moment and pressure over a limited range of angle of attack;
6. RAE 2822, surface pressure distribution, boundary layer and wake rake surveys, over a range of Mach and α (this is one of the most complete sets of data in the report);
7. NAE 75-036-13:2, Mach range from 0.5 to 0.84, α from 0 to 4° at $M = 0.75$, 2° for other Machs;
8. MBB-A3 NASA 10% supercritical, Mach from 0.6 to 0.80, α from 0.5° to 2.5°.

B.3.1.2 THREE-DIMENSIONAL TEST CASES

1. ONERA M6, pressure distributions,
2. ONERA AFV D, variable sweep wing,
3. MBB-AVA Pilot Model with supercritical wing,
4. RAE Wing A,
5. NASA Supercritical-Wing Research Airplane Model (actually the F-8, pressure distributions only).

B.3.1.3 BODY ALONE CONFIGURATIONS

1. 1.5D Ogive Circular Cylinder Body, L/D = 21.5,
2. MBB Body of revolution No. 3,
3. 10° cone-cylinder at α zero, Mach from 0.91 to 1.22,
4. ONERA calibration body model C5, Mach from 0.6 to 1.0, α zero.

B.3.2 AGARD AR-138-ADDENDUM, "ADDENDUM to AGARD AR No. 138, Experimental Data Base for Computer Program Assessment," July 1984

Five additional three-dimensional datasets were identified and included in the Addendum:

(B-6) Lockheed-AFOSR Wing A: Semi-span wing, Mach 0.62–0.84, α from −2° to 5°, Re on mac: 6 million

(B-7) Lockheed-AFOSR Wing B: Semi-span wing, Mach 0.70 to 0.94, α from −2° to +5°, Re on mac: 10 million

(B-8) ARA M100 Wing/body, full model, Mach 0.50–0.93, α from −4° to +3°, Re on mac: 3.5 million

(B-9) ARA M86 Wing/body, full model, Mach 0.50–0.82, α from 0° to +8°, Re on mac: 2.8–3.7 million

(B-10) FFA Aircraft (SAAB A32A Lansen), Mach 0.40–0.89, α from 0° to +10°, Re on mac: 10–30 million

B.3.3 AGARD R-702, "Compendium of Unsteady Aerodynamic Measurements," August 1982

Seven test cases are defined – five airfoils and two wings.

B.3.3.1 AIRFOILS

1. NACA 64006 with oscillating flap,
2. NACA 64A010 with oscillatory pitching,
3. NACA 0012 with oscillatory and transient pitching,
4. NLR 7301 airfoil with:
    (i) oscillatory pitching and oscillating flap at NLR and
    (ii) oscillating pitching (NASA Ames).

B.3.3.2 WING DATA

1. RAE Wing A with an oscillating flap
2. NORA Model with oscillation about the swept axis.

B.3.4 AGARD AR-211, "Test Cases for Inviscid Flowfield Methods," May 1985

B.3.4.1 TWO DIMENSIONAL TEST CASES

NACA 0012 airfoil at
1. $M = 0.80$, $\alpha = 1.25°$,
2. $M = 0.85$, $\alpha = 1°$,
3. $M = 0.95$, $\alpha = 0°$,
4. $M = 1.25$, $\alpha = 0°$,
5. $M = 1.25$, $\alpha = 7°$,

RAE 2822 airfoil at
6. $M = 0.75$, $\alpha = 3°$,

NLR 7301 airfoil at
7. $M = 0.720957$, $\alpha = .194°$, (theoretical data)

Chiocchia-Nocilla at
8. $M = 0.769$, $\alpha = 0°$. (sharp le)

B.3.4.2 2-D CASCADE TEST CASES

HOBSON-1   9. $M = 0.476$, $\alpha = 43.544°$, Spacing, s/c = 1.0121

HOBSON-2   10. $M = 0.575$, $\alpha = 46.123°$, Spacing, s/c = 0.5259

B.3.4.3 THREE-DIMENSIONAL CASES

ONERA M6 airfoil at   11. $M = 0.84$, $\alpha = 3.06°$,

      12. $M = 0.92$, $\alpha = 0°$,

Butler wing at   13. $M = 2.50$, $\alpha = 0°$,

Dillner wing at   14. $M = 1.50$, $\alpha = 15°$,

      15. $M = 0.70$, $\alpha = 15°$,

NASA Ames swept wing at   16. $M = 0.833$, $\alpha = 1.75°$,

AGARD B at   17. $M = 1.5$, $\alpha = 0°$,

      18. $M = 1.5$, $\alpha = 2°$,

      19. $M = 2.0$, $\alpha = 0°$,

      20. $M = 2.0$, $\alpha = 2°$.

B.3.5 AGARD AR-303, "A Selection of Experimental Test Cases for the Validation of CFD Codes," August 1994 (in two volumes)

A – Airfoil cases (13)
B – Wing-fuselage (6)
C – Bodies (6)
D – Delta wing class (5)
E – Aero-Propulsion/Pylon/Store (9)

The test data and geometry for this report are available electronically. The data will either be available on the book website or from the authors.

B.4 The NPARC Alliance

The NPARC Alliance maintains a website of test cases for CFD, including grids and results:

http://www.grc.nasa.gov/WWW/wind/valid/archive.html

B.5 Electronic Access to Workshops Related to Code Evaluation/Validation

Currently, three key activities are active online and are organized around workshops.

Drag Prediction Workshop; the fifth workshop took place in June 2012. http://aaac.larc.nasa.gov/tsab/cfdlarc/aiaa-dpw/

High Lift Prediction Workshop; the second workshop took place in June 2013. http://hiliftpw.larc.nasa.gov

High-Order CFD Methods Workshop; the second workshop took place in May 2013:

http://www.dlr.de/as/desktopdefault.aspx/tabid-8170/13999_read-35550/

and the third workshop will take place in 2015:

https://www.grc.nasa.gov/hiocfd/

These activities are ongoing, and any practicing computational aerodynamicist should make themselves aware of these valuable results.

B.6 NASA Turbulence Modeling Resource

An extensive list of verification cases for evaluating turbulence models (including grids) is readily available at: http://turbmodels.larc.nasa.gov/. The available geometries include:

Basic Cases:

    2D Zero pressure gradient flat plate
    2D Mixing Layer
    Axisymmetric subsonic jet
    Axisymmetric near-sonic jet
    Axisymmetric separated boundary layer
    2D Airfoil near-wake
    2D NACA 0012 airfoil

Extended Cases:

    2D Zero pressure gradient high Mach number flat plate
    2D Backward facing step
    2D NACA 4412 airfoil trailing-edge separation
    2D Convex curvature boundary layer
    3D Supersonic square duct

Potential Flow Review

C.1 Potential Flow Theory

Chapter 4 presented a simplified set of equations for steady, two-dimensional, incompressible, inviscid, irrotational flow of a perfect fluid with no body forces. Those equations are reproduced here for clarity for the derivations that follow. The equations derived in Chapter 4 include:

Continuity equation

$$\frac{\partial u}{\partial x} + \frac{\partial v}{\partial y} = 0. \tag{4.8}$$

Bernoulli's equation

$$p + \frac{\rho V^2}{2} = p_o. \tag{4.9}$$

Irrotational condition

$$\frac{\partial v}{\partial x} - \frac{\partial u}{\partial y} = 0, \tag{4.10}$$

where p_o is the total pressure in the flow. These equations give us a set of relations that allows for an analytic solution.

C.1.1 The Velocity Potential

Now that we have a simplified set of equations for potential flow (Equations 4.8–4.11), how will we solve them? Traditionally this has been accomplished through the introduction of a scalar function whose gradient is the velocity vector, and that is defined specifically for an irrotational flowfield, namely the *velocity potential*, Φ,

$$\vec{V} \equiv \vec{\nabla}\Phi. \tag{C.1}$$

Using the definition of the gradient in Cartesian coordinates, we obtain

$$\vec{V} = u\hat{i} + v\hat{j} = \vec{\nabla}\Phi = \frac{\partial \Phi}{\partial x}\hat{i} + \frac{\partial \Phi}{\partial y}\hat{j}. \tag{C.2}$$

Equating components of the velocity and gradient vectors yields relationships between the velocity components and the derivatives of the velocity potential

$$u = \frac{\partial \Phi}{\partial x} \qquad v = \frac{\partial \Phi}{\partial y}. \tag{C.3}$$

Substituting these relations into the irrotational condition (Equation 4.10) easily shows that the velocity potential identically defines an irrotational flowfield

$$\frac{\partial v}{\partial x} - \frac{\partial u}{\partial y} = \frac{\partial}{\partial x}\left(\frac{\partial \Phi}{\partial y}\right) - \frac{\partial}{\partial y}\left(\frac{\partial \Phi}{\partial x}\right) = 0 \tag{C.4}$$

as long as the velocity potential function is continuous and differentiable. So, we have a scalar function that is irrotational, but it must also satisfy the Continuity equation (Equation 4.8), which means

$$\frac{\partial u}{\partial x} + \frac{\partial v}{\partial y} = \frac{\partial}{\partial x}\left(\frac{\partial \Phi}{\partial x}\right) + \frac{\partial}{\partial y}\left(\frac{\partial \Phi}{\partial y}\right) = \frac{\partial^2 \Phi}{\partial x^2} + \frac{\partial^2 \Phi}{\partial y^2} = \nabla^2 \Phi = 0. \tag{C.5}$$

This is *Laplace's equation* (also discussed in Section 3.12.5), a very important partial differential equation, which represents a critical arrival point for modeling aerodynamic flowfields. Laplace's equation is a linear partial differential equation, which means that the velocity potential is a harmonic function and "superposition" of solutions is allowed; any two solutions added together create another solution. So, we have two choices at this point: (1) solve Laplace's equation numerically with appropriate boundary conditions, or (2) find basic solutions to Laplace's equation and add those solutions to create more complicated flowfields. The latter choice yields the classical area of fluid dynamics known as Potential Flow theory. This concept also serves as the basis for an important computational aerodynamic concept known as the panel method, described in Chapter 5.

C.1.2 The Stream Function

Another way to represent potential flows is using the concept of the *stream function, ψ*. The stream function is a scalar function defined to satisfy the Continuity equation (Equation 4.8):

$$u = \frac{\partial \psi}{\partial y} \quad v = -\frac{\partial \psi}{\partial x}. \tag{C.6}$$

If we check to see whether the stream function satisfies the Continuity equation, we have

$$\frac{\partial u}{\partial x} + \frac{\partial v}{\partial y} = \frac{\partial}{\partial x}\left(\frac{\partial \psi}{\partial y}\right) + \frac{\partial}{\partial y}\left(-\frac{\partial \psi}{\partial x}\right) = 0. \tag{C.7}$$

Therefore, the stream function represents a physically possible flowfield that can either be rotational or irrotational. However, if we further require the flow to be irrotational (meaning it must satisfy Equation 4.10 as well), we arrive at:

$$\frac{\partial v}{\partial x} - \frac{\partial u}{\partial y} = \frac{\partial}{\partial x}\left(-\frac{\partial \psi}{\partial x}\right) - \frac{\partial}{\partial y}\left(\frac{\partial \psi}{\partial y}\right) = \frac{\partial^2 \psi}{\partial x^2} + \frac{\partial^2 \psi}{\partial y^2} = \nabla^2 \psi = 0. \tag{C.8}$$

In other words, the stream function must also satisfy Laplace's equation. These two functions, the velocity potential and the stream function, serve as the basis for analytic fluid dynamics of incompressible flows.

C.1.3 Streamlines and Boundary Conditions

There are a number of important properties of these two functions that are essential for understanding and using potential flow for modeling aerodynamics. First of all, look at *equipotential lines* (lines along which a function is constant) for the velocity potential and stream function. Look at any stream function and take the derivative of the stream function using the chain rule:

$$\psi = \psi(x, y) \tag{C.9}$$

$$d\psi = \frac{\partial \psi}{\partial x}dx + \frac{\partial \psi}{\partial y}dy. \tag{C.10}$$

The partial derivatives can then be replaced using their velocity counterparts (see Equation C.6),

$$d\psi = -v dx + u dy. \tag{C.11}$$

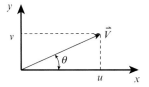

Figure C.1 Velocity components and the streamline.

Now, look what happens for ψ = constant:

$$d\psi = 0 = -vdx + udy \tag{C.12}$$

$$\left(\frac{dy}{dx}\right)_{\psi=C} = \frac{v}{u}. \tag{C.13}$$

What is this telling us? Lines of constant ψ have a slope that is equal to the ratio of the y and x components of velocity. That is also the slope of the velocity vector as shown in Fig. C.1, where $\tan\theta = v/u$. So, lines of constant ψ are aligned directly with the local flow – in other words, these lines are *streamlines* of the flowfield (lines along which the flow is always tangent). This is a very important characteristic of the stream function and also serves as an important part of the application of potential flow.

There are two basic types of surface boundary conditions in aerodynamics: inviscid flow boundary conditions and viscous flow boundary conditions. Inviscid flow at a solid surface has only one requirement – that the flow not go through the boundary, or $\vec{V} \cdot \hat{n} = 0$. This is known as the *"slip" boundary condition*. In viscous flow, we add another requirement to the boundary condition – namely, that the flow at the surface must have zero velocity, or $\vec{V}_{surface} = 0$. This is known as the *"no slip" boundary condition*.

The inviscid boundary condition requires that the flow be aligned with the surface – in other words, the boundary must be a streamline of the flowfield. This leads to important applications of potential flow, since any streamline in a potential flow can be thought of as representing a solid surface in an inviscid flow. We will use this concept frequently when we discuss potential flow applications and panel methods.

C.1.4 Relationship between Stream Function and Velocity Potential

Now look at the velocity potential (just as we did earlier for the stream function in Equation 4.20) and take the derivative of the function using the chain rule:

$$\Phi = \Phi(x, y) \tag{C.14}$$

$$d\Phi = \frac{\partial \Phi}{\partial x} dx + \frac{\partial \Phi}{\partial y} dy. \qquad (C.15)$$

The partial derivatives can then be replaced using their velocity counterparts (see Equation C.3)

$$d\Phi = udx + vdy. \qquad (C.16)$$

Now, look what happens for ϕ = constant:

$$d\Phi = 0 = udx + vdy \qquad (C.17)$$

$$\left(\frac{dy}{dx}\right)_{\Phi=C} = -\frac{u}{v} = -1 \Bigg/ \left(\frac{dy}{dx}\right)_{\psi=C}. \qquad (C.18)$$

What is this telling us? Lines of constant Φ have a slope that is equal to the negative ratio of the x- and y-components of velocity. In addition, this slope is everywhere equal to the negative reciprocal of the slope of the streamlines, which means that these two sets of lines are everywhere orthogonal. But what is the meaning of the lines of constant Φ? In order to more fully understand these two functions, it is helpful to realize that much of the development of potential flow theory was derived from electromagnetic theory. In electromagnetics there are two important field properties: the electric field and the magnetic field. These two fields portray many of the same features as potential flowfields, but while in electromagnetics both fields have a straightforward physical meaning, in fluid dynamics they do not. Yes, lines of constant ψ are streamlines and represent the flow direction, but lines of constant Φ do not lend themselves to being understood as easily – they are lines perpendicular to the flow direction. However, they can be related to something known as impulsive pressure (a type of pressure potential).[1]

C.1.5 Volumetric Flow Rate

Another important feature of stream functions is their relationship to flow rates. Since stream functions were derived to satisfy the continuity equation, they have basic properties related to flow between streamlines. Look at any two streamlines and the flow between them as shown in Fig. C.2.

Now, compute the volumetric flow rate across the control line from A to B:

$$Q = \int_A^B udy - \int_A^B vdx \qquad (C.19)$$

where the second integral is negative because a positive v velocity flowing across a positive dx will be flowing from right to left across the line. Now look

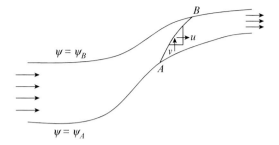

Figure C.2   Volumetric flow rate between two streamlines (from Ref. 2).

at the differential of the stream function from Equation C.11 and replace the stream function with velocity components Equation:

$$d\psi = \frac{\partial \psi}{\partial x} dx + \frac{\partial \psi}{\partial y} dy = -v dx + u dy. \tag{C.20}$$

Now, integrate this relationship from A to B and find

$$\int_A^B d\psi = \psi_B - \psi_A = -\int_A^B v dx + \int_A^B u dy. \tag{C.21}$$

But this is the same result we obtained for the volumetric flow rate (Equation C.19) – in other words:

$$Q = \psi_B - \psi_A. \tag{C.22}$$

Therefore, the volumetric flow rate between any two streamlines is given by the difference of the stream function constants for the two lines.

C.1.6 Circulation

Another concept that is crucial for understanding potential flows, and later for understanding airfoil and wing aerodynamics, is circulation. *Circulation* is defined as

$$\Gamma \equiv -\oint_C \vec{V} \cdot d\vec{l}. \tag{C.23}$$

According to the contour integration rules of mathematics, the integration should be done in a counterclockwise direction (the area inside the contour, A, should be to the left as you travel along the contour). The minus sign has been added so that circulation will be positive for a clockwise direction, something that will be important when we apply the integral to aerodynamic flows.

But what does circulation mean? Look at an arbitrary flowfield as shown in Fig. C.3. Within the flowfield we will place a closed contour, C, along which we will perform the circulation integration. At any point along the contour,

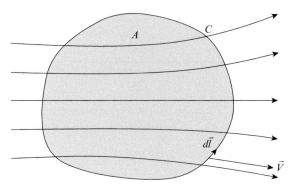

Figure C.3   Velocity components and the streamline.

there is a differential length, $d\vec{l}$, and a velocity vector, $\vec{V}$. The integral requires us to perform the scalar product of these two vectors, which leads to

$$\vec{V} \cdot d\vec{l} = \left|\vec{V}\right|\left|d\vec{l}\right|\cos\theta \tag{C.24}$$

where θ is the angle between the two vectors. What does this scalar product represent? If $\theta = 0°$ (the vectors are parallel), then $\cos\theta = 1$ and the local flow is directly in the path of the contour, contributing to the circulation. If $\theta = 90°$ (the vectors are perpendicular), then $\cos\theta = 0$ and the local flow is normal to the path of the contour, not contributing to the circulation. So, circulation happens when the flow is aligned with the contour. In other words, circulation takes place when there is a "net" turning of the flow locally.

Another way to think about circulation is on a local level, such as near the trailing edge of an airfoil. We use the symbol γ to represent local circulation, and the circulation at any point on an airfoil is $\gamma(x)$, since the local circulation can vary along the chord of the airfoil. The total circulation for the airfoil is then $\Gamma = \int_0^c \gamma(x)dx$ (where we are assuming that the airfoil is thin and the chord line is aligned with the x-axis). This local circulation can be shown to be equal to the difference in velocity between the upper and lower surfaces of the airfoil, $\gamma = V_{upper} - V_{lower}$,[2] which makes a great deal of sense. If the flow is turning, then the flow over the upper surface of the airfoil must be going faster than the flow over the lower surface, which from Bernoulli's equation also tells us that the pressures on the upper surface are lower than the pressures on the lower surface, resulting in lift. Therefore, lift is directly related to how much the flow is turned by the airfoil!

So, a flow that turns will have circulation and a flow that does not turn will not have circulation. This turning of the flow can also be related to vorticity within the flowfield, since by Green's theorem the circulation integral may be rewritten as[2]

$$\Gamma \equiv -\oint_C \vec{V} \cdot d\vec{l} = -\iint_A \left(\nabla \times \vec{V}\right) \cdot \hat{n}dA \tag{C.25}$$

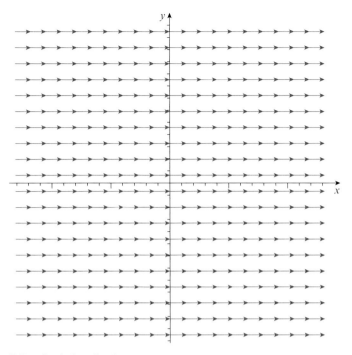

Figure C.4   Uniform flow in the x-direction.

where A is the area inside of C. This alternative method for determining circulation says that circulation can only exist if $\nabla \times \vec{V}$ is non-zero, or, in other words, the flow must have vorticity within the contour of integration. This represents another important concept for understanding aerodynamic flow – in order to have a turning flow there must be circulation, and in order for circulation to exist there must be vorticity.

C.2 Basic Potential Flow Types

There are four basic flow types that are frequently used in potential flow solutions: uniform flow, source/sink flow, vortex flow, and doublet flow. We will look briefly at each of these flows to see their characteristics and uses. A very detailed and excellent description of these flow functions is given in Currie's *Fundamental Mechanics of Fluids.*[3]

C.2.1 Uniform Flow

The first solution to Laplace's equation that we will investigate is uniform flow. First, we will look at uniform flow only in the x-direction, as shown in Fig. C.4.

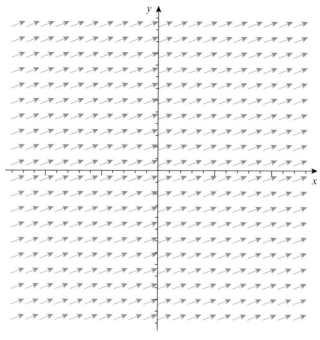

Figure C.5 Uniform flow at an angle of attack.

The velocity components for this flow are $u = V_\infty$ and $v = 0$. You can easily verify that this flowfield satisfies the Continuity equation and the Irrotational condition, meaning that the flow can be represented by a velocity potential

$$\Phi = V_\infty x + C \tag{C.26}$$

$$\psi = V_\infty y + C \tag{C.27}$$

where C represents any arbitrary constant.

This flow could be placed at an angle of attack simply by taking components of the velocity field at an arbitrary angle of attack:

$$u = V_\infty \cos\alpha \quad v = V_\infty \sin\alpha. \tag{C.28}$$

Using the basic relationships for the velocity potential and the stream function, we find

$$\Phi = V_\infty \left(x\cos\alpha + y\sin\alpha \right) \tag{C.29}$$

$$\psi = V_\infty \left(y\cos\alpha - x\sin\alpha \right) \tag{C.30}$$

which is plotted in Fig. C.5.

C.2.2 Source/Sink Flow

Another basic flow important for the development of aerodynamic theory is source/sink flow, found by defining the velocity potential and stream function as

$$\Phi = \frac{\Lambda}{2\pi}\ln r + C \tag{C.31}$$

$$\psi = \frac{\Lambda}{2\pi}\theta + C \tag{C.32}$$

where Λ is the strength of the source (if Λ is positive then the flow is called "source" flow, and if Λ is negative then the flow is called "sink" flow). We will switch to cylindrical coordinates to simply the mathematics, so $r = \sqrt{x^2 + y^2}$, and $\theta = \tan^{-1}(y/x)$. The velocity components of the flowfield are

$$u_r = \frac{\partial \Phi}{\partial r} = \frac{1}{r}\frac{\partial \psi}{\partial \theta} = \frac{\Lambda}{2\pi r} \quad u_\theta = \frac{1}{r}\frac{\partial \Phi}{\partial \theta} = -\frac{\partial \psi}{\partial r} = 0 \tag{C.33}$$

which represents a flow coming from the origin (source) or flowing into the origin (sink) along straight, radial lines, as shown in Fig. C.6. Since the streamlines are straight radial lines from the origin, and, since velocity equipotential lines are orthogonal to streamlines, the equipotential lines are circles centered at the origin.

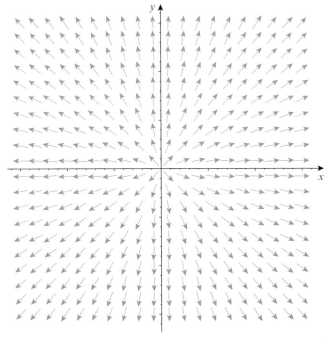

Figure C.6   Source flow centered at the origin.

C.2.3 Vortex Flow

Since source/sink flow has streamlines coming from the origin, with equipotential lines that are circles, what type of flow would we get if we exchanged the functions – what if we wanted streamlines that were circles and equipotential lines that were straight lines coming from the origin? Let's try swapping the two functions in Equations C.31 and C.32 and see that we get a vortex flow:

$$\Phi = -\frac{\Gamma}{2\pi}\theta + C \qquad (C.34)$$

$$\psi = \frac{\Gamma}{2\pi}\ln r + C \qquad (C.35)$$

where Γ is the strength of the vortex (Γ is positive for clockwise rotation). Again, use cylindrical coordinates to simplify the mathematics. The velocity components of the flowfield are

$$u_r = \frac{\partial \Phi}{\partial r} = \frac{1}{r}\frac{\partial \psi}{\partial \theta} = 0 \quad u_\theta = \frac{1}{r}\frac{\partial \Phi}{\partial \theta} = -\frac{\partial \psi}{\partial r} = -\frac{\Gamma}{2\pi r} \qquad (C.36)$$

which represents a flow rotating about the origin with equipotential lines along straight, radial lines, as shown in Fig. C.7.

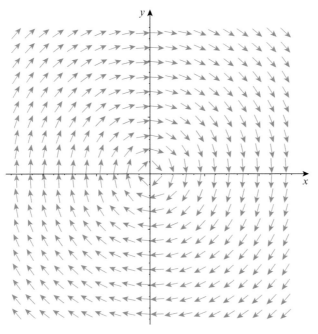

Figure C.7   Vortex flow centered at the origin.

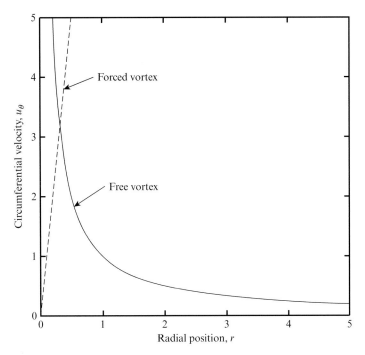

Figure C.8 Free and forced vortex velocity distributions.

The vortex creates only a circumferential velocity component (see Equation C.36) that creates a velocity distribution as shown in Fig. C.8. This simulates a *free vortex* (or irrotational vortex) but has a singularity at the center of the vortex where velocities become infinite. A "real" vortex does have a free vortex outer portion, but the inner portion is actually better modeled as a *forced vortex* (a rotational vortex) where $u_\theta = r\omega$, which is solid body rotation. An actual vortex contains a forced core, with a free outer section, which is the combination of the two vortex types.

C.2.4 Doublet Flow

Doublet flow is really a composite of a source and a sink that are very close to each other – of course, if a source and a sink of equal strengths were at the same location, there would be no flow, so this function requires a little sneaky mathematics. If the source is placed just to the left of the origin and the sink is placed just to the right of the origin, the flowfield will look as shown in Fig. C.9.

Remember, earlier we learned that potential flow functions were derived from electromagnetic theory – the source/sink combination shown previously looks like the fields around a magnet! Now, if you bring the source and sink closer and closer together, while increasing the strength of the source

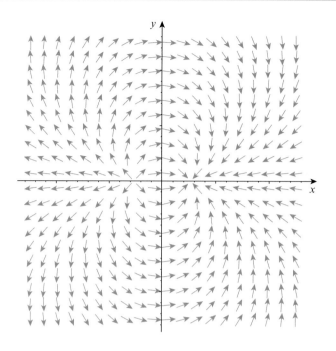

Figure C.9   Flowfield for a source located at $x = -1$ and sink located at $x = +1$.

and sink (see Ref. 3 for details), you will obtain a flow known as a doublet
with velocity potential and stream functions given by

$$\Phi = \frac{m\cos\theta}{2\pi r} + C \tag{C.37}$$

$$\psi = -\frac{m\sin\theta}{2\pi r} + C \tag{C.38}$$

where m is the strength of the doublet. The velocity components of the
flowfield are

$$u_r = \frac{\partial \Phi}{\partial r} = \frac{1}{r}\frac{\partial \psi}{\partial \theta} = -\frac{m\cos\theta}{2\pi r^2} \quad u_\theta = \frac{1}{r}\frac{\partial \Phi}{\partial \theta} = -\frac{\partial \psi}{\partial r} = -\frac{m\sin\theta}{2\pi r^2} \tag{C.39}$$

which represents a doublet (or dipole) flow as shown in Fig. C.10.

If you check, you will see that each of these basic flow functions is a solution to Laplace's equation, which means that each of these functions represents an inviscid, irrotational, incompressible flowfield. Since these functions
are all "harmonic" (solutions to Laplace's equation), they can be added
together in various ways to create a variety of more complicated flowfields.
While it is conceivable that there are an infinite number of functions that are
solutions to Laplace's equation, these are the only function we will need to
create realistic aerodynamic flowfields.

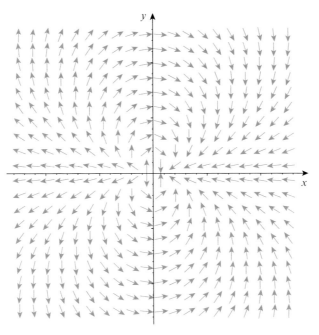

Figure C.10   Doublet flow centered at the origin.

C.3 References

1   Karamcheti, K., *Principles of Ideal-Fluid Aerodynamics*, 2nd Ed., Malabar, FL: Krieger Publishing, 1980.

2   Anderson J.D., *Fundamentals of Aerodynamics*, 5th Ed., Boston: McGraw-Hill, 2010.

3   Currie, I.G., *Fundamental Mechanics of Fluids*, 4th Ed., Boca Raton: CRC Press, 2012.

Computational Aerodynamics Programs

D.1 Codes Available from Book Website

Several types of programs are used to provide insight into aerodynamics, such as airfoil, wing, and aircraft analysis, as well as various aerodynamic design programs. The number of programs available has grown a great deal over the past few years, and a number of the programs have been discussed in Chapters 2 and 5 especially. This appendix will list and describe some of the programs that are either available from the book website (www.cambridge.org/aerodynamics) or that could be used to complete various projects listed at the end of chapters in the book. In this section we will just list and briefly describe some of the programs:

> *FOILGEN;* provides ordinates for NACA 4-digit, 4-digit modified and 5-digit airfoils
> *LADSON;* provides ordinates for NACA 6- and 6A-series airfoils
> *PANELV2:* an airfoil panel method
> *THINFOIL;* an airfoil CFD program
> *LIDRAG:* a lift-induced drag program
> *LAMDES;* a program that finds wing camber and twist to obtain a given splanload
> *FRICTION;* a skin friction and form drag estimation program
> *VLMpc;* a two-surface vortex lattice program
> *DESCAM;* an inverse design airfoil program
> *TRIDAG;* the Thomas algorithm solution approach for a tridiagonal matrix

D.2 FOILGEN

This program is used for airfoil geometry generation. For airfoils with analytically defined ordinates, this program produces airfoil definition data sets in the format required for *PANELV2*. This includes NACA 4-digit, 4-digit modified and 5-digit airfoils. In addition, the NACA 6 and 6A camber lines are available. The user can combine any combination of thickness and camber lines available within these shapes. This provides a wide range of airfoil definitions. The program runs interactively, and a sample session, with inputs and outputs, is provided at www.cambridge.org/aerodynamics.

D.3 LADSON

This is the NASA program that provides a reasonable approximation to the NACA 6 and 6A series airfoils. It was written by Charles Ladson and Cuyler Brooks. Originally it ran on the NASA CDC computer. It has been ported to run on a personal computer. Only minor modifications were made to produce a program to generate a set of ordinates in the form required as standard input by the programs described elsewhere.

The program is only an approximation to the ordinates because there is no simple algebraic formula available to describe the thickness distribution. We spoke briefly to Charles Ladson some years ago, and he said that he thought it would be impossible to generate a more accurate program. When he was doing this work he investigated the availability of more detailed notes on these airfoils and discovered that all the records have been destroyed. The only information available is that contained in the actual NACA reports. However, this program is much more accurate than attempts to simulate the 6 and 6A series thickness envelope by using a modified NACA 4-digit airfoil formula. The program was developed to handles thicknesses from 6 to 15 percent. The program runs interactively, and a sample session, with inputs and outputs, is provided at www.cambridge.org/aerodynamics.

References

Ladson, C.L., and Brooks, C.W., Jr., "Development of a Computer Program to Obtain Ordinates for NACA 6- and 6A-Series Airfoils," NASA TM X-3069, September, 1974.

Patterson, E.W., and Braslow, A.L., "Ordinates and Theoretical Pressure Distribution Data for NACA 6- and 6A- Series Airfoil Sections with Thicknesses from 2 to 21 and from 2 to 15 Percent Chord Respectively," NASA R-84, 1961.

D.4 *PANELV2*

PANELV2 is an extension of *PANEL*, which is a two-dimensional incompressible, inviscid (potential) flow program for NACA airfoils using the

Smith-Hess low-order panel method. *PANEL*, is available from Moran, J., *An Introduction to Theoretical and Computational Aerodynamics*, Mineola, NY: Dover, 2010. *PANELV2* has been extended to perform the prediction of pressure distributions over arbitrary airfoils, modification of the airfoil shape, and production of an output file for plotting or use as input to a boundary-layer analysis program. The program runs interactively, and a sample session, with inputs and outputs, is provided at www.cambridge.org/aerodynamics.

D.5 *THINFOIL*

THINFOIL solves Laplace's equation by finite differences using a variety of iteration

methods. The iteration options include SOR, SLOR, AF1 and AF2 to solve the algebraic equations. An unevenly spaced grid is used to concentrate grid points near the airfoil surface, and near the leading and trailing edges of the airfoil. Notes describing the detailed procedures and program terminology are available on the book website. A sample case shows the solution is found for the flow over a biconvex airfoil at zero angle of attack as a model problem. The program runs interactively, and a sample session, with inputs and outputs, is provided at www.cambridge.org/aerodynamics.

D.6 *LIDRAG*

This program computes the span efficiency factor e for a single planar lifting surface given the spanload. It uses the spanload to determine the "e" using a Fast Fourier Transform. Numerous other methods could be used. For reference, note that the "e" for an elliptic spanload is 1.0, and the "e" for a triangular spanload is 0.72. The original code is available on the book website in the file LIDRAG.F. The sample input is also on available and is called B2LDG.INP. The program prompts the user for the name of the input file.

The program was written by Dave Ives, and entered the public domain through the code contained in AFFDL-TR-77–122, "An Automated Procedure for Computing the Three Dimensional Transonic Flow over Wing-Body Combinations, Including Viscous Effects," February 1978. The input is the spanload obtained from any method. The output is the Trefftz plane induced drag e and the integral of the spanload, which produces the C_L. This is the "span" e. You should include a point at $\eta = 0$ and at $\eta = 1$ you should include a point with zero spanload.

D.7 *LAMDES*

This is the Lamar design program, *LamDes2*. It can be used as a nonplanar *LIDRAG* to get span *e* for multiple lifting surface cases when user supplies spanload. It has also been called the Lamar/Mason optimization code. It finds the spanload to minimize the sum of the induced and pressure drag, including canards or winglets. It also provides the associated camber distribution for subsonic flow. Since two surfaces are included, it can find the minimum trimmed drag while satisfying a pitching moment constraint.

The program will prompt you for the input file name. The program runs interactively, and a sample session, with inputs and outputs, is provided at www.cambridge.org/aerodynamics.

References

Lamar, J.E., "A Vortex Latice Method for the Mean Camber Shapes of Trimmed Non-Coplanar Planforms with Minimum Vortex Drag," NASA TN D-8090, June 1976.

Mason, W.H., "Wing-Canard Aerodynamics at Transonic Speeds – Fundamental Considerations on Minimum Drag Spanloads," AIAA Paper 82-0097, January 1982.

D.8 *FRICTION*

FRICTION provides an estimate of laminar and turbulent skin friction suitable for use in aircraft preliminary design. It is an entirely original program, but has its roots in a program by Ron Hendrickson at Grumman. The input requires geometric information and either the Mach and altitude combination, or the Mach and Reynolds number at which the results are desired. The skin friction is found by summing up the contributions of each component as discussed in Section 4.5.2. The approach uses the Eckert Reference Temperature method for laminar flow and the van Driest II formula for turbulent flow with compressibility effects. The basic formulas are valid from subsonic to hypersonic speeds, but the implementation makes assumptions that limit the validity to moderate supersonic speeds (about Mach 3). The key assumption is that the vehicle surface is at the adiabatic wall temperature (the user can easily modify this assumption). Form factors are used to estimate the effect of thickness on drag, and a composite formula is used to include the effect of a partial run of laminar flow. Because the methods are not described in detail in the text, details are provided here.

Laminar flow

The approach used is known as the Eckert Reference Temperature Method, and this particular version is the one given by F.M. White in *Viscous Fluid Flow*, New York: McGraw-Hill, 1974, pp. 589–590. In this method the incompressible skin friction formula is used, with the fluid properties chosen at a specified reference temperature, which includes both Mach number and wall temperature effects.

First, assumptions are made for the fluid properties: Prandtl number, $Pr = 0.72$, Recovery factor, $r = Pr^{1/2}$, specific heat ratio, $\gamma = 1.4$, and edge temperature, $T_e = 390$ (°R). Then, for a given edge Mach number, M_e, and ratio of wall temperature to adiabatic wall temperature T_W/T_{AW}; compute:

$$\frac{T_W}{T_e} = \frac{T_W}{T_{AW}}\left(1 + r\frac{\gamma - 1}{2}M_e^2\right). \tag{D.1}$$

Remember that

$$T_{AW} = T_e\left(1 + r\frac{\gamma - 1}{2}M_e^2\right) \tag{D.2}$$

and then compute the reference temperature:

$$\frac{T^*}{T_e} \cong .5 + .039M_e^2 + 0.5\left(\frac{T_W}{T_e}\right). \tag{D.3}$$

The Chapman-Rubesin constant based on the reference temperature and Sutherland's viscosity law is then computed from:

$$C^* = \left(\frac{T^*}{T_e}\right)^{1/2}\left(\frac{1 + K/T_e}{T^*/T_e + K/T_e}\right) \tag{D.4}$$

where $K = 200$°R for air.

Finally, the local friction coefficient (τ_w/q) is found from the standard Blasius formula, with C^* added,

$$C_f = \frac{.664\sqrt{C^*}}{\sqrt{Re_x}} \tag{D.5}$$

and

$$C_F = 2C_f \tag{D.6}$$

which comes from

$$C_F = \frac{F}{qx} = \frac{1}{x} \int_{x'=0}^{x'=x} C_f(x')dx'. \tag{D.7}$$

Recall that C_F accounts for one side of the plate only, so that if both sides are required for a drag estimate, then the skin friction coefficient, C_D, is twice C_F because the reference area is based on one side only, i.e., $S_{ref} \approx 1/2\ S_{wet}$.

Note that the results are not sensitive to the value of edge temperature for low Mach numbers, and therefore, an exact specification of T_e is not required. This method is implemented in subroutine *lamcf*.

Turbulent flow

For turbulent flow the van Driest II method is employed. This method was selected based on the recommendation of Hopkins, E.J., and Inouye, M., "An Evaluation of Theories for Predicting Turbulent Skin Friction and Heat Transfer on Flat Plates at Supersonic and Hypersonic Mach Numbers," *AIAA Journal*, Vol. 9, No. 6, 1971, pp. 993–1003. The particular algorithm is taken from NASA TN D-6945, "Charts for Predicting Turbulent Skin Friction from the Van Driest Method (II)," also by E.J. Hopkins, October 1972.

Again, assumptions are made for the fluid properties: turbulent flow recovery factor, $r = 0.88$, specific heat ratio, $\gamma = 1.4$, and edge temperature, $T_e = 222$ (°K). Then, the calculation begins for a given edge Mach number, M_e, and ratio of wall temperature to adiabatic wall temperature T_W/T_{AW} the calculation is started by computing the following constants: The analysis proceeds using barred quantities to denote "incompressible" variables, which are intermediate variables not used except to obtain the final results. Given the Reynolds number, Re_x, an iteration is used to obtain the final results. Proceed as follows, finding

$$\overline{Re}_x = F_x Re_x. \tag{D.8}$$

Now solve

$$\frac{.242}{\sqrt{\overline{C}_F}} = \log\left(\overline{Re}_x \overline{C}_F\right) \tag{D.9}$$

for $\overline{C}_F$.

Use as an initial guess

$$\overline{C}_F^0 = \frac{.074}{\overline{Re}_x^{.20}}. \tag{D.10}$$

Then, Newton's method is applied to the problem:

$$f\left(\bar{C}_F\right) = 0 \Rightarrow \bar{C}_F^{i+1} = \bar{C}_F^i - \frac{f}{f'},\tag{D.11}$$

which becomes for this equation:

$$\bar{C}_F^{i+1} = \bar{C}_F^i \left[1 + \frac{\left\{242 - \sqrt{\bar{C}_F^i}\,\log\left(Re_x \bar{C}_F^i\right)\right\}}{\left\{121 + \sqrt{\bar{C}_F^i}\,/\ln 10\right\}}\right].\tag{D.12}$$

Once this iteration is completed, and $\bar{C}_F$ is known,

$$C_F = \frac{\bar{C}_F}{F_c}.\tag{D.13}$$

Note that this value applies to one side of a plate only, so it must be doubled if the friction on both sides is desired to account for the proper reference areas. Here again, the results are not sensitive to the value of edge temperature for low Mach numbers, and the default value should be adequate for most cases. This formula is implemented in routine *turbcf*.

Composite formula

When the flow is laminar and then transitions to turbulent, an estimate of the skin friction is available from a composite of the laminar and turbulent skin friction formulas using Schlicting's formula (see Cebeci, T., and Bradshaw, P., *Momentum Transfer in Boundary Layers*, New York: McGraw-Hill, 1977, pp. 187). Given the transition position, x_c/L and Re_L, compute

$$Re_c = \left(\frac{x_c}{L}\right) Re_L\tag{D.14}$$

and compute the laminar skin friction based on Re_c and the turbulent skin friction twice, based on both Reynolds numbers and then find the value that includes both laminar and turbulent flow from:

$$C_F = C_{F_{TURB}}\left(Re_L\right) - \left(\frac{x_c}{L}\right)\left[C_{F_{TURB}}\left(Re_c\right) - C_{F_{LAM}}\left(Re_c\right)\right]\tag{D.15}$$

Several formulas are available, are all roughly equivalent, and have been evaluated extensively for incompressible flow. They are only approximate for compressible flow.

Form factors

To include the effects of thickness, it has been found that the skin friction formulas should be adjusted through the use of form factors. Two different factors are used in this code. For wing-like shapes,

$$FF = 1.0 + 1.8\left(\frac{t}{c}\right) + 50\left(\frac{t}{c}\right)^4 \qquad (D.16)$$

where t/c is the thickness ratio of a particular component. For bodies,

$$FF = 1.0 + 1.5\left(\frac{d}{l}\right)^{1.5} + 50\left(\frac{d}{l}\right)^3 \qquad (D.17)$$

where d/l is the ratio of diameter to length. This is the reciprocal of the fineness ratio. The program runs interactively, and a sample session, with inputs and outputs, is provided at www.cambridge.org/aerodynamics.

D.9 *VLMpc*

This is John Lamar's two surface vortex lattice program, developed at NASA Langley Research Center. The program treats two lifting surfaces using up to 200 panels. Vortex flows are estimated using the leading edge suction analogy. The theory and program description are available in:

Margason, R.J., and Lamar, J.E., "Vortex-Lattice FORTRAN Program for Estimating Subsonic Aerodynamic Characteristics of Complex Planforms," NASA TN D-6142, February 1971.

Lamar, J.E., and Gloss, B. B. "Subsonic Aerodynamic Characteristics of Interacting Lifting Surfaces with Separated Flow around Sharp Edges Predicted by a Vortex-Lattice Method," NASA TN D-7921, September, 1975.

The input and output from program *VLMpc* is quite lengthy and can be found at www.cambridge.org/aerodynamics.

D.10 *DESCAM*

This program provides the camber line required to obtain the user input chord loading distribution for two-dimensional incompressible flow using thin airfoil theory. Lan's quasi vortex lattice method is used, as described in Section 5.3.4. The program runs interactively, and a sample session, with inputs and outputs, is provided at www.cambridge.org/aerodynamics.

D.11 *TRIDAG*

The Thomas Algorithm is a special form of Gaussian elimination that can be used to solve tridiagonal systems of equations. The algorithm takes advantage of the large number of zeros in the matrix to start performing back substitution in order to find the unknown values. When the matrix is tridiagonal, the solution can be obtained in $O(n)$ operations, instead of $O(n^3/3)$. The form of the equation for a matrix system $[A][x]=[d]$ (as given in Equation 6.131), where $[A]$ is the coefficient matrix, $[x]$ is the vector of unknowns, and $[d]$ is the vector of constants is:

$$a_i x_{i-1} + b_i x_i + c_i x_{i+1} = d_i \quad i = 1,\ldots,k,\ldots,n \tag{D.18}$$

where a_1 and c_n are zero and b_1 and b_n can include boundary conditions. The solution algorithm, given by Conte, S.D., and deBoor, C., *Elementary Numerical Analysis*, New York: McGraw-Hill, 1972, starts with $k = 2,\ldots,n$:

$$m = \frac{a_k}{b_{k-1}} \tag{D.19}$$

$$b_k = b_k - mc_{k-1} \tag{D.20}$$

$$d_k = d_k - mc_{k-1} \tag{D.21}$$

Then

$$x_n = \frac{d_n}{b_n} \tag{D.22}$$

and finally, for $k = (n-1),\ldots,1$:

$$x_k = \frac{d_k - c_k x_{k-1}}{b_k} \tag{D.23}$$

In CFD methods this algorithm is usually coded directly into the solution procedure, unless machine optimized subroutines are employed on a specific computer. A sample FORTRAN program to implement this algorithm is given here:

```
subroutine tridag(a,b,c,d,nn)
c    solves a tridiagonal system using the Thomas Algorithm
c    there are nn equations, in the tridiagonal form:
c    a(i)*x(i-1) + b(i)*x(i) + c(i)*x(i+1) = d(i)
c    here, a(1) and c(nn) are assumed 0, and ignored
c    x is returned in d, b is altered
c    code set up to run on WATFOR-77
```

```
c   w.h. mason, April 10, 1992

    dimension a(nn),b(nn),c(nn),d(nn)

    if(nn .eq. 1)           then
                 d(1)=d(1)/b(1)
                 return
                 end if
    do 10 k = 2,nn
    km1      = k - 1
    if(b(k-1) .eq. 0.0)    then
                 write(6,100) km1
                 stop
                 end if
    xm       = a(k)/b(km1)
    b(k)     = b(k) - xm*c(km1)
    d(k)     = d(k) - xm*d(km1)
     10 continue

    d(nn)    = d(nn)/b(nn)

    k        = nn
    do 20 i = 2,nn
    k        = nn + 1 - i
    d(k)     = (d(k) - c(k)*d(k+1))/b(k)
     20 continue

    return

    100 format(/3x,'diagonal element .eq. 0 in tridag at k = ',i2/)

    end
```

A check can be made using the following main program and resulting output:

```
c  main program to check the Tridiagonal system solver
   dimension a(20),b(20),c(20),d(20)
   n        = 10
   do 10 i= 1,n
   a(i)     = -1.
   b(i)     =  2.
   c(i)     = -1.
 10 d(i)               =        0.
   d(1)     =        1.
   call tridag(a,b,c,d,n)
   write(6,610) (i,d(i), i = 1,n)
 610 format(i5,e15.7)
```

```
        stop
        end
```

The results are:

```
1          0.9090909E+00
2          0.8181819E+00
3          0.7272728E+00
4          0.6363637E+00
5          0.5454546E+00
6          0.4545454E+00
7          0.3636363E+00
8          0.2727273E+00
9          0.1818182E+00
10         0.9090909E-01
```

D.12 Freeware, Shareware, Open Source, and Other Programs

No endorsement is implied by including these programs in this appendix.
NASA Glenn Research Center educational software:
   http://www.grc.nasa.gov/WWW/k-12/freesoftware_page.htm

MIT software from Mark Drela, including a wide variety of aerodynamics
program releases via the GNU General Public License. Programs include
ASWING, AVL, DFD, DSOPT, HPA, MISES, MSES, QPROP, XFOIL,
and *XROTOR*:
   http://web.mit.edu/drela/Public/web/xfoil/
   http://web.mit.edu/drela/Public/web/avl/
   http://web.mit.edu/drela/Public/web/aswing/
   http://web.mit.edu/drela/Public/web/qprop/

CEASIOM (Computerised Environment for Aircraft Synthesis and
Integrated Optimization Methods) which was developed for the European
Union by a consortium of organizations and universities. Includes *DATCOM*,
the *Tornado* vortex lattice program, and the inviscid version of the *Edge*
CFD code:
   http://www.ceasiom.com/

PABLO: a potential flow airfoil code with a one-way coupled boundary-layer
program:
   http://www.nada.kth.se/~chris/pablo/pablo.html

Open Aerospace: a collection of open source aerospace programs, including
the *OpenFOAM* Navier-Stokes solver:
   https://www.open-aerospace.org/

OpenFOAM is also available at: http://www.openfoam.com/

Stanford University Unstructured Solver SU2: a free Navier-Stokes solver:
http://su2.stanford.edu/doxygen/group___navier___stokes___equations.
html

Mississippi State University SimCenter Software: includes *SolidMesh* and various flow solvers:
http://www.simcenter.msstate.edu/software.php

Public Domain Aeronautical Software (PDAS): a collection of public domain software packaged and documented on a DVD; includes a wide variety of programs from NASA and other sources:
http://www.pdas.com/

Desktop Aeronautics: a wide variety of airfoil, wing, and aircraft design software:
http://www.desktop.aero/index.php
Martin Hepperle Aero Tooks: includes airfoil and propeller applications:
http://www.mh-aerotools.de/airfoils/index.htm
Hanley Innovations: airfoil and wing analysis programs:
http://www.hanleyinnovations.com/
Nielsen Engineering missile aerodynamics codes:
http://www.nearinc.com/
CFD Online software list:
http://www.cfd-online.com/Links/soft.html

Software for Aerodynamics and Aircraft Design: Bill Mason's software website at Virginia Tech:
http://www.dept.aoe.vt.edu/~mason/Mason_f/MRsoft.html

engAPPLETS: a variety of fluid dynamic programs, including instructions, examples, and software from Virginia Tech:
http://www.engapplets.vt.edu/

Links to other aerospace software will be included at the book website www.
cambridge.org/aerodynamics

Structured Grid Transformations

E.1 The Role of Coordinate Systems and How Coordinate Systems Affect the Governing Equations

After a structured grid is constructed, we associate grid lines i, j, k with a coordinate system ξ, η, ζ via a transformation, as was discussed in Section 7.6. Now we have to consider the governing equations corresponding to this transformed coordinate system. You do not simply apply the governing equations defined in a Cartesian coordinate system to the new coordinate system. The coordinate system associated with an arbitrary grid is called a *generalized curvilinear coordinate system*, and the resulting governing equations are more complicated than the Cartesian coordinate system form described in Section 3.5.3. Consider the following two-dimensional example, starting with:

$$\begin{aligned} \xi &= \xi(x, y) \\ \eta &= \eta(x, y). \end{aligned} \qquad (E.1)$$

Relate derivatives along ξ, η, to x, y using the chain rule:

$$\frac{\partial}{\partial x} = \frac{\partial \xi}{\partial x}\frac{\partial}{\partial \xi} + \frac{\partial \eta}{\partial x}\frac{\partial}{\partial \eta} \qquad (E.2)$$

and now introduce the partial derivative subscript notation:

$$\frac{\partial}{\partial x} = \xi_x \frac{\partial}{\partial \xi} + \eta_x \frac{\partial}{\partial \eta}. \qquad (E.3)$$

For the y derivative, this becomes:

$$\frac{\partial}{\partial y} = \xi_y \frac{\partial}{\partial \xi} + \eta_y \frac{\partial}{\partial \eta}. \qquad (E.4)$$

We now apply this to an example equation, given in the physical plane as:

$$\frac{\partial u}{\partial x} + a\frac{\partial u}{\partial x} = 0. \tag{E.5}$$

Transforming the equation to the computational coordinates, we obtain:

$$\left(\xi_x + a\xi_y\right)\frac{\partial u}{\partial \xi} + \left(\eta_x + a\eta_y\right)\frac{\partial u}{\partial \eta} = 0 \tag{E.6}$$

which introduces a great deal of complications. So, why have we done this? These transformations are introduced to allow the boundary to be defined along a constant value of the coordinate line (which is why we constructed the grid in the manner we did), and for the finite difference approximations to be applied directly to our governing equation.

We mentioned earlier how important it might be to cast our equations in conservation law form (see Section 3.5.2), but what will happen to an equation in conservation form after applying such a transformation? Will the transformed equations still be in conservation form? We have some good luck here since there is a mathematical theorem that answers our question.

Fundamental Theorem of Lapidus[1]

Under non-singular space transformation, conservation laws are transformed into conservation laws.

In the 1970s, theoreticians derived the fluid flow conservation equations in general curvilinear coordinates, and these equations form the basis for many computer codes that are in use today. A good explanation of these equations is given by Pletcher et al.[2] and Fletcher.[3]

In considering the governing equations, the transformation used in Equation E.1 continues to be important:

$$\xi = \xi(x, y), \quad \eta = \eta(x, y). \tag{E.1}$$

This leads to the following relations:

$$\left.\begin{array}{l} d\xi = \xi_x dx + \xi_y dy \\ d\eta = \eta_x dx + \eta_y dy \end{array}\right\} \tag{E.7}$$

or

$$\begin{Bmatrix} d\xi \\ d\eta \end{Bmatrix} = \begin{bmatrix} \xi_x & \xi_y \\ \eta_x & \eta_y \end{bmatrix} \begin{Bmatrix} dx \\ dy \end{Bmatrix}. \tag{E.8}$$

The terms ξ_x, ξ_y, η_x, η_y are called the *metrics of the transformation*.

We can also relate the x- and y-coordinates to ξ, η as:

$$x = x(\xi, \eta), \quad y = y(\xi, \eta) \tag{E.9}$$

so that:

$$\begin{Bmatrix} dx \\ dy \end{Bmatrix} = \begin{bmatrix} x_\xi & x_\eta \\ y_\xi & y_\eta \end{bmatrix} \begin{Bmatrix} d\xi \\ d\eta \end{Bmatrix}. \tag{E.10}$$

We are interested in the relations between these systems since we can usually compute the metrics of x and y more easily than the other system. Thus, we use the following equality to find the metrics:

$$\begin{bmatrix} \xi_x & \xi_y \\ \eta_x & \eta_y \end{bmatrix} = \underbrace{\begin{bmatrix} x_\xi & x_\eta \\ y_\xi & y_\eta \end{bmatrix}}_{\substack{\text{usually know} \\ \text{these}}}^{-1} \tag{E.11}$$

and the metrics are given by:

$$\begin{aligned} \xi_x &= Jy_\eta \\ \xi_y &= -Jx_\eta \\ \eta_x &= -Jy_\xi \\ \eta_y &= Jx_\xi \end{aligned} \tag{E.12}$$

where

$$J = \frac{1}{x_\xi y_\eta - y_\xi x_\eta} \tag{E.13}$$

is the *Jacobian of the transformation*. The situation becomes much more complicated in three spatial dimensions. A few things should be noted, however, before the complexity of the equations becomes too daunting: (1) orthogonality means a lot of these terms are zero, and (2) metrics should be computed using a discretization consistent with the discretization of derivatives used in the flow solver (most flow solvers now read in the grid points and compute the metrics internally using finite differencing).

The general three-dimensional unsteady form of the equations is developed using the following generalized coordinate transformation:

$$\begin{aligned} \xi &= \xi(x,y,z,t) \\ \eta &= \eta(x,y,z,t) \\ \zeta &= \zeta(x,y,z,t) \\ \tau &= t \end{aligned} \tag{E.14}$$

with the Jacobian of the transformation,

$$\begin{aligned} J - \frac{\partial(\xi,\eta,\zeta)}{\partial(x,y,z)} \\ = \left[x_\xi y_\eta z_\zeta + x_\zeta y_\xi z_\eta + x_\eta y_\zeta z_\xi - x_\xi y_\zeta z_\eta - x_\eta y_\xi z_\zeta - x_\zeta y_\eta z_\xi \right]^{-1}. \end{aligned} \tag{E.15}$$

The velocities are computed in terms of the so-called contravariant velocities (introduced to simplify the following expressions):

$$U = \xi_x u + \xi_y v + \xi_z w + \xi_t$$
$$V = \eta_x u + \eta_y v + \eta_z w + \eta_t \quad \text{(E.16)}$$
$$W = \zeta_x u + \zeta_y v + \zeta_z w + \zeta_t$$

and the metric terms are:

$$\frac{\xi_x}{J} = y_\eta z_\zeta - z_\eta y_\zeta, \quad \frac{\eta_x}{J} = z_\xi y_\zeta - y_\xi z_\zeta, \quad \frac{\zeta_x}{J} = y_\xi z_\eta - z_\xi y_\eta$$

$$\frac{\xi_y}{J} = z_\eta x_\zeta - x_\eta z_\zeta, \quad \frac{\eta_y}{J} = x_\xi z_\zeta - z_\xi x_\zeta, \quad \frac{\zeta_y}{J} = z_\xi x_\eta - x_\xi z_\eta \quad \text{(E.17)}$$

$$\frac{\xi_z}{J} = x_\eta y_\zeta - y_\eta x_\zeta, \quad \frac{\eta_z}{J} = y_\xi x_\zeta - x_\xi y_\zeta, \quad \frac{\zeta_z}{J} = x_\xi y_\eta - y_\xi x_\eta$$

and

$$\xi_t = -x_\tau \xi_x - y_\tau \xi_y - z_\tau \xi_z$$
$$\eta_t = -x_\tau \eta_x - y_\tau \eta_y - z_\tau \eta_z \quad \text{(E.18)}$$
$$\zeta_t = -x_\tau \zeta_x - y_\tau \zeta_y - z_\tau \zeta_z$$

where x_τ, y_τ, z_τ, are grid speeds in the x, y, z directions, respectively.

A standard form of the governing equations is used in the literature for numerical solutions of the Navier-Stokes equations. We will provide a representative set that was used in the NASA Langley Research Center code *CFL3D* and the NASA Ames Research Center code *ARC3D*. The Navier-Stokes equations (and the other equations required in the system) are written in vector divergence form (see Section 3.5.3) for generalized curvilinear coordinates as:

$$\frac{\partial \hat{Q}}{\partial t} + \frac{\partial \left(\hat{F} - \hat{F}_v \right)}{\partial \xi} + \frac{\partial \left(\hat{G} - \hat{G}_v \right)}{\partial \eta} + \frac{\partial \left(\hat{H} - \hat{H}_v \right)}{\partial \zeta} = 0 \quad \text{(E.19)}$$

where

$$\hat{Q} = \frac{Q}{J}$$

$$\hat{F} - \hat{F}_v = \frac{|\nabla \xi|}{J} \left[\xi_x \left(F - F_v \right) + \xi_y \left(G - G_v \right) + \xi_z \left(H - H_v \right) + \xi_t Q \right]$$

$$\hat{G} - \hat{G}_v = \frac{|\nabla \eta|}{J} \left[\eta_x \left(F - F_v \right) + \eta_y \left(G - G_v \right) + \eta_z \left(H - H_v \right) + \eta_t Q \right] \quad \text{(E.20)}$$

$$\hat{H} - \hat{H}_v = \frac{|\nabla \zeta|}{J} \left[\zeta_x \left(F - F_v \right) + \zeta_y \left(G - G_v \right) + \zeta_z \left(H - H_v \right) + \zeta_t Q \right]$$

and

$$\left(\hat{k}_x, \hat{k}_y, \hat{k}_z, \hat{k}_t\right) = \left(k_x, k_y, k_z, k_t\right) / |\nabla k|$$
$$|\nabla k| = \left[k_x^2 + k_y^2 + k_z^2\right]^{1/2}. \tag{E.21}$$

The conserved variables were given by Equation 3.64 as:

$$Q = \begin{Bmatrix} \rho \\ \rho u \\ \rho v \\ \rho w \\ E_t \end{Bmatrix} = \begin{Bmatrix} \text{density} \\ \text{x-momentum} \\ \text{y-momentum} \\ \text{z-momentum} \\ \text{total energy per unit volume} \end{Bmatrix}. \tag{3.64}$$

The inviscid flux terms in curvilinear coordinates are:

$$\hat{F} = \frac{1}{J} \begin{bmatrix} \rho U \\ \rho U u + \xi_x p \\ \rho U v + \xi_y p \\ \rho U w + \xi_z p \\ (E_t + p)U - \xi_t p \end{bmatrix}$$

$$\hat{G} = \frac{1}{J} \begin{bmatrix} \rho V \\ \rho V u + \eta_x p \\ \rho V v + \eta_y p \\ \rho V w + \eta_z p \\ (E_t + p)V - \eta_t p \end{bmatrix} \tag{E.22}$$

$$\hat{H} = \frac{1}{J} \begin{bmatrix} \rho W \\ \rho W u + \zeta_x p \\ \rho W v + \zeta_y p \\ \rho W w + \zeta_z p \\ (E_t + p)W - \zeta_t p \end{bmatrix}$$

and the viscous terms are given by:

$$\hat{F}_v = \frac{1}{J} \begin{bmatrix} 0 \\ \xi_x \tau_{xx} + \xi_y \tau_{xy} + \xi_z \tau_{xz} \\ \xi_x \tau_{yx} + \xi_y \tau_{yy} + \xi_z \tau_{yz} \\ \xi_x \tau_{zx} + \xi_y \tau_{zy} + \xi_z \tau_{zz} \\ \xi_x b_x + \xi_y b_y + \xi_z b_z \end{bmatrix}$$

$$\hat{G}_v = \frac{1}{J}\begin{bmatrix} 0 \\ \eta_x\tau_{xx} + \eta_y\tau_{xy} + \eta_z\tau_{xz} \\ \eta_x\tau_{yx} + \eta_y\tau_{yy} + \eta_z\tau_{yz} \\ \eta_x\tau_{zx} + \eta_y\tau_{zy} + \eta_z\tau_{zz} \\ \eta_x b_x + \eta_y b_y + \eta_z b_z \end{bmatrix} \tag{E.23}$$

$$\hat{H}_v = \frac{1}{J}\begin{bmatrix} 0 \\ \zeta_x\tau_{xx} + \zeta_y\tau_{xy} + \zeta_z\tau_{xz} \\ \zeta_x\tau_{yx} + \zeta_y\tau_{yy} + \zeta_z\tau_{yz} \\ \zeta_x\tau_{zx} + \zeta_y\tau_{zy} + \zeta_z\tau_{zz} \\ \zeta_x b_x + \zeta_y b_y + \zeta_z b_z \end{bmatrix}$$

where b is written in indicial notation as

$$b_{x_i} = u_j\tau_{x_i x_j} - \dot{q}_{x_i}. \tag{E.24}$$

These equations should be compared with the Cartesian form presented in Section 3.5.3 to see how the transformation has added complexity to various terms. For example, the x momentum equation viscous terms in Cartesian coordinates are given by (from Equation 3.65):

$$\frac{\partial}{\partial x}(\tau_{xx}). \tag{E.25}$$

However, in the generalized curvilinear coordinate system the corresponding term is (Equation E.23):

$$\frac{\partial}{\partial \xi}\left[\frac{1}{J}\left(\xi_x\tau_{xx} + \xi_y\tau_{xy} + \xi_z\tau_{xz}\right)\right]. \tag{E.26}$$

which is much more complicated and requires more detailed coding (and remember, this is just one term of one equation!).

E.2 References

1   Lapidus, A., "A Detached Shock Calculation by Second Order Finite Differences," *Journal of Computational Physics*, Vol. 2, 1967, pp. 154–177.

2   Pletcher, R.H., Tannehill, J.C., and Anderson, D.A., *Computational Fluid Mechanics and Heat Transfer*, 3rd Ed., Boca Raton: Taylor and Francis, 2011.

3   Fletcher, C.A.J., *Computational Techniques for Fluid Dynamics*, Vol. II, Berlin: Springer-Verlag, 1988.

Commonly Used Turbulence Models

F.1 Introduction

Several RANS turbulence models, along with their empirical constants, are presented in this appendix. These models should always be validated and tested prior to being used so that you have a good understanding of their limitations and capabilities. Only RANS models are given here, since various hybrid models are usually extensions of RANS models, and LES models are beyond the scope of this book.

The effects of turbulence in the Reynolds-Averaged Navier-Stokes equations is included by an eddy viscosity μ_t which accounts for the increased mixing in a turbulent flow. This eddy viscosity is then used to define the Reynolds stress terms that are the results of the averaging process (see Chapter 8 for more details). Once the eddy viscosity and Reynolds stress terms are defined, a turbulent Prandtl number may also be specified as $Pr_t = c_p\mu_t / k_t$. The turbulent thermal conductivity coefficient, k_t, is then calculated from the turbulent viscosity and turbulent Prandtl number. For example, in the case of a two-dimensional formulation:

$$\tau_{xx} = (\mu + \mu_t)\left(\frac{4}{3}\frac{\partial u}{\partial x} - \frac{2}{3}\frac{\partial v}{\partial y}\right) \tag{F.1}$$

$$\tau_{yy} = (\mu + \mu_t)\left(\frac{4}{3}\frac{\partial v}{\partial y} - \frac{2}{3}\frac{\partial u}{\partial x}\right) \tag{F.2}$$

$$\tau_{xy} = \tau_{yx} = (\mu + \mu_t)\left(\frac{\partial u}{\partial y} + \frac{\partial v}{\partial x}\right) \tag{F.3}$$

$$q_x = -(k + k_t)\frac{\partial T}{\partial x} \tag{F.4}$$

$$q_x = -(k + k_t)\frac{\partial T}{\partial y}. \tag{F.5}$$

An experimentally determined value for the Prandtl number, $Pr_t = 0.9$, is often used. In addition, all turbulence models have constants that are determined by comparison with experimental data. The data used can lead to implicit limitations of the model.

F.2 A Sample Algebraic Model: Cebeci-Smith

One of the first useful algebraic turbulence models was proposed in 1967 for attached, turbulent boundary layers and is known as the Cebeci-Smith model.[1] It splits the turbulent viscosity into an inner (near wall) and outer (wake) region and switches to the outer viscosity at the first point away from the wall where $\mu_{t_i} = \mu_{t_o}$. The inner turbulent viscosity is given by:

$$\mu_{t_i} = \rho l_{mix}^2 \left[\left(\frac{\partial u}{\partial y} \right)^2 + \left(\frac{\partial v}{\partial x} \right)^2 \right]^{1/2}. \tag{F.6}$$

The length scale is determined by the distance from the wall with a damping function that decreases the turbulent viscosity in the viscous sublayer:

$$l_{mix} = \kappa \left(1 - e^{-y^+/A^+} \right) y. \tag{F.7}$$

where $y+$ is defined in Eqn. 8.1, and κ is the Von Karman constant defined in Table F.1.

In the outer region the viscosity is determined by:

$$\mu_{t_o} = \alpha \rho U_e \delta_v^* F_{Kleb} \tag{F.8}$$

where U_e is the velocity at the edge of the boundary layer, the Klebanoff intermittency factor is:

$$F_{Kleb} = \left[1 + 5.5 \left(\frac{C_{Kleb} y}{y_{max}} \right)^6 \right]^{-1}. \tag{F.9}$$

and δ_v^* is the velocity thickness (same as displacement thickness for incompressible flow) given by:

$$\delta_v^* = \int_0^\delta \left(1 - u/U_e \right) dy. \tag{F.10}$$

The closure coefficients used are given in Table F.1. The Cebeci-Smith model is very straightforward to code and use.

Table F.1 Cebeci-Smith Model Coefficients.

$\kappa = 0.41$	$C_{Kleb} = 0.3$	$\alpha = 0.0168$

$$A^+ = 26.0\left[1 + y\frac{dp/dx}{\rho u_\tau^2}\right]^{-1/2}$$

F.3 A Sample Algebraic Model: Baldwin-Lomax

The Baldwin-Lomax model[2] was first proposed in 1978 for RANS flow simulations with separation, yet it is still used today. Just as with the Cebeci-Smith model, it splits the turbulent viscosity into an inner (near wall) and outer (wake) region. It switches to the outer viscosity at the first point away from the wall where $\mu_{t_i} = \mu_{t_o}$. The inner turbulent viscosity is given by:

$$\mu_{t_i} = \rho l^2 |\omega| \tag{F.11}$$

where the vorticity is:

$$\omega = \frac{\partial v}{\partial x} - \frac{\partial u}{\partial y}. \tag{F.12}$$

The length scale is determined by the distance from the wall with a damping function that decreases the turbulent viscosity in the viscous sublayer:

$$l_i = \kappa\left(1 - e^{-y^+/A^+}\right)y. \tag{F.13}$$

In the outer region, the viscosity is determined by:

$$\mu_{t_o} = \alpha\rho C_{CP} F_{wake} F_{Kleb} \tag{F.14}$$

where the wake function is given by

$$F_{wake} = \min\left[y_{max}F_{max}; C_{wake}y_{max}\frac{(\Delta V)^2}{F_{max}}\right] \tag{F.15}$$

and

$$F_{max} = \max\left[\left(1 - e^{-y^+/A^+}\right)y|\omega|\right] \tag{F.16}$$

$$\Delta V = \left(u^2 + v^2\right)_{max}^{1/2}. \tag{F.17}$$

Table F.2 Baldwin-Lomax Model Coefficients.

$\kappa = 0.41$	$A^+ = 26.0$	$C_{CP} = 1.61$
$C_{wake} = 1.0$	$\alpha = 0.0168$	$C_{Kleb} = 0.3$

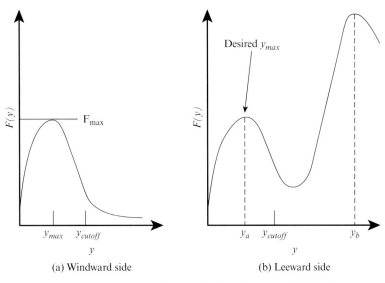

(a) Windward side   (b) Leeward side

Figure F.1   Variation of Baldwin-Lomax outer layer function, $F(y)$. (Reprinted from Ref. 3, © 1986, with permission from Elsevier).

Finally, the Klebanoff intermittency factor is:

$$F_{Kleb} = \left[1 + 5.5\left(\frac{C_{Kleb}\,y}{y_{max}}\right)^6\right]^{-1}. \tag{F.18}$$

The closure coefficients used are given in Table F.2.

While the Baldwin-Lomax model works well for attached flow, difficulties were encountered when applied to steady separated flow over forebodies at high angles of attack by Degani and Schiff.[3] The Baldwin-Lomax model contains a function, $F(y)$, that is used as part of the outer layer model:

$$F(y) = y\,|\omega|\left[1 - e^{-\left(y^+/A^+\right)}\right] \tag{F.19}$$

where y is the normal distance to a flat plate and ω is the vorticity magnitude. In an attached boundary layer, $F(y)$ increases to a maximum value and then decreases near the edge of the boundary layer. However, in a separated vortical flow layer, $F(y)$ attains a local maxima in the attached boundary layer and then reaches a global maxima in the separated layer (see Fig. F.1). The turbulence model chooses the highest $F(y)$ and its corresponding distance from the wall, which greatly overpredicts the turbulent viscosity in this region. The overpredicted turbulent viscosity creates nonphysical results when added to the laminar viscosity, altering separation locations and the proper formation of secondary and tertiary vortices.

Degani and Schiff proposed a modification to the model (and hence to all eddy-viscosity turbulence models) that obtains the correct value of $F(y)$ and gives better prediction of the flow topology in a separated flow region. This modification has led to the ability to accurately simulate steady vortical flows with RANS computations. An example of the improvements to both zero-equation and one-equation turbulence models for predicting vortical flowfields was done by Gee, Cummings, and Schiff.[4] Vortical flow modifications for the $k - \varepsilon$ turbulence model have also been suggested[5] and have been applied to flow over slender bodies at high incidence angles.[6] In spite of these improvements, however, the RANS-based equations still lead to poor predictions at high to very high angles of attack, where the flow is unsteady and the time-averaged equations are no longer capable of properly modeling the flowfield.

F.4 A Sample Half-Equation Model: Johnson-King

The Johnson-King model[7] was developed in 1984 as an eddy viscosity model with Reynolds stress model features. The model solves an ordinary differential equation, so it is known as a half-equation model. It also splits the turbulent viscosity into an inner and outer region but handles the changes in turbulence quantities with an ordinary differential equation (which is why the Johnson-King model is referred to as a "half-equation" model). The eddy viscosity is given by:

$$\mu_t = \mu_{t_o} \tanh\left(\mu_{t_i}/\mu_{t_o}\right). \tag{F.20}$$

The inner region viscosity is similar to the Cebeci-Smith and Baldwin-Lomax models, using a damped mixing length concept given by:

$$\mu_{t_i} = \rho\left[1 - \exp\left(-\frac{u_D y/v}{A^+}\right)\right]^2 \kappa u_s y \tag{F.21}$$

where

$$u_s = \sqrt{\rho_w/\rho}\, u_\tau\left(1 - \gamma_2\right) + \sqrt{\rho_m/\rho}\, u_m \gamma_2$$

$$\gamma_2 = \tanh\left(y/L_e\right)$$

$$L_e = \frac{\sqrt{\rho_w}\, u_\tau}{\sqrt{\rho_w}\, u_\tau + \sqrt{\rho_m}\, u_m}\, L_m \tag{F.22}$$

$$L_m = \begin{cases} \kappa y_m, & y_m / \delta \leq C_1 / \kappa \\ C_1 \delta, & y_m / \delta > C_1 / \kappa \end{cases}$$

$$u_m = \sqrt{\tau_m / \rho_m}$$

$$u_D = \max[u_m, u_\tau]$$

where the subscript m represents values at the point where $y = y_m$, which is where the Reynolds stress has the highest value. u_τ is the friction velocity defined in Equation 8.2, and ρ_w is the density at the wall.

The outer region viscosity is also similar to the Cebeci-Smith model, given by:

$$\mu_{t_o} = \alpha \rho U_e \delta_v^* F_{Kleb} \sigma(x) \tag{F.23}$$

where $\sigma(x)$ is the non-equilibrium parameter.

The model is comprised of three parts:

1. a non-equilibrium eddy viscosity distribution
2. a rate equation for the streamwise development of Reynolds stress
3. an equilibrium eddy viscosity distribution where production equals dissipation

The non-equilibrium eddy viscosity distribution is given by:

$$\frac{dg}{dx} = \frac{a_1}{2\bar{u}_m L_m} \left\{ \left(1 - \frac{g}{g_{eq}}\right) + \frac{C_{dif} L_m}{a_1 \delta[0.7 - (y/\delta)_m]} \left[1 - \left(\frac{v_{to}}{v_{to,eq}}\right)^{\frac{1}{2}}\right] \right\} \tag{F.24}$$

where g is the non-equilibrium eddy viscosity and this ODE needs to be solved in order to determine the distribution.

F.5 A Sample One-Equation Model: Spalart-Allmaras

The Spalart-Allmaras one-equation model[8] solves a single partial differential equation for a variable $\tilde{v}$ which is related to the turbulent viscosity. The differential equation is derived by "using empiricism and arguments of dimensional analysis, Galilean invariance and selected dependence on the molecular viscosity."[8] The model includes a wall destruction term that reduces the turbulent viscosity in the laminar sublayer and a trip term that provides a smooth transition from laminar to turbulent. The differential equation is:

$$\frac{D\tilde{v}}{Dt} = c_{b1}(1 - f_{t2})\tilde{S}\tilde{v} + \frac{1}{\sigma}\left[\nabla \cdot \left((v + \tilde{v})\nabla \tilde{v}\right) + c_{b2}\left(\nabla \tilde{v}\right)^2\right]$$
$$- \left[c_{w1} f_w - \frac{c_{b1}}{\kappa^2} f_{t2}\right]\left[\frac{\tilde{v}}{d}\right]^2 + f_{t1}\Delta U^2 \tag{F.25}$$

where the turbulent viscosity is determined by:

$$v_t = \tilde{v} f_{v1} \tag{F.26}$$

$$f_{v1} = \frac{\chi^3}{\chi^3 + c_{v1}^3} \tag{F.27}$$

$$\chi = \frac{\tilde{v}}{v}. \tag{F.28}$$

S is the magnitude of the vorticity, and the modified vorticity is:

$$\tilde{S} = S + \frac{\tilde{v}}{\kappa^2 d^2} f_{v2} \tag{F.29}$$

$$f_{v2} = 1 - \frac{\chi}{1 + \chi f_{v1}} \tag{F.30}$$

where d is the distance to the closest wall. The wall destruction function f_w is:

$$f_w = g \left[\frac{1 + c_{w3}^6}{g^6 + c_{w3}^6} \right]^{\frac{1}{6}} \tag{F.31}$$

$$g = r + c_{w2} \left(r^6 - r \right) \tag{F.32}$$

$$r = \frac{\tilde{v}}{\tilde{S}\kappa^2 d^2}. \tag{F.33}$$

Large values of r are truncated to 10. The trip functions are:

$$f_{t2} = c_{t3} \exp\left(-c_{t4}\chi^2\right) \tag{F.34}$$

$$f_{t1} = c_{t1} g_t \exp\left(-c_{t2} \frac{\omega_t^2}{\Delta U^2}\left[d^2 + g_t^2 d_t^2\right]\right) \tag{F.35}$$

where d_t is the distance from the field point to the trip (on the wall), ω_t is the wall vorticity at the trip, ΔU is the difference between the velocity at the field point and that at the trip, and $g_t \equiv \min\left(0.1, \frac{\Delta U}{\omega_t \Delta x}\right)$ where Δx is the grid spacing along the wall at the trip point. When the computations are fully turbulent, the trip functions may be set to zero (i.e., $c_{t1} = c_{t3} = 0$). The closure coefficients are given in Table F.3.

Table F.3 Spalart-Allmaras Model Coefficients

$c_{b1} = 0.1355$	$\sigma = \dfrac{2}{3}$	$c_{b2} = 0.622$	$\kappa = 0.41$
$c_{w1} = \dfrac{c_{b1}}{\kappa^2} + \dfrac{\left(1 + c_{b2}\right)}{\sigma}$	$c_{w2} = 0.3$	$c_{w3} = 2$	$c_{v1} = 7.1$
$c_{t1} = 1$	$c_{t2} = 2$	$c_{t3} = 1.1$	$c_{t4} = 2$

F.6 Spalart-Allmaras Turbulence Model with Rotation Correction

In order to more accurately model vortical flows, where the vorticity within the vortex can be quite difficult to model, a rotation correction of some sort is often applied. The Spalart-Allmaras (SA) turbulence model correction for rotating flows (Spalart-Allmaras with Rotation Correction, SARC) is based on concepts first proposed by Spalart and Shur[9] as well as Knight and Saffman.[10] The approach is "based on tracking the direction of the principal axes of the strain tensor and, thus, is both Galilean invariant and usable in a simple model."[11] The only difference between SARC and SA is that in the SARC model, the production term in the eddy viscosity transport equation is multiplied by the rotation factor f_{r1}:

$$
\begin{aligned}
f_{r1}\left(r^*, \tilde{r}\right) \\
= \left(1 + c_{r1}\right)\frac{2r^*}{1 + r^*}\left[1 - c_{r3}\tan^{-1}\left(c_{r2}\tilde{r}\right)\right] - c_{r1}.
\end{aligned}
\tag{F.36}
$$

If the variables and their derivatives are defined with respect to the rotating reference frame (rotating at rate Ω), the nondimensional quantities r^* and $\tilde{r}$ are given by:

$$
r^* = S / \omega
\tag{F.37}
$$

$$
\tilde{r} = 2\omega_{ik}S_{jk}\left[\frac{DS_{ij}}{Dt} + \left(\varepsilon_{jmn}S_{in}\right)\Omega_m\right] / D^4
\tag{F.38}
$$

where

$$
S_{ij} = 0.5\left(\frac{\partial u_i}{\partial x_j} + \frac{\partial u_j}{\partial x_i}\right)
\tag{F.39}
$$

$$
\omega_{ij} = 0.5\left[\left(\frac{\partial u_i}{\partial x_j} - \frac{\partial u_j}{\partial x_i}\right) + 2\varepsilon_{mji}\Omega_m\right]
\tag{F.40}
$$

and the constants are $c_{r1} = 1.0$, $c_{r2} = 12$, and $c_{r3} = 1.0$ (see Ref. 10 for details).

F.7 A Sample Two-Equation Model: Jones-Launder k-ε

The Jones-Launder $k - \varepsilon$ two-equation model[12] was proposed in 1973. It solves two partial differential equations for turbulent kinetic energy and turbulent dissipation. Kral et al.[13] provide the following $k - \varepsilon$ equations which can be used with other models besides the Jones-Launder:

$$\frac{D}{Dt}(\rho k) = \tau_{ij}\frac{\partial u_i}{\partial x_j} - \rho\left(1 + \alpha M_t^2\right)\hat{\varepsilon} + \frac{\partial}{\partial x_j}\left[\left(\mu + \frac{\mu_t}{\sigma_k}\right)\frac{\partial k}{\partial x_j}\right] + L_k \qquad \text{(F.41)}$$

$$\frac{D}{Dt}(\rho\hat{\varepsilon}) = c_{\varepsilon 1}f_1\tau_{ij}\frac{\partial u_i}{\partial x_j}\frac{\hat{\varepsilon}}{k} - c_{\varepsilon 2}f_2\rho\frac{\hat{\varepsilon}^2}{k} + \frac{\partial}{\partial x_j}\left[\left(\mu + \frac{\mu_t}{\sigma_\varepsilon}\right)\frac{\partial\hat{\varepsilon}}{\partial x_j}\right] + L_\varepsilon. \qquad \text{(F.42)}$$

The production term is given by:

$$\tau_{ij}\frac{\partial u_i}{\partial x_j} = \left[\mu_t\left(\frac{\partial u_i}{\partial x_j} + \frac{\partial u_j}{\partial x_i} - \frac{2}{3}\frac{\partial u_k}{\partial x_k}\delta_{ij}\right) - \frac{2}{3}\rho k\delta_{ij}\right]\frac{\partial u_i}{\partial x_j}. \qquad \text{(F.43)}$$

A compressibility term was included with a turbulent Mach number given by $M_t^2 = \dfrac{2k}{a^2}$. The turbulent viscosity is then determined from:

$$\mu_t = c_\mu f_\mu \rho\frac{k^2}{\hat{\varepsilon}}. \qquad \text{(F.44)}$$

The Jones-Launder model defines the following:

$$f_\mu = e^{\left(-\frac{2.5}{1 + 0.02\,Re_t}\right)}$$

$$f_1 = 1.0$$

$$f_2 = 1.0 - 0.3e^{-Re_t^2}$$

$$L_k = -2\mu\left(\frac{\partial\sqrt{k}}{\partial x_j}\right)^2 \qquad \text{(F.45)}$$

$$L_\varepsilon = \frac{2\mu\mu_t}{\rho}\left(\frac{\partial^2 u_i}{\partial x_j\partial x_i}\right)^2$$

$$Re_t = \frac{\rho k^2}{\mu\hat{\varepsilon}}.$$

The closure coefficients used are given in Table F.4.

Table F.4 Jones-Launder Model Coefficients

$C_\mu = 0.09$	$\sigma_k = 1.0$	$\sigma_\varepsilon = 1.3$
$C_{\varepsilon 1} = 1.44$	$C_{\varepsilon 2} = 1.92$	$\alpha = 1.0$

F.8 Another Sample Two-Equation Model: Menter's k-ω

Wilcox's $k - \omega$ model solves a transport equation for turbulent kinetic energy (k) and turbulent specific dissipation rate (ω).[14] This model is well behaved in the near wall region, where low Reynolds number corrections are not required. However, it is generally very sensitive to the freestream values of ω. On the other hand, the $k - \varepsilon$ equations are relatively insensitive to freestream values, but behave poorly in the near wall region.[15]

Menter proposed a combined $k - \varepsilon$, $k - \omega$ model that uses the best features of each model.[16] The model uses a parameter F_1 to switch from $k - \omega$ to $k - \varepsilon$ in the wake region to prevent the model from being sensitive to freestream conditions. The governing differential equations are:

$$\frac{D}{Dt}(\rho k) = \tau_{ij}\frac{\partial u_i}{\partial x_j} - \beta^* \rho \omega k + \frac{\partial}{\partial x_j}\left[(\mu + \sigma_k \mu_t)\frac{\partial k}{\partial x_j}\right] \qquad \text{(F.46)}$$

$$\frac{D}{Dt}(\rho \omega) = \frac{\gamma \rho}{\mu_t}\tau_{ij}\frac{\partial u_i}{\partial x_j} - \beta \rho \omega^2 + \frac{\partial}{\partial x_j}\left[(\mu + \sigma_\omega \mu_t)\frac{\partial \omega}{\partial x_j}\right]$$
$$+2\rho(1 - F_1)\sigma_{\omega 2}\frac{1}{\omega}\frac{\partial k}{\partial x_j}\frac{\partial \omega}{\partial x_j}. \qquad \text{(F.47)}$$

To compute the switching function F_1:

$$\text{arg}_1 = \min\left(\max\left(\frac{\sqrt{k}}{0.09\omega y};\frac{500\mu}{\rho \omega y^2}\right);\frac{4\rho\sigma_{\omega 2}k}{CD_{k\omega}y^2}\right) \qquad \text{(F.48)}$$

$$CD_{k\omega} = \max\left(2\rho\sigma_{\omega 2}\frac{1}{\omega}\frac{\partial k}{\partial x_i}\frac{\partial \omega}{\partial x_i};10^{-20}\right) \qquad \text{(F.49)}$$

$$F_1 = \tanh\left(\text{arg}_1^4\right). \qquad \text{(F.50)}$$

The production term $\tau_{ij}\dfrac{\partial u_i}{\partial x_j}$ is given by Equation F.43. The switching function also determines the value of the model constants. If φ_1 represents a generic constant of the $k - \omega$ equations, and φ_2 represents the same constant

Table F.5 Menter's Baseline Model Coefficients

Set 1 ($k-\omega$)

$\sigma_{k1}=0.5$	$\sigma_{\omega 1}=0.5$	$\beta_1=0.0750$
$\beta^*=0.09$	$\kappa=0.41$	$\gamma_1=\dfrac{\beta_1}{\beta^*}-\dfrac{\sigma_{\omega 1}\kappa^2}{\sqrt{\beta^*}}$

Set 2 ($k-\varepsilon$)

$\sigma_{k2}=1.0$	$\sigma_{\omega 2}=0.856$	$\beta_1=0.0828$
$\beta^*=0.09$	$\kappa=0.41$	$\gamma_1=\dfrac{\beta_2}{\beta^*}-\dfrac{\sigma_{\omega 2}\kappa^2}{\sqrt{\beta^*}}$

for the $k-\varepsilon$ equations, then the model constant used in Equations F.46 and F.47 is determined by:

$$\phi = F_1\phi_1 + (1-F_1)\phi_2. \tag{F.51}$$

The Baseline model uses the following relation to determine the turbulent viscosity:

$$\mu_t = \rho\frac{k}{\omega}. \tag{F.52}$$

The Baseline model constants are given in Table F.5.

Mentor also developed a Shear Stress Transport (SST) model to increase the accuracy for reversed flows. This model limits the turbulent shear stress to $\rho a_1 k$ where $a_1 = 0.31$. The turbulent viscosity is therefore given by:

$$\mu_t = \frac{\rho a_1 k}{\max\left(a_1\omega;\Omega F_2\right)} \tag{F.53}$$

where Ω is the absolute value of vorticity. The function F_2 is included to prevent singular behavior in the freestream where Ω goes to zero. F_2 is given by:

$$F_2 = \tanh\left(\arg_2^2\right) \tag{F.54}$$

$$\arg_2 = \max\left(\frac{2\sqrt{k}}{0.09\omega y};\frac{400\nu}{\omega y^2}\right). \tag{F.55}$$

The model constants were recalibrated, with the only change being in σ_{k1}, as shown in Table F.6.

Table F.6 Menter's Shear Stress Transport Model Coefficients

Set 1 $(k - \omega)$

$\sigma_{k1} = 0.85$	$\sigma_{\omega 1} = 0.5$	$\beta_1 = 0.0750$
$\beta^* = 0.09$	$\kappa = 0.41$	$\gamma_1 = \dfrac{\beta_1}{\beta^*} - \dfrac{\sigma_{\omega 1} \kappa^2}{\sqrt{\beta^*}}$

Set 2 $(k - \varepsilon)$

$\sigma_{k2} = 1.0$	$\sigma_{\omega 2} = 0.856$	$\beta_1 = 0.0828$
$\beta^* = 0.09$	$\kappa = 0.41$	$\gamma_1 = \dfrac{\beta_2}{\beta^*} - \dfrac{\sigma_{\omega 2} \kappa^2}{\sqrt{\beta^*}}$

Table F.7 Wilcox $k - \omega$ Incompressible, Fully Turbulent Model Coefficients

$\alpha_\infty = \dfrac{13}{25}$	$\alpha_\infty^* = 1$	$\sigma = \dfrac{1}{2}$	$\sigma^* = \dfrac{1}{2}$
$\beta_i = \beta_0 f_\beta$	$\beta_\infty^* = \beta_0^* f_{\beta^*}$	$\beta_0 = \dfrac{9}{125}$	$\beta_0^* = \dfrac{9}{100}$

F.9 A Third Sample Two-Equation Model: The Wilcox 1998 k-ω Model

Wilcox revised the original $k - \omega$ model to include added terms giving a more accurate spreading rate for round and radial jets, as well as planar shear layers.[14] The $k - \omega$ model solves a transport equation for turbulent kinetic energy (k) and turbulent specific dissipation rate (ω).

$$\frac{D}{Dt}(\rho k) = \tau_{ij} \frac{\partial u_i}{\partial x_j} - \beta^* \rho \omega k + \frac{\partial}{\partial x_j}\left[(\mu + \sigma^* \mu_t)\frac{\partial k}{\partial x_j}\right] \tag{F.56}$$

$$\frac{D}{Dt}(\rho \omega) = \alpha \frac{\omega}{k} \tau_{ij} \frac{\partial u_i}{\partial x_j} - \beta \rho \omega^2 + \frac{\partial}{\partial x_j}\left[(\mu + \sigma \mu_t)\frac{\partial \omega}{\partial x_j}\right] \tag{F.57}$$

where the turbulent shear stress is

$$\tau_{ij} = \alpha^*\left[\mu_t\left(\frac{\partial u_i}{\partial x_j} + \frac{\partial u_j}{\partial x_i} - \frac{2}{3}\frac{\partial u_k}{\partial x_k}\delta_{ij}\right) - \frac{2}{3}\rho k \delta_{ij}\right] \tag{F.58}$$

and the turbulent viscosity is found from

$$\mu_t = \rho \frac{k}{\omega}. \tag{F.59}$$

Incompressible, fully turbulent parameters are given in Table F.7.

Table F.8 Wilcox $k - \omega$ Low Reynolds Number Model Coefficients

$\alpha_0^* = \dfrac{\beta_i}{3}$	$\alpha_0 = \dfrac{1}{9}$	$Re_T = \dfrac{k}{\omega \nu}$
$R_\beta = 8$	$R_k = 6$	$R_\omega = 2.95$

The round jet and cross-diffusion corrections are

$$f_\beta = \frac{1 + 70\chi_\omega}{1 + 80\chi_\omega} \quad \chi_\omega = \left| \frac{\Omega_{ij} \Omega_{jk} S_{ki}}{(\beta_0^* \omega)^3} \right| \tag{F.60}$$

$$f_{\beta^*} = \begin{cases} 1; & \chi_k \leq 0 \\ \dfrac{1 + 680\chi_k^2}{1 + 400\chi_k^2}; & \chi_k > 0 \end{cases} \quad \chi_k = \frac{1}{\omega^3} \frac{\partial k}{\partial x_j} \frac{\partial \omega}{\partial x_j}. \tag{F.61}$$

If low Reynolds number corrections are not included, then:

$$\alpha^* = \alpha_\infty^* \quad \alpha = \alpha_\infty \quad \beta_i^* = \beta_\infty^*. \tag{F.62}$$

If low Reynolds number corrections are included:

$$\alpha^* = \alpha_\infty^* \frac{\alpha_0^* + Re_T / R_k}{1 + Re_T / R_k} \tag{F.63}$$

$$\alpha = \alpha_\infty \frac{\alpha_0 + Re_T / R_\omega}{1 + Re_T / R_\omega} \frac{1}{\alpha^*} \tag{F.64}$$

$$\beta_i^* = \beta_\infty^* \frac{4/15 + \left(Re_T / R_\beta \right)^4}{1 + \left(Re_T / R_\beta \right)^4} \tag{F.65}$$

where the low Reynolds number model coefficients are given in Table F.8.
The compressible values of β and β^* are

$$\beta = \beta_i \left[1 - \frac{\beta_i^*}{\beta_i} \xi^* F(M_t) \right] \tag{F.66}$$

$$\beta^* = \beta_i^* \left[1 + \xi^* F(M_t) \right] \tag{F.67}$$

with the compressibility function:

$$F(M_t) = \begin{cases} 0; & M_t \leq M_{t0} \\ M_t^2 - M_{t0}^2; & M_t > M_{t0} \end{cases} \tag{F.68}$$

where

$$M_t = \sqrt{\frac{2k}{a^2}} \tag{F.69}$$

$$M_{t0} = \frac{1}{4} \tag{F.70}$$

and a is the speed of sound.

F.10 NASA Langley Turbulence Modeling Resource

As we mentioned in Appendix B, NASA Langley Research Center maintains a collection of turbulence model codes and test cases (including grids and experimental data). The collection is done as part of the AIAA Fluid Dynamics Technical Committee working group on Turbulence Modeling Benchmarking. This is a valuable source of information if you are trying to improve your understanding of turbulence model usage in CFD. The web page is http://turbmodels.larc.nasa.gov/.

F.11 References

1 Smith, A.M.O., and Cebeci, T., "Numerical Solution of the Turbulent Boundary-Layer Equations," Douglas Aircraft Division Report DAC 33735, 1967.

2 Baldwin, B.S., and Lomax, H., "Thin Layer Approximation and Algebraic Model for Separated Turbulent Flows," AIAA Paper 78-257, January 1978.

3 Degani, D., and Schiff, L., "Computation of Turbulent Flows Around Pointed Bodies Having Crossflow Separation," *Journal of Computational Physics*, Vol. 66, No. 1, 1986, pp. 173–196.

4 Gee, K., Cummings, R., and Schiff, L., "Turbulence Model Effects on Separated Flow About a Prolate Spheroid," *AIAA Journal*, Vol. 30, 1992, pp. 655–664.

5 Dacles-Mariani, J., Zilliac, G., Chow, J., and Bradshaw, P., "Numerical/ Experimental Study of a Wingtip Vortex in the Near Field," *AIAA Journal*, Vol. 33, No. 9, 1995, pp. 1561–1568.

6 Josyula, F., "Computational Simulation Improvements of Supersonic High-Angle-of-Attack Missile Flows," *Journal of Spacecraft and Rockets*, Vol. 36, No. 1, 1999, pp. 59–66.

7 Johnson, D.A., and King, L.S., "A Mathematically Simple Turbulence Closure Model for Attached and Separated Turbulent Boundary Layers," *AIAA Journal*, Vol. 23, No. 11, pp. 1684–1692.

8 Spalart, P.R., and Allmaras, S.R., "A One-Equation Turbulence Model for Aerodynamic Flows," AIAA Paper 92-0439, January 1992.

9 Spalart, P.R., and Shur, M.L., "On the Sensitization of Turbulence Models to Rotation and Curvature," *Aerospace Science and Technology*, Vol. 1, No. 5, 1997, pp. 297–302.

10 Knight, D.D., and Saffman, P.C., "Turbulence Model Predictions for Flows with Significant Mean Streamline Curvature," AIAA Paper 78-0258, January 1978.

11 Shur, M.L., Strelets, M.K., Travin, A.K., and Spalart, P.R., "Turbulence Modeling in Rotating and Curved Channels: Assessing the Spalart-Shur Correction," *AIAA Journal*, Vol. 38, No. 5, 2000, pp. 784–792.

12 Jones, W.P., and Launder, B.F., "The Calculation of Low-Reynolds-Number Phenomena with a Two-Equation Model of Turbulence," *International Journal of Heat and Mass Transfer*, Vol. 16, 1973, pp. 1119–1130.

13 Kral, L.D., Mani, M., and Ladd, J.A., "On the Application of Turbulence Modes for Aerodynamic and Propulsion Flowfields," AIAA Paper 96-0564, January 1996.

14 Wilcox, D.C., *Turbulence Modeling for CFD*, 3rd Ed., La Cañada Flintridge: DCW Industries, Inc., 2006.

15 Menter, F.R., "Influence of Freestream Values on $k - \omega$ Turbulence Model Predictions," *AIAA Journal*, Vol. 30, No. 6, 1991, pp. 1657–1659.

16 Menter, F.R., "Two-Equation Eddy-Viscosity Turbulence Models for Engineering Applications," *AIAA Journal*, Vol. 32, No. 8, 1994, pp. 1598–1605.

Glossary

Aerodynamic center – the location on an airfoil or wing about which the pitching moment does not vary with angle of attack

Adaptive grid – a grid approach that can increase grid density in regions of high flow gradients automatically (or semi-automatically)

Algebraic grid stretching – a grid stretching approach for structured grids that uses an algebraic relation to determine spacing between grid points

Algebraic turbulence model – a turbulence model that does not use any differential equations (only algebraic relations)

Algorithm – a procedure or "recipe" for performing a numerical calculation

Amdahl's law – the maximum speedup for running a program on parallel processors based on how much of the code cannot be run in parallel mode

Architecture (computer) – the way a computer is designed to function (e.g., how a processor interacts with memory or with other processors)

Artificial viscosity – often explicitly added viscous-like terms used in a numerical algorithm to make the algorithm stable near high gradients in the flowfield like shock waves, sometime also used to describe Numerical viscosity

Backward Euler – a first-order accurate time differencing method that only uses information from the current time step to find solutions at the next time step

Banded matrix – a matrix with non-zero elements only on or near the diagonal of the matrix

Bernouli's equation – the equation that relates pressure and velocity for incompressible flow with no body forces according to momentum conservation

Blasius formula – the skin friction relation for laminar boundary layers in a flow with no pressure gradient

Block structured grid – a grid made up of several independent structured grids that intersect in order to form a grid around a complex shape

Boundary conditions – properties applied along the edges of a region where an equation is being solved

Boussinesq approximation – the assumption that turbulent viscosity behaves similar to laminar viscosity, making it possible to create a certain family of turbulence models

Boundary layer – the thin layer of flow near a surface where viscous effects dominate and a high velocity gradient is created normal to the surface

Cartesian grid – a grid that uses points aligned with a Cartesian axis system

Catalycity – the interaction of fluid particles with the material of an aircraft surface

Cell – the basic geometric shape used for numerically solving the equations of fluid motion (such as a rectangle or triangle in 2D or a cube or pyramid in 3D)

Cell aspect ratio – the ratio of the height of a cell (usually normal to a surface) and the length of the cell (usually along a surface) which is used to determine if a cell will yield good numerical results

Center of pressure – the location where the forces are assumed to act on an aircraft. Note: Technically this is a line of action, not a point.

Central processing unit – the primary calculation unit of a computer

Certification – the process of establishing the range of applicability of a verified and validated computational model

Chimera grid – a grid that is made up of multiple parts that overlap each other and provide the ability to geometrically model a complex shape

Clustering – the process for adding grid points to regions with high flow gradients

Coarse – a grid with a relatively small number of grid points or cells

Code – a computer program written to perform some set of operations

Code validation – the process of determining the degree to which a model is an accurate representation of the real world; validation is solving the right equations

Code verification – insuring that the numerical methods used to solve an equation are correct; verification is solving the equations right

Compiler – the portion of a computer that "translates" user commands in a high-level language to specific computer instructions (which are often called the machine language)

Computational aerodynamics – the use of computers to numerically solve aerodynamic problems

Computational fluid dynamics – the use of computers to numerically solve the governing equations of fluid motion directly

Conformal transformation – a geometric transformation in the complex plane that preserves angles and is used to change a simple shape (like a circle) into a more complicated shape (like an airfoil)

Continuum – the assumption that the flowfield can be modeled by the aggregate of the individual molecular motion (a macroscopic viewpoint) and ignore the motion of individual molecules (a microscopic viewpoint); this assumption is not valid for very low density flows

Conservation form – a form of a partial differential equation where all terms are inside derivatives; used in CFD to accurately model flows with shock waves and other strong flow gradients; also called Divergence form

Consistency – a finite difference equation (FDE) representation of a partial differential equation (PDE) where the difference between the two vanishes as the grid is refined

Convergence – at the most basic level, a numerical solution that approaches a single value with iterations; for a well-posed problem, a "converged" solution will be obtained if the numerical method is consistent and stable

Computer-aided design – the use of computers to model geometry, including points, lines, shapes, and at a more advanced level, entire aircraft; often used to create the surface for grids

Courant number – the relationship between the time step and the distance between grid points that will insure stability for a one-dimensional problem

Critical Mach number – the freestream Mach number where sonic flow first appears on an aircraft

D'Alembert's paradox – the disturbing result reached with potential flow theory that states there is no drag acting on a body in a flow

Delta form – a form for partial differential equations that is used for numerical solution and allows the right-hand side to go to zero as convergence takes place

Diagonally dominant matrix – for every row of the matrix, the magnitude of the diagonal value is greater than or equal to the sum of the magnitudes of all the other entries in the row

Diffusion – phase shift and Gibbs phenomenon errors in a numerical scheme due to an odd derivative error term in the modified equation

Discretization – the process of taking terms in a partial differential equation and representing the various terms on a grid of points with algebraic approximations

Dissipation – viscous-like errors in a numerical scheme due to an even derivative error term in the modified equation

Divergence form – See Conservative form

Downwash – the downward velocity field created by wing-tip vortices that changes the local velocity vector for flow over various aerodynamic surfaces like wings

Drag polar – a graph of lift coefficient vs. drag coefficient for an airfoil or aircraft; usually takes the shape of a parabolic curve

Eigenvalue – a characteristic value of a matrix that determines critical information about the solution

Elliptic PDE – a partial differential equation with coefficients that yield imaginary characteristics; solution requires boundary conditions all around the solution region

Engineering methods – prediction methods typically based on semi-empirical approaches

Empirical methods – prediction methods based entirely on experimental data

Equation of state – the thermodynamic relation describing the behavior of the specific fluid or gas

Euler's equations – the equations of fluid motion with no viscous terms included

Eulerian frame – a space-fixed reference frame used for defining the fluid equations of motion

Explicit – a numerical method that uses only known information to predict unknown information

Favre averaged – an averaging used to simplify the compressible equations of fluid motion using mass averaging (rather than Reynolds or time averaging)

Fine – a grid with a relatively large number of grid points or cells

Finite difference – an approach to numerically solving the equations of motion using a set of points that represent the fluid region; all derivatives are approximated by taking differences of values at the various points

Finite element – an approach to numerically solving the equations of motion using functions that satisfy boundary conditions

Finite volume – an approach to numerically solving the equations of motion using a set of small geometric volumes that represent the fluid region; all derivatives are approximated across the faces of the volumes

Flop – Floating-point operation

Flow solver – a computer code that solves the governing equations of fluid dynamics using various numerical schemes

Flow visualization – the field of CFD where flow solutions are visualized using various flow functions, such as scalar properties or vector properties, or other more advanced properties developed specifically for visualization purposes (such as helicity)

Formal accuracy – the theoretical accuracy of a numerical method applied to a real problem

Freestream – the flowfield far from any disturbances caused by a surface in the flow

Friction velocity – the streamwise velocity in a boundary layer nondimensionalized by the turbulent velocity and length scales

Governing equations – the partial differential or integral equations derived from the basic conservation laws of fluid dynamics that are used to find solutions for fluid flow

Graphics processing unit – a computer processor that performs calculations in order to directly output to a graphical display

Grid – a series of points used to perform finite difference calculations in order to simulate fluid flow

Grid generation – the process of creating a grid or mesh for use in fluid flow simulations

Half-equation model – a turbulence model that only uses one ordinary differential equation (as opposed to a partial differential equation)

Hardware – the physical parts of a computer, such as the processor, memory, monitor, and keypad

Helmholtz vortex theorems – the "rules" that govern how inviscid flow must behave in the vicinity of vortex filaments, developed by Hermann von Helmholtz in the nineteenth century

High-fidelity model – a prediction method that solves the governing equations of fluid dynamics to obtain results, usually at the cost of increased computational time

High- level programming language – a computer language typically created for people to learn and use, but not directly understood by the computer; a compiler is required to translate this language into a lower level programming language that the computer understands

High-performance computing – the use of world-class computers to perform advanced scientific and engineering calculations

Hyperbolic PDE – a partial differential equation with coefficients that yield real characteristics; solution requires initial conditions along a curve or surface

Hybrid grid – the combined use of semi-structured grids for viscous layers and unstructured grids away from the surface

Implicit – a numerical method that uses unknown information to predict other unknown information

Induced drag – the drag due to all vorticity shed from a wing, typically caused by downwash velocity created along the wing

Input/output device – the part of a computer that transfers information from the processor to memory or other external devices

Irrotational – a region of flow with no vorticity

Isotropic – a physical property that is not a function of axis orientation

Iteration – one sweep of a numerical algorithm through a grid or mesh

Iterative convergence – the situation as iterations are performed where changes in variables becoming small enough to consider the solution to be complete (the residual goes to zero)

Jacobian – a transformation matrix for geometry from one space to another

Kutta condition – the physical observation that fluid leaves the trailing edge of an airfoil smoothly

Kutta-Joukowski theorem – the theorem that states lift is created by circulation of the flow

Laminar separation bubble – a region of separation, typically near the leading edge of an airfoil, caused by the inability of laminar flow to withstand large pressure gradients

Laminar sublayer – also Viscous sublayer, the innermost layer of a turbulent boundary layer

Langrangian frame – a particle-fixed reference frame used for defining the fluid equations of motion

Large-scale structures – the vortical structures contained within a turbulent boundary layer, approximately the size of the boundary layer thickness

Law of the wall – the relation that the mean velocity within a turbulent boundary layer is proportional to the logarithm of the distance from the wall

Lax equivalence theorem – the statement that a numerical method will converge if it is consistent and stable

Log layer – the region in a turbulent boundary layer where the mean velocity is proportional to the logarithm of the distance to the wall

Low-fidelity model – a prediction method that uses empirical or semi-empirical approaches to obtain very quick results, usually at the cost of accuracy

Low-level programming language – also known as machine language, this is the language that the computer actually understands; all commands must be translated into this language for the computer to operate

Mach number – the ratio of velocity to the acoustic speed, or speed of sound (see M in the Nomenclature section)

Mass averaged – See Favre averaged

Massively parallel computer – the use of a large number of processors in a computer to perform calculations from a single program at the same time

Matrix – an array of numbers that represent terms in algebraic equations

Mesh   typically a set of geometric shapes that make up the solution region for the equations of fluid motion

Modified equation – the equation that is actually being solved, including truncation terms, when a numerical method is applied to a PDE

Modified Newtonian flow theory – a modification to Newtonian flow theory that uses the stagnation pressure coefficient after a normal shock to improve predictions

Moore's law – the observation that computer processor speeds double every eighteen months due to improvements in transistor technology

Multidisciplinary design optimization – a numerical optimization approach that combines the solution of multiple design aspects for an aircraft, such as aerodynamics and structures

Multigrid – a grid that is modified into multiple levels of fineness in order to accelerate convergence time

Navier-Stokes equations – the set of nonlinear PDEs that define the viscous flow of fluid in a continuum

Newton subiterations – a time integration method that decreases iteration error during calculation that take place within an iteration

Newtonian flow theory – a hypersonic flow theory that states that the pressure coefficient acting on a surface is proportional to the local slope of the surface

Newtonian fluid – a fluid where stress is proportional to the rate of strain

No-slip condition – a surface boundary condition that states that the velocity at the surface is zero

Numerical analysis – a field of mathematics that defines approaches to numerically solve mathematical problems

Numerical methods – the methods resulting from Numerical analysis

Numerical viscosity – the viscous-like error created by replacing partial derivatives with finite difference operations (or other numerical approximations); typically a combination of dissipation and dispersion

One-equation model – a turbulence model that uses one partial differential equation

Operating system – the software that manages computer hardware

Parabolic PDE – a partial differential equation with coefficients that yield a single characteristic; solution requires initial conditions and boundary conditions on the sides of the solution region

Parallel processing – using multiple computer processers in parallel (at the same time) to complete a single computation

Personal supercomputing – the concept of having high-performance computing power on a small, affordable machine

Post-processing – any use of information obtained from a fluid simulation performed on a computer, typically requiring flow visualization methods

Prandtl formula – a skin friction relation for turbulent boundary layers

Prandtl-Glauert equation – the partial differential equation that governs inviscid, isentropic compressible flow with small disturbances at subsonic or supersonic speeds

Prandtl-Glauert rule – the relation that shows how incompressible pressures change when a flow becomes compressible

Prandtl number – a nondimensional number that is the ratio of the viscous diffusion rate to the thermal diffusion rate (see Pr in the Nomenclature section)

Pressure coefficient – a nondimensional coefficient for pressure (see C_p in the Nomenclature section)

Processor – the hardware in a computer that carries out the instructions from a computer program

Program – a set of instructions used to perform calculations and provide commands to a computer processor

Pseudo-spectral method – A method for solving partial differential equations using a series of basis functions which can be represented on a quadrature grid

Reynolds averaged – a time averaging method that averages the Navier-Stokes equations to a set of equations that define the mean flowfield

Reynolds number – a nondimensional number that relates inertial effects to viscous effects (see Re in Nomenclature section)

Reynolds stress model – a turbulence model that predicts Reynolds stresses directly rather than predicting turbulent viscosity

Rotational – a region of flow with vorticity

Rotation velocity – the angular velocity of a fluid element

Scalar processing – an operation performed on a computer that can only be done in sequence (cannot be performed in parallel)

Semi-empirical methods – a method for performing engineering calculations or estimates that is based on theory and experimental data

Skew – the angular distortion of a cell

Slip condition – a fluid dynamic boundary condition that allows the flow to move relative to a solid surface

Small-scale structures – the nearly isotropic turbulence that exists throughout a boundary layer with very small length scales

Software – the coding and programming that tell a computer how to operate

Space marching – a numerical approach that iterates in a spatial direction (like the x direction) rather than in time

Sparse matrix – a matrix that has mostly zeros as elements

Spatial convergence – determining when a grid or mesh is dense enough to provide accurate predictions, typically accomplished by using a grid convergence study

Stability – a numerical scheme where errors from any source (round-off, truncation, etc.) are not permitted to grow

Structured grid – a grid with points that are placed relative to each other so that neighboring points are easily known; typically used with finite difference algorithms

Substantial derivative – a derivative that relates the time rate of change of a property to the velocity field; also known as the Material derivative

Supercomputer – a relative term describing a computer with state-of-the-art processing and memory capabilities

Surface grid – a two-dimensional grid or mesh that defines the geometry of a surface

Sutherland's law – the algebraic relation between fluid viscosity and temperature that reduces the number of independent variables in the fluid dynamic equation set

Temporal convergence – the process for insuring a steady solution has been reached within some level of accuracy

Topology – the mathematical study of shapes and spaces; often used in CA to describe the shape of a structured grid

Time marching – a numerical method that iterates in time to obtain a solution

Trailing vortex – vortices that are typically trailing behind an aircraft from the trailing edge of the wing

Tridiagonal matrix – a matrix with non-zero elements only on the main diagonal and the diagonals above and below the main element; all other elements are zero

Turbulence model – a mathematical model of the primary characteristics of turbulence (typically the "average" characteristics) used to "close" the set of equations being solved for fluid flow simulation

Two-equation model – a turbulence model that uses two partial differential equations

Unstructured grid – a grid or mesh made up of simple geometric shapes like triangles (in two dimensions) or pyramids (in three dimensions)

Validation – See Code validation

Vector form – placing a set of partial differential equations into divergence form in order to make them easier to use with numerical algorithms

Vector processing – a computer architecture that can perform a single computation on all elements of a one-dimensional array (or vector) at the same time

Verification – See Code verification

Viscous sublayer – See Laminar sublayer

Volume grid – a grid or mesh that fills a three-dimensional volume around a body surface (such as an airplane)

Von Neumann stability analysis – the process for showing the stability restrictions required for applying a numerical method to the solution of a partial differential equation

Vortex drag – the drag on an airplane caused by the vorticity shed from an airplane, typically from the vortices that form at the wing tip

Vortex filament – a three-dimensional line that is used to model a vortex

Warp – a computational cell that has been twisted

Wave drag – drag caused by the formation of shock waves on an aircraft

Zero-equation model – See Algebraic model

Index

Printed in the United States
By Bookmasters